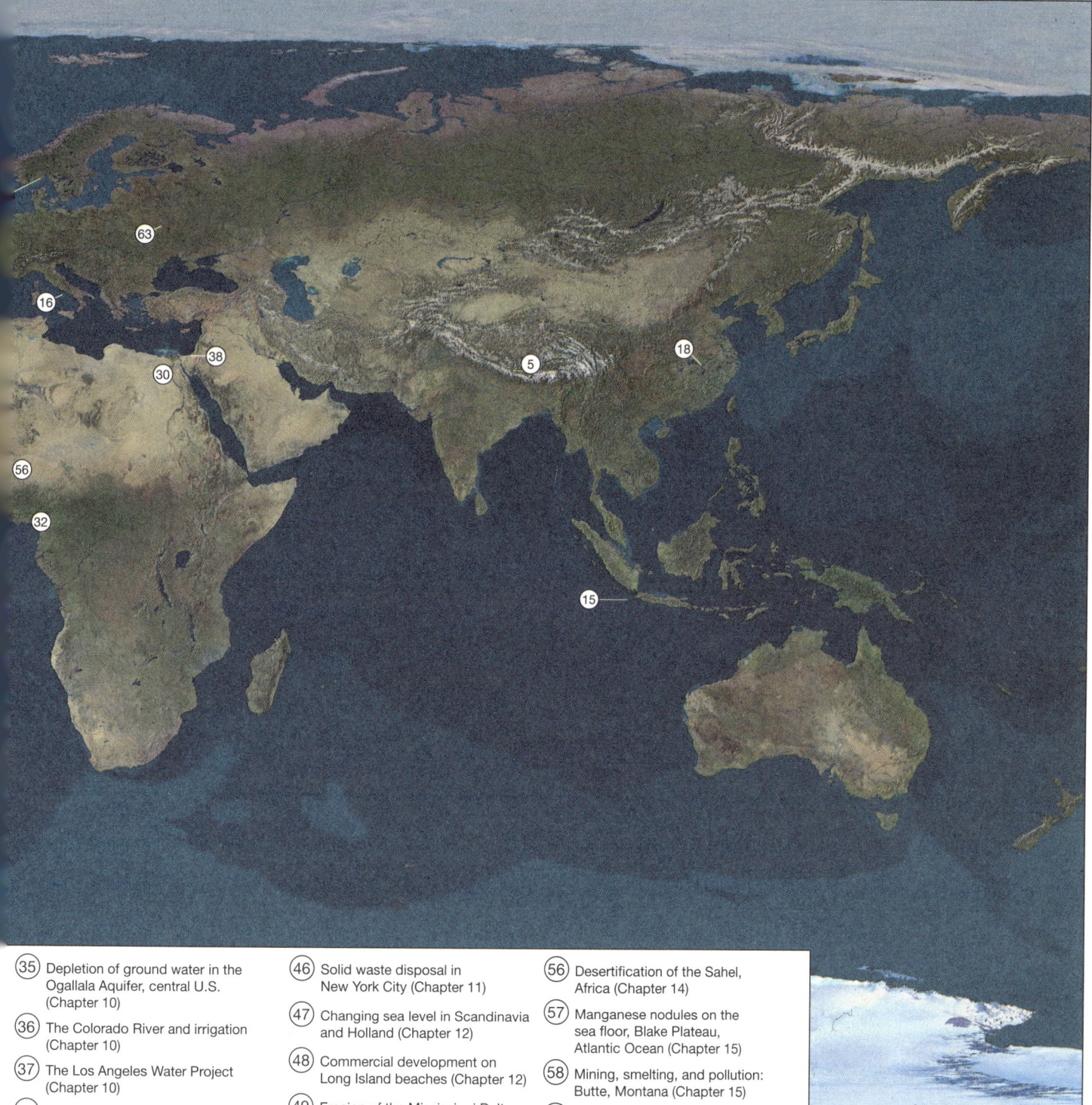

ENVIRONMENTAL Geoscience

ENVIRONMENTAL Geoscience

Jonathan Turk, PhD

Graham R. Thompson, PhD
University of Montana

Saunders Golden Sunburst Series

SAUNDERS COLLEGE PUBLISHING
Harcourt Brace College Publishers
Fort Worth   Philadelphia   San Diego   New York
Orlando   Austin   San Antonio   Toronto
Montreal   London   Sydney   Tokyo

Copyright © 1995 by Harcourt Brace & Company

All rights reserved. No part of this publication may be reproduced or transmitted in any form or by any means, electronic or mechanical, including photocopy, recording, or any information storage or retrieval system, without permission in writing from the publisher.

Requests to make copies of any part of the work should be mailed to Permissions Department, Harcourt Brace & Company, 6277 Sea Harbor Drive, Orlando, Florida 32887-6777.

Text Typeface: Goudy Old Style
Composition: Monotype Composition, Inc.
Publisher: John Vondeling
Developmental Editor: Christine Rickoff
Managing Editor: Carol Field
Project Editor: Anne Gibby
Copy Editors: Patricia M. Daly, Mary Patton
Manager of Art and Design: Carol Bleistine
Art Directors: Christine Schueler, Carol Bleistine
Art and Design Coordinator: Sue Kinney
Text Designer: Christine Schueler
Cover Designer: Lawrence R. Didona
Text Artwork: George V. Kelvin/Science Graphics
Director of EDP: Tim Frelick
Production Managers: Carol Florence, Joanne Cassetti
Marketing Manager: Sue Westmoreland

Cover Credit: Galen Rowell/Mountain Light
　Valley of the Ten Peaks, Canadian Rockies, Canada

Frontispiece: Condovivi, Bolivia

Page x: Limestone peaks, Canadian Rockies

Environmental Geoscience

ISBN: 0-03-098866-7

Library of Congress Catalog Card Number: 94-69215

5678901234 032 10 987654321

This book was printed on acid-free recycled content paper, containing **MORE THAN 10% POSTCONSUMER WASTE**

PREFACE

Ecological catastrophes contributed to the falls of several ancient civilizations. The Babylonians destroyed their farm soils by irrigating them year after year with slightly salty river water. Plato wrote about soil erosion and ground-water depletion after forests were cut from the Athenian hills. During the Industrial Revolution, governments frequently ignored the lessons of past civilizations and encouraged development regardless of environmental consequences. In many urban regions, skies turned black, rivers became polluted with chemicals and sewage, and rich soils were eroded. In 1775, Dr. Percival Pott showed that chimney soot caused cancer. In 1948, 20 people died and 6000 became ill from concentrated air pollution in Donora, Pennsylvania. In 1955, an oily film on the Cuyahoga River in Cleveland, Ohio, caught fire. These and similar incidents triggered a realization that pollution is unhealthful for humans and detrimental to the environment. People demanded action. In 1970, the United States Congress passed the National Environmental Policy Act (NEPA), committing the country to a clean environment. This act states that the United States Government must "use all practicable means, consistent with other essential considerations of national policy, to . . . fulfill the responsibilities of each generation as a trustee of the environment for succeeding generations."

Twenty-five years have passed since NEPA was signed into law. During that time, scientists of all disciplines have studied environmental degradation and searched for pollution-control strategies. Geologists contribute to this research by studying the rocks under our feet, the soil that supports plant growth, and the water that sustains us. They look backward in time to the evolution of our planet, its atmosphere, and its oceans.

Environmental Geoscience is an introductory text that explores the interaction of living organisms with the Earth. Over long expanses of geologic time, tectonic movement shapes ocean basins, raises mountains, and generates forces that cause earthquakes and volcanic eruptions. Over shorter time spans, surface processes erode mountain ranges, alter ground-water flow, and change the courses of rivers. Floods and landslides occur almost instantaneously. Both long- and short-term geologic change alters conditions for life on our planet.

Some geologic events such as earthquakes, volcanoes, and floods are beyond our control. However, if we understand these events, we can minimize their effects on human settlement. Other geologic changes, such as ground-water pollution and the impacts of mining and oil drilling, are caused by people. Since we create these disturbances, we must consider the costs and benefits of our actions.

Environmental Geoscience explains geologic processes as a background for understanding the environmental consequences of geologic change. The text highlights many environmental issues with Case Histories that relate geology to real situations. Examples include the 1993 Mississippi River flood, the California earthquake of 1994, and ground-water and nuclear waste disposal. Other topics that expand a student's understanding but are not necessary to a chapter's continuity are set off as *Focus On* boxes. Examples include "Conversion of Heat from Fuels to Work and Electricity" and "Earthquake Waves as a Tool for Investigating the Structure of the Earth."

Teaching Options

This book is divided into four parts:

Part 1: The Earth and Its Materials
Part 2: Internal Processes
Part 3: Surface Processes
Part 4: Mineral and Energy Resources

We have placed the topic of tectonics early in the book because tectonic activity creates the basic landforms that are shaped by surface processes. However, this order can be altered without much loss of continuity. Part 4, Mineral and Energy Resources, can be taught earlier in the course, or portions can be incorporated into other chapters.

We recognize that many students in an introductory geology course don't have backgrounds in chemistry and physics. Consequently, we introduce geologic processes and events in language that is readily understood by students with little or no college-level science or mathematics background.

Finally, we believe science is not a set of facts to be learned and memorized. Any study of geology would be lifeless unless students were introduced to some of the crucial experiments and thought processes that led to the development of important geological hypotheses and theories.

Special Features

As already explained, supplementary material is presented throughout the book in the form of *Focus On* boxes that are set aside and highlighted in color. These topics are not essential to the continuity of a chapter, and they can be included or omitted at the discretion of the instructor. *Case Histories* give contemporary examples that illustrate how geological events and processes affect people. The Case Histories are incorporated in the text but highlighted by color. These geological anecdotes are interesting in themselves and reinforce the main topics.

Chapter review material. Important words are highlighted in bold type within the text. Many of these key words are printed in a list at the end of each chapter for review purposes. In addition, a short summary is provided at the end of each chapter.

Questions. Two types of end-of-chapter questions are provided. The **Review Questions** can be answered in a straightforward manner from the material in the text. In contrast, **Discussion Questions** challenge students to apply what they have learned to the analysis of situations not directly presented in the text. Often there are no unique or absolute correct answers to these questions.

Appendices and Glossary. A glossary of terms is provided at the end of the book and appendices cover: A Periodic Table and International Table of Atomic Weights, Systems of Measurement, and Graphic Rock Symbols used throughout the book.

Ancillaries

This text is accompanied by extensive support materials.

Study Guide

The Study Guide, written by Vicki Harder of El Paso Community College, provides review and study aids to enhance the student's understanding of the text. The Study Guide includes an overview for each chapter, learning objectives, a detailed chapter outline, review questions, and answers.

Instructor's Manual and Test Bank

The Instructor's Manual, written by the authors, provides teaching goals, answers to discussion questions, and a short bibliography. Sample tests written by Christine Seashore are included in the manual. The test bank includes multiple-choice, true-or-false, and completion questions for each chapter.

Acknowledgments

The manuscript has been reviewed at several stages, and the numerous careful criticisms have helped shape the book and ensured accuracy.

F. William Cambray, *Michigan State University*
Susan M. Cashman, *Humboldt State University*
Edward B. Evenson, *Lehigh University*
Mark Feigenson, *Rutgers University*
Jim B. Finley, *Miami University, Ohio*
Luis A. Gonzalez, *University of Iowa*
Gilbert Hanson, *SUNY, Stony Brook*
Cindy Lampe, *San Diego State University*
Joan P. Licari, *Cerritos College*
Robert Matthews, *University of California at Davis*
Roderic A. Parnell, Jr., *Northern Arizona University*
Paul Ragland, *Florida State University*
Robert D. Shuster, *University of Nebraska at Omaha*
Duncan Sibley, *Michigan State University*
Ellen Wohl, *Colorado State University*

Geology is a visual science. We can readily observe many rocks and landforms on the Earth's surface. Although we cannot see many internal processes, these events can be visualized through the artist's eye. George Kelvin has painted most of the illustrations in this book. It has been a pleasure to work with him again.

We would never have been able to produce this book without professional support both here in Montana and from the offices of Saunders College Publishing. Thanks to Christine Seashore for finding photographs, contributing personal photographs, and for logistic collaboration. Special thanks to John Vondeling, our Associate Publisher. One of us,

Jonathan Turk, has worked with John for 25 years; we have developed a long-lasting friendship and a superb professional relationship. Christine Rickoff, our Developmental Editor, Anne Gibby, our Project Editor, Christine Schueler, Art Director, and Patricia Daly and Mary Patton, our Copy Editors, have all worked hard and efficiently to produce the finished project.

Jonathan Turk
Darby, Montana

Graham R. Thompson
Missoula, Montana

December 1994

PART 1
The Earth and Its Materials 1

CHAPTER 1
The Earth: Our Home 3

CHAPTER 2
The Earth in Time 17

CHAPTER 3
Earth Materials: Minerals, Rocks, Air, and Water 39

A CONVERSATION WITH Cindy Lee Van Dover 60

PART 2
Internal Processes 64

CHAPTER 4
Plate Tectonics 67

CHAPTER 5
Earthquakes 99

CHAPTER 6
Volcanoes 125

A CONVERSATION WITH Stephen Schneider 150

PART 3
Surface Processes 154

CHAPTER 7
Soil 157

CHAPTER 8
Mass Wasting 189

CHAPTER 9
Fresh Water: Streams, Ground Water, Lakes, and Wetlands 205

CHAPTER 10
Human Use of Water 237

CHAPTER 11
Waste Disposal and Fresh-Water Pollution 261

CHAPTER 12
Coastlines and Coastal Pollution 291

CHAPTER 13
Glaciers and Ice Ages 317

CHAPTER 14
Deserts and Desertification 337

PART 4
Mineral and Energy Resources 350

CHAPTER 15
Mineral Resources and Mining 353

CHAPTER 16
Fossil and Uranium Fuels 379

CHAPTER 17
Alternative Energy Resources 409

PART 1
The Earth and Its Materials 1

CHAPTER 1
The Earth: Our Home 3
- 1.1 Our Dynamic Planet 4
- 1.2 Ecosystems 6
- 1.3 Economics and the Environment 7
- 1.4 The Earth Summit 13

  FOCUS ON Hypothesis, Theory, and Law 10

CHAPTER 2
The Earth in Time 17
- 2.1 The Origins of the Universe, Solar System, and Earth 19
- 2.2 Formation of the Solar System and Earth 19
- 2.3 The Modern Earth 23
- 2.4 A Brief History of Life 26

  FOCUS ON Dinosaur Life in Montana 100 Million Years Ago 31

CHAPTER 3
Earth Materials: Minerals, Rocks, Air, and Water 39
- 3.1 Elements, Atoms, and Ions 40
- 3.2 The Solid Earth: Minerals 40
- 3.3 The Solid Earth: Rocks and the Rock Cycle 46
- 3.4 The Atmosphere 54
- 3.5 The Hydrosphere 56
- 3.6 Interactions Among the Spheres 57

  FOCUS ON Asbestos and Cancer 45

  FOCUS ON Radon and Rocks 54

  FOCUS ON Rocks 57

A CONVERSATION WITH Cindy Lee Van Dover 60

PART 2
Internal Processes 64

CHAPTER 4
Plate Tectonics 67
- 4.1 The Earth's Layers 69
- 4.2 Plates and Plate Tectonics 71
- 4.3 Divergent Plate Boundaries 72
- 4.4 Convergent Plate Boundaries 73
- 4.5 Transform Plate Boundaries 76
- 4.6 The Anatomy of a Tectonic Plate 76
- 4.7 The Search for a Mechanism 77
- 4.8 Supercontinents 80
- 4.9 Plate Tectonics and Mountain Building 82
- 4.10 Plate Tectonics and the Sea Floor 88

  FOCUS ON Alfred Wegener and the Origin of an Idea 79

CHAPTER 5
Earthquakes 99
- 5.1 What Is an Earthquake? **101**
- 5.2 Earthquake Waves **102**
- 5.3 Earthquake Damage **106**
- 5.4 Earthquake Danger Zones in North America **113**
- 5.5 Earthquake Prediction **119**
- 5.6 Earthquake Hazard and Human Habitation **121**

FOCUS ON Earthquake Waves as a Tool for Investigating the Structure of the Earth *108*

FOCUS ON Earthquake Control *121*

CHAPTER 6
Volcanoes 125
- 6.1 Magma **127**
- 6.2 Magma Behavior: Why Some Magma Erupts from a Volcano and Other Magma Solidifies Below the Surface **130**
- 6.3 Volcanoes and Related Landforms **130**
- 6.4 Volcanoes and Plate Tectonics **133**
- 6.5 Explosive Magma and Violent Volcanoes **137**
- 6.6 Assessing the Danger of Future Volcanism **141**
- 6.7 Volcanoes and Climate **145**

FOCUS ON The 1980 Mount St. Helens Eruption *138*

FOCUS ON Calderas in the United States *142*

FOCUS ON Disastrous Historical Eruptions *146*

A CONVERSATION WITH Stephen Schneider *150*

PART 3
Surface Processes 154

CHAPTER 7
Soil 157
- 7.1 Weathering **159**
- 7.2 Soil **160**
- 7.3 Soil Nutrients **166**
- 7.4 The Hubbard Brook Experimental Forest **167**
- 7.5 Erosion **169**
- 7.6 Agriculture and Soil Fertility **170**
- 7.7 Reducing Cropland Erosion **172**
- 7.8 Soil Erosion in Central China and the American Dust Bowl **175**
- 7.9 Croplands and World Food Supply **177**
- 7.10 Soil Engineering **179**

FOCUS ON Soil Classification *165*

FOCUS ON Organic Farming *172*

FOCUS ON Soil and Government Farm Subsidies in the United States *174*

Contents xiii

CHAPTER 8
Mass Wasting 189
- 8.1 Mass Wasting **190**
- 8.2 Factors That Control Mass Wasting **190**
- 8.3 Types of Mass Wasting **192**
- 8.4 Mass Wasting Triggered by Earthquakes and Volcanoes **197**
- 8.5 Predicting Mass Wasting **201**
- 8.6 Engineering for Landslide Avoidance **201**

CHAPTER 9
Fresh Water: Streams, Ground Water, Lakes, and Wetlands 205
- 9.1 The Water Cycle **206**
- 9.2 Streams **207**
- 9.3 Floods **211**
- 9.4 Ground Water **216**
- 9.5 Lakes **221**
- 9.6 Wetlands **227**
- 9.7 The Florida Everglades **230**

CHAPTER 10
Human Use of Water 237
- 10.1 Water in the United States: Supply and Demand **238**
- 10.2 Water Diversion and Environmental Impacts **243**
- 10.3 Salinization **250**
- 10.4 Water Use in the Western United States **250**
- 10.5 Conservation **255**
- 10.6 Water and International Politics **256**

CHAPTER 11
Waste Disposal and Fresh-Water Pollution 261
- 11.1 Water Pollution **262**
- 11.2 Water Pollutants **263**
- 11.3 Pollution by Nutrients **264**
- 11.4 Industrial Organic Wastes **266**
- 11.5 Surface Water in the United States 20 Years After the Clean Water Act **268**
- 11.6 Ground-Water Pollution **270**
- 11.7 Deep Injection Wells and Ground-Water Pollution **275**
- 11.8 Ground-Water and Nuclear Waste Disposal **275**
- 11.9 Municipal Waste Disposal **280**
- 11.10 Alternatives to Waste Disposal **285**

CHAPTER 12
Coastlines and Coastal Pollution 291
- 12.1 Emergent and Submergent Coastlines **292**
- 12.2 Erosion and Transport Along Coastlines **296**
- 12.3 Erosion of the Atlantic Coast **298**
- 12.4 Erosion of the Gulf Coast **301**
- 12.5 Erosion of the Pacific Coast **302**

12.6 Coastal Development and Public Policy **303**
12.7 Ocean Pollution **305**
12.8 Reefs **310**
12.9 Global Warming and Sea-Level Rise **312**

CHAPTER 13
Glaciers and Ice Ages *317*
13.1 Types of Glaciers **318**
13.2 Glacial Movement **319**
13.3 Glacial Erosion **322**
13.4 Glacial Deposits **325**
13.5 The Ice Ages **329**

CHAPTER 14
Deserts and Desertification *337*
14.1 Deserts of the World **339**
14.2 Water in Deserts **341**
14.3 Wind Erosion **343**
14.4 Dunes **344**
14.5 Expansion and Contraction of Deserts **345**
14.6 Land Degradation in Desert and Semiarid Lands **346**

PART 4
Mineral and Energy Resources *350*

CHAPTER 15
Mineral Resources and Mining *353*
15.1 Geologic Resources **354**
15.2 How Ore Forms **357**
15.3 Mining and Refining of Metals **364**
15.4 Future Availability of Minerals **370**

FOCUS ON Manganese Nodules on the Sea Floor *363*

FOCUS ON The 1872 Mining Law *364*

CHAPTER 16
Fossil and Uranium Fuels *379*
16.1 Fossil Fuels **380**
16.2 Coal **381**
16.3 Petroleum and Natural Gas **384**
16.4 Secondary Recovery and Synfuels **387**
16.5 Uranium and Atomic Energy **391**
16.6 The Future of Our Energy Supply **399**
16.7 The Geopolitics of Fossil Fuels **402**
16.8 Energy Strategies for the United States **403**

FOCUS ON Formation of Fossil Fuels: A Novel Idea *390*

FOCUS ON Plutonium: One of the Deadliest Environmental Contaminants *396*

CHAPTER 17
Alternative Energy Resources 409
17.1 Energy Conservation **411**
17.2 Alternative Energy Resources **415**
17.3 Solar Energy **416**
17.4 Hydroelectric Energy **419**
17.5 Geothermal Energy **420**
17.6 Energy from the Wind **422**
17.7 Biomass Energy **422**
17.8 Energy from the Seas **423**
17.9 Hydrogen Gas as a Fuel **425**
17.10 Nuclear Fusion **426**

FOCUS ON Conversion of Heat from Fuels to Work and Electricity *412*

APPENDIX A
The Elements *A-1*
International Table of Atomic Weights A-2

APPENDIX B
Systems of Measurement *A-3*

APPENDIX C
Rock Symbols *A-7*

GLOSSARY
G1

INDEX
I-1

PART 1

The Earth and Its Materials

1 The Earth: Our Home
2 The Earth in Time
3 Earth Materials: Minerals, Rocks, Air, and Water

China is the most populous nation on Earth. In recent years, many people have moved from rural areas to the cities.

CHAPTER 1

The Earth: Our Home

1.1 Our Dynamic Planet
1.2 Ecosystems
1.3 Economics and the Environment
1.4 The Earth Summit

In 1877, the Italian astronomer Giovanni Schiaparelli reported seeing a network of *channels* on Mars. An English translator mistakenly wrote that Schiaparelli had discovered *canals*. Other astronomers thought that they saw the canals as well. They argued that the canals were too straight and symmetrical to be natural and were probably irrigation conduits built by intelligent beings to transport water from the polar ice caps to deserts near the Martian equator. Since the dawn of the space age, spacecraft have studied all the planets in the Solar System except Pluto, and have learned that only Earth supports life as we know it.

The Earth is 4.6 billion years old, a time frame almost unimaginable in human terms. Communities of multicellular organisms evolved about 570 million years ago, and thus have been on the planet for about 12 percent of its history. Humans and their immediate precursors have lived on Earth for a few million years and industrial society has existed for a few hundred, both insignificant specks of Earth history. Yet within this short time—geologically speaking— humans have altered global systems so drastically that scientists are asking disturbing questions about relationships between humans and our environment.

Some factories pollute both air and water.
Hawksbury, Ontario. (SuperStock)

1.1 Our Dynamic Planet

Geology

Geology is the study of the Earth: the materials of which it is made, the physical and chemical changes that occur on its surface and in its interior, and the history of the planet and its life forms from its origin to the present.

The Earth's interior is separated into concentric layers, each with its own characteristics. As we will learn in later chapters, the outermost 100 kilometers consist of cool, rigid slabs of rock, called **tectonic plates**, that float over deeper layers of hot, plastic rock. In some places, tectonic plates collide, creating a mountain range such as the Himalayas (Fig. 1–1). In other regions, plates separate, creating or widening an ocean basin. The Atlantic Ocean is growing wider by a few centimeters each year as a result of this process. The movement of tectonic plates is an example of an **internal process**, one that is initiated deep within the Earth. Volcanic eruptions and earthquakes result from internal processes.

Geologists also study **surface processes**. Streams and glaciers erode the land, wind heaps sand into dunes, and solid bedrock weathers to soil (Fig. 1–2). Surface processes and internal processes act in concert to create familiar landscapes. Thus, as tectonic collisions raise great mountain ranges, streams and glaciers carve the uplifted rock to create peaks, canyons, and broad river valleys.

FIGURE 1–2
Big Sandy Creek has eroded a broad valley in the Wind River Mountains of Wyoming.

FIGURE 1–1
La-Moshe Peak in the eastern Himalayas.

In our everyday lives we are accustomed to thinking about time in terms of minutes, hours, days, or years. Historians study events that occurred as much as a few thousand years ago, but as mentioned previously, the Earth is 4.6 billion years old. The immense magnitude of geologic time means that processes that occur too slowly or rarely to have an impact on our daily lives are important and common in Earth history.

Slow processes are significant over geologic time. Tectonic plates creep across the surface of the Earth at a rate of a few finger widths every year. Since the first steam engine was built 200 years ago, North America has migrated 8 meters westward, a distance a sprinter can run in 1 second. Thus in human terms, continental motion is too slow to be observed except with sensitive instruments. However, if you could see a time-lapse video of a few hundred million years, a small chunk of geologic time, you would watch continents travel halfway around the Earth and great mountain ranges rise and erode away.

Events that are improbable in a human lifetime are likely over geologic time. The chances are small that the river flowing through your city will flood this spring, but if you lived to be 100 years old, you would probably see a catastrophic flood. (In fact, many residents of the Midwest saw disastrous floods in the summer of 1993.) When geologists study the 4.6 billion years of Earth history, they find abundant evidence of catastrophic events that are highly improbable in a human lifetime. Giant meteorites have smashed into our planet, vaporizing enormous volumes of rock and spreading dense dust clouds over the sky that may have driven millions of species into extinction. Similarly, huge volcanic eruptions, a few thousand times larger than that of Mount St. Helens in 1980, have changed conditions for life across the globe.

Environmental Geoscience

Environmental geoscience is the study of geology as it applies to living organisms and the quality of life on our planet.

The Earth has changed continuously since it first formed, and we are certain that it will continue to change in the future. Mountain ranges have risen and eroded away, ocean basins have opened and closed, and global climate has changed. These geologic changes affect life on Earth.

Living organisms can also affect the Earth. As you will see in a later chapter, evolving life radically altered atmospheric composition. Within the past few hundred years, humans have paved large areas of land, altered the flow of surface and ground water, extracted metals and fuels, created conditions that lead to erosion, and polluted the atmosphere and the water with solid, liquid, and gaseous wastes.

One important aspect of environmental geoscience is the interplay between natural events and human activities. For example, floods are a natural phenomenon. The floods that ravaged the Midwest during the summer of 1993 were caused by a natural event—unusually heavy rains (Fig. 1–3). But the story doesn't end here. The levees and river control projects meant to protect cities and towns from flooding increased flood severity in some places. Geologists and engineers are studying the Mississippi River valley and are searching for ways to reduce the severity of future floods. Similarly, although we cannot prevent earthquakes and volcanic eruptions, we can adapt to our dynamic planet by judicious placement and construction of human settlements.

FIGURE 1–3
Residents of Hartsburg, Missouri, boating along Main Street flooded by the Missouri River. *(Stephen Levin)*

Environmental geologists also study resources that are critical to humans and other species. A **geologic resource** is any source of raw material, including soil, surface and ground water, metal ores, coal, petroleum, building stone, sand, gravel, and a host of other rock and mineral products. Geologic resources are diminished by three processes: consumption, dispersal, and pollution.

When fossil fuels such as coal and petroleum are burned, their chemical energy is converted to heat. The fuel is **consumed** and cannot be reused or recycled.

Iron is a common element. Pick up a rock or dig a shovelful of soil in your back yard, and your sample almost certainly contains iron. But the concentration of iron in the soil in your back yard is so low that it would be prohibitively expensive to extract it. Iron ore is valuable because the iron is concentrated. If you mine and refine this ore, manufacture cars, paper clips, and other objects, and then throw these objects away, the iron becomes dispersed. Not one atom is lost, but it is expensive to collect the iron again. In this way, **dispersal** depletes resources.

If a factory spills toxic chemicals onto the soil and these poisons seep into a river, lake, or ground water, the water is neither consumed nor dispersed. But its quality is degraded so that it is no longer fit for human use. **Pollution** is an undesirable change in the quality of the environment that is caused by wastes or by excessive heat, noise, or harmful radiation. Many different factors can cause a waste product to pollute a resource. One factor is quantity. In a natural ecosystem, plant and animal wastes are not pollutants because they are consumed by decay organisms. However, if thousands of cattle are penned together in a feedlot or if millions of people crowd into a city, then their wastes can overwhelm natural decay mechanisms and become pollutants.

Another factor is toxicity. **Toxic substances** are acute poisons at specific concentrations. Carbon monoxide is a minor component of the atmosphere, and we inhale some with every breath, but if you breathe air with a high carbon monoxide concentration, you will die quickly. Therefore, carbon monoxide is not toxic at low concentrations but is highly toxic at high concentrations.

Some substances that are not acute poisons can damage plants and animals in other ways. For example, small quantities of the pesticide DDT won't poison you. However, chronic exposure to low doses of DDT (or many other substances) might cause cancer or birth defects. A cancer-producing substance is a **carcinogen**, and a substance that causes inheritable birth defects, or **mutations**, is called a **mutagen**. As discussed in Section 1.3, it is often difficult to prove whether or not a substance is carcinogenic or mutagenic.

Many substances are not harmful to people but are detrimental to the environment. For example, carbon dioxide is harmless to people but may alter global climate.

Fossil fuels, metal ores, and some ground-water reservoirs are **nonrenewable resources** because the geological processes that created them were much slower than the span of human history. Once they are gone, we must learn to live without them. Other resources are **renewable** because they are replenished within human lifetimes. If you drink the water from a stream or lake, the spring rains will replace what you removed. If you cut a tree (and do not destroy the soil), a new tree will grow within one or two lifetimes.

Decay organisms and chemicals break down most dead plants, animals, and their waste products in the environment. Thus, these materials are **biodegradable**. Many substances produced in industrial processes are so different from naturally occurring chemicals that they do not degrade rapidly; some persist for decades or centuries. **Nonbiodegradable** pollutants can be especially troublesome because they accumulate in the environment.

1.2 Ecosystems

If you look around a forest, a city park, or your own back yard, you may wonder why animals and plants live in the ways they do. You may ask yourself questions like these:

> Grass grows in the open but not under that tree. What does grass need that it finds in one place but not in another?
>
> A single housefly may lay several thousand eggs at one time. Why haven't flies overpopulated the Earth?
>
> If this forest is cut down, can crops grow on the land, or would the soil and climate be unsuitable for agriculture?

All these questions are about **ecology**, the science that studies interactions between living organisms and their environments. Ecologists have discovered that these questions cannot be answered by studying a single organism or even a single species. Instead, we must study entire **systems**. We hear the word *system* used in a variety of contexts. People speak of a digestive system, a sound system, a system of government, or an ecosystem. Systems may be quite different from one another, yet they all share a common concept:

> **A system is a group of elements that interact and function together as a whole.**

Thus, a digestive system is more than a disconnected assembly of stomach, liver, intestines, and pancreas; the individual parts work together in an integrated manner. If one of the components is removed or damaged, the whole system malfunctions.

The entire Earth with its oceans and atmosphere is too complex to study as a whole, so ecologists work with manageable units such as hillsides, forested valleys, lakes, or fields. In 1887, Stephen Forbes, a biologist for the Illinois Natural History Survey, wrote that

> a lake . . . forms a little world within itself—a microcosm within which all the elemental forces are at work and the play of life goes on in full, but on so small a scale as to bring it easily within the mental grasp. . . .
>
> If one wishes to become acquainted with the black bass, for example, he will learn but little if he limits himself to that species. He must evidently study also the species upon which it depends for its existence and the various conditions upon which these depend.

Today we would call Forbes's lake, or any other manageably small unit, an **ecosystem** (Fig. 1–4). Ecosystems are nearly self-contained. Think of an island in the middle of the ocean. Plants grow, die, and decay. Animals eat and in turn are eaten. Yet there is little exchange of life or raw materials with the ocean or other islands. Similarly, think of a valley in the mountains. Although seeds blow from one valley to another and many animals, such as birds and deer, roam far from their birthplaces, most of the exchanges occur within the valley. Of course, no ecosystem is truly isolated and independent. Wind blows soil and pollen over great distances, and animals migrate from system to system. Sunlight is absorbed, atmospheric gases are exchanged, rain falls, and streams flow toward the sea. Thus, each ecosystem is just one part of the Earth.

> **An ecosystem is a system formed by the interactions of a variety of individual organisms with one another and with their environment. Ecosystems are nearly self-contained, so that the exchange of nutrients within the system is much greater than exchanges with other systems.**

Ecosystems resist change. For example, during drought, some trees in a forest may die and be replaced by drought-resistant shrubs, but the system continues despite shifts in species abundance. Yet ecosystems have changed dramatically throughout Earth history. Dinosaurs wallowed in humid, tropical swamps in central Montana, and then about 65 million years ago, the dinosaurs and half of all other land animals became extinct. Twenty thousand years ago, massive continental glaciers

1.3 Economics and the Environment

Cost–Benefit Analysis

We can minimize the negative environmental impacts of mining, agriculture, logging, and industry if we are willing to pay the price. For example, mining creates huge amounts of waste dirt and rock. If the waste is simply piled adjacent to mines and smelters, sediment and toxic pollutants contaminate surface and ground water. Mine wastes can be stabilized and pollution reduced, but these processes cost money.

Limited environmental protection is relatively inexpensive, but costs rise as we seek a more pristine environment. Some people suggest that the environment should be protected only when the protection produces a positive economic return. This approach, known as **cost-benefit analysis**, balances the monetary cost of environmental protection against the monetary benefits of the protection.

Environmental degradation costs money, as shown on the left side of Figure 1–5. The cost of living in a degraded environment includes direct costs such as medical bills, lost work, and damage to waterways, crops, and livestock. It also includes indirect costs resulting from

FIGURE 1–4
(A) Sparse vegetation characterizes the eastern Arizona desert. (B) A temperate rainforest in southeastern Alaska. (C) East African savannah. (*Amos Turk*)

covered the northern part of the state. Until a few hundred years ago, bison roamed the lush Montana prairies.

Today, we are living in a period of rapid environmental change. Within the past 500 years, one third of the world's forests have been cut, the atmospheric carbon dioxide concentration has increased by 27 percent, and the human population has increased from 500 million to 5.5 billion. We ask, "How much change can occur before our life support systems become disrupted?"

In our search for answers to this disturbing question, we look at the Earth as an ecosystem that is composed of many smaller interconnected ecosystems. Our planet has been compared to a spaceship: Although we receive a continuous supply of energy from the Sun, we are otherwise isolated and depend entirely on the resources contained in our planet.

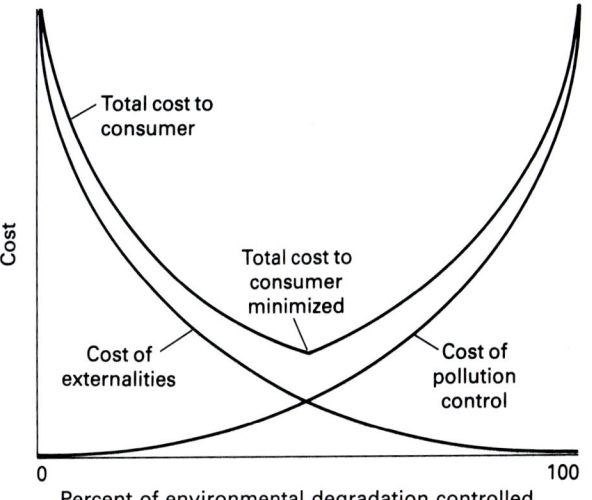

FIGURE 1–5
The relationship between the cost of environmental protection and the cost of externalities. The actual numbers vary with the situation.

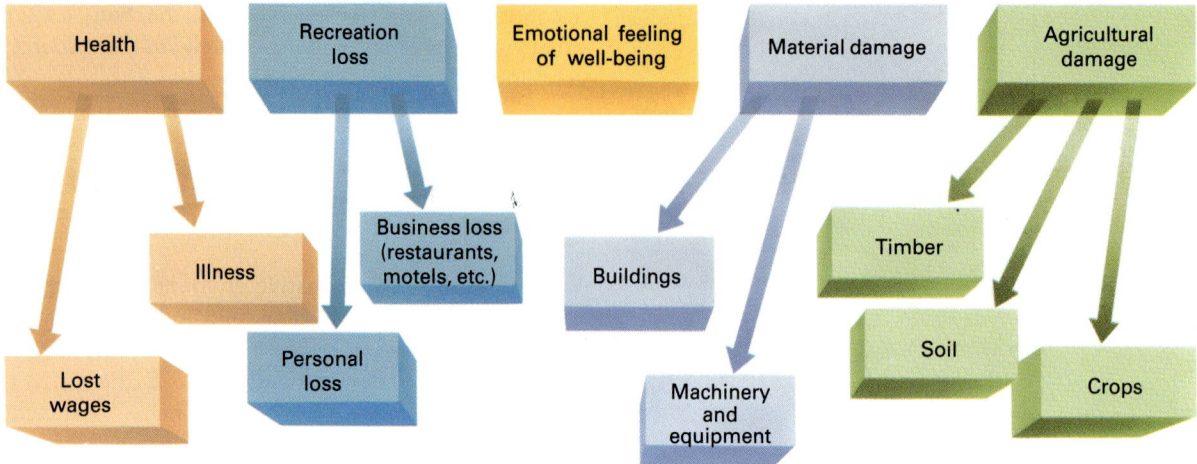

FIGURE 1–6
Some external and human costs of pollution.

effects such as reduced tourism and lowered land values if people no longer want to visit or live in a degraded area (Fig. 1–6). All of these costs are called **externalities**.

Measures to protect the environment cost money, but they also save money because the cost of externalities decreases. The uppermost curve in Figure 1–5 shows the total cost of a hypothetical polluting process, which is the sum of the costs of environmental protection and externalities. In many cases, it is relatively inexpensive to decrease environmental degradation by 50 percent, and when this is done, the cost of externalities drops dramatically. Thus the total cost decreases. As pollution is further reduced, however, the control cost rises rapidly and the cost of externalities decreases more slowly. In this scenario, shown by the right side of the graph in Figure 1–5, a pristine environment becomes expensive.

So how much environmental protection is desirable? Some people suggest that we should minimize the total cost and thus accept considerable degradation and pollution. Others argue that cost–benefit analysis is fundamentally flawed because it ignores the **quality of life**. How, they ask, can you place a dollar value on the annoyance of a vile odor, a persistent cough, polluted streams, dirty air, or industrial noise? Such annoyances damage our sense of well-being. What about recreational opportunities? One hundred years ago, most rivers in the United States were relatively unpolluted. People fished, boated, swam, and explored along the shores and marshes. Today, swimming in natural waterways is unhealthy in many parts of the country, the number of edible game fish has declined, and many of the marshes have disappeared. Alternative recreational opportunities are now available. Swimmers can find swimming pools, children can play in parks, and people who want to see wildlife can drive to a zoo or watch educational programs on television. For many, these changes involve poignant losses, but their dollar value is difficult to assess (Fig. 1–7). If all of these noneconomic costs of pollution are considered, then a higher level of pollution control becomes desirable despite its higher dollar cost.

Habitats for many organisms are destroyed when forests are cut, wetlands drained, and natural prairies plowed. As a result, millions of species are being driven to extinction today. This loss of species and the resultant loss of diversity of organisms on the planet is known as the

FIGURE 1–7
If rivers are polluted or dammed, people lose irreplaceable recreational resources.

decline in **biodiversity**. Biodiversity loss, like a decline in the quality of life, does not affect short-term economic balance sheets. However, species extinction alters ecosystems in ways that may ultimately affect all of the Earth's organisms.

Long-Term versus Short-Term Considerations

One problem with cost–benefit analysis lies in choosing the appropriate time period for the analysis. Suppose that a portion of a tropical rainforest is cut and the land is planted with grain. The logger makes a profit from selling the wood, and the farmer makes an additional profit from growing wheat. There may be no immediate losses. However, many tropical soils deteriorate quickly when they are farmed. As a result, in the long term both the local economy and the natural environment become impoverished. In this example, an immediate gain leads to a future loss. When analyzing the costs and benefits of environmental protection, it is important to specify the length of time under consideration.

Tragedy of the Commons

In the preceding example, if one individual clears a small area of rainforest to plant a garden, the rainforest ecosystem is not affected. But if many people clear small plots (or if one big corporation clears a large plot), then serious problems can arise. Deforestation of tropical rainforests causes species extinction and may affect local and global climates. "**The Tragedy of the Commons**" describes the cumulative effect of many people's actions. Ecologist Garrett Hardin used the example of the old English commons, grazing land shared by all the farmers in a region.[1] Each farmer, as a rational human being, chooses to maximize his or her personal gain. If a farmer adds one cow to the commons, that person gains one cow. The cow degrades the pasture by eating more grass, but the reduction in the quality of the pasture is shared by all the farmers. Because the gain is individual but the loss is shared, a farmer profits by placing more animals on the commons. The tragedy arises when everyone acts in the same manner; then, too many cows overgraze the pasture and all the cows starve (Fig. 1–8). To continue the argument in Hardin's words,

> In a reverse way, the tragedy of the commons reappears in problems of pollution. Here it is not a question of taking something out of the commons, but of putting something in—sewage, or chemical, radioactive, and heat wastes into water; and noxious and dangerous fumes into the air. ... The rational person finds that his or her share of the costs of the wastes discharged into the commons is less than the cost of purifying the wastes before releasing them. Since this is true for everyone, we are locked into a system of "fouling our own nest," so long as we behave only as independent, rational, free-enterprisers.[2]

FIGURE 1–8
If too many cows graze a common pasture, the range is depleted and all the cows die. In recent decades, overgrazing and natural drought have led to repeated famines in North Africa. *(Jason Laure)*

The tragedy of the commons operates on many levels. The dilemma of an individual farmer on a commons is mirrored by local, regional, and national activities. On a much larger scale, the current economy of the United States depends on extensive fossil fuel use, but release of carbon dioxide when these fuels are burned may contribute to global climate change.

Risk Assessment

If someone buries a carcinogenic chemical in a landfill, it may leach into the soil and pollute ground water. If people drink the water, some will develop cancer but many will remain healthy. Thus, the pollutant poses a risk, but not a certainty, of causing cancer in an individual. This element of probability has made the assessment of risk a significant and controversial issue in public policy.

---

[1] Garrett Hardin, "The Tragedy of the Commons," *Science, 162* (1968), 1243.

[2] Ibid.

Hypothesis, Theory, and Law

On an afternoon field trip you may find several different types of rocks or watch a river flow by. But you can never see the rocks or river as they existed in the past or as they will exist in the future. Yet a geologist could tell you how the rocks formed millions or even a few billion years ago and could predict how the river valley might change in the future.

Scientists not only study events that they have never observed and never will observe, but they also study objects that can never be seen, touched, or felt. In this book we examine the core of the Earth 6400 kilometers beneath our feet, even though no one has ever visited it and we are certain that no one ever will.

Much of science is built on inferences about events and objects outside the realm of direct experience. An inference is a conclusion based on thought and reason. How certain are we that a conclusion of this type is correct?

In science, inferences are called laws, theories, or hypotheses, depending on the degree of certainty. A **law** is a formal statement of the way in which events always occur under given conditions. It is considered to be factual and always correct. A law is the most certain of scientific statements. For example, the law of gravity states that all objects are attracted to one another in direct proportion to their masses. We cannot conceive of any contradiction of this principle, and none has been observed. Hence, the principle is called a law.

A **theory** is less certain than a law. It is an interpretation or explanation of some aspect of the world that is supported by experimental or factual evidence but is not so conclusively proved that it is accepted as a law. For example, the theory of plate tectonics states that the outer layer of the Earth is broken into a number of plates that move horizontally relative to each other. As you will see, this theory is supported by many observations and seems to have no major inconsistencies.

A **hypothesis**, or a **model**, is weaker than a theory. It is a tentative explanation of observations that can be tested by comparing it with other observations and experiments. Thus, a hypothesis or model is a rough draft of a theory that is tested against the facts. If it explains some of the facts but not all of them, it must be altered, or if it cannot be changed satisfactorily, it must be discarded and a new hypothesis developed.

Scientists develop hypotheses and theories according to a set of guidelines known as the **scientific method**, which involves three basic steps: (1) observation, (2) forming a hypothesis, and (3) testing the hypothesis and developing a theory.

Observation

All modern science is based on observation. Suppose that you observed an ocean wave carrying and depositing sand. If you watched for some time, you would see that the sand accumulates slowly, layer by layer, on the beach. You might then visit Utah or Nevada and see cliffs of layered sandstone hundreds of meters high. Observations of this kind are the starting point of science.

Forming a Hypothesis

Simple observations are only a first step along the path to a theory. A scientist tries to organize observations to recognize patterns. You might notice that the sand layers deposited along the coast look just like the layers of sand in the sandstone cliffs. Perhaps you would then infer that the thick layers of sandstone had been deposited on an ancient beach. You might further conclude that, since the ocean deposits layers of sand slowly, the thick layers of sandstone must have accumulated over a long time.

If you were then to travel, you would observe that thick layers of sandstone are abundant all over the world. Because thick layers of sand accumulate so slowly, you might infer that a long time must have been required for all that sandstone to form. From these observations and inferences you might form the hypothesis that the Earth is old.

Testing the Hypothesis and Forming a Theory

Theories differ widely in form and content, but all obey four fundamental criteria.

1. A theory must be based on a series of confirmed observations or experimental results.
2. A theory must explain all relevant observations or experimental results.
3. A theory must not contradict any relevant observations or other scientific principles.
4. A theory must be internally consistent. Thus, it must be built in a logical manner so that the conclusions do not contradict any of the original premises.

Most theories can be used to predict events that have not yet been observed, and if the theory is a good one, the predictions will be correct. When first proposed in the late 1700s, the hypothesis that the Earth is old was based only on the observation that sand layers accumulate slowly. In the past 200 years, results of many different measurements and experiments have proved consistent with this hypothesis. Today the idea that the Earth is old is a firmly grounded theory.

It is impossible to live without risk. People drown in their bathtubs, are swept away by tornados, or are struck by lightning. However, each of us makes personal decisions on the magnitude of the risks that we choose to accept. Some people live cautiously and seldom leave their homes, others climb rock cliffs or ski avalanche-prone mountains, and still others elect cosmetic surgery despite the risks inherent in any operation. Other risks are imposed on us by society. Nuclear power plants, the threat of nuclear war, and the introduction of industrial poisons into the environment are imposed on society as a whole.

To understand the factors involved in environmental risk assessment, consider one experiment and two observations.

Experiment. Scientists suspect that many synthetic compounds in our food, air, and water may cause cancer in humans. Imagine that you want to test the carcinogenic properties of red food dye used to give the filling in jelly doughnuts that bright, day-glo color. Because experiments with humans are unethical, scientists frequently study animals, often laboratory rats. In the initial experiment scientists feed the rats a concentrated diet of the dye, perhaps a quantity equivalent to eating 10,000 jelly doughnuts a day. Imagine that in this experiment one out of every five rats develops cancer (Fig. 1–9). Then the dose is reduced to the equivalent of 1000 jelly doughnuts per day, and on this dose one rat in 50 develops a tumor. At a dose equivalent to 100 jelly doughnuts per day, one rat in 500 gets a tumor. Now the research is becoming more difficult. The experiment is using too many rats, and the dose is still too high—no one eats 100 jelly doughnuts per day. Ultimately, the important question is whether it is dangerous to eat one a day. The data obtained thus far are plotted in Figure 1–9. In this example, the three points lie in a straight line.

To estimate the danger of eating one jelly doughnut per day, we make two assumptions: (1) It is valid to extrapolate the straight line to the one-doughnut-per-day equivalent dose. This assumption shows that if 50,000 rats each ate one doughnut per day, one would develop cancer. (2) After correction for differences in body weight, the chances of a tumor in rats and in people are equal. If this assumption is correct, then 20 cancers will develop for every million people who eat a jelly doughnut every day.

These two assumptions may or may not be valid and, as a result, conclusions based on this type of study are uncertain. Yet decisions must be made. Someone offers you a jelly doughnut. Do you eat it or politely refuse? Government regulators must analyze the data and decide when to ban a food additive and when to permit its use.

First Observation. In 1974, the Stauffer Chemical Company opened a factory near Cleveland, Ohio, to manufacture agricultural fungicides. Engineers at the plant drilled a 2000-meter-deep well into bedrock and pumped hazardous wastes into the well. Such injection wells are legal because the wastes are discharged below ground-water reserves and thus are isolated from the biosphere. In 1986, a moderate-sized earthquake occurred near the well, although the rock in the region had previously been stable and no earthquakes had been observed in recent times (Fig. 1–10). Geologists suspect that the waste fluids were forced into small cracks in the underground rock, thereby enlarging the cracks and creating stress. The earthquake occurred when the stress became greater than the strength of the rock. This suspicion is supported by the observation that earthquakes have occurred around injection wells near Denver, Colorado, and in other locations. Furthermore, the earthquakes near Denver stopped when engineers stopped injecting fluids into the wells. However, another possibility must be considered. Perhaps these earthquakes were natural events and were not triggered by the injection wells.

Second Observation. During the latter half of the nineteenth century, chemists discovered several synthetic fabric dyes. In 1890, a company in Germany started producing large quantities of one of these dyes, 2-naphthylamine. For 5 years, people worked in the factory, prospered, and remained healthy. Then one of the workers developed bladder cancer. During the following 10

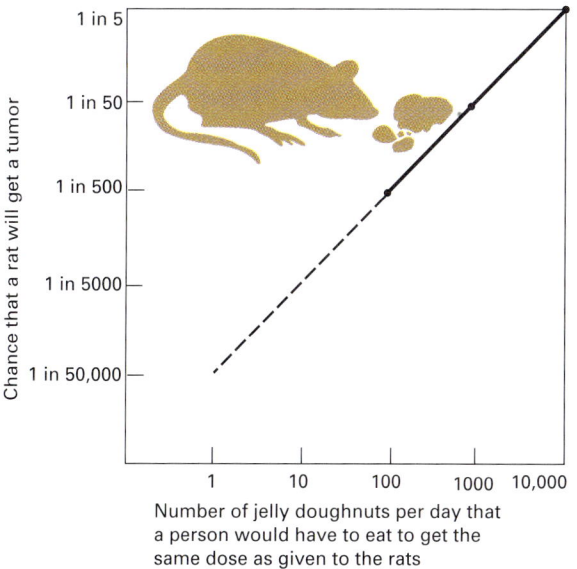

FIGURE 1–9
An attempt to extrapolate a laboratory study on rats to determine the probability that a food additive will cause cancer.

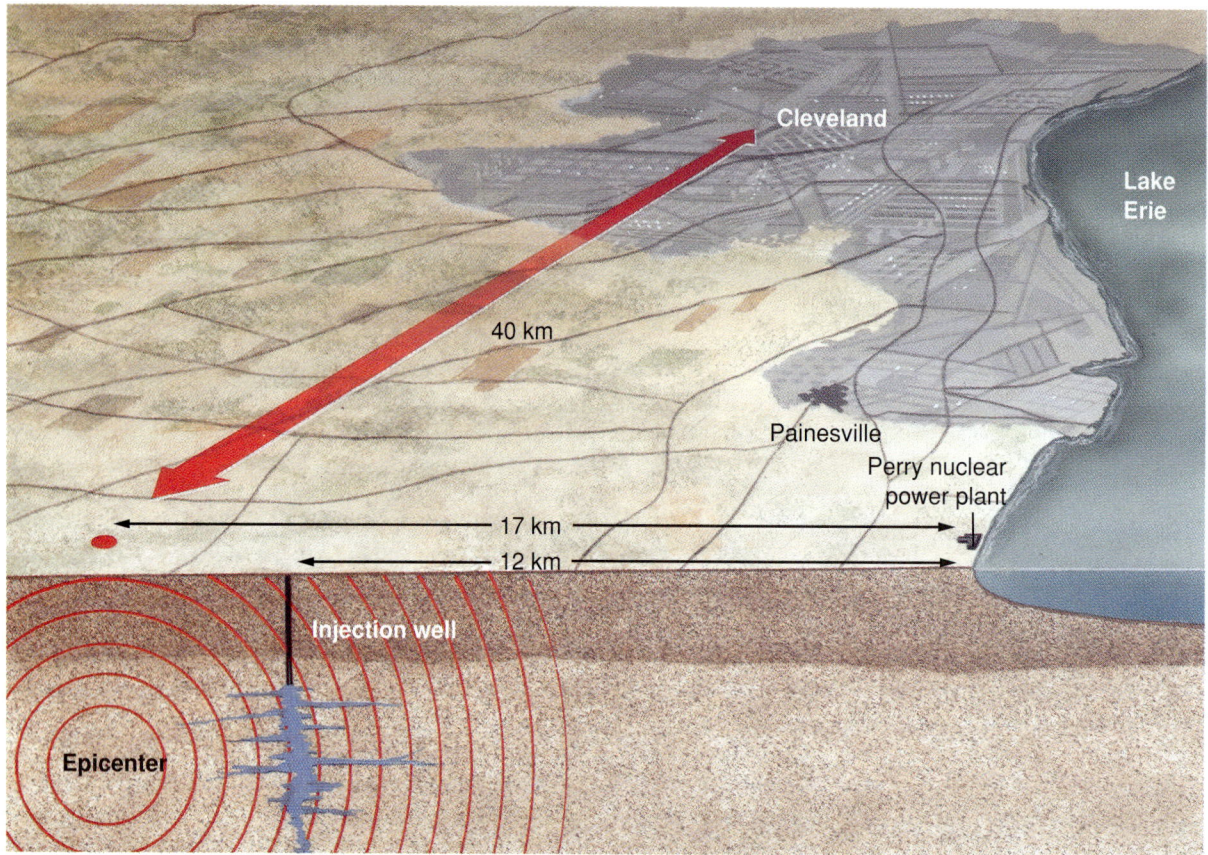

FIGURE 1–10
Locations of the Stauffer Chemical Company deep injection well and an adjacent earthquake. The area had experienced no historical earthquake activity, and geologists suspect that the injected fluids may have initiated the quake. However, this hypothesis is difficult to prove.

years, 10 percent of the factory workers contracted bladder cancer. Thirty years after the plant had begun operations, 95 percent of the original workers had died of bladder cancer. Bladder cancer is a rare disease. The overwhelming mortality convinced health officials that 2-naphthylamine causes cancer, although the disease may not appear for a decade or more. Similar observations of delayed response to carcinogens have been made for chimney soot, asbestos, and other materials. Thus, exposure to a carcinogen may be lethal to a large proportion of the people exposed, even though no effects are immediately discernible.

Because it is often difficult and frequently impossible to perform controlled experiments on the Earth or on people, many hypotheses in environmental geoscience are difficult to test. Yet people need to respond to problems or to potential problems. One argument, called the **precautionary principle**, is that "it is better to be safe than sorry." Proponents of the precautionary principle argue that people commonly must act on the basis of incomplete proof. If a mechanic told you that your brakes were faulty and likely to fail within the next 1000 miles, you would recognize this as an opinion, not fact. Yet would you wait for proof or replace the brakes now?

In the case of the Stauffer Chemical Company injection well, a nuclear power plant is situated 5 kilometers from the well. Proponents of the precautionary principle argue that because there is good evidence that injection wells have caused earthquakes at several sites, and because an earthquake might trigger a nuclear accident, waste disposal in this well in Ohio should be prohibited. In the case of human exposure to chemicals, proponents of the precautionary principle argue that animal tests are uncertain and health effects may not become apparent for decades, after many people have died. Therefore, public safety demands regulation even if the initial data do not prove a link between a compound and disease in humans.

On the other hand, if we pursued the precautionary principle to the extreme, we would regulate human activity so severely that agriculture and industry would grind to a halt. People would suffer from lack of essential commodities and ultimately starve.

Threshold Effects

One problem that arises when we seek a balance between caution and risk is that the relationships between environmental disturbances and the effects on humans and global ecosystems are neither simple nor linear. Often, soils are plowed, deep wells dug, and pollutants released, and the environment is barely altered. But many systems exhibit **threshold effects**. Once the threshold is reached, a small additional disturbance may cause a large change.

As an example of a threshold effect, consider the potential rise in sea level from melting of glaciers. Ice melts at 0°C. Imagine that during the summer, the average temperature on a continental glacier is −2°C. If the temperature increases by 1.5°C, nothing much happens because the temperature rises to only −0.5°C, and the glacier remains frozen. If you did not know the melting point of ice and observed that a temperature rise of 1.5°C caused little effect, you might extrapolate that another 1.5°C temperature rise would also cause little change. However, this prediction would be wrong because it ignores the fact that the first temperature rise placed the ice at a threshold. Another 1.5°C rise would melt the ice. Then sea level would rise and flood coastal cities.

Scientists know that thresholds exist in many systems. In a recent study, climatologists interpreted the climatic history of Greenland by studying a long cylinder of ice, called a core, taken by drilling into the thickest part of the Greenland ice sheet (Fig. 1–11).[3] They learned that significant climate changes have occurred in periods of time as short as 3 to 10 years. The author of the study concluded that "what this shows us is that there are big thresholds or instabilities and we don't know what those are yet. So maybe there are some surprises out there."

1.4 The Earth Summit

In June 1992, delegates from 170 nations met in Rio de Janeiro, Brazil, to attend the United Nations Conference on Environment and Development, commonly called the **Earth Summit**. For the first time in history, national governments acknowledged that environmental degradation is a global problem and that solutions require international cooperation. The largest gathering of world leaders ever—118 heads of state—attended the final ceremonies (Fig. 1–12).

The meeting raised global environmental awareness and produced an unprecedented agreement that humans must work together to protect the environment. The **Rio Declaration** stated that economic growth must follow the guidelines of **sustainable development**. Sustainable development requires that people look at both the short- and long-term consequences of their actions. Although we recognize the need to build strong economies today, we must also be careful not to squander our children's resources.

However, representatives at the Rio meeting failed to agree on many specific issues and at many times the meeting was argumentative and fractious. A fundamental

[3]Kendrick Taylor, "Ice core shows speedy climate change," as quoted in *Science News*, 142 (1992), 404.

FIGURE 1–11
Scientists remove a portion of the ice core from a glacier in Greenland. Studies of ice cores provide information about past climate. (*R. Gaillarde/Gamma-Liaison*)

FIGURE 1–12
George Bush addresses the United Nations Conference on Environment and Development in Rio de Janeiro, 1992.
(*Wide World Photos*)

disagreement arose between the rich and poor countries. The rich countries are often called the **developed** or **industrial countries**, and the poorer nations are called the **developing** or **less developed countries**, depending on their economic growth. Leaders of developing and less developed countries insisted that environmental protection is meaningful only if the living standards of the inhabitants improve. Brazilian President Collor de Mello said, "You can't have an environmentally healthy planet in a world that is socially unjust." The developed countries recognized the problem of unequal distribution of wealth but refused to give sufficient economic support to resolve the inequity. They also refused to commit to specific target dates for reducing their own industrial pollution.

Agenda 21

Despite these disagreements, the delegates compiled **Agenda 21**, an 800-page document intended to set up a strategy for reducing environmental degradation. It outlines 115 global projects to preserve forests, decrease poverty, provide safe drinking water, and respond to many other urgent needs. However, it establishes no mechanism to fund the proposals. More specific treaties on climate, biodiversity, and use of forest resources also failed to meet environmentalists' hopes.

Most of us who live in the developed nations enjoy unprecedented wealth. Heat and lighting are available at the flick of a switch, sanitary drinking water is piped directly to our homes, we drive or fly across the country in upholstered luxury, and our diet is richer and more varied than that available to royalty a century ago. Yet billions of people, some in our own country, live in poverty and die from malnutrition, poor drinking water, or inadequate medical attention. In addition, problems of resource depletion, pollution, and climate change endanger everyone's well-being.

The Rio Conference has taught us that there is no easy consensus. But surely one step forward is to study the nature of environmental problems and to understand the choices that we face.

SUMMARY

Geology is the study of the Earth: the materials of which it is made, the physical and chemical changes that occur on its surface and in its interior, and the history of the planet and its life forms from its origin to the present. Geologists study both **internal** and **surface processes**. **Environmental geoscience** is the study of geology as it applies to living organisms and the quality of life on our planet. It includes interactions between human activities and geologic phenomena and loss of resources by consumption, dispersal, and pollution.

An **ecosystem** is a system formed by the interactions of individual organisms with each other and with their environment. Ecosystems are nearly self-contained, so that the exchange of nutrients within the system is much greater than exchanges with other systems.

Cost–benefit analysis balances the cost of environmental protection against the cost of living in a degraded environment. Costs of environmental degradation, called **externalities**, are an important component of cost–benefit analysis. One problem with cost–benefit analysis is balancing long-term and short-term considerations. The **tragedy of the commons** arises when an individual or group benefits by an act that harms the general population and therefore eventually harms the individual.

Environmental geoscientists conduct risk assessment because it is often difficult to prove that an environmental change causes a specific effect. **Threshold effects** occur when a small additional perturbation causes a large change. The **Earth Summit** achieved a consensus that global cooperation is needed to solve environmental problems, but the final agreements contained few binding treaties.

KEY TERMS

Geology 4
Environmental geoscience 5
Geologic resource 5
Pollution 5
Toxic substances 5
Carcinogen 5
Mutagen 5

Systems 6
Ecosystem 6
Cost–benefit analysis 7
Externalities 8
Tragedy of the Commons 9
Precautionary principle 12
Threshold effects 13

Earth Summit 13
Sustainable development 13
Developed countries 14
Industrial countries 14
Developing countries 14
Less developed countries 14

REVIEW QUESTIONS

1. Define geology and environmental geoscience.
2. Differentiate between a surface process and an internal process in geology. Give an example of each.
3. Explain how resources are diminished by consumption, dispersal, and pollution.
4. What is an ecosystem?
5. Explain the basic concepts of cost–benefit analysis.
6. Discuss the tragedy of the commons.
7. Explain why risk assessment is necessary in analysis of environmental problems.
8. What is the precautionary principle? How does it relate to risk assessment?
9. Discuss the achievements and failures of the Earth Summit.

DISCUSSION QUESTIONS

1. Explain how an ecosystem meets the definition of a system given in the text.
2. Identify an ecosystem in your area. Show how it meets the definition given in the text.
3. Identify a source of pollution or environmental degradation in your community. Discuss laws that might alleviate the problem. Describe the costs and benefits of your proposed law.
4. The Natural Resources Defense Council (NRDC) is a private organization that protects environmental quality by filing lawsuits against companies or individuals who are allegedly causing excess pollution. Former Environmental Protection Agency (EPA) administrator William Ruckelshaus has criticized the NRDC for failing to account for the costs of the regulation that their suits have mandated. D. Doniger, a lawyer for the NRDC, replied, "We take the view that there are rights involved here . . . rights to be protected from threats to your health regardless of the cost involved." Prepare a class debate. Have one side argue Mr. Ruckelshaus's position and the other argue Mr. Doniger's position.
5. On February 26, 1972, heavy rains destroyed a dam across Buffalo Creek in Logan County, West Virginia. The resulting flood killed 75 people and rendered 5000 homeless. The dam had been built from unstable coal mine refuse, or "spoil." A United States Geological Survey report had warned of the instability of many dams in Logan County, especially the one at Buffalo Creek. Soil stabilization and reclamation programs directed at mine spoil dams would have added to the cost of producing coal. (a) How would you relate the concept of economic externalities to this tragedy? (b) Suggest legislation designed to prevent future dam disasters. (c) How would your legislation affect the price of coal?
6. If you were the mayor of a small town, and if a prosperous factory, the largest single employer in the town, were illegally dumping untreated wastes into a stream, what action would you recommend? What if the factory were barely profitable?
7. If the burden of environmental control falls heavily on particular segments of society, do you think it would be fair for the government to provide compensation? Would you favor granting such compensation to workers who lose their jobs? To companies whose costs for environmental clean-up are high? To stockholders whose equities are reduced in value? For each of these segments, what kinds of abuses are possible? What benefits to the environment would such compensations provide?
8. Suppose that you were a reporter assigned to compare the cost of stopping the pollution of a local river with the economic cost of the pollution. What sources of information would you seek out? Which cost would be more difficult to estimate? Defend your answer.
9. Suggest one action that you personally might take to help preserve the environment. How could you assess the importance of your own action? How would you estimate the number of people who would have to join you before the action would have a significant environmental impact?
10. Explain why the need for environmental laws arises out of the tragedy of the commons.
11. In the United States, about 55 percent of aluminum beverage cans are recycled. In 1993, scrap aluminum was worth $0.20 per pound. Estimate the percentage of aluminum cans that would be recycled if the scrap value were zero, $0.50 per pound, $1.00 per pound, and $5.00 per pound. Discuss your estimates with your classmates.
12. In the United States today, many farmers are losing soil to erosion. Yet agricultural yields remain high because the losses can be balanced by increased application of fertilizers. Is this an example of the tragedy of the commons? If your answer is no, defend your reasoning. If your answer is yes, what is the commons, and what is the nature of the tragedy?
13. Do you think that it would be reasonable to eliminate all risks generated by technology? Defend your answer.

CHAPTER 2
The Earth in Time

2.1 The Origins of the Universe, Solar System, and Earth
2.2 Formation of the Solar System and Earth
2.3 The Modern Earth
2.4 A Brief History of Life

Historians look at past events to understand human history. In a similar manner, geologists look back in time to the earliest days of our planet, before continents, oceans, or even a solid crust had evolved. From that beginning they reconstruct the Earth's evolution: the growth of mountain ranges, oceans, atmosphere, and climate. By studying its history, we learn that the Earth has changed continuously and that we can therefore expect change in the future. By understanding the events that cause change, we hope to control and modify our activities to minimize the negative impacts on our own environment.

Exposed rocks in the walls of Grand Canyon record events that occurred during the latter half of Earth history.

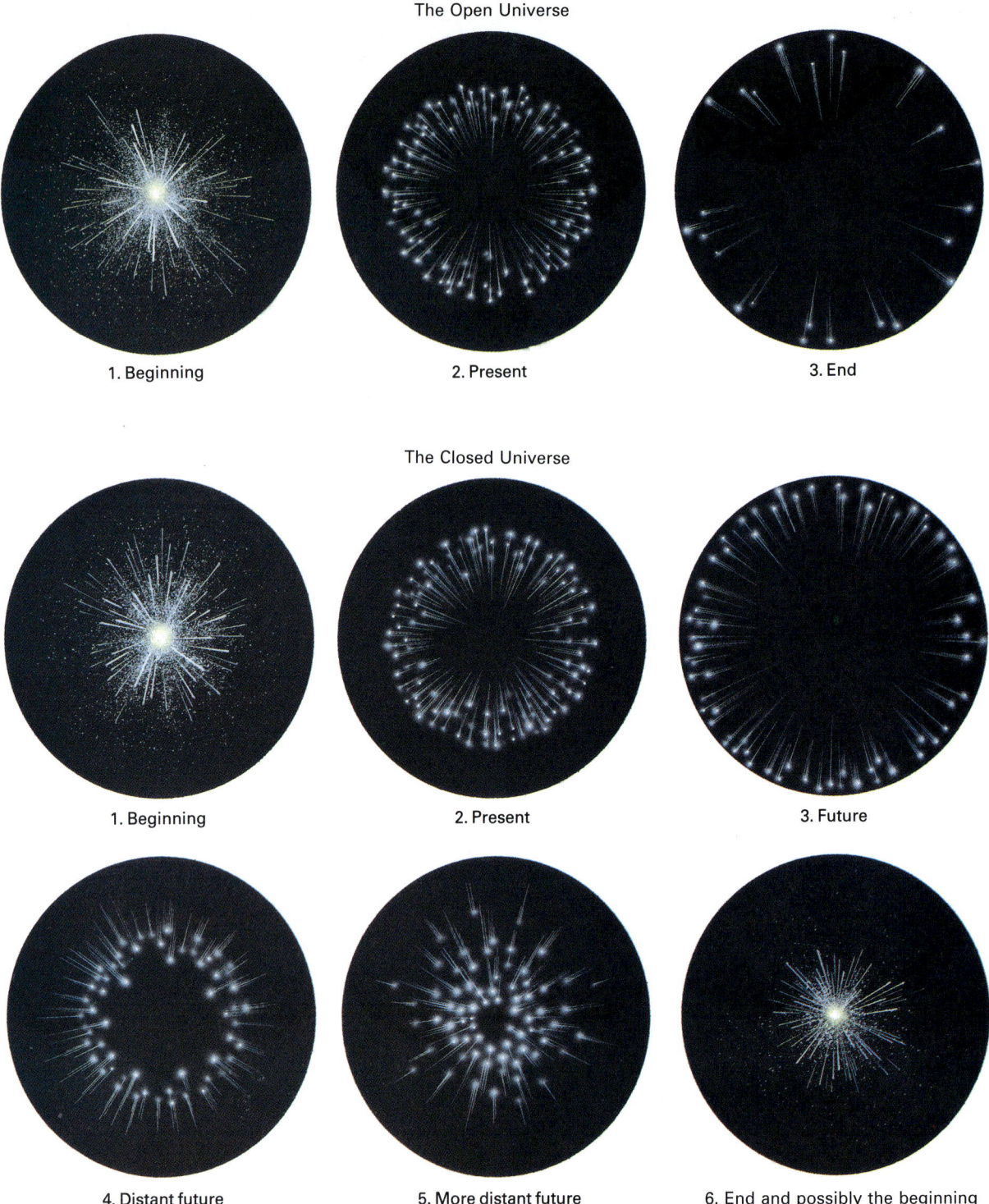

FIGURE 2–1
A schematic representation of two possible cosmologies, the open Universe and the closed Universe.

2.1 The Origins of the Universe, Solar System, and Earth

In 1929, the astronomer Edwin Hubble observed that all galaxies in the Universe are speeding away from each other. He measured their speeds and the distances between them and retraced their paths back to a time when the entire Universe was packed into a single, infinitely dense point. Most modern cosmologists think that the Universe began when this point exploded in a cataclysmic event called the **big bang**. The big bang was the origin of the Universe and the start of time. It was no ordinary explosion. It is not comparable to a hydrogen bomb or the explosive death of a giant star. This explosion instantaneously created the Universe. Matter, energy, space, and even time came into existence with this single event. Most astronomers date the big bang between 15 and 18 billion years ago.

As the Universe was born, it was so hot that atoms did not exist and matter consisted of a sea of subatomic particles. This expanding sea of matter cooled, and after 1 million years the particles coalesced into atoms, mostly hydrogen and helium. With time, gravitational forces drew the atoms into galaxies and stars.

At present all galaxies are flying away from one another, spreading farther and farther into space. What will happen in the future? Think of a rocket ship taking off from a planet. If the planet's gravitational field is weak enough, the rocket will escape into space and never return. However, if the planet has a large mass and therefore a strong gravitational field, the rocket will fall back to the surface. In the same manner, if the gravitational force of the Universe is sufficient, all the galaxies will eventually slow down, reverse direction, and fall back to the center. This possibility is called the **closed Universe**. This point may then explode again to form a new universe. The process may repeat indefinitely, creating a continuous series of universes. It is possible, however, that the gravitational force of the Universe is not sufficient to stop the expansion, and the galaxies will continue to fly apart forever. Within each galaxy, stars will eventually consume all their nuclear fuel and stop producing energy. As the stars fade and cool, the galaxies will continue speeding outward into the cold void. This scenario is called the **open Universe** (Fig. 2–1). Astronomers are attempting to measure the mass of the Universe in an effort to determine which of the two models is more realistic. However, the measurements are uncertain and the answer remains elusive. If the mass is great enough, then the force of gravity will eventually cause all matter to collapse back to a single point. But if the mass is less than a certain value, the Universe will continue to expand forever.

2.2 Formation of the Solar System and the Earth

The Milky Way is one of billions of galaxies that formed after the big bang. Five billion years ago, matter in a small part of the Milky Way Galaxy drew closer together, forming an immense cloud of dust and gas rotating slowly in space. Small gravitational attractions drew the dust and gas into a sphere (Fig. 2–2). Some astronomers suggest that a nearby star may have exploded and that the shock wave triggered the condensation (Fig. 2–3). As gravity pulled the cloud into a smaller, denser mass, its rotation accelerated, and the sphere spread into a disk, as shown in Figure 2–2C.

More than 90 percent of the matter in the disk coalesced at the center to form the **protosun**, the earliest form of the Sun. As matter sped inward, particles smashed together, producing heat. However, the protosun was not a true star because it did not yet generate energy by nuclear fusion.

Formation of the Sun and the Planets

Heat from the protosun warmed the inner region of the disk. Then, as gravitational collapse became nearly complete, the disk cooled. Gases condensed to form small aggregates, much as raindrops or snowflakes form when moist air cools in the Earth's atmosphere. Over a time span of 10,000 to 100,000 years, the aggregates grew into small rocky spheres, which, in turn, collided and grew into mini-planets called **planetesimals**, ranging in size from a few kilometers to about 100 kilometers in diameter. The planetesimals then coalesced into a few large planets, including the Earth. The Earth formed about 4.6 billion years ago.

The protosun was changing at the same time that planets were evolving. Gravitational attraction pulled its gases inward, creating extremely high pressure and temperature. The core of the protosun became so hot that hydrogen nuclei combined in a process called **nuclear fusion**. The solar fusion released vast amounts of nuclear energy in a process comparable to the continuous explosion of millions of hydrogen bombs. The onset of nuclear fusion was the birth of the modern Sun.

At first, the cloud that evolved into the Solar System was homogeneous; that is, it was the same throughout. However, as the Sun became hotter, many light gases, such as hydrogen and helium, boiled away from the inner Solar System and collected in the frozen outer regions. As a result, the four planets closest to the Sun—Mercury, Venus, Earth, and Mars—are now rocky with metallic centers (Fig. 2–4a). They lost most of their gases during

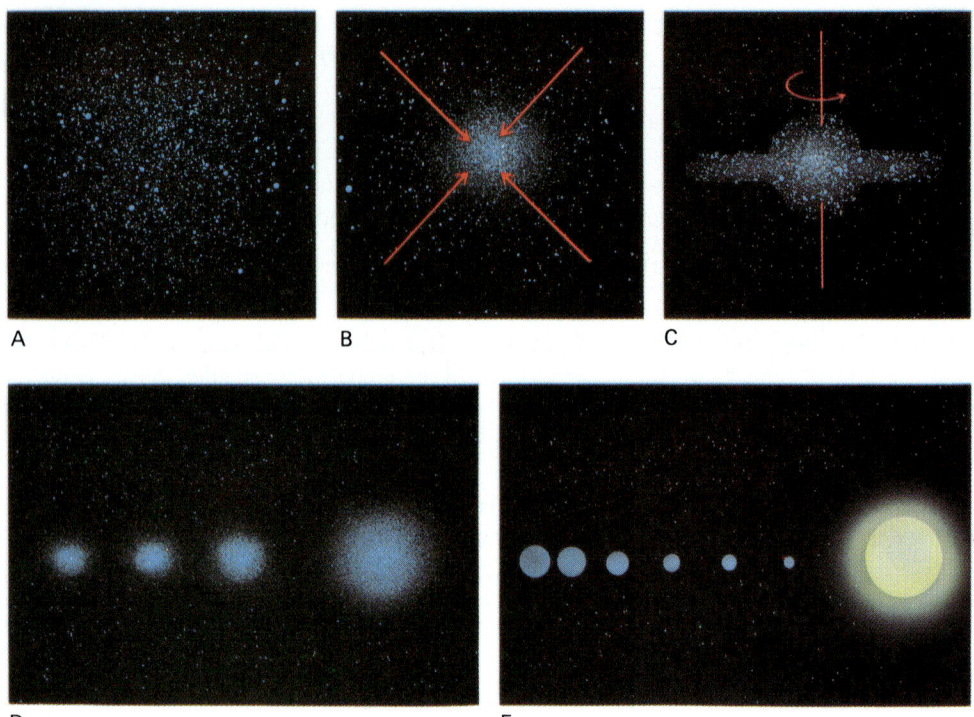

FIGURE 2-2
Formation of the Solar System. (A) The Solar System was originally a diffuse cloud of dust and gas. (B) The dust and gas began to coalesce due to gravity. (C) The shrinking mass began to rotate and formed a disk. (D) The mass broke up into a discrete protosun orbited by large protoplanets. (E) The Sun heated up until fusion temperatures were reached. The heat from the Sun drove most of the hydrogen and helium away from the closest planets, leaving small, solid cores behind. The massive outer planets remain composed mostly of hydrogen and helium.

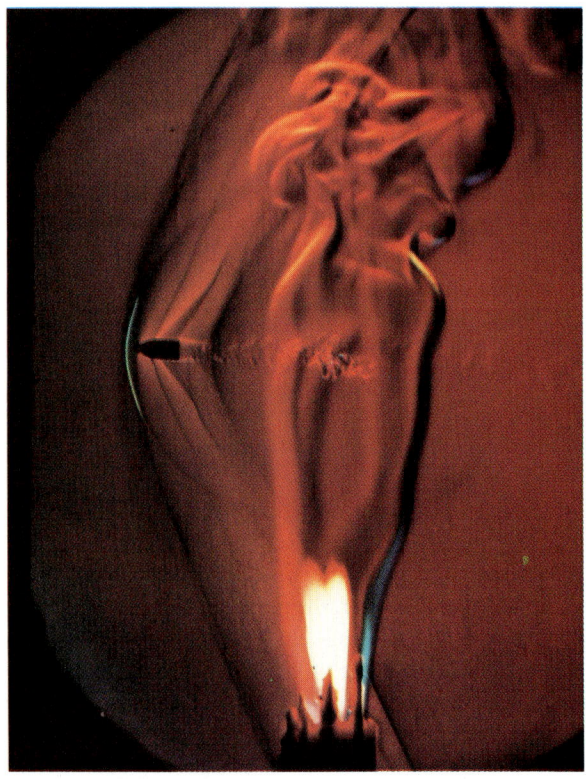

FIGURE 2-3
A shock wave forms as a bullet passes through hot gases generated by a burning candle. The leading edge of the shock wave is on the left. According to one theory, a shock wave created by an exploding star initiated the collapse of the cloud of dust and gas and thereby was the first step in the formation of the Solar System. *(Harold E. Edgerton, MIT; courtesy of Palm Press)*

A   B

FIGURE 2–4
(A) Mercury is a small planet close to the Sun. Because of this proximity, most of its lighter elements have long since been boiled off into space, and today its surface is solid and rocky. (B) Jupiter, on the other hand, is composed mainly of gases and liquids, with a small solid core. This image is a close-up of its turbulent atmosphere. *(NASA)*

this early event. These four are called the **terrestrial planets** because they are all "Earthlike."

In contrast, the four outer planets—Jupiter, Saturn, Uranus, and Neptune—are composed mostly of liquids and gases such as hydrogen, helium, water, ammonia, and methane (Fig. 2–4B). These planets are called **Jovian planets** because they are all similar to Jupiter. Pluto, the outermost planet, is anomalous. Figure 2–5 is a schematic representation of the modern Solar System.

The Early Earth

The early Earth grew by multiple collisions of smaller bodies drawn together by gravity. The collisions among the speeding planetesimals and other bits of matter generated heat. As the Earth grew larger, its surface was further heated by an intense bombardment of rock, comets, and other debris floating about in the early Solar System. At the same time, Earth's natural radioactivity released additional heat. A thick layer of surface rocks insulated inner parts of the Earth, and the heat accumulated.

As a result, the Earth became so hot that it began to melt a few hundred million years after it formed. Many geologists now think that the early Earth melted completely, forming a sphere of molten liquid rock. Others believe that only inner portions melted, whereas an outer layer remained solid although it became very hot. In either case, **in the Earth's molten or near-molten state, heavy iron and nickel sank toward its center while lighter material floated upward toward its surface.** At the same time, volatile elements and compounds such as hydrogen, helium, and lesser amounts of water, carbon

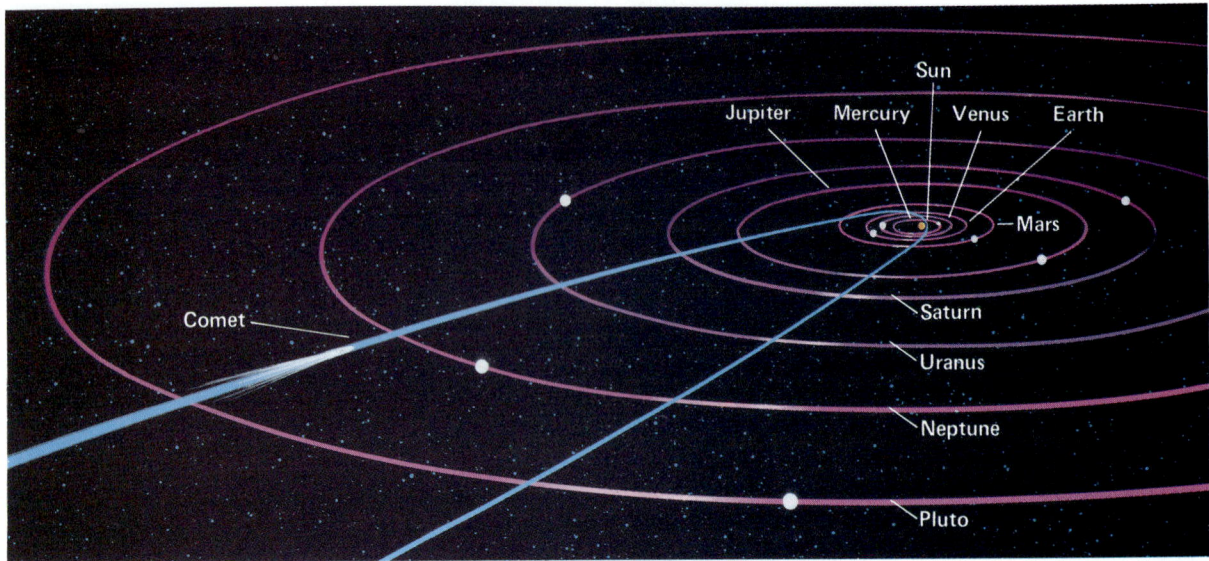

FIGURE 2–5
A schematic view of the Solar System.

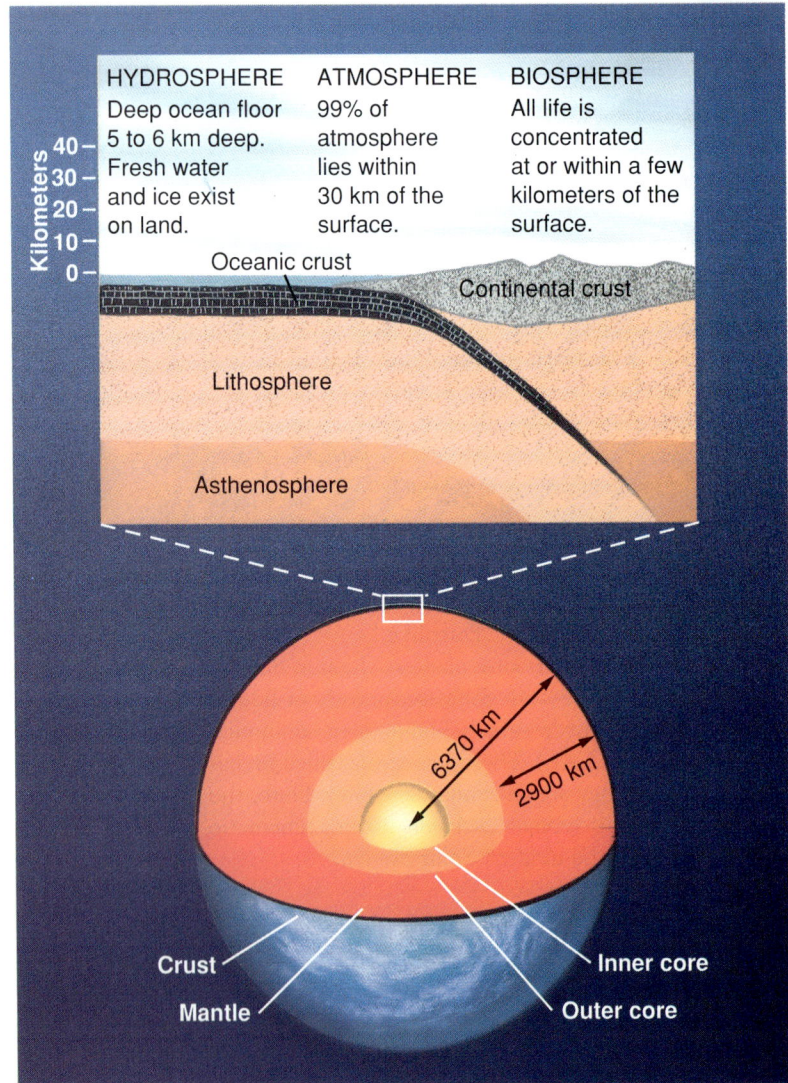

FIGURE 2–6
A schematic view of the Earth's layers. The inset is a view of the outer layers.

dioxide, molecular oxygen, and nitrogen escaped from the molten rock to form Earth's earliest atmosphere.

2.3 The Modern Earth

As a result of its early separation into layers, the modern Earth is a layered planet (Fig. 2–6). The innermost zone of the **solid Earth** is a dense, hot **core** consisting mostly of iron and nickel. The inner core is solid while a larger outer core is liquid. The light rocks of the **crust** form a thin surface layer. The thick rock layer lying between the crust and the core is denser than the crust but less so than the core, and is called the **mantle**.

The outermost layer of the solid Earth, including both the crust and the upper mantle, consists of cool, solid, strong, and relatively rigid rock. This layer is called the **lithosphere** (Greek for "rock layer"). It extends from the Earth's surface to a depth of about 100 kilometers. Immediately beneath the lithosphere, at a depth of about 100 kilometers, the mantle's physical properties suddenly change; here the mantle rock becomes hot, plastic, and weak. This layer of hot, plastic rock extends from a depth of 100 kilometers to about 350 kilometers beneath the Earth's surface and is called the **asthenosphere** (Greek for "weak layer"). Below the asthenosphere, mantle rock again becomes relatively strong and rigid. **The asthenosphere is so weak and plastic that the lithosphere floats on it much as blocks of wood float in a tub of honey.** The asthenosphere is solid, not liquid like honey, but because of its plastic nature, it can flow.

According to the theory of **plate tectonics**, the lithosphere is segmented into seven major plates and several smaller ones. These lithospheric plates are packed tightly together like the segments of a turtle shell (Fig. 2–7). Each plate is about 100 kilometers thick and may be thousands of kilometers in width and length. **The plates move across the Earth's surface at rates up to 18 centimeters per year by floating on and gliding over the hot, plastic asthenosphere.** Commonly two adjacent plates move in different directions. As you might expect, a boundary between two plates is geologically active. When two plates rub, jerk, or jump past one another, the Earth shakes. This vibration is called an **earthquake**. In some regions near plate boundaries, parts of the lithosphere and the underlying asthenosphere melt. The resulting liquid rock is called **magma**. Large quantities of magma rise through the lithosphere and pour out onto the Earth's surface in **volcanic eruptions**.

The hydrosphere, atmosphere, and biosphere lie in

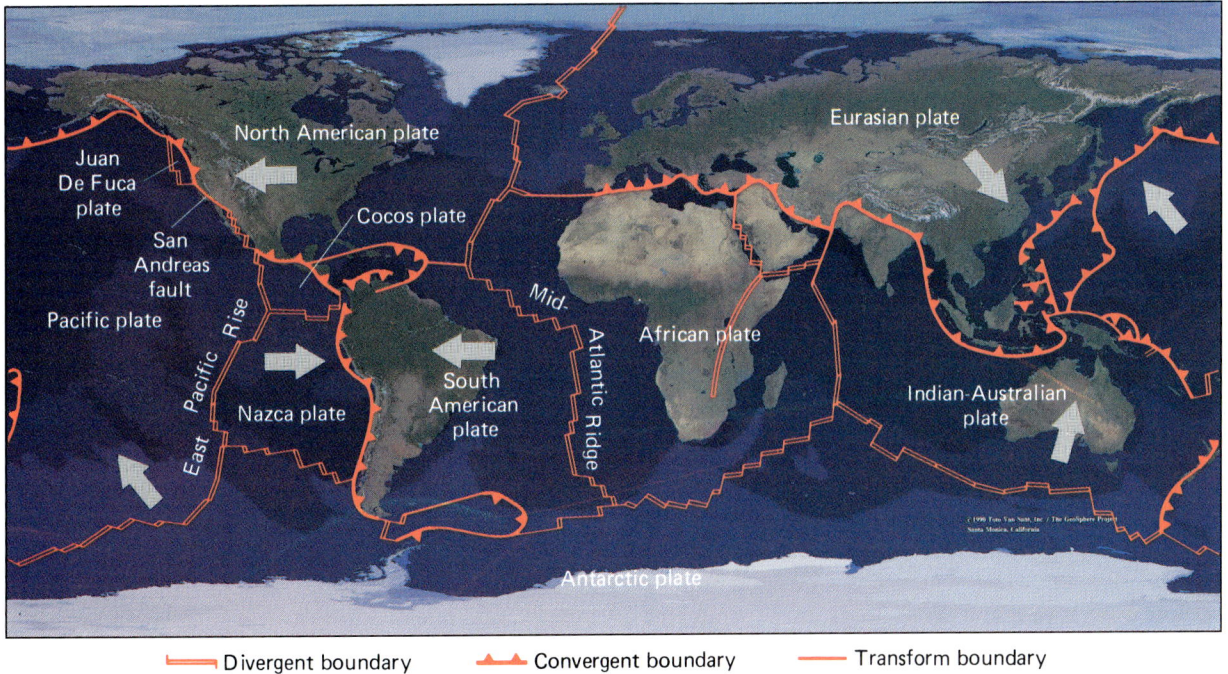

FIGURE 2–7
Major plates of the world.

a thin zone near the surface of the solid Earth. The **hydrosphere** includes water in streams, lakes, and oceans; in the atmosphere; and frozen in glaciers. It also includes ground water, which soaks soil and rock to a depth of 2 or 3 kilometers.

The **atmosphere** is a mixture of gases, mostly nitrogen and oxygen. It is held to the Earth by gravity and thins rapidly with altitude. Ninety-nine percent is concentrated in the first 30 kilometers, but a few traces remain even 10,000 kilometers above the Earth's surface.

The **biosphere** is the thin zone of life. Organisms inhabit the uppermost solid Earth, the hydrosphere, and the lower parts of the atmosphere. Although some bacteria live in rock to depths of several kilometers, specialized organisms live in the deepest oceanic trenches, and a few windblown microorganisms are found at heights of 10 kilometers or more, most creatures live in a very thin layer at the Earth's surface. Land plants grow on the surface, their roots penetrating a few meters into the soil. Most animals live on the surface, fly a kilometer or two above it, or burrow a few meters underground. Sea life also concentrates near the surface of the oceans, where sunlight is available.

If you could drive a magical vehicle from the center of the solid Earth to the outer fringe of the atmosphere at 100 kilometers per hour, you would be within the Earth for 64 hours. In another 20 minutes you would pass through nearly all of the atmosphere and would enter the rarefied boundary between Earth and space. You would pass most living organisms in a few seconds and the entire biosphere in 6 minutes.

The separation of the Earth into layers was only one step in the evolution of the planet. Changes in the interior and on the surface created the Earth's earliest environments and then altered them continuously over the great length of geologic time. We can trace three separate stories: evolution of the oceans, atmosphere, and life. Yet these stories aren't separate, for changes in one realm affect the others—rock, water, air, and life are all intertwined.

Evolution of the Oceans

Recall that the solid lithosphere floats on the plastic asthenosphere. The crust beneath ocean basins is different from that of continents. Regions of the Earth covered with oceanic crust float at low elevations because oceanic crust is thin and dense. In contrast, continents float at higher elevations because they are thicker and less dense. Because water runs downhill, the low ocean basins fill with water and the continents protrude above sea level.

The Earth's water comes from three sources:

1. As the early Earth formed, its gravity attracted large amounts of water remaining in the inner Solar System.
2. Comets are composed mainly of water and other light compounds. During the early history of the Solar System, comets and meteorites collided with Earth, adding water.
3. Recall that the early Earth was hotter than the modern Earth. As a result, extensive volcanic activity occurred early in Earth history. The eruptions carried huge amounts of steam to the surface with the fiery magma. Most of this volcanic water collected on the surface during the first half-billion years of Earth's history, although volcanoes continue to emit water today.

Co-evolution of the Atmosphere and Life

All planets and moons of the Solar System evolved from the same mixture of dust and gas, but today, no two have identical atmospheres. Mercury and our Moon have no atmospheres at all. The atmosphere of Venus is mainly carbon dioxide, and that of Jupiter is mostly hydrogen and helium.

The Earth's present-day atmosphere is dramatically different from its original one. Hydrogen and helium are the most abundant elements in the Universe, and the Earth's earliest atmosphere consisted mainly of these two elements. Most of the hydrogen and helium escaped into space shortly after the Earth formed, 4.6 billion years ago. A second atmosphere then accumulated as volcanoes emitted gases trapped within the planet, and the Earth's gravitation attracted other gases from outer space. Scientists have deduced that this second atmosphere consisted mainly of carbon dioxide (CO_2), nitrogen (N_2), and water vapor (H_2O). Oxygen in the form of O_2 was present in trace quantities only. This atmosphere formed within the first 500 million years of Earth history. Most modern organisms would rapidly suffocate and die in such an environment.

The oldest fossils are tiny imprints of bacteria that date back to about 3.5 billion years ago, although life probably originated earlier (Fig. 2–8). Sometime during the next 500 million years, organisms developed the ability to conduct photosynthesis, combining carbon dioxide and water in the presence of sunlight to form glucose (sugar) and oxygen (Fig. 2–9). They used glucose for food and released the oxygen into the atmosphere. In this way, photosynthesizers convert solar energy to chemical energy. As early organisms evolved and multiplied, they released more oxygen into the atmosphere than they consumed.

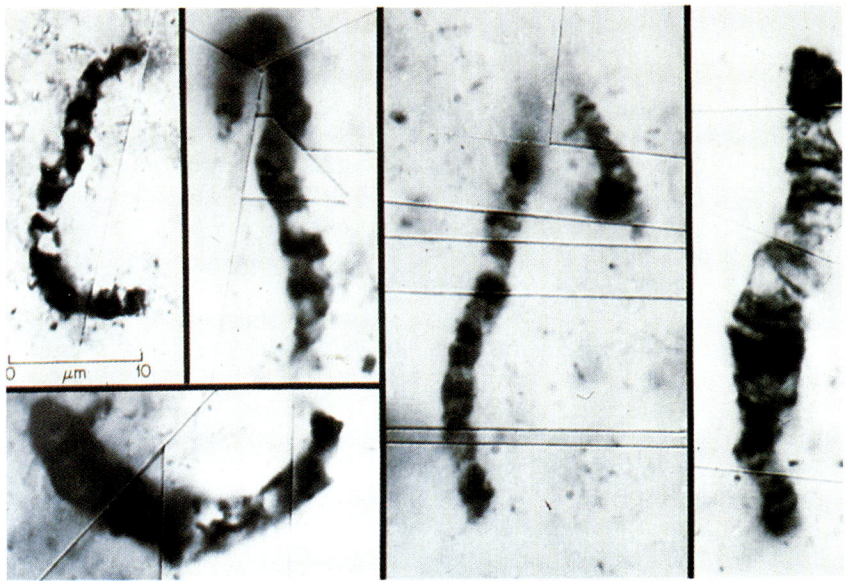

FIGURE 2–8
Tiny, 3.5-billion-year-old, bacteria-like fossils from western Australia. (*J.W. Schopf and B.M. Packer*)

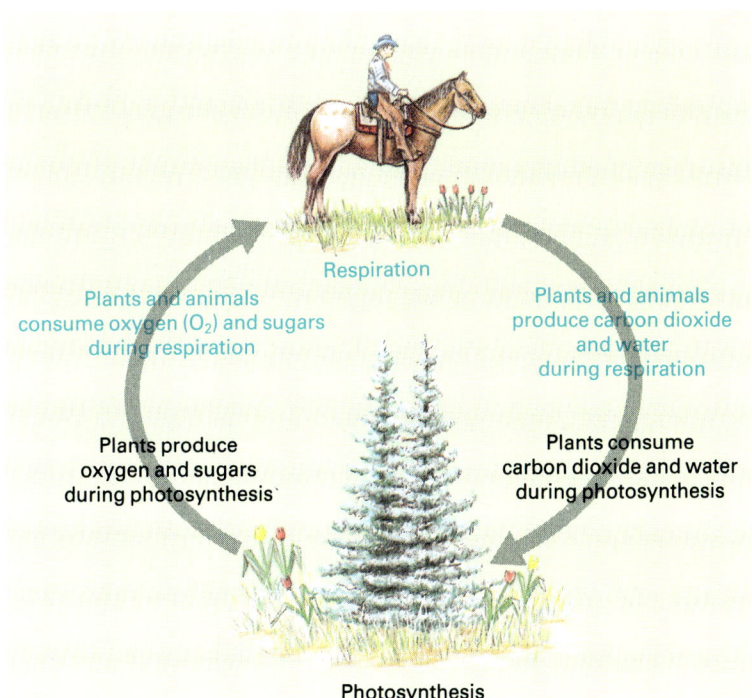

FIGURE 2–9
Early life forms conducted photosynthesis by combining carbon dioxide and water in the presence of sunlight to form glucose (sugar) and free molecular oxygen.

For the next 3 billion years, single-celled organisms dominated the Earth, and the oxygen content of the atmosphere increased slowly. Then about 570 million years ago, multicellular plants and animals suddenly evolved and thrived. According to one theory, multicellular organisms could not have evolved earlier because the oxygen concentration was too low to support them.

Today, the Earth's atmosphere contains about 21 percent oxygen (Fig. 2–10). Fires burn rapidly if oxygen is abundant, so if its concentration were to increase by even a few percent, fires would burn uncontrollably across the planet. If the oxygen concentration were to decrease appreciably, most modern plants and animals would not survive. If the carbon dioxide concentration were to

 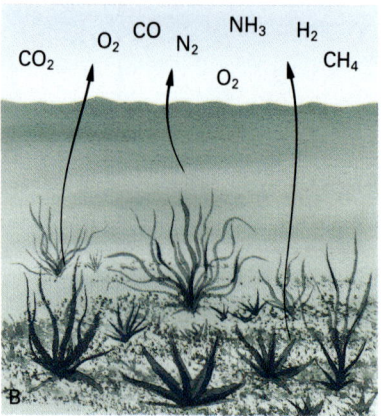 

FIGURE 2–10
Evolution of the atmosphere. (A) The primitive atmosphere contained carbon dioxide, nitrogen, and other gases. (B) As plants evolved, the composition of the atmosphere began to change. Oxygen, released during photosynthesis, began to accumulate. (C) The modern atmosphere is composed mainly of nitrogen and oxygen, with smaller concentrations of water, carbon dioxide, and other gases. The ratio of oxygen to carbon dioxide is maintained by dynamic exchange among plants and animals.

increase by a small amount, the average temperature of the Earth would rise because carbon dioxide traps heat.

Are we merely fortunate to have inherited such a nearly perfect atmosphere? The answer is no, luck is not responsible for the compatibility of Earth's atmosphere with its organisms. Organisms did not simply adapt to an existing atmosphere; they partially created it by photosynthesis and respiration. Thus, not only is our delicate oxygen–carbon dioxide balance biologically maintained, but the very presence of oxygen in our atmosphere is the result of biological activity. If all life on Earth were to cease, the atmosphere would revert to its primitive, oxygen-poor composition and become poisonous to modern plants and animals.

2.4 A Brief History of Life

Geologists have constructed the **geologic time scale,** shown in Table 2–1. Geologic time is subdivided into **eons,** which are divided into **eras.** Eras are subdivided in turn into **periods** (the most commonly used time unit), and periods are further subdivided into **epochs.**

The **Hadean Eon** (Greek for "beneath the Earth"; 4.6 to 3.8 billion years ago) is the earliest time in the Earth's history. Geologists infer that the Earth existed during Hadean time from a few Earth rocks and from extraterrestrial rocks. No fossils of Hadean age are known. It may be that geologic processes have destroyed traces of Hadean life over this long time span or that life had not yet evolved in Hadean time.

Nearly all of the oldest known Earth rocks formed during the **Archean Eon** (Greek for "ancient"; 3.8 to 2.5 billion years ago). Some Archean rocks contain microscopic fossils of bacteria, and a few contain fossil algae. Thus, life had become well established during this eon.

The evolution of living organisms from nonliving molecules is improbable. No one has reproduced this transition in the laboratory, and no one has found evidence that it has occurred elsewhere in the Solar System. Yet life evolved early in Earth history. Our planet formed about 4.6 billion years ago. For 100 to 900 million years, the surface was too hot for life to exist. The earliest fossils date to about 3.5 billion years ago, early in Archean time. Thus, life originated within 200 million to 1 billion years after the Earth's surface cooled. During the following 3 billion years, organisms remained predominantly single celled and microscopic.

Rocks of the **Proterozoic Eon** (Greek for "earlier life"; 2.6 to 0.57 billion years ago) contain abundant fossils of algae and other simple plants. Animal fossils have been found in the youngest Proterozoic rocks. The animals lacked hard parts such as shells and skeletons. Consequently, the fossils consist of imprints of their soft bodies in shale. Preservation of such delicate remains requires a gentle environment in which the soft bodies are not consumed by scavengers, lost to decay, or destroyed by sedimentary processes. Such an environment must have been uncommon. Therefore, similar animals may have lived in other places and even at earlier times, but their remains were not preserved.

TABLE 2-1 The Geologic Column and Time Scale *

Time Units of the Geologic Time Scale					Distinctive Plants and Animals	
Eon	Era	Period		Epoch		
Phanerozoic Eon (*Phaneros* = "evident"; *Zoon* = "life")	Cenozoic Era	Quaternary		Recent or Holocene	"Age of Mammals"	Humans
				Pleistocene —2—		
		Tertiary	Neogene	Pliocene —5—		Mammals develop and become dominant
				Miocene —24—		
			Paleogene	Oligocene —37—		
				Eocene —58—		Extinction of dinosaurs and many other species
				Paleocene —66—		
	Mesozoic Era	Cretaceous —144—			"Age of Reptiles"	First flowering plants, greatest development of dinosaurs
		Jurassic —208—				First birds and mammals, abundant dinosaurs
		Triassic —245—				First dinosaurs
	Paleozoic Era	Permian —286—			"Age of Amphibians"	Extinction of trilobites and many other marine animals
		Carboniferous	Pennsylvanian —320—			Great coal forests; abundant insects, first reptiles
			Mississippian —360—			Large primitive trees
		Devonian —408—			"Age of Fishes"	First amphibians
		Silurian —438—				First land plant fossils
		Ordovician —505—			"Age of Marine Invertebrates"	First fish
		Cambrian —570—				First organisms with shells, trilobites dominant
Proterozoic		Sometimes collectively called Precambrian				First multicelled organisms
—2500—						
Archean						First one-celled organisms
—3800—						Approximate age of oldest rocks
Hadean —4600 ±—						Origin of the Earth

*Absolute ages of boundaries are based on radiometric dating.

Then, within a short time at the end of the Proterozoic Eon and the start of the Phanerozoic Eon, large, multicellular plants and animals evolved abruptly, proliferated, and spread over the Earth. This sudden explosion of life is a puzzle. Why did it take much longer for multicellular organisms to evolve from single-celled organisms than it did for life to evolve from nonliving molecules? Why have multicellular organisms existed for only the most recent 15 percent of the time that life has existed on Earth? At least part of the answer to these questions may be that the Earth's early atmosphere contained only a small amount of oxygen. Although single-celled organisms could thrive in this environment, multicellular ones could not evolve until the oxygen concentration increased to suitable levels at the end of Proterozoic time.

The Phanerozoic Eon: The Last 570 Million Years

Four changes occurred at the beginning of the **Phanerozoic Eon**.

1. Animals evolved shells and skeletons.
2. The total number of individual organisms increased greatly.
3. The total number of species increased greatly.
4. The sizes of individual organisms increased from microscopic to macroscopic.

Apparently something happened about 570 million years ago that resulted in the sudden appearance of animals with shells and other hard parts. No one knows why hard parts developed, but it is easy to speculate that a shell allowed an animal to survive in places where soft organisms could not—for example, in the surf zone of a beach or in a habitat surrounded by carnivorous predators. In this scenario, animals with shells had such a survival advantage that they rapidly dominated the biosphere. As discussed above, their rise was probably related to an increase in atmospheric oxygen.

The Phanerozoic Eon is subdivided into the Paleozoic, Mesozoic, and Cenozoic Eras on the basis of abundant plants and animals of each era.

Paleozoic Life

Sedimentary rocks that formed during the **Paleozoic Era** (Greek for "old life"; 570 to 245 million years ago) contain fossils of animals and plants that evolved early, including invertebrates, fish, amphibians, reptiles, ferns, and cone-bearing trees. In early Paleozoic time, life was almost completely confined to the oceans. The beds of warm, shallow seas were populated with snail-like gastropods, worms, brachiopods that looked like clams and oysters, colonies of corals, and crab-like trilobites. Algae and other simple plants shared the sea floor with these animals. Fish and sharks swam the oceans, increasing in diversity and numbers until they dominated the seas in late Paleozoic time. About 450 million years ago, plants and animals spread onto the land. By late Paleozoic time, amphibians and reptiles were abundant, and scale trees, ferns, ginkgoes, and conifers formed great coal swamps (Fig. 2–11).

The Paleozoic Era ended with a **mass extinction**, in which half of all families of organisms disappeared suddenly. At least four mass extinctions have taken place since the beginning of Paleozoic time. The causes of extinctions are discussed in a following section.

Mesozoic Life

The extinction of most life at the end of the Paleozoic Era left a depopulated Earth. As the survivors crawled out from under the wreckage left by whatever force caused the extinction, they found a world with great ecological imbalance and disrupted food chains. Prey, suddenly without predators, bred and multiplied, exploring the new **Mesozoic** world with rapid evolutionary experiments. Many new families of plants and animals evolved quickly to fill ecological niches suddenly vacated by the extinction.

As a result, rapid and spectacular evolutionary radiations dominate the Mesozoic fossil record. On land, birds, mammals, and dinosaurs appeared early in Mesozoic time and diversified rapidly into many shapes and sizes. Flowering plants and insects followed later in the Mesozoic. New types of **phytoplankton**, plants that float at or near the sea surface, evolved to form the base of the marine food chain (Fig. 2–12). New families of fish and corals appeared to replace their deceased Paleozoic ancestors. Beautiful swimming cephalopods called **ammonoids** became abundant (Fig. 2–13).

More than 70 percent of amphibian and reptilian families perished in the terminal Paleozoic extinction. In early Mesozoic time, however, one surviving variety of reptile evolved into the dinosaurs, a group of animals that captures our fancy like no other prehistoric beast.

Evolving from small, two-legged Triassic reptiles, dinosaurs developed rapidly into many species of all sizes and shapes and came to rule the Mesozoic landscape (Fig. 2–14). Conventional wisdom portrays them as green, scaly, cold-blooded, and reptilian. However, modern evidence suggests that many were warm-blooded, brightly colored, and either hairy or covered with feathers like modern birds, and that they cared for their young after they hatched.

(Text continues on page 31)

FIGURE 2-11
(A) The Paleozoic seas teemed with a wide variety of life. *(Smithsonian Institution)* (B) The land was inhabited by reptiles and amphibians and was covered by ferns, gingkoes, and conifers. *(Ward's Natural Science Establishment)*

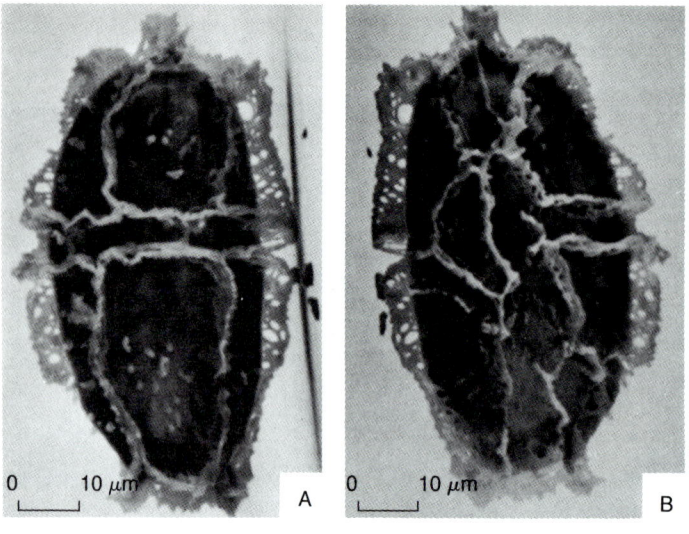

FIGURE 2-12
Fossil Cretaceous phytoplankton from Alaska.

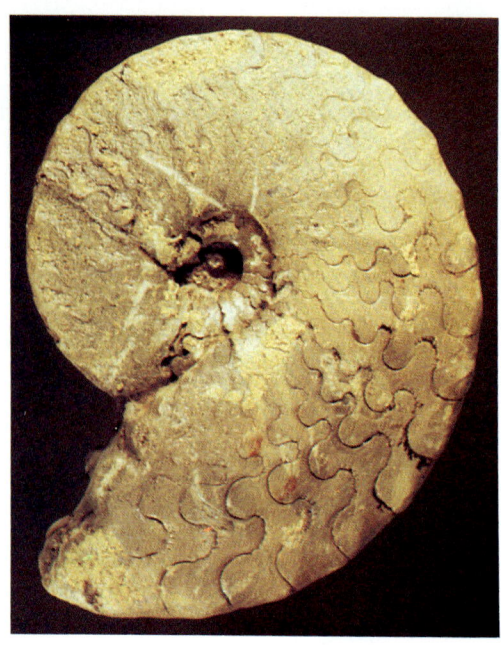

FIGURE 2-13 A Triassic ammonoid from Germany.

FIGURE 2-14
Evolution of the dinosaurs. (From E. H. Colbert, *Evolution of the Vertebrates*, New York: John Wiley & Sons, 1969)

FOCUS ON

Dinosaur Life in Montana 100 Million Years Ago

Jack Horner, a paleontologist at the Museum of the Rockies in Bozeman, Montana, has reconstructed a scene of dinosaur life 100 million years ago in Montana. A shallow inland sea flooded much of North America east of the newly rising Rocky Mountains. Erosion of the mountains filled the seaway with mud and sand, creating a vast, swampy plain lying between the mountains and the sea and crossed by streams meandering from the rising highlands.

Fossils of more than 70 species of dinosaurs, including some of the largest land animals known, have been found in rock formed from the mud and sand deposited on the plain. Bones, eggs, nests, and footprints of the animals contain evidence of their habits and life styles (Fig. 1). Horner envisions herds of thousands of duck-billed dinosaurs up to 30 feet long. Following warmth and blooming vegetation, the herds migrated seasonally a thousand kilometers or more north and south along the margins of the sea. They stopped along the way to build nests and raise their young, remaining in the nesting area until the babies were able to travel with the herd. Solitary carnivores followed the herds, preying on the young and infirm. Occasionally, great volcanic eruptions from the mountains buried an entire herd of dinosaurs in ash, preserving eggs, baby dinosaurs in their nests, and adults.

FIGURE 1

In this reconstruction, a mother duckbill dinosaur nurtures her babies in their nest. Duckbills lived adjacent to shallow seas and swamps in eastern Montana from 175 to 100 million years ago. (Museum of the Rockies)

At nearly the same time that one variety of reptile evolved into the dinosaurs, another gave rise to mammals. Mesozoic mammals were small, furtive, rodent-like animals that spent most of their lives worrying about dinosaurs. During much of Mesozoic time, they quietly improved their evolutionary abilities to adapt, survive, and reproduce. Mammals developed an efficient system for maintaining body temperature that allowed them to thrive in both cold and warm climates and in climates with wide daily or seasonal temperature ranges. Near the end of Mesozoic time, the numbers of dinosaurs and other reptiles were declining even before the next great extinction. As reptiles became less numerous, mammals increased rapidly in both number of individuals and number of species. New species also grew much larger in size as they replaced the great saurians.

Mass Extinctions

Sixty-five million years ago, all remaining species of dinosaurs suddenly became extinct. This event greatly resembled the terminal Paleozoic extinction because it included not only dinosaurs but many other plants and animals. At least one fourth of all animal species on Earth, both marine and terrestrial, vanished in this terminal Mesozoic extinction, which defines the end of Mesozoic time and the beginning of the Cenozoic Era. Why do mass extinctions occur?

In 1977, a father-and-son team, Walter and Luis Alvarez, found a sooty, thin, 65-million-year-old clay layer containing 50 to 100 times more of the element iridium than is normal in such rock. Iridium is rare in the Earth's crust but abundant in meteorites. This same 65-million-year-old, iridium-rich, sooty clay has now been found at several locations around the Earth.

Alvarez and Alvarez suggested that, 65 million years ago, a giant meteorite 10 to 25 kilometers in diameter hit the Earth. Astronomers calculate that an object of that size would have struck Earth with the energy of 10 million to 100 million hydrogen bombs. The impact vaporized both the meteorite and the Earth's crust at the point of impact, igniting massive fires. Soot from the fires and iridium-rich meteorite dust rose into the upper atmosphere, circling the globe. The thick, dark cloud blocked out the Sun. Surface waters froze. Many plant and animal species froze and starved to death. Then the sooty, iridium-rich dust settled to Earth to form the distinctive clay layer.

Other scientists have found evidence of meteorite impacts that occurred exactly 65 million years ago in both western India and the Caribbean Sea. Geologists continue to find evidence of giant meteorite impacts that coincide with other mass extinctions, including the one that marks the end of Paleozoic time. In July 1994, a fragmented comet smashed into Jupiter. Each impact provided a visual reminder that such collisions do occur and alter atmospheric and surface conditions on a planet.

Many scientists disagree with the meteorite hypothesis of mass extinctions, and several other hypotheses have been offered to explain the sudden and simultaneous disappearances of large numbers of plants and animals. The extinctions occurred on a global scale, and hence a global explanation must be sought. Most other theories invoke episodes of intense volcanic activity, sudden changes in global climate, or changes in sea level.

FIGURE 2–15
The 2-meter-tall flightless Eocene bird *Diatryma*, from Wyoming. *(Levin)*

Cenozoic Life

With the dinosaurs eliminated and reptiles and amphibians depleted by the terminal Mesozoic extinction, mammals and birds evolved rapidly to dominate land in the **Cenozoic Era**. The available evidence indicates that the basic features of birds have remained the same since the beginning of Cenozoic time. Those features include skeletal structures, feathers, lightweight porous bones, a toothless horny beak, and constant body temperature (Fig. 2–15). Birds resemble dinosaurs more closely than they do any other Mesozoic animals, and many paleontologists think that birds are the evolutionary descendants of dinosaurs.

Hair and mammary glands characterize mammals. Neither is commonly preserved in fossil remains, however. As a result, paleontologists characterize mammal fossils on the basis of their skeletal structures. The fossil record shows that mammalian evolution involved rapid changes in bone structures, particularly those of the skull and lower jaw (Fig. 2–16).

Grasses evolved in the Miocene Epoch. They were so successful that lush grassy prairies and savannahs spread widely over the globe. The grasses were a new and nourishing kind of food. In response, large herds of grazing animals, such as horses, began a rapid evolution to take advantage of grass (Fig. 2–17). Escape from predators is difficult on the open plains, but predators can be seen from a distance. Horses and other grazers became larger,

FIGURE 2–16
Early Miocene mammals. *(National Museum of Natural History, J.H. Matternes)*

taller, and faster to escape from carnivores, which evolved to prey on the expanding herds.

In the seas, many types of phytoplankton had become extinct at the end of the Mesozoic Era. Those that survived evolved rapidly in early Cenozoic time to fill empty ecological niches. All of the beautiful ammonoids had also vanished at the end of the Mesozoic. Because ammonoids competed with fish for habitat and food, the number of fish expanded rapidly in early Cenozoic time. Reef-forming corals also proliferated throughout regions of warm, shallow water.

About 5 million years ago, near the end of Miocene time, creatures that resembled modern humans more than apes appeared in India. Later, between 3 and 1 million years ago, several separate human-like lineages developed and are preserved as fossils in East Africa (Fig. 2–18). Primitive men and women had evolved between a million and 500,000 years ago and had spread across Europe and Africa by 400,000 years ago. They used fire, made tools, and lived in caves. Modern humans (*Homo sapiens*) evolved in Africa between 140,000 and 100,000 years ago. Biochemists, using DNA evidence, suggest that present-day humans have descended from these ancestors. Neanderthal people roamed Europe, Africa, and the Middle East by 100,000 years ago. They made fine tools and had brains larger than those of modern humans. By 30,000 years ago, early modern humans, called Cro-Magnon people, had replaced the Neanderthal people in Europe. They crafted well-made weapons and tools, developed religions, and created art. They are our immediate ancestors.

For 90 percent of their time on Earth, humans hunted game and gathered wild plants for their nourishment. Then, about 10,000 years ago, people in many parts of the world domesticated plants and animals. Agriculture became increasingly important over the following centuries, gradually displacing hunting and gathering. About 5000 years ago, Sumerian scholars developed a written language, enabling people to communicate over time and distance. Civilization grew rapidly and engineers built great cities and monuments. The human population expanded. But the expansion came with a price. Irrigation canals brought salty water to the Middle Eastern desert, ruining farmland. Deforestation in Greece led to depletion of ground-water reserves; wells dried up. Destruction of forest soils caused starvation and dislocation in the great Mayan cities of the Western Hemisphere.

Then, about 250 years ago, great advances in science, technology, and medicine accelerated the pace of human development. Machines brought unprecedented wealth and luxury. Medicine and hygiene decreased infant mortality and increased life expectancy. Populations boomed; cities grew; people logged entire continents to establish farms. These changes rapidly altered many of the Earth's systems. This book examines these changes and how they affect the Earth's geological systems and its inhabitants.

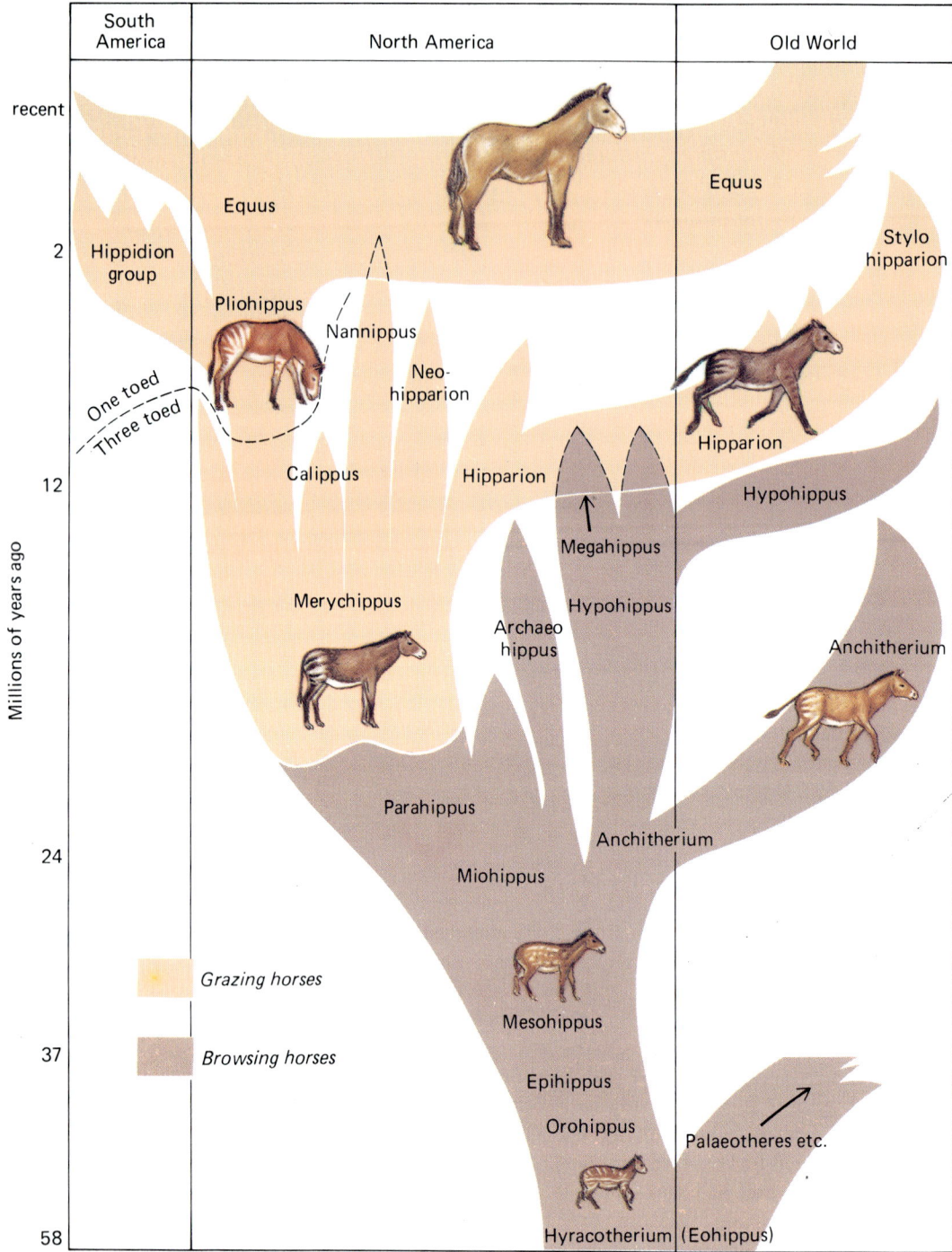

FIGURE 2-17
Evolution of the horses. (Redrawn from Stephen Jay Gould, "Life's Little Joke," *Natural History*, April 1987, p. 16)

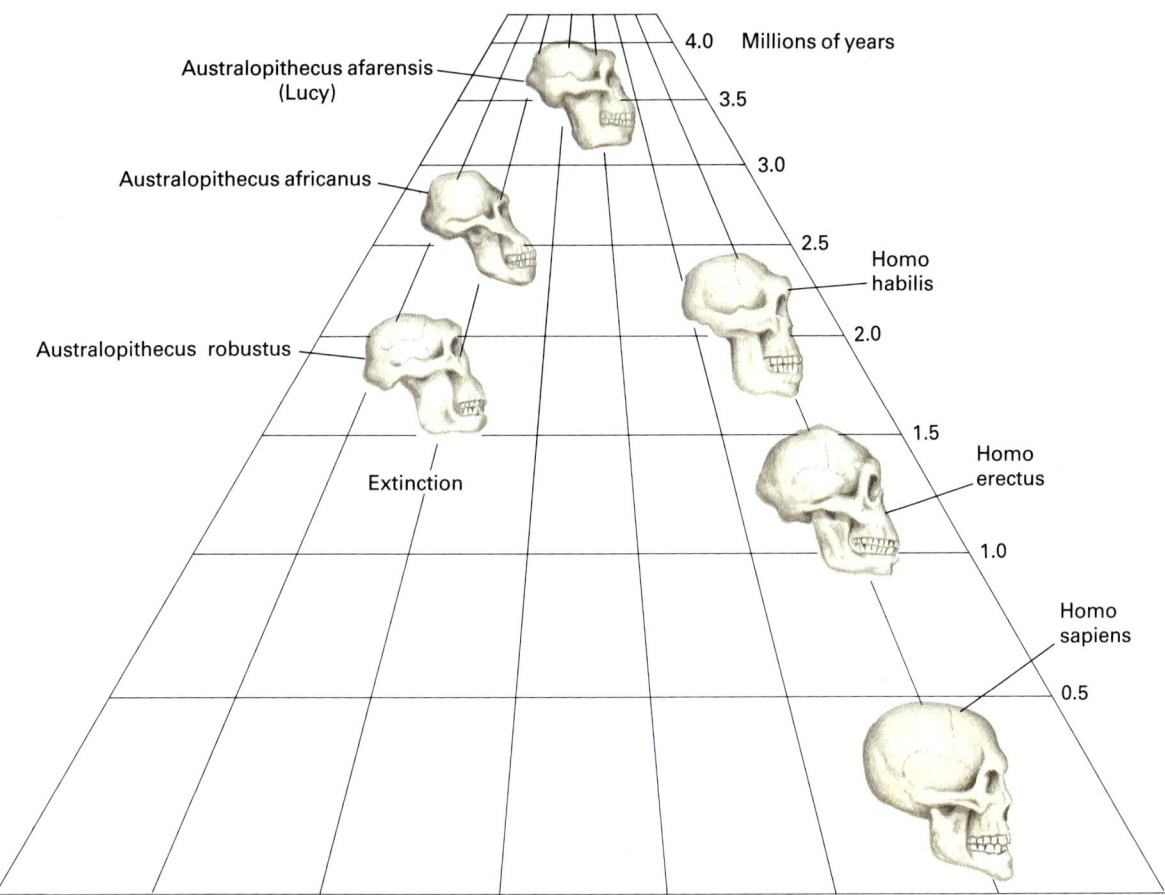

FIGURE 2–18
One interpretation of human evolution.

SUMMARY

The **big bang theory** states that originally all matter was compressed into a single point and that this point exploded between 15 and 18 billion years ago to form the Universe and mark the beginning of time.

The Solar System formed from a mass of dust and gases that rotated slowly in space. Within its center, gravitational attraction was so great that the gases were pulled inward with enough velocity to initiate nuclear fusion and create the Sun. Planets formed from coalescing dust and gases. Most of the lighter elements escaped from the terrestrial planets but remained on the outer giants.

The primordial Earth was heated by energy released from radioactive decay and by bombardment from outer space. It has since cooled so that most of it is solid, although the inner layers remain hot. The modern **solid Earth** is made up of a dense **core** of iron and nickel, a **mantle** of lower density, and a **crust** of yet lower density. The **lithosphere** contains the uppermost portion of the mantle and the crust. According to the **theory of plate tectonics**, the lithosphere is broken into several plates that float on the underlying **asthenosphere** and move about relative to one another. The **hydrosphere**, **atmosphere**, and **biosphere** occupy a thin layer near the Earth's surface.

The Earth's oceans originated from water trapped in the Earth's interior, water from the inner Solar System, and water from comets and meteorites that originated in the outer Solar System and later collided with the Earth. The earliest atmosphere consisted mainly of hydrogen, with smaller amounts of helium and other gases. These gases escaped into space and were replaced by gases released during volcanic eruptions. This secondary atmosphere was then modified by living organisms.

The **geologic time scale** is a calendar of Earth history from the Earth's origin 4.6 billion years ago through present time. The earliest fossils date to about 3.5 billion years ago. Single-celled organisms dominated the Earth for nearly 3 billion years following life's origin. The primitive organisms gradually released oxygen, changing the atmosphere.

Multicellular plants and animals evolved rapidly about 570 million years ago to mark the start of the **Phanerozoic Eon** (meaning "evident life"). Early **Paleozoic** oceans were dominated by snail-like gastropods, worms, brachiopods that looked like clams and oysters, colonies of corals, and horseshoe-crablike trilobites. Algae and other simple plants shared the sea floor with these animals. Gradually plants and animals moved onto land. Trees, amphibians, and reptiles evolved.

Mass extinctions occurred at the ends of both the Paleozoic and Mesozoic Eras, and rapid evolutionary developments followed each of the two extinction events. At least four such mass extinctions have occurred in Earth history and may have been caused by collisions of large meteorites with Earth.

Birds, mammals, dinosaurs, and flowering plants evolved on land during **Mesozoic** time. Dinosaurs dominated the land, and mammals were small, elusive, and uncommon. Ammonoids were abundant in the seas and were more common than fishes. The mass extinction that ended the Mesozoic Era eliminated many dominant Mesozoic animals and plants, including dinosaurs and ammonoids.

After the terminal Mesozoic extinction, mammals and birds experienced evolutionary explosions and rapidly dominated the land in **Cenozoic** time. With the ammonoids extinct, fishes and corals increased greatly, in variety and numbers, in the seas.

Human precursors evolved about 5 million years ago, and primitive men and women had appeared by 500,000 years ago.

KEY TERMS

Big bang 19
Closed Universe model 19
Open Universe model 19
Protosun 19
Planetesimal 19
Nuclear fusion 19
Terrestrial planets 21
Jovian planets 21
Solid Earth 23
Core 23
Crust 23
Mantle 23

Lithosphere 23
Asthenosphere 23
Plate tectonics 23
Earthquake 23
Magma 23
Volcanic eruptions 23
Hydrosphere 24
Atmosphere 24
Biosphere 24
Geologic time scale 26
Eons 26
Eras 26

Periods 26
Epochs 26
Hadean Eon 26
Archean Eon 26
Proterozoic Eon 26
Phanerozoic Eon 28
Paleozoic Era 28
Mass extinction 28
Mesozoic Era 28
Phytoplankton 28
Ammonoids 28
Cenozoic Era 32

REVIEW QUESTIONS

1. Briefly outline the formation of the Universe.
2. Briefly outline the evolution of the planets.
3. How did the Sun form? How is its composition different from that of the Earth? Explain the reasons for this difference.
4. Compare and contrast the properties of the terrestrial planets with those of the Jovian planets.
5. The entire Earth was molten soon after its formation. Explain how it became hot enough to melt, and why it cooled.
6. Briefly outline the layered structure of the modern Earth.
7. Describe how the modern atmosphere evolved.
8. Describe how the modern seas evolved.
9. Construct a simple table listing the four eons of geologic time. Make the space allotted to each eon proportional to its length. Which eon is longest? Which is shortest?
10. On what basis have geologists defined the boundary between the Paleozoic and Mesozoic Eras?
11. On what basis have geologists defined the boundary between the Mesozoic and Cenozoic Eras?
12. From what major group of organisms did dinosaurs evolve?
13. When did mammals first appear?
14. When did dinosaurs become extinct?
15. What other events coincided with the extinction of dinosaurs?
16. Describe the evidence that indicates that a meteorite impact caused the terminal Mesozoic mass extinction.

DISCUSSION QUESTIONS

1. Explain how the theory of the evolution of the Solar System explains the following observations: (a) All the planets in the Solar System are orbiting in the same direction. (b) All the planets in the Solar System except Pluto are orbiting in the same plane. (c) The chemical composition of Mercury is similar to that of Earth. (d) The Sun is composed mainly of hydrogen and helium but also contains all the elements found on Earth. (e) Venus has a solid surface, whereas Jupiter is mainly a mixture of gases and liquids with a small, solid core.

2. The radioactive elements that are responsible for the heating of the Earth decompose very slowly over billions of years. How would the Earth be different if these elements decomposed much more rapidly—say, over a period of a few million years? Defend your answer.

3. Explain how the size of a terrestrial planet can affect its surface environment.

4. What conditions are necessary for the formation of continents on an evolving planet?

5. What kinds of information regarding the physical characteristics and habits of a particular species of dinosaur might be obtained from its tracks?

6. Why would a mass extinction be followed by a sudden increase in rates of evolutionary change in surviving life forms?

7. Discuss the possibilities of an evolutionary connection between dinosaurs and birds.

8. Discuss the implications of dinosaurs as warm-blooded animals who nurtured their young in contrast with the traditional view of them as cold-blooded animals who laid and then abandoned their eggs (as modern reptiles do).

CHAPTER 3

Earth Materials: Minerals, Rocks, Air, and Water

3.1 Elements, Atoms, and Ions
3.2 The Solid Earth: Minerals
3.3 The Solid Earth: Rocks and the Rock Cycle
3.4 The Atmosphere
3.5 The Hydrosphere
3.6 The Earth as a System

Imagine walking along a rocky coast as a storm blows in from the sea. Wind whips the ocean into whitecaps, gulls hurtle overhead, and waves crash onto shore. Blowing spray and rain soak your clothes as you scramble over the last rocks to your car. During this adventure, you have interacted directly with the four major realms of the Earth. The rocks and soil underfoot are part of the solid Earth. The rain and sea are parts of the hydrosphere, the watery part of our planet. The wind and air are portions of the atmosphere. Finally, you, the gulls, the beach grasses, and all other forms of life in the sea, on land, and in the air are parts of the biosphere, the thin layer of living organisms.

These four realms overlap and interact near the Earth's surface to create and modify our environment. Water, for example, evaporates from the ocean to become part of the atmosphere. It then falls onto the land and sea as rain and snow. The snow accumulates in the high mountains to form glaciers, which erode deep valleys in the solid Earth as they flow downhill. As another example, water, air, and plants interact with solid rock, decomposing it to soil. Flowing water erodes the soil, carrying sand and clay to a seacoast where the material accumulates. Eventually, the sand and clay are buried and become cemented together to form a new rock layer. Along seacoasts, shells of animals such as corals, clams, and oysters may become cemented to form a type of rock called limestone.

The nature and distribution of Earth materials directly affect everyone. People congregate near seacoasts because the coasts provide access to transportation and an aesthetically pleasing living environment. Our most productive agricultural regions lie in areas of the country where rainfall is abundant or where irrigation water is available. On a smaller scale, a building foundation or roadbed in a gravel-rich soil may remain stable for centuries, whereas soil rich in certain types of clay may swell

The red sandstone of Arches National Park, Utah, is composed of sand that accumulated in windblown desert dunes about 200 million years ago. Erosion formed the arches themselves during the past few million years.

and heave during the first rainy season, destroying the structures.

Minerals and rocks are the fundamental materials of geology. Thus, an understanding of the relationships between geology and our environment involves an understanding of the natures of minerals and rocks.

3.1 Elements, Atoms, and Ions

In the third century B.C., the Greek philosopher Aristotle defined an **element**, saying that "everything is either an element or composed of elements." Although Aristotle's definition is still correct, a more complete modern definition is that **an element is a fundamental form of matter that cannot be broken into simpler substances by ordinary chemical processes.** Elements are the building blocks of which rocks, minerals, air, water, and all other substances are composed. Thus, elements are the fundamental materials of the Earth.

A total of 88 elements occur naturally in the Earth's crust. **Of those 88, only eight elements—oxygen, silicon, aluminum, iron, calcium, magnesium, potassium, and sodium—make up more than 98 percent of the Earth's crust.** All of the elements are listed in Appendix 1. Each element is assigned a one- or two-letter symbol. The symbols for the eight most abundant elements are given in Table 3–1.

An **atom** is the basic unit of an element. It consists of a small, dense, positively charged center called a **nucleus** surrounded by a cloud of negatively charged **electrons** (Fig. 3–1). An electron is a fundamental particle; as far as we know, it is not made up of smaller components. The nucleus, however, is made up of two different kinds of particles: (1) positively charged **protons**, and **neu-**

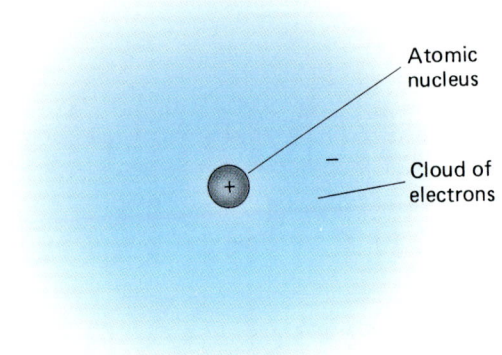

FIGURE 3–1
An atom consists of a small, dense, positive nucleus surrounded by a much larger cloud of negative electrons.

trons, which have no charge. In a neutral atom the number of protons equals the number of electrons. Therefore, the positive and negative charges balance each other so that neutral atoms have no overall electrical charge.

However, in most elements, including all eight of the most abundant ones, the neutral atoms easily lose or gain electrons. When an atom loses one or more electrons, its positive charges then outnumber its negative ones. The atom therefore becomes positively charged. If an atom gains one or more extra electrons, it becomes negatively charged. Atoms with a positive or negative charge are called **ions**.

3.2 The Solid Earth: Minerals

Below a thin layer of soil on land and a thin layer of mud on the sea floor, the outer layers of the Earth are composed entirely of **rock**. Even a casual observer sees that rocks are different from one another: Some are soft, others hard, and they come in many colors. Most rocks are composed of tiny, differently colored grains, each of which is a **mineral** (Fig. 3–2).

Minerals are the substances that make up rocks. **A mineral is a naturally occurring, inorganic solid with a characteristic chemical composition and a crystalline structure.** Thus, a mineral has five characteristics: (1) It is natural in origin, (2) it is inorganic, (3) it is solid, (4) it has a characteristic chemical composition, and (5) it has a crystalline structure. The most important properties of a mineral are its chemical composition and its crystalline structure. They distinguish any mineral from all others.

The Chemical Composition of Minerals

When we say that a mineral has a characteristic chemical composition, we mean that **a mineral is made up of**

TABLE 3–1 The Eight Most Abundant Chemical Elements in the Earth's Crust

Element	Chemical Symbol	Common Ion(s)
Oxygen	O	O^{2-}
Silicon	Si	Si^{4+}
Aluminum	Al	Al^{3+}
Iron	Fe	Fe^{2+} and Fe^{3+}
Calcium	Ca	Ca^{2+}
Magnesium	Mg	Mg^{2+}
Potassium	K	K^{1+}
Sodium	Na	Na^{1+}

FIGURE 3–2
Each of the differently colored grains in this granite is a different mineral. The pink grains are feldspar, the white ones are quartz, and the black ones are amphibole.

elements bonded together in specific proportions. Therefore, its composition can be expressed with a chemical formula. A few minerals consist of only a single element. Gold and silver are single-element minerals. Most minerals, however, are made up of two to six elements. For example, the formula of quartz is SiO_2, meaning that it consists of one atom of silicon (Si) for every two of oxygen (O).

The Crystalline Nature of Minerals

A crystal is any substance whose atoms are arranged in a regular, orderly, periodically repeated pattern. All minerals are crystals. Halite has the composition NaCl: one sodium ion (Na^+) for every chlorine ion (Cl^-). Figure 3–3 includes a photo of halite crystals and two sketches showing halite's arrangement of sodium and chlorine ions. Figure 3–3A is an "exploded" view that allows you to see into the structure. Figure 3–3B is more realistic, showing the ions in contact. They lie in orderly rows and columns of alternating sodium and chlorine ions. This orderly arrangement is the **crystalline structure** of halite. All minerals have their atoms in orderly arrangements, although the pattern is not always as obvious as in halite. Any solid with such an orderly, repetitive arrangement of atoms is a crystal.

Crystal faces are flat surfaces that form if a crystal grows without obstructions. The halite in Figure 3–3C has well-developed crystal faces. In nature, crystal growth is often hindered by other minerals. For this reason, minerals rarely show perfect crystal faces.

The Rock-Forming Minerals

The 88 elements that occur naturally in the Earth's crust can combine in many different ways—in fact, more than 2500 different minerals are known! As a result, we might expect to see an overwhelming variety of minerals on

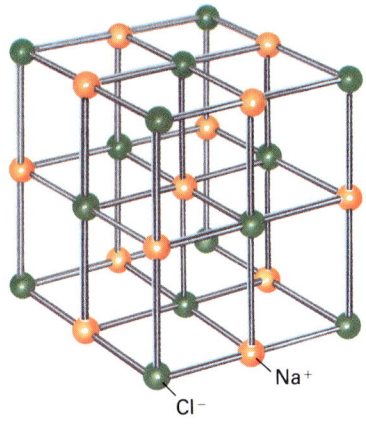

A

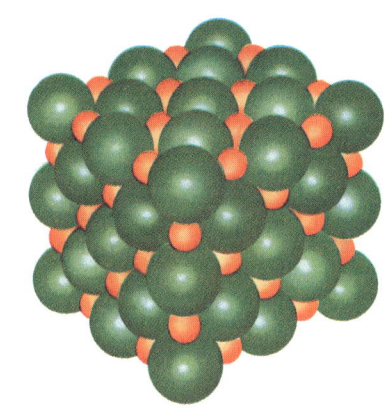

B

C

FIGURE 3–3
(A and B) The orderly arrangement of sodium and chlorine ions in halite. (C) Halite crystals. The crystal model in (A) is exploded so that you can see into it; the ions are actually closely packed as in (B). Note that the arrangement of ions in (A) and (B) is that of a cube, and the crystals in (C) are also cubes. *(C, American Museum of Natural History)*

a half-day field trip. However, only eight elements are abundant, and they commonly combine in only a few ways. Thus, only nine **rock-forming minerals** make up most rocks of the Earth's crust (Table 3–2). Because they are so common, they are the minerals you are most likely to find and identify.

Geologists classify minerals according to their anions (negatively charged ions). A simple anion is a single negatively charged ion such as O^{2-}. Minerals whose principle anion is O^{2-} are called **oxides.** Alternatively, two or more atoms can bond firmly together and acquire a negative charge to form a complex anion. Two common examples of complex anions are **silicate,** $(SiO_4)^{4-}$,

TABLE 3–2 The Rock-Forming Minerals

Silicates	Carbonates
Feldspar	Calcite
Quartz	Dolomite
Pyroxene	
Amphibole	
Mica	
Clay Minerals	
Olivine	

TABLE 3–3 Important Mineral Groups

Group	Member	Formula	Economic Use
Oxides	Hematite	Fe_2O_3	Ore of iron
	Magnetite	Fe_3O_4	Ore of iron
	Corundum	Al_2O_3	Gemstone, abrasive
	Ice	H_2O	Solid form of water
	Chromite	$FeCr_2O_4$	Ore of chromium
Sulfides	Galena	PbS	Ore of lead
	Sphalerite	ZnS	Ore of zinc
	Pyrite	FeS_2	Fool's gold
	Chalcopyrite	$CuFeS_2$	Ore of copper
	Bornite	Cu_5FeS_4	Ore of copper
	Cinnabar	HgS	Ore of mercury
Sulfates	Gypsum	$CaSO_4 \cdot 2H_2O$	Plaster
	Anhydrite	$CaSO_4$	Plaster
	Barite	$BaSO_4$	Drilling mud
Native elements	Gold	Au	Electronics, jewelry
	Copper	Cu	Electronics
	Diamond	C	Gemstone, abrasive
	Sulfur	S	Sulfa drugs, chemicals
	Graphite	C	Pencil lead, dry lubricant
	Silver	Ag	Jewelry, photography
	Platinum	Pt	Catalyst
Halides	Halite	$NaCl$	Common salt
	Fluorite	CaF_2	Used in steel making
	Sylvite	KCl	Fertilizer
Carbonates	Calcite	$CaCO_3$	Portland cement
	Dolomite	$CaMg(CO_3)_2$	Portland cement
	Aragonite	$CaCO_3$	Portland cement
Hydroxides	Limonite	$FeO(OH) \cdot nH_2O$	Ore of iron, pigments
	Bauxite	$Al(OH)_3 \cdot nH_2O$	Ore of aluminum
Phosphates	Apatite	$Ca_5(F,Cl,OH)(PO_4)_3$	Fertilizer
	Turquoise	$CuAl_6(PO_4)_4(OH)_8 \cdot 4H_2O$	Gemstone
Silicates	(See Fig. 3–5 for the silicate minerals.)		

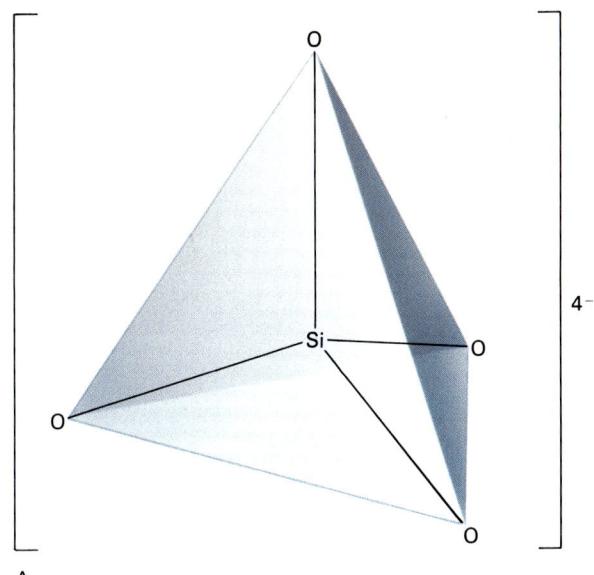
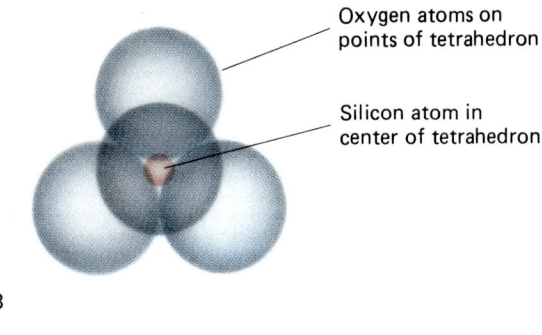

FIGURE 3–4
The silicate tetrahedron consists of one silicon atom surrounded by four oxygens. It is the fundamental building block of all silicate minerals. (A) A schematic representation. (B) A proportionally accurate model.

and **carbonate**, $(CO_3)^{2-}$. Common and useful mineral groups and important minerals in each group are listed in Table 3–3.

Seven of the nine rock-forming minerals are silicates. The silicate group makes up more than 95 percent of the Earth's crust. Silicate minerals are so abundant for two reasons. First, they are made up principally of the two most plentiful elements in the crust, silicon and oxygen. Second, silicon and oxygen bond together readily, forming the **silicate tetrahedron**, the fundamental building block of all silicate minerals (Fig. 3–4).

Feldspar (Fig. 3–5A), making up more than 50 percent of the Earth's crust, is the most abundant mineral in the crust. It is a major component of nearly all common rocks. Feldspar is actually a group of minerals with similar crystal structures and compositions. The feldspar minerals are named according to whether they contain potassium, sodium, or calcium. **Orthoclase** is a common type of potassium feldspar. Feldspar containing calcium and sodium is called **plagioclase**.

Quartz (Fig. 3–5B) is pure SiO_2. It is widespread and abundant in continental rocks but rare in oceanic crust and the mantle.

Pyroxene (Fig. 3–5C), like feldspar, is a group of similar minerals. It is a major component of oceanic crust and the mantle and is abundant in some rocks of the continents. **Amphibole** (Fig. 3–5D) also is a group of minerals with similar properties. It is common in many rocks of the continents. Pyroxene and amphibole can resemble each other so closely that they are difficult to tell apart.

Mica (Fig. 3–5E) is a plate-shaped mineral from which layer after layer can be peeled, like the layers of an onion. Mica is common in continental rocks. The **clay minerals** (Fig. 3–5F) are similar to mica but are so small that they can barely be seen with a good optical microscope. Most clay forms when other minerals weather at the Earth's surface. Thus, clay is abundant at and near the Earth's surface and is an important component of soil.

Olivine (Fig. 3–5G) occurs in small quantities in both continental and oceanic rocks. However, olivine and pyroxene make up most of the upper mantle.

Calcite and **dolomite** are the only two rock-forming minerals that are not silicates. They are abundant carbonate minerals in near-surface rocks of the continents (Figs. 3–6A and B). Calcite and dolomite make up the rocks called carbonate rocks, or sometimes simply limestone.

Other Important Minerals

We study the rock-forming minerals because they make up most of the Earth's crust. A small number of other minerals are important for economic reasons or because they are commonly found in small quantities. The most notable of these minerals are described in Appendix 2.

Ore minerals are minerals from which metals or other elements can be recovered profitably. Thus, they are minerals that contain commercially valuable elements or compounds. Native gold and native silver are ore minerals composed of pure metals. Most other metals exist in nature as compounds. The industrially important metals copper, lead, and zinc are obtained from chalcopyrite, galena, and sphalerite, respectively. Halite is mined for table salt, and gypsum is mined for the manufacture of plaster and sheetrock.

FIGURE 3–5
The seven rock-forming silicate minerals. (A) Feldspar, represented here by a variety of potassium feldspar. *(Jeffrey Scovil)* (B) Quartz. *(Jeffrey Scovil)* (C) Pyroxene, variety diopside. *(Jeffrey Scovil)* (D) Amphibole, variety actinolite. *(Jeffrey Scovil)* (E) Black biotite is one common type of mica. White muscovite is the other. *(Geoffrey Sutton)* (F) Clay. *(Geoffrey Sutton)* (G) Olivine. *(Geoffrey Sutton)*

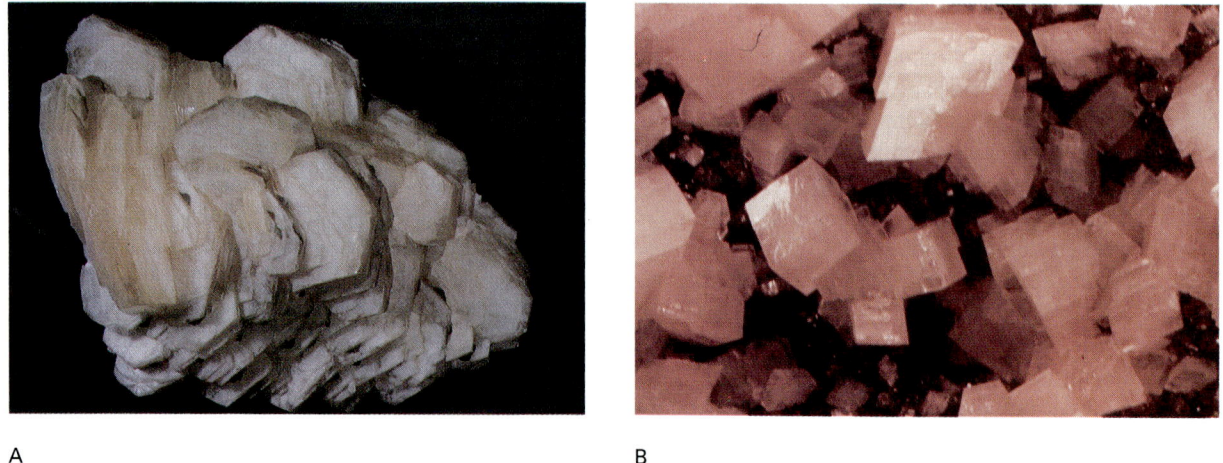

FIGURE 3–6
(A) Calcite *(Harvard University Mineralogical Museum)* and (B) dolomite *(Ward's Natural Science Establishment)* are the only two rock-forming minerals that are not silicates.

Mining and refining of fossil fuels, ore minerals, and rock and sediment, are necessary to produce most of the energy, all of the metals, and many other materials—such as sand, gravel, and concrete—that are fundamental to a modern culture. But mining, quarrying, and smelting also change the Earth's surface; they add gases such as sulfur dioxide and carbon dioxide to our atmosphere, and they pollute surface and ground water. Those effects have changed our environment in ways that affect humans and all other organisms of the biosphere. Many scientists think that those environmental changes threaten the health of the biosphere itself, and some feel that they threaten its very survival.

A **gem** is a mineral that is prized for its rarity and beauty rather than for industrial use. Depending on its value, a gem can be either precious or semiprecious. Precious gems include diamond, emerald, ruby, and sapphire. Several varieties of quartz, including amethyst, agate, jasper, and tiger's-eye, are semiprecious gems. Garnet, olivine, topaz, turquoise, and many other minerals sometimes occur as semiprecious gems.

Accessory minerals are minerals that are seen commonly but usually only in small amounts. They are not abundant enough to classify as rock-forming minerals. Chlorite, garnet, limonite, magnetite, and pyrite are among the most common accessory minerals.

FOCUS ON

Asbestos and Cancer

Many minerals, such as quartz, are relatively inactive in biological systems. Others, such as apatite (a calcium phosphate mineral mined to produce phosphate fertilizer), provide nutrients that are necessary for plant growth. A few, such as lead and arsenic minerals and asbestos, are toxic. In nature, these minerals are usually buried beneath rock and soil, but they can become environmental hazards when they are mined and humans are exposed to them.

Asbestos is an industrial name for a group of minerals that crystallize as long, thin fibers (Fig. 1). The two most common types are fibrous varieties of the minerals **chrysotile** and **amphibole**. Chrysotile has a crystal structure and a composition similar to those of the micas. The fibers of chrysotile form tangled, curly bundles, whereas amphibole asbestos occurs as straight, sharply pointed needles.

Asbestos fibers are commercially valuable because they are flameproof, chemically inert, and extremely strong. For example, a chrysotile fiber is eight times stronger than a steel wire of equivalent diameter. Asbestos fibers have been used to manufacture brake linings, fireproof clothing, insulation, shingles, tile, pipe, and gaskets but now are allowed only in brake linings, shingles, and pipe.

In the early 1900s, asbestos miners and others who worked with asbestos learned that prolonged exposure to the fibers caused **asbestosis**, an often lethal lung disease. Later, in the 1950s and 1960s, studies showed that asbestos also causes lung cancer and other forms of cancer. One reason that so much time passed before scientists recognized the cancer-causing properties of asbestos is that cancer commonly does not develop until decades after the first exposure to asbestos.

Experiments have shown that lung diseases are caused by the fibrous nature of asbestos, not by its chemical composition. For example, amphibole with identical composition can occur in both fibrous and nonfibrous forms. In a laboratory study, a group of rodents was exposed to fibrous amphibole and another group to identical amounts of nonfibrous amphibole. The group exposed to the fibrous type developed cancers, but the other group did not.

Another experiment with rodents showed that amphibole asbestos is a more effective cause of lung cancer than is chrysotile. Apparently the curly chrysotile fibers are more easily expelled from the lungs, whereas the sharp amphibole needles remain in the lungs. Additionally, the incidence of cancer among chrysotile workers is proportionally lower than among those working with amphibole asbestos. Although it is not clear how asbestos causes cancer, it is clear that the fibers are important and that the sharp needles of the less common amphibole asbestos are more dangerous than chrysotile.

In response to growing awareness of the health effects of asbes-

(Continued)

FOCUS ON (continued)

tos, the Environmental Protection Agency (EPA) banned the use of the material in construction in 1978. However, the ban did not address the issue of what should be done with the asbestos already installed. In 1986 Congress passed a ruling called the Asbestos Hazard Emergency Response Act, requiring that all schools be inspected for asbestos. Public response has resulted in hasty programs to remove asbestos from schools and other buildings (Fig. 1). The EPA estimates that removal of asbestos from schools and public and commercial buildings will cost between $50 billion and $150 billion. But what is the real level of hazard?

Most asbestos is the less dangerous chrysotile. More important, most asbestos in buildings is woven into cloth or glued into a tight matrix, and often the surface has been further stabilized by painting. Therefore, the fibers are not free to blow around. The lev-

FIGURE 1 *Asbestos.*

els of airborne asbestos in most buildings are no higher than that in outdoor air. Many scientists argue that asbestos insulation poses no health danger if left alone, but as the material is removed it is disturbed and asbestos dust escapes. Not only are the removal workers endangered, but airborne asbestos persists in the building for months after completion of the project.

Thus, when assessing the health effects of asbestos, we must understand how it is transported and incorporated into living tissue. Asbestos is unquestionably deadly in a mine where rock is drilled and blasted and dust hangs heavy in the air. However, in a school or commercial building it may be harmless until, in the interest of public safety, workers release fibers as they disturb the insulation during removal.

FIGURE 2
Workers removing asbestos from a commercial building in New York City. (Andrew Savulich/Wide World Photo)

3.3 The Solid Earth: Rocks and the Rock Cycle

Geologists separate rocks into three groups according to how they form: igneous rocks, sedimentary rocks, and metamorphic rocks.

Igneous Rocks

Under certain conditions, rocks of the upper mantle and lower crust melt, forming a hot liquid called **magma**. **Igneous rock forms when magma cools and solidifies.** Igneous rock is the most abundant kind of rock. It makes up much more than half of the Earth's crust. Granite and basalt are the two most common igneous rocks.

The temperature of magma varies from about 600° to 1400°C, depending on its chemical composition and the depth at which it forms. As a comparison, an iron bar turns red-hot at about 600°C and melts at slightly over 1500°C. Blacksmiths easily heat iron to redness on the glowing embers of a coal forge.

When rock melts to form magma, it expands by about 10 percent. Therefore, magma is of lower density than the solid rock around it. Because of its lower density, magma rises toward the Earth's surface as soon as it forms, just as a hot balloon rises in the atmosphere. As the magma rises, it enters the cooler, lower-pressure environment near the Earth's surface. When temperature and pressure drop sufficiently, the liquid solidifies to form solid igneous rock.

Igneous rocks are divided into two groups on the basis of where the magma solidifies as it rises. **Intrusive igneous rocks** form when magma solidifies within the Earth, before it can rise all the way to the surface. Intrusive rocks are sometimes called **plutonic rocks** after Pluto, the ancient Roman god of the underworld. An intrusive rock mass is called a **pluton** (Fig. 3–7).

Extrusive igneous rocks form when magma erupts and solidifies on the Earth's surface. Because extrusive rocks are so commonly associated with volcanoes, they are also called **volcanic rocks** after Vulcan, the Roman god of fire. Magma flowing over the Earth's surface, as

FIGURE 3–7
The granite peaks of Sam Ford Fjord, Baffin Island, Canada, are part of a large pluton.

FIGURE 3–8
Molten lava erupting from Mauna Loa, Hawaii, in 1984. (*Bob Seibert, Hawaii Volcanoes National Park*)

well as the volcanic rocks formed when this magma solidifies, is also called **lava** (Fig. 3–8).

The emplacement of plutons deep within the Earth has little effect on the Earth's surface and consequently little effect on our environment. Volcanic eruptions, on the other hand, can be extremely violent and have killed tens of thousands of humans within a few seconds. In addition, many geologists now think that massive volcanic eruptions alter climate.

Volcanic rocks are usually fine grained, whereas plutonic rocks are medium or coarse grained (Table 3–4). The difference exists because crystals usually grow slowly as magma solidifies. On the Earth's surface, volcanic magma cools rapidly and solidifies before crystals have time to grow large. In contrast, plutonic magma cools slowly within the crust, and crystals have a long time to grow to greater sizes.

Geologists use both minerals and texture to name igneous rocks. For example, **granite** is mostly feldspar and quartz and is medium or coarse grained. **Rhyolite** contains the same minerals but is very fine grained (Fig. 3–9B). The same magma that erupts onto the Earth's surface to form rhyolite also forms granite if it solidifies slowly within the crust.

A

B

FIGURE 3–9
Although granite (A) and rhyolite (B) contain the same minerals, they have very different textures because granite cools slowly and rhyolite cools rapidly. (*Geoffrey Sutton*)

TABLE 3–4 Igneous Rock Textures Based on Grain Size	
Grain Size	**Name of Texture**
No mineral grains (obsidian)	Glassy
Too fine to see with naked eye	Very fine grained
Up to 1 millimeter	Fine grained
1–5 millimeters	Medium grained
More than 5 millimeters	Coarse grained
Relatively large grains in a finer-grained matrix	Porphyry

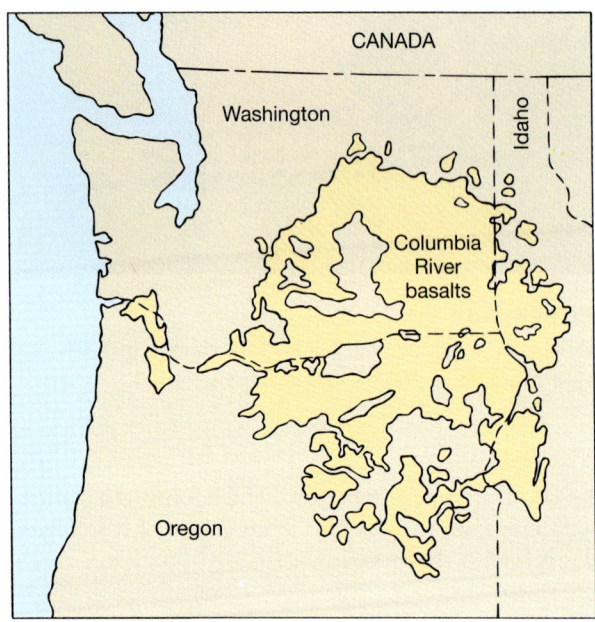

on continents. In some cases the fluid magma may flood thousands of square kilometers of land, forming a large **basalt plateau** (Fig. 3–10). **Gabbro** is mineralogically identical to basalt but has larger mineral grains because it is a plutonic rock.

Andesite is a volcanic rock intermediate in composition between basalt and granite. It is commonly gray or green and consists of plagioclase and dark minerals (usually biotite, hornblende, or pyroxene). It is named for the Andes Mountains, the volcanic chain on the western edge of South America, which is made up largely of this rock. Because it is volcanic, andesite typically has a very fine-grained texture. **Diorite** is the medium- to coarse-grained plutonic equivalent of andesite. Diorite plutons commonly underlie large areas of volcanic andesites, such as the Andes. They formed from the same magmas that produced the andesite but solidified in the crust beneath the volcanoes.

Sedimentary Rocks

Rocks of all kinds decompose, or **weather**, at the Earth's surface. Weathering breaks large rocks into smaller fragments such as gravel, sand, silt, and clay. All such weathered fragments are collectively called **sediment**. Streams, wind, glaciers, and gravity pick up and remove this sediment from the place where it formed by a process called **erosion**. The same agents then **transport** the sediment downhill and **deposit** it at lower elevations. The sand on a beach and mud on a mud flat accumulate in this way.

Lithification is a collection of processes that convert loose sediment to solid rock. The most common of these processes occurs when a mineral such as calcite or quartz precipitates from ground water seeping through loose sediment. The mineral forms a cement that glues the loose sediment together, creating a hard, solid rock.

With time, sand, mud, and other kinds of sediment become lithified to form **sedimentary rock.** The most common sedimentary rocks are **sandstone**, **shale**, and **limestone**. When beach sand is lithified, it turns to sandstone; mud becomes shale. Lithified shells of corals, clams, and other marine organisms become limestone. Because sediment accumulates in layers at the Earth's surface, sedimentary rocks show a distinctive layered structure called **bedding.** Sedimentary rocks make up less than 5 percent of the Earth's crust. However, because they form a thin veneer over about 80 percent of the continental land, it is easy to get the impression that sedimentary rocks are more abundant than they really are.

Sedimentary rocks are broadly divided into three categories based on the types of sediment of which they are made: clastic, organic, and chemical.

FIGURE 3–10
(A) The Columbia River basalt plateau covers much of Washington, Oregon, and Idaho. (B) Columbia River basalt in eastern Washington. Each layer is a separate lava flow.
(Larry Davis)

Thus, igneous rocks are classified in pairs. The members of each pair contain the same minerals but have different textures. The texture depends mainly on whether the rock is volcanic or plutonic in origin.

Basalt is a dark, very fine-grained volcanic rock. It is about half plagioclase and half pyroxene. Most oceanic crust is basalt. Basalt magma also erupts in great volumes

TABLE 3-5 Sizes and Names of Sedimentary Particles and Clastic Rocks			
Diameter (mm)	Sediment		Clastic Sedimentary Rock
256— 64— 2—	Boulders Cobbles Pebbles	Gravel (rubble)	Conglomerate (rounded particles) or breccia (angular particles)
1/16—	Sand		Sandstone
1/256—	Silt Clay	Mud	Siltstone Claystone } Mudstone or shale

Clastic Sedimentary Rocks

Clastic sedimentary rocks are composed of cemented clay, silt, sand, and gravel. The clastic particles, or **clasts**, are rock and mineral fragments or broken shells that have been physically transported and deposited (Table 3–5). Clastic rocks make up more than 80 percent of all sedimentary rocks.

Conglomerate (Fig. 3–11) is cemented gravel. **Sandstone** consists of cemented sand grains (Fig. 3–12). Of the nine rock-forming minerals, quartz is the most resistant to weathering. Feldspar and the other common minerals succumb to chemical attack and physical abrasion during weathering and transport. In contrast, about all that happens to quartz grains during weathering and transport is that they become rounded. Consequently, most sandstones consist of rounded quartz grains.

A

B

FIGURE 3–12
Sandstone is lithified sand. (A) A close-up of sandstone. Notice the well-rounded sand grains. (B) A sandstone cliff in Zion National Park, Utah.

FIGURE 3–11
Conglomerate is lithified gravel.

A　　　　　　　　　　　　　　　　　　　　　B

FIGURE 3–13
Shale is made up mostly of platy clays. Therefore, it shows very thin layering called fissility. (A) An outcrop of shale. (B) A close-up of the same outcrop.

Shale is a fine-grained clastic sedimentary rock (Fig. 3–13A). It consists of clay minerals and small amounts of quartz and feldspar. Shale has thin bedding called **fissility**, along which the rock splits easily (Fig. 3–13B). Clay minerals, like micas, have platy shapes. The plates stack like dishes or sheets of paper. The fissility of shales results from parallel alignment of the clay plates. Shale makes up about 70 percent of all sedimentary rocks (Fig. 3–14). Its abundance reflects the vast quantities of clay produced by weathering. Shale is usually gray to black due to the presence of decayed remains of plants and animals that are commonly deposited with clay. This organic material in shale is the source of most oil and natural gas.

Organic Sedimentary Rocks

Organic sedimentary rocks consist of the remains of plants and animals. **Chert** is pure quartz (Fig. 3–15). Microscopic examination of most chert shows that it is made up of the remains of tiny marine organisms that make their skeletons of silica.

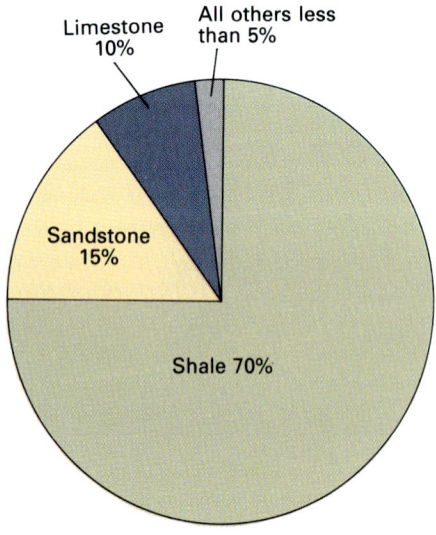

FIGURE 3–14
The relative abundances of sedimentary rocks.

FIGURE 3–15
Red nodules of chert in light-colored limestone.

Coal is composed of plant remains. When plants die, they usually decompose by reaction with oxygen. However, in warm swamps and other environments where plants grow rapidly, dead plants can accumulate so quickly that the available oxygen is used up before decay is complete. The partly decayed plant remains form **peat**. As peat is buried and compacted by overlying sediment, it converts to **coal,** a hard, black, combustible rock (Fig. 3–16).

Chemical Sedimentary Rocks

When rocks weather, some elements such as calcium, sodium, potassium, and magnesium dissolve. Those dissolved ions are transported by ground water and streams to the oceans or to saline lakes such as Great Salt Lake in Utah. When the concentration of the dissolved ions increases, or when other ions are added, some ions may bond together to form a solid mineral that **precipitates** from the solution. **Chemical sedimentary rocks** form by direct precipitation of minerals from solution. **Evaporites** are sedimentary rocks formed when evaporation of water concentrates dissolved ions to the point that they precipitate from solution. The most common minerals found in evaporite deposits are **gypsum**[1] ($CaSO_4 \cdot 2H_2O$) and **halite** (NaCl). Evaporites constitute only a small proportion of all sedimentary rocks but can be important sources of salt and other materials (Fig. 3–17).

A

B

FIGURE 3–16
(A) Coal forms from partly decayed plant remains. (B) This sample of organic-rich shale contains fossil ferns; it was taken between two coal beds.

[1] The $2H_2O$ in the chemical formula of gypsum means that water is incorporated into the mineral structure.

FOCUS ON

Rocks

The terms *basement rock, bedrock, parent rock,* and *country rock* are commonly used by geologists.

Basement rock is the igneous and metamorphic rock that lies beneath the thin layer of sediment and sedimentary rocks covering much of the Earth's surface, and thus it forms the basement of the crust.

Bedrock is the solid rock that lies beneath soil or unconsolidated sediments. It can be igneous, metamorphic, or sedimentary.

Parent rock is any original rock before it is changed by metamorphism or any other geologic process.

The rock enclosing or cut by an igneous intrusion or by a mineral deposit is called **country rock**.

FIGURE 3–17
The Bonneville Salt Flats, Utah.

Carbonate Rocks: Limestone and Dolomite

Limestone and dolomite can form by clastic, organic, and chemical processes. They are made up primarily of the minerals **calcite** and **dolomite**. They are called **carbonates** because both minerals contain the carbonate ion, CO_3^{2-}. Calcite-rich carbonate rocks are called **limestone**, whereas carbonate rocks rich in the mineral dolomite are called **dolomite**. Some geologists use the term **dolostone** for the rock to distinguish it from the mineral.

Seawater contains much dissolved calcium carbonate. Clams, corals, some types of algae, and many other marine organisms make their shells and other hard body parts of calcium carbonate. After the organisms die, waves or ocean currents break up and transport the shell fragments. Limestone formed by lithification of such sediment is called **bioclastic limestone**. The name indicates that both biological and clastic processes contributed to its origin. Most limestones are bioclastic. The bits and pieces of shells appear as fossils in the rock (Fig. 3–18). Some limestone also forms by chemical precipitation of calcium carbonate.

Metamorphic Rocks

Tectonic activity can force a portion of the Earth's crust downward. As the crust sinks, rocks that once were at the surface become buried by a thick pile of sediment accumulating in the depression. When rock is buried in this way, both temperature and pressure increase. The higher temperature and pressure change both the minerals and the texture of the rock. These changes are called **metamorphism**, and the rock is termed a **metamorphic rock**. Metamorphic changes can also occur when tectonic forces deform a rock during mountain building.

Two types of metamorphic reactions occur. As a general rule, when a **parent rock** (the original rock) contains only one mineral, metamorphism forms a rock composed of the same mineral but with a coarser texture. For example, fossils give fossiliferous limestone its texture

A B

FIGURE 3–18
Most limestone is lithified shell fragments and other remains of marine organisms. (A) A limestone mountain in Alberta, Canada. (B) A close-up of shell fragments in limestone.

(Fig. 3–19A). Both the fossils and the cement between them are made of small calcite crystals. If the limestone is buried and heated, the calcite grains grow larger. In the process, the fossils are destroyed. The resulting metamorphic rock, called **marble** (Fig. 3–19B), is still made of calcite, but its texture is now one of large interlocking grains. In a similar manner, metamorphism of quartz sandstone forms **quartzite**, a rock composed of recrystallized quartz grains.

In contrast, metamorphism of a parent rock containing several minerals usually forms new and different minerals. Shales commonly contain clay minerals as well as quartz and feldspar. During metamorphism, shales always grow entirely new minerals.

Radon and Rocks

Radon is a radioactive gas that commonly forms in some igneous and sedimentary rocks. Invisible, odorless, and tasteless, it occurs naturally in bedrock and soil. It seeps from the ground into homes and other buildings, where it concentrates and causes an estimated 5000 to 20,000 cancer deaths per year among Americans. The risk of dying of radon-caused lung cancer in the United States is about 0.4 percent over a lifetime, much greater than the risk of dying from cancer caused by asbestos, pesticides, or other air pollutants and nearly as high as the risk of dying in an auto accident or from a fall or fire at home.

However, Americans are not all exposed to equal amounts of radon. Some homes contain very low concentrations of the gas; others have high concentrations. The variation in concentration is due to two factors: geology and home ventilation.

Radon is one of a series of radioactive elements formed by the radioactive decay of uranium. Thus, radon forms wherever uranium occurs. Uranium occurs naturally in all types of rock in tiny amounts, but it concentrates in some types and occurs in only minuscule amounts in others. It concentrates in granite and shale. For this reason, radon concentrations are highest in poorly ventilated homes built on granite or shale bedrock or on soil derived from these rocks. In other cases, building materials contain rocks with high uranium contents. The highest home radon concentrations ever measured have occurred in houses built on the Reading Prong, a uranium-rich granite pluton that extends from Reading, Pennsylvania, through northern New Jersey and into New York. The air in one home in this area contained 700 times as much radon as the EPA "action level"—the concentration at which the Environmental Protection Agency recommends that measures be taken to reduce the amount of radon in indoor air.

Thus, a homeowner may ask two questions: "What is the radon concentration in my house?" and "If it is high, what can be done about it?" Because radon is radioactive, it can be measured with a simple detector that is available at most hardware stores and from local government agencies for about $25.

If the detector indicates excessive radon in a home, measures can be taken to remove it. Radon gas forms by slow radioactive decay of uranium in building materials, soil, or bedrock beneath a house. After radon seeps from the ground or foundation into the basement, it rises and accumulates in indoor air. An unventilated house seals out the fresh outdoor air and therefore accumulates radon. Three types of solutions have proved effective. The first is simply to extend a ventilation duct from the basement directly outside the house. In this way, air from the basement does not circulate throughout the entire house, and the basement air is changed frequently so that radon does not accumulate. The second solution is to ventilate the entire house so that indoor air is refreshed continually. Essentially, an open window suffices. However, an open window allows hot air to escape and thereby increases fuel consumption. A third solution is to pump outside air into the house to keep indoor air at a slightly higher pressure than the outside air. This positive pressure prevents gas from seeping from soil or bedrock into the basement.

It is impossible to avoid exposure to radon completely because it is everywhere, in outdoor air as well as in homes and other buildings. But it is relatively easy and inexpensive to minimize exposure and thus avoid a major cause of cancer.

FIGURE 3–19
Metamorphism converts fossiliferous limestone (A) to marble (B), which has a very different texture, although both are made of the mineral calcite.

Tectonic forces commonly deform rocks at the same time that the rocks are metamorphosed by rising temperature. When deformation and metamorphism occur together, the metamorphic rocks develop a layered texture called **foliation**. The metamorphic layers range from a fraction of a millimeter to a meter or more in thickness. This metamorphic layering results from alignment of platy minerals, such as mica, that grow during metamorphism. Metamorphic foliation superficially resembles sedimentary bedding, but is quite different.

When metamorphism of shale begins, clay minerals decompose and new minerals grow, forming **slate**. With increasing heat and pressure, more new minerals form and foliation becomes well developed. Rock of this type is called **schist** (Fig. 3–20). At high metamorphic grades, light- and dark-colored minerals often separate into bands 1 centimeter or more thick, to form a rock called **gneiss** (pronounced "nice").

The Rock Cycle

Igneous, sedimentary, and metamorphic rocks seem to be permanent features of the Earth over a human life span and even over the range of human history. Archaeologists have used biblical descriptions of rocky peaks to locate ancient ruins. But historical records go back only a few thousand years, whereas geologic time extends back 4.6 billion years. Over this much greater length of time, rocks change. In geologic time it is common for Earth processes to convert an igneous rock to a sedimentary rock or a sedimentary rock to a metamorphic rock. In turn, when a metamorphic rock becomes hot enough, it melts to form magma. The magma then cools to form a new igneous rock. Thus, no rock is permanent over geo-

FIGURE 3–20
When shale (A) is metamorphosed to schist (B), both a new texture and new minerals form. *(A-Geoffrey Sutton)*

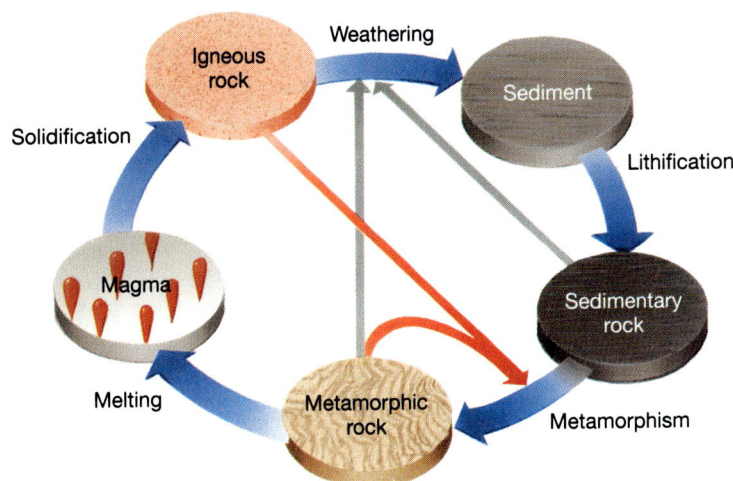

FIGURE 3-21
The rock cycle shows that rocks of the crust change continuously over geologic time. The arrows show paths that rocks can follow as they change.

logic time; instead, all rocks change slowly from one of the three types to another. This continuous transformation is called the **rock cycle** (Fig. 3-21).

Although the term *rock cycle* implies an orderly progression from one type of rock to another, such a regular sequence does not necessarily occur. Shortcuts are common, as shown by the arrows cutting across the circle of Figure 3-21. For example, a sedimentary or metamorphic rock may weather to form sediment. An igneous rock may be metamorphosed. The rock cycle simply expresses the concept that rocks are not permanent, but change continuously over geologic time.

3.4 The Atmosphere

The Earth's atmosphere is mostly gas with small quantities of water droplets and dust. The gaseous composition of dry air is roughly 78 percent nitrogen (N_2), 21 percent oxygen (O_2), and 1 percent other gases (Fig. 3-22, Table 3-6). Nitrogen, the most abundant gas, does not react readily with other substances. Oxygen, on the other hand, reacts chemically as fires burn, iron rusts, and plants and animals respire.

TABLE 3-6 Gaseous Composition of Natural Dry Air*	
Gas	Concentration (in percent)
Nitrogen, N_2	78.09
Oxygen, O_2	20.94
Inert gases, mostly argon, with much smaller concentrations of neon, helium, krypton, and xenon	0.93
Carbon dioxide, CO_2	0.03
Methane, CH_4, a natural part of the carbon cycle	0.0001
Hydrogen, H_2	0.00005
Oxides of nitrogen, mostly N_2O and NO_2, both produced by solar radiation and by lightning	0.00005
Carbon monoxide, CO, from oxidation of methane and other natural sources	0.00003
Ozone, O_3, produced by solar radiation and by lightning	Trace

*Natural dry air is defined as air without water or industrial pollutants. Carbon dioxide, methane, oxides of nitrogen, carbon monoxide, and ozone are all components of natural air, but they are also industrial pollutants. Therefore, the concentrations of these gases may vary, especially in urban areas.

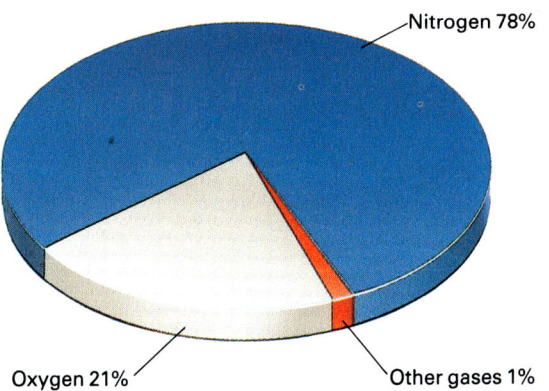

FIGURE 3-22
The gaseous composition of natural dry air. See Table 3-4 for a list of the 1 percent "other gases."

FIGURE 3–23
A thundercloud forming over the desert.

In addition to gases, the atmosphere contains water vapor, water droplets, and dust. The types and quantities of these components vary with both location and altitude. In a hot, steamy jungle, air may contain 5 percent water vapor by weight, whereas in a desert or cold polar region, only a small fraction of a percent may be present (Fig. 3–23).

If you sit in a house on a sunny day, you may see a sunbeam passing through a window. The visible beam is made up of light reflected from tiny specks of suspended dust. Clay, salt, pollen, bacteria, viruses, bits of cloth, hair, and skin are all components of dust. People travel to the seaside to enjoy the "salt air." Visitors to the Great Smoky Mountains in Tennessee view the bluish, hazy air formed by sunlight reflecting from pollen and other dust particles.

Within the past century, humans have altered the chemical composition of the atmosphere by releasing large quantities of pollutants (Fig. 3–24). Some are gases such as sulfur dioxide, carbon dioxide, and nitric oxide; others, such as soot and smoke, are dust particles.

If air in one region is heated above the temperature of surrounding air, this warm air becomes less dense and rises. As the warm air rises, cooler and denser air in another portion of the atmosphere sinks. Air then flows along the surface to complete the cycle (Fig. 3–25). The steady winds that blow across the tropical oceans, a tornado that ravages a city in the Midwest, and a thunderstorm that drops rain and hail on a Sunday picnic are all caused by air movement resulting from temperature differences.

3.5 The Hydrosphere

About 1.3 billion cubic kilometers of water exist at the Earth's surface. Of this huge quantity, 97.5 percent is salty seawater: The oceans cover about 71 percent of our planet. An additional 1.8 percent is frozen in the great ice sheets of Antarctica and Greenland. Only about 0.65 percent is fresh water in streams, lakes, and underground reservoirs (Fig. 3–26).

Fresh water is most conspicuous in streams and lakes. But about 30 times more fresh water is hidden just below the Earth's surface, saturating soil and bedrock as **ground water**. Ground water pumped from wells becomes drinking water for more than half of the population of North America and is a major source of water for irrigation and industry.

All land-dwelling organisms, including humans, need fresh water to survive. In addition, we use rivers and streams for transportation, irrigation, and industry, and we dam them to produce energy. We obtain protein from fresh-water fisheries.

FIGURE 3–24
Air pollution emitted from a wood products mill in Lewiston, Idaho.

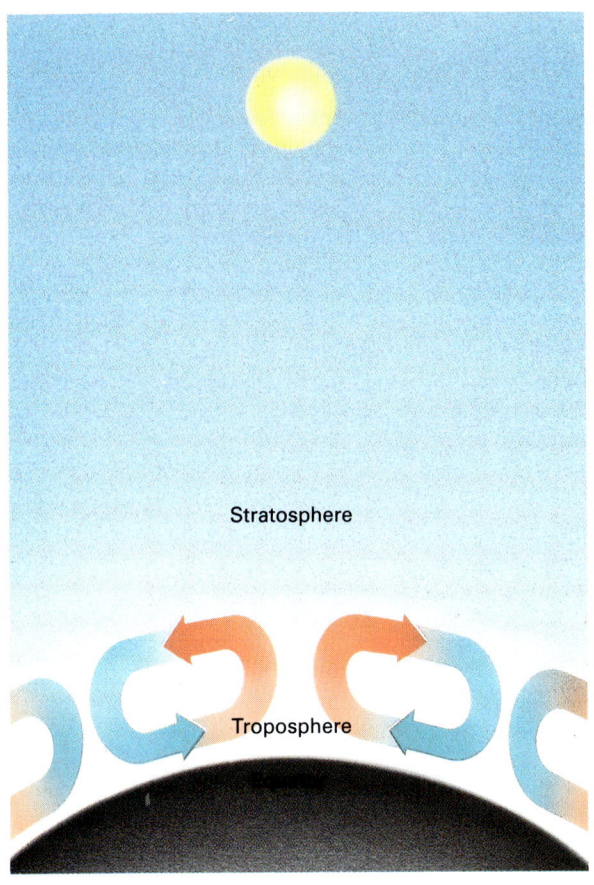

FIGURE 3–25
The Sun heats air near the Earth's surface, causing it to rise. Cooler air in other parts of the atmosphere sinks and flows along the surface to replace the rising air. We feel the air flowing over the surface as wind.

FIGURE 3–26
A lake in the Beartooth Mountains, Montana.

3.6 The Earth as a System

Tectonic forces, which move the solid Earth, may raise a mountain range, exposing fresh rock at the Earth's surface. Soon thereafter, rainwater attacks the rock chemically. Sulfur dioxide from volcanic eruptions and factories dissolves in cloud droplets to make the rain strongly acidic and even more corrosive. In addition, plant roots secrete chemical agents that hasten weathering. In this way, the atmosphere, hydrosphere, and biosphere act directly on the solid Earth to convert a barren, rocky surface to productive soil capable of sustaining an ecosystem, including humans.

But erosion removes part of the soil as it forms, and running water, glaciers (parts of the hydrosphere), wind (the atmosphere), and gravity carry the sediment downhill, eventually depositing it near a seacoast. Over time, the sediment is buried and heat and pressure convert it to sedimentary rock, returning it to the realm of the solid Earth. Eventually, tectonic events may force the rock to depths of several kilometers, where heat and pressure convert it to metamorphic rock or even melt it to form magma.

Much of environmental geology is a study of such interactions and the ways in which they affect humans. In the following chapters, we will learn how geologic change affects humans and how, in turn, humans alter the Earth.

SUMMARY

Our geologic environment includes the **solid Earth**, the **atmosphere**, the **hydrosphere**, and the **biosphere**. **Rocks** and **minerals** make up the solid Earth. A mineral is a naturally occurring inorganic solid with a definite chemical composition and a crystalline structure. The **crystalline structure** of a mineral is the orderly, periodically repeated arrangement of its atoms. Every mineral is distinguished from others by its chemical composition and crystal structure.

Common minerals are easily recognized and identified visually. Although more than 2500 minerals are known in the Earth's crust, only the nine **rock-forming minerals** are abundant in most rocks. They are **feldspar, quartz, pyroxene, amphibole, mica,** the **clay minerals, olivine, calcite,** and **dolomite.** The first seven on this list are **silicates.** The silicates are the most abundant minerals because silicon and oxygen are the two most abundant elements in the Earth's crust and bond together readily to form the **silicate tetrahedron.**

Ore minerals and **gems** are important for economic reasons. **Accessory minerals** are found commonly but in small amounts. A few minerals, such as asbestos and those containing lead, arsenic, and other poisonous elements, are dangerous to human health.

Geologists divide rocks into three groups depending on how they formed. **Igneous rocks** solidify from **magma**. **Sedimentary rocks** form by **lithification** of **sediment** that collects at the Earth's surface. **Metamorphic rocks** form when any rock is changed by temperature, pressure, or deformation. The **rock cycle** summarizes the processes by which rocks continuously recycle in the outer layers of the Earth, with new rocks forming from old ones.

The atmosphere is mostly gas. The gaseous composition of dry air is about 78 percent nitrogen (N_2), 21 percent oxygen (O_2), and 1 percent other gases. In addition to gases, air contains water vapor, water droplets, and dust. Some human activities pollute the atmosphere and modify its composition in ways that may cause climatic changes.

Of the 1.3 billion cubic kilometers of water at the Earth's surface, 97.5 percent is salty seawater, 1.8 percent is frozen in the ice sheets of Antarctica and Greenland, and only 0.65 percent is fresh water in streams, lakes, and **ground water**. As is the case with the atmosphere, human activities pollute water supplies and, in some cases, deplete water resources.

Common Igneous Rocks in the Earth's Crust

Plutonic rocks	Granite	Diorite	Gabbro
Volcanic rocks	Rhyolite	Andesite	Basalt

Common Sedimentary Rocks

Sandstone	Shale	Limestone	Dolomite

Common Metamorphic Rocks

Marble	Slate	Schist	Gneiss

KEY TERMS

Element 40
Atom 40
Ions 40
Mineral 40
Crystal 41
Crystalline structure 41
Crystal faces 41
Rock-forming minerals 41
Oxides 42
Silicate tetrahedron 43
Ore mineral 43

Gem 45
Accessory minerals 45
Magma 46
Igneous rocks 46
Intrusive igneous rocks 46
Plutonic rock 46
Extrusive igneous rocks 46
Volcanic rocks 46
Lava 47
Weathering 48
Sediment 48

Lithification 48
Sedimentary rock 48
Bedding 48
Clastic sedimentary rock 49
Conglomerate 49
Organic sedimentary rocks 50
Chemical sedimentary rocks 51
Bioclastic limestone 52
Metamorphic rock 52
Foliation 53
Rock cycle 53

REVIEW QUESTIONS

1. What properties distinguish minerals from other substances?
2. What does the chemical formula for quartz, SiO_2, tell you about its chemical composition? What does $KAlSi_3O_8$ tell you about orthoclase feldspar?
3. List the eight most abundant chemical elements in the Earth's crust. Are any unfamiliar to you? List familiar elements that are not among the eight. Why are they familiar?
4. What is an atom? An ion? What roles do they play in minerals?
5. Every mineral has a crystalline structure. What does this mean?
6. List the rock-forming minerals. Why are they called "rock-forming"? Which are silicates? Why are so many of them silicates?
7. Why is asbestos considered to be an environmental hazard?
8. Explain the differences between the two types of asbestos.
9. What are the three main types of rock in the Earth's crust?
10. How do the three main types of rock differ from each other?
11. What is magma, and where does it originate?
12. Describe and explain the differences between a plutonic rock and a volcanic rock.
13. What is sediment? How does it form?
14. Describe how sediment becomes sedimentary rock.
15. What are the differences and similarities among shale, sandstone, and limestone?
16. Explain why almost all sedimentary rocks are layered, or bedded.
17. What is metamorphism? What factors cause metamorphism?
18. What kinds of changes occur in a rock as it is metamorphosed?
19. What is metamorphic foliation? How does it differ from sedimentary bedding?
20. Explain what the rock cycle tells us about Earth processes.
21. Why do radon concentrations in indoor air vary widely, both regionally and among individual buildings?

DISCUSSION QUESTIONS

1. Diamond and graphite are two minerals with identical chemical compositions of pure carbon (C). Diamond is the hardest of all minerals, and graphite is one of the softest. If their compositions are identical, why do they have such profound differences in physical properties?
2. Silicon and oxygen together make up nearly 75 percent by weight of the Earth's crust. But silicate minerals make up more than 95 percent of the crust. Explain the difference.
3. Would you expect minerals found on the Moon, Mars, or Venus to be different from those of the Earth's crust? Explain your answer.
4. Make a list of minerals, besides asbestos, that present health hazards or other kinds of environmental threats. Discuss the reasons that each mineral on your list is hazardous.
5. Magma usually begins to rise toward the Earth's surface as soon as it forms. It rarely accumulates as large pools in the upper mantle or lower crust, where it originates. Why?
6. In the San Juan Mountains of Colorado, parts of the range are made up of granite, and other parts are volcanic rocks. Explain why these two types of igneous rock are likely to be found together.
7. Suppose you are given a fist-size sample of igneous rock. How can you tell whether it is volcanic or plutonic in origin?
8. How can you tell whether another rock sample is igneous, sedimentary, or metamorphic in origin?
9. Why is shale the most common sedimentary rock?
10. One sedimentary rock is composed of rounded grains, and the grains in another are angular. What can you tell about the histories of the two rocks from these observations?
11. What happens to bedding when sedimentary rocks undergo regional metamorphism?
12. How can granite form during metamorphism?

INTERVIEW

A CONVERSATION WITH
Cindy Lee Van Dover

Cindy Lee Van Dover received a B.S. in environmental science from Rutgers University (1977), followed by an M.S. in ecology from the University of California, Los Angeles (1985). In 1980 she published the first of three articles on decapod crustaceans from the Indian River region of Florida. Then, in 1982, she participated in the Oasis expedition to hydrothermal vents in the East Pacific rise. This was only the beginning of her relationship with the deep ocean and specifically hydrothermal vent communities. She took part in several other expeditions—aboard the research vessel Atlantis II and, as a scientist, aboard the submersible ALVIN—to places such as the Rose Garden and Gorda ridge.

In 1989 Van Dover completed a Ph.D. in biological oceanography in the Massachusetts Institute of Technology and Woods Hole Oceanographic Institution Joint Program. Since then she has held the position of postdoctoral investigator with the Biology Department at Woods Hole Oceanographic Institution in Woods Hole, Massachusetts. Also in 1989, Van Dover began her pilot training, and by 1990 she had piloted ALVIN to a variety of underwater locations throughout the Pacific Ocean. Later that year, she participated in a joint United States–Soviet Union dive aboard the MIR-2 submarine in Monterey Bay, off Baja California, Mexico, and in the nearby Guaymas basin. At the controls of ALVIN, Van Dover explores hydrothermal vents with dexterity and thoroughness, combining her expert knowledge of sea life with her Navy-certified piloting skills. Most recently, she has employed data gathered from ALVIN expeditions to study the effects of sludge from deep-sea waste dumping on the biota of hydrothermal vent communities.

Van Dover is a member of the Deep Submersible Pilots' Association and the American Geophysical Union. In addition to her considerable scholarly contributions, she has published several nontechnical articles on the undersea world as seen from her unusual perspective. In 1988 she was honored as one of Ms. magazine's "women of the year." Her commitment to science and her pioneering spirit have melded to help unlock the mysteries of the deep ocean, one of the few remaining earthly frontiers.

Where were you born and where did you attend school?

I was born and grew up in New Jersey, about 5 miles from the coast, in Eatontown; I went to a regional high school there, which happened to have a summer marine biology program and excellent teachers. That's how I got into marine science to begin with. As an undergraduate, I went to Rutgers University. After that, I spent some years working up and down the East Coast at various marine labs, as a technician. Eventually, I traveled west to earn my master's degree at UCLA. Then, back once more on the East Coast, I worked as a technician at the Marine Biological Lab in Woods Hole before starting work on my Ph.D. in the Woods Hole Oceanographic Institution–Massachusetts Institute of Technology Joint Program in biological oceanography.

So, from an early time in your career, you've been involved in marine biology or biological oceanography.

I always liked animals—especially invertebrates, because they come in so many different forms. Bugs and spiders I still think are creepy, but the animals that live in water are wonderful; some look more like plants than animals, others have fantastic features—feather-like appendages or multiple pairs of tiny eyes. Marine animals can be the stuff of science fiction, and I wanted to know more about them.

Then, too, I always wanted to be on a boat when I was little, even though I would get terribly seasick. I don't know where the sense of the romance of the sea comes from, but I certainly had it. Partly the animals fascinated me. But I admit that the adventure, the sense of independence or strength, the chance to prove myself against the sea, lies behind my wanting to sail.

What did you do your Ph.D. on?

My Ph.D. was on the ecology of deep-sea hydrothermal vents. I started off my dissertation work studying food webs at vents—who eats whom. Geologists had just discovered hydrothermal vents on the Mid-Atlantic ridge, back in 1985. Thousands of shrimp swarm over the sulfide chimneys there. One scientist described them as looking like maggots swarming on a hunk of rotten meat—a not very poetic, but very accurate, description. My advisor was given some specimens; he passed them on to me: "Find out what they are doing, how they are making a living."

As part of my dissertation work, I joined an ALVIN expedition to vent sites off the coast of Washington. I had learned that the chief scientist was planning to use an electronic camera developed here in Woods Hole. At the local bar on a Friday afternoon, I talked with the engineers who built the camera and learned that it ought to be able to detect the sort of light we supposed might be generated at a high-temperature vent. The chief scientist was willing to configure the camera appropriately, position it in front of a black smoker, turn out the submersible lights, and collect a time exposure of the ambient light. He and one of his colleagues were able to do this successfully, capturing a spectacular image of a glow emitted by the hot water.

Can you tell us some more about what these black smokers are, and something about the animal or plant life associated with them?

Black smokers occur along mid-ocean ridge spreading centers, which are amazing places. Along the spreading centers, the Earth's plates are pulling apart and new ocean crust is forming. It is a very black, high-contrast environment with incredible terrain, basically a linear volcano; pillowed and ponded lavas cover the sea floor. It is a very fluid-looking, dynamic landscape, with lava in frozen pools and swirls, buckles and fractures. Fissures there are often big enough for the submersible to drive into; exposed along the walls are the histories of eruptions, flows on top of flows.

Where the sea floor is moving apart, there is a lot of earthquake activity, a lot of cracking of the basalt. Seawater fills those cracks and actually reaches, if not the magma chamber, then very, very hot rock. It is heated, reacts with the rock, and becomes modified. The heated water is buoyant and rises up through the crust to exit at the sea floor with a very interesting chemistry—enriched in metal sulfides and depleted in magnesium and oxygen. It is the chemical characteristics of the vent water that support microorganisms, the bacteria at the base of the food web. Life at vents is based *not* on photosynthesis, like we are used to in shallow-water and terrestrial systems, but on chemosynthesis. Instead of using light energy, the microorganisms use energy in reduced compounds to fix carbon. Instead of plants, the bacteria are the "grasses" of the hydrothermal vent community. The bacteria live both as free-living organisms suspended in the water column and on surfaces of rocks, sulfides, and animals; they also live as endosymbionts in tube worms and bivalves, supplying nutrients to their invertebrate hosts.

Can you describe the macrofauna around a black smoker a little bit more? What, in addition to tube worms, lives there?

In some places tube worms are the dominant organism. They are beautiful; they have bright red plumes and very white tubes. Living with them are a variety of crabs and other crustaceans. There can also be large golden-brown mussels and giant white clams with blood-red bodies occurring in large numbers at the bases of black smokers, in the cracks and crevices where warm, sulfide-laden vent fluids are flowing.

How did you originally become interested in learning to pilot the ALVIN?

I think I published something like a dozen papers on the biology or ecology of vent fauna, but I'd only dived once to one vent. I was writing about animals I never saw alive, in their natural setting. I would go on lots of cruises, out at sea for a month or more each time, yet felt very lucky to get one dive. A simple look around me showed me who dove the most: the pilots. They could dive every other day, every three days, depending on the number of pilots. If you are a scientist, you dive only at the specific site you wrote the proposal for—you can't go all over the globe, diving anywhere. But the pilots get to go to all the different sites. As an ecologist, it seemed important to get a broad perspective of what the sea floor and hydrothermal vents are like. So that was the scientific motivation. I also wanted to do it just because of the adventure of it.

How did you first get involved with the ALVIN group?

ALVIN is Navy-certified, which

means they have to get recertified at intervals and prove that they are operating a safe sub. Traditionally, repairs on ALVIN were supervised by a shop manager, and he knew how to maintain the thing. There was little in the way of procedures written down. But the Navy insisted they put together a maintenance manual. I came along at the right time with the talent to write and organize a document—I'd just finished my Ph.D. dissertation as proof of that ability. I began as a volunteer and was eventually hired.

What sort of training do you need to drive, or pilot, the submarine?

The training is all on the job, and when I trained, it was all done independently; the schematics and instrument manuals were in files and I was encouraged to work my way through them all. In order to qualify, you have to pass a qualifying dive with the chief pilot, proving you can competently operate the sub. You also have to pass a series of qualifying boards, with the two most important being an Engineering Board and a Navy Board. During the Engineering Board, you stand in front of the engineers who designed and built the submarine. They have you draw and explain the schematics of every major system in the submarine. It was, at my exam, an unfriendly board. I passed. The Navy Board was mostly related to safety procedures and was a thoroughly professional exchange.

Have you ever had to troubleshoot beneath the surface?

All the time. We constantly work with untested scientific gear that needs troubleshooting. Some of the failures are more scary than others. You learn there are certain failures you cannot continue a dive with; when one of those happens, you really pay attention.

As an example, once I had a 10K leak indication, which meant that I had water in one of the outside electrical boxes. These boxes are oil-filled; water in a box could short out the system and cause serious damage. When I saw that 10K indication, my pulse shot up rapidly. You see, at first you don't know where the 10K leak is. It could be in a battery box—bad news; these are 120-volt batteries and they don't like salt water. There are a series of toggle switches to go through that will locate where the problem is. You go through them one at a time, systematically, fast. It is tense. You know once you find out where the problem is, you'll know what to do. Still, it's tense. In this particular case, the problem was in the variable ballast system. I just shut that system down and continued the dive. On deck we discovered that a single drop of water had closed the circuit in the leak detector. But there was no way of knowing the problem was minor under water.

The moment of a failure is always a challenge; you have to be on top of things, know your submarine; you have to know those schematics to know what to do next.

What do you see as your professional direction in the next decade?

I hope to continue to be a research scientist and to continue to study the ecology of the sea floor. It is an important field if we are to appreciate the world in which we live—not the small proportion that we ourselves occupy and modify, but the vast ocean environment that buffers our world and makes it livable. Continuing work on hydrothermal vents is an exciting prospect, pure scientific adventure, where the rules have not even begun to be defined.

Tube worms and other organisms living on sulfide chimneys at the Juan de Fuca ridge. *(Cindy Lee Van Dover)*

I'm also involved in some of the sewage sludge disposal issues off New Jersey. I'm interested in trying to trace the entry of sewage sludge into the benthic community of the deep sea. Dumping in deep water was initially started with the belief that none of the discharged materials would be detectable on the sea floor—the particles were so fine, they would be diluted and dispersed in the surface waters. My advisor and some of his colleagues thought maybe that wasn't the case and obtained funding to investigate.

My contribution to the project is to use stable isotopes as tracers of the sewage sludge. The technique requires an hypothesis that the organisms have two potential, isotopically distinct food sources. In this case, the sewage sludge has a very distinctive sulfur isotopic composition compared to normal sources of deep-sea organic sulfur. We collected benthic invertebrates—sea stars, sea urchins—from within the dump site and from a reference area upstream along the same depth contour. From the isotopic data for one species, an urchin, there was a strong sewage-sludge signal in the dump-site specimens. In fact, as much as 25 to 30 percent of the organic sulfur in dump-site urchins may be derived from sewage sludge.

What do you think that means? Is it good or bad?

Well, that's a little hard to say. It is a perturbation. In a way, it is good in that the sewage sludge provides a source of nutrition for the urchins in a food-limited environment. But the urchins may pay a price in accumulating contaminants. There is the chance for magnification in the food web. It might also select for opportunistic species and change the structure of the benthic community. What strikes me as significant is the fact that the original thought was "Oh, we don't need to worry about this; it can't possibly get there." But sewage sludge is reaching the sea floor. If we can still be so naive about what we are doing to the environment, there is a lesson: we need to assess what we are doing a little more carefully, we need to know more about the deep sea.

What would you tell a student who might be interested in further investigating or even professionally pursuing a field of biological oceanography or marine biology, your field?

If you make it in this field, you will have a life that is exciting and adventurous. I believe there truly are unimaginable discoveries waiting for us in the oceans. But you only get there through hard work and creative thinking. You don't do this for money or prestige or glory. Each scientist has a philosophy, a way of viewing the world and tackling questions. My approach is probably more unorthodox than most, and I pay my dues for that. I don't hesitate to cut across boundaries, to venture into fields that I know nothing about in order to get at some aspect of my work. Becoming an *ALVIN* pilot is a good example of that; I had no technical skills appropriate for that job. My first cruise with the *ALVIN* group, I sailed as an electrician, of all things! But I learned, and learned fast. Have the confidence in yourself to go after what you want; have the consciousness of your abilities to know how far you can go.

PART 2

Internal Processes

4 Plate Tectonics
5 Earthquakes
6 Volcanoes

A cascade of molten lava pours from Mauna Loa volcano during the July 1993 eruption. (J. D. Griggs/USGS)

CHAPTER 4

Plate Tectonics

4.1   The Earth's Layers
4.2   Plates and Plate Tectonics
4.3   Divergent Plate Boundaries
4.4   Convergent Plate Boundaries
4.5   Transform Plate Boundaries
4.6   The Anatomy of a Tectonic Plate
4.7   The Search for a Mechanism
4.8   Supercontinents
4.9   Plate Tectonics and Mountain Building
4.10 Plate Tectonics and the Sea Floor

Science usually creeps forward by innumerable little discoveries, each won by months or years of hard work in the field or laboratory. Occasionally, however, scientists gather all the little advances into a new idea or a new way of looking at old ideas to initiate a major scientific revolution. Such a revolution occurred in geology in the 1960s, and changed the science forever. Its effects are as exciting and important to geologists as Einstein's theory of relativity was to physicists earlier in this century. It allows us to comprehend many of the geologic events that create and change our environment, including earthquakes, volcanic eruptions, mountain building, and some climatic changes. Furthermore, because geologists now understand the forces that cause these events, we are better able to predict and respond to many such hazards and changes before they affect human lives.

The theory that has evolved from this revolution is called the **plate tectonics theory**. Briefly, this theory describes the Earth's outer layer as a 100-kilometer-thick shell of rigid, brittle rock called the **lithosphere**, which is broken into independent segments called **plates** (Fig. 4–1). The plates float on a layer of hot, soft rock called the **asthenosphere**. They move horizontally across the Earth's surface by gliding over the asthenosphere, like sheets of ice floating back and forth on a pond.

A **fault** is any fracture in rock along which movement has occurred. A **plate boundary** is a fault separating one plate from another. Because the plates move relative to one another as they float over the asthenosphere, they collide at some of the plate boundaries. At other boundaries, two plates move apart from each other; and at yet other boundaries, two plates move horizontally past each other. California's famous San Andreas fault is an example of this last type of plate boundary.

A time exposure shows Izalco volcano in El Salvador erupting at night.

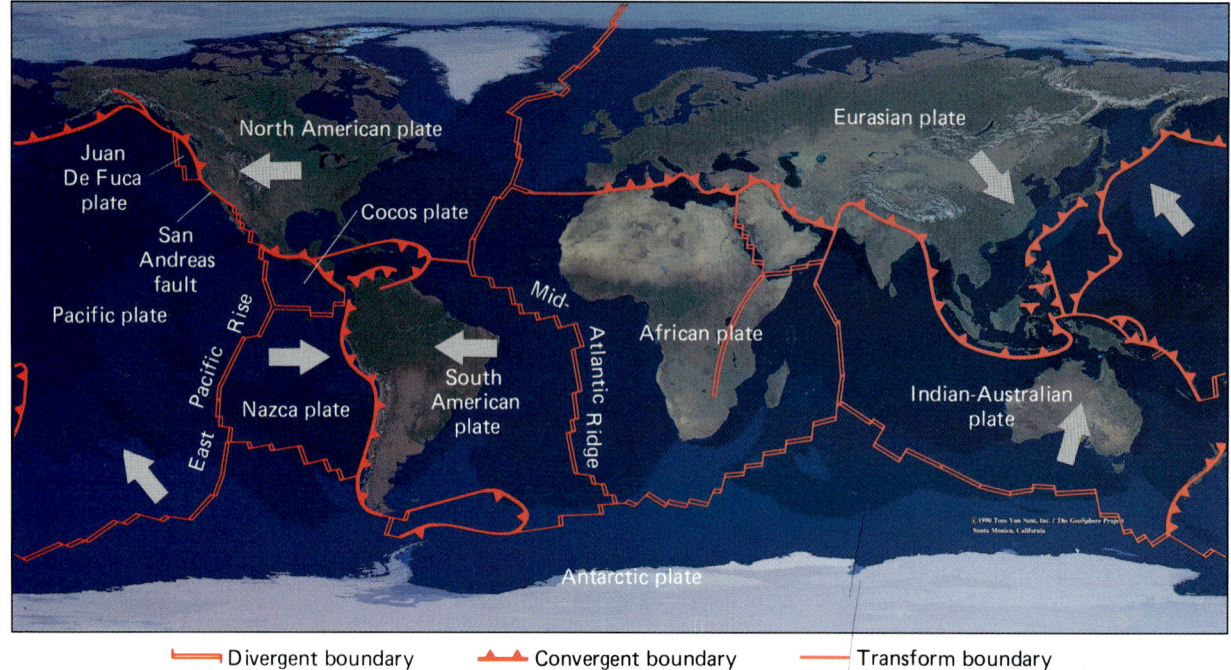

FIGURE 4-1

Lithospheric plate boundaries are shown in red. The major plates are the African, Eurasian, Indian-Australian, Antarctic, Pacific, North American, and South American. A few of the smaller plates are also shown. Gray arrows indicate directions of plate movement. *(Tom Van Sant, Geosphere Project)*

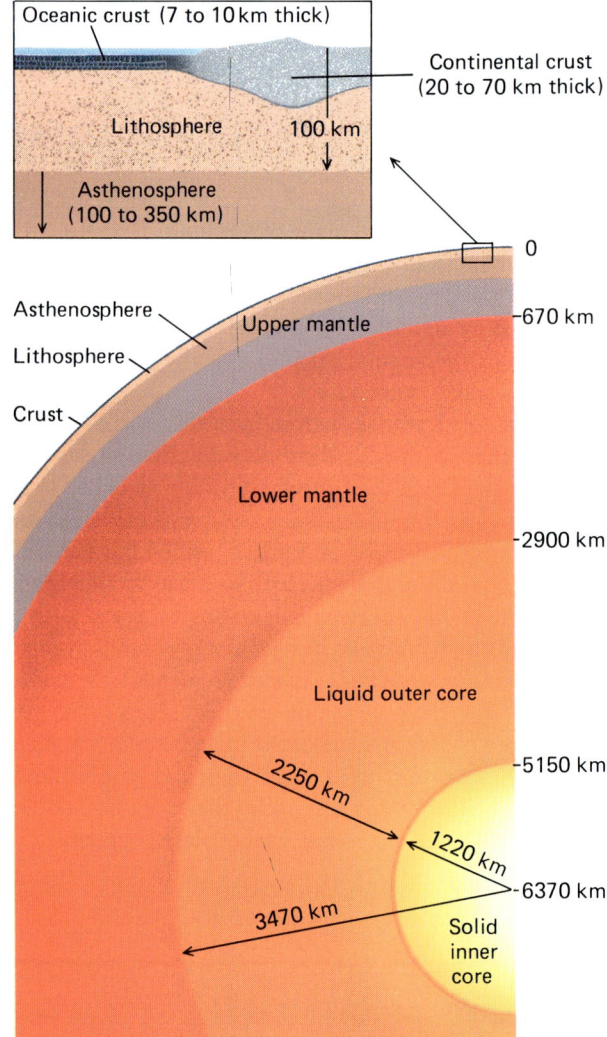

FIGURE 4-2

The Earth is a layered planet. The insert is drawn on an expanded scale to show near-surface layering.

Because 100-kilometer-thick plates bump and grind together at a plate boundary, you can easily imagine that a boundary is a geologically active place. As tectonic plates move, continents migrate around the globe, ocean basins open and close, mountain ranges rise, earthquakes shake the ground, and volcanoes erupt dust and gas into the atmosphere. These processes affect the Earth's climate; they also remind us of the interrelationships among the solid Earth, oceans, atmosphere, and life.

4.1 The Earth's Layers

The Earth is a layered planet, and the theory of plate tectonics depends on an understanding of the Earth's layers.

The Crust

Figure 4–2 shows a cross-sectional view of the Earth, and Table 4–1 summarizes properties of its layers. The **crust** is the thinnest and outermost layer. Both the thickness and the composition of the crust vary. Oceanic crust ranges from 7 to 10 kilometers thick and is composed mostly of basalt. Continental crust is much thicker. Continents not only rise above the ocean floor, but they have roots that extend downward into the mantle. In mid-continent regions, the crust is about 20 to 40 kilometers thick; under mountain ranges its thickness is as much as 70 kilometers. Most continental crust is of granitic composition.

Rocks of the crust are rigid and brittle. It is important to understand that rigidity and brittleness are relative. Imagine that you could cut out the state of Kansas from the surface to a depth of 35 kilometers and place it on top of a neighboring state. It could neither support its own weight nor hold its shape. The edges would crumble and Kansas would flow like honey over the Great Plains.

The Mantle

The **mantle** lies directly below the crust. It is almost 2900 kilometers thick and makes up about 80 percent

TABLE 4–1 The Layers of the Earth

	Layer	Composition	Depth	Properties
Crust	Oceanic crust	Basalt	7–10 km	Cool, rigid, and brittle
	Continental crust	Granite	20–70 km	Cool, rigid, and brittle
Lithosphere	Lithosphere includes the crust and the uppermost portion of the mantle	Varies; the crust and the mantle have different compositions	About 100 km	Cool, rigid, and brittle
Mantle	Uppermost portion of the mantle included as part of the lithosphere	Entire mantle is ultramafic rock; its mineralogy varies with depth		
	Asthenosphere		Extends from 100 to 350 km	Hot and plastic, 1% or 2% melted
	Remainder of upper mantle		Extends from 350 to 670 km	Hot, under great pressure, rigid, and brittle
	Lower mantle		Extends from 670 to 2900 km	High pressure forms minerals different from those of the upper mantle
Core	Outer core	Iron and nickel	Extends from 2900 to 5150 km	Liquid
	Inner core	Iron and nickel	Extends from 5150 km to the center of the Earth	Solid

of the Earth's volume. Its chemical composition is nearly constant throughout. However, Earth's temperature and pressure increase with depth, and these changes cause the minerals and the physical properties of the mantle to vary with depth. The upper part of the mantle consists of two layers.

The Lithosphere

The outer 100 kilometers of the Earth, including both the uppermost mantle and the crust, make up a layer called the **lithosphere** (Greek for "rock layer"). The lithosphere is close enough to the surface that it is relatively cool. As a result, the rock in this layer is rigid relative to the rocks below. Although the compositions of the crust and upper mantle are different, this 100-kilometer-thick zone behaves as a single layer. Most earthquakes originate in the rigid rock of the lithosphere.

The Asthenosphere

At a depth of about 100 kilometers, the cool, rigid rock of the lithosphere suddenly gives way to the hot, soft, plastic rock of the **asthenosphere**. The rock of this layer is so hot and plastic that it flows readily, even though it is solid. (To visualize a solid that can flow, think of Silly Putty® or road tar on a hot day.) The asthenosphere extends from the base of the lithosphere to a depth of about 350 kilometers. At the base of the asthenosphere, increasing pressure causes the mantle to become more rigid and less plastic, and it remains in this state all the way down to the core.

Isostasy

The lithosphere is less dense than the asthenosphere. Therefore, it floats on the asthenosphere much as an iceberg or a block of wood floats on water. The concept of a floating lithosphere is called **isostasy**.

To illustrate isostasy, imagine three icebergs of different sizes floating in the ocean. When ice floats in sea water, approximately 10 percent of it lies above water, whereas the remaining 90 percent is under water. Of the three bergs in Figure 4–3A, the largest has the highest peak but it also extends to the greatest depth. The lithosphere behaves in a similar manner. Continents stick up above sea level, but they also have "roots" that extend into the mantle. High mountain ranges have deeper roots than low plains, just as the bottom of a large iceberg is deeper than the base of a smaller one.

If the theory of isostasy is correct, it should explain why the ocean floor is lower than the continents. Even if both the ocean floor and the continents were made of the same material, the thinner oceanic lithosphere should float lower in the asthenosphere, just as a thin iceberg sits lower in the water than a thick one does. In addition, oceanic crust is basalt, which is denser than granite. The denser oceanic lithosphere settles down even farther in the asthenosphere, enhancing the difference in elevation between ocean basins and continents.

If you have ever loaded equipment onto a small boat, you may have noticed that the boat settles lower in the water as you add more cargo, whereas the boat rises when it is unloaded. The lithosphere behaves in a similar manner. But how is "cargo" added to or subtracted from

A

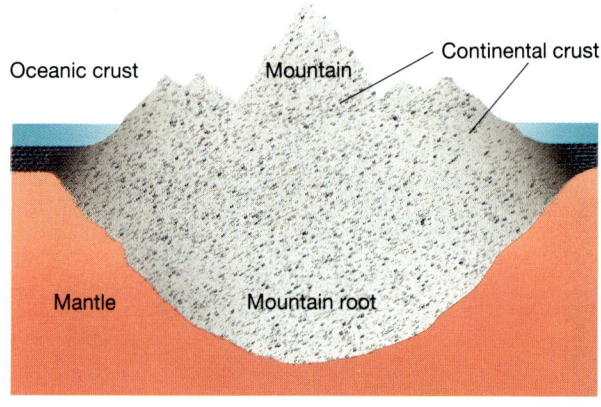
B

FIGURE 4–3
The principle of isostasy. (A) The largest of the three icebergs has the most material under water and also the most above. (B) In an analogous manner, continental crust extends more deeply into the mantle beneath high mountains than it does under lower areas of the continents.

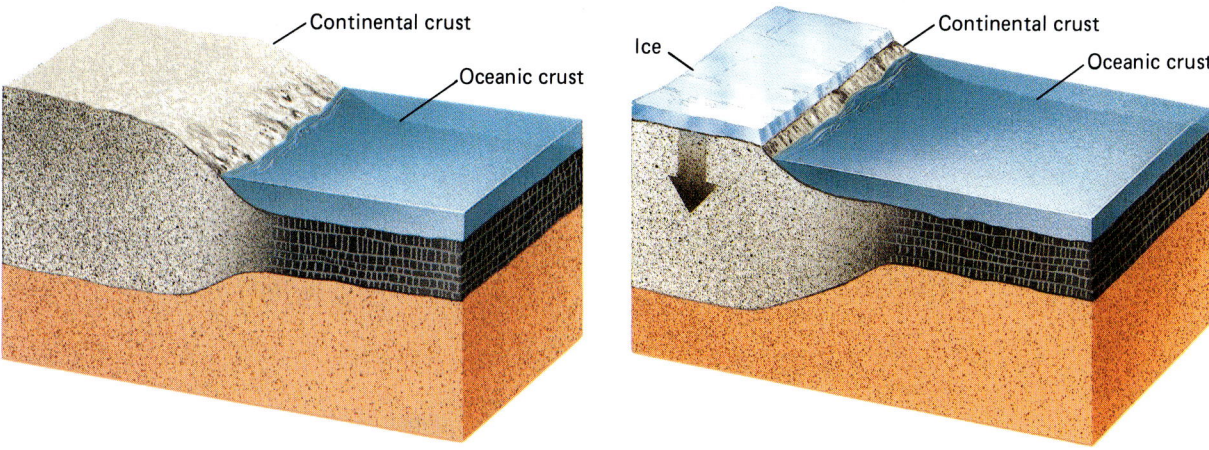

FIGURE 4–4
The weight of an ice sheet causes continental crust to sink isostatically.

the lithosphere? One mechanism that adds and removes weight is the growth and melting of large glaciers. When a glacier grows, the added weight of ice forces the continent to sink. The central portion of Greenland lies under a 3000-meter-thick ice sheet. The ice has depressed the continental crust below sea level. On the other hand, when a glacier melts, the continent rebounds, or rises up. Geologists have discovered ice age beaches tens of meters above modern sea level because the land rose as glaciers melted. This vertical movement in response to a changing burden is called **isostatic adjustment** (Fig. 4–4). Mountains also rise isostatically as rain, glaciers, and landslides erode their tops. Artificial structures can also trigger isostatic adjustment. When a dam is built, the new lake may depress lithosphere by a measurable amount. This sinking often causes swarms of small earthquakes.

The Core

During the early history of the Solar System, Earth was a homogeneous mixture of dust and gas. As the planet evolved, it became hotter until it began to melt. Once the Earth melted, heavy iron and nickel sank toward the Earth's center whereas lighter material floated toward the surface. As a result, the **core** consists mostly of the heavy metals iron and nickel. It is a sphere with a radius of about 3470 kilometers, larger than the planet Mars. The outer core is molten because of the high temperatures near the Earth's center. The inner core is even hotter, but it is solid because the higher pressures in the inner core overwhelm the temperature effect. Near its center the core has a temperature of about 6000°C, as high as that at the surface of the Sun. The pressure exceeds 1 million atmospheres.

4.2 Plates and Plate Tectonics

The plate tectonics theory is a model of the Earth in which the rigid lithosphere floats on the hot, plastic asthenosphere. The lithosphere is broken into seven large plates and several small ones, resembling segments of a turtle's shell. The plates move across the Earth's surface, each in a different direction from its neighbors. They glide slowly over the weak, plastic asthenosphere at rates ranging from less than 1 to about 18 centimeters per year, about as fast as a fingernail grows.

Because the plates move in different directions, they bump and grind together at their boundaries. Just imagine two 100-kilometer-thick slabs of rock colliding along a boundary a few thousand kilometers in length. No convenient analogy exists to describe such a collision. If a billion large bulldozers were to drive into a giant city made of all the buildings on Earth, the force would be tiny compared with that resulting from a collision of two tectonic plates. Because of the great forces generated at plate boundaries, mountains rise, volcanoes erupt, and earthquakes shake the land where two plates meet. These events are called tectonic activity, from the ancient Greek word for "construction." Tectonic activity constructs and modifies the Earth's crust. In contrast with plate boundaries, interior portions of lithospheric plates are usually tectonically quiet because they are far from the zones where two plates interact.

Simple logic tells us that one plate can move relative

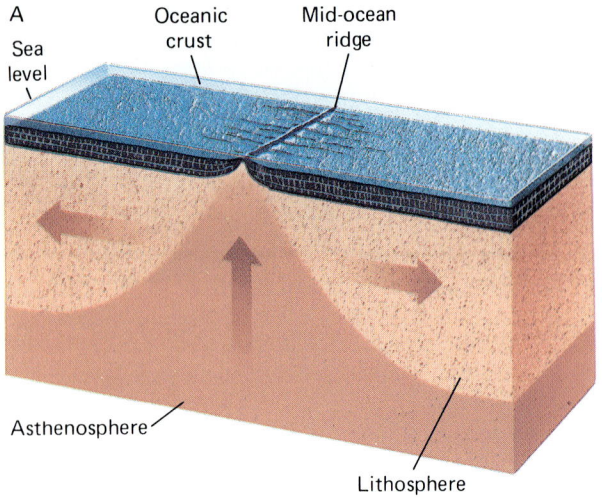
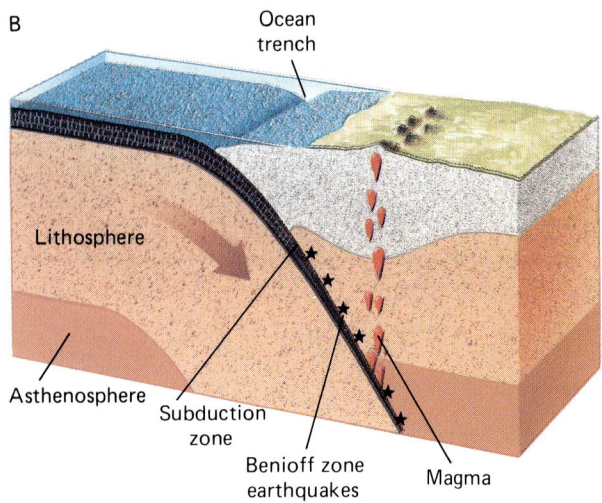
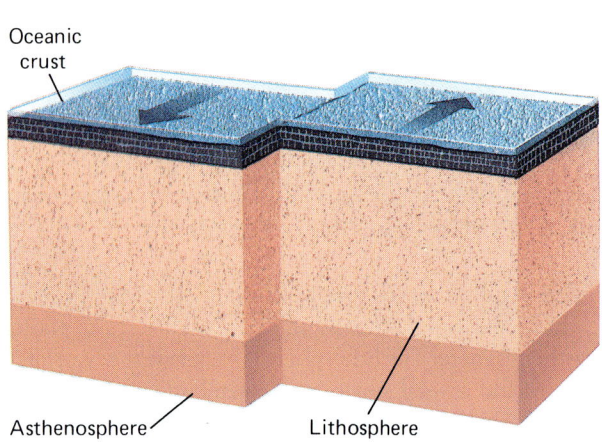

FIGURE 4–5

Three types of plate boundaries. (A) Two plates separate at a divergent boundary. New lithosphere forms as hot asthenosphere rises to fill the gap where the two plates spread apart. The lithosphere is relatively thin at this type of boundary. (B) Two plates collide at a convergent boundary. If one or both plates carry oceanic crust, the dense oceanic plate sinks into the mantle in a subduction zone. Here an oceanic plate is sinking beneath a less-dense continental plate. Magma rises from the subduction zone, and a trench forms where the subducting plate sinks. The stars mark Benioff Zone earthquakes. (C) At a transform plate boundary, rocks on opposite sides of the fracture slide horizontally past each other.

to an adjacent plate in three different ways (Fig. 4–5). At a **divergent plate boundary**, two plates move apart, or separate. At a **convergent plate boundary**, two plates move toward each other and collide. At a **transform plate boundary**, two plates slide horizontally past each other. Each of these three types of boundaries creates different tectonic features. Table 4–2 summarizes characteristics of each type of plate boundary and lists modern examples of each.

4.3 Divergent Plate Boundaries

A divergent boundary, also called a **spreading center** and a **rift**, occurs where two plates move apart horizontally (Fig. 4–6). As the two plates separate, hot, plastic asthenosphere rock flows upward to cool and form new lithosphere in the gap left by the diverging plates. The rising asthenosphere partly melts, forming basalt magma that oozes to the surface to form new crust. Rifts occur in both oceanic and continental crust.

Divergent Boundaries in Oceanic Crust: The Mid-Oceanic Ridge

Newly formed lithosphere at a divergent boundary is hot and therefore of low density. If a divergent boundary occurs in an ocean basin, the new, submarine lithosphere rises isostatically to form an undersea mountain chain called the **mid-oceanic ridge**. The mid-oceanic ridge circles the globe, like the seam on a baseball, to form the Earth's longest mountain chain. The basalt magma that oozes out onto the sea floor at the ridge forms oceanic crust on top of the new lithosphere. Approximately 6.5×10^{18} (6,500,000,000,000,000,000) tons of new oceanic crust form in this way each year. However, it is

TABLE 4–2 Plate Boundaries				
Type of Boundary	Types of Plates Involved	Topography	Geologic Events	Modern Examples
Divergent	Ocean–ocean	Mid-oceanic ridge	Sea-floor spreading, shallow earthquakes, rising magma, volcanoes	Mid-Atlantic ridge
	Continent–continent	Rift valley	Continents torn apart, earthquakes, rising magma, volcanoes	East African rift
Convergent	Ocean–ocean	Island arcs and ocean trenches	Subduction, deep earthquakes, rising magma, volcanoes, deformation of rocks	Western Aleutians
	Ocean–continent	Mountains and ocean trenches	Subduction, deep earthquakes, rising magma, volcanoes, deformation of rocks	Andes
	Continent–continent	Mountains	Deep earthquakes, deformation of rocks	Himalayas
Transform	Ocean–ocean	Major offset of mid-oceanic ridge axis	Earthquakes	Offset of East Pacific rise in South Pacific
	Continent–continent	Small deformed mountain ranges, deformations along fault	Earthquakes, deformation of rocks	San Andreas fault

not merely oceanic crust that spreads outward from the mid-oceanic ridge, but the entire lithosphere spreads, carrying the sea floor on top of it in piggyback fashion. The mid-oceanic ridge and other features of the sea floor are described further in Section 4.10.

Divergent Boundaries in Continental Crust: Continental Rifting

A divergent plate boundary can rip a continent in half; the process is called **continental rifting**. Continental rifting is occurring today along a north–south fault zone in eastern Africa called the East African rift. If the rifting continues, eastern Africa will eventually separate from the main portion of the continent. Basalt magma will rise to fill the growing gap, forming a new ocean basin between the separating portions of Africa. Continental rifting may also be occurring in North America along the Snake River plain, which extends westward from Yellowstone National Park to southeastern Oregon, and along the Rio Grande rift, extending from southern Colorado to El Paso, Texas. Elongate depressions called **rift valleys** develop along continental rifts because continental crust becomes stretched, fractured, and thereby thinned as it is pulled apart. The East African rift, the Snake River plain, and the Rio Grande rift are all great valleys in continental crust.

4.4 Convergent Plate Boundaries

A convergent boundary develops where two plates are moving toward each other and therefore are colliding (Fig. 4–5B). Collisions can occur (1) between a plate carrying oceanic crust and another carrying a continent, (2) between two plates carrying oceanic crust, and (3) between two continental plates.

FIGURE 4–6
The outer few hundred kilometers of the Earth. In the center of the drawing, new lithosphere forms at a spreading center. At the sides of the drawing, old lithosphere sinks into the mantle at subduction zones. The lithospheric plates move away from the spreading center by gliding over the weak, plastic asthenosphere.

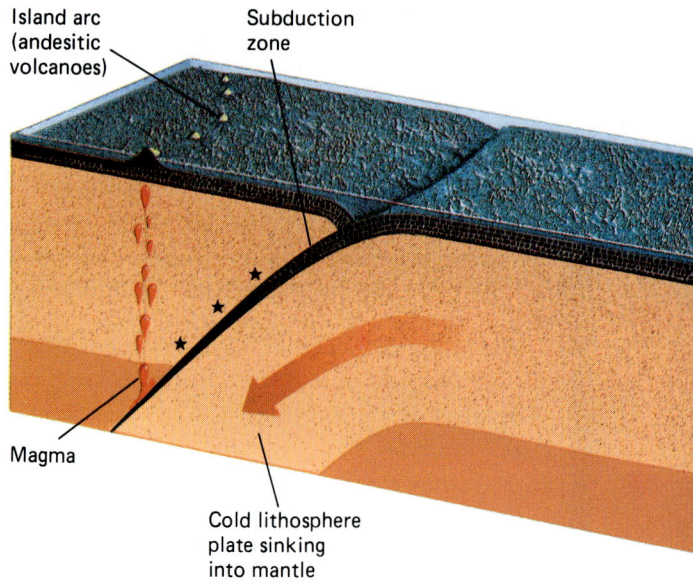

Convergence of Oceanic Crust with Continental Crust

Oceanic crust is composed of basalt, which is denser than the granitic crust of a continent. Differences in density determine what happens in a collision. Think of a boat colliding with a floating log. The log is denser than the boat, so it sinks beneath the boat. When a continental plate collides with an oceanic plate, the denser oceanic plate sinks beneath the continental plate and dives into the mantle. This process is called **subduction**.

A **subduction zone** is a long, narrow belt where a lithospheric plate dives into the mantle. On a worldwide scale, the rate at which old lithosphere sinks into the mantle at subduction zones is equal to the rate at which new lithosphere forms at spreading centers. In this way, a perfect global balance is maintained between the creation of new lithosphere and the destruction of old lithosphere.

Only lithosphere covered with oceanic crust can sink into the mantle. Continental crust is of lower density than oceanic crust and cannot sink far. Attempting to stuff a continent down a subduction zone would be like trying to flush a marshmallow down a toilet: It won't go because it is too light. The oldest sea-floor rocks on Earth are only about 200 million years old because oceanic crust is continuously recycled back into the mantle at subduction zones. Far older rocks are found on continents because subduction consumes little or no continental crust.

Today, oceanic plates are subducting beneath continents along the western edge of South America; along the coasts of Oregon, Washington, and British Columbia; and around most of the rest of the margin of the Pacific Ocean (see Fig. 4–1). Earthquakes, active volcanoes, and rising mountains all characterize these regions.

Subduction and Earthquakes

As a lithospheric plate sinks toward the mantle, it slips and jerks beneath the opposite plate, causing numerous earthquakes. These quakes trace the path of the subducting plate as it sinks into the mantle (Fig. 4–6). This zone of earthquakes is called a **Benioff zone**, after one of the geologists who first recognized it. The deepest earthquakes known occur in Benioff zones at a depth of about 700 kilometers. Below 700 kilometers, subducting plates become so hot that they flow plastically rather than fracture. Earthquakes are common along all subduction zones.

Subduction and Volcanoes

As you learned earlier in this chapter, the asthenosphere is so hot that it is soft and plastic—in fact, it is very close to melting. The descending lithosphere generates huge quantities of magma in the subduction zone. The most important mechanism for magma production in this environment is addition of water. Because oceanic crust is covered by the sea, the upper part of a subducting plate consists of water-soaked mud and basalt. As a subducting plate sinks, the water rises into the asthenosphere where it melts the hot rock to form magma. (A more detailed discussion of magma production is found in Chapter 6.) The magma then rises through the overlying lithosphere. Some solidifies within the crust, whereas some erupts onto the Earth's surface to form volcanic rocks (Fig. 4–7). Igneous activity is common in the Cascade Range of Oregon, Washington, and British Columbia; in western South America; and in most of the rest of the Pacific margin.

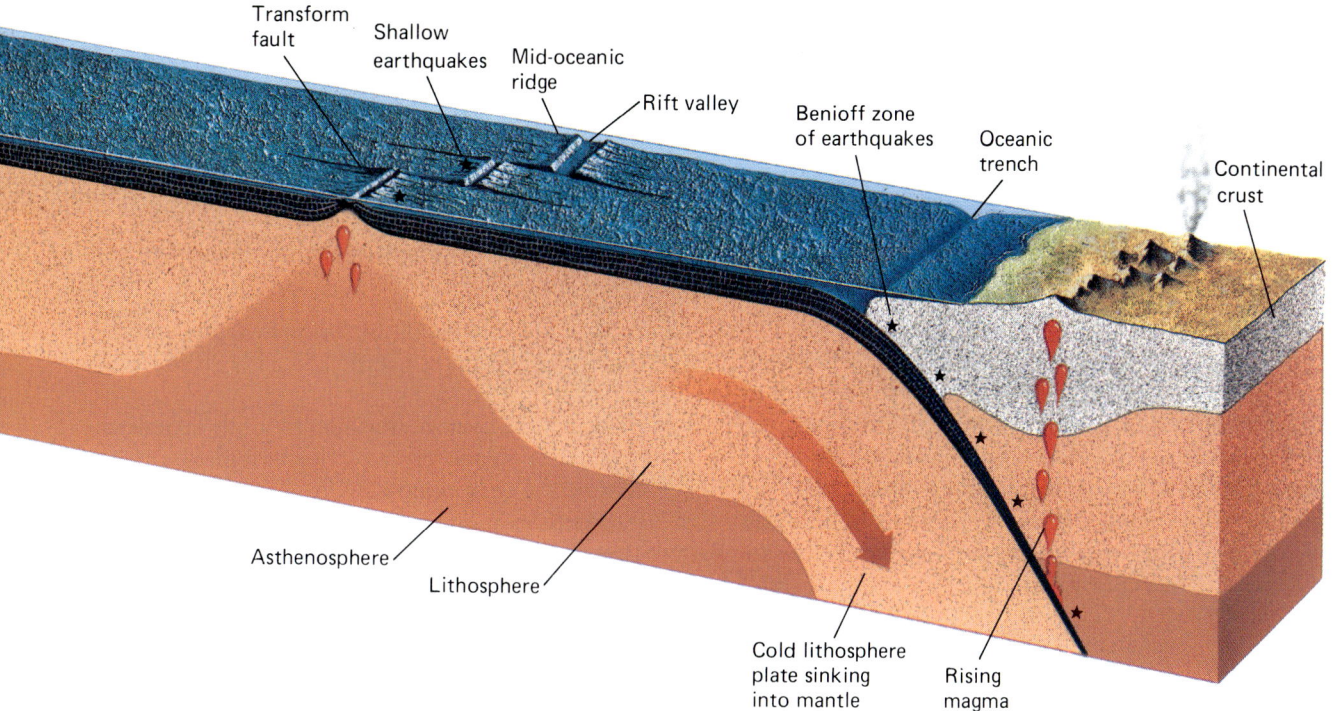

Subduction and Mountain Building

Many of the world's great mountain chains, including the Andes and parts of the mountains of western North America, formed at subduction zones. Several factors contribute to growth of mountain chains at subduction zones. The great volume of magma rising into the Earth's crust thickens the crust, causing mountains to rise. Volcanic eruptions pour huge amounts of lava onto the surface, constructing chains of volcanoes. Additionally, the Earth's crust crumples and buckles into mountain ranges where two lithospheric plates collide. The tectonic causes of mountain building are described further in Section 4.9.

Oceanic Trenches

An **oceanic trench** is a long, narrow trough in the sea floor formed where a subducting plate bends downward to sink into the mantle (Fig. 4–6). A trench can form wherever subduction occurs—where oceanic crust sinks beneath either continental or oceanic crust. Trenches are the deepest parts of the ocean basins. The deepest point on Earth is in the Mariana trench in the southwestern Pacific Ocean north of New Guinea, where the sea floor is as much as 10.9 kilometers below sea level (compared with the average sea-floor depth of about 5 kilometers).

FIGURE 4–7
Ollagüe volcano, in Chile near the Bolivian border, erupted magma generated by subduction off the west coast of South America.

Convergence of Two Plates Carrying Oceanic Crust

Subduction also occurs where two oceanic plates collide. Recall that new oceanic lithosphere is hot and therefore of low density when it first forms at the mid-oceanic ridge. It cools and becomes denser as it ages and spreads away from the ridge. When two oceanic plates collide, the older, cooler, and denser plate sinks into the mantle. Oceanic trenches, volcanic island arcs, and earthquakes mark mid-oceanic subduction zones and are further discussed in Section 4.10.

Convergence of Two Plates Carrying Continental Crust

If two colliding plates are both covered with continental crust, subduction cannot occur because continental crust is too light to sink into the mantle. In this case, the two continents collide and crumple against each other, forming huge mountain chains in the collision zone. The Himalayas, the Alps, and the Appalachians all formed as results of continental collisions. These processes are discussed further in Section 4.9.

4.5 Transform Plate Boundaries

A **transform plate boundary** forms where two plates slide horizontally past each other (Fig. 4–5C). California's San Andreas fault is a transform boundary between two major lithospheric plates, the North American plate and the Pacific plate. Although earthquakes are common at this type of boundary, igneous activity is not.

4.6 The Anatomy of a Tectonic Plate

The nature of a tectonic plate can be summarized as follows:

1. A plate is a segment of the lithosphere; thus, it includes the uppermost mantle and all of the overlying crust.
2. Although the average thickness of a lithospheric plate is 100 kilometers, a portion of a plate with continental crust composing its uppermost layer is thicker than one bearing oceanic crust (Fig. 4–8). The average thickness of the part of a plate carrying oceanic crust is about 75 kilometers,

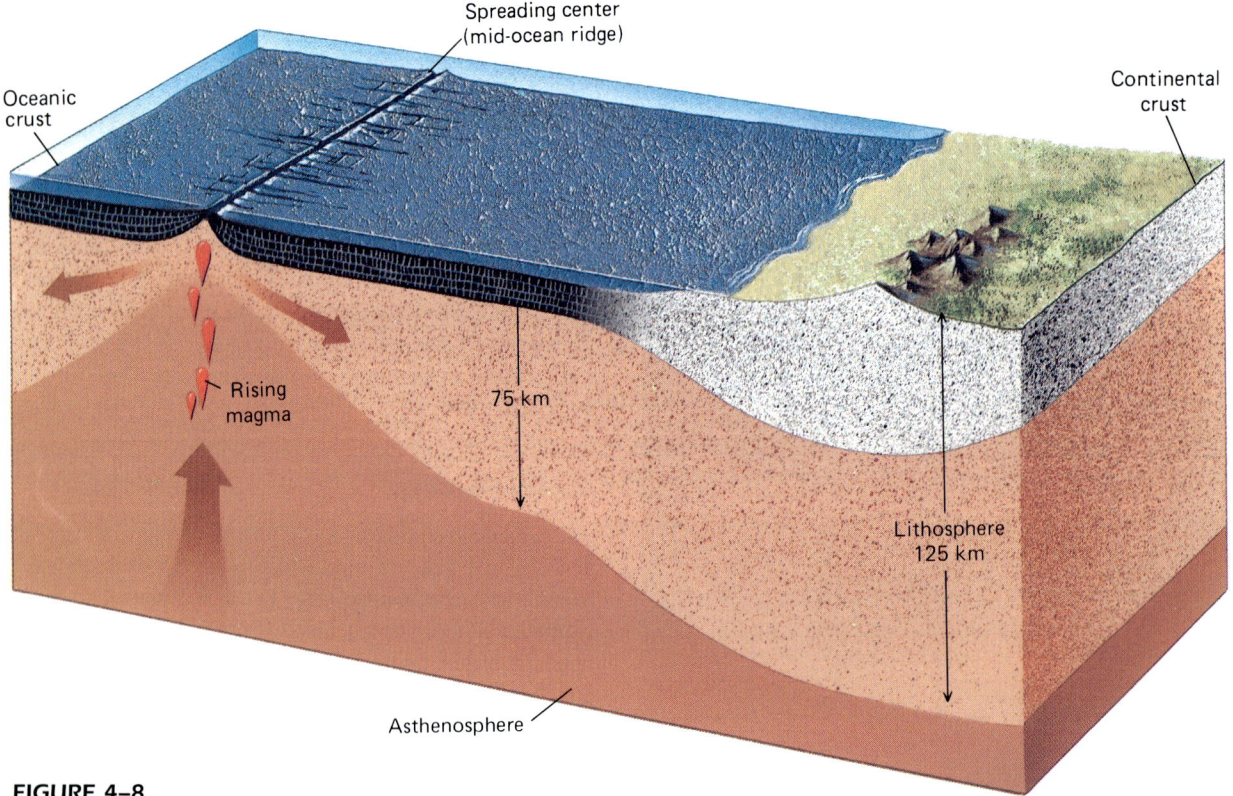

FIGURE 4–8
Lithosphere carrying continental crust is typically thicker than lithosphere bearing oceanic crust.

whereas that of a plate bearing continental crust is about 125 kilometers. Lithosphere may be as little as 20 kilometers thick at an oceanic spreading center.
3. A plate is hard, rigid or nearly rigid rock.
4. A plate floats on the underlying hot, plastic asthenosphere and glides horizontally over it.
5. A plate behaves like a large slab of ice floating on a pond. It may flex slightly, as thin ice does when a skater goes by, allowing minor vertical movement such as isostatic adjustment. In general, however, each plate moves as a single, large, intact sheet of rock.
6. A plate margin is tectonically active. Earthquakes and faulting are common at all plate boundaries. Volcanic eruptions and intrusion of magma are common at subduction zones and spreading centers. The Earth's crust deforms at transform boundaries and subduction zones. In contrast, the interior of a lithospheric plate is normally tectonically stable.
7. Tectonic plates move at rates that vary from less than 1 centimeter to 18 centimeters per year.

4.7 The Search for a Mechanism

Geologists have accumulated ample evidence that lithospheric plates move and can even measure how fast they move. But what force moves the plates?

Mantle Convection

Although the mantle is solid rock (except for small partially melted zones in the asthenosphere), it is so hot that over geologic time it flows slowly as a stiff fluid. Mantle rock flows in elliptical or circular patterns. Hot rock from deep in the mantle rises to the base of the lithosphere. At the same time, cooler upper-mantle rock sinks. This flow of solid rock is called **mantle convection**.

Mantle convection is thought to be closely related to the movement of lithospheric plates. However, it is not clear whether convection of the mantle causes the plates to move or, conversely, movement of the plates causes mantle convection.

Mantle Convection as the Cause of Plate Movement

Convection occurs when a fluid is heated from below, as in a pot of soup on a stove. The soup at the bottom of the pot expands as it is heated and becomes less dense than the soup at the top. Because it is less dense, it rises.

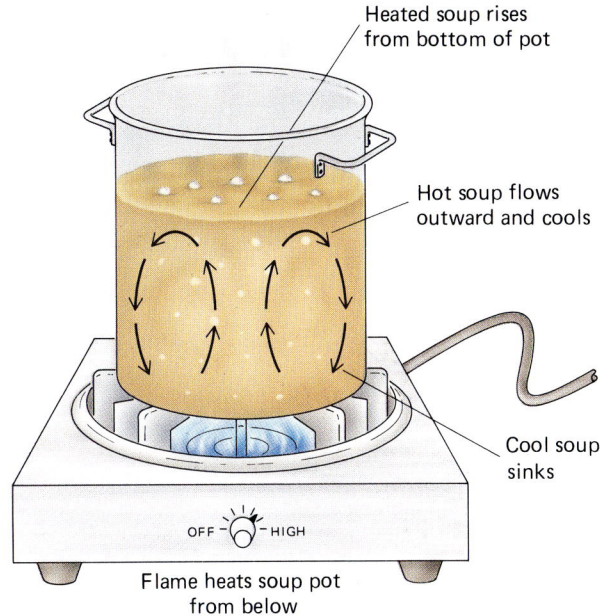

FIGURE 4–9
Soup convects when it is heated from the bottom of the pot.

When the hot soup reaches the top of the pot, it flows along the surface until it cools and sinks (Fig. 4–9). If the heat source persists, this cool, sinking soup is then warmed. It rises, and the convection continues.

Mantle convection might occur in a manner similar to that in the soup pot. In this model, the base of the mantle is heated from below, perhaps by the hot core. In turn, the heating causes mantle convection. Imagine a block of wood floating in a tub of honey. If you heated the honey from below so that it started to convect, the horizontal flow of honey along the surface would drag the block of wood along with it. Some geologists suggest that lithospheric plates are dragged along in a similar manner by a convecting mantle (Fig. 4–10).

Plate Movement as a Cause of Mantle Convection

Other geologists have suggested that movement of lithospheric plates might be the cause, rather than the effect, of mantle convection. Return to our analogy of the block of wood and the tub of honey. If you dragged the block of wood across the honey, friction between the block and the honey would make the honey flow. Similarly, if some force caused the plates to move, their motion might cause the mantle to flow. But what force might cause the plates to move?

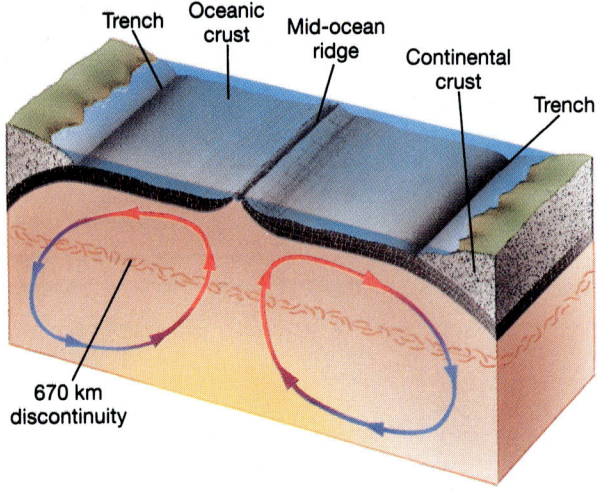

FIGURE 4-10
According to one explanation, lithospheric plates are dragged along by mantle convection.

Gravitational Sliding as a Cause of Plate Movement

A plate may simply glide downhill, away from a spreading center. Because new lithosphere at a spreading center is hot and of low density, it is thin and floats isostatically to a relatively high level. That is why the mid-oceanic ridge rises above the rest of the sea floor. As the lithospheric plate spreads away from the ridge, it cools and thickens. Therefore, both the surface and the base of the lithosphere slope downward from the spreading center.

The average slope of the surface of the mid-oceanic ridge is only about 0.6 percent. However, because the lithosphere thickens as it spreads away from the ridge, the slope of the base of the lithosphere beneath the ridge is about 8 percent, steeper than almost any paved road in North America (Fig. 4–11). Calculations show that if the slope is as slight as 0.03 percent, a plate should glide away from a spreading center at a rate of a few to several centimeters per year.

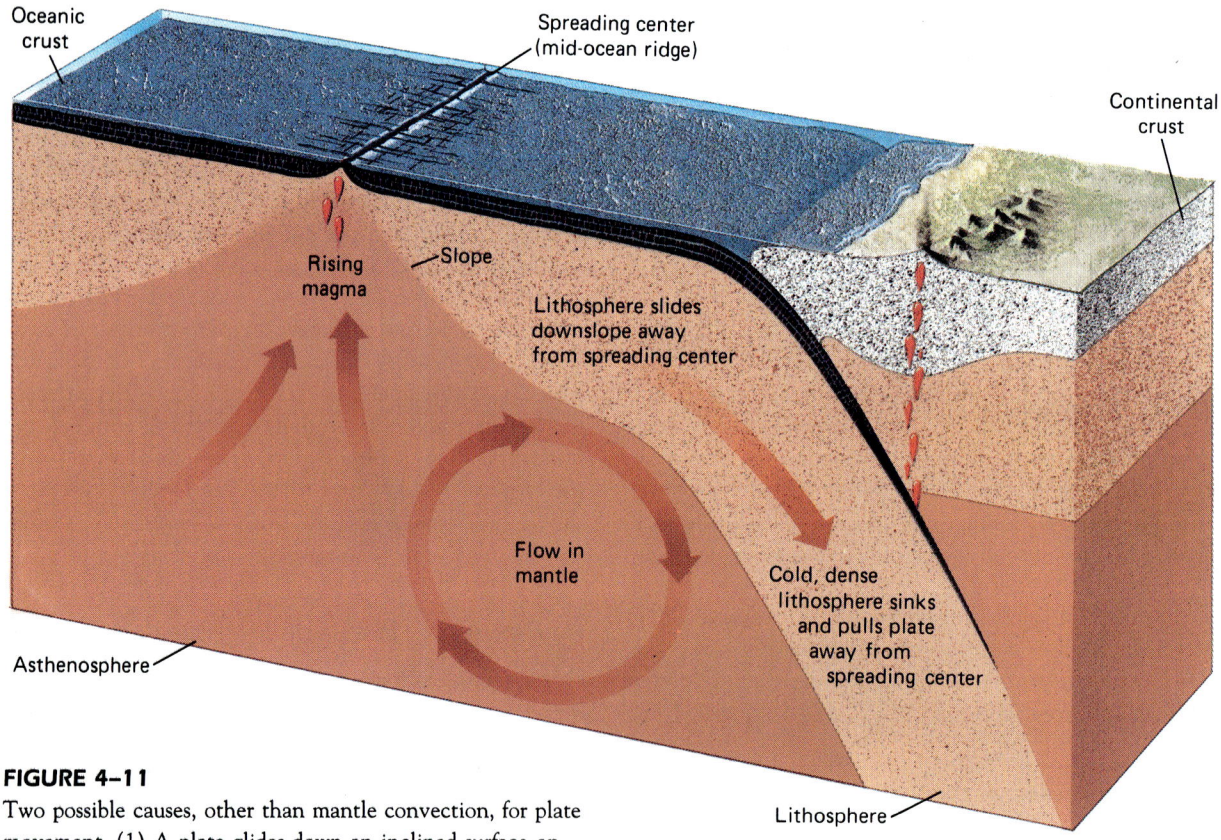

FIGURE 4-11
Two possible causes, other than mantle convection, for plate movement. (1) A plate glides down an inclined surface on the asthenosphere. (2) A cold, dense plate sinks at a subduction zone, pulling the rest of the plate along with it. In this drawing, both mechanisms are operating simultaneously.

As two plates separate at a spreading center, subduction must also occur; if a plate is growing at one end, it (or some other lithospheric plate) must sink back into the mantle elsewhere. As the old, cold lithosphere sinks into the mantle, it pulls on the rest of the plate, like a weight on the edge of a tablecloth. Many geologists now think that plates move because they glide downslope from a spreading center and, at the same time, are pulled along by their sinking ends—a combined mechanism called the **push–pull model** of plate movement.

FOCUS ON: Alfred Wegener and the Origin of an Idea

In the early twentieth century, a young German scientist named Alfred Wegener noticed that the African and South American coastlines on opposite sides of the Atlantic Ocean seemed to fit as if they were adjacent pieces of a jigsaw puzzle. He realized that the jigsaw-like fit suggested that the continents had once been joined together and had later split and drifted apart. Studying world maps, Wegener noticed that not only did the continents on both sides of the Atlantic fit together well, but other continents, when moved properly, also fit like additional pieces of the same jigsaw puzzle. He constructed a map of the Earth based on the fit of continents. On his map, all the continents were joined together, forming one supercontinent that he called **Pangaea** from the Greek root words for "all lands." The northern part of Pangaea is commonly called **Laurasia** and the southern part **Gondwanaland**.

Wegener then mapped the locations of fossils of several species of animals and plants that could neither swim well nor fly. Fossils of the same species are now found in Antarctica, Africa, Australia, South America, and India. Why would a single species be found on continents separated by thousands of kilometers of ocean? To solve this dilemma, Wegener plotted the fossil localities on his Pangaea map. All of them lie in the same region of Pangaea (Fig. 1), suggesting that each species evolved and spread over a part of Pangaea rather than mysteriously migrating across thousands of kilometers of open ocean.

The fossil evidence led Wegener to an additional interpretation of Pangaea's history. He observed that rocks of Ordovician through Triassic age (505 through 208 million years ago) from Africa and South America contained identical fossils. Rocks younger than Triassic age, however, showed development of different species on the two continents. Wegener reasoned that the Atlantic Ocean basin had begun to open and tear Pangaea apart at the end of the Triassic Period. Rocks that had formed before this separation contained identical fossils because the plants and animals had evolved and spread throughout Pangaea. Rocks formed after Pangaea split up contained different fossils because evolution had followed different paths on the separating continents.

Certain types of sedimentary rocks form in specific climatic zones of the Earth. Glaciers and glacial sediment, for example, concentrate at high latitudes and high altitudes. Deserts and the rocks that form in deserts cluster around latitudes 30° north and south. Coral reefs and coal swamps thrive in near-equatorial tropical climates. Thus, sedimentary rocks reflect the latitudes at which they formed.

Wegener plotted sedimentary rocks that indicate climate and latitude on maps showing the modern distribution of continents. Figure 2 shows his map of 300-million-year-old glacial deposits. The light blue area shows how large the ice mass would have been if the continents had been in their present positions. Notice that the glacier would have crossed the equator, and glacial deposits would have formed in tropical and subtropical zones. Figure 2B shows the same glacial deposits plotted on Wegener's Pangaea map. Here they are neatly clustered about the South Pole.

Wegener noticed several instances in which an uncommon rock type or a distinctive sequence of rocks on one side of the Atlantic Ocean was identical to rocks on the other side. When he plotted the rocks on a Pangaea map, those on the east side of the

(Continued)

FOCUS ON (continued)

Atlantic were continuous with their counterparts on the west side. For example, the Cape Fold Belt of South Africa consists of a sequence of deformed rocks similar to rocks found in the Buenos Aires Province of Argentina. Plotted on a Pangaea map, the two sequences of rocks appear as a single, continuous belt of folded rocks.

Wegener's concept of a single supercontinent that broke apart to form the continents as we know them today is called the theory of **continental drift**. This theory is similar to the plate tectonics theory in that both ideas involve continental movements. But Wegener never postulated the existence of lithospheric plates or many of the ramifications of the plate tectonics theory.

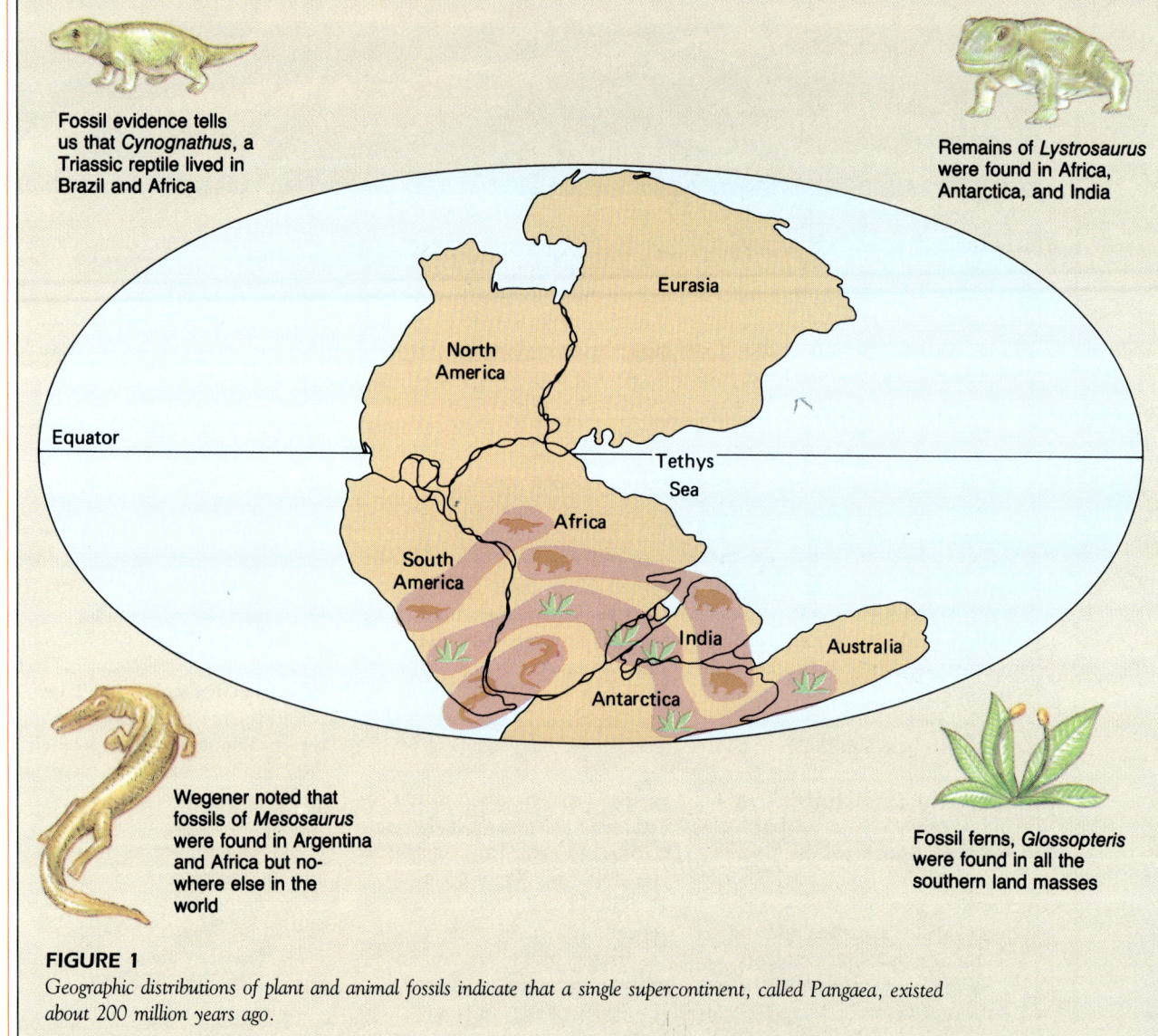

Fossil evidence tells us that *Cynognathus*, a Triassic reptile lived in Brazil and Africa

Remains of *Lystrosaurus* were found in Africa, Antarctica, and India

Wegener noted that fossils of *Mesosaurus* were found in Argentina and Africa but nowhere else in the world

Fossil ferns, *Glossopteris* were found in all the southern land masses

FIGURE 1
Geographic distributions of plant and animal fossils indicate that a single supercontinent, called Pangaea, existed about 200 million years ago.

4.8 Supercontinents

Recently, Paul Hoffman of the Geologic Survey of Canada has suggested that at least three times during the Earth's history all continents were joined together in a single **supercontinent** and then split apart. Hoffman estimated that it takes about 500 million to 700 million years for a supercontinent to assemble, split apart, and then reassemble.

In this model, rifting breaks up a supercontinent and the fragments begin to separate. But because the Earth is a sphere, the continental fragments migrate halfway

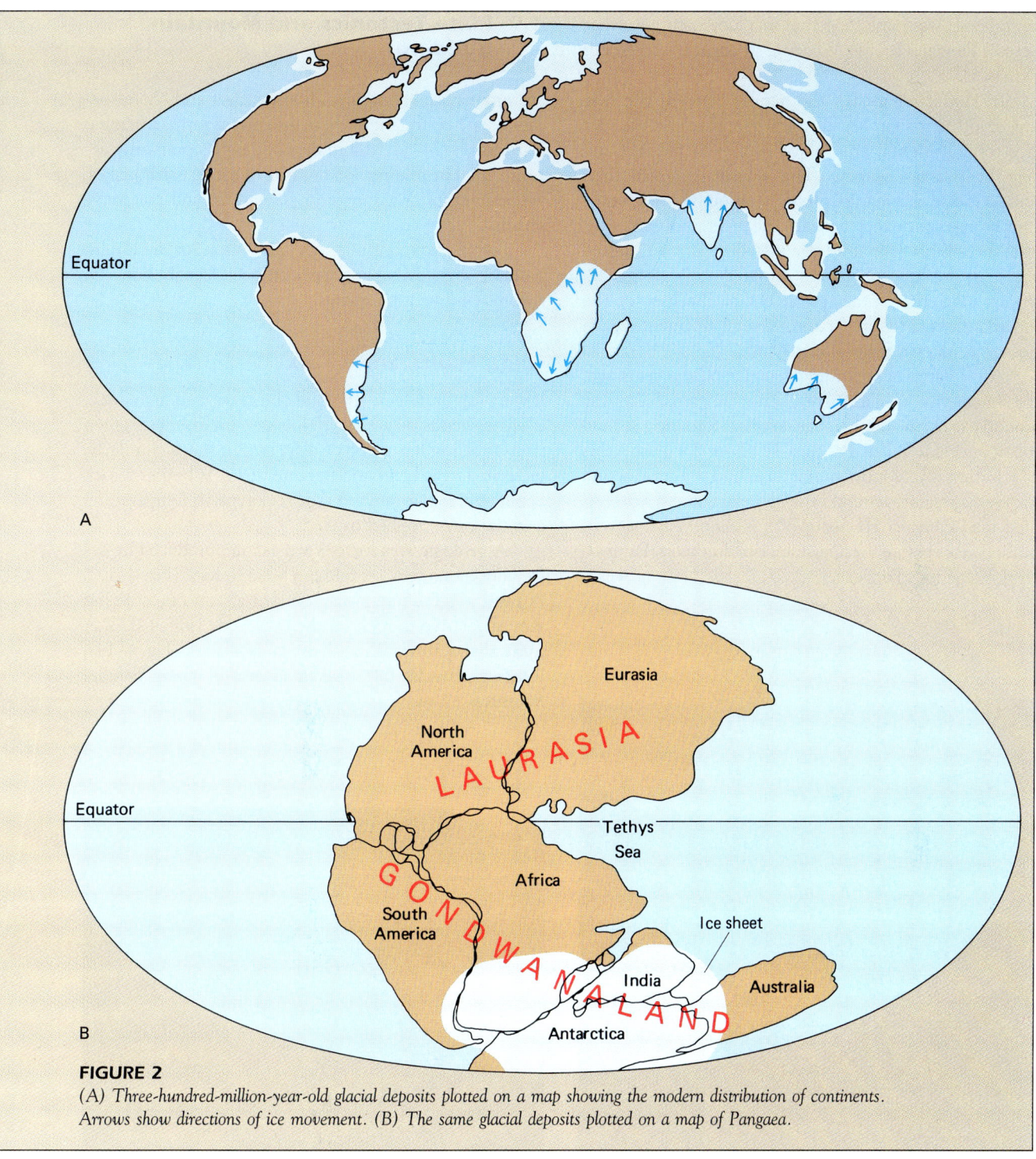

FIGURE 2
(A) Three-hundred-million-year-old glacial deposits plotted on a map showing the modern distribution of continents. Arrows show directions of ice movement. (B) The same glacial deposits plotted on a map of Pangaea.

around the globe and then collide on the far side to reassemble as a new supercontinent. Thus, the breakup of one supercontinent leads to the assembly of a new one.

Prior to about 2 billion years ago, the landmasses of the Earth consisted mainly of small island chains and microcontinents (small masses of continental crust, such as Japan and New Zealand) scattered about the globe and separated by ocean basins. Then movements of tectonic plates swept all of the islands and microcontinents together into a single great landmass. This accretion took about 200 million years to complete. Thus, Hoffman suggests, by 1.8 billion years ago, all of the Earth's continental crust was joined together in a supercontinent that we call **Pangaea I**, after Wegener's Pangaea (see "Focus On" box).

In Hoffman's model, the mantle beneath the new supercontinent soon began to warm up because the vast layer of continental crust acted as a giant blanket and kept heat from escaping. The warming mantle then started to rise as a **mantle plume**, a column of hot, vertically rising mantle rock beneath the supercontinent. Another possibility is that a large meteorite impact blasted a hole in the lithosphere and initiated a mantle plume. In any case, the plume spread out at the base of the supercontinent, tearing it into several fragments, each riding on its own lithospheric plate.

After Pangaea I split up about 1.3 billion years ago, the fragments of continental crust migrated halfway around the Earth and then reassembled, forming a second supercontinent called **Pangaea II**, about 1 billion years ago. In turn, this continent fractured and the continental fragments migrated around the globe again and reassembled into **Pangaea III** about 300 million years ago, 70 million years before the appearance of dinosaurs. Pangaea III is Wegener's Pangaea.

4.9 Plate Tectonics and Mountain Building

Mountains usually occur in linear ranges and chains because plate boundaries, where mountains form, are linear features (Fig. 4–1). Mountain building is commonly accompanied by folding and faulting of rocks, earthquakes, volcanic eruptions, intrusion of plutons, and metamorphism.

Before we continue our discussion of mountain building, consider what happens to rocks caught between two moving tectonic plates.

Folds, Faults, and Joints

If you push or pull on any solid object with enough force, it either bends or breaks. If you step on a supple, green stick, it bends; if you step on a dry twig, it snaps. Similarly, tectonic forces bend and fracture rocks. Whether a rock bends or breaks depends on the rock type, its temperature, and the force applied to it.

A **geologic structure** is any feature produced by deformation of a rock. A **fold** is a bend in rock (Fig. 4–12). A **fault** is a fracture along which rock on one side has moved relative to rock on the other side (Fig. 4–13). A **joint** is a fracture without movement of rock (Fig. 4–14). Joints and faults often occur as sets of many parallel fractures.

A   B

FIGURE 4–12
(A) Folds in limestone, Canadian Rockies, Alberta. (B) Small-scale folds, McCarty's Mountain, Montana.

FIGURE 4–13
A small fault in sedimentary rocks near Kingman, Arizona.
(Ward's Natural Science Establishment)

The Building of Two Mountain Chains: The Andes and the Himalayas

Imagine that you have the opportunity to spend a season mountaineering and trekking through the Andes in western South America and, shortly thereafter, another season traveling through the Himalayas between Asia and India. Initially you might be struck by similarities between the two mountain chains. Both contain towering, snow-covered peaks. The Andes have 49 peaks with elevations above 6000 meters (nearly 20,000 feet). The highest is Aconcagua, at 6962 meters. The Himalayas have 14 peaks above 8000 meters (26,000 feet), including Mount Everest, the highest peak on Earth, at 8848 meters. Furthermore, both mountain chains are deeply eroded by glaciers that were once much more extensive than they are today. Both are still capped by immense alpine glaciers.

Thus, the two mountain chains might seem nearly identical at first. However, as a student of geology, you can distinguish among igneous, sedimentary, and metamorphic rocks. After a month in the Andes, you would have seen all three, with igneous rocks, particularly volcanic rocks, being the most abundant. In fact, if you hiked the entire length of the Andes and were then asked to summarize the geology in a single phrase, you might reply, "It's a pile of volcanic rocks!" After a month or two of trekking through the Himalayas, you would also have seen igneous, sedimentary, and metamorphic rocks. But folded and faulted sedimentary rocks dominate Himalayan geology.

The fact that the Andes are predominantly a chain of volcanic rocks, whereas the Himalayas contain mostly sedimentary rocks, suggests that the two mountain chains must have been built by different geologic processes. Both chains formed at convergent plate boundaries. However, the Andes rose where a tectonic plate carrying oceanic crust collided with another carrying continental crust, whereas the Himalayas developed where two continents collided.

The Andes: Subduction at a Continental Margin

During Triassic time, from 245 to 208 million years ago, all of the Earth's continents were gathered in the supercontinent called Pangaea III. When Pangaea III broke apart about 200 million years ago, it first split into two large pieces, a southern portion called **Gondwanaland** and a northern piece called **Laurasia**. During the next 60 million years, each of those two fragments also broke apart. At this time, the lithospheric plate that included South America started moving westward. To accommodate the westward motion, oceanic lithosphere began to subduct and dive into the mantle beneath the west coast of South America (Fig. 4–15A).

By about 130 million years ago, the diving plate had reached the lithosphere–asthenosphere boundary. Melting then began in the asthenosphere above the descending plate, forming vast amounts of basaltic magma (Fig. 4–15B). The magma rose to the base of the South American continent where it, in turn, melted the granitic

FIGURE 4–14
Vertical joints in sandstone, Indian Creek, Utah.

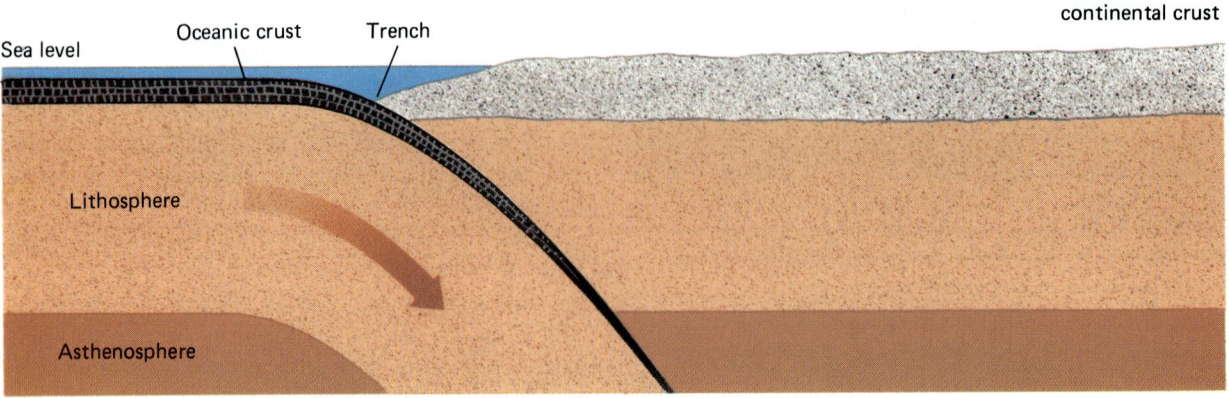

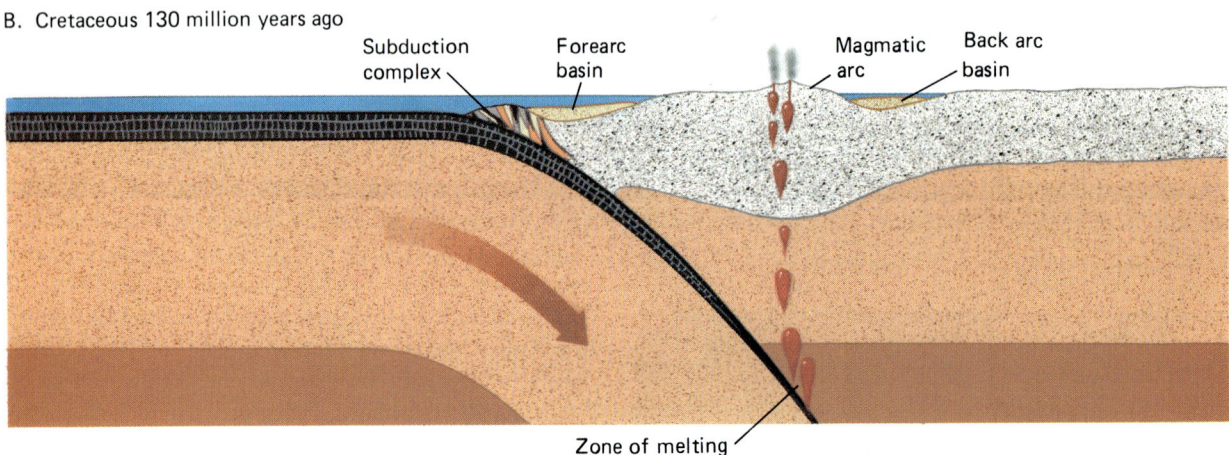

FIGURE 4–15
Development of the Andes, seen in cross section looking northward. (A) As the South American lithospheric plate began to move westward in early Cretaceous time, about 140 million years ago, subduction began and a trench formed at the west coast of the continent. (B) By 130 million years ago, the subducting plate crossed the lithosphere–asthenosphere boundary. Igneous activity began and a subduction complex formed.

crust. The basaltic and granitic magmas then rose into the continent. Some solidified within the crust to form plutons. However, vast quantities of magma continued upward to erupt at the surface, forming volcanoes. This intrusive and volcanic activity occurred along the entire length of western South America, but only in a narrow band directly over the zone of melting.

Thus, the Andes are a relatively narrow mountain chain of volcanic and plutonic rocks produced by subduction at a continental margin. The chain also contains extensive sedimentary rocks on both sides of the mountains. These rocks formed from sediment eroded from the rising peaks. The Andes are a good general example of mountains formed by subduction at a continental margin, and this type of plate margin is called an **Andean margin**.

The Himalayan Mountain Chain: A Collision Between Continents

The Himalayan mountain chain separates the Earth's two most populous nations, China and India (Fig. 4–16). The world's highest mountains, including Mount Everest and K2, are in the Himalayas. Today, if you were to stand in southern Tibet and look southward, you would see the high peaks of the Himalayas. Beyond this great mountain chain are the rainforests and hot, dry plains of India. If you had been able to look south from the same place 100 million years ago, you would have seen only ocean. At that time, India lay south of the equator, separated from Tibet by thousands of kilometers of open ocean. The Himalayas had not yet begun to rise.

FIGURE 4–16
The Himalayas separate the Indian subcontinent from southern Asia.

When Pangaea III split into Laurasia and Gondwanaland, the two were separated by open ocean. About 120 million years ago, a large triangle-shaped piece of lithosphere split off from Gondwanaland. It began drifting northward toward Laurasia at high speed, geologically speaking—perhaps as fast as 20 centimeters per year (Fig. 4–17A). The northern part of this lithospheric plate carried oceanic crust, and the Indian subcontinent lay on its southern corner (Fig. 4–17B).

Figure 4–18A shows that India and southern Asia were separated by oceanic crust before India began moving northward. As the Indian plate started to move, oceanic crust subducted at Asia's southern margin (Fig. 4–18B). As a result, magma formed, volcanoes erupted, and granite plutons were emplaced in southern Tibet. At this point, southern Tibet was an Andean-type continental margin, and it continued to be so from about 120 to 50 million years ago, while India drew closer to Asia.

By about 50 million years ago, all of the oceanic lithosphere between India and Asia had been consumed by subduction (Fig. 4–18C). Then the two continents collided. Because both are continental crust, neither could subduct deeply into the mantle. The collision did not stop the northward movement of India, but it did slow it down. India had been speeding northward at 20 centimeters per year and suddenly slowed to about 5 centimeters per year.

Although India had collided with southern Asia, it continued to move northward in two ways. The leading edge of India began to slide under Tibet in a process called **underthrusting**. As a result, the thickness of continental crust in the region doubled. As India slid beneath Tibet, the edge of Tibet scraped the soft Indian sedimentary rocks from harder basement rock and pushed them into folds and faults. These deformed sedimentary rocks make up the high parts of the Himalayas (Fig. 4–18D).

The second way in which India continued moving northward after the collision was by crushing Tibet and wedging China out of the way along huge faults. India has pushed southern Tibet 1500 to 2000 kilometers northward since the beginning of the collision. The tectonic forces have created major mountain ranges and basins north of the Himalayas.

The underthrusting of India beneath Tibet and the squashing of Tibet have produced unusually thick continental crust in the entire region. Consequently, the Himalayas and the Tibetan Plateau isostatically float at high elevation. Even the valley bottoms lie at elevations of 3000 to 4000 meters, and the Tibetan Plateau has an average elevation of 4000 to 5000 meters. One reason the Himalayas contain all of the Earth's highest peaks is simply that the bases of the peaks are at such high elevations. From its base to its summit, Mount Everest is actually smaller than Alaska's Denali (Mount McKinley), North America's highest peak. Mount Everest rises about 3300 meters from base to summit, whereas Denali rises about 4200 meters. The difference in elevation of the respective summits lies in the fact that the base of Mount Everest is at about 5500 meters, but Denali's base is at 2000 meters.

The Himalayan chain is only one example of mountain building by continent–continent collision. The Appalachian Mountains formed when eastern North America collided with Europe, Africa, and South America between 400 and 250 million years ago. The European Alps formed during repeated collisions between northern Africa and southern Europe beginning about 30 million years ago. The Urals of northwestern Asia formed by a similar process about 250 million years ago.

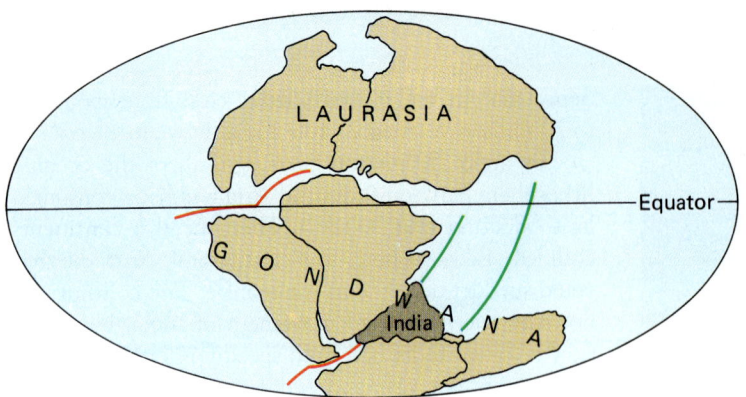

A 200 Million years ago

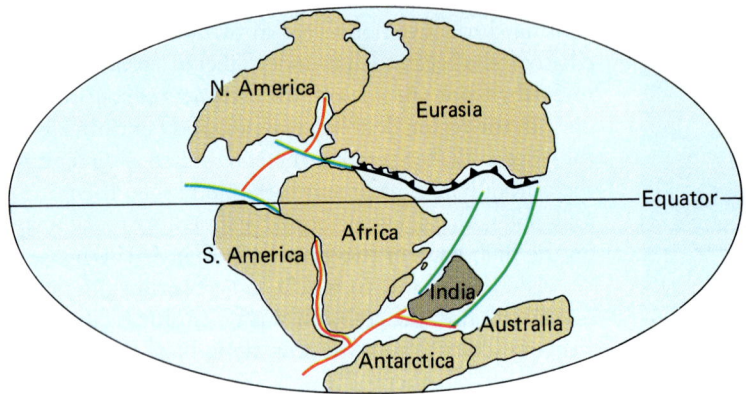

B 120 Million years ago

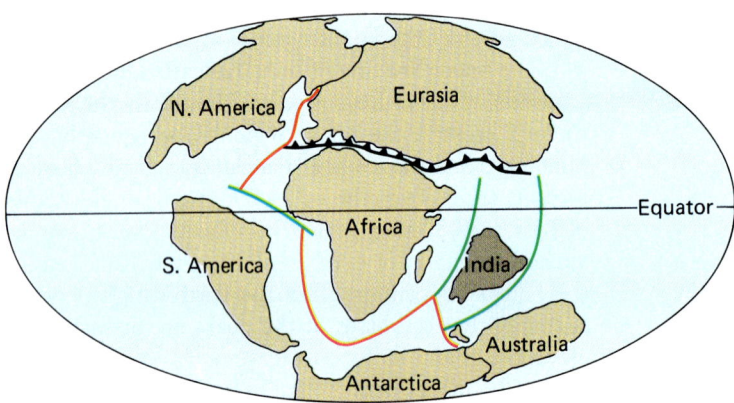

C 80 Million years ago

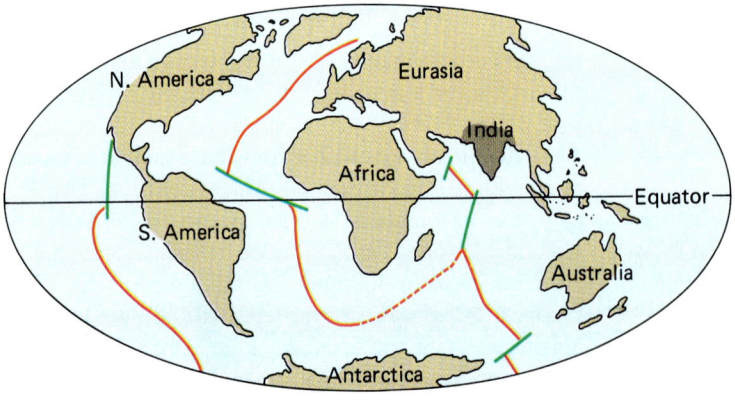

D 40 Million years ago

FIGURE 4–17
(A) Gondwanaland and Laurasia formed shortly after 200 million years ago as a result of the early breakup of Pangaea. Notice that India was initially part of Gondwanaland. (B) About 120 million years ago, India broke off from Gondwanaland and began drifting northward. (C) By 80 million years ago, India was isolated from other continents and was approaching the equator. (D) By 40 million years ago, it had moved 4000 to 5000 kilometers northward and collided with Asia.

FIGURE 4–18 ▶
These cross-sectional views show the Indian and Asian plates before and during the collision between India and Asia. (A) Shortly before 120 million years ago, India, southern Asia, and the intervening ocean basin were parts of the same lithospheric plate. (In this figure, the amount of oceanic crust between Indian and Asian continental crust is abbreviated to fit the diagram on the page.) (B) When India began moving northward, the plate broke and subduction began at the southern margin of Asia. By 80 million years ago, an oceanic trench and subduction complex had formed. Volcanoes erupted, and granite plutons formed in the region now called Tibet. (C) By 40 million years ago, India had collided with Tibet. The leading edge of India was underthrust beneath southern Tibet. (D) Continued underthrusting and collision between the two continents has crushed Tibet and created the high Himalayas by folding and thrust-faulting the sedimentary rocks. India continues to underthrust and crush Tibet today.

4.9 Plate Tectonics and Mountain Building 87

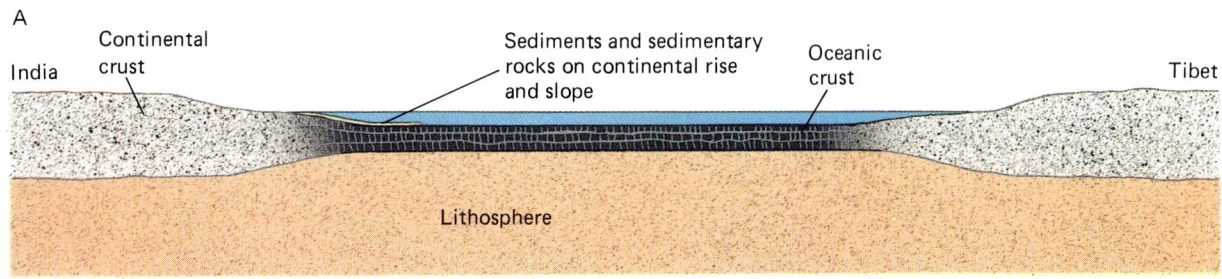
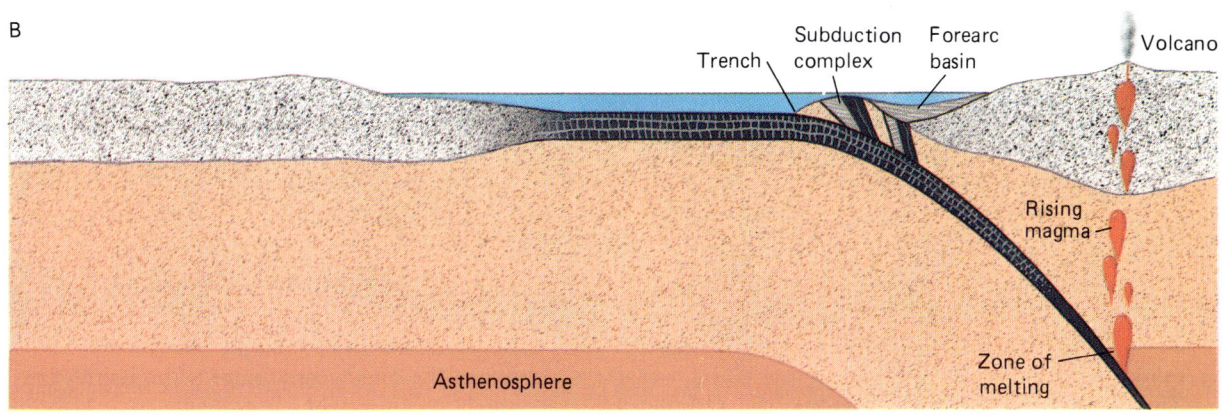
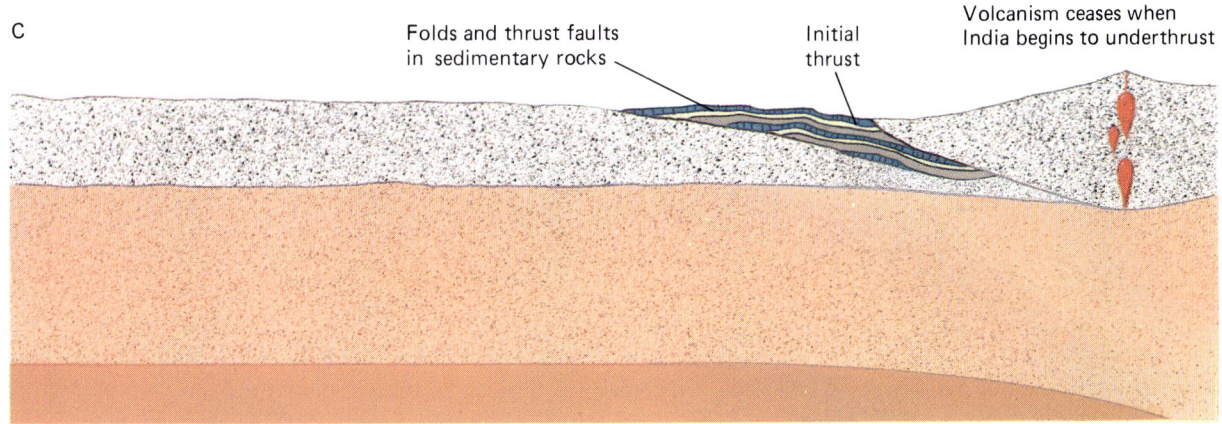
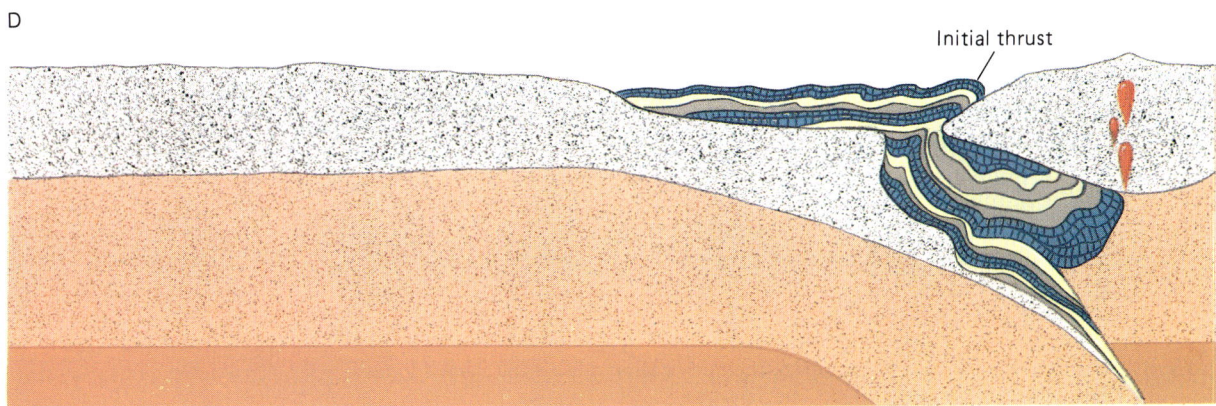

4.10 Plate Tectonics and the Sea Floor

If you were to ask most people to describe the difference between continents and oceans, they might show surprise and reply, "Why, obviously oceans are water and continents are land!" This is true, of course, but to a geologist a more important distinction exists. He or she would explain that the rocks beneath the oceans are very different from those of the continents. The accumulation of seawater in the world's ocean basins is a result of that difference.

Recall that the lithosphere floats on the asthenosphere. The concept of isostasy tells us that thin, dense lithosphere should sink to a low elevation, whereas thicker, lighter lithosphere ought to float to a higher level. Oceanic crust is dense basalt and is only about 10 kilometers thick. In contrast, continental crust is granite, which is of lower density and averages about 30 kilometers thick. As a result of these differences, continents float at high elevations, and the ocean basins sink to low elevations. The Earth's oceans collect in the huge depressions formed by oceanic crust.

The size of each ocean basin changes over geologic time because new oceanic crust forms at spreading centers such as the Mid-Atlantic ridge, and subduction consumes old sea floor. At present, the Atlantic Ocean is growing while the Pacific is shrinking. These changes affect ocean currents and the transport of heat across the globe. Therefore, such tectonic changes affect climate and life on Earth.

The Mid-Oceanic Ridge

Recall that a divergent plate boundary, or spreading center, runs through all of the Earth's ocean basins. This spreading center forms the mid-oceanic ridge. At the spreading center, the new lithosphere is hot and therefore of low density. Then it cools and becomes more dense as it spreads outward from the ridge axis. The new hot, low-density rock at the spreading center floats at a higher elevation than the older, cooler, and denser rock on both sides. Thus, the mid-oceanic ridge is elevated above surrounding sea floor because it is made of the newest, hottest, and lowest-density lithosphere.

Although the mid-oceanic ridge is the Earth's longest mountain chain, we do not normally see it because it lies below sea level (Fig. 4–19). It is more than 80,000

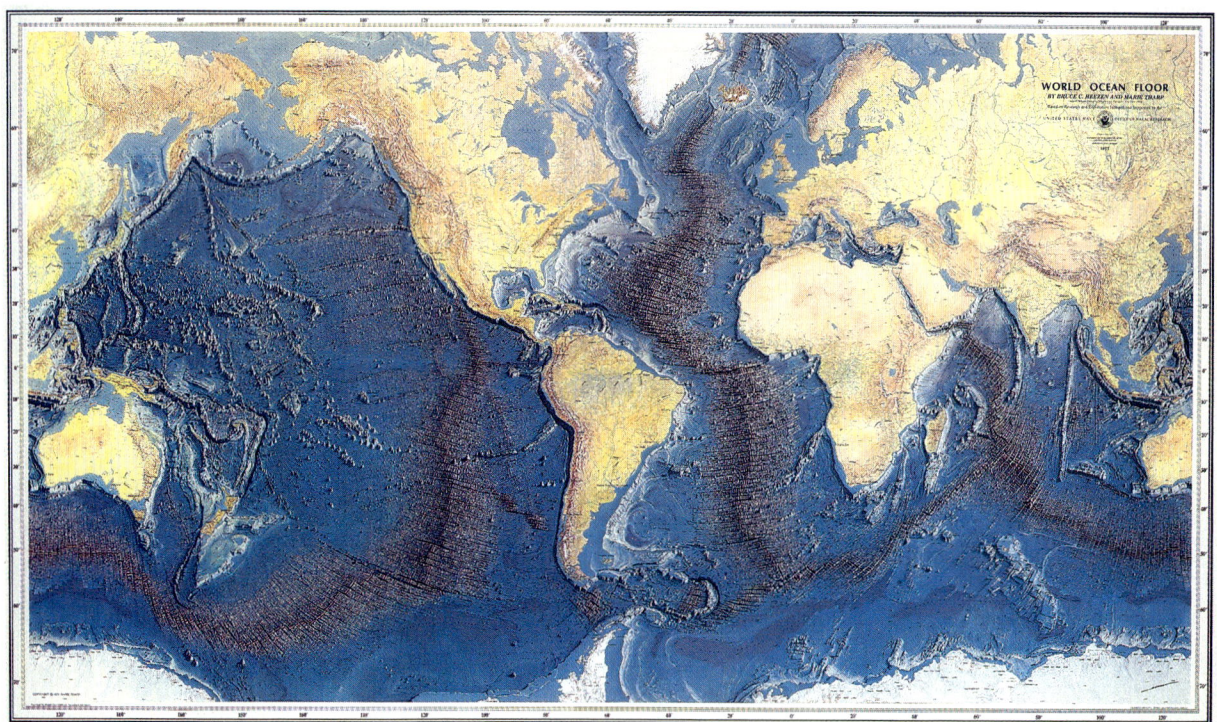

FIGURE 4–19
Although it lies almost entirely below sea level, the mid-oceanic ridge is the Earth's longest mountain chain. *(Marie Tharp)*

kilometers long, and is 1500 kilometers wide in most places. The ridge rises an average of 3 kilometers above the surrounding deep-sea floor and covers more than 20 percent of the Earth's surface, nearly as much as all continents combined. It winds through all of the Earth's ocean basins, much like the seam on a baseball. Occasionally it rises above sea level to form islands such as Iceland.

A rift valley 1 to 2 kilometers deep and several kilometers wide splits the crest of the mid-oceanic ridge. In 1974, French and American scientists used a small research submarine to dive into the rift valley in the Atlantic Ocean. They saw gaping vertical cracks up to 3 meters wide on the floor of the rift. Nearby were basalt flows so young that they were not covered by any sediment. The cracks form when oceanic crust separates at the ridge axis. Basalt magma then rises through the cracks and flows onto the floor of the rift valley. This basalt becomes new oceanic crust as two lithospheric plates spread outward from the ridge axis. Thus, the rift valley is the boundary between two diverging plates. Shallow earthquakes commonly occur at the mid-oceanic ridge as a result of fracturing and faulting of oceanic crust as the two plates separate (Fig. 4–20). Blocks of new oceanic crust drop downward along the faults, forming the rift valley.

This submarine mountain chain displaces a huge volume of seawater. Therefore, the very existence of the mid-oceanic ridge causes sea level to be higher than if the ridge did not exist. If the mid-oceanic ridge were smaller, it would displace less seawater and sea level would fall. If it were larger, sea level would rise. Therefore, one possible cause of global sea-level rise is an increase in the volume of the mid-oceanic ridge.

As described earlier, the mid-oceanic ridge stands highest at the spreading center, where new rock is hottest and has the lowest density. The elevation of the ridge decreases on both sides of the spreading center because the lithosphere cools and shrinks with time as it moves outward. Now consider a spreading center where spreading is very slow—say, 1 to 2 centimeters per year. At such a slow rate of spreading, the newly formed lithosphere would become quite old and thus would cool before it migrated far from the spreading center. This slow rate of spreading would produce a narrow, low-volume ridge, as shown in Figure 4–21A. In contrast, rapid sea-floor spreading, on the order of 10 to 20 centimeters per year, would create a high-volume ridge because the newly formed, hot lithosphere would be carried a considerable distance away from the spreading center before it began to cool and shrink (Fig. 4–21B). This high-volume ridge would displace considerably more seawater than a low-volume ridge and produce a global rise in sea level.

Close examination shows that the mid-oceanic ridge is cut and offset by numerous fractures called **transform faults** (Fig. 4–22). Transform faults extend through the entire thickness of the lithosphere. They develop because the mid-oceanic ridge is not a single, continuous spreading center, but rather it consists of many short segments. Each segment is slightly offset from adjacent segments by the cross-cutting transform faults. Thus, transform faults are original features of a mid-oceanic ridge; they begin to form at the same time sea-floor spreading begins.

The Earth is 4.6 billion years old, and rocks as old as 4.1 to 4.2 billion years have been found on continents. However, no oceanic crust is older than about 200 million years. Oceanic crust is so young because it forms continuously at spreading centers and recycles into the mantle at subduction zones. Thus, oceanic crust is youngest at the mid-oceanic ridge and becomes older with increasing distance from the ridge. In contrast, once continental crust forms, little or none can return into the mantle because of its low density.

Continental Margins

Continental margins are regions where continental crust meets oceanic crust. Two types of continental margins exist. A **passive continental margin** is characterized by a firm connection between continental and oceanic crust. Little tectonic activity occurs at this type of boundary. Continental margins on both sides of the Atlantic Ocean

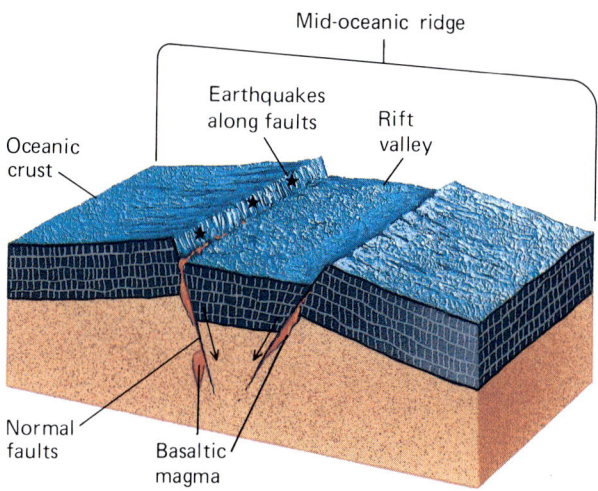

FIGURE 4–20
A cross-sectional view of the central rift valley of the mid-oceanic ridge. As the plates separate, blocks of rock drop down along the fractures to form the rift valley. The moving blocks cause earthquakes.

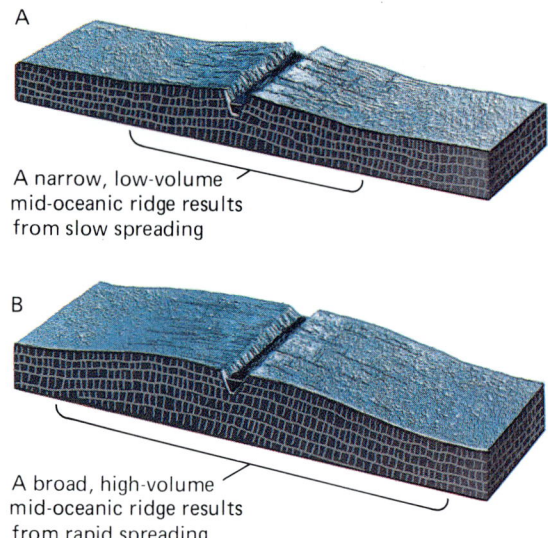

FIGURE 4–21
(A) A slow rate of sea-floor spreading produces a narrow, low-volume mid-oceanic ridge and results in low sea level. (B) Rapid sea-floor spreading creates a high-volume ridge and high sea level.

FIGURE 4–22
Transform faults offset segments of the mid-oceanic ridge. Adjacent segments of the ridge may be separated by steep cliffs 3 or 4 kilometers high. Note the flat abyssal plain far from the ridge.

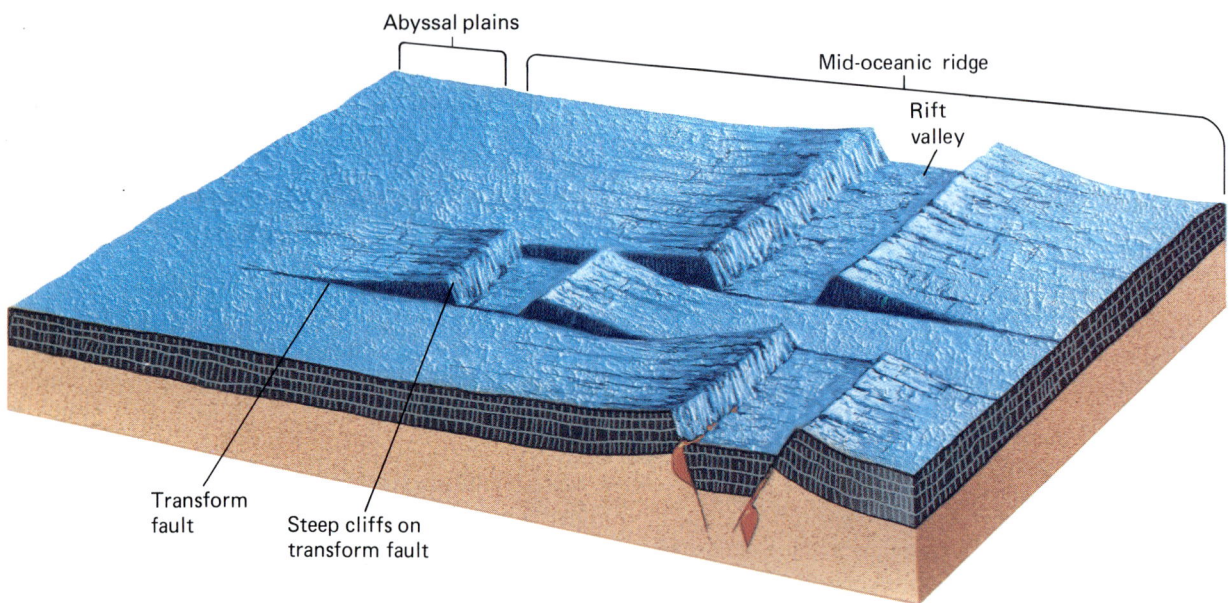

are passive margins. In contrast, an **active continental margin** is characterized by subduction of oceanic lithosphere beneath a continent. At an active margin, subduction typically occurs very close to the edge of the continent, and the subducting plate descends at an angle for hundreds of kilometers beneath the continent. The west coast of South America is an example of an active margin.

Passive Continental Margins

Consider the passive margin of eastern North America. Recall that about 200 million years ago, all of the Earth's continents were joined, forming the supercontinent Pangaea III. Shortly thereafter, Pangaea broke into the continents as we know them today. As Pangaea was heated and pulled apart, its continental crust stretched and thinned (Fig. 4–23A). Eventually the lithosphere fractured and separated where the crust was thinnest. Basaltic magma rose through the cracks and flowed out onto the splitting continent. Continued eruption of basalt formed new oceanic crust between the separating fragments of Pangaea as they drifted apart (Fig. 4–23B). However, no further tectonic activity occurred along the ocean–continent boundary; hence the term passive continental margin.

Continental Shelf

Streams and rivers carry sediment from land to sea and deposit it on coastal deltas. Then ocean currents redistribute the sediment along the coast. As sediment accumulated on the passive margin of the east coast of North America, it built a shallow, gently sloping submarine surface on the submerged edge of the continent. This surface is a **continental shelf** (Fig. 4–24). Its depth increases gradually from the shoreline to about 200 meters at the outer shelf edge. A continental shelf on a passive margin is often a large feature. The shelf off the coast of southeastern Canada is about 500 kilometers wide. Parts of the shelves of Siberia and northwestern Europe are even wider.

Most continental shelves are covered by young sediment carried to the continental margin by modern rivers. Thick layers of shale, limestone, and sandstone lie beneath the younger sediment. Many of these older sedimentary rocks contain oil. Some of the world's richest offshore petroleum reserves occur in the North Sea between England and Scandinavia, in the Gulf of Mexico, and in the Beaufort Sea on the northern coast of Alaska and western Canada. In recent years, extensive exploration and development of offshore petroleum reserves have taken place on continental shelves. Deep drilling for oil has revealed that granitic crust lies beneath the sedimentary rocks of the shelves, confirming that the continental shelves are truly parts of the continents even though they are covered by sea water.

Continental Slope and Rise

At the outer edge of a continental shelf, the sea floor suddenly steepens as its depth increases from 200 meters to about 5 kilometers. This steep region averages about 50 kilometers wide and is called the **continental slope** (Fig. 4–24). It is formed by sediment much like that of the shelf. Its steeper angle is due primarily to rapid thinning of underlying continental crust as it approaches the junction with oceanic crust.

The steepness of a continental slope decreases as it gradually merges with the deep ocean floor. This region, called the **continental rise**, consists of an apron of sediment that was transported across the continental shelf and down the slope. It then came to rest on the deep ocean floor at the foot of the slope. The continental rise averages a few hundred kilometers wide. Typically, it joins the deep sea floor at a depth of about 5 kilometers.

In essence, then, the continental shelf–slope–rise complex is formed by accumulation of sediment near the continental margin. The sediment is derived by erosion of the continent and creates a smooth, sloping surface at the junction of continental and oceanic crust.

Active Continental Margins

An active continental margin forms where an oceanic plate subducts beneath a continental plate. As described previously, a long, narrow, steep-sided depression called a trench forms on the sea floor where the oceanic plate dives into the mantle (Fig. 4–25). An active continental margin commonly has a much narrower continental shelf and a considerably steeper continental slope than does a passive margin. The continental rise is absent because sediment flows into the trench instead of accumulating on the ocean floor. The landward wall of the trench forms the continental slope of an active margin. Subduction at an active continental margin causes earthquakes, mountain building, and volcanic eruptions. The west coast of South America is an example of such an environment.

Island Arcs

In many parts of the Pacific Ocean and elsewhere, two oceanic plates are colliding. When oceanic plates converge, one subducts beneath the other, diving into the

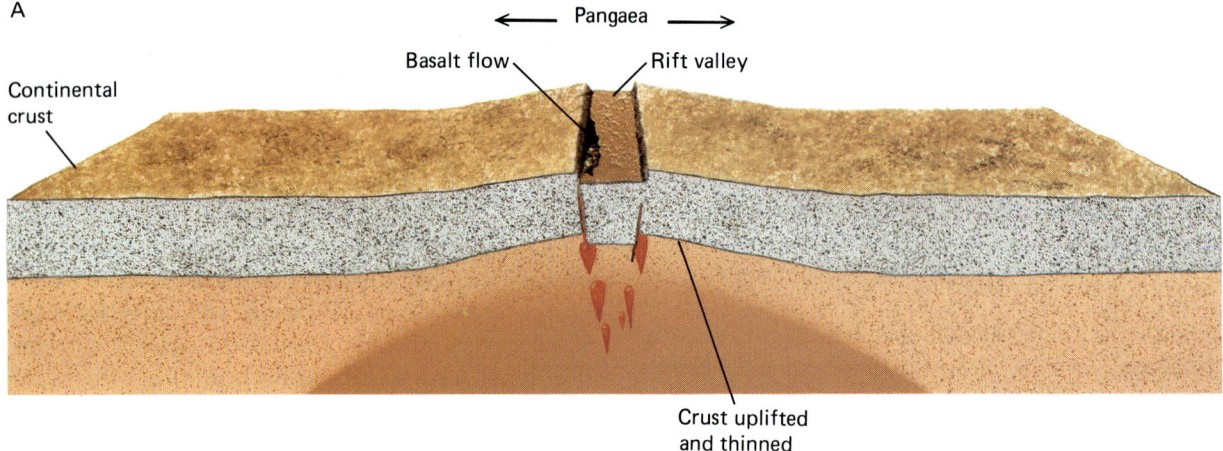
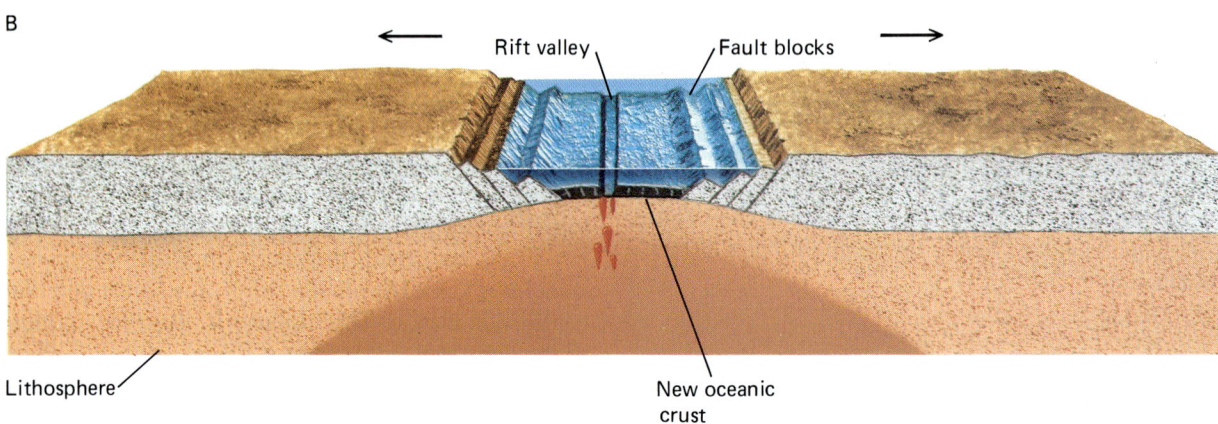
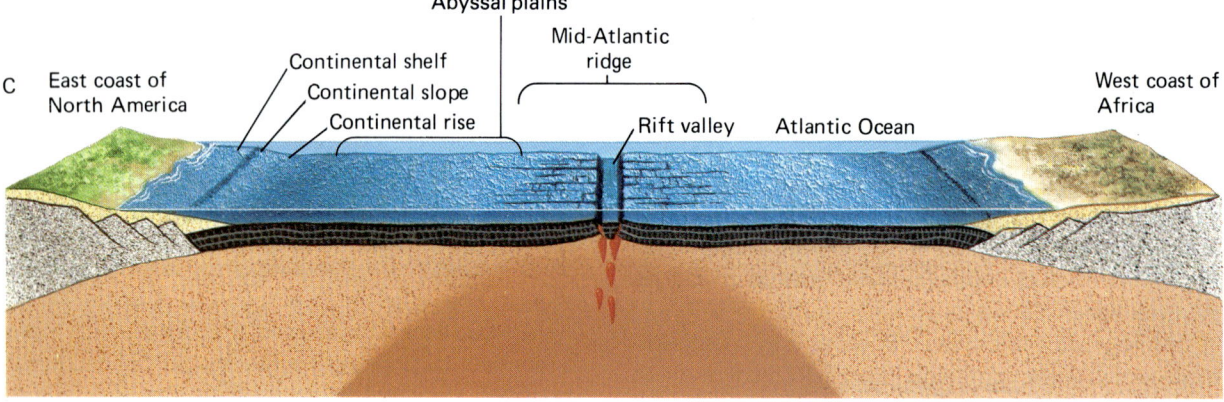

FIGURE 4-23
Development of the passive continental margin of eastern North America. The Atlantic Ocean basin formed as Pangaea rifted apart. (A) Pangaea is elevated over a rising mantle plume. (B) The crust thins due to faulting and erosion as a rift valley forms. Pangaea tears apart, and rising basalt magma forms new oceanic crust between the two halves of continental crust. (C) As the new Atlantic Ocean basin widens, sediment from the continents accumulates to form a broad continental shelf–slope–rise complex and buries the faults.

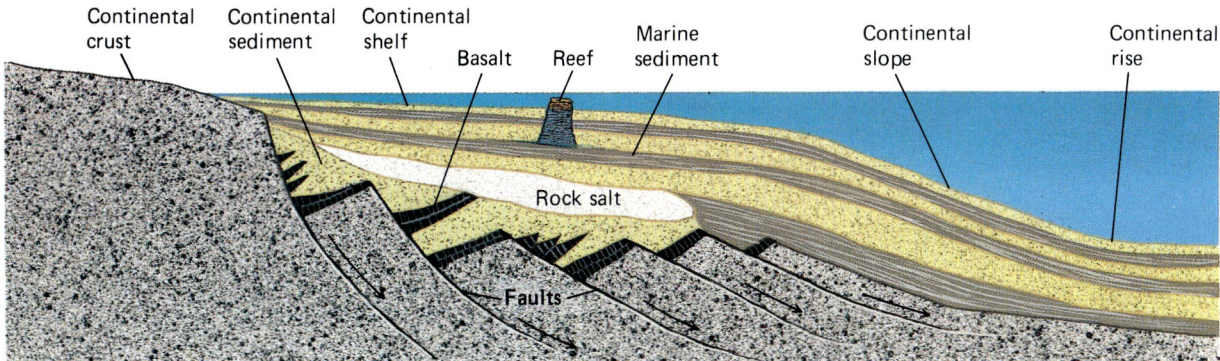

FIGURE 4–24
A passive continental margin is characterized by a broad continental shelf, slope, and rise formed by accumulation of sediment eroded from the continent. In some areas, salt deposits form when the land is uplifted or sea level falls. In tropical areas, reefs may also grow on the continental shelf.

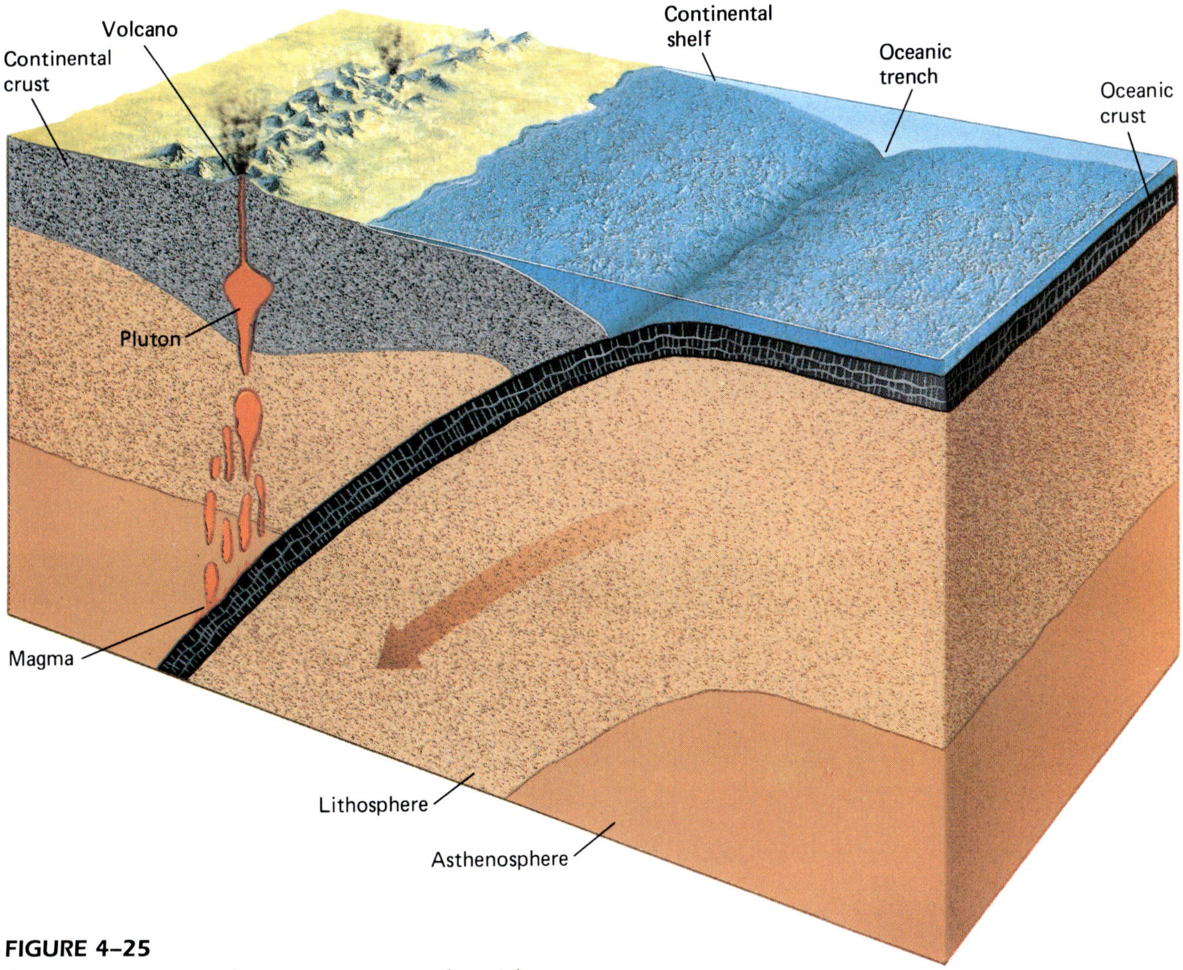

FIGURE 4–25
At an active continental margin, an oceanic plate sinks beneath a continental plate, forming an oceanic trench.

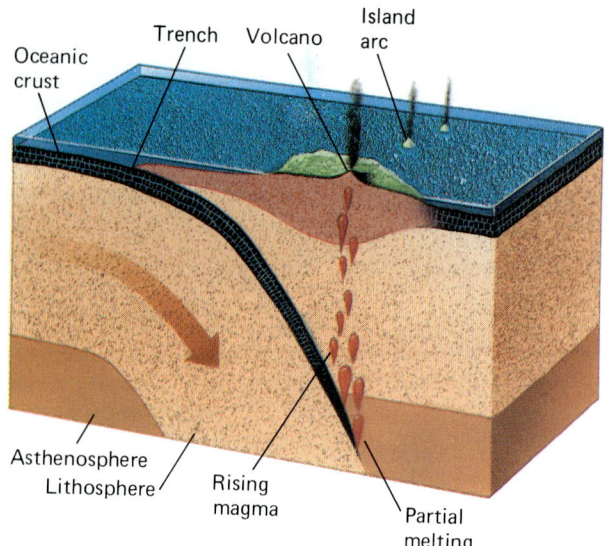

mantle and forming a mid-oceanic trench (Fig. 4–26). At the same time, great amounts of magma form directly above the subducting plate. The magma rises and erupts onto the sea floor to form submarine volcanoes next to the trench. The volcanoes eventually grow to become a chain of islands, called an **island arc**. The western Aleutian Islands are an example of an island arc. Many others are found in the southwestern Pacific.

FIGURE 4–26
An island arc forms at a convergent boundary between two oceanic plates. One of the plates sinks, generating magma that rises to form the islands.

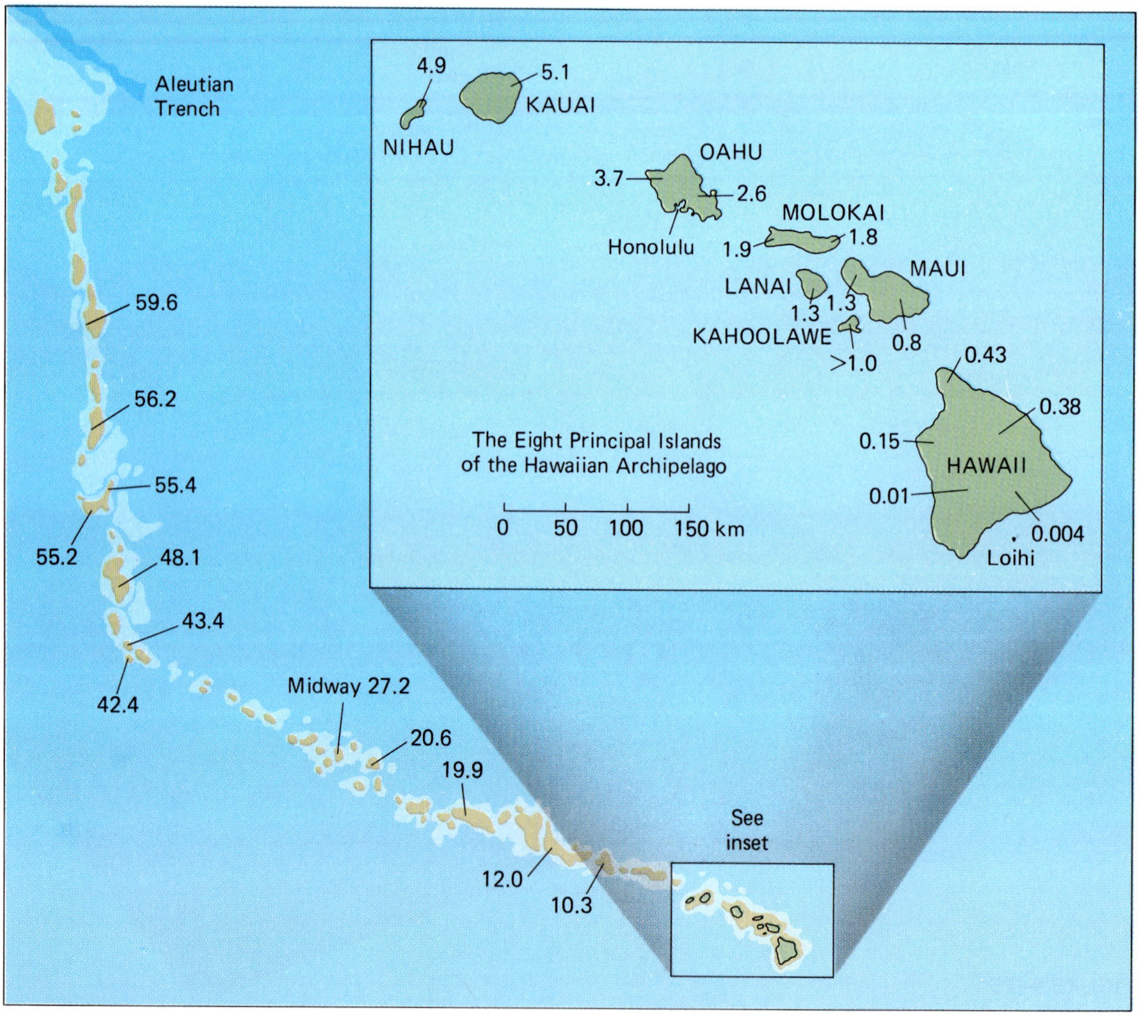

FIGURE 4–27
The Hawaiian Island–Emperor Seamount Chain. The ages, in millions of years, are for the oldest volcanic rocks found on each island or seamount. The islands become progressively older as they migrate away from the mantle plume.

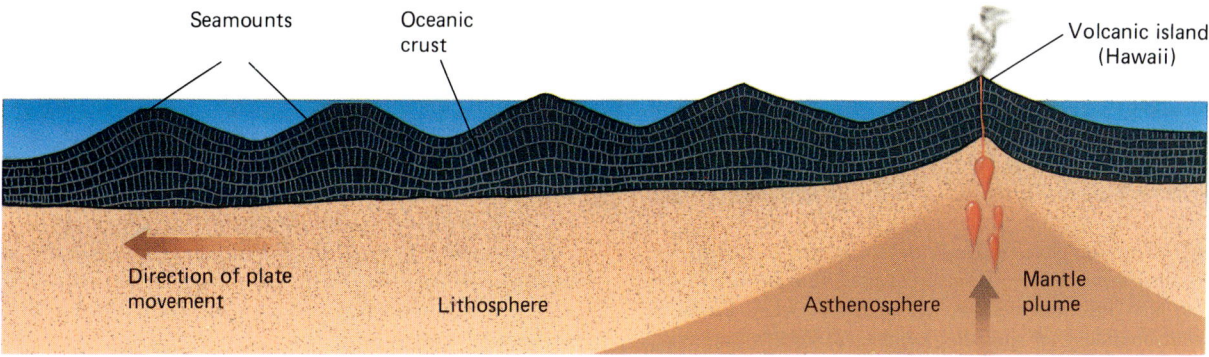

FIGURE 4–28
The Hawaiian Islands and Emperor Seamounts sink as they move away from the mantle plume.

Seamounts, Oceanic Islands, and Aseismic Ridges

A **seamount** is a submarine mountain that rises 1 kilometer or more above the surrounding sea floor. An **oceanic island** is a seamount that protrudes above sea level. Both seamounts and oceanic islands are common in all ocean basins, but they are particularly abundant in the southwestern Pacific Ocean. Seamounts and oceanic islands occur most commonly as isolated peaks on the ocean floor. However, some occur in chains of mountains called **aseismic ridges**. *Aseismic* means "no earthquake activity," as opposed to the mid-oceanic ridge and island arcs, where earthquakes are common. The Hawaiian Island–Emperor Seamount Chain is an example of an aseismic ridge. Dredge samples show that seamounts, like oceanic islands and the ocean floor itself, are made of basalt.

Seamounts and oceanic islands are submarine volcanoes probably formed at **hot spots**, areas of localized volcanic activity that lie above plumes of rising mantle rock. Isolated seamounts and islands must have formed over plumes that persisted for only a short time. In contrast, aseismic ridges, such as the Hawaiian Island–Emperor Seamount Chain, formed over long-lasting plumes. In this case, the lithospheric plate on which the volcanoes formed migrated over the plume as the magma continued to rise. Each volcano formed directly over the plume and then became extinct as the moving plate carried it away from the plume. As a result, the seamounts and islands become progressively younger toward one end of the chain (Fig. 4–27).

After a volcanic island or seamount forms, it begins to sink. Three factors contribute to the sinking:

1. If the hot spot feeding the volcanic eruptions cools and stops supplying magma, the lithosphere beneath the island cools, becomes denser, and contracts. Alternatively, the island migrates away from the hot spot if the lithospheric plate moves. This also causes cooling, contraction, and sinking of the island.
2. The weight of the newly formed volcano causes isostatic sinking.
3. Erosion lowers the top of the volcano.

These three factors commonly result in the gradual transformation of a volcanic island into a seamount (Fig. 4–28). If the Pacific Ocean plate continues to move at its present rate, the island of Hawaii may sink beneath the sea within 10 to 15 million years.

SUMMARY

The **plate tectonics theory** is the concept that the **lithosphere**, the outer, 100-kilometer-thick layer of the Earth, is segmented into seven major **plates**, which move relative to one another by gliding over the **asthenosphere**. Most of the Earth's major geological activity occurs at huge fractures called plate boundaries.

The Earth is a layered planet. The **crust** is its outermost layer and varies from 7 to 70 kilometers in thickness. The **mantle** extends from the base of the crust to a depth of 2900 kilometers, where the core begins. The **lithosphere** is the cool, brittle outer 100 kilometers of the Earth, which includes all of the crust and the uppermost mantle. The lithosphere floats on the hot, plastic **asthenosphere**, which extends from 100 to 350 kilometers in depth. The concept that the lithosphere floats on the asthenosphere is called **isostasy**. When weight is added to or subtracted from portions of the crust, it rises or falls. This vertical movement in response to changing burdens is called **isostatic adjustment**. The **core** is mostly iron and nickel and consists of a liquid outer layer and a solid inner sphere.

Tectonic plates move at rates that vary from 1 to 18 centimeters per year. Three types of plate boundaries exist. (1) New lithosphere forms and spreads outward at a **divergent boundary**, or **spreading center**. (2) Two lithospheric plates collide at a **convergent boundary**, which develops into a **subduction zone** if at least one plate carries oceanic crust. (3) Two plates slide horizontally past each other at a **transform plate boundary**. Interior parts of lithospheric plates are tectonically stable. The cause or causes of plate motion are not well understood at present. Mantle convection may cause plate movement or plate movement may cause mantle convection. A plate may move because it slides downhill from a spreading center as its cold leading edge sinks into the mantle. Supercontinents may assemble, split apart, and reassemble every 500 million to 700 million years.

Mountains and ranges form by tectonic activity at boundaries between lithospheric plates. Volcanic eruptions, intrusion of granite, metamorphism, earthquakes, and folding and faulting of rocks commonly accompany growth of a mountain chain.

Folds, faults, and **joints** develop as mountains are built. The Andes formed where the Pacific oceanic plate subducted beneath the continental plate of western South America. The Himalayas grew at a tectonic plate boundary where India collided with Asia.

Continents are composed of relatively thick, low-density granite, whereas oceanic crust is mostly thin, dense basalt. Thin, dense oceanic crust lies at low topographic levels and forms ocean basins. The **mid-oceanic ridge** is a submarine mountain chain that extends through all of the Earth's major ocean basins. A **rift valley** runs down the center of the ridge, and the ridge and rift valley are both offset by numerous **transform faults**. The mid-oceanic ridge forms along a spreading center, where new oceanic crust is added to the sea floor.

The age of sea-floor rock increases regularly away from the mid-oceanic ridge. No oceanic crust is older than about 200 million years because it recycles into the mantle at subduction zones.

A **passive continental margin** includes a **continental shelf**, a **slope**, and a **rise** formed by accumulation of sediment transported from the continents. An **active continental margin**, where oceanic crust subducts beneath the margin of a continent, usually includes a narrow continental shelf and a continental slope that steepens rapidly into a **trench**. A trench is an elongate trough in the ocean floor formed where oceanic crust dives downward at a subduction zone. Trenches are the deepest parts of ocean basins.

Island arcs are common features of some ocean basins, particularly the southwestern Pacific. They are chains of volcanoes formed at subduction zones where two oceanic plates collide. **Seamounts, oceanic islands**, and **aseismic ridges** form in oceanic crust as a result of volcanic activity over mantle plumes.

KEY TERMS

Plate tectonics theory 67
Lithosphere 67
Plates 67
Asthenosphere 67
Fault 67
Plate boundary 67
Crust 69
Mantle 69
Isostasy 70
Isostatic adjustment 71
Core 71
Divergent plate boundary 72
Convergent plate boundary 72
Transform plate boundary 72
Spreading center 72
Mid-oceanic ridge 72

Continental rifting 73
Rift valleys 73
Subduction 74
Subduction zone 74
Benioff zone 74
Oceanic trench 75
Transform plate boundary 76
Mantle convection 77
Supercontinent 80
Pangaea 82
Mantle plume 82
Geologic structure 82
Fold 82
Fault 82
Joint 82
Gondwanaland 83

Laurasia 83
Andean margin 84
Underthrusting 85
Transform fault 89
Continental margins 89
Passive continental margin 89
Active continental margin 91
Continental shelf 91
Continental slope 91
Continental rise 91
Island arc 94
Seamount 95
Oceanic island 95
Aseismic ridges 95
Hot spots 95

REVIEW QUESTIONS

1. Draw a cross-sectional view of the Earth. List all the major layers and the thickness of each.
2. Describe the physical properties of each of the Earth's layers.
3. What properties of the asthenosphere allow the lithospheric plates to glide over it?
4. How is it possible for the solid rock of the mantle to flow and convect?
5. Summarize the important aspects of the plate tectonics theory.
6. How many major tectonic plates exist? List them.
7. Describe the three types of tectonic plate boundaries.
8. Explain why tectonic plate boundaries are geologically active and the interior regions of plates are geologically stable.
9. Describe a reasonable model for a mechanism that causes movement of tectonic plates.
10. In Paul Hoffman's model, why do supercontinents break up within a few hundred million years after they form?
11. How many supercontinents have formed in Hoffman's model?
12. What is the difference between a fault and a joint?
13. How are faults related to earthquakes?
14. Describe the similarities and differences between the Andes and the Himalayan chain. Why do the differences exist?
15. Sketch a cross section of an Andean-type plate boundary to a depth of several hundred kilometers. Show the positions of the subducting plate, trench, subduction complex, volcanoes and plutons, and earthquakes.
16. Draw a series of cross-sectional sketches showing the evolution of a Himalayan-type plate boundary. Why does this boundary start out as an Andean-type margin?
17. Describe the Mid-Atlantic ridge and the mid-oceanic ridge.
18. Why are the oldest sea-floor rocks only about 200 million years old, whereas some continental rocks are more than 3 billion years old?
19. Why do oceanic islands sink after they form?

DISCUSSION QUESTIONS

1. Central Greenland lies below sea level because the crust is depressed by the ice cap. If the glacier were to melt, would Greenland remain beneath the ocean? Why or why not?
2. At a rate of 5 centimeters per year, how long would it take for a continent to drift the width of your classroom? The distance between your apartment or dormitory and your classroom? The distance from New York to London?
3. Although most earthquakes occur at plate margins, occasionally very large earthquakes occur within lithospheric plates. How might this happen?
4. Discuss how microcontinents might have formed.
5. How and why does an oceanic trench form?
6. The east coast of South America has a wide continental shelf, whereas the west coast has a very narrow shelf. Discuss and explain this contrast.
7. Seismic data indicate that continental crust thins where it joins oceanic crust at a passive continental margin, such as on the east coast of North America. Other than that, we know relatively little about the nature of the junction between the two types of crust. Speculate on the nature of that junction. Consider rock types, geologic structures, ages of rocks, and other features of the junction.
8. Why do most major continental mountain chains form at convergent plate boundaries? What topographic and geologic features characterize divergent and transform plate boundaries in continental crust? Where do these types of boundaries exist in continental crust today?
9. Compare and explain the similarities and differences between the Andes Mountains and the Himalayan chain. How would the Himalayas 60 million years ago compare with the modern Andes?
10. If you were studying photographs of another planet, what features would you look for to determine whether the planet is or has been tectonically active?
11. The largest mountain in the Solar System is Olympus Mons, a volcano on Mars. It is 25,000 meters high, nearly three times the elevation of Mount Everest. Speculate on the factors that might permit such a large mountain on Mars.

CHAPTER 5
Earthquakes

5.1 What Is an Earthquake?
5.2 Earthquake Waves
5.3 Earthquake Damage
5.4 Earthquake Danger Zones in North America
5.5 Earthquake Prediction
5.6 Earthquake Hazard and Human Habitation

About a million earthquakes occur worldwide every year. Most are too mild to be felt and are detected only with sensitive instruments. Many shake houses and rattle windows but cause little damage. A few topple buildings, cause landslides, and fracture dams, roadways, pipelines, and bridges. An average of about 10,000 earthquake fatalities occur every year (Table 5–1).

As we learned in Chapter 4, the outer Earth is segmented into 100-kilometer-thick plates that migrate slowly across the globe. The plates separate, collide, and slip past one another at their boundaries. As a result of these motions, most earthquakes occur at the plate boundaries.

The 1989 Loma Prieta earthquake in California fractured the ground surface near the epicenter.

TABLE 5-1 Major Historical Earthquakes

Year	Date	Region	Deaths	Magnitude*	Comments
1556		China, Shensi	830,000		More deaths than in any other natural disaster in history
1663	Feb. 5	Canada, St. Lawrence River	?		Chimneys broken as far away as Massachusetts
1811	Dec. 16	Missouri, New Madrid	Several	8.7†	Three shocks; largest historical earthquake sequence in the U.S.
1857	Jan. 9	California, Fort Tejon	?	8.0†	San Andreas fault rupture
1868	Aug. 16	Ecuador and Colombia	Ecuador, 40,000 Colombia, 30,000		
1886	Aug. 31	South Carolina, Charleston–Summerville	About 60		Most recent major quake in eastern United States
1896	June 15	Japan, Riku-Ugo	22,000		Giant tsunami
1906	April 18	California, San Francisco	650	8.3†	San Francisco fire
1923	Sept. 1	Japan, Kwanto	200,000	8.2†	Great Tokyo fire
1960	May 21–30	Southern Chile	5700	8.5–8.7	Largest sequence of earthquakes ever recorded
1964	March 28	Alaska	131	8.6	Damaging tsunami
1970	May 31	Peru	66,000	7.8	Great rockslide destroyed the town of Yungay
1971	Feb. 9	California, San Fernando	65	6.5	$550 million damage
1976	July 28	China, Tangshan	250,000	7.6	Great economic damage; not predicted
1985	Sept. 19	Mexico, Michoacán	9500	7.9	More than $3 billion damage; 30,000 injured; small tsunami
1988	Dec. 7	Armenia	55,000+	6.9	Death toll high due to poor construction
1989	Oct. 17	California, Loma Prieta	65	6.9	May be harbinger of additional Bay Area quakes
1994	Jan. 17	California, San Fernando	55	6.6	Occurred on secondary faults in the San Andreas fault zone.

*Magnitudes were not calculated prior to 1935. †Estimated magnitudes based on reported damage.

5.1 What Is an Earthquake?

Although tectonic plates move continuously at rates from less than 1 to 18 centimeters per year, slippage at a plate boundary is not smooth and continuous. Commonly, as two plates move past one another, rocks in the fracture zone are held tightly by friction and no slipping occurs at the fault. **Two plates can glide past each other without motion on the fault because rock can stretch or compress like a spring.** We do not commonly think of rock as elastic, but if you drop a rock onto a cement floor, it bounces. Although the magnitude of the bounce is small, the rock behaves like any other elastic object, such as a tennis ball.

When two tectonic plates move past each other without motion on the fault, rock near the plate boundary stretches or compresses. As it does so, it stores **elastic energy**, energy stored by elastic deformation of the rock. It is comparable to the energy in a stretched elastic band. When the elastic energy overcomes the friction that is keeping the rocks from moving, the rocks break loose along the fault and spring back to their original shapes (Fig. 5–1). This rapid motion sets up vibrations that travel through the Earth like the vibrations of a bell struck by a hammer. An **earthquake** is the vibration generated by sudden slippage of rocks along a fault. Figure 5–2 shows that most earthquakes happen along tectonic plate boundaries, where two plates slip past one another.

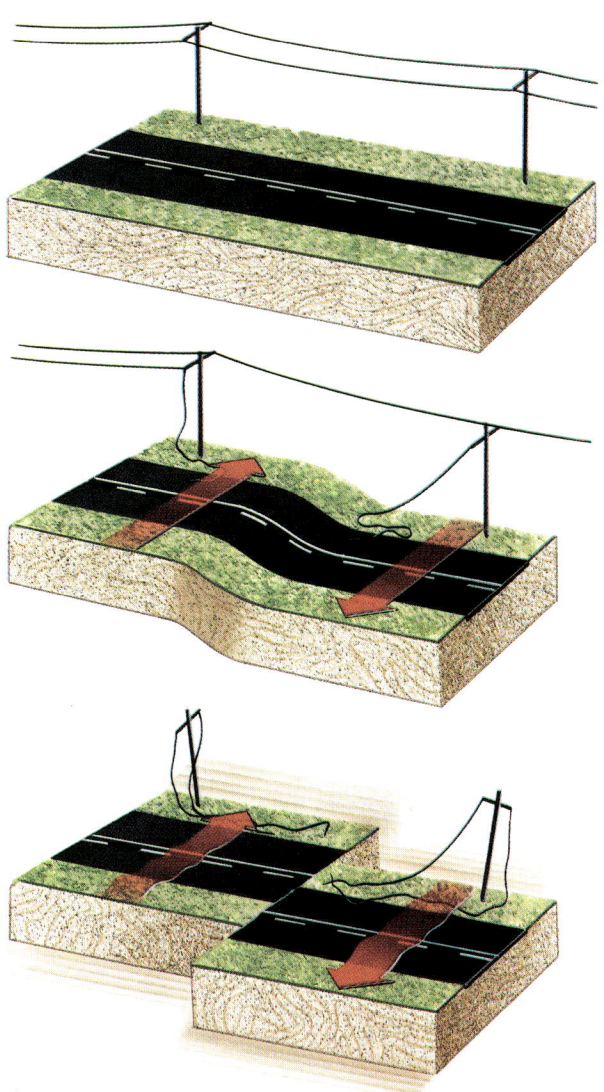

B

FIGURE 5–1
(A) When rock is stretched by a tectonic force, it stores elastic energy. Eventually the rock fractures and snaps back to its original shape but in a new position, creating an earthquake. (B) Fractures in a roadway in Santa Cruz following the 1989 quake.

A

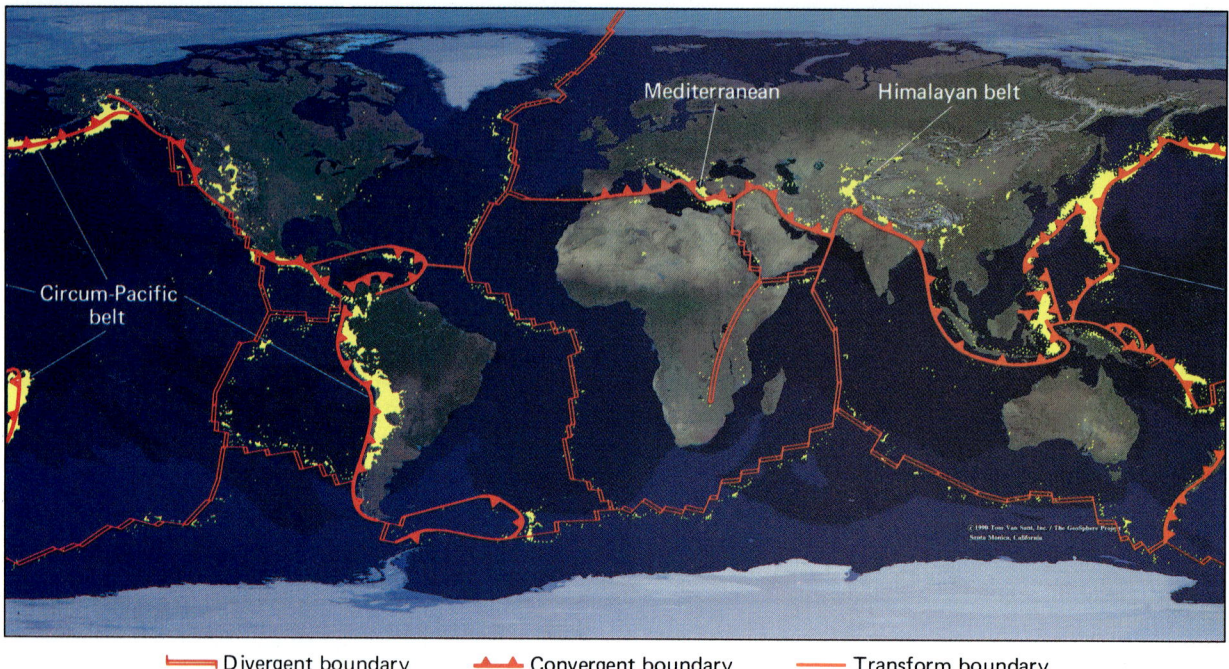

Divergent boundary   Convergent boundary   Transform boundary

FIGURE 5–2
The Earth's major earthquake zones coincide with tectonic plate boundaries. Each yellow dot represents an earthquake that occurred between 1961 and 1967. Note the concentration of earthquakes along the San Andreas fault. *(Tom Van Sant, Inc./Geosphere Project, Santa Monica, CA)*

5.2 Earthquake Waves

If you have ever bought a watermelon, you know the challenge of trying to pick out a ripe, juicy one without being able to look inside. One trick is to tap the melon gently with your knuckle. If you hear a sharp, clean sound, it is probably ripe; a dull thud indicates that it may be overripe and mushy. The watermelon illustrates two points that can be applied to the Earth: (1) The energy of your tap is transmitted through the melon, and (2) the nature of the melon's interior affects the quality of the sound.

A wave transmits energy from one place to another. Thus, a drumbeat travels through air to your ear as a sequence of waves, and the Sun's heat travels to Earth as waves. Similarly, a tap travels through a watermelon in waves. Waves that travel through rock are called **seismic waves**. Earthquakes and explosions produce seismic waves. **Seismology** is the study of earthquakes and of the nature of the Earth's interior based on evidence from seismic waves.

An earthquake produces several different types of seismic waves. **Body waves** travel through the Earth's interior. They start at the earthquake's **focus**, where rocks move along a fault, and radiate outward in concentric

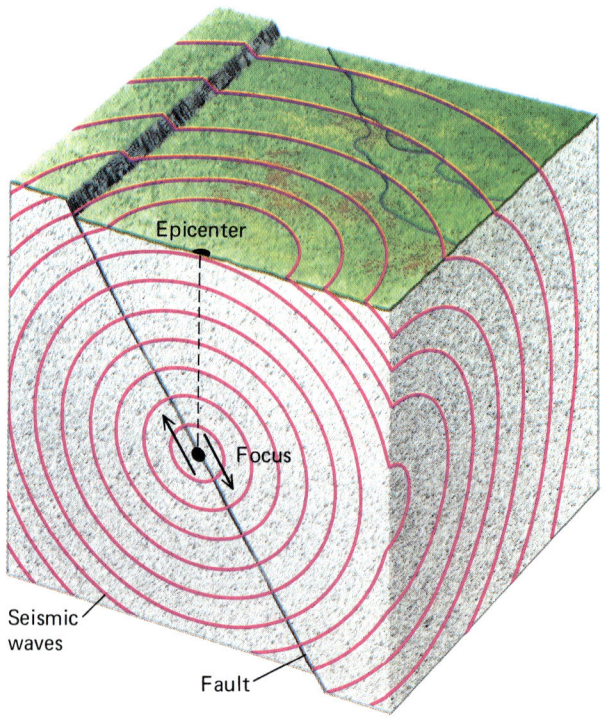

FIGURE 5–3
Body waves radiate outward from the focus of an earthquake.

spheres (Fig. 5–3). An earthquake's focus may be as deep as 700 kilometers below the surface, but most earthquakes start at 100 kilometers or less.

The point on the Earth's surface directly above the focus is the **epicenter**. During an earthquake, **surface waves** radiate away from the epicenter along the surface of the Earth like the waves that ripple across the water when you throw a rock into a calm lake.

Body Waves

Two main types of body waves travel through the Earth's interior. A **primary wave**, or **P wave**, forms by alternate compression and expansion of the rock (Fig. 5–4A). Consider a long spring such as the popular Slinky® toy. If you stretch a Slinky and strike one end, a compressional wave travels along its length. Sound also travels as a compressional wave. A ringing bell produces a sound wave in air, which is a type of P wave. Liquids and solids also transmit P waves. Next time you take a bath, immerse your head until your ears are under water and listen as you tap the sides of the tub with your knuckles. In a similar manner, the music from the radio in the apartment next door travels easily through the walls when you are trying to read this chapter.

P waves travel at speeds between 4 and 7 kilometers per second in the Earth's crust and at about 8 kilometers per second in the upper mantle. As a comparison, the speed of sound in air is only 0.34 kilometer per second, and the fastest jet fighters fly at about 0.85 kilometer per second. Therefore, even the slowest P waves in the Earth travel more than ten times faster than the speed of sound in air, and five times faster than a jet fighter. P waves are called *primary* waves because they are so fast that they are the first waves to reach an observer.

A second type of body wave, called an **S wave**, is a **shear wave**. An S wave can be illustrated by tying a rope to a wall, holding the end, and giving it a sharp up-and-down jerk (Fig. 5–4B). S waves are slower than P waves, traveling at speeds between 3 and 4 kilometers per second in the crust. As a result, they arrive after P waves and are the *secondary* waves to reach an observer on Earth.

Unlike P waves, S waves move only through solids. Because molecules in liquids and gases are only weakly bound together, they slip past each other and thus cannot transmit a shear wave. Geologists infer that the Earth's

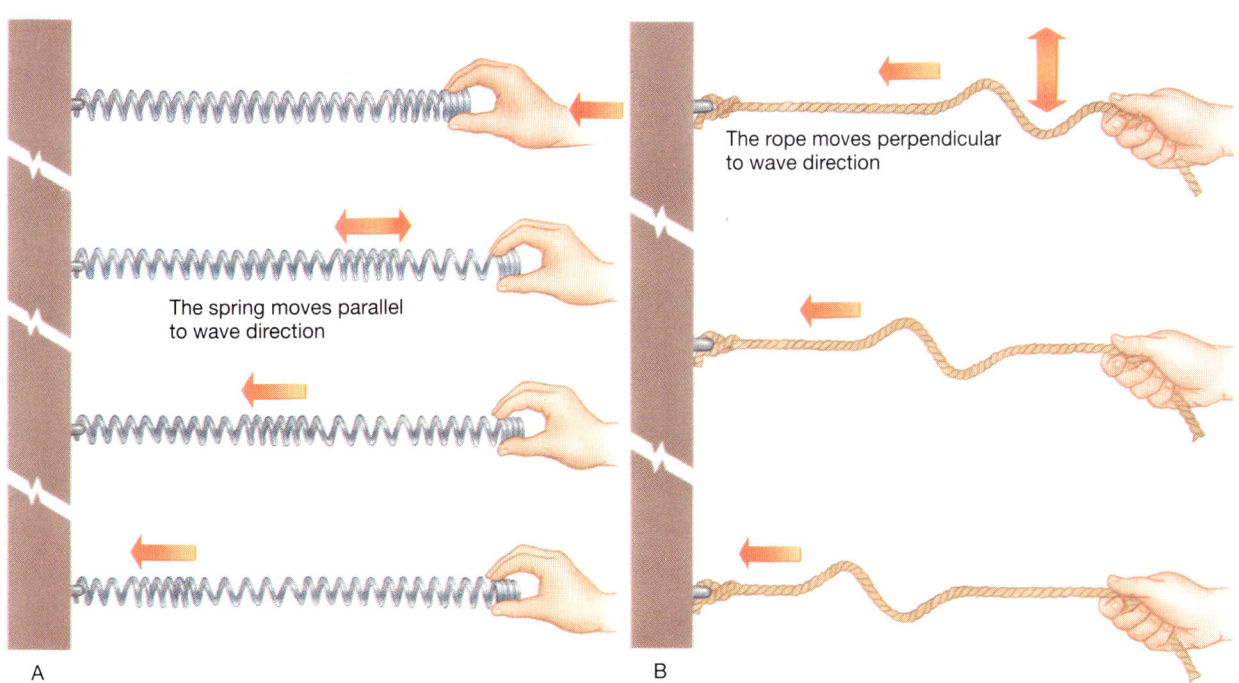

FIGURE 5–4
Two different types of body waves travel through the Earth. Their respective characteristics are shown by a spring and a rope. (A) A compressional, or P, wave travels along the spring. The particles in the slinky move parallel to the direction in which the wave itself travels. (B) An S wave travels along a rope, but the particles in the rope move perpendicular to the direction in which the wave travels.

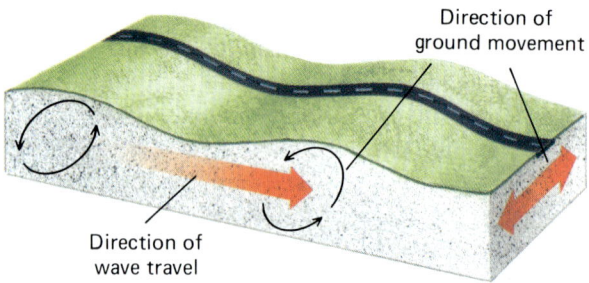

FIGURE 5–5
Surface waves move up and down, like ocean waves, and also from side to side.

outer core is liquid because P waves travel through it but S waves do not.

Surface Waves

Surface waves, called **L waves**, travel more slowly than either type of body wave. Two types of surface waves occur simultaneously (Fig. 5–5). One is an up-and-down motion and the other is a side-to-side vibration. Thus, during an earthquake, roadways and pipelines roll like ocean waves and writhe from side to side like snakes (Fig. 5–6).

Measurement of Seismic Waves

A **seismograph** is a device that records seismic waves. To understand how a seismograph works, consider the act of writing a letter while riding in an airplane. If the plane hits turbulence, your handwriting becomes wiggly. Because the paper is on a tray that is connected to the frame of the aircraft, the paper moves when the plane bounces. But your hand is connected to your body by a series of movable joints that flex when the plane lurches. Inertia keeps your hand relatively stationary as the plane moves back and forth beneath it, so your hand does not move as erratically as the plane. The paper jiggles back and forth beneath a relatively motionless hand, and your handwriting becomes erratic.

Early seismographs worked on the same principle. Imagine a weight suspended from a spring. A pointer attached to the weight is aimed at the zero mark on a scale (Fig. 5–7). The scale is attached firmly to bedrock by solid metal braces, but the weight and pointer hang

FIGURE 5–6
Vehicles stranded by multiple failures of an overpass on Highway 14 near Los Angeles, resulting from the 1994 earthquake. *(Douglas C. Pizac/Wide World Photo)*

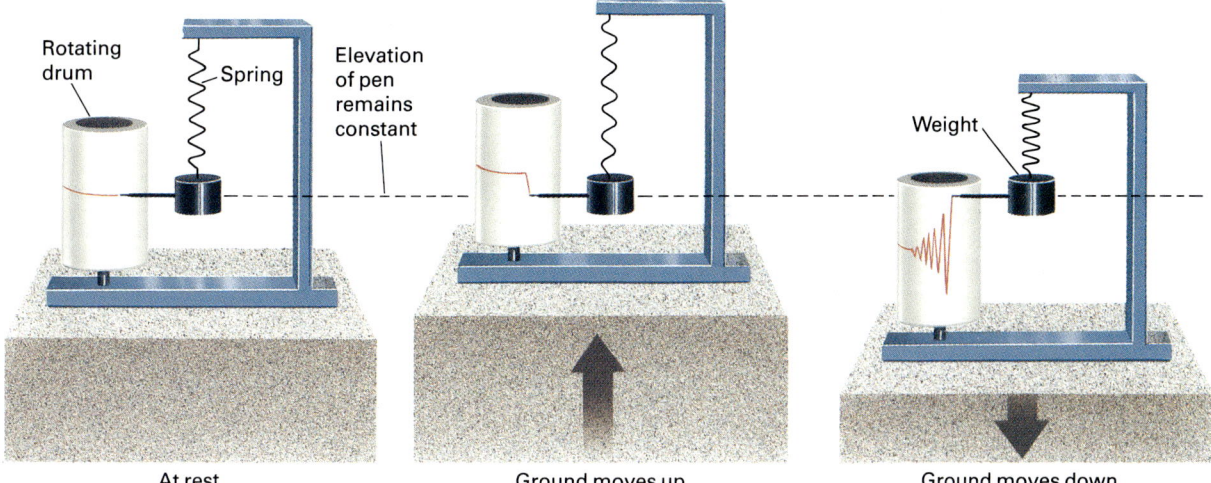

FIGURE 5–7
A seismograph records ground motion during an earthquake. When the ground is stationary, the pen draws a straight line across the rotating drum. When the ground rises abruptly during an earthquake, it carries the drum up with it. But the spring stretches so that the weight and pen hardly move. Therefore, the pen marks a line lower on the drum. Conversely, when the ground sinks, the pen marks a line higher on the drum. During an earthquake, the pen traces a jagged line as the drum rises and falls.

from the flexible spring. During an earthquake, the scale jiggles up and down, but inertia keeps the suspended weight stationary. As a result, the scale moves up and down beneath the pointer. If the scale is replaced by a rotating drum and a pen is mounted on the pointer, then the pen records earthquake motion on the rotating drum. This record of Earth vibration is called a **seismogram**.

A modern seismograph uses electronic motion detectors.

Measurement of Earthquake Strength

Earthquakes vary from gentle tremors that cannot be detected without a seismograph to destructive giants that create large-scale movements of the Earth's surface and catastrophic losses of life and property.

The **magnitude** of an earthquake is proportional to the amount of energy released by the quake. From about 1935, when it was first refined by Charles Richter, to recently, the **Richter scale** was the most common means of indicating the strength of a quake, and it is still widely used in the popular press. The Richter magnitude is calculated from the height of the largest earthquake wave recorded on a seismograph. However, the relationship between the recorded wave height and earthquake energy depends on many complex factors including distance between the epicenter and the seismograph, the depth of the quake, and the rock types through which the waves passed. Seismologists often spent the days following a large quake on the telephone with other seismologists, adjusting their numbers so that they all agreed on a single Richter magnitude.

Modern equipment and methods enable seismologists to measure the amount of rock displacement during a quake. This value allows them to calculate the total surface area of movement. This number is then used to calculate the **moment magnitude**, which is directly proportional to the energy released during an earthquake. The moment magnitude is now used by most seismologists because it is a more accurate indicator of the amount of energy released by a quake than the Richter magnitude.

Seismologists have calibrated the moment magnitude scale so that the number for any earthquake is identical to an accurately calculated Richter magnitude. Thus, the main difference between the two scales lies in the way in which the magnitude is calculated. Both the moment magnitude and Richter scales are logarithmic; an increase of one unit—for example, from 7 to 8— on both scales corresponds approximately to a 30-fold increase in energy released during the quake. Thus, a magnitude 8 quake releases 30 times as much energy as a magnitude 7 quake and 900 times as much as a magnitude 6 quake. An earthquake with a magnitude of 6.5 has an energy of about 10^{21} (10 followed by 21 zeros)

TABLE 5-2 Frequency and Energy Equivalence of Earthquakes

Magnitude	Effects	Energy Equivalence	Average Number per Year
-2, -1, 0, 1, 2	Not felt	100-watt light bulb left on for 1 week / 0.5 kg of conventional explosive	900,000
3	Felt but only minor damage	1000 tons of conventional explosive	100,000
4			15,000
5			3,000
6	Significant, depending on quality of construction	Small atomic bomb	100
7	Major earthquake; very destructive		20
8	Great earthquake; some communities totally destroyed	8.7 was largest earthquake ever recorded; it released about 900 times as much energy as the atomic bomb dropped on Hiroshima at the end of World War II	One every 5 to 10 years

ergs.[1] The atomic bomb dropped on the Japanese city of Hiroshima at the end of World War II released about that much energy. The largest possible earthquake is determined by the strength of rocks. A strong rock can store more elastic energy before it fractures than a weak rock. The largest earthquakes ever measured had magnitudes of 8.5 to 8.7, about 900 times greater than the energy released by the Hiroshima bomb (Table 5–2).

5.3 Earthquake Damage

Ground Motion

During an earthquake, waves travel both along the Earth's surface and through subterranean rock. Under proper conditions, the earthquake may produce permanent displacement of the Earth's surface (Fig. 5–8). The New Madrid, Missouri, earthquake of 1811 changed the course of the Mississippi River. A cliff created by an earthquake is called a **scarp** (Fig. 5–9). During the 1964 Alaskan earthquake, some beaches rose 12 meters, leaving harbors high and dry, whereas other beaches sank 2 meters, causing coastal flooding.

Most of the people killed during an earthquake die because falling structures crush them. Structural damage, injury, and death depend on the magnitude of the quake, its proximity to population centers, rock and soil types, topography, and the quality of construction in the region.

Structural damage also depends in part on the depth of the focus. A magnitude 8.2 quake occurred about 300 kilometers northeast of La Paz, Bolivia, on June 8, 1994. Although it was one of the strongest earthquakes ever recorded and was felt as far away as Toronto, Canada, it caused almost no damage, even at the epicenter. In contrast, an 8.1 magnitude quake described later in this section caused extensive damage and death in Mexico City, 500 kilometers from its epicenter. Part of the reason for the difference in the damage caused by the two quakes lies in the fact that the Bolivian quake occurred in a

---

[1] An erg is the standard unit of energy in scientific usage. One erg is a small amount of energy. Approximately 3×10^{12} ergs are needed to light a 100-watt light bulb for 1 hour. However, 10^{21} is a very large number and 10^{21} ergs represents a considerable amount of energy.

FIGURE 5–8
The 1964 Alaska earthquake destroyed much of Anchorage.
(Ward's Natural Science Establishment, Inc.)

FIGURE 5–9
A scarp formed by the Loma Prieta, California, earthquake of 1989.

relatively unpopulated region, whereas the Mexican quake occurred near one of the world's largest cities. However, the amount of ground shaking at the epicenter of the Bolivian quake was much less than that during the Mexican quake because the Bolivian quake was also one of the deepest earthquakes recorded, with a focus 625 kilometers beneath the Earth's surface.

The deep earthquake caused little damage and only moderate earth movement at the epicenter because the source of the energy was so far from the Earth's surface. The energy of the waves had weakened by the time they arrived at the epicenter. At the same time, the waves caused minor shaking as far away as Toronto because the focus was so deep, where the rocks are more elastic than those at lesser depths.

How Rock and Soil Influence Structural Damage

In many regions, bedrock lies at or near the surface and buildings are anchored directly to the rock. Bedrock vibrates during an earthquake, and if the motion is violent enough, buildings may fail. However, most bedrock returns to its original shape when the earthquake is over, so if the structures can withstand the shaking, they will survive. Because rock fractures in only a few places, bedrock forms a desirable foundation in earthquake hazard areas.

In other regions, thick layers of sediment lie over bedrock and structures are built on sand, clay, or silt. If you pour sand into a coffee can and then tap the can lightly on a tabletop, the sand settles. Sand and sandy

FOCUS ON

Earthquake Waves as a Tool for Investigating the Structure of the Earth

Several properties of seismic waves enable geologists to study the Earth's interior.

1. The speed with which a seismic wave travels depends on the kinds of rock that it travels through. In addition, temperature, density, rigidity, and other properties of rock affect wave speed.

2. A wave refracts (bends) and sometimes reflects (bounces back) as it passes from one medium into another. If you place a pencil in a glass half filled with water, the pencil appears bent. Of course, the pencil doesn't really bend; the light rays bend. Light rays slow down when they pass from air to water, and as their speed changes they refract. A mirror reflects light from your face when you look into it. Seismic waves refract and reflect as they pass from one medium into another.

3. P waves travel through gases, liquids, and solids, whereas S waves travel only through solid rock.

Discovery of the Core

If the entire Earth were composed of one type of rock, then P and S waves would travel everywhere and would be detectable anywhere on the planet. However, S waves do not pass through the Earth's outer core (Fig. 1). Because S waves travel through solids but not through liquids, the Earth's outer core must be liquid.

Neither S nor P waves arrive in a shadow zone between 105° and 140° from an epicenter. Beyond 140°, direct P waves arrive, but direct S waves do not. The shadow zone, too, results from the change from solid rock to molten liquid at the core–mantle boundary. Earthquake waves curve gently as they pass through the Earth. But P waves refract sharply as they pass from the mantle into the core. As a result, no P waves arrive in the shadow zone.

P waves refract sharply again when they pass from the outer core to the inner core, indicating another radical change in physical properties of the Earth's interior. In this case, the change in direction results from an abrupt transition from the molten outer core to the solid inner core. Thus, seismic data tell us that the core is composed of an inner solid sphere surrounded by an outer liquid shell. The entire core is composed mostly of iron and nickel.

Discovery of the Crust–Mantle Boundary

In 1909 a seismologist, Andrija Mohorovičić, discovered that seismic wave velocities increase sharply at a depth varying from 7 to 70 kilometers beneath the Earth's surface. This sudden velocity change indicates an abrupt transition in rock type at that depth. The upper layer is the Earth's crust, and the lower layer is the mantle. The boundary between the two is called the **Mohorovičić discontinuity** or the **Moho**, in honor of its discoverer.

The Lithosphere–Asthenosphere Boundary

Recall that the lithosphere is the 100-kilometer-thick brittle outer shell of the Earth. It includes both the crust and the uppermost part of

---

soils settle during earthquakes, causing permanent displacement of the land. This motion tilts buildings, breaks pipelines and roadways, and fractures dams. To avert structural failure in sandy soils, engineers drive steel or concrete pilings through the sand to the bedrock below. These pilings anchor and support the structures even if the ground beneath them settles.

Clay soils can also create trouble during an earthquake. Clay particles absorb water and shake like jello during an earthquake. In the process, they can amplify earthquake waves. To observe this effect, place some jello in a bowl and shake the bowl gently and rhythmically. The waves will grow until the jello shakes more than the bowl.

Mexico City is built on a high plateau ringed by even higher mountains. When the Spaniards first invaded central Mexico, lakes dotted the plateau and the Aztec capital lay on an island at the end of a long causeway in the middle of one of the lakes. Over the following centuries, European settlers drained the lake and built the mantle and floats on the hot, plastic asthenosphere. Geologists discovered the boundary between the two layers when they observed that both the speed and the strength of seismic waves decrease abruptly at a depth of 100 kilometers. The sudden changes occur because the cool, brittle rock of the lithosphere transmits seismic waves more effectively than the hot, soft rock of the asthenosphere.

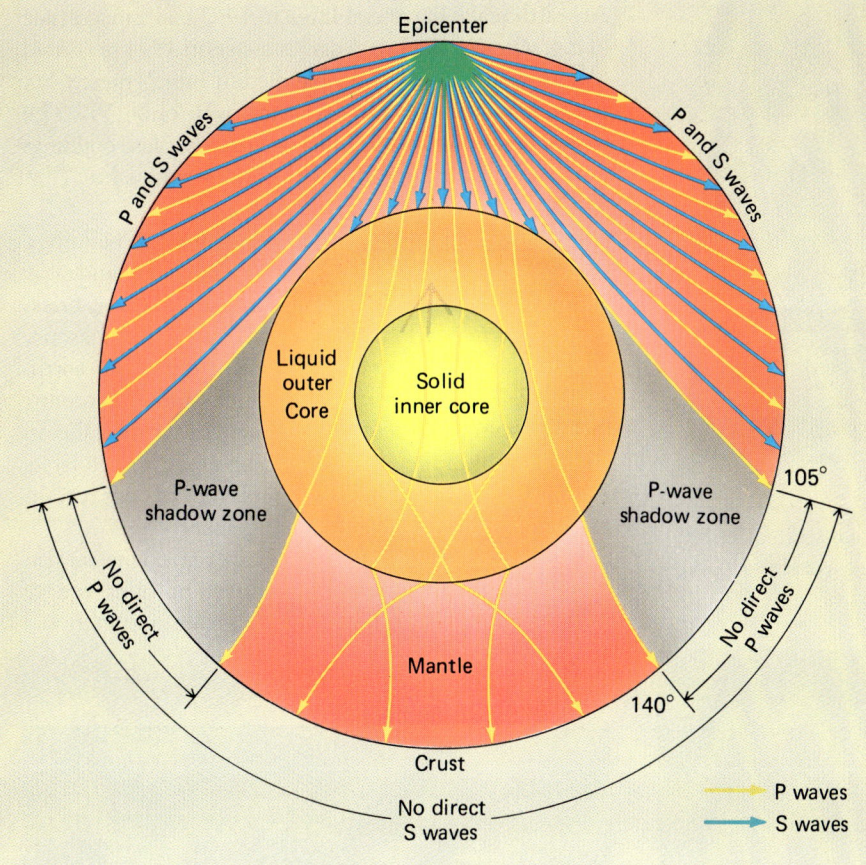

FIGURE 1
Seismic waves curve gently as they pass through the Earth. They also bend sharply where they cross major layer boundaries in the Earth's interior. The blue S waves do not travel through the liquid outer core, and therefore direct S waves are observed only within 105° of the epicenter. The yellow P waves bend sharply at the core–mantle boundary to create a shadow zone of no direct P waves from 105° to 140°.

the modern city on the water-soaked, clay-rich sediment. On September 19, 1985, an earthquake with a magnitude of 8.1 occurred about 500 kilometers west of the city. The earthquake waves shook the wet clay beneath the city and bounced back and forth between the bedrock sides of the basin, just as waves in the bowl of jello bounced off the side of the bowl. The reflections amplified the waves until structures collapsed. The shaking destroyed more than 500 buildings and killed 8000 to 10,000 people (Fig. 5–10). Meanwhile, comparatively little damage occurred in Acapulco, which was much closer to the epicenter but is built on bedrock.

Clay soils can be particularly unstable on hillsides. Recall from Chapter 2 that clay is composed of tiny, platelike minerals that soak up large amounts of water. If the ground is jiggled, the clay can release some of the water, creating a liquid slurry like wet concrete. This process is called **liquefaction**. When a clay soil liquefies, the slurry slides downslope, carrying structures along with it. During the 1964 earthquake near Anchorage, Alaska, a clay-rich bluff 2.8 kilometers long, 300 meters wide, and 22 meters high liquefied. The slurries carried houses into the ocean and buried some so deeply that bodies were never recovered (Fig. 5–11).

Pilings are less effective in clay soils than in sandy soils. When sand moves, individual particles flow past one another. The particles also flow around the pilings, like water in a river. But clay is so cohesive that it shakes and even destroys the pilings.

Construction Design and Earthquake Damage

An earthquake in central India in 1993 had a magnitude of 6.4, whereas the Los Angeles quake in 1994 measured 6.6. More than 30,000 people died in India's quake, but the death toll in Los Angeles was only 55. The tremendous mortality in India occurred because buildings were not engineered to withstand earthquakes (Fig. 5–12).

Some common framing materials used in buildings, such as wood and steel, are flexible; they bend and sway during an earthquake and resist failure. However, brick, stone, concrete, adobe (dried mud), and other masonry products are brittle and likely to fail during an earthquake. Falling concrete, adobe, brick, and clay roofing tiles crush unlucky residents and cause heavy mortality in many regions of the world. If enough money is available, masonry can be reinforced with steel, but in poor regions people cannot afford such reinforcement.

Given adequate resources, however, engineers can build earthquake-resistant structures. In one type of earthquake-resistant design, buildings are not anchored

FIGURE 5–10

The 1985 Mexico City earthquake had a magnitude of 8.1. It killed between 8000 and 10,000 people. *(Wide World Photos)*

FIGURE 5–11

A landslide in a suburban area of Anchorage, Alaska, triggered by the 1964 earthquake. When the ground shook, marine clay and silt became unstable and slid downslope. *(USGS)*

FIGURE 5–12

A family sits forlornly near the remains of their home destroyed by the 1993 earthquake in India's Maharashtra State. Structural damage and mortality were high because buildings were not designed or built to withstand earthquakes. *(Sherwin Crasto/Wide World Photos)*

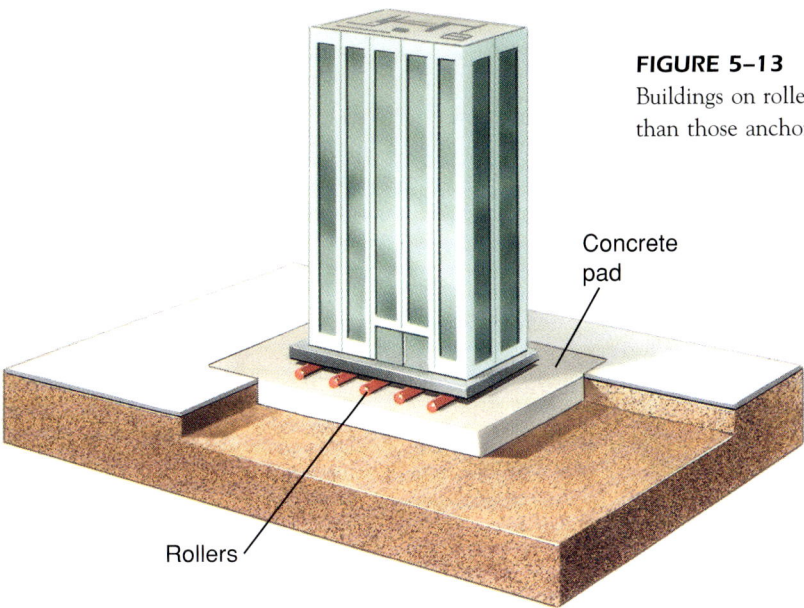

FIGURE 5–13
Buildings on rollers suffer less damage during an earthquake than those anchored to bedrock or sediment.

to bedrock but are set on rollers or wheels mounted on rails like a railroad car (Fig. 5–13). When the ground sways, the wheels roll as the Earth slips beneath them. These dynamic structures are less likely to fail than buildings that are firmly secured to rock or soil. In an even more advanced design, seismographs in the basement detect ground motion and relay signals to motors on the roof. The motors move heavy counterweights to dampen the structural sway (Fig. 5–14).

Fire

Moving rock may rupture buried gas mains and electrical wires, leading to fire, explosions, and electrocutions (Fig. 5–15). Water mains may also rupture, so fire fighters may not have water. Most of the damage from the 1906 San Francisco earthquake resulted from fires.

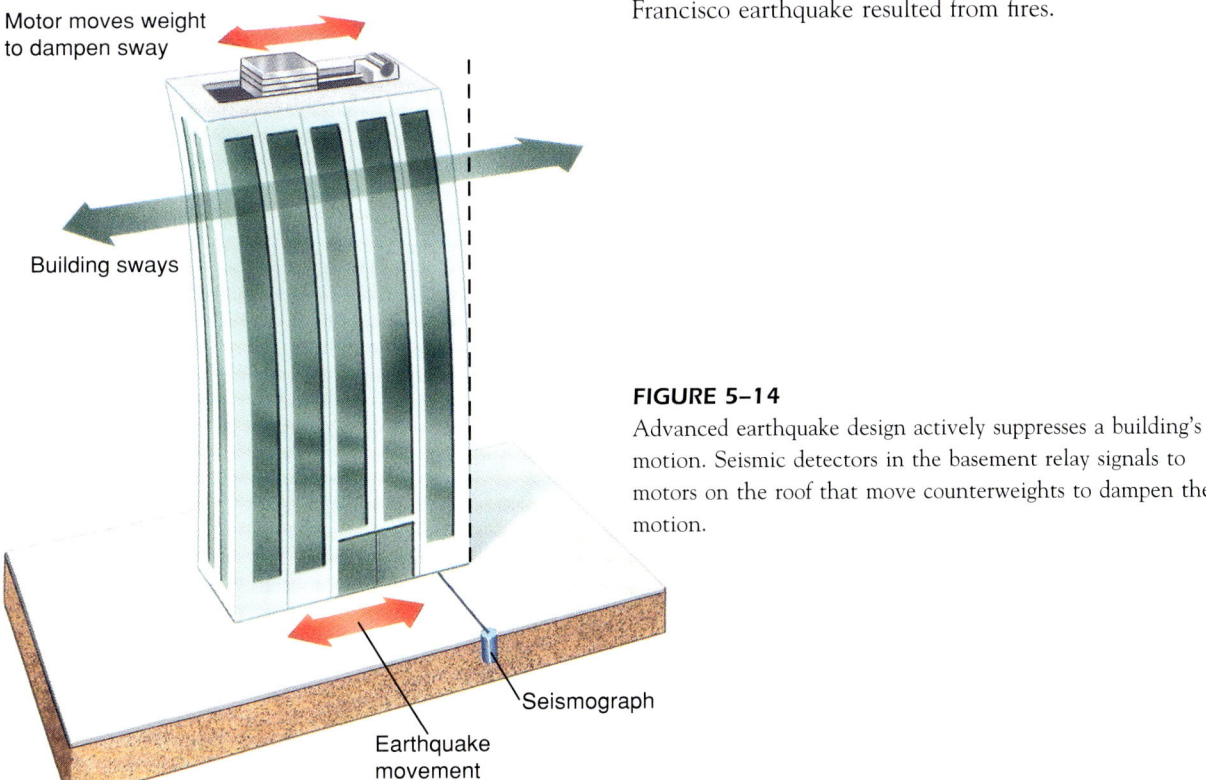

FIGURE 5–14
Advanced earthquake design actively suppresses a building's motion. Seismic detectors in the basement relay signals to motors on the roof that move counterweights to dampen the motion.

FIGURE 5–15
A fire caused by the 1989 San Francisco earthquake. *(Michael Williamson/Sygma)*

Landslides

Landslides may occur when the Earth trembles. As mentioned earlier, landslides in water-soaked clays caused extensive damage and loss of life in the 1964 Alaskan earthquake. Earthquake-related landslides are discussed in more detail in Chapter 8.

Tsunamis

When an earthquake occurs beneath the sea, part of the sea floor drops, as shown in Figure 5–16. As water level drops with the rock beneath, water from the surrounding area quickly rushes in to fill the depression, forming a wave. Sea waves produced by an earthquake are often called tidal waves, but because they have nothing to do with tides, geologists prefer to call them by their Japanese name, **tsunami**.

A. Normal state, before earthquake

B. Earthquake! Sea floor drops, sea level falls with it

C. Water rushes into low spot, and overcompensates, creating a bulge

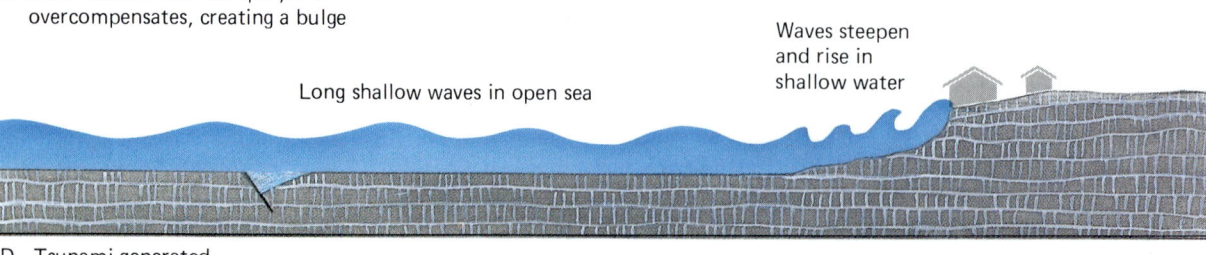

D. Tsunami generated

FIGURE 5–16
A tsunami develops when part of the sea floor drops during an earthquake. Water rushes in to fill the low spot, but the inertia of the rushing water forces too much water into the area, creating a bulge in the water surface. The long, shallow waves can build up into destructive giants when they reach shore.

In the open sea, a tsunami is so flat and spread out that it is barely detectable. Typically, the crest may be only 1 to 3 meters high, and successive crests may be more than 100 to 150 *kilometers* apart. However, tsunamis travel rapidly and are very destructive when they reach land. A tsunami may attain speeds of 750 kilometers per hour (450 miles per hour). When the wave approaches shore, its base drags against the ocean floor and slows down. The wave farther out to sea rapidly catches up with the one near shore. As the wave crests compress, the water between them stacks up, thus increasing the height of the wave. The rising wall of water then flows inland. A tsunami can flood the land for as long as 5 to 10 minutes before it withdraws (Fig. 5–17).

One of the worst tsunamis in history struck the eastern coast of Japan in 1896. The wave was probably formed by a submarine earthquake in the western Pacific Ocean. As it approached shore, the water rose about 30 to 35 meters (about 100 feet) above high-tide level, flooding villages and killing 26,000 people. The same wave was 3 meters high when it reached Hawaii, where it destroyed buildings near the coast; minor effects were felt on the west coast of North America. The wave then bounced off the North American coast and sped westward, back across the Pacific Ocean, until it reached New Zealand and Australia. By then it had lost enough energy that it was no longer destructive.

FIGURE 5–18
The San Andreas fault slices the Earth's surface in San Luis Obispo County, California. *(R. E. Wallace, USGS)*

FIGURE 5–17
An earthquake in Chile in May 1960 created a tsunami that flooded Hilo and other parts of the Hawaiian coast, halfway across the Pacific Ocean. *(Ward's Natural Science Establishment, Inc.)*

5.4 Earthquake Danger Zones in North America

Earthquakes at Transform Plate Boundaries: The San Andreas Fault

About 10 percent of the U.S. population and industrial resources are located in California, and 85 percent of the people and industry in California are concentrated along the west coast from San Francisco to San Diego. This area straddles the great San Andreas fault zone, the boundary between the North American tectonic plate to the east and the Pacific plate to the west (Fig. 5–18). At present, the Pacific plate is moving northwest relative to the North American plate at about 3.5 centimeters per year. This motion has produced hundreds of thousands of earthquakes in the past few centuries. Geologists of the United States Geological Survey recorded 10,000 earthquakes in 1984 alone, although most could be detected only with seismographs. Severe quakes occur periodically. One shook Los Angeles in 1857, and another destroyed San Francisco in 1906. A large quake in 1989 was centered south of San Francisco, and another rocked Los Angeles in January 1994. The fact that the San Andreas fault zone is part of a major plate boundary tells us that more earthquakes are inevitable in the future.

Three types of motion occur along the San Andreas fault:

1. Along some portions of the fault, rocks on opposite sides slip past one another at a continuous, snail-like pace. This type of movement is called **fault creep**. Many years ago in Hollister, California, houses were inadvertently built directly over the fault. Slowly, millimeter by millimeter, fault creep has torn the houses in two. The movement occurs without violent and destructive earthquakes because the rocks move continuously and slowly.
2. In other segments of the fault, the plates pass one another in a series of small hops, causing numerous small, nondamaging earthquakes.
3. Along the remaining portions of the San Andreas fault, friction binds rocks together, keeping opposite sides of the fault motionless for decades. As the plates continue to move past one another, the rock along the fault deforms and stores elastic energy. Because the plates move past one another at 3.5 centimeters per year, 3.5 meters of elastic deformation accumulate over a period of 100 years. When the accumulated energy exceeds friction, the rock suddenly slips along the fault and snaps back to its original shape, producing a large, destructive earthquake.

CASE HISTORY

Earthquake Activity in the San Francisco Bay Area

San Francisco lies on a segment of the San Andreas fault. To the east, across San Francisco Bay, the Hayward fault extends past San Jose and through Oakland. Both faults are active. A devastating earthquake struck San Francisco on April 18, 1906. The moment magnitude and Richter scales did not exist in 1906, but geologists estimate that the quake was of magnitude 8.3. The maximum movement along the fault was horizontal and exceeded 6 meters. Although the ground motion toppled many buildings, fires caused the greatest damage (Fig. 5–19). The moving Earth severed underground gas lines, and flames spread through the city, destroying the downtown area. At least 500 people died, and 250,000 were left homeless.

FIGURE 5–19
A portion of downtown San Francisco following the 1906 earthquake and fire. *(USGS)*

On October 17, 1989, a magnitude 6.9 earthquake left 65 people dead in San Francisco and surrounding areas. Its epicenter was in Loma Prieta, east of Santa Cruz and about 90 kilometers south of San Francisco (Fig. 5–20). Ground motion destroyed

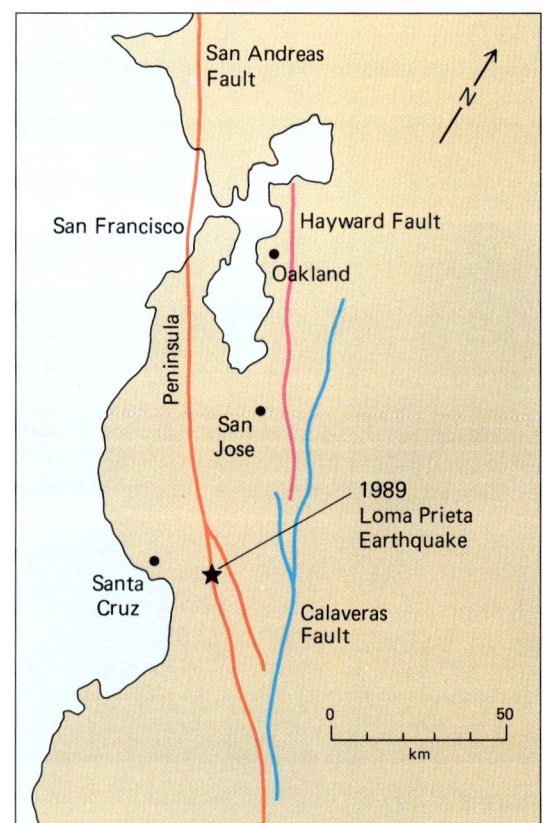

FIGURE 5–20
Active faults in the San Francisco Bay area.

FIGURE 5–21
The collapse of Interstate Highway 880 during the 1989 earthquake. *(Paul Scott/Sygma)*

much of the Santa Cruz business district, segments of both the Bay Bridge and Interstate Highway 880 collapsed, and damage was heavy in the Marina district, which had been built on an old landfill. The total damage was estimated at more than $4 billion (Fig. 5–21).

The Loma Prieta quake caused death and destruction, but it was not a major catastrophe. Two factors reduced its impact. First, the epicenter was located in a sparsely populated area, not in one of the densely populated Bay Area cities. Second, although a magnitude 6.9 quake is substantial, it is not a great earthquake. It released only one sixtieth as much energy as the magnitude 8.3 quake in 1906. If a magnitude 7 or 8 earthquake occurred in San Francisco or Oakland, the damage would eclipse that of the 1989 quake. Many homes in the San Francisco Bay Area are wood-frame structures that resist collapse. Therefore, if a quake were to strike at night or during the weekend, experts predict a low mortality, perhaps 1000 out of a total population of 2 million in the Bay Area. However, if the quake struck late on a summer afternoon, when bridges and subways were jammed with commuters and the streets were packed with late shoppers, the death toll would be higher.

Earthquake Activity in Southern California

In 1857, an earthquake with an estimated magnitude of 8.0 struck just north of Los Angeles. This segment of the San Andreas fault has not moved since 1857, but the same fault to the north and south has moved 4.6 meters since 1857 by fault creep and frequent small quakes. Thus, the accumulated elastic deformation of rock in the locked segment must be about 4.6 meters. This elastic energy is stored in the

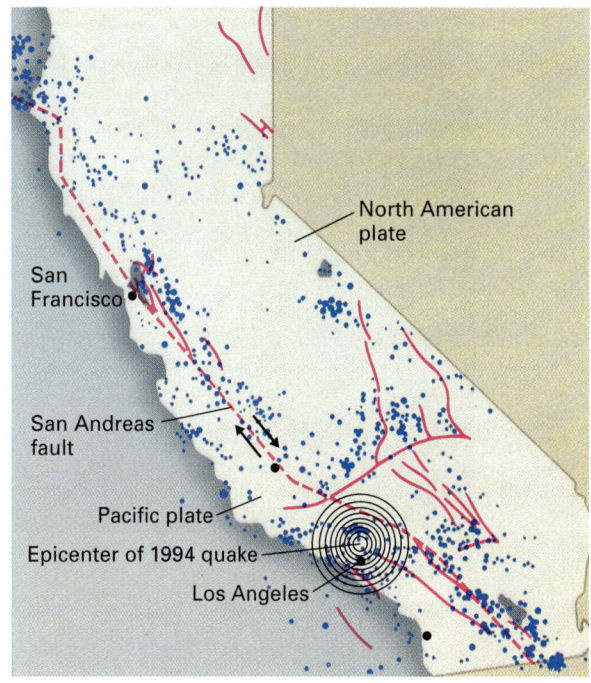

FIGURE 5–22
The location of the January 1994 earthquake near Los Angeles. The dashed line is the San Andreas fault; dots are epicenters of other recent earthquakes.

rocks, waiting for a sudden fracture to release it.

In April 1992, a magnitude 6.1 earthquake occurred in the Mojave Desert, northeast of Palm Springs, California. Then another earthquake with a magnitude of 7.3 struck in late June near Landers. Numerous additional quakes were recorded in the vicinity within the following few weeks. The Landers quake had a magnitude greater than that of the Loma Prieta quake, but it occurred in a sparsely inhabited desert and led to only one death and little property damage. Thus, in practical terms, the impacts of these earthquakes depended not only on magnitude but also on location.

The Landers earthquake occurred on a series of faults that intersect the San Andreas fault at an angle. In 1992, two geophysicists from Columbia University, Steven Jaume and Lynn Sykes, calculated that motion on the faults near Landers stretched rock along the San Andreas fault and increased the probability of an earthquake near Los Angeles.[2] Then, in January 1994, a magnitude 6.6 earthquake struck the San Fernando Valley just north of Los Angeles (Fig. 5–22). Fifty-five people died and property

---

[2]Steven C. Jaume and Lynn R. Sykes, "Changes in State of Stress on the Southern San Andreas Fault Resulting from the California Earthquake Sequence of April to June 1992," *Science*, 258 (November 20, 1992), 1325.

FIGURE 5–23

Commercial buildings damaged by the 1994 Los Angeles earthquake. *(Wide World Photos)*

damage was estimated at $8 billion (total damage, which included lost work and business revenues, was much higher) (Fig. 5–23).

A cross section of the Los Angeles area shows that numerous smaller faults cut through the region and intersect each other (Fig. 5–24). During an earthquake, energy is transmitted through the rock in complex ways due to interactions among the faults. Thus, geologists are not certain whether the most recent quake will relieve stress on the San Andreas fault or increase it, leading to a higher probability of a more disastrous quake in the near future. Although many disastrous and expensive earthquakes have affected southern California in the past few decades, none of them has been "the big one" that seismologists still fear.

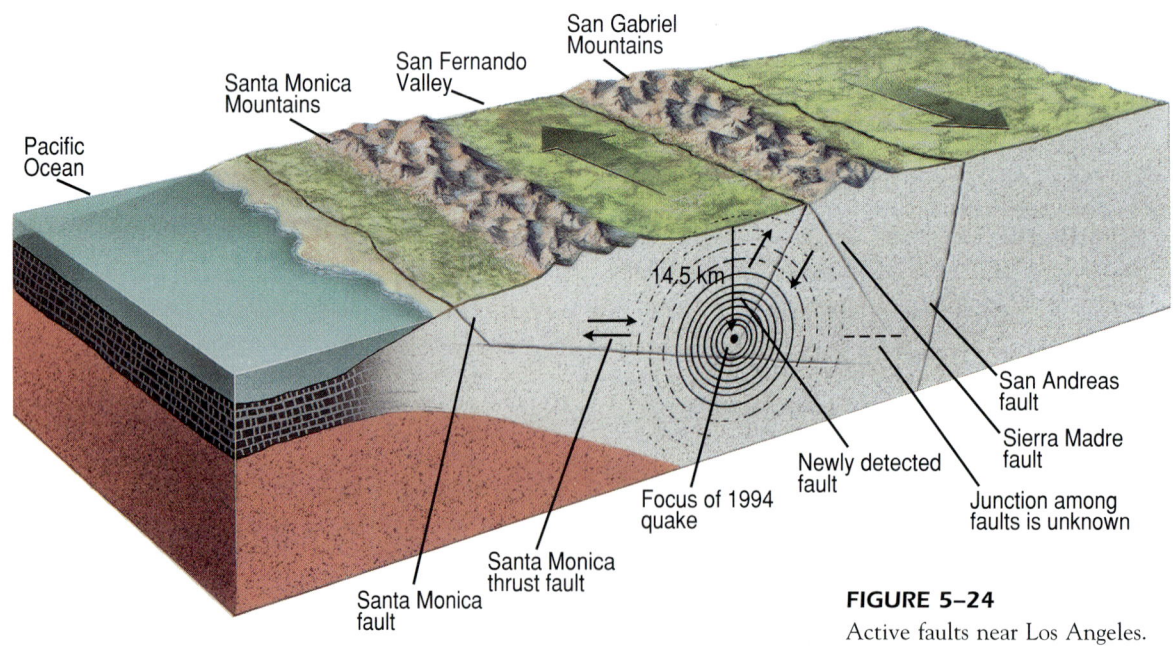

FIGURE 5–24

Active faults near Los Angeles.

5.4 Earthquake Danger Zones in North America 117

FIGURE 5–25
(A) A descending lithospheric plate generates magma and deep earthquakes along subduction zones. (B) The 1964 Alaskan earthquake was caused by subduction. The quake fractured the ground surface and (C) caused a tsunami that carried fishing boats inland and deposited them on city streets. *(A and C, USGS)*

Earthquakes at Subduction Zones

In a subduction zone, a relatively cold, brittle lithospheric plate dives beneath another plate and slowly sinks into the mantle. In most places, the subducting plate sinks with intermittent slips and jerks, giving rise to numerous earthquakes. The earthquakes concentrate at the Benioff zone along the upper part of the sinking plate, where it scrapes past the opposing plate (Fig. 5–25). Earthquakes of this type occur frequently along the west coast of South America, in Japan, in Alaska, in the Aleutian Islands, and in other places where subduction is active today. Many of the world's strongest earthquakes occur in subduction zones. Both the 1964 Alaskan earthquake and the 1994 Bolivian quake occurred at subduction zones.

CASE HISTORY

Earthquake Activity in the Pacific Northwest

Subduction is occurring today along the coasts of Oregon, Washington, and southern British Columbia in the Pacific Northwest. In those places, an oceanic plate is diving beneath the continent at a rate of 3 to 4 centimeters per year. Magma generated in the subduction zone rises to erupt from Mount St. Helens, Mount Rainier, and other Cascade Range volcanoes (Fig. 5–26). One would also expect many earthquakes at such an active plate boundary. Small earthquakes occasionally shake the region, but no large ones have occurred in the past 150 to 200 years.

Why are earthquakes relatively uncommon in this active subduction zone? The answer to this question is important because the Pacific Northwest is densely populated and heavily industrialized. Two possible answers exist. Subduction may be occurring slowly and continuously by fault creep. Thus, elastic energy is not accumulating in nearby rocks, and large-magnitude earthquakes are unlikely. Alternatively, rocks

FIGURE 5–26
Both earthquakes and active volcanoes characterize regions near a subduction zone. Mount Adams, Washington.

along the fault may be locked together by friction, accumulating a huge amount of elastic energy that will be released in a giant, destructive quake sometime in the future.

Recently geologists have discovered evidence of great prehistoric earthquakes in the Pacific Northwest. A major coastal earthquake commonly creates violent sea waves, which deposit a layer of mud and sand along the coast. Six such layers have formed within the past 7000 years in the Pacific Northwest. Other evidence indicates that the coastline dropped suddenly by 0.5 to 2 meters as each of the sediment layers formed. This information suggests that subduction is not occurring by creep or frequent small earthquakes. Instead, friction has locked the plates together for an average of 1150 years between large, devastating earthquakes. Thus, many geologists anticipate another major, destructive earthquake in the Pacific Northwest.

On April 25, 1992, a magnitude 6.9 earthquake rocked northern California and southern Oregon. Centered near Eureka, California, it caused damage locally and was felt as far away as Reno, Nevada. Two smaller quakes with magnitudes of 6.5 and 6.0 followed on the next day. These earthquakes occurred in the area where the Pacific Northwest subduction zone joins the San Andreas fault system. Although they caused some damage, these quakes were not the destructive major events that are possible in the region.

Earthquakes at Plate Interiors

No major earthquakes have occurred in the central or eastern United States in the past 100 years, and no lithospheric plate boundaries are known in these regions. Therefore, one might conclude that earthquake danger is insignificant. Indeed, building codes in most major central and eastern cities do not require earthquake-resistant construction.

Today, seismologists are questioning this complacency:

> We were told by theory that plate interiors should be quiet areas. Then people started building nuclear power plants and started to look at the details of data and discovered, in fact, that there is a lot of seismicity in interior plates and that we know very little about it.[3]

The largest historical earthquake sequence in the contiguous 48 states occurred not in California, but near New Madrid, Missouri. In 1811 and 1812, three shocks with estimated magnitudes between 7.3 and 7.8 altered the course of the Mississippi River and made church bells ring 1500 kilometers away in Washington, D.C. Another large quake with an estimated magnitude of 7.0 occurred in Charleston, South Carolina, in 1886.

Geologists have remeasured distances between old survey pins near New York City and found that the pins have moved significantly during the past 50 to 100 years. This motion indicates that the crust in this region is being deformed. Historical reviews show that, although earthquakes are infrequent in the Northeast, they do occur (Fig. 5–27). If a major quake were to occur near New York City today, the consequences could be disastrous.

Earthquakes in plate interiors are not as well understood as those at plate boundaries, but modern research is revealing some clues. The New Madrid region lies in an ancient continental rift zone bounded by deep parallel faults. The rift may have formed over a mantle plume or as a result of some other force that began to tear the continent apart. It was active during Paleozoic and Mesozoic time between 570 and 65 million years ago, but failed to develop into a divergent plate boundary. Even though spreading did not occur, the deep faults remain a weakness in the lithosphere. As the North American plate glides horizontally over the asthenosphere, it may undergo some vertical motion above irregularities, or "bumps," in that plastic zone. The motion may cause slippage along weak parts of the lithosphere, such as the deep faults near New Madrid. Alternatively, under special circumstances, stress from a plate boundary may be transmitted hundreds of kilometers into the interior of a plate. When stress builds in a plate interior, it is likely to be released as an earthquake along old faults.

Some intraplate earthquakes occur where thick piles of sediment have accumulated on great river deltas. The

---

[3]Leonardo Seeber of Columbia University, quoted in the *New York Times*, March 1, 1988.

underlying lithosphere cannot support the weight of sediment, and the lithosphere fractures as it settles.

Earthquakes Triggered by Human Activity

Shortly after Hoover Dam was built near Las Vegas, numerous earthquakes rocked the region, although few earthquakes had been recorded previously. The quakes occurred along shallow faults on the east side of Lake Mead, behind the dam. The mass of water in the new lake had stressed the underlying crust. In addition, water from the lake had seeped into the faults, lubricating them and allowing the rock to slip. However, the area stabilized after 5 years, and earthquake activity stopped. Although earthquakes along Lake Mead caused no damage, earthquakes induced by a newly formed lake nearly destroyed the Nurek Dam in the Soviet Union. Deep injection wells can also cause earthquakes. This topic will be discussed in Chapter 11.

5.5 Earthquake Prediction

Long-Term Prediction

Most earthquakes start on old faults that have moved many times in the past and will move again in the future. Earthquakes occur over and over in the same places because it is easier for rocks to move along an old fracture than for a new fault to form in solid rock. Many of these faults lie along tectonic plate boundaries.

In 1979, William McCann and coworkers at Columbia University developed the **seismic gap hypothesis**. This hypothesis starts with the simple premise that all segments of a fault, such as the San Andreas, eventually must move by the same amount because the fault is the boundary between two tectonic plates moving past each other. McCann proposed that an earthquake releases the accumulated elastic energy in rocks adjacent to the fault. Thus, the probability is low that a second large quake will occur in the same place in the near future. However,

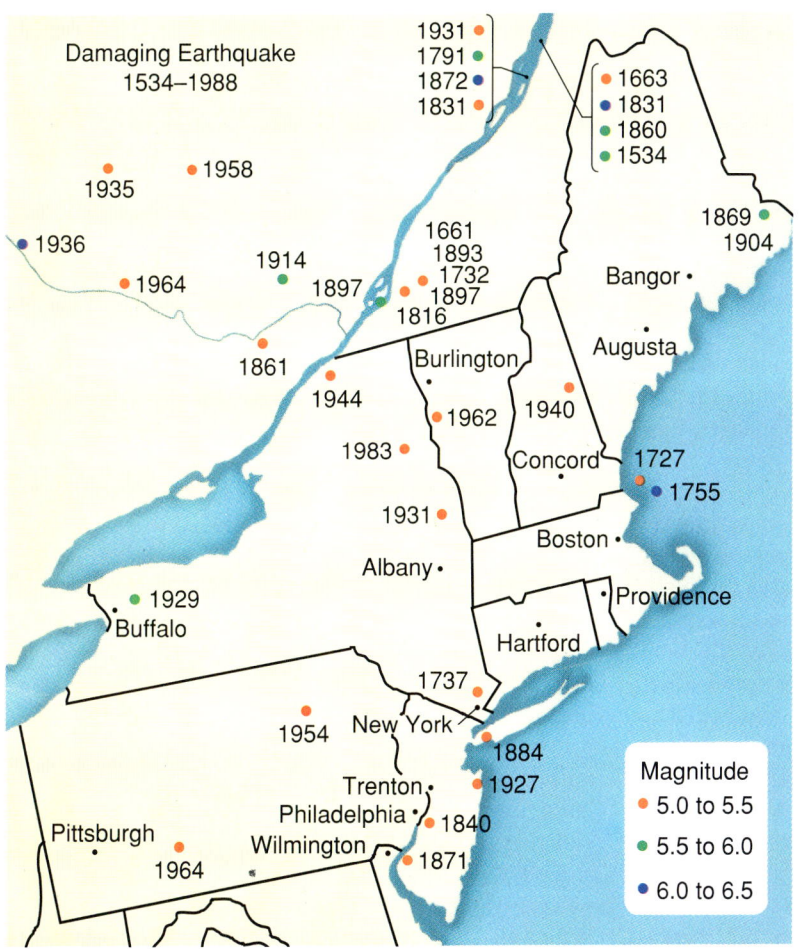

FIGURE 5-27

Earthquake activity in the northeastern United States and southeastern Canada between 1534 and 1988. (Redrawn from *Geotimes*, May 1991, p. 6)

strain continues to build near other parts of the fault where earthquakes have not occurred. As a result, portions of the fault that have not experienced a recent major quake should be at high risk of an earthquake in the near future. During the 1980s, geologists used the seismic gap hypothesis to draw maps of earthquake probabilities in California and other seismically active regions.

In 1989, David Jackson and Yan Kagan of UCLA tested the seismic gap hypothesis by reviewing all the recorded earthquakes of magnitude 7 or higher along both coasts of the Pacific Ocean. They learned that many earthquakes occurred in the same region in quick succession, and they did not detect a higher frequency of earthquakes in regions that had been quiet for a long time. Why did the seismic gap hypothesis fail? The seismic gap hypothesis explains motion along a single fault. However, the San Andreas fault zone contains many related faults in the same area. If many faults intersect at different angles, motion along one may increase or relieve stress on nearby faults.

Geologists know that earthquakes are common along tectonic plate boundaries and less frequent in plate interiors. Thus, earthquake hazard is high in parts of California and low in Iowa. However, long-term forecasts of precisely when and where the next quake will strike remain unreliable.

Short-Term Prediction

Short-term prediction depends on a reliable early warning system—a signal or group of signals that immediately precedes an earthquake. **Foreshocks** are small earthquakes that precede a large quake by an interval ranging from a few seconds to a few weeks. The cause of foreshocks can be explained by a simple analogy. If you try to break a stick by bending it slowly, you may hear a few small cracking sounds just before the final snap. If foreshocks consistently preceded major earthquakes, they would be a reliable tool for short-term prediction. However, foreshocks preceded only about half of a group of recent major earthquakes. At other times, swarms of small shocks that could have been foreshocks were not followed by a large quake.

Another approach to short-term earthquake prediction is to measure changes in the shape of the land surface. Seismologists monitor rising bulges and other unusual Earth movements with tiltmeters and laser surveying instruments (Fig. 5–28). The concept behind these studies is that distortions of the crust may precede major earthquakes. Some earthquakes have been predicted successfully with this method, but in other instances predicted quakes did not occur, or quakes occurred that had not been predicted.

Another prediction method is based on the observa-

FIGURE 5–28
Geologists measuring land surface distortions on the flanks of Mount St. Helens prior to the 1980 eruption. (C. D. Miller/USGS)

tion that, when rock is deformed to near its rupture point, microscopic cracks and pores open between the mineral grains. These tiny openings may produce measurable effects. For example, radon gas trapped in rocks and minerals may be released before an earthquake. In addition, the formation of cracks and pores increases the volume of the rock, which, in turn, may cause the water table to rise or fall. Thus, the water levels in some wells have fluctuated just prior to an earthquake. Furthermore, air-filled holes do not conduct electricity as well as solid rock, so the electrical conductivity of rock decreases as pores open up.

Chinese scientists reported that, just prior to the 1975 quake in the city of Haicheng, snakes crawled out of their holes, chickens refused to enter their coops, cows broke their halters and ran off, and even well-trained police dogs became restless and refused to obey commands. Some researchers in the United States have attempted to quantify the relationship between animal behavior and earthquakes. In one study near San Francisco, scientists asked animal trainers and zoo keepers to report unusual animal behavior. The scientists received a flurry of calls—*after* a magnitude 5.7 earthquake—from people who reported that they had neglected to call earlier, but now that they thought about it, they remembered that some animals had behaved strangely before the earthquake. However, there had been no significant increase in calls before the quake, so no predictions had been made.

In the early 1970s, Chinese geophysicists used the seismic gap hypothesis and historical data to issue a long-range warning for a major earthquake near the city of Haicheng. Seismic stations were established in and around the city. In January 1975, scientists recorded swarms of foreshocks and unusual bulges in the land. When the foreshocks became intense on February 1, authorities ordered an evacuation of portions of the city. The evacuation was completed on the morning of February 4, and in the early evening of the same day, a large earthquake destroyed houses, apartments, and factories but caused few deaths.

Immediately after that success, geologists hoped that a new era of earthquake prediction had begun. But a year later, Chinese scientists failed to predict an earthquake in the adjacent city of Tangshan. This major quake was not preceded by a swarm of foreshocks, no warning was given, and at least 250,000 people died.[4] Shortly after that failure, a quake was predicted in a third city and the city was evacuated, but the earthquake did not occur.

Earthquake prediction involves many political, social, and economic issues. Imagine that geologists observe many signs indicating that a quake is imminent in a city and issue a warning, as the Chinese have done. The scientists might be right or wrong. What are government officials to do—ignore the warning, or shut down businesses, close trade and commerce, and evacuate millions of people? Such draconian measures would cost billions of dollars, and what if no quake occurred? On the other hand, if a major quake were predicted, nothing were done, and the earthquake did occur, many people would die needlessly. Either way, the consequences of the wrong choice are severe.

5.6 Earthquake Hazard and Human Habitation

All of these examples tell us that at present, neither long-term nor short-term earthquake prediction is reliable.

[4]Accurate reports of the death toll are unavailable. Published estimates range from 250,000 to 650,000.

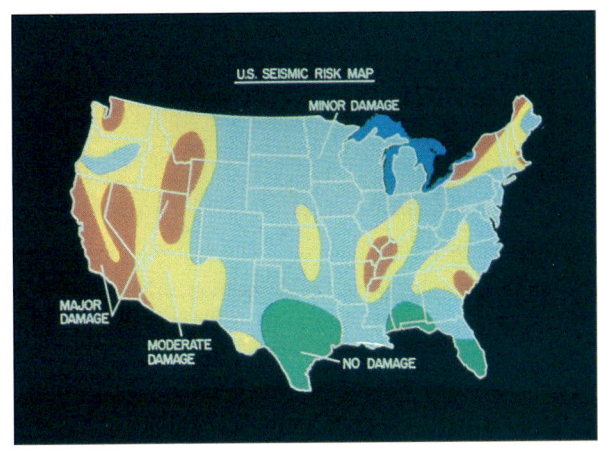

FIGURE 5–29

This map shows potential earthquake damage in the United States. The predictions are based on records of the frequencies and magnitudes of historical earthquakes. (*Ward's Natural Science Establishment, Inc.*)

However, as explained earlier, geologists can identify high-hazard areas. Even though many faults occur within tectonic plates, the largest and most active faults are the boundaries between tectonic plates (Fig. 5–29).

Engineers can build structures that will withstand even the most severe earthquakes, but only at great expense. As a result, building codes are a compromise between safety and cost. Requirements for nuclear power plants are stricter than those for bridges because if a nuclear power plant were to fail, thousands of people could die, whereas only a few would die if a bridge failed (Fig. 5–30). In the late 1970s, the California state government authorized the Seismic Safety Commission to recommend upgrading state-owned structures to meet

FOCUS ON

Earthquake Control

Is it possible to control earthquakes to minimize the damage they cause to human-made structures?

Recall from Chapter 1 that wastes injected into the Earth through deep wells have caused earthquakes. Some geologists suggest that we could control earthquakes by lubricating known active faults with fluids. According to this hypothesis, the lubrication would trigger numerous minor earthquakes, which would relieve the stress accumulation that precedes a major quake.

To test the hypothesis, geologists injected water under high pressure into old faults near Rangely, Colorado. As predicted, small earthquakes resulted. However, no one has used this technique along a tectonic plate boundary such as the San Andreas fault, where the accumulated energy is presumably much greater. One problem is that the risk is too great. Fluid injection could release enough stress to trigger a major and catastrophic earthquake.

FIGURE 5–30
The remains of a bridge at the intersection of Highways 14 and 5 that collapsed during the 1994 earthquake near Los Angeles. *(Douglas C. Pizac/Wide World Photos)*

stricter engineering criteria. The commission evaluated soil and bedrock at probable earthquake zones and then recommended construction upgrades based on the probable number of lives saved per reconstruction dollar. This cost–benefit analysis reduces economic disruption. However, during a large earthquake, people would die as substandard structures collapsed.

Why do people live in high-earthquake-hazard areas? One reason is that many plate tectonic boundaries lie along coastlines, which are attractive for many reasons. The entire west coast of the United States lies within earthquake zones, but people live there anyway. Some like the weather, some find favorable business opportunities, and others like to be near friends and family. After the 1994 Los Angeles quake, journalists asked local citizens why they chose to live in an area where earthquakes are almost certain to occur again. Many Californians argued that they would rather accept earthquake risks than face severe winters. Some people pointed out that during the winter of 1994, cold and snow in the northeastern United States led to more deaths than were caused by the Los Angeles quake.

SUMMARY

An **earthquake** is a sudden motion or trembling of the Earth caused by the release of energy stored when rocks stretch or compress. About a million earthquakes occur each year, but only a small fraction are strong enough to be felt. Most earthquakes occur along the boundaries between moving lithospheric plates. Earthquake energy is released at the **focus** of the earthquake, where rocks move abruptly. The **epicenter** of a quake is the point on the Earth's surface directly over the focus. Earthquake energy travels through rock as **seismic waves**, which occur as **body waves** and **surface waves**. Two main types of body waves transmit earthquake energy. A **P wave** (compression wave) travels through solids, liquids, and gases, but an **S wave** (shear wave) is transmitted only through solids and travels more slowly than a P wave. Surface waves called **L waves** travel along the Earth's surface.

An earthquake is detected and measured by a **seismograph**. The **moment magnitude** measures the energy released by a quake. Structural damage is related to the magnitude of the quake, its proximity to population centers, rock and soil types, and the quality of construction in the region. Earthquakes also cause fires, landslides, and tsunamis.

The most active earthquake region in North America is the San Andreas fault zone in California, a transform plate boundary. Recent earthquakes have occurred all along this fault zone, and more are expected in the future. Subduction is occurring in the Pacific Northwest, and earthquakes are likely in this region as well. Earthquakes also occur occasionally in plate interiors. Small quakes may be triggered by injection wells or lakes that form behind dams.

The earthquake history of an area and identification of **seismic gaps** along a fault zone form the basis of long-term earthquake predictions. However, recent studies question the seismic gap hypothesis. Short-term prediction is based on **foreshocks**, changes in the shape of the land, changes in ground-water levels and electrical conductivity, release of radon gas, and animal behavior.

KEY TERMS

Elastic energy 101
Earthquake 101
Seismic waves 102
Seismology 102
Body waves 102
Focus 102
Epicenter 103
Surface waves 103

Primary or P wave 103
Shear or S wave 103
L wave 104
Seismograph 104
Seismogram 105
Magnitude 105
Richter scale 105
Moment magnitude 105

Scarp 106
Mohorovičić discontinuity 108
Liquefaction 109
Tsunami 112
Fault creep 114
Seismic gap hypothesis 119
Foreshocks 120

REVIEW QUESTIONS

1. Explain how energy is stored in rocks and then released during an earthquake.
2. Why do most earthquakes occur at the boundaries between tectonic plates?
3. Describe the differences among P waves, S waves, and L waves.
4. Explain how a seismograph works. Sketch a seismogram before and during an earthquake.
5. Compare and contrast the Richter and moment magnitude scales. Explain their logarithmic nature.
6. Discuss mechanisms for structural damage during an earthquake for buildings set on rock, sand, and clay.
7. Discuss the fundamental elements of earthquake-proof building design.
8. List and discuss three events besides direct building collapse that are triggered by earthquakes and that destroy cities and towns.
9. Explain why fault creep along a segment of a fault provides evidence that an earthquake may soon occur in a nearby segment where creep is *not* occurring.
10. Describe current earthquake activity along the Pacific coast of the United States.
11. Explain two mechanisms whereby human activity can trigger earthquakes.
12. Discuss the scientific reasoning behind long-term and short-term earthquake prediction.
13. Review the scientific argument between the geologists who support and criticize the seismic gap theory.

DISCUSSION QUESTIONS

1. If rock is caught between two lithospheric plates moving in different directions, it may bend plastically or fracture. Which type of movement is more likely to cause an earthquake? Explain.
2. New York City is a low-earthquake-hazard area, but rock deformation has been recorded by measuring distances between old survey pins. Discuss the trade-offs of writing laws that require better earthquake-resistant design for buildings in New York City.
3. Evaluate the safety of your home, school building, or dormitory in the event of a large earthquake. What is the probability of a large earthquake in your area?
4. If you were the mayor of San Francisco, would you encourage or discourage the injection of fluids into the San Andreas fault? Defend your stance.
5. Significant earthquakes occurred in Parkfield, California, in 1857, 1881, 1901, 1922, 1934, and 1966. Draw a graph with the dates on the vertical (Y) axis and the numbers of the events (simply 1, 2, 3, and so on) spaced evenly on the X axis. Use your graph to predict when the next earthquake might occur in Parkfield.
6. Imagine that geologists predict a major earthquake in a densely populated region. The prediction may be right or it may be wrong. City planners may heed it and evacuate the city, or they may ignore it. The possibilities lead to four combinations of prediction and response, which can be set out in a grid as follows:

	Does the predicted earthquake really occur?	
	Yes	No
Is the city evacuated? Yes		
Is the city evacuated? No		

For example, the space in the upper left corner of the grid represents the situation in which the predicted earthquake occurs and the city is evacuated. For each space in the square, outline the consequences of that sequence of events.

7. Supply a plausible mechanism to support the seismic gap theory. Supply a plausible mechanism to refute it.
8. In a recent article, seismologist Bruce Bolt noted that the 1989 Loma Prieta earthquake occurred in a region of the Santa Cruz Mountains where geologists had thought earthquake probability was low. He wrote,

> In recent years, the concept of "characteristic earthquake" on a particular fault has been formulated. If faulting processes repeat, there is hope that the prediction of future earthquake behavior can be reduced to a known set of basic earthquake types. Yet the source of the earthquake in the Santa Cruz Mountains presents, at least for the time being, a number of problems that cast doubt on earthquake invariance along even such geologically well-exposed structures as the San Andreas fault.[5]

Discuss whether this statement is relevant to the 1994 Los Angeles earthquake. Discuss this statement in relationship to the argument about seismic gap hypothesis.

---

[5]Bruce Bolt, "Balance of Risks and Benefits in Preparation for Earthquakes," *Science*, 251 (January 1991), 169ff.

CHAPTER 6
Volcanoes

6.1 Magma
6.2 Magma Behavior: Why Some Magma Erupts from a Volcano and Other Magma Solidifies Below the Surface
6.3 Volcanoes and Related Landforms
6.4 Volcanoes and Plate Tectonics
6.5 Explosive Magma and Violent Volcanoes
6.6 Assessing the Danger of Future Volcanism
6.7 Volcanoes and Climate

In the spring of 1980, 1 cubic kilometer of rock, lava, and ash exploded from Mount St. Helens in western Washington (Fig. 6–1). The blast flattened and burned surrounding forests, killing at least 63 people. The sky grew so dark with pulverized rock that motorists 150 kilometers from the mountain had to use their headlights at noon. But this eruption was minor compared with others in history.

In the past 2000 years, volcanic eruptions have killed more than 1 million people. Because the greatest disasters obliterated entire cities, they destroyed population records along with the populations, so the exact numbers of deaths will never be known. During the past 100 years, eruptions killed approximately 100,000 people and caused about $10 billion in damage. A few eruptions caused most of this death and destruction, and the damage was more closely related to the distance between the erupting volcano and human habitation than to the size of the eruption. In some cases, death and damage resulted directly from hot lava or volcanic ash burying a town or city. The 1902 eruption of Mount Pelée on the Caribbean island of Martinique buried the city of Saint Pierre in a glowing cloud of volcanic ash that killed 29,000 (Fig. 6–2). However, many deaths also result from events triggered by the eruptions. The 1893 eruption of Krakatoa in the southwestern Pacific Ocean generated a tsunami (tidal wave) that killed 36,000 people. A mudflow triggered by a small eruption in 1985 of Nevado del Ruiz in Colombia buried the town of Armero, killing more than 22,000 people. In 1986, volcanic gases—mostly carbon dioxide—that had accumulated beneath Lake Nyos in a volcanic crater in Cameroon, Africa, seeped into the town of Lower Nyos, killing 1700 people. The five or six survivors told of sitting and talking with their friends only to watch them drop dead as the invisible cloud flowed through the village. To this day, no one knows why

The March 1969 eruption from the East Rift of Kilauea volcano, Hawaii.
(J. Judd/USGS)

FIGURE 6–1
The May 1980 Mount St. Helens eruption. (*USGS/R. P. Hoblitt*)

FIGURE 6–2
The smouldering ruins of Saint Pierre on May 14, 1902, following the May 8 eruption of Mount Pelée. (*Institute of Geological Sciences, London*)

the gases suddenly escaped from the lake water. The Lower Nyos disaster is described further in Chapter 9.

Even the greatest historical eruptions have been small compared with some prehistoric ones. An eruption 1.9 million years ago in Yellowstone Park, Wyoming, ejected approximately 2500 cubic kilometers of rock, lava, and ash, 2500 times more than the 1980 Mount St. Helens eruption. The red-hot volcanic ash buried thousands of square kilometers of surrounding countryside, killing all living things in its path. An older eruption in the San Juan Mountains of Colorado blasted out 3000 cubic kilometers of similar material. In contrast with these violent explosions, Hawaiian volcanoes erupt gently enough that tourists approach closely to photograph flowing lava and to watch fire fountains erupting into the sky.

Volcanoes are common in some regions and unheard of in others. Eighteen recently active volcanoes from high peaks in the Cascade Range in northern California, Oregon, and Washington, but no volcanoes exist in the central or eastern United States. Why are some parts of the Earth volcanically active, whereas other parts are not? Why do some volcanoes explode violently but others erupt gently? Can we predict volcanic eruptions and warn people of impending disasters? To answer these questions, consider how magma forms and behaves.

6.1 Magma

Magma forms in three geologic environments: subduction zones, spreading centers, and mantle plumes. Thus, volcanic activity and volcanic threats to human habitation concentrate in regions of the Earth where these features are active today.

Magma Production in a Subduction Zone

At a subduction zone, one 100-kilometer-thick plate sinks hundreds of kilometers into the mantle as it dives beneath an adjacent plate. Three processes combine to form magma in a subduction zone (Fig. 6–3).

Heating

Everyone knows that a solid melts when it becomes hot enough. Butter melts in a frying pan, and snow melts under the spring sun. Recall that the asthenosphere is

FIGURE 6–3
Three factors contribute to the melting of the asthenosphere and the production of magma at a subduction zone: (1) Friction heats rocks in the subduction zone; (2) water rises from oceanic crust on top of the subducting plate; and (3) circulation in the asthenosphere decreases pressure on hot rock.

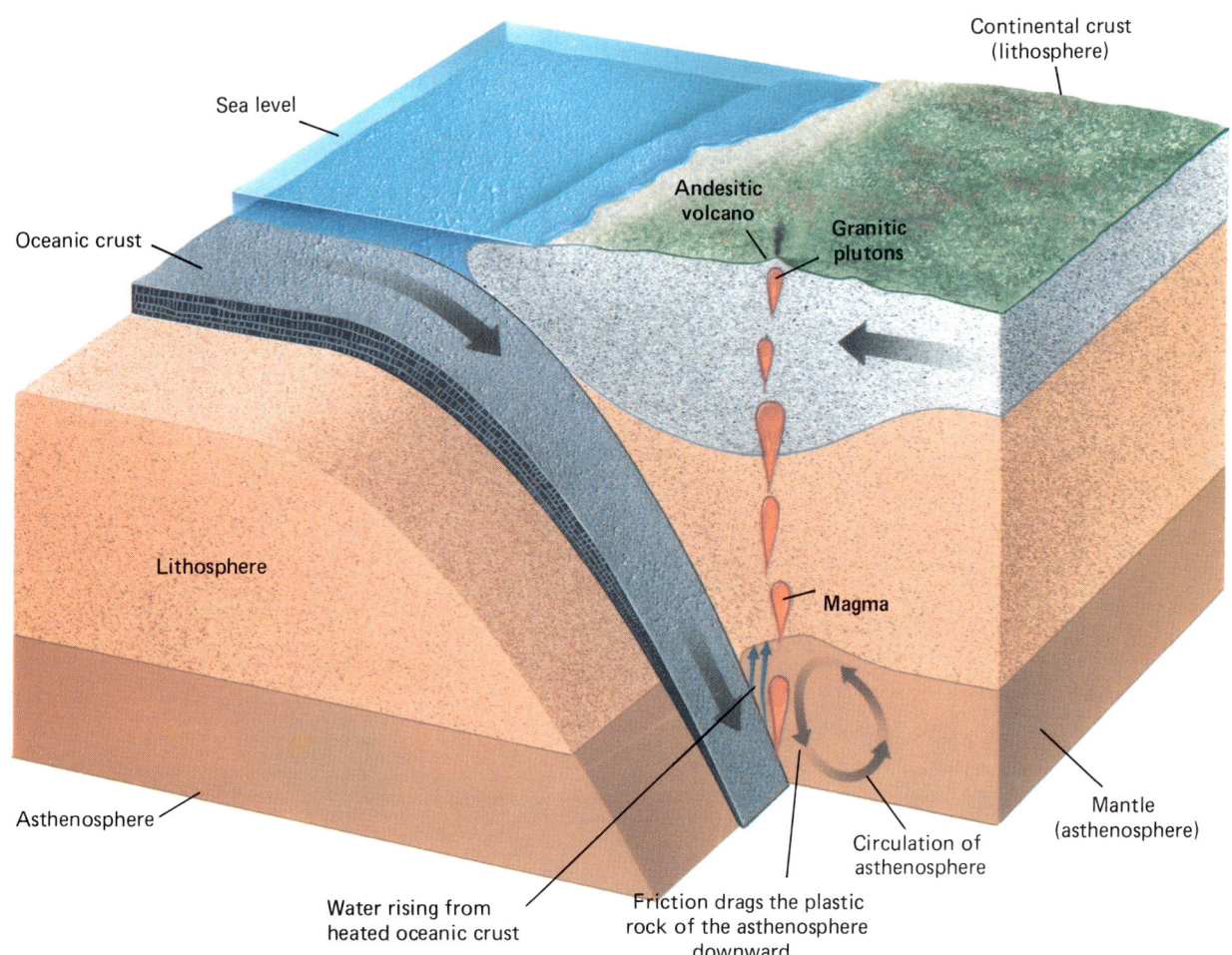

very hot—so hot that it is plastic. When a tectonic plate sinks into the mantle at a subduction zone, it enters hotter zones of the Earth. It also generates frictional heat as it scrapes past the opposite plate. Both sources of heat bring the rock closer to its melting point. Oddly, however, increasing temperature is the least important of the three causes of melting in a subduction zone.

Addition of Water to the Asthenosphere

In general, a wet rock melts at a lower temperature than an identical dry rock. Therefore, addition of water can melt a rock that is already close to its melting point.

A subducting plate is covered by oceanic crust, and the mud and fractured basalt of oceanic crust are soaked with seawater. As wet oceanic crust dives into the mantle, it becomes hotter and the water escapes, rising into the hot asthenosphere directly above the sinking plate. This addition of water melts portions of the asthenosphere, forming huge quantities of magma. This process is the most important cause of magma formation in a subduction zone.

Pressure-Relief Melting

In Chapter 3 you learned that a rock expands by about 10 percent when it melts to form magma. If the rock is near the Earth's surface, it can expand and melt easily because there is little pressure preventing it from doing so. However, in the asthenosphere, pressure is so great from the weight of the overlying 100 kilometers of lithosphere that expansion and melting are more difficult. However, if the pressure should somehow decrease, the hot asthenosphere rock could expand and melt. In other words, a drop in pressure can melt a hot rock. Melting caused by decreasing pressure is called **pressure-relief melting**.

A subducting plate drags plastic asthenosphere rock down with it, as shown by the arrows in Figure 6–3. Rock from deeper in the asthenosphere then flows upward to replace the sinking rock. As this rock rises, pressure decreases and pressure-relief melting occurs.

To summarize, three processes contribute to melting of the asthenosphere above a subducting plate. (1) The most important is addition of water from subducting oceanic crust; (2) pressure-relief melting is second; and (3) rising temperature is least important (Fig. 6–4). The result is that huge amounts of magma form in subduction zones at depths of about 100 kilometers, where the subducting plate passes from the lithosphere into the asthenosphere. This magma rises to form abundant igneous rocks. The volcanoes of the Pacific Northwest and the granite cliffs of Yosemite are examples of igneous rocks formed at subduction zones.

Magma Production in a Spreading Center

Where two lithospheric plates separate at a spreading center, soft, hot asthenosphere oozes upward to fill the

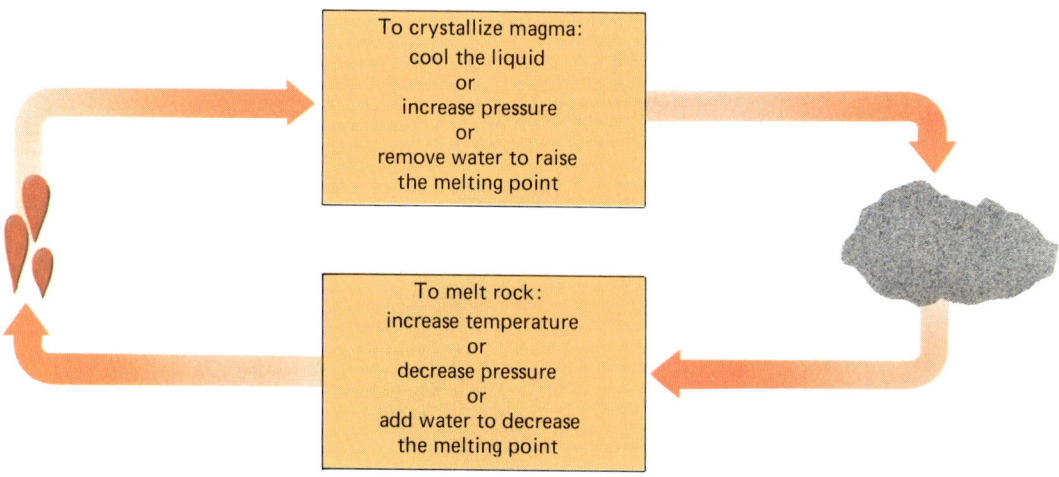

FIGURE 6–4
Increasing temperature, addition of water, and decreasing pressure all melt rock to form magma.

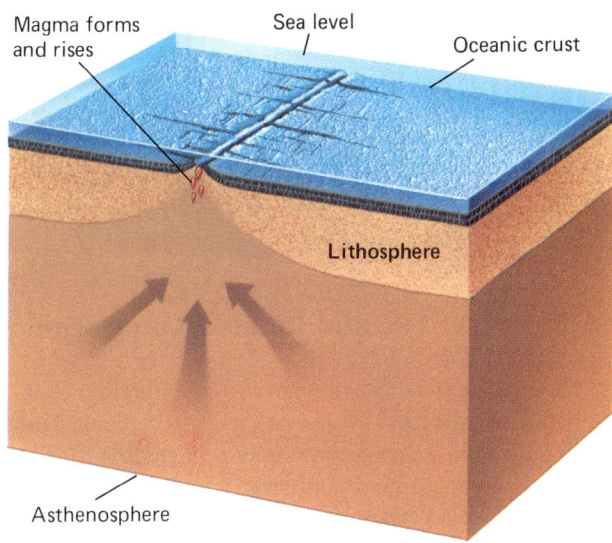

FIGURE 6–5
Pressure-relief melting occurs where hot asthenosphere rises beneath a spreading center.

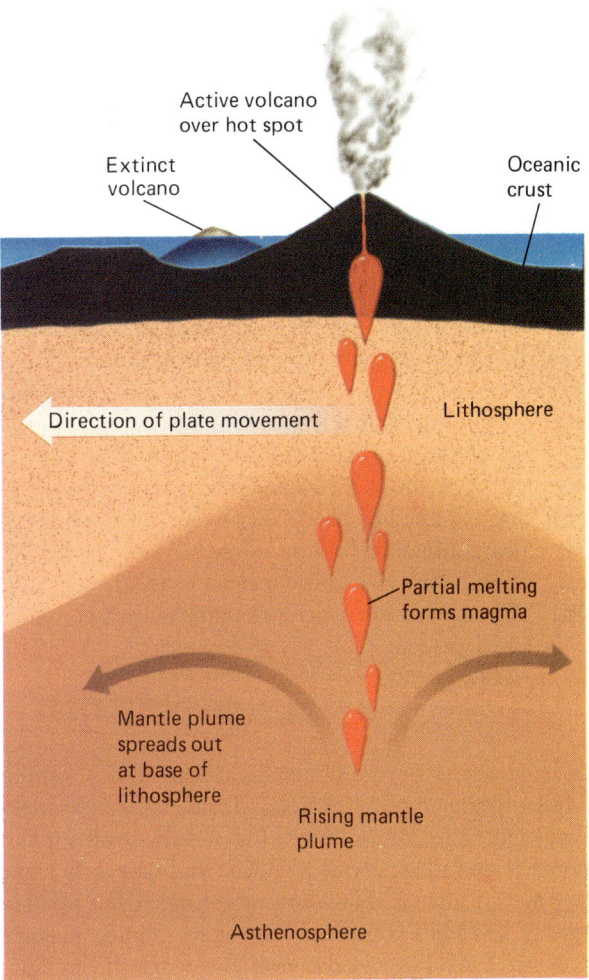

FIGURE 6–6
Pressure-relief melting occurs in a rising mantle plume, and magma rises to form a volcanic hot spot.

gap (Fig. 6–5). As the hot asthenosphere rises, pressure-relief melting forms vast quantities of magma. Melting below the mid-oceanic ridge forms basaltic magma that rises to become new oceanic crust. The basalt crust forms at the mid-oceanic ridge and then spreads outward. Thus, all of the Earth's oceanic crust is created at the mid-oceanic ridge. Some rifts occur in continents, and there, too, great amounts of basaltic magma erupt onto the Earth's surface.

Magma Production at a Hot Spot

A **hot spot** is a volcanic region at the Earth's surface directly above a rising plume of hot, plastic mantle rock. As a mantle plume rises, pressure-relief melting forms magma that flows upward to erupt at the Earth's surface (Fig. 6–6). If a mantle plume rises beneath the sea, volcanic eruptions build submarine volcanoes and volcanic islands. As a lithospheric plate continues to move over the relatively stationary asthenosphere, a mantle plume may generate magma continuously under the migrating plate. This process forms a chain of volcanic islands that becomes progressively younger toward one end, such as the Hawaiian Islands.

Granitic and Basaltic Magmas

Melting of the asthenosphere normally produces basaltic magma. Many volcanoes, such as those of the Hawaiian Islands, erupt basaltic magma directly from this source. But some volcanoes erupt granitic magma, and most of the world's plutons are granite. How does granitic magma form?

Granitic magma forms when basaltic magma rises into continental crust. This occurs where subduction, rifting, or a mantle plume forms basaltic magma beneath a continent. Recall that continental crust is mostly granite. Granite melts at a temperature 200° to 300°C lower than the temperature of basaltic magma. Thus, when basaltic magma rises into continental crust, it melts portions of the crust, forming granitic magma. Most of the granite plutons of the world, including those of the Appalachians, the Rocky Mountains, Yosemite Valley, and the Alps, formed by this two-step process.

6.2 Magma Behavior: Why Some Magma Erupts from a Volcano and Other Magma Solidifies Below the Surface

Once magma forms, it rises toward the Earth's surface because it is less dense than surrounding rock. Basaltic magma commonly rises all the way to the surface to erupt from a volcano. In contrast, granitic magma usually solidifies within the crust, although under special conditions it, too, can rise to erupt from a volcano. These special conditions, described in Section 6.5, cause the most violent and destructive volcanic eruptions.

The contrasting behavior of granitic and basaltic magmas is a result of their different compositions. Granitic magma contains about 70 percent silica, whereas the silica content of basaltic magma is only about 50 percent. In addition, granitic magma generally contains up to 10 percent water, whereas basaltic magma contains only 1 to 2 percent.

Effects of Silica on Magma Behavior

In magma, the silicate tetrahedra described in Chapter 3 link together, forming chains and other large molecules. Large molecules formed by linking of many smaller components are called **polymers**. Silica tetrahedra form large polymers if silica is abundant in the magma, but smaller ones if less silica is present. Therefore, granitic magma contains much larger polymers of silicate tetrahedra than does basaltic magma. The large polymers become tangled, making the magma stiff and resistant to flow. Basaltic magma, with its smaller polymers, flows easily. Because of its fluidity, it rises rapidly to erupt at the Earth's surface. Granitic magma, in contrast, rises slowly because of its stiffness. Therefore, it commonly solidifies within the crust before reaching the surface.

Effects of Pressure and Water on Magma Behavior

Granitic magma contains more water than basaltic magma. Water keeps magma liquid to lower temperatures. Thus, if dry granitic magma solidifies at 700°C, the same magma with 10 percent water may not become solid until it cools to 600°C.

At high temperatures, water tends to escape from magma as steam. Deep in the crust where granitic magma forms, high pressure prevents the water from escaping. But when magma rises, pressure decreases and water escapes. Because the water escapes, the solidification temperature of the magma rises. This loss of water causes rising granitic magma to solidify within the crust. For this reason, most granitic magmas become solid and stop rising at depths of 5 to 20 kilometers beneath the Earth's surface. This process forms granitic plutons.

Because basaltic magmas have only 1 to 2 percent water to begin with, any loss is relatively unimportant. As basaltic magma rises, it remains liquid all the way to the surface. Thus, basalt volcanoes are common.

6.3 Volcanoes and Related Landforms

Volcanic eruptions create many different landforms, including several types of volcanoes and lava plateaus. They also have built many oceanic islands, including the Hawaiian Islands, Iceland, and most islands in the southwestern Pacific Ocean. Occasionally a violent eruption destroys a volcanic peak, as happened in the 1980 eruption of Mount St. Helens, which blew away most of the mountaintop.

Flood Basalts

The gentlest type of volcanic eruption occurs when magma is so fluid that it oozes from cracks and flows over the land like water. When a large flow of this type solidifies, it forms a broad plain or plateau. Basaltic magma commonly pours out as a **flood basalt**, so named because it spreads out like a flood. The flat landscape created by such a lava flow is called a **lava plateau**. The Columbia Plateau in eastern Washington, northern Oregon, and western Idaho is a large lava plateau (Fig. 6–7). This sequence of basalt flows contains 350,000 cubic kilometers of rock, is 3000 meters thick in places, and covers 200,000 square kilometers. It formed rapidly about 15 million years ago as layer upon layer of basaltic magma oozed from fissures in eastern Washington and Oregon. Each flow formed a layer between 15 and 100 meters thick. Similar large basalt plateaus are found in western India, northern Australia, Iceland, Brazil, Argentina, and Antarctica.

Volcanoes

If lava is too viscous to spread out as a lava plateau, it builds up into a hill or mountain called a **volcano**. Volcanoes differ widely in shape, structure, and size (Table 6–1). Lava, rocks, and gas erupt from an opening called a **vent** or from **fissures**, linear cracks in the sides of the volcano. In many volcanoes the vent is located in a **crater**, a bowl-like depression at the summit of the volcano.

FIGURE 6-7
(A) The Columbia Plateau covers much of Washington, Oregon, and Idaho. (B) Columbia Plateau basalt in eastern Washington. Each layer is a separate lava flow.
(Larry Davis)

TABLE 6-1 Characteristics of Types of Volcanoes

Type of Volcano	Form of Volcano	Size	Type of Magma	Style of Activity	Examples
Basalt plateau	Flat to gentle slope	100,000 to 1,000,000 km² in area; 1 to 3 km thick	Basalt	Gentle eruption from long fissures	Columbia Plateau
Shield volcano	Slightly sloped, 6° to 12°	Up to 9000 m high	Basalt	Gentle; some fire fountains	Hawaii
Cinder cone	Moderate slope	100 to 400 m high	Basalt or andesite	Ejections of pyroclastic material	Parícutin, Mexico
Composite volcano	Alternate layers of flows and pyroclastics	100 to 3500 m high	Variety of types of magmas and ash	Often violent	Vesuvius, Mount St. Helens, Aconcagua
Caldera	Cataclysmic explosion leaving a circular depression called a caldera	Less than 40 km in diameter	Granite	Very violent	Yellowstone, San Juan Mountains

FIGURE 6–8
Mount Skjoldbreidier in Iceland shows the typical low-angle slopes of a shield volcano.
(Science Graphics, Inc./Ward's Natural Science Establishment, Inc.)

Shield Volcanoes

Basaltic magma commonly forms a gently sloping volcanic mountain called a **shield volcano** (Fig. 6–8). The sides of a shield volcano generally slope away from the vent at angles between 6° and 12° from horizontal. As a reference, a ski slope for beginning to intermediate skiers is about 10°, and an expert slope is commonly 30°. Unless erosion has formed deep gullies, you will find no challenging ski runs on a shield volcano; it isn't steep enough.

When a shield volcano erupts, the fluid lava usually flows gently through vents or fissures. Although the Hawaiian shield volcanoes erupt regularly, the eruptions are normally gentle and only rarely life threatening. Lava flows occasionally overrun homes and villages, but the flows advance slowly enough to give people time to evacuate (Fig. 6–9).

FIGURE 6–9
This automobile was partially buried by a lava flow on the island of Hawaii. *(Kenneth Neuhauser)*

FIGURE 6–10
Cinder cones of Wupatki Sunset Crater National Monument, Arizona. *(Wupatki National Monument)*

Cinder Cones

A **cinder cone** is a small volcano, as high as 400 meters, made of volcanic cinders blasted out of a central vent at high velocity. A cinder cone forms when large amounts of gas accumulate within rising magma. When the gas pressure builds up sufficiently, the magma erupts explosively, hurling cinders, ash, and molten magma into the air. All such material erupted explosively from a volcano is called **pyroclastic** material, or **tephra**. The particles then fall back around the vent to accumulate as a small mountain.

Volcanic ash and **cinders** are not the same as the ashes and cinders produced by a conventional fire; they are, rather, fine particles of solidified magma. In December 1989, Mount Redoubt volcano, southwest of Anchorage, Alaska, erupted. The pilot of a Boeing 747 carrying 231 passengers ignored warnings about the rising ash cloud and flew into it. The abrasive particles were sucked into the jet engines and quickly ground the sharp blades of the compressor into small stubs. The intense heat in the combustion chamber melted the volcanic ash particles. Moments later, the liquid solidified to form glass on the cooler turbine blades. As a result, all four engines stalled, and the jet fell 4000 meters in 8 minutes. Finally the pilot was able to restart the engines and land the disabled craft in Anchorage.

As the name implies, a cinder cone is symmetrical. It also can be as steep as 30° (Fig. 6–10). A cinder cone is usually active for only a short time because, once the gas escapes, the driving force behind the eruption is removed. Because the pyroclastic fragments are not cemented together, a cinder cone erodes easily and quickly. Therefore, in geologic time it is a transient feature of the landscape.

About 350 kilometers west of Mexico City, many extinct cinder cones rise from a broad plain. A small hole had existed in the flat ground of the plain for as long as anyone could remember, and farmers grew corn just a few meters away. In February 1943, as two farmers were preparing their field for planting, the hole started to emit smoke and sulfurous gases. As night fell, hot, glowing rocks flew skyward from the hole, creating arcing flares like a giant fireworks display. By morning, a 40-meter-high cinder cone had grown in the middle of the cornfield. For the next 5 days, pyroclastic material erupted 1000 meters into the sky and the cone grew to 100 meters in height. After a few months, a fissure opened at the base of the cone, extruding a lava flow that buried the town of San Juan Parangaricutiro. Two years later, the cone had grown to a height of 400 meters. After 9 years the eruptions ended, and today the volcano, called El Parícutin, is dormant.

Composite Cones

Some of the most beautiful and violent volcanoes in the world are **composite cones**, sometimes called **stratovolcanoes**. They form by a series of alternating lava flows and pyroclastic eruptions. As explained previously, pyroclastic eruptions form steep but unconsolidated slopes. However, when lava flows cover the loose cinders, the hard lava rock protects them from erosion (Fig. 6–11).

A composite cone grows as a result of many eruptions occurring over a long time. Many of the highest mountains of the Andes are composite cones, as are many of the most spectacular mountains of the Cascade Range of the Pacific Northwest. Two examples are Mount St. Helens and Mount Adams (Fig. 6–12). Mount St. Helens erupted in 1980. Mount Adams has been dormant in recent times but could become active at any time; repeated eruptions over hundreds of thousands or even millions of years is a trademark of a composite volcano.

After an eruption, the vent of a volcano may fill with magma that later cools and solidifies. Commonly this **volcanic neck** is harder than surrounding rock. Given enough time, the slopes of the volcano may erode, leaving only the tower-like neck exposed (Fig. 6–13).

6.4 Volcanoes and Plate Tectonics

Subduction Zone Volcanoes

We learned in Section 6.1 that magma forms abundantly in subduction zones. It then rises to form both volcanoes and plutons directly over the subducting tectonic plate. Figure 6–14 shows that most of the world's volcanoes are located near subduction zones. The "ring of fire"

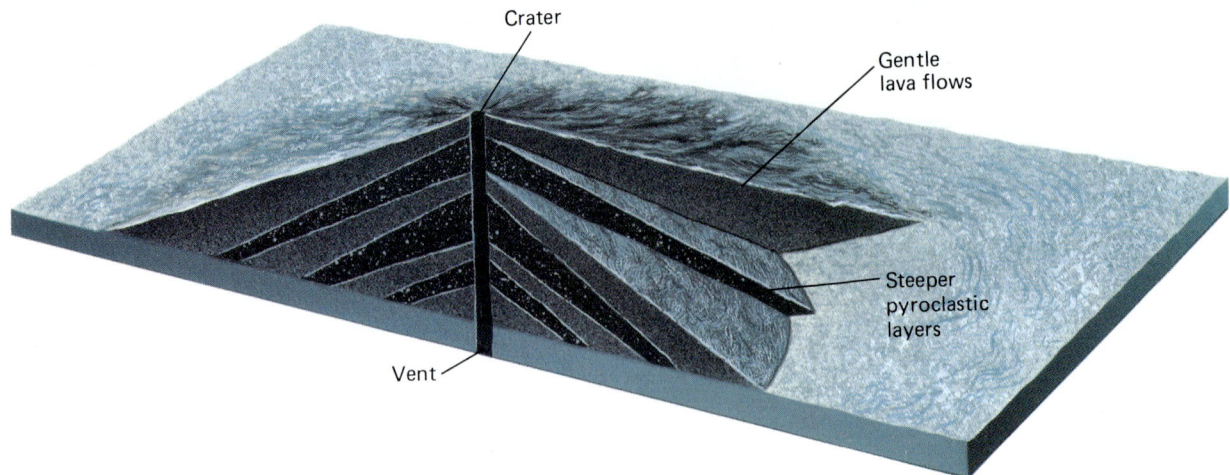

FIGURE 6-11
(A) A schematic cross section of a composite cone showing alternating layers of lava and pyroclastic material. (B) Steam and ash pouring from Mount Ngauruhoe, a composite cone in New Zealand. *(Don Hyndman)*

FIGURE 6-12
Mount Adams is a composite cone.

FIGURE 6-13

Shiprock, New Mexico, is a volcanic neck. The great rock was once the core of a volcano. The softer flanks of the cone have now eroded away. A dike several kilometers long extends to the left.
(Dougal McCarty)

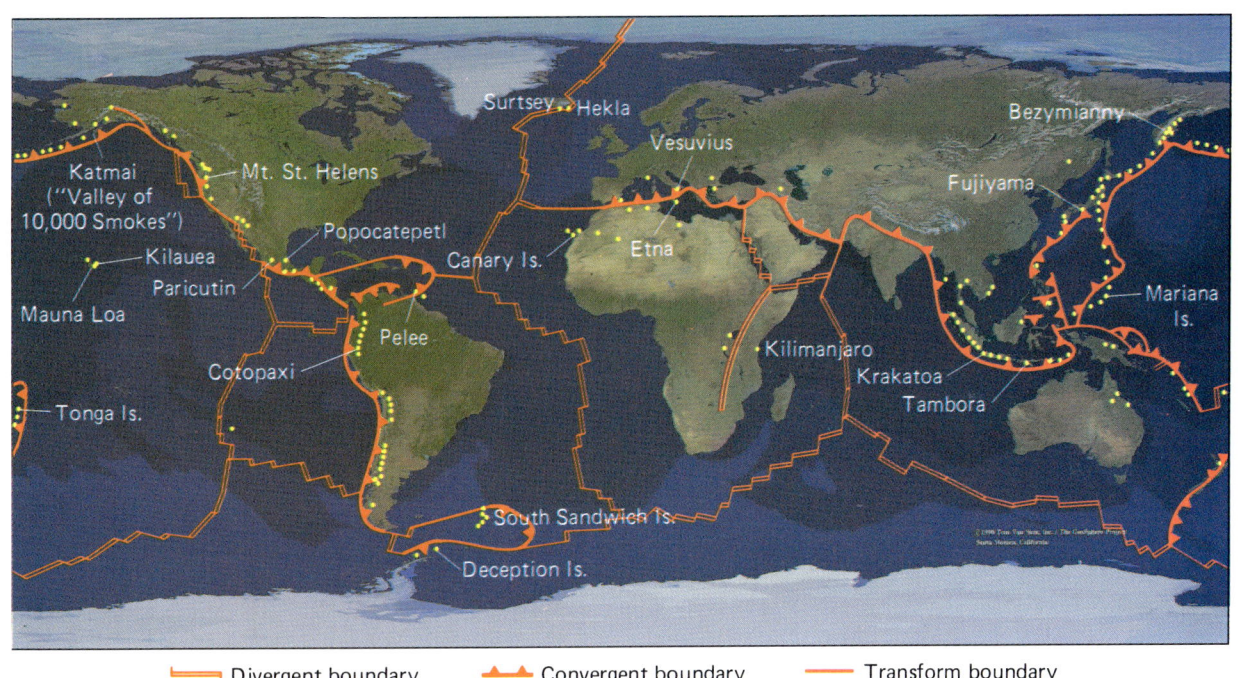

FIGURE 6-14

This map shows the close relationship between subduction zones (convergent boundaries) and major volcanoes, indicated by the yellow dots.

is a zone of concentrated volcanic activity that traces subduction zones encircling the Pacific Ocean basin. About 75 percent of the Earth's active volcanoes lie in the ring of fire. Most of the volcanoes that threaten human life are in this region.

All 18 volcanoes of the Cascade Range, from Mount Baker near the Canadian border to Mount Lassen in northern California, have been active in the past 2 million years, fired by subduction of oceanic plates beneath the continent. Mount St. Helens erupted in 1980 and Mount Lassen in 1915. About 7000 years ago, a Cascade volcano, posthumously named Mount Mazama, erupted 10 cubic kilometers of rock and ash. The eruption blew some of the mountain away, and the remainder collapsed

FIGURE 6-15
Crater Lake, Oregon. *(Crater Lake National Park Administration)*

into the hole formed by the blast, leaving what is now Crater Lake, Oregon (Fig. 6–15). The ash is found in soils over much of western and central North America.

Some volcanoes that have not erupted recently show signs of activity today. Mount Baker has recently experienced earthquake swarms, and gas regularly escapes from its crater. Steam caves lie beneath the summit glaciers of Mount Rainier. It is reasonable to predict that the Cascades will see additional eruptions, although it is difficult to predict where, when, or how large. It is important to remember that, despite the loss of life and tremendous damage, the 1980 Mount St. Helens eruption was small compared with other known eruptions from similar volcanoes.

On June 9, 1991, Mount Pinatubo in the Philippines began ejecting a gray-green cloud of ash, rock, and smoke 15 kilometers into the sky (Fig. 6–16). By the end of June, 338 people had died and several towns were evacuated because of continued explosions. As many as 200 earthquakes a day were felt. Both the U.S. Clark Air Base, 15 kilometers from the volcano, and Subic Bay Naval Station sustained heavy damage. Clark was evacuated because of the threat to personnel and aircraft. Mount Mayon, also in the Phillipines, erupted violently, killing several people in February 1993.

Rift Zone Volcanoes

The island of Iceland in the North Atlantic Ocean is about the size of Virginia and supports a quarter of a million people. It lies directly over the Mid-Atlantic ridge and thus is on a spreading boundary between two

FIGURE 6-16
An American soldier guards the entrance to Clark Air Base in the Philippines as Mount Pinatubo erupts steam and ash on June 29, 1991. The base had been evacuated a week before this picture was taken, and was destroyed by later eruptions. *(Wide World Photos)*

tectonic plates. The island formed by repeated volcanic eruptions at the ridge.

Iceland has experienced numerous eruptions since its settlement by Vikings before A.D. 1000. Some have covered villages with ash and cinders, although human injury and death have been rare. In 1973, a newly formed volcanic vent on the island of Heimaey just off the south coast of Iceland began spewing cinders and lava. Eventually, the cinders buried the town of Vestmannaeyjar, a prosperous fishing village, but no one died. The lava threatened to block the harbor used by the fishermen. Government workers sprayed great amounts of seawater on the front of the flow, hoping to cool it and thus stop its advance toward the harbor. The cold water did chill the margins of the flow and slow it down, but it was not clear that the cooling altered the final extent of the flow. Between 1975 and 1981, an 80-kilometer section of the Mid-Atlantic ridge in northeastern Iceland spread apart by 5 meters, erupted eight times, and was the source of innumerable earthquakes. Today, hot springs and hot rock resulting from recent volcanism are used to generate electrical power for Iceland, and the spectacular volcanic scenery is a major tourist attraction.

Typically, rift-zone eruptions in oceanic crust such as those in Iceland are gentle and not life threatening because volcanoes in this environment erupt basaltic magma. This type of magma is very fluid because of its low silica content and is not violently explosive because of its low water content. As a result, the eruptions eject cinders and lava flows that move slowly enough that people can escape from the path of an oncoming flow.

Rifting can also occur in continental crust. When it does, pressure-relief melting in the mantle forms basaltic magma below the continent. This basaltic magma rises into lower granitic continental crust and melts it to form granitic magma. Thus both types of magma erupt in continental rifts. Continental rifting is occurring today along a north–south trend through eastern Africa in a zone called the East African rift. It may also be occurring in Idaho's Snake River plain and along the Rio Grande rift in central Colorado and New Mexico along the Rio Grande River. Granitic eruptions of this type can be extremely violent, as explained in Section 6.5.

Mantle-Plume and Hot-Spot Volcanoes

The island of Hawaii is composed of several overlapping volcanoes that lie above a mantle plume. The youngest, Kilauea, frequently erupts basaltic lava, commonly for weeks or months. As in Iceland, lava flows occasionally destroy homes and agricultural land, but they rarely cause injury or death because eruptions are comparatively gentle and the flows advance slowly enough that people can evacuate threatened areas. Because the eruptions are relatively safe, at least from a distance, and because they continue for long periods of time, tourists flock to the island to see the eruptions.

Mantle plumes also rise beneath continents. As in the case of continental rifting, basaltic magma forms in a rising mantle plume beneath the continent. The magma melts lower continental crust as it ascends, forming granitic magma, which in turn rises to create a continental hot spot, such as Yellowstone National Park. Continental hot spots often give rise to explosive volcanic eruptions, as discussed below.

6.5 Explosive Magma and Violent Volcanoes

Some granitic magma rises all the way to the Earth's surface, where it characteristically erupts with great violence. Why is the volcanic behavior of such magma so violent and dangerous?

Most granitic magma contains as much as 10 percent water and large amounts of carbon dioxide. At the high pressure of the great depths where the magma forms, the water and carbon dioxide are dissolved in the magma. When the magma rises toward the Earth's surface and pressure decreases, the dissolved gases separate to form bubbles in the liquid magma (Fig. 6–17A). As an analogy, think of a bottle of beer or soda pop. When the cap is on and the contents are under pressure, the carbon dioxide gas is dissolved in the liquid. Because the gas is dissolved, there are no visible bubbles. If you remove the cap, the pressure is reduced. As a result, the gas escapes from solution, and bubbles form and rise to the surface. If the conditions are favorable, the frothy mixture of gas and liquid erupts through the opening. In a similar manner, gas bubbles form in rising magma, creating a frothy, expanding mixture of gas and liquid magma. The temperature of this mixture may be as high as 900°C.

As the magma rises to within a few kilometers of the Earth's surface, it lifts and fractures overlying rocks. The highly pressured, foamy mixture of magma and gas may then explode through the cracks in the roof rock (Fig. 6–17B). Alternatively, the mixture may just ooze from the cracks, like root-beer foam oozing from a newly opened bottle.

Recall that the 1980 eruption of Mount St. Helens blasted out 1 cubic kilometer of ash and rock and that, in contrast, the 1.9-million-year-old Yellowstone eruption ejected 2500 cubic kilometers of pyroclastic material. In a large eruption of the latter type, the explosion may blast a column of ash and rocks 12 or 15 kilometers above

The 1980 Mount St. Helens Eruption

Mount St. Helens erupted violently in 1857 and lay dormant for over a century. In March 1980, the mountain experienced several small to moderate earthquakes. Puffs of steam and volcanic ash emanated from a newly formed crater on the summit. The activity convinced geologists to install seismographs and tiltmeters near and on the volcano. Seismographs record earthquakes caused by moving magma that might precede an impending eruption. Tiltmeters detect tilting of the Earth's surface caused by magma moving upward in the volcano and swelling the flanks.

In early April, the emissions of steam and ash ceased, although earthquakes continued. The tiltmeters detected swelling on the north side of the volcano. By early May the swelling had grown to an ominous bulge that rose as fast as 1.5 meters per day. Eventually it measured 1 by 2 kilometers and extended outward more than 100 meters. Warned by the geologists, officials of the United States Forest Service closed the volcano and surrounding area to the public and evacuated most local inhabitants.

Two strong earthquakes rocked the mountain at 8:27 and 8:31 on Sunday morning, May 18. The bulge had grown so steep that the second earthquake caused it to break away from the mountain, forming an immense landslide that carried rock, earth, and glacial ice from the summit. The landslide roared down the mountain and, in doing so, relieved the pressure on the gas-charged magma that had been causing the bulge. The magma then exploded

FIGURE 1
The eruption of Mount St. Helens. (USGS, R. P. Hoblitt)

out through the side of the mountain where the bulge had been, blowing away the entire north side and the top 410 meters of the volcano (Fig. 1). The horizontal blast flattened trees as far as 25 kilometers away and destroyed 400 square kilometers of forest (Fig. 2). The landslide, combined with large volumes of volcanic ash from the eruption, poured down the Toutle River and into the Columbia River, filling the channels with ash and debris.

A vertical plume of ash, rocks, and gas rose to a height of 18 kilometers within a few minutes of the initial eruption. Static electricity generated by the billowing plume caused lightning strikes in the surrounding forest. Mountaineers on Mount Adams, 50 kilometers to the east, saw sparks jump from their ice axes, were pelted with falling debris, and felt intense heat from the blast.

An airborne ash cloud spread eastward on the prevailing winds at 60 kilometers per hour. It darkened the sky and for several days turned the air gritty and unpleasant all the way to western Montana. Some ash drifted to Minnesota and Oklahoma. The ash made breathing difficult, destroyed crops, and ruined automobile engines. Damage from the eruption was estimated in the hundreds of millions of dollars. Sixty-three people are known or presumed to have been killed.

Five months after the eruption, hot lava forced its way upward, forming a new dome in the crater. During the following decade, the lava dome grew to 350 meters in height. This dome reminds us that Mount St. Helens is likely to erupt again.

FIGURE 2
Burned and uprooted trees 13 kilometers from Mount St. Helens. (Larry Davis)

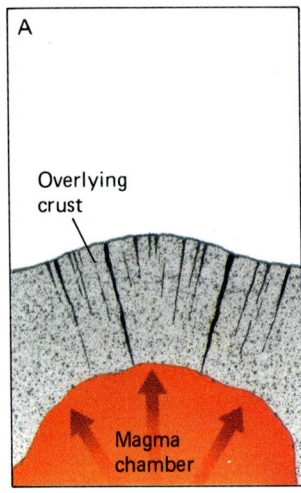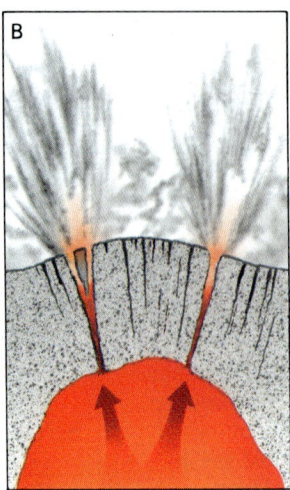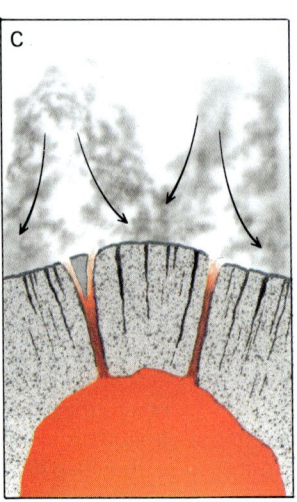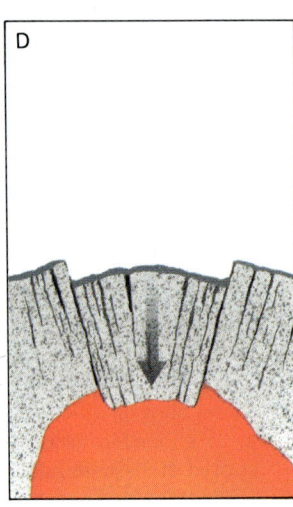

FIGURE 6–17
(A) When granitic magma rises to within a few kilometers of the Earth's surface, it stretches and fractures overlying rock. Gas separates from the magma and rises to the upper part of the magma body. (B) The gas-rich magma explodes through fractures, rising as a vertical column of hot ash, rock fragments, and gas. (C) When the gas is used up, the column collapses and spreads outward as a high-speed ash flow. (D) Because so much material has erupted from the top of the magma chamber, the roof collapses to form a caldera.

the Earth's surface; the column may be several kilometers in diameter. A cloud of fine volcanic dust may rise to much greater heights, into the upper atmosphere. Continued eruption can hold the column up for hours or even days as additional material streams from the magma chamber.

Ash Flows

Eventually, when the gas pressure in the upper part of the magma chamber is exhausted, the eruption stops and the frothy column falls back to the Earth's surface (Fig. 6–17C). The material in the column is mostly gas consisting of water and carbon dioxide from the magma plus trapped atmospheric gases, glassy magma bubbles, frothy liquid magma, crystals from the magma, and rock fragments that were ripped away from the roof during the explosion. Although the collapsing column contains solid particles, it behaves as a fluid called an **ash flow**. When it reaches the Earth's surface, the falling mixture flows outward from its point of impact. Small ash flows have been seen traveling at speeds up to 200 kilometers per hour. Larger flows have traveled distances exceeding 100 kilometers. One large flow jumped over a 700-meter-high ridge as it crossed from one valley into another.

When an ash flow comes to a stop, much of the gas escapes into the atmosphere, leaving a chaotic mixture of volcanic ash, crystals, and rock fragments picked up along the way. **Tuff** is a rock that forms when this mixture solidifies (Fig. 6–18).

FIGURE 6–18
Tuff forms when an ash flow comes to a stop. The fragments in the tuff are pieces of rock that were carried along with the volcanic ash and gas. *(Geoffrey Sutton)*

Calderas

When such an immense volume of material suddenly erupts, nothing is left to hold up the overlying roof rock. Therefore, the roof of the magma chamber collapses, forming a circular depression called a **caldera** (Fig. 6–17D). A large caldera may be 20 to 40 kilometers in diameter and 1 kilometer deep. We usually think of volcanic landforms as gracefully symmetrical cones. Calderas, as topographic depressions, are interesting exceptions to this notion. Oregon's Crater Lake (Fig. 6–15) is a small caldera.

Magmas that produce ash-flow tuffs and calderas commonly erupt more than once. Following an eruption, the remaining upper part of the magma is depleted in gases and thus has lost its explosive potential. However, lower portions of the magma continue to release gases, which rise and accumulate in the upper portion of the magma until pressure increases enough to begin a new cycle of eruption. The time interval between successive eruptions typically varies from a few thousand to about a half million years.

6.6 Assessing the Danger of Future Volcanism

An **active volcano** is one that is erupting or is expected to erupt. A **dormant volcano** is one that is not now erupting but has erupted in the past and will probably do so again. Thus, no clear distinction exists between active and dormant volcanoes. In contrast, an **extinct volcano** is one that is expected never to erupt again.

There are approximately 1300 active volcanoes in the world, and 5564 known eruptions have occurred in the past 10,000 years (Table 6–2). Many volcanoes have erupted recently, and we are certain that others will erupt soon. Is there any way to predict when and where an eruption will occur? How can we reduce the risk of volcanic disaster?

Prediction and risk reduction of volcanic eruptions involve evaluation of both frequency and potential violence. Policy decisions are then made on the basis of geology, politics, and economics.

Long-Term Prediction

As explained earlier in this chapter, certain tectonic environments are volcanically active, and others are free of volcanic hazard. The environments that are prone to frequent volcanic eruptions are (1) those in which tectonic plates are rifting apart, such as East Africa and Iceland; (2) those in which subduction is occurring, such as in the Pacific Northwest and western South America; and (3) those in which a mantle plume creates a volcanic hot spot, such as Hawaii and Yellowstone. In contrast, most interior areas of tectonic plates, such as the central and eastern United States, have no volcanic activity. Thus, the first level of assessing the volcanic hazard of a particular region is to understand its place in the global tectonic framework.

You have also learned that certain types of volcanoes are much more dangerous than others. Mount St. Helens exploded on May 18, 1980, killing 63 people. Mount Pelée on the Caribbean island of Martinique erupted in 1902, destroying the town of Saint Pierre and killing

TABLE 6–2 Some Notable Volcanic Disasters Since the Year A.D. 1000 Involving 5000 or More Fatalities

			Primary Cause of Death and Number of Deaths				
Volcano	Country	Year	Pyroclastic Flow	Debris Flow	Lava Flow	Posteruption Starvation	Tsunami
Kelut	Indonesia	1586		10,000			
Vesuvius	Italy	1631			18,000		
Etna	Italy	1669			10,000		
Lakagigar	Iceland	1783				9,340	
Unzen	Japan	1792					15,190
Tambora	Indonesia	1815	12,000			80,000	
Krakatoa	Indonesia	1883					36,000
Pelée	Martinique	1902	29,000				
Santa Maria	Guatemala	1902	6,000				
Kelut	Indonesia	1919		5,110			
Nevado del Ruiz	Colombia	1985		>22,000			

Calderas in the United States

Figure 1 shows that calderas and ash-flow tuffs are of major importance in the geology of mountainous regions in the western United States. Yellowstone National Park and the Long Valley caldera in California are two famous and important calderas; both have the potential for future eruptions.

Yellowstone National Park

Yellowstone National Park in Wyoming and Montana is the oldest national park in the United States. Its geology is dominated by three large overlapping calderas and the tuffs that erupted from them (Fig. 2). The oldest of the three eruptions occurred 1.9 million years ago and ejected the 2500-cubic-kilometer Huckleberry Ridge tuff. Roof collapse accompanying the eruption formed the Big Bend Ridge caldera, an elongate depression about 25 by 40 kilometers with walls up to 1 kilometer high. The next major eruption, 1.3 million years ago, ejected about 280 cubic kilometers of ash and produced the smaller Henry's Fork caldera. The most recent eruption, 0.6 million years ago, blew out the 1000-cubic-kilometer Lava Creek tuff and formed the Yellowstone caldera in the center of the park.

Yellowstone Park is famous for its numerous geysers and hot springs, features that often form where a shallow magma chamber heats ground water. This activity, coupled with seismographic detection of magma movement within the Earth's crust, confirms that liquid magma exists below the surface at Yellowstone.

Consider the periodicity of the three Yellowstone eruptions. They occurred at 1.9, 1.3, and 0.6 million years ago. Intervals of 0.6 to 0.7 million years have separated the eruptions. It has been 0.6 million years since the last eruption. The periodicity of the eruptions, the presence of magma at shallow depths, and the well-known tendency of magmas of this type to erupt multiple times all suggest that a fourth Yellowstone eruption may be due at any time. However, a forecast based on periodicity is only approximate. With a periodicity of 0.6 million (or 600,000) years, a 1 percent error

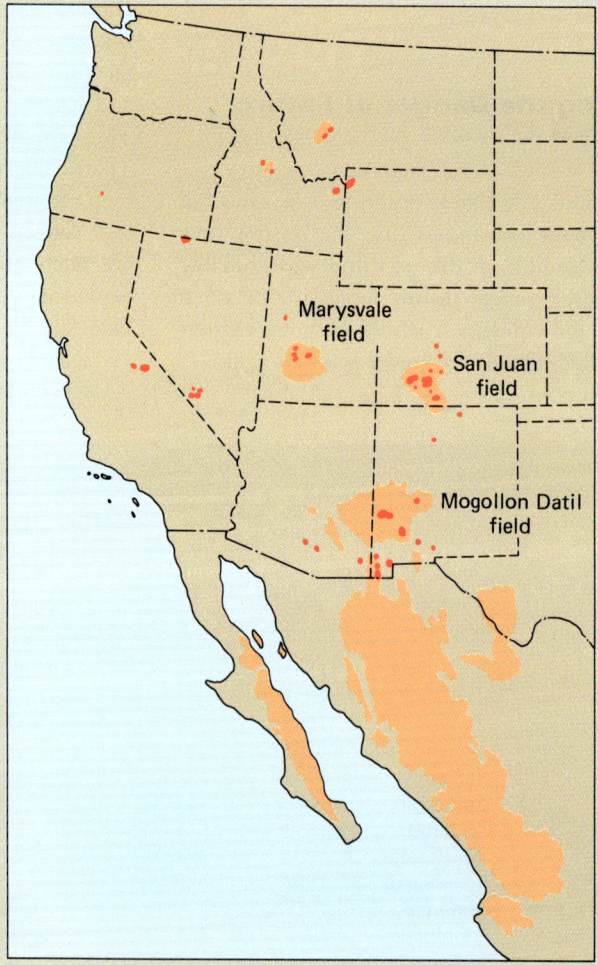

FIGURE 1
Calderas (red dots) and ash-flow tuffs (orange areas) in western North America.

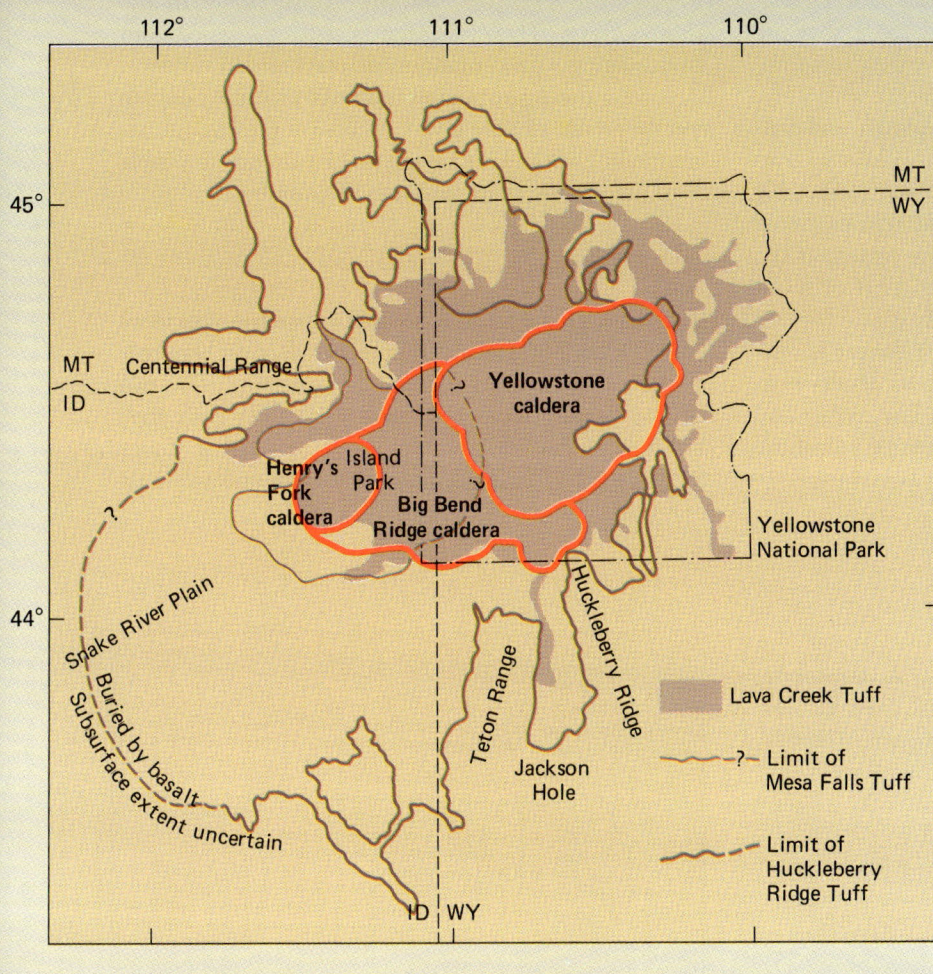

FIGURE 2
Calderas and ash-flow tuffs of Yellowstone Park. The Big Bend Ridge caldera formed during the eruption of the Huckleberry Ridge Tuff 1.9 million years ago. The Henry's Fork caldera formed during the eruption of the Mesa Falls Tuff 1.3 million years ago. The Yellowstone caldera formed during the eruption of the 0.6-million-year-old Lava Creek Tuff. Portions of older calderas are obliterated by younger calderas. Dashed boundaries of the Huckleberry Ridge and Mesa Falls Tuff are covered by younger rocks. (Figure modified from Hildreth and others, *Journal of Geophysical Research* 89, 1984)

FIGURE 3
The popular Mammoth Mountain ski area lies on the edge of the Long Valley caldera.

is 6000 years. Thus, it is conceivable that the next eruption will occur within our lifetimes. Alternatively, it may not occur for several thousand years, or perhaps Yellowstone has seen its last eruption.

The Long Valley Caldera, California

A situation somewhat similar to that in Yellowstone is found near Yosemite National Park in eastern California. Here the 170-cubic-kilometer Bishop tuff erupted from the Long Valley caldera 0.7 million years ago. Only one major eruption has occurred in this area to date, but seismic monitoring indicates that active magma lies beneath Mammoth Mountain, a popular California ski area on the southwest edge of the Long Valley caldera (Fig. 3).

29,000 people. The Indonesian volcano Tambora exploded in 1815, directly killing 12,000 people and destroying crops, which led to death by starvation of another 80,000. In contrast with these examples of death and annihilation, large populations coexist peacefully with active volcanoes in Hawaii and Iceland.

As explained previously, the difference between relatively safe volcanic environments and those that are dangerous is closely related to magma composition. Granitic magmas tend to be viscous and charged with water and carbon dioxide and are therefore explosive and dangerous. Basaltic magmas are fluid and have low gas contents and thus usually erupt in a gentle and predictable manner. The nature of volcanic activity can be predicted from a knowledge of the geology of the region. If the cause of volcanism is rifting of oceanic crust or a hot spot under oceanic crust, then relatively gentle basaltic volcanism should predominate, and the danger is relatively low. In contrast, if volcanic activity results from subduction or from a hot spot or rift below continental crust, then the much more dangerous granitic volcanism is likely, and the hazard level is high.

Another important hazard factor is the probability of recurrence of eruptions. Cinder cones usually erupt only once from a vent. Subsequent eruptions form new vents elsewhere. In contrast, shield and composite volcanoes and calderas of the Yellowstone type commonly erupt many times.

Figure 6–19 shows volcanic danger zones in the United States. The evaluation of danger is based both on frequency of past eruptions and on potential violence. However, the figure merely outlines probabilities. No specific predictions are given, because at present our understanding of the Earth is not precise enough to enable us to predict the times and sizes of eruptions.

Short-Term Prediction

Long-term prediction tells us that composite volcanoes such as Mount St. Helens erupt repeatedly over hundreds of thousands or even millions of years. In 1978, two United States Geological Survey (USGS) geologists, Dwight Crandall and Don Mullineaux, noted that Mount St. Helens had erupted more frequently and violently

FIGURE 6–19

The United States Geological Survey lists U.S. volcanoes in three groups according to potential for future eruptions. This assessment is based on frequency of past eruptions. The first group is the most likely to erupt, and the third group is the least likely. Within each group, volcanoes with the greatest potential for destructive eruptions are listed first. The first group consists of volcanoes that have erupted on average every 200 years or less, have erupted in the past 300 years, or both. This group includes (1) Mount St. Helens, (2) Mono-Inyo Craters, (3) Lassen Peak, (4) Mount Shasta, (5) Mount Rainier, (6) Mount Baker, and (7) Mount Hood. The second group consists of volcanoes that have erupted less frequently than every 1000 years and last erupted more than 1000 years ago. They include (A) Three Sisters, (B) Newberry volcano, (C) Medicine Lake volcano, (D) Crater Lake (Mount Mazama), (E) Glacier Peak, (F) Mount Adams, (G) Mount Jefferson, and (H) Mount McLoughlin. The third group consists of volcanoes that last erupted more than 10,000 years ago but overlie large magma chambers. This group includes (AA) Yellowstone caldera, (BB) Long Valley caldera, (CC) Clear Lake volcanoes, (DD) Coso volcanoes, (EE) San Francisco Peak, and (FF) Socorro. *(Adapted from United States Geological Survey Open-file Report 83-400)*

during the past 4500 years than any other volcano in the contiguous 48 states. They predicted that the volcano would erupt again before the end of the century.

In March 1980, about 2 months before the great May eruption, Mount St. Helens showed signs of renewed activity. Puffs of steam and volcanic ash rose from the crater, and swarms of small to moderate earthquakes occurred beneath the mountain. This activity convinced USGS geologists that Crandall and Mullineaux's prediction might be about to come true. In response, they installed networks of seismographs, tiltmeters, and surveying instruments on and around the mountain.

Short-term volcanic prediction methods monitor a known volcano to detect signals that the volcano is about to erupt. The signals may include changes in the shape of the mountain and surrounding land, earthquake swarms indicating movement of magma beneath the mountain, increased emissions of ash or gas from the vent, increasing temperatures of nearby hot springs, and any other signs that magma is approaching the surface. Short-term predictions are forecasts that a specific volcano may erupt in the near future.

Education and Public Policy

A short-term prediction of a volcanic eruption is never 100 percent certain. In some cases, magma rises beneath a volcano, giving all signs that an eruption is imminent. The volcano swells, the ground tilts, earthquake activity increases, temperatures of nearby vents and hot springs rise, but the volcano doesn't erupt. On the other hand, another volcano may erupt without warning. Given this uncertainty, consider the dilemma of a politician or civil defense official told by geologists that a volcano looming over the city is showing increased earthquake activity and swelling and tilting of its flanks, signs of an eruption in the near future. He or she faces the same uncertainty that characterizes response to earthquake forecasts. Should he or she order an expensive evacuation of the city and risk the embarrassment of being wrong if the volcano doesn't erupt? Or should he or she ignore the warning and risk mass death if it does erupt? The social, ethical, and financial aspects of public policy response can be more troublesome than the geology.

Prior to 1985, geologists warned Colombian officials that the town of Armero rested on top of an 1845 mudflow that had come from the flanks of Nevado del Ruiz volcano, and that if a similar mudflow occurred, it would probably follow the path of the older one. Officials took no action in response to the warning. Then, on November 13, 1985, a small eruption melted ice and snow near the top of the volcano. The meltwater saturated the volcanic ash on the flanks of the peak, starting a mudflow that buried Armero and killed 22,000 people.

In 1975, Soufrière volcano on the Caribbean island of Guadeloupe began shaking from small swarms of earthquakes and minor explosive eruptions. Recalling the 1902 eruption of nearby Mount Pelée that killed 29,000 people, authorities evacuated 72,000 people from the island. People were kept from their homes and businesses for 3 months, but a major eruption never occurred. Scientists and politicians argued long afterward about the wisdom of the expensive evacuation.

In July 1983, Colo volcano on an island in Indonesia experienced earthquake swarms and small explosive eruptions. This type of behavior had preceded major eruptions of nearby volcanoes, so officials moved 7000 people off the island. On July 23, a major eruption blasted hot ash flows across the island, destroying the coconut plantations, but all of the inhabitants survived to return and replant the coconut groves.

Thus, although forecasting of volcanic eruptions is imperfect, expensive, and even embarrassing when wrong, it does save lives.

6.7 Volcanoes and Climate

When Mount St. Helens erupted, the ash darkened the sky over a wide area. In Yakima, Washington, 140 kilometers from the mountain, people had to turn their car headlights on at noon. In Missoula, Montana, 620 kilometers from Mount St. Helens, an eerie darkness, or dry fog, was observed. Clouds of volcanic ash reflect light and heat from the sun out into space, which causes cooling of the Earth's surface. The immediate effects are easy to document. However, it is more difficult to assess the long-term climatological effects of major volcanic eruptions.

The largest volcanic eruption in recent history occurred in 1815, when Mount Tambora in the southwestern Pacific Ocean exploded, ejecting approximately 100 times more magma and ash into the atmosphere than Mount St. Helens would. The following year, 1816, was one of the coldest years in history worldwide and has been called the Year Without a Summer. Crop failures, compounded by devastation caused by the Napoleonic Wars, led to widespread famine in Europe, a period called the "last great subsistence crisis in the Western world." The question remains: Was this cold period caused by the Mount Tambora eruption, or was it merely a coincidence that the two events occurred together?

Figure 6–20 shows a plot of global temperatures before and after eight recent major volcanic eruptions. This graph clearly shows a correlation between global cooling and volcanic eruptions. The correlation agrees with me-

FOCUS ON: Disastrous Historical Eruptions

Volcanic eruptions have caused much loss of life and destruction of property throughout history. When Mount St. Helens erupted in 1980, it exploded with the force of 500 atomic bombs of the size used on Hiroshima in World War II. The volcanic blast ejected nearly 1 cubic kilometer of rock and ash, leveled forests, and—despite adequate warning that an eruption was imminent—killed 63 people. Yet, compared with other historical eruptions, Mount St. Helens was a small event. Table 6–2 summarizes the major known volcanic disasters since A.D. 1000.

Mount Pelée

On May 2, 1902, the coastal town of Saint Pierre on the Caribbean island of Martinique was destroyed by an ash flow that erupted from the nearby volcano of Mount Pelée. All but two of the 29,000 residents of the town died instantly as an 800°C cloud of gas and volcanic ash roared down the side of the volcano and through town at speeds up to 100 kilometers per hour. (Ironically, one of the survivors was a convicted murderer who was imprisoned in a dungeon when the eruption occurred.) Only one of 18 ships in the harbor at the time of the explosion escaped, and it lost many crew members. The magma responsible for the eruption was viscous, silica-rich, and charged with gas.

Krakatoa

Before August of 1893, Krakatoa was an 800-meter-high island lying between Java and Sumatra in the southwestern Pacific Ocean. It consisted of three volcanic cones lying within an old caldera. In the days before August 27, several small volcanic explosions occurred on the island. On August 27, the entire island exploded with an amount of energy equivalent to 100 million tons of TNT. Although Krakatoa itself was unpopulated, the volcanic explosion formed an immense tsunami, between 35 and 40 meters high, that radiated outward, inundating coastal villages on nearby islands and killing about 36,000 people.

Most of the island of Krakatoa disappeared in the explosion, leaving in its place a basin about 100 meters beneath the sea. The sound of the explosion was heard in Australia, more than 4000 kilometers away. Dust hurled into the upper atmosphere circled the globe for years afterward, turning sunsets red in London and blocking the sun to the extent that the Earth's mean temperature fell by a few degrees.

It is thought that the explosion occurred because seawater poured into cracks that had been formed in the overlying bedrock by the rising magma. The magma rapidly heated the seawater, causing a steam explosion.

Mount Vesuvius

In A.D. 79, Mount Vesuvius erupted and destroyed the Roman cities of Pompeii and Herculaneum and several neighboring villages near what is now Naples, Italy. Prior to that eruption, the volcano had been inactive for more than 2000 years, so long that vineyards had been cultivated on the sides of the mountain all the way to the summit. During the eruption, a hot ash flow streamed down the flanks of the volcano, burying the cities and towns under 5 to 8 meters of hot ash.

Pompeii was located and excavated 17 centuries later. The excavations revealed molds of the inhabitants, both people and animals, trapped by the ash flow as they attempted to flee or find shelter (Fig. 1). Some of the molds of humans even appear to preserve facial expressions of terror.

Mount Vesuvius returned to relative quiescence, only to become active again in 1631. It was frequently active from 1631 to 1944, and in this century it erupted in 1906, 1929, and 1944.

FIGURE 1
Molds of Pompeiians killed during the A.D. 79 eruption of Mount Vesuvius. (UPI/Bettmann)

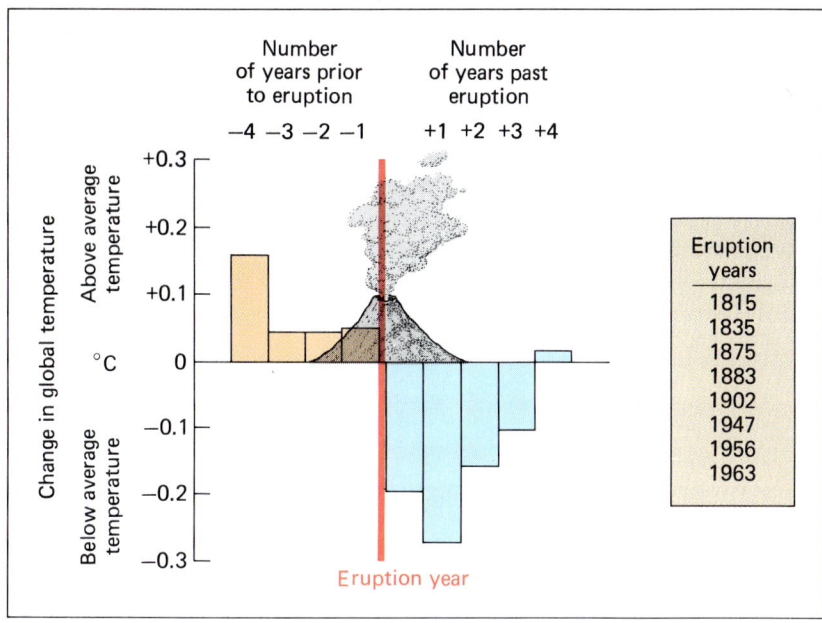

FIGURE 6–20
Temperature changes in the Northern Hemisphere in the 4 years immediately before and after eight large recent eruptions. (Michael Rampino, *Annual Review of Earth and Planetary Science* 16 (1988): 73–99)

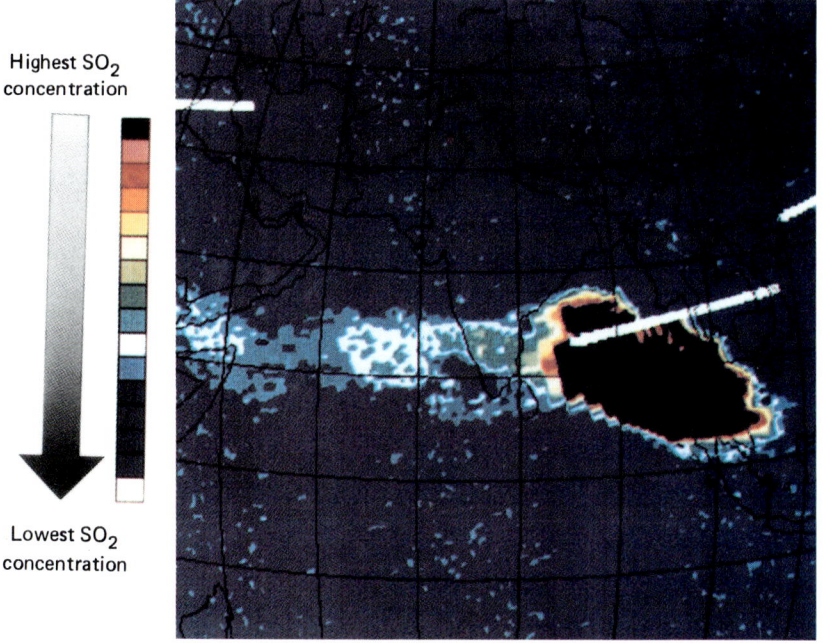

FIGURE 6–21
A false color image of the sulfur dioxide cloud from the eruption of Mount Pinatubo. The volcano is at the lower right side of the photograph. The image was taken on June 18, 1991, 2 days after the initial eruption. The cloud had already drifted 8000 kilometers. The white lines represent missing data, and the light blue specks scattered throughout the image are caused by "noise" in the experimental data. (NASA)

teorological models showing that high-altitude dust acts as an umbrella to prevent sunlight from reaching the Earth.

The 1991 series of eruptions of Mount Pinatubo produced the greatest ash and sulfur clouds since the beginning of the space age (Fig. 6–21). Satellite measurements show that the total solar radiation reaching the Earth's surface declined by 2 to 4 percent after the Pinatubo eruptions. The following two years, 1992 and 1993, were a few tenths of a degree Celsius cooler than the first half of 1991. Many meteorologists believe that the ash and sulfur clouds from Pinatubo caused the cooling. About one third of the atmospheric sulfur compounds from a volcano wash out every year. Thus, the cooling effect will remain with us, but with diminishing intensity, for a few years.

Historical eruptions have been small compared with some in the more distant past. What would happen if a true monster like one of the eruptions that created the Yellowstone calderas were to occur? As a worst-case scenario, dust and sulfur aerosols would block out so much sunlight that daytime would be just a little brighter than a full-moon night for nearly a year after the eruption. However, other calculations indicate that a bright, sunny day a year later would be about as bright as a normal overcast day. In either case, the effects on climate and agriculture would be significant, perhaps catastrophic. But on the positive side, the effects would only last a few years before normal climatological conditions returned.

Many scientists think that global climatic changes caused by major volcanic eruptions, or times of unusually frequent eruptions, were responsible for major extinctions, including that of the dinosaurs 65 million years ago. It may be that the abnormal volcanic episodes resulted from meteorite impacts or from the Earth's internal processes. Nevertheless, most scientists agree that volcanic activity can cause major climatic changes on Earth.

SUMMARY

Volcanic eruptions have the potential to alter our environment, kill humans, and destroy property more rapidly and profoundly than any other geological event. Eruptions occur when molten magma flows or explodes from the Earth's interior onto its surface.

Magma forms in three geologic environments: subduction zones, spreading centers, and mantle plumes. Heat, addition of water, and **pressure-relief melting** cause rocks to melt and form magma. Granitic magma typically solidifies within the Earth's crust, whereas basaltic magma usually erupts from a **volcano** at the surface. This contrast in behavior of the two types of magma is due to differences in silica and water contents.

Magma may flow onto the Earth's surface as **lava flows** or erupt explosively as **pyroclastic material**, or **tephra**. Fluid lava forms **lava plateaus** and **shield volcanoes**. A pyroclastic eruption may form a **cinder cone**. Alternating eruptions of fluid lava and pyroclastic material from the same vent form a **composite cone**. About 75 percent of the Earth's volcanic activity occurs along a circle of subduction zones in the Pacific Ocean called the **ring of fire**. When granitic magma rises to the Earth's surface, it usually erupts explosively, forming **ash-flow tuffs** and **calderas**. Eruptions of this type have caused widespread death and destruction throughout history.

Prediction of eruptions and reduction of risk of death and damage from eruptions fall into three categories: **long-term prediction** based on the eruptive history of a region or a particular volcano, **short-term prediction** based on monitoring, and **public education and development of policies** for governmental response to scientific predictions of eruptions.

Volcanic eruptions may have caused major long-term climatic changes in the geologic past and have the potential to do so again.

KEY TERMS

Pressure-relief melting 128
Polymer 130
Flood basalt 130
Lava plateau 130
Volcano 130
Vent 130
Fissure 130
Crater 130

Shield volcano 132
Cinder cone 133
Pyroclastic 133
Tephra 133
Volcanic ash 133
Volcanic cinders 133
Composite cone 133
Stratovolcano 133

Ash flow 140
Tuff 140
Caldera 141
Active volcano 141
Dormant volcano 141
Extinct volcano 141

REVIEW QUESTIONS

1. Describe several different ways in which volcanoes and volcanic eruptions can threaten human life and destroy property.

2. What has been the death toll from volcanic activity during the past 2000 years? During the past 100 years?

3. Why is it difficult to assess human death and property damage caused by major historical volcanic eruptions that occurred prior to about 100 years ago?

4. List and describe the major tectonic environments in which magma forms.

5. List and explain the most important factors in the melting of rock to form magma.

6. What happens to most basaltic magma after it forms?

7. What happens to most granitic magma after it forms?

8. Explain why basaltic magma and granitic magma behave differently as they rise toward the Earth's surface.

9. How much silica does the average granitic magma contain? How much does basaltic magma contain?

10. How much water does the average granitic magma contain? How much does basaltic magma contain?

11. Why does magma rise soon after it forms?

12. How do a shield volcano, a cinder cone, and a composite cone differ from one another? How are they similar?

13. Which type of volcanic mountain has the shortest life span? Why is this structure a transient feature of the landscape?

14. How does a composite cone form?

15. What is a volcanic neck? How is it formed?

16. What is the ring of fire? Why do most of the Earth's volcanoes occur in this zone?

17. Explain why and how granitic magma forms ash-flow tuffs and calderas.

18. Explain why some eruptions of granitic magma are extremely violent and threaten human life and property.

19. How does tuff form?

20. How does a caldera form?

21. How much pyroclastic material can erupt from a large caldera?

22. Explain why additional eruptions in Yellowstone Park seem likely. Describe what such an eruption might be like.

23. Describe how volcanic eruptions can cause major global climatic change.

DISCUSSION QUESTIONS

1. Discuss the various ways in which volcanic eruptions and other volcanic events create conditions that threaten life and property.

2. Many mountains are composed of granite that is exposed at the Earth's surface. Does this observation prove that granite forms at the Earth's surface?

3. Are basalt plateaus made of extrusive or intrusive rocks? How could you distinguish between a basalt plateau and an uplifted batholith?

4. Much of the surface of the Moon is scarred by meteor craters. However, some regions, called seas or maria, are flat expanses of rock. Outline a plausible geologic explanation for these maria. Include a chronology of events in your sequence.

5. Sometimes gases dissolve in a liquid, and at other times they form bubbles within it. Discuss the difference between these two conditions and its relevance to the geology of volcanoes.

6. Imagine that you detect a volcanic eruption on a distant planet, but you have no other data. What conclusions could you draw from this single bit of information? What types of information would you search for to expand your knowledge of the geology of the planet?

7. Explain why some volcanoes have steep, precipitous faces, but many do not.

8. Parts of the San Juan Mountains of Colorado are composed of granite plutons, and other parts are volcanic rock. Explain why these two types of rock are likely to occur in proximity.

9. Smith Rock in central Oregon is composed of tuff that rises above the Columbia Basalt Plateau. Explain how tuff and basalt could form in such proximity.

10. Compare and contrast the danger of living 5 kilometers from Yellowstone National Park with the danger of living an equal distance from Mount St. Helens. Would your answer differ for people who live 50 kilometers or 500 kilometers from the two regions?

11. Evaluate the potential for a destructive volcanic eruption in the place where you live.

12. Fill in the following table and use it to evaluate the danger to humans of various volcanic regions.

Region	Past Frequency of Eruptions	Past Violence of Eruptions	Proximity to Human Habitation	Type of Danger
Hawaii				
Yellowstone				
Long Valley				
Cascades				

INTERVIEW

A CONVERSATION WITH
Stephen Schneider

Stephen Henry Schneider was born in 1945 in New York City. He received his B.S. and M.S. in mechanical engineering and his Ph.D. in mechanical engineering and plasma physics at Columbia University. Following a brief stint with NASA's Goddard Space Center, Schneider in 1972 signed on with the National Center for Atmospheric Research in Boulder, Colorado, and from 1987 to 1992 served as the head of the Interdisciplinary Climate Systems Section there. In 1992 he became a professor of biological sciences and international studies at Stanford University.

The author of several books on climate and climate change, and coauthor of many others, Schneider has focused his research on climate modeling and the forecasting of the implications of climate change on our environment. He applies his comprehensive knowledge of the field as an editor of the international journal Climatic Change and of The Encyclopedia of Climate and Weather. He also serves as a science advisor and editorial board member for many other organizations. As a result of his nearly 100 articles in publications as diverse as Scientific American and Good Housekeeping, and his appearances on several television programs, in 1991 he was given the American Association for Advancement of Science/Westinghouse Award for Public Understanding of Science, and in 1992 was named a MacArthur Foundation Fellow for creativity.

Schneider has been engaged as an expert speaker at many universities, including the University of Virginia, Stanford University, University of Wisconsin–Madison, and Northwestern University. He has also served as consultant to environmental and government agencies around the world on energy policies and the implications of nuclear weapons. Included as one of Science Digest's "One Hundred Outstanding Young Scientists in America" in 1984, Schneider has proven worthy of the honor through his many contributions to both the scientific and global communities.

Where were you born and raised?

I grew up on the south shore of Long Island in a town called Woodmere. What I remember enjoying a lot about Long Island was going to a square-mile acre of woods, where I would run around and just enjoy streams and nature. One day a hurricane came by, and I went back to the forest, and half the trees were knocked down and it all had been disturbed. Even at the age of nine, I realized that ecology and climate and soils were all connected systems. Later on, I suppose, this was an emotional driver for getting involved in Earth sciences.

At what point did you become interested in, and then commit yourself to, atmospheric research?

I liked racing cars when I was in high school. So I went to Columbia University's engineering school to learn how to build the fastest race car. I ended up a mechanical engineering student studying fluid mechanics, then called engineering physics. So my initial practical notion to build race cars was completely dashed by my own choice to go into more theoretical parts of engineering.

My interest in atmospheric research began around 1970, when I attended the first Earth Day celebrations at Columbia. I was a little bored working on a plasma physics problem. I wanted to do something environmentally useful. At one of the Earth Day presentations, somebody (I think it was Barry Commoner) said, "What if pollution could change the climate—it could either heat it up if it's greenhouse gases or cool it down if it's sulfur injections," and I didn't believe it.

There was an atmospheric sciences course taught at Columbia by Ichtiaque Rasool. He went over the difference between Mars, Earth, and Venus. He said Venus is very hot with its very thick atmosphere, a super greenhouse effect. Mars is very cold with a very thin atmosphere, a weak greenhouse effect. Earth is right in the middle; water is what makes us different, and pollution could, in fact, dirty the greenhouse window. This was fascinating to me. Rasool said, "I will give you your postdoctoral fellowship if you'll leave plasma physics and convert to atmospheric science—to mathematically modeling the climate." So I took up Rasool's offer and became a postdoc at the Goddard Institute for Space Studies, a NASA laboratory at Columbia.

What did you do next?

In 1972 I went to the National Center for Atmospheric Research (NCAR) to help them start what we dubbed the Climate Project, and began doing climate research in Boulder. Climate involves the integration of materials from many disciplines—from oceanography, ecology, geography, meteorology, chemistry, and physics. Critics argued that such an integration was premature because each of these subtopics is not yet understood to the satisfaction of the practitioners. And my answer is that the world has to have the answer whether the disciplines are ready or not, so why don't we take halting steps to try to see how the system is integrated and connected—what we now call Earth systems science?

Much of your work in atmospheric research is based on computer modeling of the atmosphere. Can you tell us how that works?

Climate theory is not a perfect replica of nature; it's a model. And it's not a physical model in the laboratory, because you cannot make a physical model in a laboratory that includes enough of the important complexity. For example, the single most important component of the Earth's climate—the evaporation of water at the surface, and the recondensation in clouds—can't be done meaningfully in the lab. So the lab is very limited. Thus, the lab that you have to have for your model, literally, is a computer Earth. It's an Earth that we can pollute—it's an Earth we can modify by just changing something in the computer, and then resimulate the climate under these new conditions.

Tell us more about your modeling technique.

You break the atmosphere up into a bunch of boxes, or what we call grid squares. You break it up into a latitude–longitude grid of 4.5 degrees latitude and 7.5 degrees longitude. There are 20,000 of these grid squares around the world if this grid is piled into ten vertical layers.

Now, if you have a box 500 by 500 kilometers—say, the size of the state of Colorado typically—you can't, obviously, resolve an individual cloud. Nobody's ever seen a cloud the size of Colorado. So we've got a problem, because clouds are the venetian blinds of the Earth. They control, more than any other elements, the amount of solar energy absorbed, which is how much the planet is heated and the amount and distribution of infrared radiative energy that escapes back to space—the so-called greenhouse effect. They are more important radiatively than water vapor, carbon dioxide, chlorofluorocarbons, methane, and all those things that we argue about for the human-induced greenhouse effect.

Yet our models cannot explicitly resolve clouds, because they are too small. They are not 500 by 500 kilometers square. They are 5 by 5, maybe, or less. So what do we do? Some people say, "Well, your models are no darn good. Throw them away." But you don't need to know every detail in order to make a prediction about how something works. I mean, you don't have to predict what happens in every play to know that the 49ers would beat Stanford in a football game.

So the question, then, is: How can we get the *average effects* of clouds at the grid scale, even though we're not explicitly calculating individual clouds? We use a technique called parameterization, which is a short contraction for parametric representation.

Can you give an example of this so we can get an idea about the strengths and weaknesses of models?

Say the model produces humidity, temperature, and wind at the grid box. We know the relative humidity of, say, Colorado even if we don't know how many clouds there are. So we can make a rule, a parametric representation which says if the air is humid it's more likely to be cloudy than if the air is dry. Every farmer knows that. You don't need a Ph.D. in atmospheric science to know that. So then you have a parametric form which says cloudiness is equal to a number, a parameter, times the relative humidity.

But it's more sophisticated than that. Is it more likely to be cloudy if the air is rising or sinking? Think about it for a second. If air is rising, it would be drawing in the humid air from below. If air is sinking it's starting from a high place, which is dry, and coming down. Plus it's heating when it's going down, which tends to cause evaporation. And when it's rising, the air is expanding, which means it's cooling, which means it would tend to make condensation. So then you can say, "All right, my cloud parameterization will get more sophisticated. The cloudiness in the grid box is equal to a parameter times the relative humidity plus another parameter times the vertical velocity, both of which are calculated at the grid box average." The point is, you can get more and more sophisticated without resolving the explicit nature of the details. You are able to get the averages.

The problem is, how well are we doing this? And that's why there's an endless debate among the practitioners as to whether it's been done well, or medium well, or terribly. And how do we validate it? The Earth warms up somewhere between 1.5 and 4.5°C for almost all of the models that have been run for carbon dioxide doubling. Now, there are some that are a little hotter and some that aren't quite as hot. But the bulk of them fall in that range. So there's an uncertainty factor of 3 (between 1.5 and 4.5). So our models could be off by a factor of 3 global average temperatures. And this factor of 3 is largely because of differences in models' cloudiness parameterizations.

We're talking about clouds being the critical factor in changing the atmospheric temperature in one way or another. Why, then, are we all so interested in carbon dioxide and possibly other greenhouse gases?

The reason we worry about CO_2 is, we know beyond a doubt that if you double CO_2 you're going to trap something like 4 watts of energy over every square meter of Earth. Now, since the Industrial Revolution, we've added 25 percent more CO_2 and doubled the methane; that's known beyond doubt. The point is that, because we've added all these chemicals, we've trapped about 3 extra watts of energy [per square meter] in the last 100 years in the Earth's surface layers. That's not the debate. The debate is: Should 3 watts of energy warm up the Earth a quarter of a degree, 2 degrees, 4 degrees? And in order to answer that question, you've got to then say, "Now, what happens to the natural system? Do the trees get darker and greener, which would accelerate the warming? Or do they get slightly lighter and brighter, which would retard the warming? What do clouds do?" All these so-called "feedback processes" are endlessly debated and cause that uncertainty factor of 3 or so that we argue about.

The point is, there's positive and negative feedback in nature and in a physical and chemical world. And the sum of the positives and negatives, as to who wins, is not known. That's the debate. So the fact that clouds are a very large factor doesn't mean that that factor will necessarily be a positive or a negative feedback. We simply don't know. If you double CO_2, as is forecast sometime in the next century, from growing populations using fossil fuel at a higher and higher rate and continuation of deforestation, then what's going to happen is, we're going to trap another couple of watts of energy.

The debate, then, is translating that into temperature change. Is it going to end up just a half a degree to a degree more warming? Or are we going to get 3 or 4 degrees? In other words, are we going to have a mild change or a catastrophic change—because that's the range. And the debate goes on between mild and catastrophic without any clear and obvious answer right now, because we do not have either the theory or the measurements to validate the overall effect on the globe to much better than a factor of 3.

We know, based on your work and that of many other people, that CO_2 concentration in the atmosphere has increased. We know the absorptive qualities of CO_2. Basically it boils down to the fact that we know that increased CO_2 in the atmosphere is going to lead to global warming. Tell us more about the rate and magnitude of global warming.

The rate at which nature changes is on the order of 5°C. That's how much colder an ice age is than a so-called interglacial period, on a global average basis. Say it takes nature 5000 years to end the ice age, then we end up with a rate that's about 1°C per millennium as natural average rate of change.

The conservative forecast for global warming is a degree a century, or ten times faster than the natural average rate of change. The radical forecast says we could warm up 10 degrees in a century. That's a hundred times faster than natural rates. Pick a middle number, and it's something like 2 to 5 degrees' warming projected to occur over the next century. That would be something like 20 to 50 times faster than natural rates of change.

Why do you think there's such difficulty in producing a social, political, and human response to this problem even though the logical conclusion is that, if the global climate warms significantly in a short time, the effects will be more negative than positive?

It's not like chlorofluorocarbons and the ozone hole, because we all admit we can do without the spray cans and fluorocarbon refrigerators and change what's in the air conditioners in our cars. It might cost a few percent more, but very few people are willing to risk skin cancer and disruption of nature for a few chemicals that are substitutable.

When we're talking about global warming, we're talking about methane produced by agriculture, coal mining, natural gas, and landfills. We're talking about CO_2 produced by coal, oil, and gas, which the Third World is expecting to use to power their industrial revolution just the way the Western countries did in the Victorian era. And people do not want to hear that these mainstays of economic growth that permit growing populations to increase standards of living have side effects that might be dangerous.

So the problem is, it's not easy to get political agreement to slow down emissions, because it could be painful. Thus, people use the honest and legitimate scientific debate as an excuse to wait and see. They'll say, "Well, we're really not sure."

No honest scientist says that we know the answer. The degree of uncertainty ranges from mild to catastrophic, and we don't know where in this range the actual outcome is going to happen. And we're not going to know in the next 5, 10, or probably even 20 years. Whether to take the chance is not a scientific question per se, but a personal value judgment about which you fear more—investing *present* resources as the hedge against potentially catastrophic change, even though that change may not be so bad if you're lucky, or *not* investing present resources to find that you might be unlucky and, in not trying to slow it down, you have gotten a really whopping big dose that will be impossible to stop without irreversible damage.

So what should we do?

Study after study has shown that, depending upon how much we're willing to invest to buy these new materials, we can cut somewhere between 10 and 40 percent of our greenhouse emissions by replacing existing inefficient technology, and do it at below zero net cost. In other words, the amount of money you'll save in reduced energy costs actually will pay for itself without even counting the free extras, like reducing acid rain and the threat of disrupted ecosystems.

So my personal opinion is, why don't we do those things first that are freestanding and make sense anyway? Get rid of the fluorocarbons, because they not only trap 25 percent of the heat, but they also help cause ozone depletion. That's the easiest and the first thing to do. Let's use energy efficiently, because not only does it reduce global warming, but it also reduces acid rain and reduces dependency on foreign supplies, which are expensive and sometimes militarily dangerous to protect. It reduces local air pollution. It also makes our products more competitive in the long run. All of these things have to take place, and can, and are, but we need to push them harder. We can also pursue vigorously research and development on non-fossil-fueled energy systems. And then in the time frame of 5 to 15 years, while we're slowing down the rate at which we're making the system change, and therefore slowing down the rate at which nature will have to adapt, we can buy time to have the scientists determine what is likely to happen. To use the excuse that the thing is now very uncertain and therefore we shouldn't act is to say we should never have insurance, the police, or the military.

What advice would you give to students taking Earth science?

What students need to recognize is, we're not talking about the environment *or* the economy. What we're talking about is, there are trade-offs between the two, but that if we use our brains and hands cleverly we can find ways to have the economy grow in a much less environmentally destructive way.

Problems are increasingly cross-cutting. Environmental problems are only one example. Health problems are another example. And the way we're organized is not set up to deal with problems that crosscut. In universities we learn disciplines; in governments we deal with departments. Real issues, like global warming, involve solutions that have a little bit in population, forestry, agriculture, energy, and foreign policy. In order to manage that problem, each group has to do a little, and *it has to be coordinated.* And therefore an Earth systems science approach, such as this book is trying to take by integrating the disciplines, is the only way to be in tune with the way the world's real problems are.

PART 3

Surface Processes

7 Soil
8 Mass Wasting
9 Fresh Water: Streams, Ground Water, Lakes, and Wetlands
10 Human Use of Water
11 Waste Disposal and Fresh-Water Pollution
12 Coastlines and Coastal Pollution
13 Glaciers and Ice Ages
14 Deserts and Desertification

The amount of unspoiled natural land on the Earth is rapidly diminishing. A desert landscape in Arizona.

CHAPTER 7

Soil

7.1 Weathering
7.2 Soil
7.3 Soil Nutrients
7.4 The Hubbard Brook Experimental Forest
7.5 Erosion
7.6 Agriculture and Soil Fertility
7.7 Reducing Cropland Erosion
7.8 Soil Erosion in Central China and in the American Dust Bowl
7.9 Croplands and World Food Supply
7.10 Soil Engineering

Plant roots reach into the soil to draw up water and nutrients. Most animals are herbivores and eat plants. Others are predators and eat herbivores. When plants and animals die, their remains fall to the Earth and decay, thus recycling nutrients back to the soil. Plants and animals also rely on soil for shelter and support. Roots anchor plants, gophers burrow into the root zone, birds nest in trees that grow in soil, deer find shelter in the cool forest, and people use trees for building materials. People also build their homes, factories, and roadways on and below the ground surface. Despite its importance to terrestrial life, soil is a thin, fragile, and vanishing resource (Fig. 7–1).

Human activity is responsible for the fact that soil is disappearing more rapidly than it is being replenished. Soil is eroding more rapidly than it is forming on 35 percent of the world's croplands. About 23 trillion kilograms of soil (about 25 billion tons) are lost every year. The soil lost annually would fill a train of freight cars long enough to encircle our planet 150 times.

In the United States, we have lost about one third of the topsoil that existed when the first European settlers arrived. The American Farmland Trust estimates that the current average soil erosion rate in the United States is about 30 tons per hectare[1] per year. To put this erosion rate in perspective, it will lead to complete loss of topsoil from the United States in about 120 years. A total of more than 6 billion tons of soil per year is lost from the United States, according to the U.S. Department of Agriculture.

In this chapter we will consider how soil forms, how human activity increases erosion, and what factors make soil a stable or unstable platform for construction.

[1]One hectare equals 2.47 acres.

Farmers cut terraces into the steep hillsides surrounding Kangding in western China.

FIGURE 7–1
Soil is a thin, fragile, and vanishing resource. McKenzie River, Yukon Territory.

FIGURE 7–2
In these Roman ruins, mechanical weathering has broken and toppled large building stones, and chemical weathering has rounded and pitted the stones. *(Italian State Tourist Office)*

7.1 Weathering

Weathering is the decomposition and disintegration of rocks and minerals at the Earth's surface by both mechanical and chemical processes. This process converts solid rock to gravel, sand, clay, and soil but involves little or no movement of the decomposed rocks and minerals away from the places where they form. **Erosion** is the removal of the weathered rocks by flowing water, wind, glaciers, or gravity. Thus, weathering and erosion are different but related processes. After water, wind, ice, or gravity erodes weathered material, those same agents **transport** it downslope and finally **deposit** it in a stream bed, a beach, or some other new environment.

The physical and chemical environment at the Earth's surface is corrosive to most materials. A knife rusts when it is left out in the rain. For similar reasons, rocks decompose naturally. Thus, over the centuries, stone cities have fallen into ruin. Two types of changes destroy ancient cities and weather rocks in natural environments. First, large stones break into smaller fragments (Fig. 7–2). **Mechanical weathering** is the physical disintegration of rock into smaller pieces. For example, plant roots may grow in cracks in building stones, enlarging the cracks and eventually toppling the walls. In cold climates, water expands as it freezes in cracks, fracturing rocks and reducing stone buildings or cliffs to rubble. Such mechanical processes break rocks into smaller pieces, but they do not alter the chemical compositions of the rocks and minerals.

Abrasion is the mechanical wearing away of rock by wind or water. Wind and water are not abrasive, but they transport sand and other abrasive sediment. Thus, a stream loaded with sediment or a desert dust storm can be thought of as flowing sandpaper. Many streams are lined with rounded rocks whose edges have been chipped away by abrasion (Fig. 7–3).

If you look closely at a building stone in an ancient city, you may see a second type of disintegration. The stone's surface may be pitted and discolored, and the rock may be soft and earthy rather than hard and solid, as it was when quarried. **Chemical weathering** occurs when air and water attack rocks chemically. The changes are similar to rusting because a chemically weathered rock contains different minerals and has a different chemical composition from the original rock. Solution, oxidation, and hydrolysis are the three most important kinds of chemical weathering.

Solution

Natural water is never pure. Atmospheric carbon dioxide reacts with rainwater to form a weak acid called carbonic acid. This acid chemically attacks rocks and minerals.

FIGURE 7–3

Abrasion rounded these boulders in a stream bed in Yellowstone National Park, Wyoming.

In addition, some air pollutants react in moist air to form strong acids. Sulfur dioxide, released mainly when coal is burned, reacts to form sulfuric acid. Atmospheric nitrogen reacts in hot automobile exhaust to form nitric acid. Both of these acids fall to Earth as **acid precipitation** and accelerate natural chemical weathering.

When water flows across or beneath the surface of the Earth, it dissolves ions from minerals. In some instances, these ions make the water acidic; in other cases, the water becomes basic. In either case, the resulting **solution** has a greater ability to dissolve rocks and minerals.

Oxidation

Oxygen, one of the most common atmospheric gases, also causes chemical weathering. Many elements react rapidly with molecular oxygen, O_2. In our everyday experience, iron reacts with water and oxygen to form rust. Rusting is one manifestation of a more general process called **oxidation**.[2]

About 21 percent of the Earth's atmosphere is oxygen, and as a result, oxidation is common. It usually turns useful material to waste. Wood burns (oxidizes) to form ashes, iron oxidizes to rust, and so on. Rocks and minerals oxidize when their iron reacts with oxygen.

Hydrolysis

Feldspar and quartz are two of the most abundant minerals in the rocks of continental crust. Water reacts with

---

[2]Oxidation is properly defined as the loss of electrons from a compound or element during a chemical reaction. In the weathering of common minerals, this usually occurs when the mineral reacts with molecular oxygen.

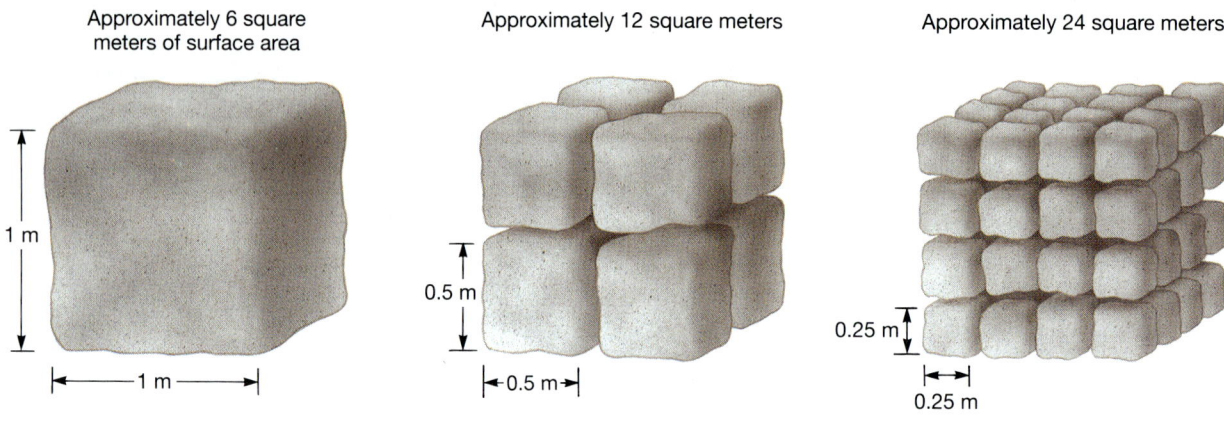

FIGURE 7–4
Mechanical weathering produces more surface area for chemical weathering to attack.

feldspar. In the process, the feldspar decomposes and its decomposition products combine with some of the water to form clay minerals. This process is called **hydrolysis** because it involves the chemical addition of water. Feldspar makes up about half of continental rocks, and, as a result, clay is an abundant weathering product. In contrast with feldspar, quartz resists chemical weathering. When bedrock weathers, the quartz grains are released from the solid rock but remain mostly unchanged. As a result of these processes, bedrock breaks into clay and quartz-rich sand as it weathers, and thus clay and sand are two important components of soil.

Mechanical and chemical weathering reinforce each other. For example, chemical processes generally act on the surface of a solid object. Therefore, a chemical process speeds up if the surface area increases. Think of a burning log; the fire starts on the outside and works its way inward. If you want the log to burn faster, split it in half to increase its surface area. Mechanical weathering cracks rocks, exposing more surface area for chemical agents to work on (Fig. 7–4).

7.2 Soil

As a result of weathering, a layer of loose rock fragments mixed with abundant clay and sand overlies bedrock nearly everywhere on land. This material is called **regolith**. In engineering and construction, the words *soil* and *regolith* are interchangeable. However, soil scientists define **soil** as upper layers of regolith that support plant growth. That is the definition we use here.

Components of Soil

Soil is a mixture of mineral grains, organic material, water, and gas. The mineral grains include clay, silt, sand, and rock fragments. Clay grains are so small and closely packed that water and gas do not flow through them readily. Pure clay is so impermeable that plants growing in clay soils often become waterlogged and suffer from lack of oxygen. In contrast, water flows rapidly through sand.

If you walk through a forest or prairie, you will find bits of leaves, stems, and flowers on the surface. This material is called **litter**. Litter is any dead plant or animal matter that has retained its original form (Fig. 7–5). When litter decomposes so that the origins of its individual pieces can no longer be determined, it becomes **humus**. Humus is an essential component of most fertile soils. Scoop up some forest soil or rich garden soil with your hand. Soil rich in humus is light and spongy and readily absorbs water. Its ability to soak up water and then dry out causes it to swell after a rain and shrink during a dry spell. This alternate shrinking and swelling keeps the soil loose, allowing roots to grow easily. A rich layer of humus also insulates deeper soil from heat and cold and retains water. The most fertile soil is **loam**, a mixture of sand, clay, silt, and generous amounts of humus.

Soil Profiles

Undisturbed soil consists of several layers, or **soil horizons** (Fig. 7–6). The uppermost layer of mature soil is the **O horizon**, named for its organic component. This layer is mostly litter and humus with a small proportion

FIGURE 7–5
Litter is organic matter that has fallen to the ground and started to decompose, but still retains its original form.

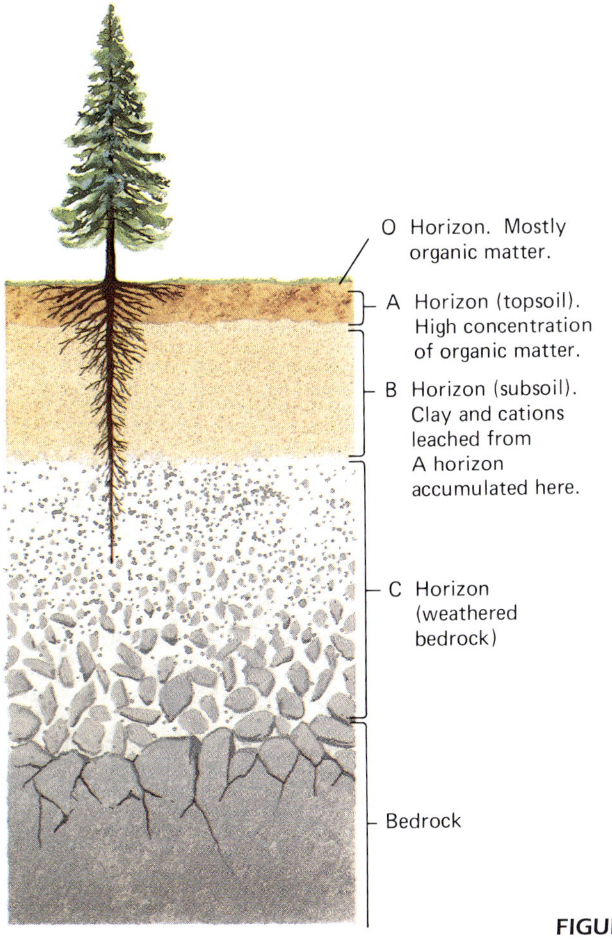

FIGURE 7–6
(A) A well-developed soil commonly shows several distinct horizons. (B) Soil horizons are often distinguished by color and texture. The dark upper layer is the A horizon; the white lower layer is the B horizon.

of minerals. The next layer down, called the **A horizon**, is a mixture of humus, sand, silt, and clay. The upper part of soil, including both O and A horizons, is often called **topsoil**. A kilogram of average fertile topsoil contains about 30 percent by weight organic matter, including 2 trillion bacteria, 400 million fungi, 50 million algae, 30 million protozoa, and thousands of larger organisms such as insects, worms, and mites.

The third layer, the **B horizon**, or subsoil, is a transitional zone between topsoil and weathered parent rock below. Roots and other organic material occur in the B horizon, but the proportion of organic matter is low. The deepest layer, called the **C horizon**, is partially weathered bedrock with little or no organic matter. It lies directly on unweathered parent rock.

When rain falls on soil, it sinks into the O and A horizons. As it seeps downward through the topsoil, it partially dissolves minerals and carries the dissolved ions to lower layers. This downward movement of dissolved material is called **leaching**. The A horizon is commonly sandy because the water also carries clay downward but leaves the sand grains behind. Because materials are removed from the A horizon, it is called the **zone of leaching**.

The dissolved ions and clay leached from the A horizon accumulate in the B horizon, which is therefore called the **zone of accumulation**. This layer retains moisture because of its high clay content. Although moisture retention may be beneficial, too much clay may form a dense, waterlogged soil.

Soil-Forming Factors

Why is one soil rich and another poor, or one sandy and another loamy? What factors contribute to the character of a specific soil? Five **soil-forming factors** control how soil develops as parent material weathers: parent rock, time, development of organic matter, slope angle and aspect, and climate.

Parent Rock

The type of parent rock exerts a strong influence on soil. Granite, the most abundant rock type on continents, contains mostly feldspar and quartz. When it weathers, its feldspar converts to clay, but the quartz grains resist chemical weathering and become sand. Thus, sandy soil commonly forms on granite. In contrast, basalt contains much feldspar but no quartz, and soil formed from basalt is likely to be clay rich but not sandy. In addition to texture, the parent material provides nutrients to soil, so nutrient abundance or deficiency depends in part on the chemical composition of the parent rock.

Time

In a geologically young soil, weathering of feldspar and other minerals may be incomplete, and so the soil is likely to be sandy. As a soil matures and more feldspar decomposes, the soil's clay content increases. Thus many young soils are sandy and consist mostly of slightly weathered minerals inherited from bedrock.

Addition of new material over time also changes the character of soil. A stream may deposit layers of sand or mud on its floodplain, or wind may blow in dust. The sediment may mix with the soil, changing its composition and texture, or may bury it completely.

Development of Organic Material

The most fertile soils are those of prairies and forests in temperate latitudes where plant growth and decay are balanced so that thick layers of humus form. In the tropics, organic material decays so rapidly that little humus accumulates (Fig. 7–7). The Arctic is so cold that plant growth and decay are slow. Therefore, litter and humus form slowly and Arctic soils contain little organic matter.

Slope Angle and Aspect

Soil generally migrates downslope in response to gravity. Therefore, if all other factors are equal, soil is thinner and poorer on a hillside, and the valley floor has the deepest and richest soil.

Aspect is the orientation of a slope with respect to compass direction. Exposure of a slope to the Sun affects soil formation. For example, in the semiarid western United States, thick soils and dense forests cover the north-facing hill slopes, but thin soils and grass dominate southern exposures. The reason for this difference is that, in the Northern Hemisphere, more water evaporates from the sunny southern slopes. Therefore, fewer plants grow, weathering occurs more slowly, and soil development is retarded. The moister northern slopes weather more deeply to form thicker soils.

Climate

Rain seeps downward through soil, but other processes pull the water back upward. Roots suck soil water toward the surface, and water near the surface evaporates. In addition, because water is electrically attracted to soil particles, **capillary action** draws water upward if the pores in the soil are small enough. Capillary action can be

FIGURE 7-7
(A) The tropical soil of Costa Rica supports lush growth but organic material decays so rapidly that little humus accumulates. (B) The Arctic soil of Baffin Island, Canada, supports sparse vegetation and contains little organic matter.

demonstrated by placing the corner of a paper towel in water and watching the water rise (Fig. 7–8).

During a rainstorm, water percolates down through the A horizon, dissolving soluble ions such as calcium, magnesium, potassium, and sodium. In regions where rainfall is moderate or high, the water continues downward, carrying the leached ions away from the soil. However, in deserts and semiarid regions, rainstorms are usually brief and followed by sunshine, which dries out the upper soil layers. In this environment, capillary action and plant roots then draw the water back up toward the surface, where it evaporates. As it evaporates, many of its dissolved ions precipitate in the B horizon, encrusting the soil with salts. A soil of this type is a **pedocal** (Fig. 7–9). Such a process often deposits enough calcium carbonate to form a rock-hard cement called **caliche** in the soil. The Greek word *pedon* means soil; the word *pedocal* is a composite of *pedon* and the first three letters of calcite, *cal*.

In a wet climate, ground water leaches soluble ions from both the A and B horizons. The less soluble elements such as aluminum, iron, and some silicon remain behind, accumulating in the B horizon to form a soil type called a **pedalfer** (Fig. 7–9). The subsoil in a pedalfer commonly is rich in clay, which is mostly aluminum and silicon. Iron and aluminum accumulate in the lower layers of a pedalfer; hence the prefix *ped* is followed by the chemical symbols Al for aluminum and Fe for iron. Iron oxides commonly color pedalfers red.

In regions of very high rainfall, such as a tropical rainforest, the great amount of water seeping through the soil leaches away nearly all the soluble elements. Only insoluble aluminum and iron minerals remain (Fig. 7–9). Soil of this type is called **laterite**. Laterites are often colored rust-red by iron oxide (Fig. 7–10). A highly aluminous laterite, called **bauxite**, is the world's main source of aluminum ore.

FIGURE 7-8
Capillary action causes colored water to soak upward through the pores in a paper towel.

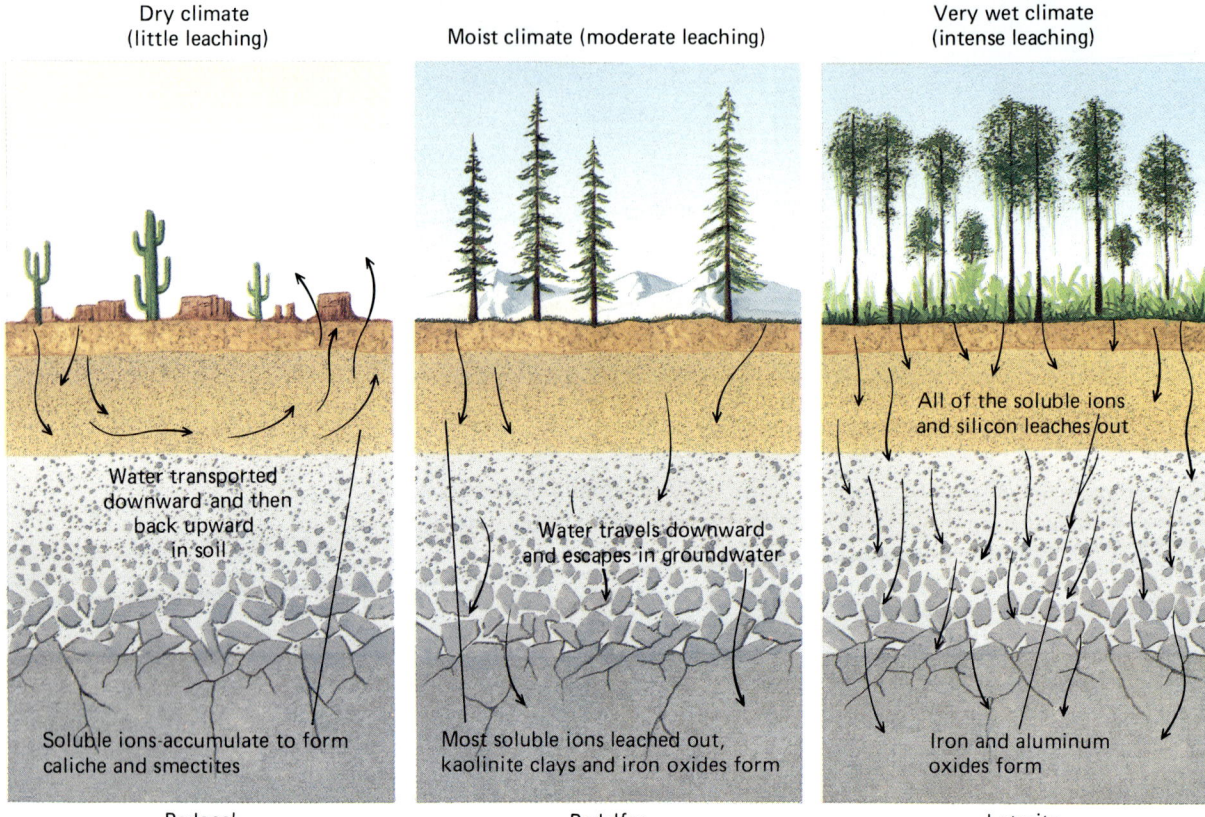

FIGURE 7–9
Pedocals, pedalfers, and laterites form in different climates.

FIGURE 7–10
Iron oxide colors this Georgia laterite soil red.

Tropical rainforests support a rich and varied community of plants and animals, but paradoxically their laterite soils are poor in both humus and nutrients, and thus are unsuitable for farming. The hot, moist environment promotes rapid decay of litter, minimizing the amount of humus that forms. In addition, plant growth is so rapid that the growing plants absorb most of the nutrients released by the decaying litter. The great amount of water seeping through the soil leaches the remaining nutrients from the root zone. As a result, the soil contains little humus and nutrients.

Since 1945, more than half of the Earth's tropical rainforests have been cut for timber and to clear land for farming. When the rainforest is cut, tropical rains quickly leach the small amount of nutrients from the soil. Farmers can grow crops for a few years, but on most cleared land leaching and crop removal soon exhaust the nutrients and the soil becomes useless for agriculture. Because the nutrients have been lost, it is then impossible for the natural rainforest to regenerate, and the abandoned soil lies bare and vulnerable to erosion (Fig. 7–11).

FIGURE 7–11
This deforested hillside in the tropics of Costa Rica is vulnerable to erosion.

Soil Classification

The classification of soils into pedalfers, pedocals, and laterites is based primarily on climate. However, as discussed in the text, many other factors contribute to soil characteristics. For example, limestone bedrock contains no iron or aluminum, so it will never produce a soil rich in these elements, no matter how much rain falls and leaching occurs. In 1965, the Soil Conservation Service adopted the U.S. Comprehensive Soil Classification System based on the following ten categories.

Soil Order	Description
Oxisols	Tropical soils (laterites) that have been intensely leached; found in tropical rainforests.
Ultisols	Soils that have formed under warm, moist conditions but not as extreme as for oxisols; found in southeastern U.S.
Alfisols	Soils with low humus content and a clay-rich B horizon, indicating extensive leaching but less than in ultisols; found in the Mississippi Valley.
Mollisols	Prairie soils with a thick layer of humus. The grain-growing breadbasket of the Midwest lies on mollisols, as do some forests.
Spodosols	Acidic soils that form in cool, moderately moist climates. The A horizon is generally sandy, and the B horizon contains leached organic matter and clay.
Aridosols	Alkaline soils, frequently with caliche layers (pedocals).
Vertisols	Clay-rich soils that expand and contract with moisture content. The Lakewood subdivision was built on vertisols.
Histosols	Boggy soils that form in regions such as the Florida Everglades and the Mississippi River delta.
Entisols	Poorly developed soils with little vertical profile. They often form in regions with little vegetation, such as deserts, tundra, and areas recently disturbed by glaciers or volcanic eruptions.
Inceptisols	Very young soils with weakly developed layering and little mineral alteration from the parent rock.

Rates of Soil Formation

The rate at which soil forms varies greatly from place to place and over time. The variations result mostly from geographic and temporal differences in the soil-forming factors described earlier. As a result, estimates of soil formation rates vary widely and are uncertain. Nevertheless, many geologists and soil scientists use 0.8 millimeters per year as a value for the average global rate of soil development. This amount translates to slightly more than 12 tons per hectare per year.

7.3 Soil Nutrients

Soil nutrients are chemical elements needed by plants. Only 16 of the 88 elements that occur naturally in the Earth's crust are essential to plants (Table 7–1). Of these nutrients, the three needed in the largest quantities are phosphorus, nitrogen, and potassium. Animals require all of the elements listed in Table 7–1 except boron. In addition, they need sodium, iodine, selenium, and cobalt.

Nutrients occur in many different forms in the atmosphere, the hydrosphere, the biosphere, and the solid Earth. The movement of a nutrient through these realms is called a **biogeochemical cycle**. This name recognizes that biological, geological, and chemical processes affect nutrient transport. It also recognizes that nutrients are recycled within ecosystems.

The Phosphorus Cycle

Animals need phosphorus to make shells, bones, teeth, and cell membranes and to regulate cell functions. Plants require phosphorus for cell membranes and metabolism.

TABLE 7–1 Nutrients Essential for Plant Growth

Macronutrients from air and water	
Carbon	
Oxygen	The basic building blocks of all organic tissue
Hydrogen	
Macronutrients from soil	
Nitrogen	Most critical element for growth of plant proteins
Phosphorus	Important for metabolism and development of cell membranes
Potassium	Maintains plant cell permeability
Calcium	Important for plant cell walls
Magnesium	Important for production of chlorophyll
Sulfur	Required for synthesis of plant vitamins and proteins
Micronutrients from soil	
Iron	
Copper	
Manganese	
Boron	
Zinc	
Chlorine	
Molybdenum	

Even though many rocks and minerals contain phosphorus, plants and animals cannot use this element when it is bound in a mineral. When a phosphorus mineral weathers chemically, it releases phosphorus into the soil, usually in the form of phosphates, $(PO_4)^{3-}$. Plants absorb phosphates through their roots. Then animals eat the plants, predators and scavengers eat the herbivores, and decay organisms decompose organic tissue after plants

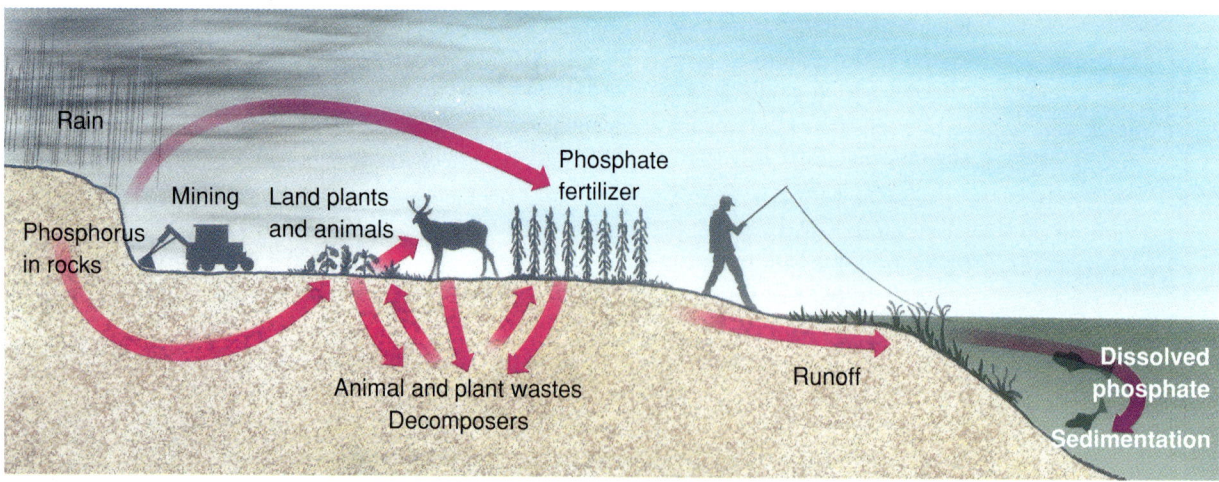

FIGURE 7–12
The phosphorus cycle.

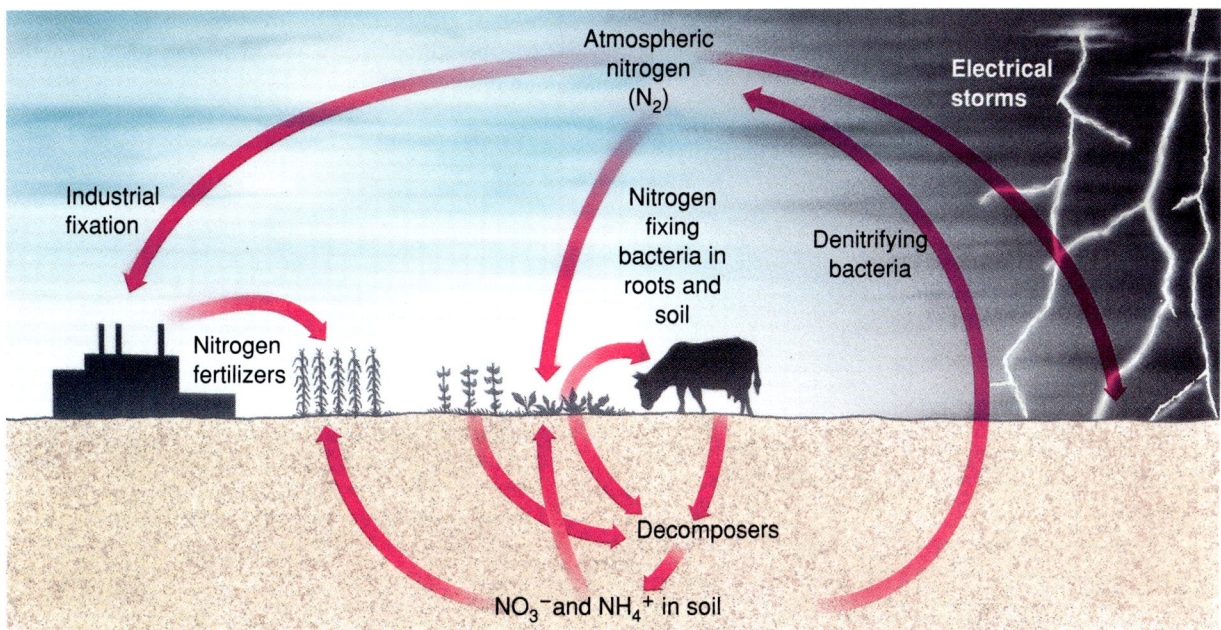

FIGURE 7-13
The nitrogen cycle.

and animals die. Thus phosphorus moves through the ecosystem, changing forms many times (Fig. 7-12). Small amounts of phosphorus escape from the ecosystem by dissolving in water and flowing to the ocean. Trivial amounts of phosphorus return to land in the droppings of ocean birds or in fish caught by terrestrial animals. In most natural ecosystems, biological transfers of phosphorus are much faster than weathering and erosion. As a result, large quantities of phosphorus recycle in the day-to-day exchanges among living organisms and only small amounts enter or leave the ecosystem. Thus an ecosystem thrives by conserving and recycling its phosphorus.

The Nitrogen Cycle

Even though our atmosphere is about 78 percent nitrogen, few organisms can use atmospheric nitrogen (N_2) directly. Most plants can absorb this vital nutrient only when it is in the form of nitrites (NO_2^-), nitrates (NO_3^-), or ammonium ions (NH_4^+). Soil bacteria convert atmospheric nitrogen to the forms usable by plants (Fig. 7-13). This process is called **nitrogen fixing**. Terrestrial plants either absorb the nitrogen compounds directly from the nitrogen-fixing bacteria, or indirectly from wastes that they excrete into the soil. Lightning also converts atmospheric nitrogen and oxygen to nitrogen compounds that are usable by plants. Weathering of nitrogen minerals releases some nitrogen to the soil, although this process is not as important to the nitrogen cycle as weathering is to the phosphorus cycle. Once nitrogen is converted to biologically usable forms, plants and animals conserve and recycle it within the ecosystem.

Denitrifying bacteria break up nitrogen-containing organic molecules for their own food. In the process, they convert the nitrogen compounds back to N_2 gas, which returns to the atmosphere. These bacteria live in anaerobic (without oxygen) conditions in mud at the bottom of some lakes, in bogs and estuaries, and in parts of the sea floor. Terrestrial ecosystems also lose some nitrogen to erosion.

CASE HISTORY

7.4 The Hubbard Brook Experimental Forest

Ecologists have studied water and nutrient cycling in the Hubbard Brook Experimental Forest in New Hampshire (Fig. 7-14). The ecosystem consists of a series of small valleys, each drained by a single creek. The bedrock beneath the valleys is impermeable, and water cannot escape by seeping underground. Researchers built concrete dams across several of the creeks and anchored the dams on bedrock. They then measured the amount of water flowing from each valley and the amounts of sediment and dissolved nutrients carried by each creek.

In an undisturbed forest, some of the rainwater soaks into the soil, some evaporates from the soil surface, and some evaporates from plant leaves in a process called **transpiration**. As shown in Figure 7-15, in a healthy forest only 25 percent of the rainwater soaks into the soil or is lost to runoff, whereas 75 percent returns to the atmosphere through transpiration and evaporation. Thus a forest recycles much of its water back into the atmosphere.

FIGURE 7–14
One forest watershed was completely devegetated, while adjacent watersheds were left in their natural states, to measure the effects of forests on soils and streams in the Hubbard Brook Experimental Forest, New Hampshire.
(Hubbard Brook Experimental Forest)

Scientists cut all the trees and shrubs in one Hubbard Brook valley and sprayed its soil with herbicides to prevent regrowth. With no plants to absorb water, most of the rainwater flowed downslope quickly and stream flow increased by 40 percent over that in an adjacent undisturbed drainage. In addition, the amounts of dissolved nutrients and sediment carried by the stream draining the devegetated valley increased.

The scientists at Hubbard Brook learned that the soils of the deforested valley lost nutrients six to eight times faster than the adjacent undisturbed watershed did (Fig. 7–16). The great increase in water runoff carried large amounts of nutrients from the soil that otherwise would have been conserved by the vegetation. Scientists have recorded similar results for other forests, grasslands, and wetlands. These experiments demonstrate the interrelationships between soil and vegetation. Plants maintain healthy soil, which in turn, sustains them.

FIGURE 7–15
The water cycle in a natural forest ecosystem.

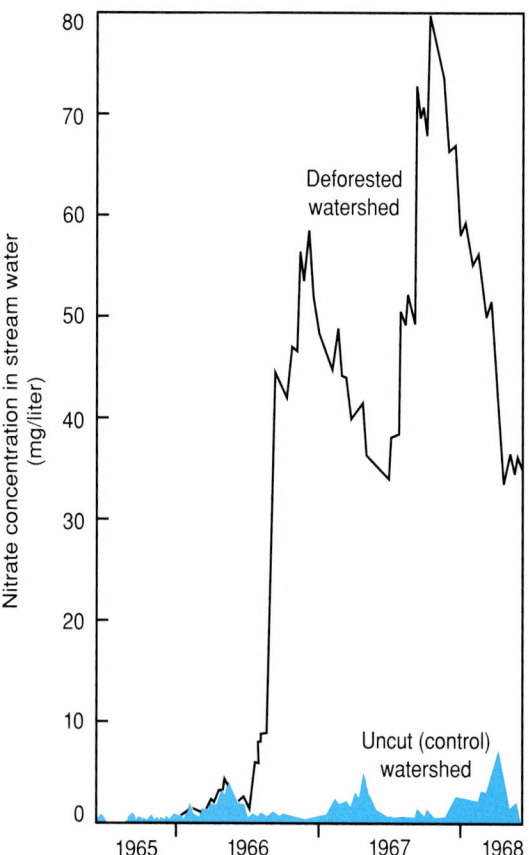

FIGURE 7–16
The Hubbard Brook experiment showed that a completely devegetated forest valley lost its soil nutrients up to six times faster than a comparable undisturbed adjacent watershed.

7.5 Erosion

Weathering decomposes bedrock to create soil at the Earth's surface. The process has gone on throughout most of geologic time. However, soil has not accumulated and thickened throughout geologic time. If it had, the Earth would be covered by a mantle of soil hundreds or thousands of meters thick, and rocks would be unknown at the Earth's surface. Instead, flowing water, wind, and glaciers erode soil as it forms (Fig. 7–17). In addition, some soil and bedrock simply slide downhill under the influence of gravity. All forms of erosion combine to remove soil about as fast as it forms on land undisturbed by human activities. For this reason, soil is only a few meters thick or less in most parts of the world.

Once soil erodes, the sediment begins a long journey as it is carried downhill by streams, glaciers, wind, and gravity. During the journey, the sediment may come to rest in a stream bed, a sand dune, or a lake bed, but only for a short time. Sooner or later most sediment erodes again and is carried farther downhill, until finally it is deposited where the land meets the sea. There it remains and is buried by younger sediment until it lithifies to form sedimentary rocks.

Human Activity and Erosion

In most ecosystems, vegetation and litter cover the soil. (Deserts, seacoasts, and land recently disturbed by glaciers and fire are exceptions.) However, human activity frequently disturbs the surface and exposes bare soil to the

FIGURE 7–17
Flowing water has eroded gullies in this hillside.
(Don Hyndman)

FIGURE 7–18
Logging near the China–Tibet border has exposed forest soil to erosion.

elements. Farmers plow the soil, removing all protective cover. Logging machinery digs ruts on steep hillsides, and logging roads often create open scars (Fig. 7–18). Miners leave huge, unstable piles of waste rock and soil. Construction machinery cuts roads and levels homesites on hillsides.

Whenever rain falls on bare soil, each raindrop knocks loose a tiny grain, moving it slightly downslope. Later drops continue to bump the grain farther downhill. Much of the rain soaks into the soil, but if the storm lasts long enough, water begins to flow over the bare soil in a thin, broad sheet that continues to move grains downslope. Eventually the sheet of water flows into a streambed, carrying the sediment with it. This process, called **rainwash**, can move great amounts of sediment from bare soil into a stream channel. In addition, when soil is exposed, its humus oxidizes more rapidly than it does in an undisturbed state. This oxidization destroys humus and thus reduces the soil's ability to retain moisture and nutrients.

Although human activities greatly accelerate soil erosion, soil continues to form by weathering at its usual slow, natural pace. However, accelerated erosion caused by human activity is frequently much faster than soil formation, leading to net soil loss.

The World Resources Institute estimates that, in the past 45 years, human abuse has degraded 11 percent of the Earth's soil to the point that its ability to support plants has been at least partly destroyed (Table 7–2).

TABLE 7-2 Human-Induced Land Degradation Worldwide, 1945 to Present

Region	Overgrazing	Deforestation	Agricultural Mismanagement	Other[1]	Total	Degraded Area as Share of Total Vegetated Land
	(million hectares)					(percent)
Asia	197	298	204	47	746	20
Africa	243	67	121	63	494	22
South America	68	100	64	12	244	14
Europe	50	84	64	22	220	23
North and Central America	38	18	91	11	158	8
Oceania	83	12	8	0	103	13
World	679	579	552	155	1965	17

[1]Includes exploitation of vegetation for domestic use (133 million hectares) and bioindustrial activities, such as pollution (22 million hectares).

SOURCE: Worldwatch Institute, based on "The Extent of Human-Induced Soil Degradation," Annex 5 in L.R. Oldeman et al., *World Map of the Status of Human-Induced Soil Degradation* (Wageningen, Netherlands: United Nations Environment Programme and International Soil Reference and Information Centre, 1991).

The affected area is as large as India and China combined, about 1.2 billion hectares. Another 0.7 billion hectares of soil have been "lightly degraded." A total of 35 percent of the world's cropland and 17 percent of all vegetated land on Earth has suffered some human-caused soil deterioration. The deterioration has resulted mainly from erosion, nutrient depletion, loss of soil humus, and salt accumulation in soils (described in Chapter 10). The same study estimates that agricultural misuse caused 28 percent of the damage, logging caused 30 percent, overgrazing was responsible for 35 percent, and mining and construction led to the remaining 7 percent (Fig. 7-19).

FIGURE 7-20
Plowed soil is unprotected by vegetation.

7.6 Agriculture and Soil Fertility

A bare, plowed field is vulnerable to erosion (Fig. 7-20). During the growing season, the soil is protected if cover crops such as alfalfa and wheat are grown, but if a row crop such as corn or vegetables is planted, the rows of crop plants alternate with strips of unprotected soil (Fig. 7-21). If weeds sprout in the strips, the farmer roots them out with a tractor or kills them with herbicides. Similarly, intensive grazing strips away plants that normally protect the soil (Fig. 7-22). Runoff from bare agri-

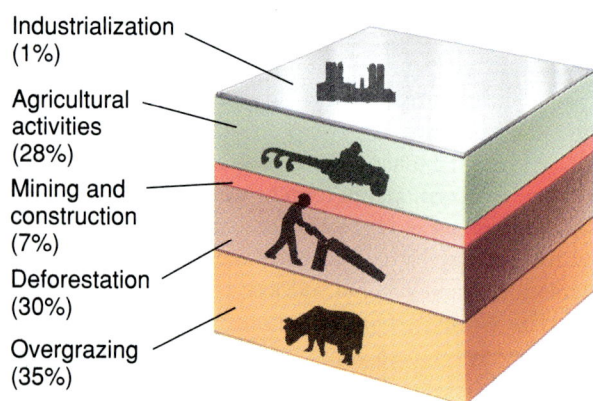

FIGURE 7-19
Causes of world soil degradation.

7.6   Agriculture and Soil Fertility   171

FIGURE 7–21
Row crops leave strips of bare soil.

FIGURE 7–23
In China, human waste is applied directly to crops from carts such as this one.

cultural soil removes nutrients, just as runoff from the deforested valley in Hubbard Brook removed nutrients.

All crops contain nutrients derived from the soil. When a farmer harvests a crop and ships it away, the nutrients are lost from the soil. In a natural ecosystem, new plant growth and decay replenish soil humus as old humus oxidizes. But when crops are exported to market, the new plant growth is removed and soil humus is gradually depleted.

As a result, croplands lose soil particles, nutrients and humus faster than they are replaced by natural mechanisms. Farmers must replace the lost nutrients by fertilizing. In traditional agriculture, farmers fertilized with manure, crop residues, or other plant and animal matter (Fig. 7–23). These fertilizers contain both nutrients and organic matter that decomposes to humus. However, plant and animal matter is often heavy and expensive to transport and spread on fields. As a result, farmers in developed nations use chemical fertilizers. To produce chemical fertilizers, nitrogen is extracted from the atmosphere by an industrial process. Phosphorus and potassium are mined from mineral deposits.

Chemical fertilizers have increased global food production and thus help feed the expanding human population. Today, average yields are higher than they have ever been. For this reason, modern farmers all over the world use fertilizers in ever-increasing amounts. In types of intensive agriculture where chemical fertilizers are used, crop residues such as straw are removed, and soil humus is thus depleted continuously. When humus is lost, soil retains less water, nutrients leach from the soil, and the agricultural system becomes heavily dependent on irrigation and chemical fertilizers.

The most obvious effect of soil degradation and erosion is the loss of the soil itself and the resulting loss of productivity. This effect occurs at the site where the degradation or erosion has happened and is called the **on-site impact**. However, the eroded soil washes into streams, causing siltation and other problems; nutrients leached from the eroding soils contaminate both stream water and ground water. These latter effects occur away from the site of erosion and are called **off-site impacts**.

Off-site impacts can be either nontoxic or toxic. Soil particles such as silt and sand are not toxic, but they settle to the bottoms of streams and fill in the small spaces between the rocks on the stream bed. In a healthy stream, these nooks and crannies provide protected places for fish to lay their eggs. Once the eggs hatch, the baby fish hide in the small, restricted passageways where

FIGURE 7–22
Intensive grazing exposes soil to erosion in this fragile Andean ecosystem.

FOCUS ON

Organic Farming

Most farmers in the United States use synthetic fertilizers, herbicides, and pesticides. In contrast, an organic farmer uses only organic fertilizer such as manure and waste plant products and applies no synthetic pesticides or herbicides.

In 1990, the National Academy of Sciences emphatically endorsed organic farming. Although the report recognized the undeniable production success of technological agriculture in the United States, it also described many shortcomings of farming based on synthetic fertilizers and pesticides. For example, since World War II, pesticide applications on corn have increased by a factor of 1000, but crop losses to pests have risen from 3.5 percent to 12 percent. A generation ago, farmers rotated crops and planted smaller fields, so pest populations were limited by a changing food supply. Modern agribusiness, however, commonly plants the same crops over and over again in the same fields, so pests that require a constant supply of a single crop can become established. They often develop a resistance to pesticides. Many farmers claim that they must plant large fields of high-priced crops such as corn to stay in business, and thus use of chemicals is necessary.

According to the U.S. Department of Agriculture, if all the farms in the United States used organic techniques, erosion and off-site pollution would diminish. Because many organic farms produce lower yields than those using chemicals, less food would be available for export. On the other hand, organic farming would reduce or eliminate many of the environmental problems associated with conventional agriculture. Today, pesticide-related sickness leads to loss of life and $2 billion in economic losses every year. Continued soil depletion will cause crop loss in the future, and energy dependency is potentially dangerous at a time when fossil fuel reserves are being depleted.

---

the bigger fish cannot catch them. If the river is silted, these habitats are physically blocked, resulting in sharply increased mortality and decreased fish populations. The increased sediment supply also silts reservoirs behind dams, fills irrigation and drainage ditches, chokes navigable river channels so that they must be dredged more frequently, and increases the severity of floods.

Some leached fertilizers are also nontoxic. Phosphorus doesn't poison aquatic communities; it enhances the growth of aquatic plants. However, as discussed in Chapter 11, uncontrolled growth can convert a sparkling trout stream to a waterway choked with aquatic weeds. Nitrate fertilizers are also not acute poisons, but they are suspected carcinogens, and water contaminated with nitrogen fertilizers is not fit to drink.

On the other hand, pesticide residues are toxic. In high concentrations they kill both fish and people. In the lower concentrations common in streams that drain agricultural regions, many pesticides are suspected carcinogens and mutagens. Heavy metals leached from mine dumps are also toxic.

A report by the Conservation Foundation estimates that damage caused by off-site impacts costs more than $6 billion, and maybe as much as $13 billion, per year.

We need food, wood products, metals, and places to build homes and businesses, but soil erosion and the off-site impacts of water-polluting practices reduce our quality of life. Fortunately, solutions exist. With a little extra care, erosion can be reduced drastically.

7.7 Reducing Cropland Erosion

Tillage is the way in which a farmer prepares a field for planting crops and controlling weeds. The type of tillage used affects the erosion potential of the tilled land. Improved tillage methods are the mainstay of erosion control on farmland.

Contouring

If a farmer plows furrows that run uphill and downhill, water runs down them, eroding soil as it flows. In contrast, if the farmer plows furrows that follow the contours of the hillsides, the furrows trap and hold water, preventing it from flowing over the surface (Fig. 7–24). The water seeps into the ground and carries off little or no soil. Studies by the Environmental Protection Agency and

7.7 Reducing Cropland Erosion

FIGURE 7-24
The gently curved furrows created by contour plowing follow the topography and prevent soil erosion. (USDA)

pletely exposed until the new crop grows sufficiently to protect it. The methods range from no-till soil preparation for planting, in which seeds are planted directly through the previous crop's stubble and roots, to mulch-till and reduced-till preparation, in which some crop residue is plowed under but 20 to 30 percent of it is left on the surface to protect the soil (Fig. 7–25). Most of these methods require the use of herbicides and pesticides because the mulch provides a medium for weed growth and attracts insects and vermin.

The effectiveness of conservation tillage is proportional to the amount of crop residue left on the surface to protect the soil. Studies reported by the Conservation Foundation show that water runoff decreases by 10 to 60 percent, soil erosion decreases by 15 to 90 percent, and nutrient losses diminish by 15 to 70 percent when conservation tillage is used instead of normal plowing.

Cropping Patterns

The erosion rate on farmland is strongly affected by the type of crop grown. Row crops, such as corn, normally leave strips of bare soil exposed between the rows. In contrast, field crops such as hay, legumes (peas, beans, clover, alfalfa, and so on), and grains provide more complete soil covers. In addition, some crops such as corn deplete soil nutrients, whereas others, such as legumes and some grasses, actually return nitrogen and organic matter to the soil.

other groups estimate that contouring reduces erosion of plowed fields by 20 to 70 percent and loss of nitrogen and phosphorus by 25 to 65 percent.

In many cases, particularly on family farms where relatively small tractors and plows are used, contouring is a cheap, easy, and effective conservation technique. In recent years, however, large industrial farms operated by corporations have replaced many family farms. The larger farms use much larger agricultural machinery, which makes contouring considerably more difficult and expensive. As a result, the amount of contour-plowed land in the United States has decreased.

Conservation Tillage

Conservation tillage consists of a group of methods that leave vegetation on the land surface to protect soil from rainwash erosion. These methods contrast strongly with normal tillage, in which plowing leaves bare soil com-

FIGURE 7-25
Old stubble prevents soil erosion between the rows in one type of conservation tillage. (USDA)

FOCUS ON

Soil and Government Farm Subsidies in the United States

In the United States, the federal government has established a complex set of laws designed to assure the public of a plentiful food supply at reasonable prices. The core of this program is payments to farmers for agricultural practices deemed beneficial to the nation. In 1993, these payments, called **subsidies**, exceeded $17 billion. Subsidies vary according to the crop grown, the yield, and the land area under production. To understand how these policies affect soil erosion and off-site pollution, consider the following:

1. The government pays subsidies mainly on seven crops: corn, sorghum, wheat, barley, oats, rice, and cotton. Imagine that a farmer realizes that he or she can conserve soil by rotating corn with a cover crop such as alfalfa. That farmer is penalized because a neighboring farmer who plants corn continually receives more subsidies.
2. Some land is so hilly or the soil so poor that it is uneconomical to farm. Without subsidies, this land is left unplowed for pasture. However, because subsidies are linked to total land area in production, farmers often plow and plant the marginal areas. Soil erosion from marginal farmlands is generally high. If enough soil is lost, the land is no longer suitable even for pasture.
3. Fertilizers and pesticides increase yields, but they are expensive and pollute waterways and ground water. If a farmer reduces fertilizer and pesticide applications, yields may decrease but profits increase because production costs are lower. However, because government subsidies are related to yields, the farmer who uses more fertilizer and pesticides to boost yields is rewarded with a larger government check.

Thus, farm subsidies tend to increase off-site pollution that results from overuse of fertilizer and pesticides.

In 1990, the United States Department of Agriculture (USDA) initiated the **Integrated Farm Management Program Option** to reward farmers who practice soil and water conservation. However, payments under this plan are often less than the payments under other policies.

Crop rotation is the practice of changing the type of crop grown on a field from year to year. One Missouri study showed a decrease in the annual soil erosion rate from 8 tons per hectare to 1.1 tons per hectare when continuous corn cropping was replaced by rotation of corn, wheat, and clover. Another study showed that crop rotation decreased phosphorus loss by 30 to 75 percent and nitrogen loss by 55 to 80 percent. Although crop rotation is effective in reducing erosion and loss of nutrients, the practice can reduce farm income because soil-depleting row crops such as corn often sell for higher prices and produce greater yields than do the soil-building cover crops.

Other cropping patterns that conserve soil include **cover-cropping, strip-cropping**, and **intercropping**. In cover-cropping, farmers plant crops such as grasses or legumes when a field would otherwise be left plowed and bare, waiting for the next planting of the main crop. The cover crops protect the soil from erosion, return nitrogen and organic matter for humus improvement, and provide crop residue for conservation tillage when the main crop is planted. In strip-cropping, cover crops such as grasses and legumes are planted in the strips between row crops. The cover crops trap the water that runs over the field, thereby preventing erosion and loss of nutrients. This technique reduces soil loss by as much as 85 percent. Intercropping is similar to cover-cropping but involves planting cover crops as undergrowth between trees in orchards and similarly widely spaced crops. As in the aforementioned techniques, the cover crops greatly reduce erosion and add nutrients to the soil.

7.8 Soil Erosion in Central China and in the American Dust Bowl

Soil Erosion in Central China

Agriculture has been practiced in China for over 6000 years. Today, most farmers use hand tools, draft animals, or small, two-wheeled tractors to till small plots. Because land is scarce and population high, soil is a cherished resource. In many places the land is more fertile today than it was several thousand years ago. However, some soils are so prone to erosion that it is difficult to stabilize them, even with the utmost care.

Windblown silt can accumulate in thick deposits called **loess**. Loess is porous and typically forms a single homogeneous bed. Often the angular silt particles interlock with one another. As a result, even though the loess is not cemented, it often forms vertical cliffs and bluffs (Fig. 7–26).

The largest loess deposits in the world are found in central China. There the loess beds cover 800,000 square kilometers and reach a thickness of more than 300 meters. The silt was blown from the Gobi and Takla Makan deserts of central Asia. The silt particles are so cohesive that people dig caves in the loess cliffs and make their homes there.

If properly cultivated, loess is highly productive soil, but it is also vulnerable to erosion. When it is dry, it becomes a fluffy powder that is easily lifted by the wind. Alternately, raindrops dislodge the tiny grains and transport them as suspended matter in rainwash and streams. Therefore, the Yellow River, which drains the central Chinese highlands, is the muddiest river in the world; on an average it contains 34 percent by weight of sediment. Much of the eroded loess flows to the ocean, but large quantities settle to the bottom of the river. Two thousand years ago, Chinese engineers built earthen levees along the sides of the river to keep it confined to its channel. As more sediment accumulated in the stream bed, the river bottom rose above the surrounding plain (see Chapter 9 for a discussion of this effect). To prevent flooding, people built the levees higher. Today, the Chinese raise the level of the levees by 1 meter each decade. Periodic floods caused by torrential rains occasionally burst these structures and inundate the plains.

The Yellow River Plain is the most productive region in China and contains more than 17.5 percent of all the cultivated land in that country. As China's population continues to increase, more food must be produced in this area. But the combined effect of erosion and river dynamics threatens this vital breadbasket.

FIGURE 7–26
Villagers in Askole, Pakistan, have dug caves in these vertical loess cliffs.

CASE HISTORY

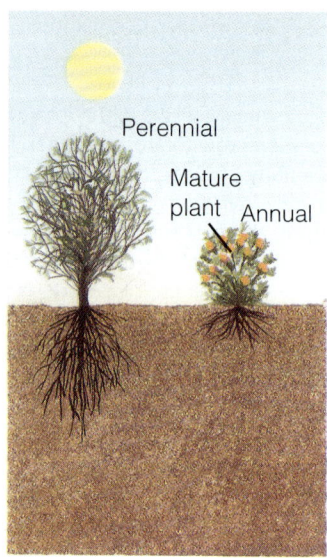  

FIGURE 7–27

Root systems of prairie plants. (A) Both annual and perennial plants grow in a prairie. (B) In a dry year, the annuals die but the perennials hold the soil. (C) In years of high rainfall, annuals sprout quickly and hold the soil.

The Dust Bowl

Periodic droughts have afflicted the semiarid and arid plains of the United States for centuries. However, the natural prairie ecosystem is drought-resistant. The perennial native bushes and grasses have long roots that collect water and nutrients from deep beneath the surface. When rainfall is adequate, shallow-rooted annual plants sprout from seed every spring, grow quickly, and then die in the fall. In years of high rainfall, the annuals cover the soil and prevent erosion. During dry years, so little water is available that many of the annuals fail to sprout. However, the deep-rooted perennials survive and protect the soil from wind erosion. Thus, the natural prairie plants protect the soil in both dry and wet years (Fig. 7–27).

Cultivated land is less resistant to changing weather. Farmers remove the protective armor of natural vegetation by tilling the land before they plant. At this point, the field is vulnerable to erosion. If a drought occurs, the unprotected soil dries up and blows away. If spring rains are too heavy, rainwash erodes the exposed soil before the crops begin to grow. Many crops, such as corn, cotton, and vegetables, are cultivated in rows, and weeds growing between the rows are removed by herbicides or by further tilling. As a result, much of the soil is exposed throughout the growing season. In addition, most crops are shallow-rooted annuals that cannot withstand drought.

During the early 1900s, poor farming practices caused a decline in soil fertility in the south and western United States. As a result, much soil lay bare and exposed. Wind and rain had begun to erode it. Then a prolonged drought occurred; the seeds failed to sprout, and the dry, exposed soil eroded rapidly. In 1934, a summer wind stripped the topsoil from entire counties and blew some of the dust more than 1500 kilometers eastward into the Atlantic Ocean. The erosion destroyed 3.5 million hectares of farmland and reduced productivity on an additional 30 million hectares (Fig. 7–28). The same drought had little effect on the natural prairies.

FIGURE 7–28

During the 1920s and 1930s, windblown dust devastated agriculture in the United States. *(National Archives)*

7.9 Croplands and World Food Supply

Recall that in 1789 the Reverend Thomas Malthus predicted that global population would grow faster than food supply. This prediction has not been fulfilled in the past 200 years because of tremendous advances in agriculture unforeseen by Malthus. In the nineteenth century, food production increased largely because huge areas of forests or prairies were converted to farmland (Fig. 7–29). During this time Europeans migrated to North and South America and Australia and cleared forests on those continents. In the first half of the twentieth century, fertilizers, pesticides, and irrigation increased crop yields dramatically. But shortly after World War II, the rapidly expanding world population threatened to catch up with food production. Then, in the mid-1960s, plant scientists used selective breeding techniques to develop new varieties of wheat and rice capable of producing higher yields than older strains. These varieties have upright leaves so that plants can grow close together without shading one another. In addition, their stalks are short and thick so that they will not bend and break under the weight of heavy seed clusters. These new varieties were so productive that many people heralded their introduction as the **Green Revolution**. In the 1940s, Mexican wheat fields produced an average of 750 kilograms per hectare. The new strains of seeds increased yields to 3200 kilograms per hectare in the 1970s. New wheat varieties raised the harvest in India and Pakistan 50 to 60 percent during a period of two growing seasons in the late 1960s.

Despite the success of the Green Revolution, the "wonder seeds" do not miraculously produce large quantities of food. They only carry the genetic potential for high yield. To fulfill that potential, the plants must be fertilized and watered heavily. The new grain varieties planted in impoverished soil and dependent on inconsistent rainfall may produce smaller yields than native grains that have been cultivated in such poor areas for centuries. In addition, the new varieties are less resistant to insects and fungi than the traditional plants. As a result, farmers

FIGURE 7–29
Nineteenth-century agriculture.

FIGURE 7–30
Intensive use of fertilizer and pesticides increases crop yields but causes long-term environmental problems. *(Cold-Flo/Seeds)*

who invest in the seed, the fertilizer, and the irrigation systems must also invest in pesticides (Fig. 7–30).

During the 1970s and 1980s, global fertilizer production, pesticide use, the amount of farmland, and the use of Green Revolution seeds all increased. As a result, global grain production once again increased faster than population.

However, in the first few years of the 1990s, several of these trends reversed. World fertilizer production peaked in 1990 and then declined (Fig. 7–31). A farmer can increase crop yields by fertilizing impoverished soil, but once enough nutrients are added, greater quantities of fertilizer do no further good. Consider the analogy of feeding children. An adequate diet will help a malnourished child realize his or her potential, but more-than-adequate food will not produce a giant. In the past, increases in fertilizer led to increased yields, but this trend has reached a plateau.

The total land area in grain production peaked in 1981 after several millennia of increase. Because population continued to rise, the amount of land per person devoted to growing grain declined (Fig. 7–32). As farmers have pushed into deserts, the Arctic, and tropical rainforests, they have gradually plowed most of the high- and medium-quality land in the world. At present, farmers are converting tropical rainforests to farmland, but that land is being lost to erosion, nutrient loss, and soil degradation at a nearly equal rate.

Total global grain production has continued to increase. However, the increase has not kept up with population growth, and world per-capita grain production

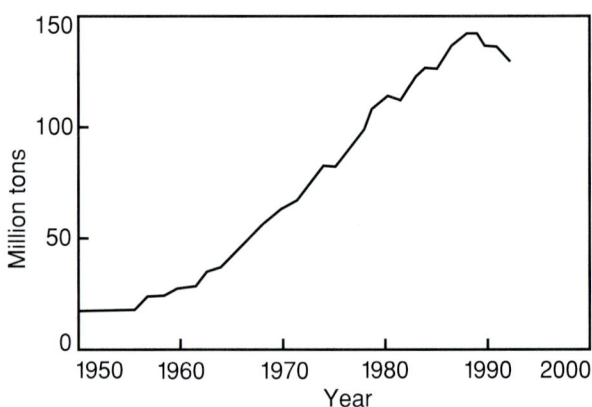

FIGURE 7–31
World fertilizer use from 1950 to 1992. (From Lester Brown, Hal Kane, and Ed Ayres, *Vital Signs 1993*, New York, W. W. Norton)

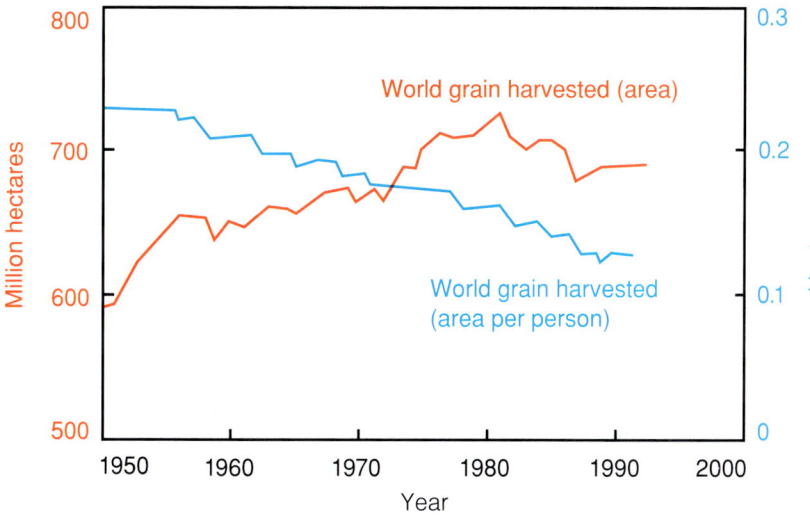

FIGURE 7–32
The red line shows the global area planted to grain between 1950 and 1992. Note that the curve rose from 1950 to 1981, then declined slightly and stabilized. However, because population rose during this period, the grain area per person (blue line) decreased.

leveled off or fell slightly between 1980 and 1992 (Fig. 7–33). Even though per-capita grain production has begun to decline, there is still enough food to feed the world's people. Yet news telecasts show starving people around the world. According to Worldwatch Institute, 700 million people, 12.7 percent of the world's population, do not eat enough to live and work at their full potential.[3] People are starving today primarily because of war and unequal distribution of food.

[3]Lester Brown et al., *State of the World 1994*. New York, W.W. Norton and Co., 1994, p 6.

In 1990, the wealthiest 20 percent of the world population consumed 83 percent of the resources. In contrast, the poorest 20 percent consumed only 1.4 percent. Rich people eat large amounts of meat. Because several kilograms of grain are required to produce one kilogram of meat, rich people use a disproportionate share of the world's grain. For example, per-capita grain consumption in the United States is about five times that in India. The average American doesn't eat five times as much wheat and rice as the average Indian, but Americans eat much more meat.

Warfare destroys crops and interrupts food shipments. Many of the recent famines in Africa and other parts of the world have been caused or at least intensified by war.

7.10 Soil Engineering

A foundation for a single-family dwelling is usually a concrete pad dug a meter or two into the soil. Skyscrapers require more substantial foundations, which are frequently built on pilings anchored to bedrock. At most building sites, soil and rock are stable and foundation engineering is routine. However, some sites require special engineering, and mistakes can be expensive or even fatal. One example of an unstable site is a steep hillside where rock and soil may slide downslope under the influence of gravity. These problems are discussed in Chapter 8. In the case histories that follow, we discuss soil engineering problems in four different geological environments.

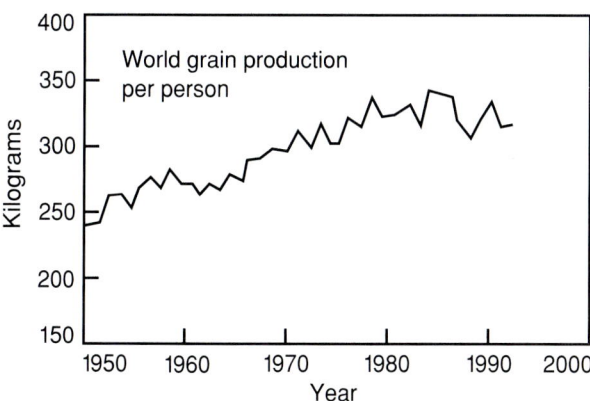

FIGURE 7–33
World grain production per person between 1950 and 1992. (From Lester Brown, Hal Kane, and Ed Ayres, *Vital Signs 1993*, New York, W. W. Norton)

CASE HISTORY

Housing Subdivisions in Lakewood, Colorado

When you buy a home, you expect it to last for decades at least. However, houses in some subdivisions in Lakewood, Colorado, began to fail when they were only a few years old. Foundations cracked, the houses tilted so that doors jammed, and floors buckled (Fig. 7–34). Geologists learned that the subdivisions were built on shale. The rock had expanded unevenly a few years after the houses were built, shifting foundations and damaging houses.

Two mechanisms have been proposed for this expansion. Recall that shale contains a high proportion of clay, which absorbs large amounts of water. According to one hypothesis, Lakewood is a naturally dry environment, but suburbanites flush large amounts of water into the ground as they irrigate their lawns. Clays expand when they absorb water, and the expanding clays cause the damage. An alternative explanation suggests that the weight of overlying rock and soil compresses the shale. When a foundation is dug and a house built, the weight of rock removed is greater than the weight of the house. The underlying shale then expands, just as a spring expands when you remove weight from it.

Whatever the mechanism, if a geologist had been consulted before construction, he or she would have recognized that shales are unstable and would have recommended either not building on the site or using proper engineering to accommodate the unstable base.

FIGURE 7–34
Uneven expansion of shale distorted the land surface in the Lakewood subdivision. *(David Noe, Colorado Geological Survey.)*

CASE HISTORY

Construction in Mexico City

In A.D. 1325, Aztec Indians built a small city of temples and palaces on an island in a shallow lake in central Mexico. The lake lay in a mountain basin and was surrounded by Mexico's highest snowcapped peaks. The Spanish explorer Hernán Cortés destroyed the city in 1521 and built his own capital on the ruins. Over the next few centuries, settlers expanded the city by filling the lake with rock and soil excavated from the nearby mountains. The population expanded from 250,000 in the sixteenth century to 20 million by 1991 (Fig. 7–35).

The mountains surrounding the Mexico City basin are largely volcanic, and settlers used volcanic ash and sand to fill the lake. Volcanic ash decomposes to clay. In 1985, an earthquake destroyed many buildings in the city, although the epicenter was hundreds of kilometers away. The seismic waves reflected back and forth within the wet mud beneath Mexico City, causing extensive property damage (Fig. 7–36). Additional damage to the buildings of the city occurs because clay swells when it is wet and the swelling tilts buildings and cracks foundations.

To make the situation even worse, residents of Mexico City obtain much of their water from wells dug into the old lake beds underlying the city. As they pump water from the sediment, the particles shift to fill the space left by the lost water. As a result, the sediment settles and the overlying ground sinks. This process is **subsidence**, the sinking or settling of the Earth's surface. Parts of Mexico City have subsided up to 3 meters (nearly 10 feet). Thus subsidence and expansion are occurring simultaneously, resulting in an unstable base for the city.

FIGURE 7–35
As Mexico City's population continues to expand, streets become crowded and the government struggles to maintain basic services. A Sunday afternoon gathering at the Shrine of Guadalupe. *(SuperStock)*

FIGURE 7–36

The 1965 Mexico City earthquake damaged many buildings. *(David Tenenbaum/Wide World Photos)*

Mexico City should never have been built on the site, but you can't move a multibillion-dollar city and 20 million people (Fig. 7–37). You can, however, design new structures to make the best of a bad situation. The Tower Latino Americano is a modern skyscraper. When engineers drilled 70-meter-deep test holes at the construction site, they found no rock, but they located a stable sand layer 33.5 meters below the surface. They drove pilings into the sand and poured a concrete slab on the pilings. The slab is a fixed platform that lies on the unstable surface soil but is supported by the pilings. The tower was then built on this concrete slab. So far, the tower has been stable.

FIGURE 7–37

Tower Latino Americano rises above the Mexico City landscape. *(SuperStock)*

CASE HISTORY

Building a Foundation Beneath a River

New York City's World Trade Center gained tragic notoriety on February 26, 1993, when Muslim fundamentalists attempted to topple one of its two 110-story towers by detonating a huge car bomb in a parking garage beneath the tower. The bomb killed six people, injured more than 1000 others, and blasted a crater 60 meters in diameter and several stories deep in the basement. But the massive explosion caused little structural damage and came nowhere near toppling the tower.

The World Trade Center consists of four low buildings and the two giant towers, located on 6.5 hectares of land on the west side of Manhattan Island (Fig. 7–38). Most of that land lay beneath the waters of the Hudson River when George Washington was president. As New York City grew, workers dumped rubble and trash in the river, burying abandoned wharves, shacks, and even old sailing ships as they pushed the shoreline more than 200 meters westward into the river.

Engineers faced the problem of designing a foundation for the 1,250,000 tons of buildings, including the two 400-meter-tall towers, in the loose, rotting jetsam that lay beneath the site. To complicate the problem, the site is only a few meters above the river, and thus the soil is water-saturated down to bedrock. Clearly the mud and trash would not support the buildings, and the foundation had to extend all the way down to bedrock. But how could they keep the river out as workers excavated a 6.5-hectare hole nearly 21 meters below river level?

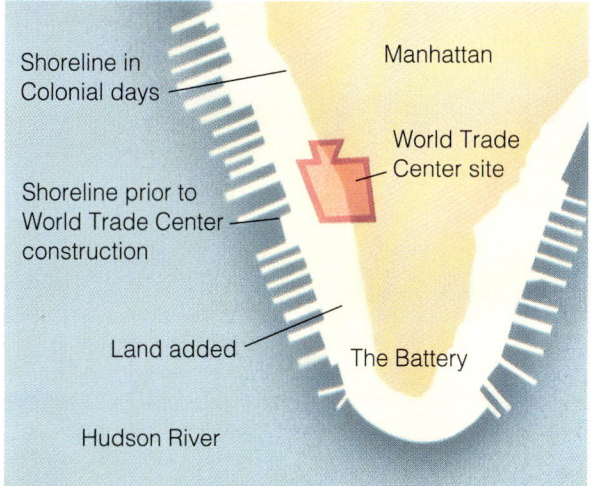

FIGURE 7–38

The World Trade Center occupies 6.5 hectares of Manhattan Island, most of which lay beneath the Hudson River in George Washington's day.

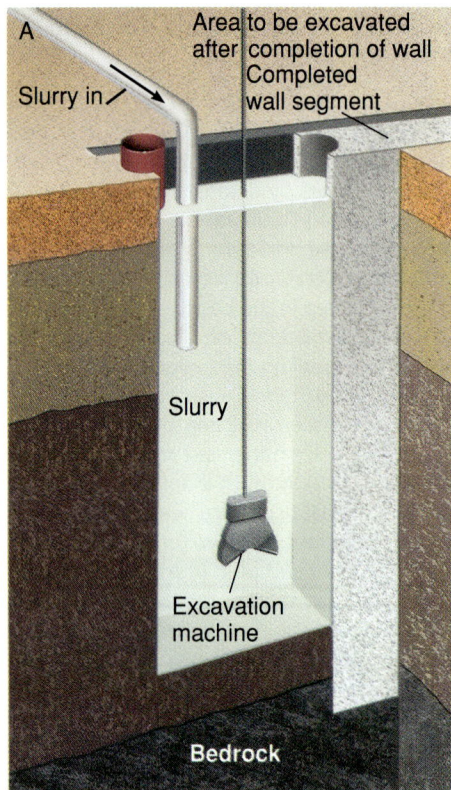

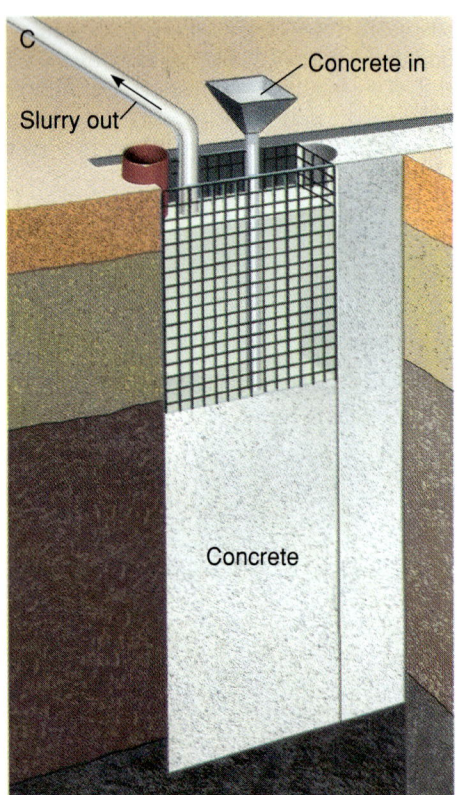

FIGURE 7–39
(A) As excavating machines cut each 1-meter-wide, 6.6-meter-long trench segment downward to bedrock, a clay–water slurry was pumped into the trench to keep river water out and maintain the trench walls. (B) After each trench segment was completed and filled with slurry, a 25-ton cage of reinforcing steel was lowered into the trench. (C) Finally, concrete was pumped into the trench, forcing the slurry upward. The slurry was pumped to the next segment of trench being excavated.

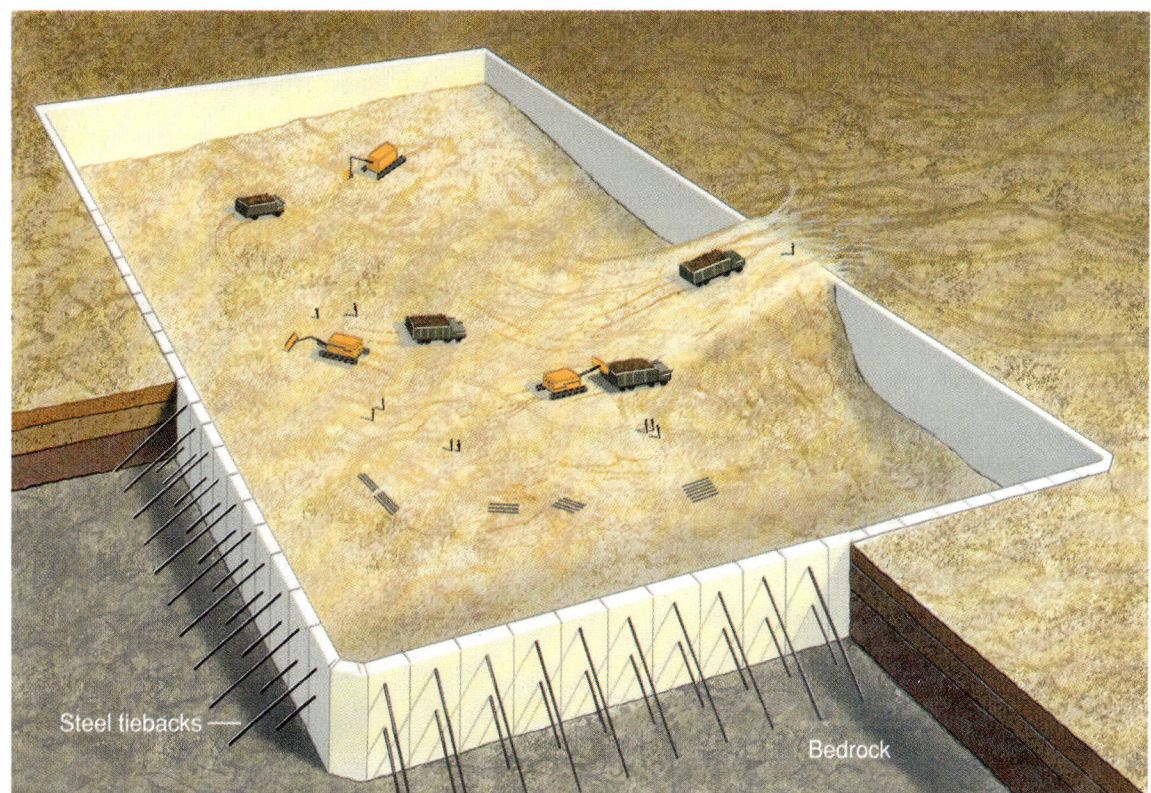

FIGURE 7–40
When the wall was completed, workers excavated the muck from its interior and began building the foundations for the buildings of the World Trade Center.

The engineers decided to build a "bathtub"; the bottom of the tub would be the bedrock, and the sides would be a 1-meter-thick steel-reinforced concrete wall surrounding the entire 6.5 hectares. After the walls were finished, the muck would be excavated from the tub, and, instead of holding water in as most bathtubs do, the tub would hold out the Hudson River. Workers could then build the foundation directly on the bedrock. The engineers had to devise a method to construct 940 meters of 21-meter-high, 1-meter-thick steel-reinforced concrete walls below ground and below the river before they began excavating for the foundation.

Common methods for building underground walls in saturated soil were unsatisfactory for a project of this scale. Instead, engineers chose a process called the slurry trench method that had been used in Europe and Canada but never before in the United States. Excavating machinery cut a trench 1 meter wide and 6.6 meters long all the way to bedrock. As the machinery pumped muck out of the trench, pumps forced a slurry of clay and water into the trench to replace the muck. The slurry kept river water from filling the trench and, at the same time, kept the trench walls from collapsing (Fig. 7–39A). When each segment was finished and filled with slurry, a 25-ton cage of reinforcing steel was pushed through the slurry until it rested on bedrock (Fig. 7–39B). Finally, concrete was pumped in through pipes that carried it to the base of the trench. As the concrete filled the trench from the bottom, the clay–water slurry floated upward and was pumped to the next section of trench to be built (Fig. 7–39C). When all 152 segments were completed, the bathtub was finished. Steel tiebacks connected the wall to bedrock, providing tension to prevent water pressure from collapsing the walls inward (Fig. 7–40). Workers then excavated the muck from inside the completed wall down to bedrock and began constructing the foundations for the buildings in a dry hole nearly 21 meters below the river.

CASE HISTORY

Arctic Soils and the Alaska Pipeline

Due to its harsh climate, the population of the Arctic has always been low. However, in recent years, with the discovery of oil and ore in the far north and the desire of indigenous people to increase the comfort of their lives, a rapid increase in construction has occurred in previously pristine northern environments. A permanent layer of ice and frozen soil, called **permafrost**, lies just below the surface in most Arctic regions.

A delicate balance exists between Arctic permafrost and the stability of the soil. During summer, when air temperature is above freezing, vegetation and surface soil insulate the deeper soil layers. This insulation keeps the permafrost frozen. When oil was discovered on the North Slope of Alaska, engineers started to plan a pipeline to transport the oil southward, and built a temporary road for initial survey studies. The road was built in the winter when the ground was frozen. Operators simply bulldozed the vegetation and surface soil away and trucks drove on the frozen ground beneath. The following summer, with the insulation removed, the permafrost beneath the road melted. The meltwater created a swampy quagmire where the road had been. Vegetation could not reestablish itself in this narrow swamp, and today the swamp appears to be a permanent feature of the landscape. Without expensive reconstruction, natural regeneration of the land will take decades or even centuries.

With this lesson in mind, engineers realized that they had to devise a way to build the pipeline itself without melting more permafrost. If the pipeline were buried or laid on the surface, the oil, which is heated to improve flow, would melt the permafrost and weaken the structural support of the pipe. Therefore, the pipeline was set on a trellis to keep it off the ground (Fig. 7–41).

Rather than build a straight pipeline, as is done in temperate regions, portions of the Alaska pipeline were laid out in a zig-zag pattern. If the ground shifted due to movement of soil over the ice or if an earthquake occurred, engineers believe that a straight pipe would break like a taut string. The zig-zag pipe, however, is expected to flex like an accordion.

FIGURE 7–41

The Alaska pipeline was set on trellises so the warm oil wouldn't melt the permafrost. The pipe was laid out in a zig-zag pattern so ground movement and earthquakes wouldn't snap it. *(Alyeska Pipeline Company)*

SUMMARY

Although all terrestrial life depends on soil, human activity is eroding and degrading soil more rapidly than soil is being renewed. **Weathering** is the decomposition and disintegration of rocks and minerals by **mechanical** and **chemical** processes. **Erosion** is the physical removal of weathered rock or soil by moving water, wind, glaciers, or gravity. After rock or soil has been eroded, it may be **transported** great distances and is eventually **deposited**.

Mechanical weathering occurs when physical processes break rocks and minerals into smaller grains. Chemical weathering occurs when rocks and minerals react with water, acids, oxygen, and other chemicals. Mechanical and chemical weathering processes reinforce one another. Natural rainwater is slightly acidic as a result of reactions between water and atmospheric carbon dioxide. Air pollutants make rainwater even more acidic. Chemical weathering of feldspar and other common minerals, except quartz, produces clay. Both chemical and physical weathering release quartz grains from rock to form sand. Clay and sand are two of the most abundant components of soil.

Soil is the layer of weathered material overlying bedrock. **Sand, silt, clay**, and **humus** are common components of soil. Water **leaches** soluble ions downward through soil. Clay is also transported downward by water. The uppermost layer of soil, called the **O horizon**, consists mainly of litter and humus. The **A horizon** is the **zone of leaching**, and the **B horizon** is the **zone of accumulation**.

Five major factors control the character of soil: **parent rock, time, climate, vegetation**, and **slope angle and aspect**.

In dry climates, **pedocals** form. In pedocals, leached ions precipitate in the B horizon, where they accumulate and may form **caliche**. In moist climates, soluble ions are removed from the soil, and **pedalfer** soils develop with high concentrations of less soluble aluminum and iron. **Laterite** soils form in very moist climates, where leaching removes all of the soluble ions. Rates of soil formation vary greatly, but a rate of 0.8 millimeters per year is commonly taken as a global average.

Sixteen elements, called **soil nutrients**, are essential for plant growth. The movement of these nutrients through the atmosphere, hydrosphere, biosphere, and solid Earth is called **biogeochemical cycling**. Logging and agriculture are examples of human intervention that disrupt biogeochemical cycles.

In nature, flowing water, wind, glaciers, and mass wasting erode soil about as rapidly as it forms. Plants protect soil from erosion. Human activities, such as logging, farming, and overgrazing by livestock accelerate erosion rates by destroying protective plant cover. **Off-site** impacts of erosion and soil degradation include nontoxic effects, such as siltation of streams and reservoirs, rapid growth of aquatic weeds, and increased severity of floods. Toxic off-site impacts include nitrate, pesticide, and heavy metal pollution.

Building foundations must lie on bedrock or stable soil. Structures built on an unstable base require special engineering.

KEY TERMS

Weathering 159
Erosion 159
Mechanical weathering 159
Abrasion 159
Chemical weathering 159
Acid precipitation 159
Oxidation 159
Hydrolysis 160
Regolith 160
Litter 160
Humus 160
Loam 160
Soil horizons 160
O horizon 160
A horizon 162

Topsoil 162
B horizon 162
C horizon 162
Leaching 162
Zone of leaching 162
Zone of accumulation 162
Capillary action 162
Pedocal 163
Caliche 163
Pedalfer 163
Laterite 163
Bauxite 163
Soil nutrients 166
Biogeochemical cycle 166
Nitrogen fixing 167

Denitrifying bacteria 167
Transpiration 167
Rainwash 169
On-site impact 171
Off-site impact 171
Tillage 174
Crop rotation 174
Cover-cropping 174
Strip-cropping 174
Intercropping 174
Loess 175
Green Revolution 177
Subsidence 180

REVIEW QUESTIONS

1. Explain the differences among weathering, erosion, transportation, and deposition.
2. Explain the differences between mechanical weathering and chemical weathering. Give examples of each.
3. List and explain three types of chemical weathering. Explain how chemical weathering can cause mechanical weathering.
4. Explain how mechanical weathering can speed up chemical weathering.
5. Both water and oxygen are abundant at the Earth's surface. Explain how each causes chemical weathering.
6. Explain why clay and sand are abundant components of many soils.
7. What are the components of healthy soil? What is the function of each component?
8. Characterize the four major horizons of a mature soil.
9. List the five soil-forming factors and briefly discuss each one.
10. Imagine that soil forms on granite in two regions, one wet and the other dry. Will the soil in the two regions be the same or different? Explain.
11. Explain how a soil formed from granite bedrock might change with time.
12. What is a laterite soil? How does it form? Why is it unsuitable for agriculture?
13. How rapidly does soil form?
14. What is a soil nutrient? List three major soil nutrients.
15. Use phosphorus or nitrogen to describe the nature of a biogeochemical cycle.
16. Describe the effects of vegetation destruction as illustrated by the Hubbard Brook study.
17. Compare the water cycle in a forest with a nearby deforested region.
18. Describe the ways in which agriculture can disrupt biogeochemical cycles.
19. Explain how rainwash erodes soil.
20. Explain the differences between on-site impacts and off-site impacts of soil erosion and degradation.
21. Why are croplands more prone to erosion than natural forests and grasslands?
22. Discuss five tillage practices that conserve soil.
23. Briefly discuss trends in food production from 1800 to 1992.
24. What is the Green Revolution? What are its promises and limitations?

DISCUSSION QUESTIONS

1. Discuss the long-term consequences of the fact that current rates of soil erosion exceed those of soil formation.
2. Describe a scenario of terrestrial life in a world without soil.
3. What process is responsible for each of the following observations and phenomena? Is the process mechanical or chemical? (a) A board is sawn in half. (b) A board is burned. (c) A cave forms when water seeps through limestone. (d) Calcite precipitates from a hot underground spring. (e) Ice forms in cracks in granite, expanding the cracks and fracturing the rock. (f) Rockfall from a steep cliff in the Canadian Rockies is more common in the spring than in mid-summer.
4. Arctic regions are cold most of the year, and summers are short. Therefore, decomposition of organic matter is slow. In temperate regions, decay is much more rapid. How does this difference affect the character of soils formed in the two regions?
5. How might a soil formed on quartz sandstone bedrock differ from one formed on granite or basalt?
6. In some regions subsoil lies on top of a layer of topsoil. Suggest a plausible process to explain this.
7. Show how time interacts with the other soil-forming factors listed in the text.
8. Discuss the roles that climate plays in biogeochemical cycling.
9. How might the results of the Hubbard Brook study have been different if herbicides were not used to suppress regeneration of vegetation?
10. Discuss ways in which chemical fertilizers affect soil differently from natural fertilizers.
11. Discuss the relative importance of rainwash erosion in three different climatic zones: temperate, desert, and Arctic. Which climatic zone does plowed farmland resemble most closely?
12. List the potential off-site impacts of a very large, severely eroding farm in the middle of the United States.
13. In a pamphlet titled, *Agricultural Policy and Sustainability*, Paul Faeth and Robert Repetto of the World Resources Institute argue that, "U. S. agriculture's exemption from the 'polluter pays principle' should be dropped. Farmers should pay fines and fees for their off-site pollution just as mining and construction industries do..." Argue for or against this opinion.
14. Sand shifts underfoot when you walk on it, whereas dry clay is firm. Which sediment would make a more stable foundation for a house? Explain.
15. Portions of some interstate highways repeatedly break apart, while other portions remain smooth for decades. If you were an engineer on a road-building project, what data would you obtain before building an interstate? What soil types would require special engineering?
16. Compare and contrast the engineering problems encountered for the Tower Latino Americano and those of the World Trade Center.

CHAPTER 8
Mass Wasting

8.1 Mass Wasting
8.2 Factors That Control Mass Wasting
8.3 Types of Mass Wasting
8.4 Mass Wasting Triggered by Earthquakes and Volcanoes
8.5 Predicting Mass Wasting
8.6 Engineering for Landslide Avoidance

Every year, small landslides destroy homes and farmland. Occasionally, an enormous landslide buries a town or city, killing thousands or even tens of thousands of people. Landslides cause billions of dollars in damage every year. The total global property damage from landslides in a single year equals that caused by earthquakes in 20 years. In many instances, losses occur because people do not recognize danger zones that are obvious to a geologist. In other cases, damage results from poor planning and construction.

Consider three recent landslides that have affected humans:

1. A movie star builds a mansion on the edge of a picturesque California cliff. After a few years, the cliff collapses and the house slides into the valley (Fig. 8–1A).

2. A ditch carrying irrigation water across a hillside in Montana leaks water into the ground. After years of seepage, the muddy soil slides downslope and piles against a house at the bottom of the hill (Fig. 8–1B).

3. Excavations for roads and high-rise buildings undercut the base of a steep hillside in downtown Hong Kong. Suddenly, the slope slides, destroying everything in its path (Fig. 8–1C).

The word *landslide* is a general term for all kinds of mass wasting and the landforms created by such movements. The terms discussed in this chapter, such as *debris flow*, *mudflow*, and *slump*, are more specific than the generic word *landslide*.

Landslides on the edge of a steep cliff destroyed these expensive homes in San Clemente, California. (Super Stock)

FIGURE 8–1
Examples of mass wasting. (A) A few days after this photo was taken, the corner of the house hanging over the gully fell in. *(J.T. Gill, USGS)* (B) A landslide, triggered by a leaking irrigation ditch, threatens a house in Darby, Montana. (C) An expensive landslide in Hong Kong. *(Hong Kong Government Information Services)*

8.1 Mass Wasting

Mass wasting is the downslope movement of Earth material, primarily under the influence of gravity. Look at a hill or mountain and think about the bedrock and soil near the top. Gravity constantly pulls them downward, but on any given day the rock and soil are not likely to slide or tumble down the slope. Their own strength and friction keep them from sliding downhill. Even so, a slope can become unstable and mass wasting can occur. For example, stream erosion can undermine a rock cliff so much that it collapses. Water can add weight and lubricate soil on a slope and cause it to slide. Mass wasting occurs naturally in any hilly or mountainous terrain. Steep slopes are especially vulnerable, and scars from recent movement of rock and soil are common in mountainous country.

In recent years, the human population has increased dramatically. As land has become overpopulated, the character of human settlements has changed. In poor countries, more and more people try to scratch out a living in mountains once considered too harsh for homes and farms. In rich nations, people have moved into the hills to escape congested cities. As a result, permanent settlements have grown in previously uninhabited steep terrain. Many slopes are naturally unstable. Construction and agriculture destabilize others.

Why does mass wasting occur? Is it possible to avoid or predict landslides and similar catastrophes to reduce property damage and loss of life?

8.2 Factors That Control Mass Wasting

Imagine that you are a geological consultant on a construction project. The developers want to build a road at the base of a hill, and they wonder whether landslides will threaten the road. What factors should you consider?

Steepness of the Slope

Obviously, the steepness of a slope is a factor in mass wasting. If ice dislodges a rock from a vertical cliff, the rock tumbles to the valley below. However, a similar rock is less likely to roll down a gentle hillside. The relationship between slope steepness and mass wasting can be illustrated by placing a block of wood on a board and slowly lifting one end of the board. When the board is slightly tilted, friction holds the block in place. However, when the slope exceeds a critical angle, the block slides or tumbles.

Type of Rock and Orientation of Rock Layers

The block of wood is a coherent mass; either the entire block slides, or none of it moves. A hillside does not behave in the same way. Any portion of a slope can move. For example, if sedimentary rock layers dip in the same direction as a slope, the upper layers may slide over the lower ones. Imagine a hill underlain by shale, sandstone, and limestone layered parallel to the slope, as shown in Figure 8–2A. If the base of the hill is undermined (Fig. 8–2B), the upper portion is left hanging and may slide over the layer of weak shale. In contrast, if the rock layers dip at an angle to the hillside, the slope may be stable even if it is undercut (Figs. 8–2C and 8–2D).

Several processes can undermine a slope. A stream or ocean waves can erode its base. Road cuts and other types of excavation can create instability in the same manner. Therefore, geologists and engineers must consider not only a slope's stability before construction, but how the project might alter its stability.

The Nature of Unconsolidated Materials

The **angle of repose** is the maximum slope or steepness at which loose material remains stable. When chunks of rock break from a cliff, they fall to the bottom, where they collect to form **talus**. Because rocks in talus are angular and irregular, they interlock and jam together, allowing talus to maintain a high angle of repose, up to 45°. The slope cannot be any steeper because, if it were, the rock would slide. In contrast, rounded sand grains do not interlock as well as angular talus and therefore have a lower angle of repose (Fig. 8–3).

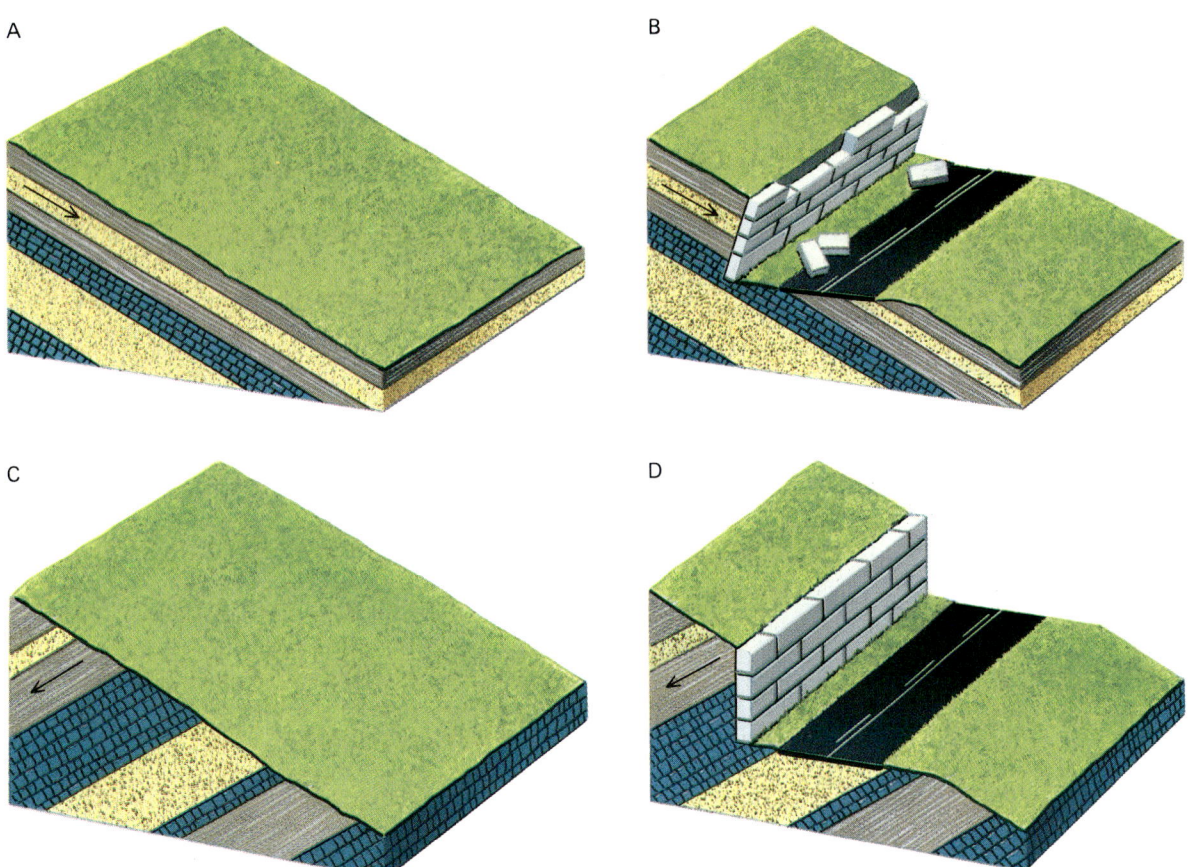

FIGURE 8–2
(A) Sedimentary rock layers dip parallel to this slope. (B) If a road cut undermines the slope, the dipping rock provides a good sliding surface, and the slope may fail. (C) Sedimentary rock layers dip at an angle to this slope. (D) The slope may remain stable even if it is undermined.

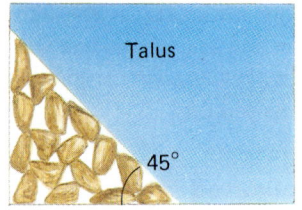 

FIGURE 8–3
The angle of repose is the maximum slope that can be maintained by a specific material.

Water and Vegetation

To understand how water affects slope stability, think of a sand castle. Even a novice sand-castle builder knows that it is impossible to build steep-sided towers and walls with dry sand; the sand must be moistened first (Fig. 8–4). But if you add too much water, the castle collapses. Small amounts of water bind sand grains together because electrical charges on water molecules attract sand grains. However, excess water adds too much weight to a slope and lubricates the sand. When some soils become saturated with water, they flow downslope, just as the sand castle collapses. In addition, if water collects on impermeable clay or shale, it may provide a weak, slippery layer that enables overlying rock or soil to move easily.

FIGURE 8–4
The angle of repose depends on both the type of material and its water content. Dry sand forms only low mounds, but if you moisten the sand, you can build steep, delicate towers with it.

Roots hold soil together and plants absorb water; therefore, a highly vegetated slope is more stable than a similar bare one. Many forested slopes that were stable for centuries have slid when the trees were removed during logging, agriculture, or construction.

Mass wasting is common in regions with intermittent rainfall and in deserts. For example, southern California has dry summers and occasional heavy winter rain. Vegetation is sparse because of summer drought and wildfires. When winter rains fall, bare hillsides often become saturated and slide. For similar reasons, mass wasting occurs during infrequent but intense storms in deserts.

Earthquakes and Volcanoes

Earthquakes and volcanic eruptions have triggered many devastating landslides. An earthquake can cause mass wasting by shaking the ground, causing an unstable slope to slide. A volcanic eruption may melt snow and ice near the top of a volcanic mountain. The water then soaks into the slope to release a landslide.

8.3 Types of Mass Wasting

Mass wasting can occur slowly or rapidly. In some cases, rocks fall freely down the face of a steep mountain. In other instances, rock or soil creeps downslope so slowly that the movement is unnoticed by casual observers. Mass wasting falls into three categories: flow, slide, and fall (Fig. 8–5). To understand these categories, think again of building a sand castle. If you added too much water to the sand, it would flow like warm honey down the face of the structure. During **flow**, loose, unconsolidated regolith moves as a fluid. Flow can be slow; some slopes flow at a speed of 1 centimeter per year or less. On the other hand, mud with a high water content can flow almost as rapidly as water.

If you undermined the base of a sand castle, a block of sand might break away from the rest of the castle and move downslope as a coherent unit. This type of movement is called **slide**. Slide is usually faster than flow, but it still may take several seconds for the block to slide down the face of the castle.

If you took a huge handful of sand out of the bottom of a sand castle, the whole tower would topple. This rapid, free-falling motion is called **fall**. Fall is the most rapid type of mass wasting. In extreme cases, rock can fall at a speed dictated solely by the force of gravity and air resistance.

Table 8–1 outlines the characteristics of flow, slide, and fall. Details of these three types of mass wasting are explained in the following sections.

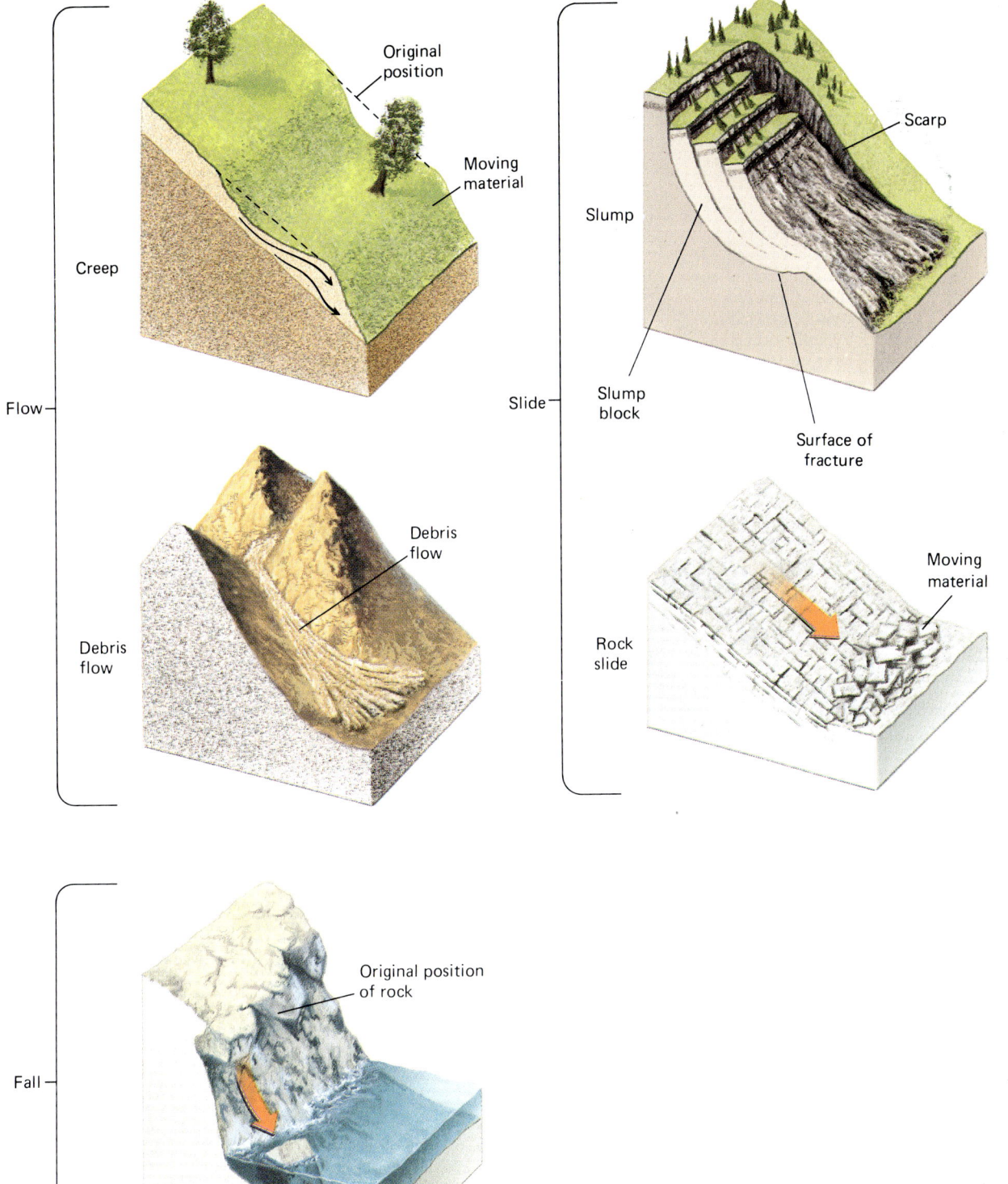

FIGURE 8–5
Flow, slide, and fall are the three categories of mass wasting.

TABLE 8-1 Some Categories of Mass Wasting

Type of Movement	Description	Subcategory	Description	Comments
Flow	Individual particles move downslope independently of one another, not as a consolidated mass. Typically occurs in loose, unconsolidated regolith.	Creep	Slow, visually imperceptible movement.	Often occurs in conjunction with slump.
		Debris flow	More than half the particles larger than sand size; rate of movement varies from less than 1 m/year to 100 km/hr or more.	Common in arid regions with intermittent heavy rainfall, or can be triggered by volcanic eruption.
		Earthflow and mudflow	Movement of fine-grained particles with large amounts of water.	
		Solifluction	Movement of waterlogged soil over permafrost.	Can occur on very gradual slopes
Slide	Material moves as discrete blocks; can occur in regolith or bedrock.	Slump	Downward slipping of a block of Earth material, usually with a backward rotation on a concave surface	Often triggers flow; trees on slump blocks remain rooted
		Rockslide	Usually rapid movement of a newly detached segment of bedrock	
Fall	Material falls freely in air; typically occurs in bedrock.	—	—	Occurs only on steep cliffs

Flow

Types of flow include creep, debris flow, mudflow, earthflow, and solifluction.

Creep

As the name implies, **creep** is a slow downhill movement of rock or soil. Individual particles move independently of one another, and the slope does not move as a consolidated mass. Typically, movement occurs at a rate of about 1 centimeter per year—so slowly that the motion cannot be detected without careful observation. In a creeping slope, the surface moves more rapidly than deeper layers (Fig. 8-6). As a result, anything with roots or a foundation tilts downhill. For example, if you look at a hillside cemetery, you may note that older headstones are tilted,

FIGURE 8-6

Creep has bent the layering in sedimentary rocks in a downslope direction. *(Ward's Natural Science Establishment)*

FIGURE 8–7
During creep, the soil surface moves more rapidly than deeper layers, so the tombstone embedded in the soil tilts downhill.

whereas newer ones are vertical (Fig. 8–7). Over the years, soil creep has tipped the older monuments, but the newer ones have not yet had time to tilt.

Creep also tilts fences and telephone poles, and in some instances it may tear entire buildings apart. When soil creep tilts trees, they develop curved trunks as a result of their natural tendency to grow straight upward. The result is a J-shaped configuration called pistol butt (Fig. 8–8). If you ever contemplate buying hillside land for a home site, examine the trees. If they have pistol-butt bases, the slope is probably creeping.

Debris Flows, Mudflows, and Earthflows

Debris flows, mudflows, and earthflows all involve downslope flow of wet soil or regolith as a plastic or semifluid mass. Think of what can happen when heavy rain falls on an unvegetated slope, as in a desert. The rain rapidly mixes with soil to form a slurry of mud and rocks. A slurry consists of water and solid particles, but it flows as a liquid. Wet concrete is a familiar example of a slurry. It flows easily and is routinely poured or pumped from a trunk.

The advancing front of a flow often forms a tongue-shaped lobe (Fig. 8–9). A slow-moving flow can travel at a rate of about 1 meter per year, but others can move as fast as a car speeding along an interstate highway. The destructive potential of a flow depends on its speed and the consistency of the slurry. Flows can pick up boulders and automobiles and smash houses, filling them with mud or even dislodging them from their foundations.

Different types of flows are characterized by the sizes of the solid particles. A **debris flow** consists of a mixture of clay, silt, sand, and rock fragments in which more than half of the particles are larger than sand. In contrast, mudflows and earthflows are predominantly sand and mud. Some **mudflows** have the consistency of wet concrete, and others are more fluid. Because of their high water content, mudflows may race down stream channels at speeds up to 100 kilometers per hour. **Earthflows** have less water than mudflows and are therefore less fluid.

FIGURE 8–8
If a hillside creeps as a tree grows, the tree develops pistol butt.

FIGURE 8-9
A debris flow in the Cascade Range of Washington has a characteristic lobe-shaped form.

FIGURE 8-10
Arctic solifluction is characterized by lobes and a hummocky surface. *(R.B. Colton, USGS)*

Solifluction

In temperate regions, soil moisture freezes in winter and thaws in summer. However, in very cold regions such as the Arctic and the Antarctic, and even in high tropical mountain ranges such as the Bolivian Andes, a layer of permanently frozen soil or subsoil, called **permafrost**, lies about a half meter to a few meters beneath the surface. Because ice is impermeable, summer meltwater cannot percolate downward as it does in warmer regions, and it therefore collects on the ice layer. This leads to two unique characteristics in these soils:

1. Water cannot penetrate the ice layer, so it collects near the surface. As a result, even though many Arctic regions receive little annual precipitation, bogs and marshes are common.
2. Ice, especially ice with a thin film of water on top, is extremely slippery. Therefore, permafrost regions are particularly susceptible to mass wasting.

Solifluction is a type of mass wasting that occurs when water-saturated soil moves over permafrost (Fig. 8-10). Solifluction can occur even on very gentle slopes.

Slide

In many cases, a large block of rock or soil—sometimes an entire hillside—fractures and moves. The material does not flow as a fluid but rather **slides** downslope as a coherent mass or as several blocks that remain intact. There are two types of slide: slump and rockslide.

A **slump** occurs when a gently curved fracture forms in rock or regolith. Overlying blocks of material slide downhill along the fracture, as shown in Figure 8-11. Trees remain rooted in the moving blocks. However, because the blocks rotate on the concave fracture, trees on the slumping blocks are tilted backward. Thus, you can distinguish slump from creep by the orientation of the trees. Slump tilts trees uphill, whereas creep tilts them downhill. At the lower end of a large slump, the blocks often pile up and break apart to form a jumbled, hummocky topography.

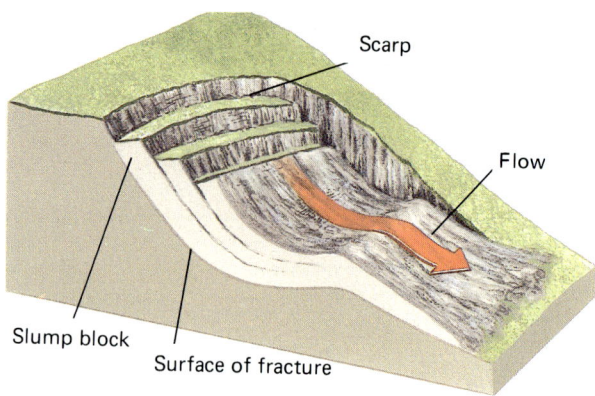

FIGURE 8-11
In slump, blocks of soil or rock remain intact as they move downslope.

It is useful to identify slump because it often recurs in the same place or on nearby slopes. Thus, a slope that shows evidence of past slump is not a good place to build a house. Figure 8–1B shows a slump that almost destroyed a home in Darby, Montana. In the western United States, irrigation water commonly flows through unlined ditches. If the ground is porous, water seeps from the ditch into soil and bedrock. The slump in Figure 8–1B resulted from water seeping through an irrigation ditch cut across an unstable slope. Alternatively, excessive irrigation can trigger slump. A Los Angeles man left his sprinkler on when he went away for a vacation. His property was on a hillside, and when he returned he found that not only was his lawn gone, but his house had slid downslope as well.

The Himalayas are the highest mountain chain on Earth and lie in subtropical latitudes. The flanks of the tall peaks rise steeply out of narrow valleys. In recent years, the population in this region has expanded rapidly, and this expansion has led to increased needs for food and fuel. The most inexpensive short-term solution to these needs has been to cut hillside forests for fuel, construct level terraces on the steep slopes, and plant grain on the terraces. However, many slopes are only semistable at best, and where the forest has been cut and steep slopes terraced, slumping has become common (Fig. 8–12). Entire hillsides are frequently lost, and slopes that were once useful for growing timber and grazing livestock have become useless. In many cases the landslides have crushed houses and killed their unlucky inhabitants.

During a **rockslide**, or **rock avalanche**, a fracture occurs in bedrock and the overlying rock slides down-slope. Characteristically, the rock breaks into small pieces of rubble, and a turbulent mass of broken rock tumbles down the hillside.

> ## CASE HISTORY
>
> ### Rock Avalanche near Kelly, Wyoming
>
> A classic rock avalanche occurred in 1925 in the hills above the Gros Ventre River near the town of Kelly, Wyoming. Before the slide, a layer of sandstone rested on shale, which in turn was supported by a thick bed of limestone (Fig. 8–13). The sedimentary layers dipped 15° to 20° toward the river and parallel to the slope. Over time, the Gros Ventre River had undermined the sandstone, leaving the slope unsupported above the river. Only a trigger was needed to release the hillside.
>
> Snowmelt and heavy rains provided the trigger in the spring of 1925. Water seeped into the ground, saturating the soil and bedrock and increasing their weight. The water collected on the shale, forming a slippery surface. Finally the sandstone layer broke loose and began to slide over the shale. In a few moments, approximately 38 million cubic meters of rock tumbled into the valley. The sandstone crumbled into small blocks, forming a 70-meter-high natural dam across the Gros Ventre River. (For comparison, a 20-story building is about 70 meters high.) But a dam made of rockslide debris is generally unstable. Two years later the lake overflowed the dam, washing it out and creating a flood downstream that killed several people.

Fall

If a rock dislodges from a steep cliff, it falls rapidly under the influence of gravity. When water freezes and thaws, the alternate expansion and contraction can dislodge rocks from cliffs and thus cause rockfall. Rockfall also occurs when a cliff is undermined. For example, if ocean waves undercut a steep shoreline, rock above the waterline may tumble (Fig. 8–14).

8.4 Mass Wasting Triggered by Earthquakes and Volcanoes

Mass wasting often occurs when a geological event releases an unstable slope. Water commonly triggers landslides because it adds weight, lubricates, and reduces shear strength. Earthquakes and volcanic eruptions also initiate mass wasting. In fact, in many cases, an earthquake or eruption itself causes comparatively little damage, but the resultant mass wasting is devastating. Consider the following case histories.

FIGURE 8–12
Slumping is common in the steep terrain of the Himalayas. Much terraced farmland is lost to landslides.

198   Chapter 8   Mass Wasting

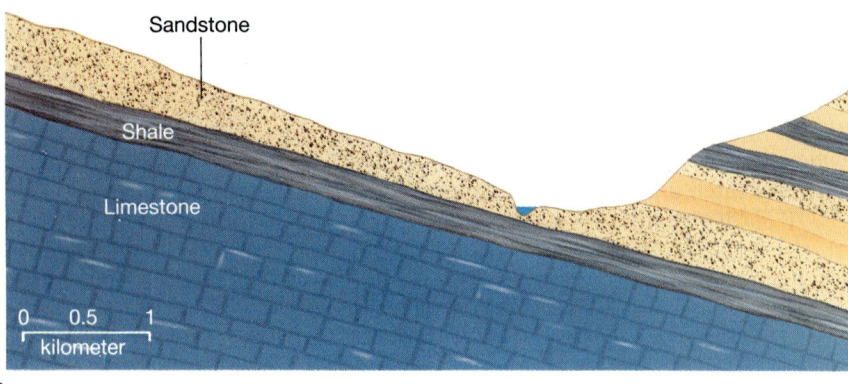
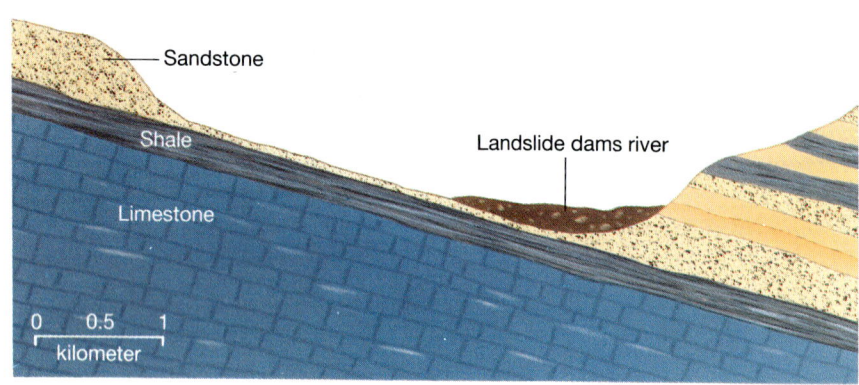

FIGURE 8–13
A profile of the Gros Ventre hillside (A) before and (B) after the slide. (C) About 38 million cubic meters of rock and soil broke loose and slid downhill during the Gros Ventre slide.

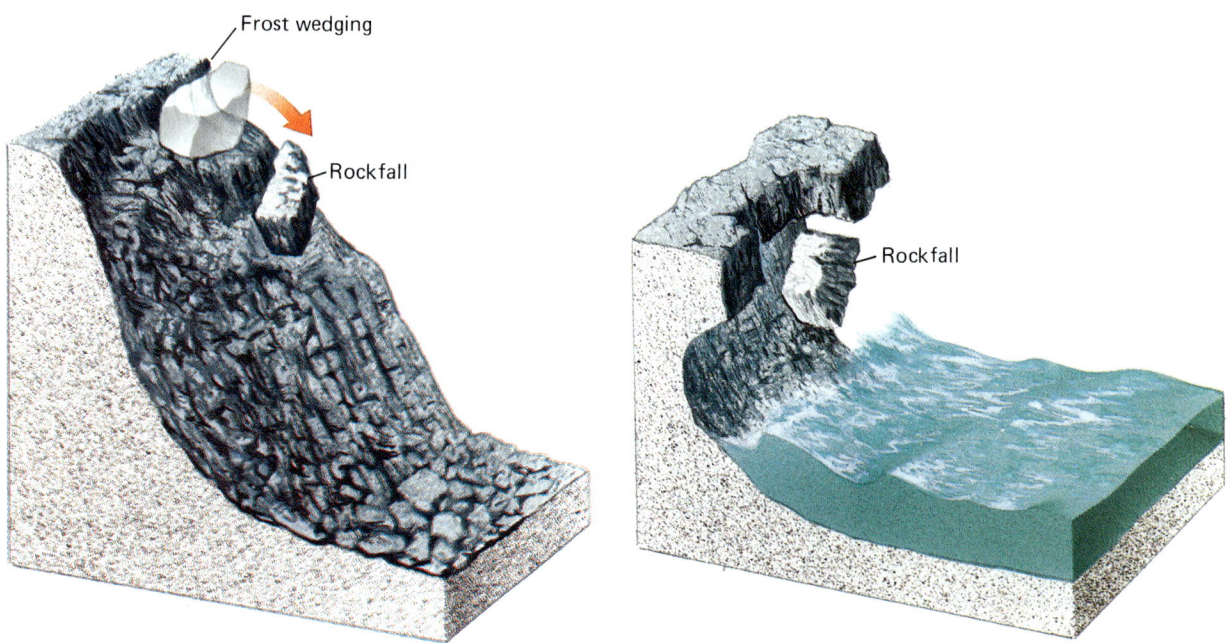

FIGURE 8–14
Rockfall commonly occurs in spring or fall when freezing water dislodges rocks from cliffs. Undercutting of cliffs by waves, streams, or construction can also cause rockfall.

FIGURE 8–15
This landslide near Yellowstone Park buried a campground, killing 26 people. (*Donald Hyndman*)

The Madison River Slide, Montana

In August 1959, an earthquake occurred just west of Yellowstone National Park. This region is sparsely populated, and most of the buildings in the area are wood-frame structures that can withstand quakes. Thus, the earthquake itself caused little property damage and no loss of life. However, the ground motion triggered a massive rockslide from the top of Red Mountain above the Madison River. About 30 million cubic meters of rock broke loose and slid into the valley below, burying a campground and killing 26 people. The mass of falling rock compressed large quantities of air, creating intense winds as the air escaped from beneath the falling debris. The winds lifted a car off the ground and carried it more than 10 meters into nearby trees. The slide's momentum carried it more than 100 meters up the mountain on the opposite side of the valley. The slide debris dammed the Madison River, forming a lake behind the dam that was later named Quake Lake. Figure 8–15 shows the pile of debris and some of the damage caused by this slide.

Nevado del Ruiz, Colombia

In November 1985, the volcanic mountain Nevado del Ruiz in central Colombia erupted. The eruption caused only minor damage, but heat from the ash and lava melted large quantities of ice and snow that lay on the mountain. The rushing water mixed with ash and debris on the mountainside, forming a mudflow that raced down gullies and stream valleys to the town of Armero, 48 kilometers from the mountain. The mudflow buried and killed 20,000 people in Armero and caused additional loss of life and property damage in a dozen other villages in nearby valleys (Fig. 8–16).

FIGURE 8–16
The eruption of Nevado del Ruiz in Colombia triggered a mudflow that buried the town of Armero, killing 20,000 people. (*Wide World Photos/Associated Press*)

Yungay, Peru

The Cordillera Blanca, the White Mountains of central Peru, contains some of the highest peaks in the Andes and therefore some of the higher peaks in the world. The summit of Nevado Huascarán, the loftiest peak in the region, is 6663 meters (21,860 feet) above sea level, and the heavily glaciated mountain rises steeply above the surrounding valleys. Because the Cordillera Blanca lies near the equator, the lowland valleys are fertile and support thriving farm communities. Prior to 1970, the town of Yungay was situated in a valley about 20 kilometers from the summit of Huascarán. Because the west coast of South America is a boundary between colliding tectonic plates, the entire region is geologically active and earthquakes are common.

In 1970, an earthquake shook loose an 800-meter-wide slab of ice from a glacier near the top of Huascarán. The ice initially tumbled about 1 kilometer over steep ice faces and rock cliffs. At the foot of this precipice was a pile of loose rocks and soil. When the ice hit this material, it mobilized tremendous quantities of the loose material, and the growing, icy landslide continued downslope. As the landslide accelerated, it compressed a cushion of air beneath it that buoyed it up like a puck in an air hockey game. Lubricated in this manner, it flowed into a valley at the foot of the mountain, racing at a speed exceeding 400 kilometers per hour (240 miles per hour). When it reached a curve in the valley, it jumped over a ridge and completely buried the town of Yungay, killing an estimated 17,000 people (Fig. 8–17). Within a few weeks, the ice melted and the mud solidified, leaving the church steeple as the only visible remnant of the town.

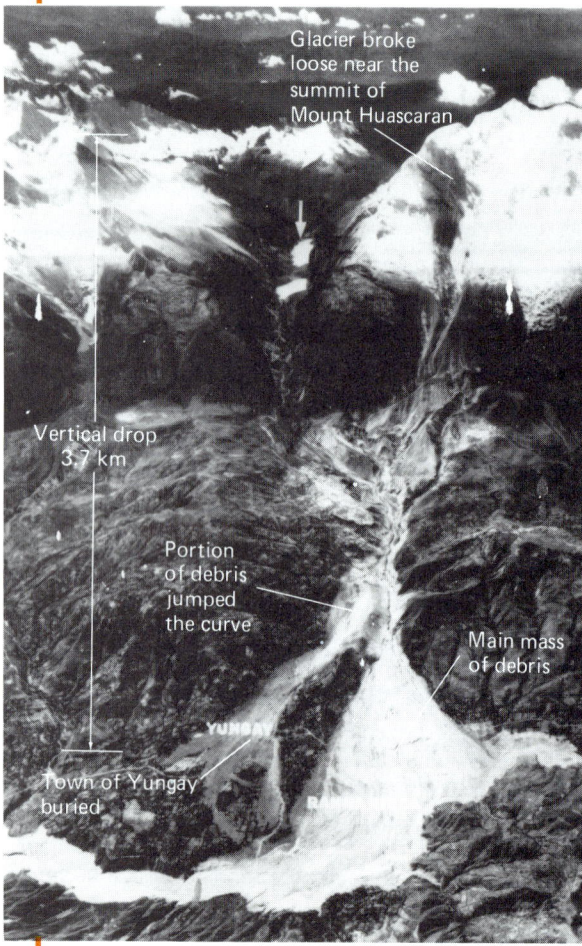

FIGURE 8–17
(A) The landslide that buried the town of Yungay, Peru. *(USGS)*

CASE HISTORY

Mass Wasting in Washington

Several volcanoes near the west coast of Washington State have been active in recent geologic history. The 1980 eruption of Mount St. Helens blew away the entire north side of the mountain and ejected about 1 cubic kilometer of rock and ash. The heat of the eruption melted glaciers and snowfields near the summit, and the water mixed with volcanic ash and surface soil to create mammoth mudflows. Although the eruption and mudflows killed 63 people, the total loss of life was small compared with the tragedies in Armero and Yungay.

Why was the death toll from the Mount St. Helens eruption so much lower? Is a catastrophic mass-wasting event possible or likely elsewhere in Washington State?

The answer to the first question is twofold. First, Mount St. Helens provided ample warning of an impending eruption. When geologists detected increases in earthquakes and volcanic rumblings on and near the mountain during the spring of 1980, the United States Forest Service evacuated many residents and withdrew water from reservoirs so that the dams would partially contain the anticipated mudflows. Second, the region around the mountain is a forested park, and there are no cities in the immediate vicinity.

Could other eruptions in Washington and nearby locations lead to much greater disasters? Unfortunately, the answer is maybe. Mount Baker, an active, glacier-covered volcano north of Seattle, is situated just 20 kilometers upriver from the town of Glacier, Washington, and less than 50 kilometers upriver from the city of Bellingham, which has a population of about 50,000. (Recall that the city of Armero was 48 kilometers from Nevado del Ruiz.) Mount Baker has been active recently; steam and gases still escape periodically from its crater (Fig. 8–18). It would be no surprise if it erupted again. If an eruption did occur, it is conceivable that it would melt the large glaciers on the mountain to create mudflows that would follow the valley leading to the town of Glacier or even to Bellingham.

We are not predicting an eruption of Mount Baker or a disaster in Glacier or Bellingham; we are simply stating that both are geologically reasonable. So the question remains: What should we do? It is impractical to move an entire city. The only alternative is to monitor the mountain carefully and continuously and either hope that it will not erupt or hope that, if it does, it will provide enough warning that urban areas can be evacuated in time.

FIGURE 8–18
A wisp of steam rises near the summit of Mount Baker (lower center).

8.5 Predicting Mass Wasting

Earlier in this chapter we asked you to imagine that you were evaluating the potential for mass wasting at a proposed construction site. Unfortunately, this danger is often ignored. Throughout the world, homes, roadways, and commercial buildings are built on unstable slopes and in the paths of landslides. As explained previously, construction projects themselves can also create instability. For example, consider a hillside supported by weak rock and stabilized by vegetation. If a highway is cut at the base of the hill, the road cut may weaken the slope. Houses built above the road, where occupants have a nice view, add weight to the hillside. If homeowners water their lawns excessively, the bedrock and soil may become saturated. Under these conditions, the slope could become unstable and begin to move. This scenario is neither imaginary nor uncommon. Loss of life and property due to human-caused mass wasting occurs regularly.

What lessons can be learned from disasters of this kind? One rule about mass wasting is that it commonly recurs in the same area or even precisely in the same location. The reason for this recurrence is that the geological conditions that cause mass wasting are often constant over a large area and remain constant for long periods of time. If a cliff overlooking the ocean has slumped, adjacent cliffs are likely to slump as well.

If older landslides have followed a particular path on the slopes of a steep mountain, the next slide is likely to follow the same route. Yungay was built on the debris of an older rock avalanche. Similarly, Nevado del Ruiz was known to be an active volcano, and geologists were aware that an eruption could cause a mudflow. Furthermore, they had predicted that a mudflow from Nevado del Ruiz was likely to inundate Armero. It is relatively easy to predict the path of a mudflow. It will follow gullies and stream valleys, and Armero lay in a stream valley beneath Nevado del Ruiz.

If geologists' warnings had been heeded, mass wasting at Yungay and Armero might not have caused loss of life. However, the situation is not always as simple in real life as it is on paper. Many towns were founded decades or centuries ago, before geological disasters were understood. Often the original choice of a town site was not dictated by geological considerations but by factors related to agriculture, commerce, or industry, such as proximity to rivers and ocean harbors and the quality of the farmland. For example, Yungay was located on a pleasant, well-drained, gravelly slope above the valley. Armero was located on fertile floodplain along a river. The original settlers did not realize that Yungay's gravelly slope had been created by landslides and that Armero's valley was fertile because of fresh soils formed on earlier mudflows. Once a city is established, moving it would be prohibitively expensive and therefore impractical. Furthermore, geologists' warnings that a disaster might occur are often ignored. After all, predictions of earthquakes and volcanic eruptions are commonly incorrect. Even in areas known to be active, a quake or eruption may not occur for decades or even centuries. The new town of Yungay has been built away from the paths of future landslides, but many other towns and cities remain on sites vulnerable to geological disasters.

8.6 Engineering for Landslide Avoidance

The first defense against mass-wasting damage is awareness and avoidance. By combining data on soil, bedrock, slope angle, and past history, geologists construct maps of slope stability. Building codes then regulate or prohibit construction in unstable areas. For example, recall from Figure 8–3 that the angle of repose of sand is 35°. According to the *United States Uniform Building Code*, you may not build a house on a sandy slope steeper than 27°. Thus, the law leaves a safety margin of 8°. Architects can obtain permission to build on more precipitous slopes if they anchor the foundation to stable rock.

In some instances, engineers cannot avoid construction on steep hillsides and must stabilize the site. Some common stabilization techniques are discussed in the following case history.

CASE HISTORY

Lost Trail Pass

In 1803, President Thomas Jefferson appointed Meriwether Lewis and William Clark to explore the newly acquired Louisiana Purchase and to find an overland route to the Pacific Ocean. A year later, Lewis, Clark, and their small band of explorers had crossed the plains to search for a gap through the formidable Rocky Mountains. They encountered a large river that headed westward, but found the canyon so narrow and the rapids so treacherous that they believed passage was impossible. They named it the "River of No Return" and turned north to find another route. (Today, the River of No Return is called the Salmon River.) Hungry and confused, they crossed a mountain pass that seemed to lead nowhere and named it Lost Trail Pass. Today, U.S. Highway 93 crosses Lost Trail Pass, linking Montana's Bitterroot Valley with the Salmon River valley to the south. In the summer of 1994, construction crews were widening and improving the road.

Any road built across a mountain pass necessarily crosses potentially unstable slopes. The most common technique to reduce the threat of mass wasting is **excavation**. Builders decrease the slope angle above the road by removing soil and rock with heavy equipment (Fig. 8–19). In the road cut shown here, workers cut a sequence of small **terraces** into the newly formed slope. The terraces provide level surfaces to reduce both mass wasting and erosion. Terraces also form stable platforms for vegetation, which stabilizes the slope still further.

Since it is impossible to excavate the entire mountain to reduce slope angle, engineers constructed a **retaining wall** to support the roadway. The retaining wall on Lost Trail Pass is built of rocks bound together with heavy wire mesh. The wall itself is stabilized by large steel rods drilled into bedrock. A wall constructed in this manner is porous, so water seeps through it readily and does not accumulate behind it. Walls of other projects are built of concrete, rock, or steel.

How effective are these measures? How long will the roadway at Lost Trail Pass remain? No one knows, but the answer is probably measured in centuries—not forever. Slowly, inexorably, the stream bed in the canyon below cuts deeper, undermining the base of the slope. Someday, spring rains may fall on an unusually deep winter snowpack, weighting and lubricating the hillside. Perhaps an earthquake will shake the region. The retaining wall may buckle and the roadway collapse. Engineers can stabilize a slope, but they can't stop geologic change.

FIGURE 8–19
Excavating on Lost Trail Pass.

SUMMARY

Mass wasting is the downhill movement of rock and soil under the influence of gravity. The stability of a slope and the severity of mass wasting depend on (1) steepness of the slope, (2) orientation and type of rock layers, (3) nature of unconsolidated materials, (4) climate and vegetation, and (5) earthquakes or volcanic eruptions.

Mass wasting falls into three categories: flow, slide, and fall. During **flow**, a mixture of rock, soil, and water moves as a viscous fluid. **Creep** is a slow type of flow that occurs at a rate of about 1 centimeter per year. A **debris flow** consists of a mixture in which more than half the particles are larger than sand. **Earthflows** and **mudflows** are mass movements of predominantly fine-grained particles mixed with water. Earthflows have less water than mudflows and are therefore less fluid. **Solifluction** is a type of flow that occurs when water-saturated soil moves over permafrost.

Slide is the movement of a coherent mass of material. **Slump** is a type of slide in which the moving mass travels on a concave surface. In a **rockslide**, a newly detached segment of bedrock slides along a tilted bedding plane or fracture.

Fall occurs when particles fall or tumble down a steep cliff.

Earthquakes and volcanic eruptions trigger devastating mass wasting. Damage to human habitation can be averted by proper planning, which involves avoidance, excavation, terracing, construction of retaining walls, and adequate drainage.

KEY TERMS

Mass wasting 190
Angle of repose 191
Talus 191
Flow 192
Slide 192
Fall 192
Creep 194
Debris flow 195
Mudflow 195
Earthflow 195
Permafrost 196
Solifluction 196
Slide 196
Slump 196
Rockslide 197
Rock avalanche 197

REVIEW QUESTIONS

1. List and describe each of the factors that control slope stability.
2. What is the angle of repose? Why is the angle of repose different for different types of materials?
3. Explain how a small amount of water might increase slope stability, whereas a landslide might occur on the same slope during heavy rainfall or rapid snowmelt.
4. How does vegetation affect slope stability?
5. Why is mass wasting common in deserts and semiarid lands?
6. How do volcanic eruptions cause landslides?
7. How do earthquakes cause landslides?
8. Discuss the differences among flow, slide, and fall. Give examples of each.
9. Compare and contrast creep, debris flow, and mudflow.
10. What does a pistol-butt tree trunk tell you about slope stability?
11. Why is solifluction more likely to occur in the Arctic than in temperate or tropical regions?
12. Compare and contrast slump and rockslide.
13. Explain how trees are bent but not killed by slump. How are trees affected by rockslide?
14. How do landslides reach and destroy towns and villages many kilometers from the steep slopes where the slides originate?

DISCUSSION QUESTIONS

1. The Moon is considerably less massive than the Earth, and therefore its gravitational force is less. It has no atmosphere and therefore no rainfall. The interior of the Moon is cool, and thus it is geologically inactive. Would you expect mass wasting to be a common or an uncommon event in mountainous areas of the Moon? Defend your answer.
2. Explain how wildfires affect slope stability and mass wasting.
3. What types of mass wasting (if any) would be likely to occur in each of the following environments? (a) A very gradual (2 percent) slope in a heavily vegetated tropical rainforest. (b) A steep hillside composed of alternating layers of conglomerate, shale, and sandstone, in a region that experiences distinct dry and rainy seasons. The dip of the rock layers is parallel to the slope. (c) A hillside similar to that of (b) in which the rock layers are oriented perpendicular to the slope. (d) A steep hillside composed of clay in a rainy environment in an active earthquake zone.
4. Identify a hillside in your city or town that might be unstable. Using as much data as you can collect, discuss the magnitude of the potential danger. Would the landslide be likely to affect human habitation?
5. Explain how the mass wasting triggered by earthquakes and volcanoes can have more serious effects than the earthquake or volcano itself. Is this always the case?
6. How do mudflows and debris flows transport automobile-sized boulders?
7. If you were asked to choose locations for new campgrounds in and near Yellowstone National Park, what criteria would you use to avoid a reenactment of the 1959 disaster?
8. Develop a strategy for minimizing loss of life from mass wasting if Mount Baker should show signs of an impending eruption similar to those shown by Mount St. Helens in the spring of 1980. How would your strategy apply to towns such as Armero and Yungay?

CHAPTER 9

Fresh Water: Streams, Ground Water, Lakes, and Wetlands

9.1 The Water Cycle
9.2 Streams
9.3 Floods
9.4 Ground Water
9.5 Lakes
9.6 Wetlands

Fresh water lies at the base of the terrestrial food web; neither plants nor animals would survive on land without fresh water. Not only do organisms need water to live, most are largely composed of water. For example, a tree is about 60 percent water by weight. Humans and most other animals are about 65 percent water.

Fresh water serves humans and shapes our environment in many ways. At home we drink water, cook with it, and wash our bodies and clothes in it. Crops need water from rain or irrigation. Organisms and ecosystems are nurtured by rainfall. Industry uses large amounts of water as a coolant, solvent, chemical reagent, or for washing. Rivers and lakes are transportation arteries. Fresh-water fisheries yield protein for human consumption. Wetlands provide habitat for millions of waterfowl and other aquatic organisms. In addition, although tectonic processes initially create mountains, valleys, and plains, flowing water erodes and shapes them into the familiar landforms that surround us.

About 1.3 billion cubic kilometers of water exist at the Earth's surface. If the surface were perfectly level, water would form a layer about 3 kilometers thick surrounding the entire planet. Of this huge quantity, however, 97.5 percent is salty seawater, and another 1.8 percent is frozen into the great ice caps of Antarctica and Greenland. Only about 0.65 percent is fresh water in streams, underground reservoirs, lakes, and wetlands. Thus, although the hydrosphere contains a great amount of water, only a tiny fraction is fresh. In this chapter we will examine fresh water in streams, ground water, lakes, and wetlands. Chapters 10 through 12 describe human use and pollution of water, and Chapter 13 discusses the vast amount of water frozen in glaciers.

Streams commonly flood and damage human structures built on flood plains. The 1993 Mississippi River flood caused the greatest damage of any flood in U.S. history.
(Wide World Photos)

9.1 The Water Cycle

Water moves constantly through the hydrosphere. It evaporates from the sea, falls as rain, and flows over land and underground as it returns to the ocean. The constant circulation of water among sea, land, and the atmosphere is called the **hydrologic cycle**, or the water cycle (Fig. 9–1). As water cycles through different parts of the hydrosphere, natural processes purify it, removing contaminants and pollutants that it may have acquired.

Water **evaporates** from sea and land to become invisible water vapor and visible clouds in the atmosphere. Water also evaporates directly from plants as they breathe in a process called **transpiration**. Atmospheric moisture then returns to the Earth's surface as **precipitation**: rain, snow, hail, and sleet.

Water that falls onto land can follow three different paths:

1. Water that flows back to the oceans in streams and rivers is called **runoff**. This water may become temporarily trapped in a lake, pond, reservoir, or wetland—an area of land that is under water for at least part of the year. All such water is called **surface water**.
2. Some seeps into the ground to become part of a vast subterranean reservoir known as **ground water**. Although surface water is much more conspicuous, 30 times more water is stored as ground water than in all streams, lakes, and wetlands combined. Ground water also seeps through bedrock and soil back to the sea, although it flows much more slowly than surface water.
3. The remainder of water that falls onto land evaporates back into the atmosphere.

If we could watch a single water molecule as it fell in a raindrop, entered a stream, and flowed to the sea, we would learn that water spends more time in some places than in others. The average time spent by all water molecules in a portion of the hydrologic cycle is called the **residence time**. Any water molecule in a large basin such as the Pacific Ocean may evaporate tomorrow, but, on the average, water has the longest residence time in the oceans and in deep ground water and the shortest in the atmosphere (Table 9–1). Thus, water-borne pollutants move through the ocean and ground-water reservoirs slowly, while they move quickly through the atmosphere and streams.

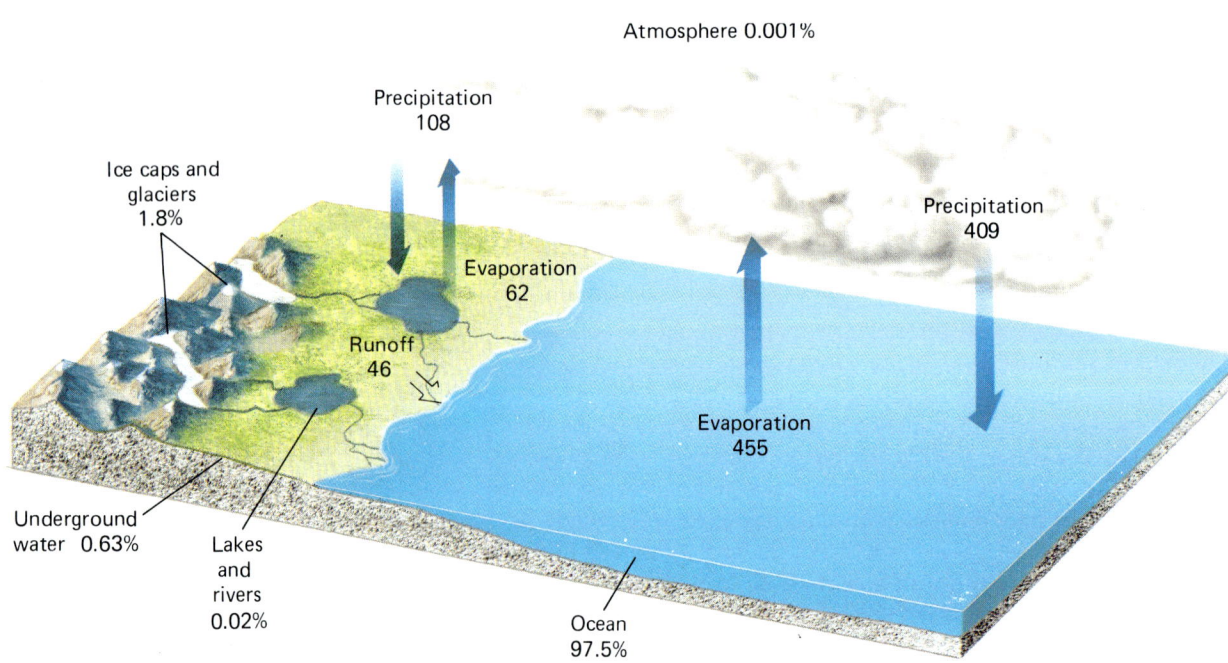

FIGURE 9–1
The hydrologic cycle shows that water circulates constantly among the sea, the atmosphere, and the land. Numbers indicate thousands of cubic kilometers of water transferred each year. Percentages show proportions of total global water in different portions of the Earth's surface.

TABLE 9-1 Average Residence Times of Water in Different Parts of the Hydrologic Cycle

Location	Average Residence Time
Atmosphere	9–10 days
Ocean	
Shallow layers	100–150 years
Deepest layers	30,000–40,000 years
World ocean average	3000 years
Continents	
Rivers	2–3 weeks
Lakes	10–100 years
Ice caps and glaciers	10,000–15,000 years
Shallow ground water	Up to hundreds of years
Deep ground water	Up to millions of years

9.2 Streams

Water flowing in channels has a variety of names, such as creek, brook, rivulet, stream, and river. To avoid confusion, geologists use the term **stream** for all water flowing in a channel, regardless of size. The term **river** is commonly used for any large stream fed by smaller streams called **tributaries**. Most streams run year round, even when it is not raining, because they are fed by ground water that seeps into the stream bed.

Normally a stream flows in its **channel**. The floor of the channel is called the **bed**, and the sides of the channel are the **banks**. When rainfall is heavy or when snow melts rapidly, a flood may occur. During a **flood,** a stream overflows its banks and water flows over adjacent land called a **flood plain**.

Stream Flow

A slow stream flows at 0.25 to 0.5 meter per second (1 to 2 kilometers per hour), whereas a steep, flooding stream may race along at about 7 meters per second (25 kilometers per hour). The gradient of a stream bed and the flow, or discharge, of the stream both affect stream velocity.

Gradient

Obviously, if all other factors are equal, water flows more rapidly down a steep slope than a gradual one. **Gradient** is the vertical drop of a stream within a certain horizontal distance. The lower Mississippi River has a shallow gradient and drops only 10 centimeters per kilometer. In contrast, a tumbling mountain stream may drop 40 meters or more per kilometer (Fig. 9–2).

FIGURE 9–2
The Selway River in Idaho has a gradient of about 5 meters per kilometer.

Discharge

Discharge is the volume of water flowing downstream per unit of time. It is expressed in cubic meters per second (m^3/sec). The velocity of a stream increases when its discharge increases. Thus, a stream flows faster during flood, even though its gradient is unchanged.

The largest river in the world is the Amazon, with a discharge of 150,000 m^3/sec. In contrast, the Mississippi River, the largest river in North America, has a discharge of about 17,500 m^3/sec, approximately one ninth that of the Amazon.

A stream's discharge can change dramatically from month to month or even during a single day. For example, the Selway River, a mountain stream in Idaho, has a discharge of 100 to 130 m^3/sec during early summer, when mountain snow is melting rapidly. During the dry season in late summer, the discharge drops to about 10 to 15 m^3/sec (Fig. 9–3). In extreme cases, discharge can vary almost instantaneously. A desert stream bed may be completely dry in mid-summer. But a sudden thunderstorm can send a wall of water rushing violently down the stream bed. After a few hours, the flow may die off to a gentle trickle as the stream dries up again.

The Ability of a Stream to Erode and Transport Sediment

A rapidly flowing stream has more energy to erode and transport sediment than a slow stream. The **competence** of a stream is a measure of the largest particle it can carry. A fast-flowing stream can pick up and carry cobbles and boulders in addition to small particles. A slow stream erodes and carries only silt and clay.

The **capacity** of a stream is the total amount of sediment it can carry past a point in a given amount of time. Capacity is proportional to both current speed and discharge. Thus, a large, fast stream has a greater capacity than a small, slow one. Because the ability of a stream to erode and carry sediment is proportional to both velocity and discharge, most erosion and sediment transport occur during the few days each year when the stream is in flood. Relatively little erosion and sediment transport occur during the remainder of the year. To see this effect for yourself, look at any stream during low water. It will most likely be clear, indicating little erosion or sediment transport. Look at the same stream later, when it is flooding. It will probably be muddy and dark, indicating that much sediment is being eroded from the bed and banks and carried by the current.

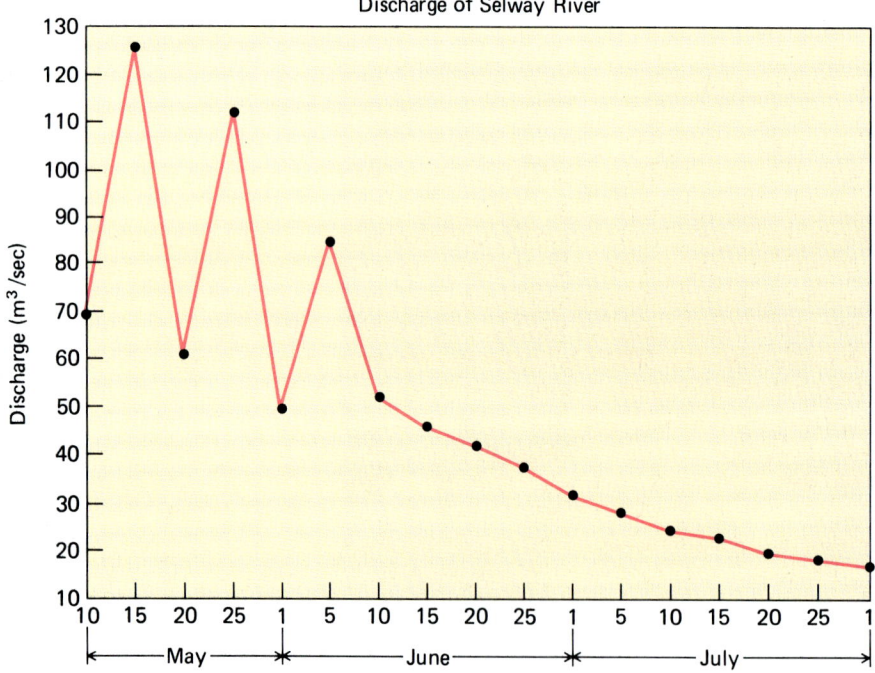

FIGURE 9–3
Discharge for northern Idaho's Selway River during the spring and early summer of 1988. The discharge drops to about 10 to 15 m^3/sec during the dry season. (U.S. Forest Service)

Downcutting and Base Level

A stream creates a valley by eroding bedrock and carrying off the sediment. The flowing water erodes both downward into the stream bed and laterally against the banks. Downward erosion is called **downcutting**. How deeply can a stream erode its channel? The **base level** of a stream is the deepest level to which it can erode its bed. For most streams,[1] the lowest possible level of downcutting is sea level, which is called the **ultimate base level**. This concept is straightforward. Water can only flow downhill. If a stream were to cut its way down to sea level, it would stop flowing and hence would no longer erode its bed or banks.

In addition to ultimate base level, a stream may have a number of **local** or **temporary base levels**. For example, where a stream flows into a lake, its current stops and erosion ceases because the stream has reached a temporary base level. A layer of rock that resists erosion may also establish a temporary base level because it resists downcutting and flattens the stream gradient. Thus, the stream slows down and erosion decreases. The top of a waterfall is an example of a temporary base level established by resistant rock (Fig. 9–4). Beneath the waterfall, less resistant rock is more easily eroded. Niagara Falls is held up by a resistant layer of dolomite over softer shale. The shale erodes first, then the harder dolomite cap is undermined and collapses. Eventually, both the shale

[1] We say "for most streams" because a few empty into valleys that lie below sea level.

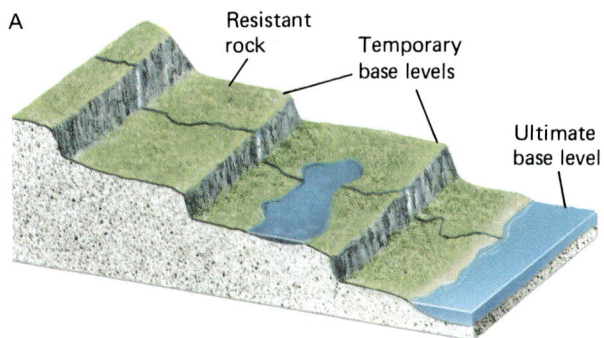

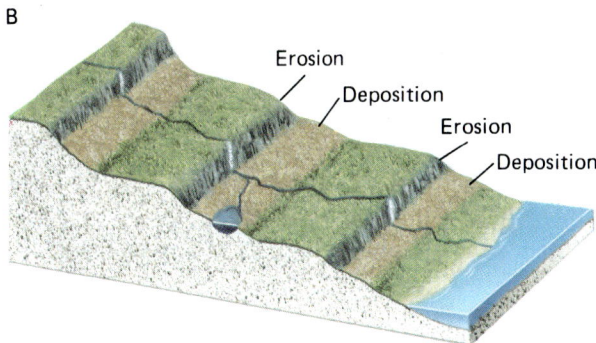

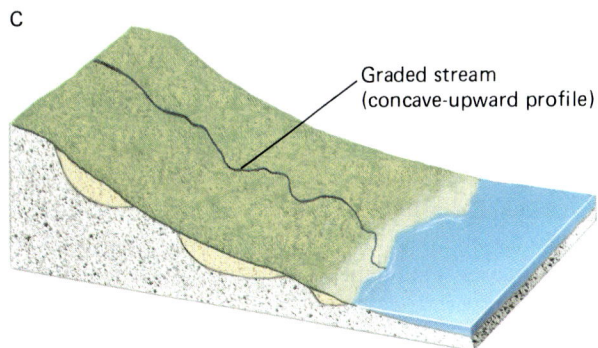

FIGURE 9–5
(A) An ungraded stream has many temporary base levels.
(B) With time, the stream smooths out the irregularities to develop a graded profile (C).

FIGURE 9–4
The pool at the top of this waterfall forms a temporary base level for the stream feeding the falls. White Mountains, New Hampshire.

and dolomite erode and the cliff face retreats upstream. Niagara Falls has retreated 11 kilometers since its formation.

If a stream has numerous temporary base levels, it erodes its bed in the steep places where flow is rapid, and it deposits sediment in the low-gradient stretches where it flows more slowly (Fig. 9–5). Over time, erosion and deposition smooth out the irregularities in the gradient. The resultant **graded stream** has a smooth, concave

FIGURE 9–6
A V-shaped valley formed by the Yellowstone River.
(Larry Davis)

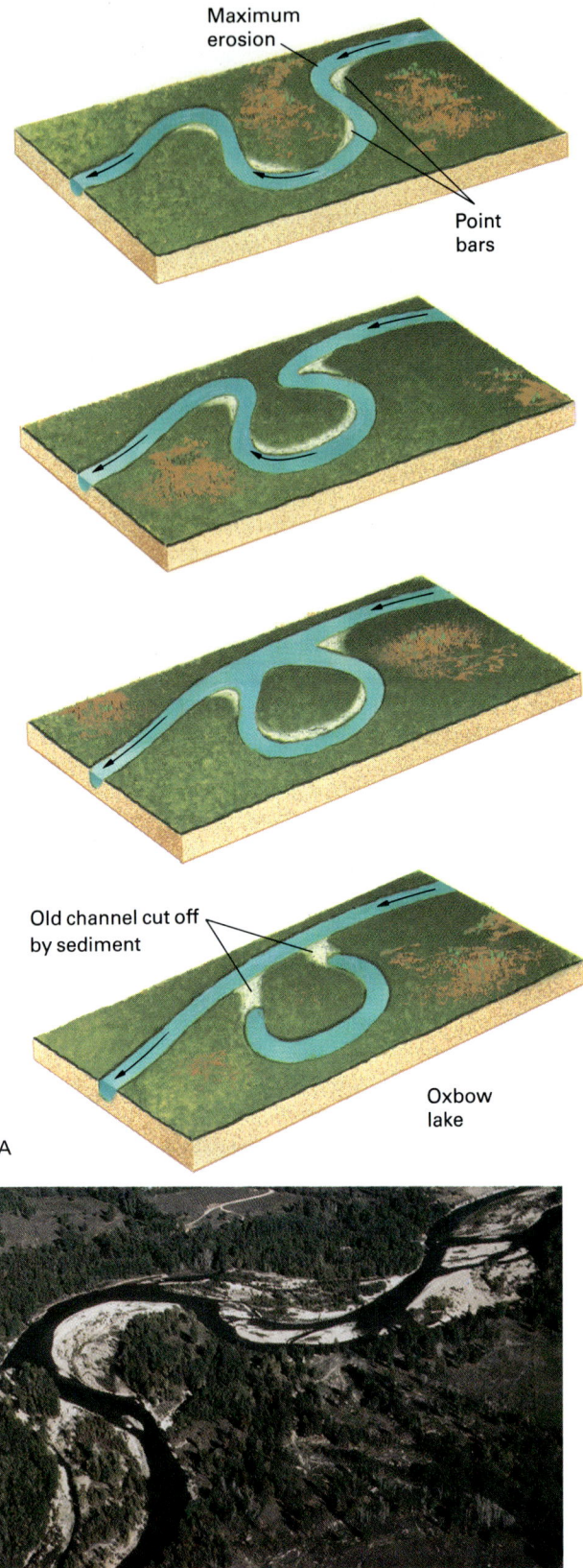

FIGURE 9–7
(A) An oxbow lake forms where a stream erodes through a meander neck. (B) Meanders in the Bitterroot River, western Montana.

profile. Once a stream becomes graded, there is no net erosion or deposition and the stream profile no longer changes. An idealized graded stream such as this does not actually exist in nature, but many streams come close.

A steep mountain stream usually downcuts rapidly compared with the rate of lateral erosion. As a result, it cuts a relatively straight channel with a steep-sided, V-shaped valley (Fig. 9–6). The steep valley walls erode by mass wasting and the stream removes the fallen sediment, enlarging the valley. In contrast, a low-gradient stream is closer to base level, and its downcutting is therefore slower. It mainly erodes laterally, wandering back and forth across the land in broad loops called **meanders** (Fig. 9–7). Occasionally, an **oxbow lake** forms when the stream cuts across the neck of a meander and isolates an old meander loop. Such a stream forms a wide valley with a flat bottom.

According to a model popular in the first half of this century, streams erode mountain ranges and create landforms in a particular sequence. At first, the streams cut steep, V-shaped valleys. Over time, mass wasting and lateral erosion widen the valleys into broad flood plains. Eventually, the entire landscape flattens, forming a large, featureless plain. However, we now know that the Earth is 4.6 billion years old; more than enough time has passed for all landforms to have eroded to a flat, low surface. Why, then, do mountain ranges, valleys, and high plateaus still exist?

This model of continuous leveling of the Earth's surface is invalid because it tells only half the story. Streams do continuously erode the landscape, flattening mountains and widening flood plains. But at the same time, tectonic activity may uplift the land and interrupt the simple, idealized sequence. Consider the Himalayas. Today, streams, glaciers, and mass wasting are eroding deep valleys. The huge amounts of sediment carried from the Himalayas to the sea by the Indus and Ganges Rivers are evidence of this erosion. However, at the same time, tectonic forces continue to raise the peaks. As a result, the Himalayas are rising more rapidly than erosion is wearing them down.

Drainage Basins

Only a dozen or so major rivers flow into the sea along the coastlines of the United States (Fig. 9–8). Each is fed by a number of tributaries, which in turn are fed by smaller streams. Mountain ranges or other raised areas called **drainage divides** separate adjacent river systems. The region drained by a single river is called a **drainage basin**. It can be large or small and is bounded by drainage divides. For example, the Rocky Mountains separate the Colorado and Columbia drainage basins to the west from the Mississippi and Rio Grande basins to the east.

9.3 Floods

A stream creates its channel by eroding and transporting sediment. A large stream creates a deep, wide channel to accommodate a large volume of water, and a small stream creates a small channel. However, as discussed earlier, stream discharge varies from season to season and year to year. When the volume of water in a stream is greater than the volume of its channel, water overflows onto the flood plain, creating a flood. Disastrous floods occur every year and are common and costly geologic hazards. In the United States floods cause an average of more than $1 billion in property damage and kill 85

FIGURE 9–8
Major drainage basins of the United States.

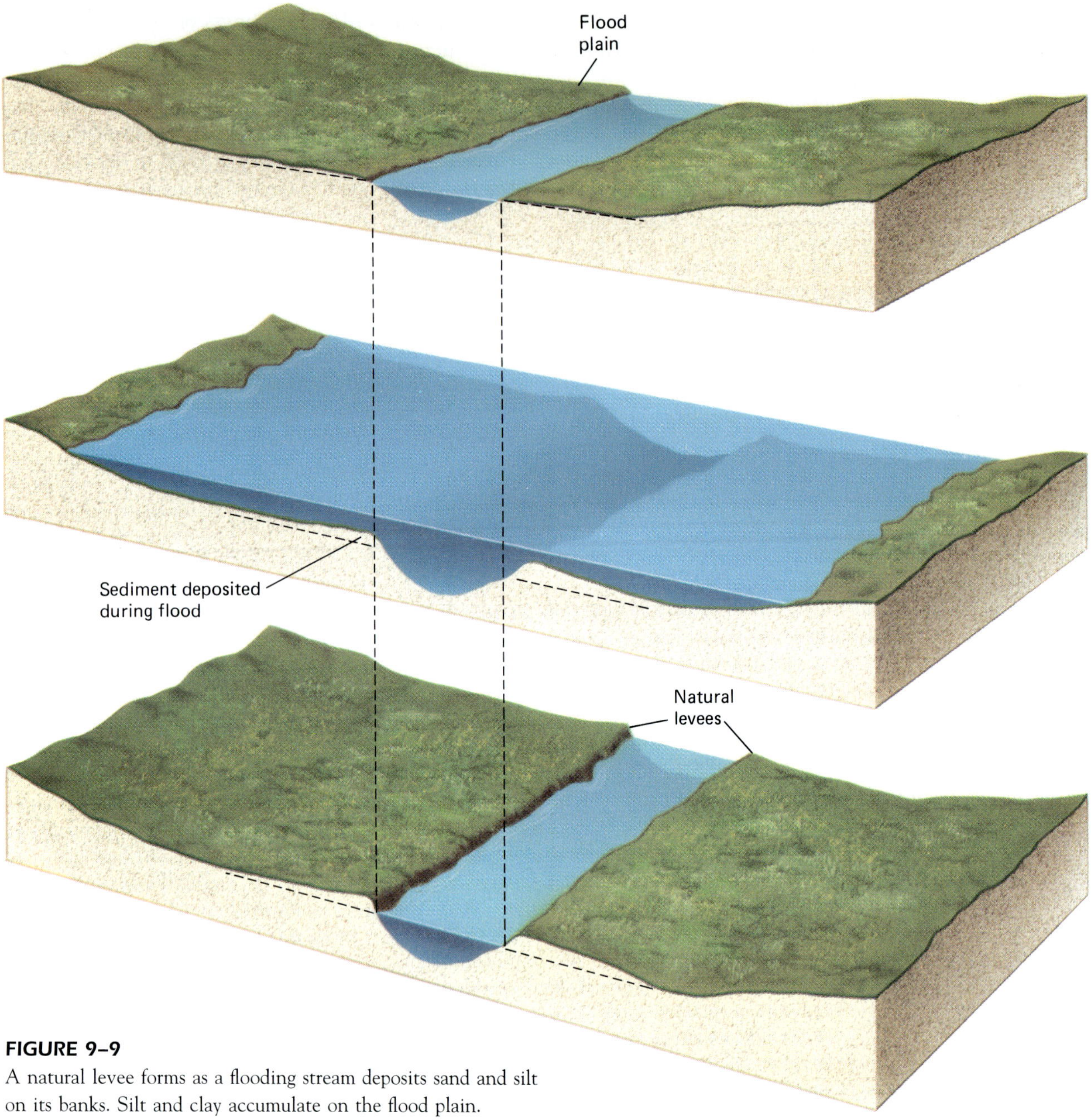

FIGURE 9–9

A natural levee forms as a flooding stream deposits sand and silt on its banks. Silt and clay accumulate on the flood plain.

people annually. In the summer of 1993, raging rivers killed more than 150 people in southern China; flash floods and related landslides killed 1800 people in Nepal, and 1350 people died in monsoon floods in northern India and Bangladesh. The Mississippi River flood in the summer of 1993 cost about $12 billion.

While flooding is a natural event, flood frequency and severity are often augmented by logging, farming, and urbanization. Recall from Chapter 7 that stream discharge increased when portions of the Hubbard Brook Experimental Forest were deforested. Without trees and shrubs to retain and absorb surface moisture, rainfall runs over the land and seeps through the ground to collect in streams. Similar conditions develop when prairies are plowed and farmed. Urbanization also increases runoff. Nearly all rainwater runs off pavement directly into nearby streams.

Floods are costly because people choose to live in flood plains. Paradoxically, flooding creates the rich soils and flat plains that make these regions so desirable.

As a stream rises to flood stage, both its discharge and its velocity increase. Therefore, it erodes its bed and banks. When a flooding stream overflows its banks, the current slows abruptly where water leaves the channel to flow onto the flood plain. Because it slows down so suddenly, the current deposits sand and silt on the banks of the stream. This sediment forms ridges called **natural levees** at the margins of the channel (Fig. 9–9). Farther out on the flood plain, the flood water carries mostly clay and silt. This finer sediment accumulates on flood plains to form fertile soils.

CASE HISTORY

Flooding and Agriculture on the Nile River

The oldest agricultural settlements along the Nile River in Egypt were established about 5000 B.C. However, farming communities thrived 3000 years earlier in the Sahara to the west and in Syria to the east. Why did the Egyptians develop agriculture so much later than their neighbors? Twenty thousand years ago, much of the Earth's water was incorporated into great ice sheets, and sea level was lower than it is today. With its large discharge and lowered base level, the Nile flowed more rapidly than it does today. The rapidly flowing river deposited coarse gravel along its flood plain and delta, while finer sediment washed out to sea. Then, as glaciers receded, rising sea level raised base level and reduced the stream's gradient, lowering its velocity. As velocity decreased, the coarse sediments were deposited farther upstream and floods spread out onto the lower flood plain and delta. As a result of these geologic changes, the Nile deposited rich silt on its flood plain and delta. Nomads settled and farmed. Once food became abundant, people had the time and energy to build the great Egyptian civilization.

In the late 1960s, the Egyptian government began construction of the Aswan Dam to control flooding and provide electricity. However, the dam also blocked the flow of silt, depriving downstream regions of their yearly infusion of soil. Today, deprived of the replenishing silt, the Nile delta is shrinking, a victim of coastal erosion and sediment deprivation. Thus, efforts to control flooding have led to loss of farmland.

Floods and Stream Size

Rapid snowmelt or even a single intense thunderstorm can cause flooding in a small stream. For example, in 1976, a series of summer storms saturated soil and bedrock near Rocky Mountain National Park, northwest of Denver. Then a large thunderhead dropped 19 centimeters of rain in the headwaters of Big Thompson Canyon in 1 hour. The Big Thompson River flooded, filling its narrow valley with a deadly, turbulent wall of water. Some people tried to escape by driving toward the mouth of the canyon, but traffic clogged the two-lane road and drowned trapped motorists in their cars. A few residents tried to escape by scaling the steep canyon walls; some of them were caught by the rising waters. Within a few hours, 139 people had died and five were missing. By the next day, the flood was over (Fig. 9–10).

In contrast, the Mississippi River drainage basin covers 3.2 million square kilometers and the river itself has a normal discharge of 17,500 m^3/sec. A thunderstorm in Big Thompson Canyon or any of its other small tributaries would have no effect on the Mississippi. A flood on the Mississippi is caused by large amounts of rainfall over a broad area. (A rainfall of only 1 centimeter [less than 1/2 inch] over the entire Mississippi basin amounts to 3.2 cubic kilometers of water.) The Mississippi River flood of 1993, discussed in the following Case History, plagued the area for 2 months.

Flood Frequency

Many streams flood regularly, some every year. A given stream floods to higher levels in some years than in others. A 100-year flood is the largest flood that occurs in a given stream an average of once every 100 years. A

FIGURE 9–10

The 1976 Big Thompson Canyon flood in Colorado lifted this house from its foundation and carried it downstream.

10-year flood is the largest that occurs on an average of once every 10 years. In any stream, small floods are more common than large ones. For example, a stream may rise 7 meters above its banks during a 100-year flood, but only 2 meters during a 10-year flood. Thus, a 100-year flood is higher and larger, but less frequent, than a 10-year flood.

Does a 100-year flood mean that if a large flood occurs in 1995, one of equal size will occur again in 2095? No, not at all. The 100-year cycle is an expression of probability: The chance of a 100-year flood occurring in any given year is 1 in 100. Think of a roulette wheel with 100 slots. Imagine that one is marked F for a 100-year flood, and 99 are marked NF for no flood (or just a small flood). Now you spin the wheel. The chance on any spin that you will land on F is 1 out of 100. If you land on F on one spin, you might land on F again on the very next try, or you might spin 200 times before it happens again. It is a game of chance.

FIGURE 9–11
Residents of Hartsburg, Missouri, boating along Main Street flooded by the Missouri River. *(Stephen Levin)*

CASE HISTORY

Flood Control and the 1993 Mississippi River Floods

During the late spring and summer of 1993, unusually heavy rains soaked the upper Midwest, raising the levels of the Mississippi River and its tributaries. In mid-July, 2.5 centimeters of rain fell in 6 minutes in Papillion, Nebraska. Thirteen centimeters fell in already saturated central Iowa in a day. In Fargo, North Dakota, the Red River, fed by a day-long downpour, rose 1.2 meters in 6 hours, flooding the town and backing up sewage into homes and Dakota Hospital. In St. Louis, Missouri, the Mississippi crested 14 meters above normal and 1 meter above the highest previously recorded flood level. At its peak, the flood inundated nearly 44,000 square kilometers in at least a dozen states. Damage to homes and businesses on the flood plain reached $12 billion, and 45 people died (Fig. 9–11).

A flood plain is a desirable place for human habitation. Floods deposit rich, fertile soil on the flood plain, creating excellent agricultural land. In addition, rivers are important transportation corridors, and as a result cities and towns have grown on the banks of many rivers. But floods are part of the natural cycle of a stream. Therefore, a conflict arises. People want to live and work along streams to take advantage of their benefits, but they don't want the inconvenience and economic loss associated with floods.

As a result, people attempt to control floods through massive engineering projects. During the 1993 Mississippi River flood, control projects saved entire towns and prevented millions of dollars in damage. However, control measures increased damage in some locations. Geologists, engineers, and city planners are studying these conflicting results to plan for the next flood, which may be years or decades away.

Artificial Levees

An **artificial levee** is a wall built of earth, rocks, or concrete along the banks of a stream to prevent rising water from spilling out of the stream channel onto the flood plain. In the past 70 years, the U.S. Army Corps of Engineers has spent billions of dollars building flood control structures, including 11,000 kilometers of levees along the banks of the Mississippi and its tributaries (Fig. 9–12).

As the Mississippi River crested in July 1993, flood waters surged through low areas of Davenport, Iowa, built on the flood plain. However, the business district of nearby Rock Island, Illinois, on the same flood plain, remained mostly dry. In 1971,

FIGURE 9–12
Floodwaters pouring from the channel of the Missouri River through a broken levee onto the flood plain, Boone County, Missouri. *(Stephen Levin)*

Rock Island had built levees to protect low-lying areas of the town, whereas Davenport had not built levees. Hannibal, Missouri, had just completed levee construction when the flood struck. The $8 million project is credited with saving the Mark Twain home and museum as well as much surrounding land.

Unfortunately, two major problems plague flood control projects that rely on artificial levees: Levees are temporary solutions to flooding, and in some cases they cause higher floods along nearby reaches of the river.

In the absence of levees, when a stream floods, it deposits mud and sand on the flood plain. When artificial levees are built, the stream cannot overflow, so it deposits the sediment in its channel, raising the level of the stream bed. After several floods the entire stream may rise above its flood plain, contained only by the levees (Fig. 9–13). This configuration creates the potential for a truly disastrous flood because if the levee should be breached, the entire stream then flows out of its channel and onto the flood plain. As a result of levee building and channel sedimentation, portions of the Yellow River in China now lie 10 meters above the flood plain. Thus, levees may solve flooding problems in the short term, but in a longer time frame they may cause even larger and more destructive floods.

Levees protecting one portion of a flood plain may also cause higher flood levels in unprotected portions of the flood plain just upstream. A flooding river spreads horizontally over its flood plain, temporarily forming a wide path of flowing water. But when levees constrict the river into its narrow channel, they form a partial dam, causing the waters to rise to even higher flood levels upstream from the levees (Fig. 9–14).

Channelization

Engineers have tried to solve the problem of channel sedimentation by building artificial channels across meanders. When the stream is thus straightened, its velocity increases, and therefore it scours more sediment from its channel. This solution, however, has its drawbacks. When a stream bed is straightened, the total volume of the channel is reduced. Therefore, the channel cannot contain excess water, and flooding is likely to increase downstream.

Flood Plain Management

When a river floods naturally, a large amount of water escapes from the channel and flows over the flood plain, and the total discharge downstream is reduced. On the other hand, if levees prevent the river from overflowing onto its flood plain, then more water flows downstream. Thus, levees that save one city may endanger another.

One approach to flood control is to abandon some of the flood control projects and let the river spill out onto its flood plain. Of course, the problem is which land to sacrifice. Every farmer or homeowner on the river will want to maintain the levees that protect his or her land and allow the flooding to occur elsewhere. Currently, federal and state governments are establishing wildlife reserves in strategic flood-plain areas. Since no development is allowed in these reserves, they will flood during the next high water. However, a complete river management plan involves complex political and economic considerations.

A Normal flow

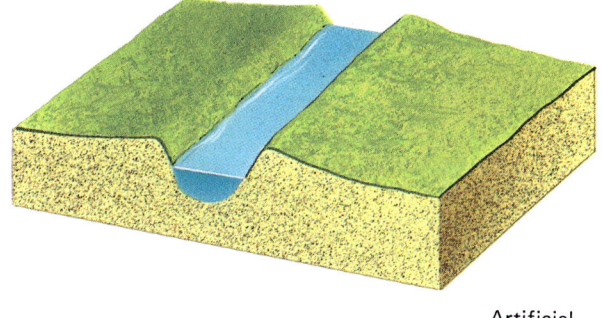

B Flood level

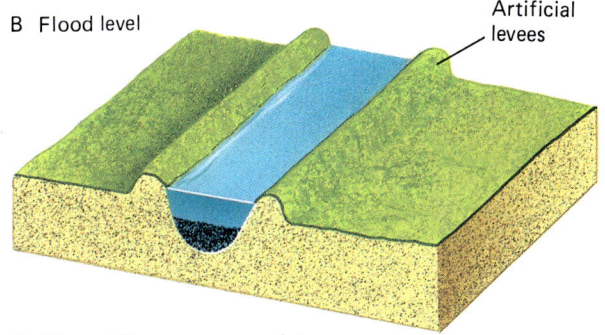

C Normal flow many years later

FIGURE 9–13

Artificial levees cause sediment to accumulate in a stream channel, eventually raising the channel above the level of the flood plain and creating the potential for a disastrous flood.

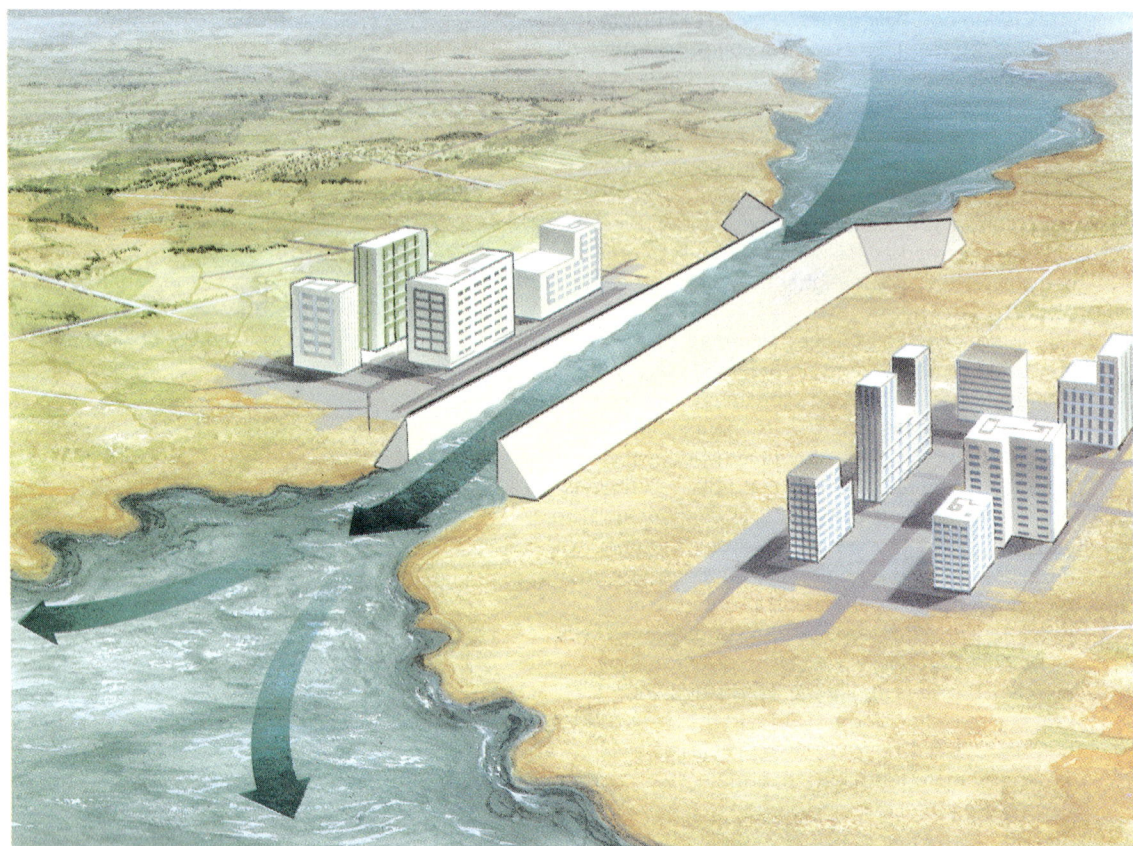

FIGURE 9–14
Levees force a flooding river into a restricted channel, forming a partial dam that raises the flood level upstream from the restriction.

9.4 Ground Water

If you drill a hole into the ground in most places, after a few days its bottom fills with water. The water appears even if no rain falls and no streams flow nearby. The water that seeps into the hole is part of the vast reservoir of subterranean ground water that saturates the Earth's crust in a zone between a few meters and a few kilometers below the surface (Fig. 9–15).

Globally, 30 times more water is stored as ground water than in all streams and lakes combined, as mentioned earlier in this chapter. We use this resource by digging wells and pumping the water to the surface. Ground water provides drinking water for more than half of the population of North America and is a major source of water for irrigation and industry. Human use and pollution of ground water are described in Chapters 10 and 11.

Characteristics of Ground Water

Ground water fills small cracks and voids in soil and bedrock. **Porosity** is the proportional volume of rock or soil that consists of open spaces. Igneous and metamorphic rocks, such as granite, gneiss, and schist, have low porosities unless they are fractured. However, many sedimentary rocks are porous. Sandstone can have 5 to 20 percent porosity (Fig. 9–16). The porosity of loose sediment and soil can be even greater, reaching 50 percent in sand and even 90 percent in clay.

Although porosity tells us how much water rock or soil can retain, it tells us nothing about the rate at which water can flow through the pores. **Permeability** is a measure of the speed at which water can travel through porous soil or bedrock. For a well to supply adequate amounts of water, it must be drilled into soil or bedrock that is both porous and permeable.

FIGURE 9–15
A spring forms where ground water flows onto the land surface. Thunder River originates at this spring flowing from bedrock in Grand Canyon.

Soil and loose sediment such as sand and gravel are both porous and permeable. Thus, they can hold a lot of water, and it flows easily through them. Sandstone and conglomerate can also have high permeabilities. Although clay and shale are porous and can hold a large amount of water, the pores in these fine-grained materials are so small that water flows very slowly through them. Thus, clay and shale have low permeabilities, and wells dug into these materials have poor flow.

The Water Table

When rain falls on dry soil, much of it soaks into the ground. Water does not descend into the crust indefinitely, however. Below a depth of a few kilometers, rock is at a high enough temperature and pressure to be plastic. The weight of overlying rock closes the pores, and rock from this depth on down is both nonporous and impermeable. Water accumulates on this impermeable barrier, filling the pores in the rock and soil above it. This completely wet layer of soil and bedrock is called the **zone of saturation**. The **water table** is the top of the zone of saturation (Fig. 9–17). Above the water table lies the **unsaturated zone**, or **zone of aeration**. In this layer, the rock or soil may be moist but not saturated. The **soil moisture belt** is the soil layer. It holds more water than

A

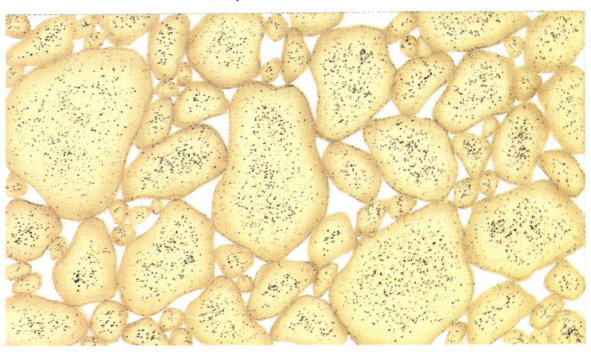

B

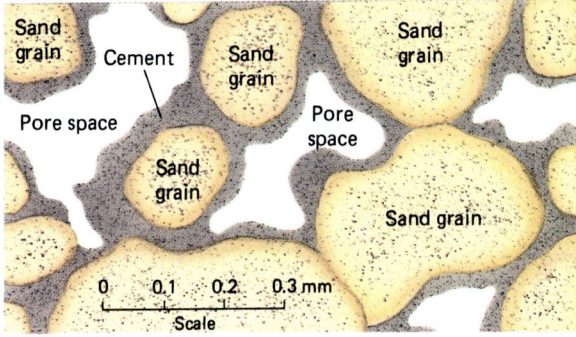

C

FIGURE 9–16
Different materials have different amounts of open pore space between grains. (A) Well-sorted sediment consists of equal-size grains and has a high porosity, about 30 percent in this case. (B) In poorly sorted sediment, small grains fill the spaces among the large ones, and porosity is lower. In this drawing it is about 15 percent. (C) Cement partly fills pore space in sedimentary rock, lowering the porosity.

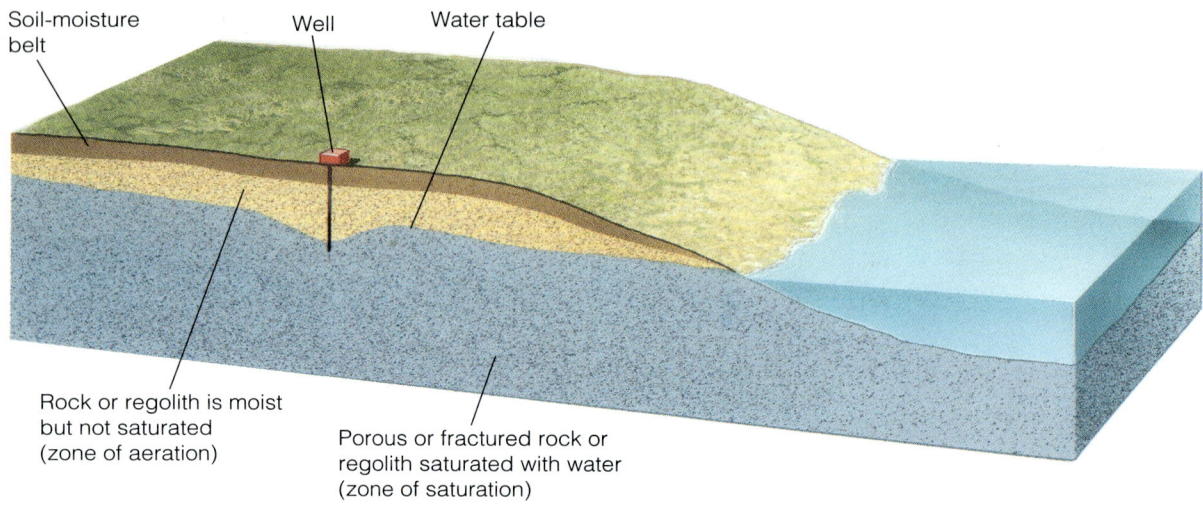

FIGURE 9–17
The water table is the top of the zone of saturation near the Earth's surface. It intersects the land surface at lakes and streams and is the level of standing water in a well.

the unsaturated zone below and supplies much of the water needed by plants.

If you dig into the unsaturated zone, the hole does not fill with water. However, if you dig below the water table into the zone of saturation, you have dug a **well**, and the water level in the well is at the water table.

An **aquifer** is any body of rock or soil that can yield economically significant quantities of water. An aquifer must be both porous and saturated. Thus, it must contain water and must also be permeable so that water flows into a well to replenish water that is pumped out. High-quality aquifers commonly occur in sand and gravel, sandstone, limestone, and highly fractured igneous and metamorphic rock.

Movement of Ground Water

In a few regions, underground streams flow through caverns, but they are the exception, not the rule. Nearly all ground water seeps slowly through interconnected pores in bedrock and soil. Think of an aquifer as a sponge through which water seeps, not as an underground pool or stream. Typically, ground water flows at about 4 centimeters per day (about 15 meters per year), although flow rates may be much faster or slower. The rate depends on the permeability of soil or rock or on the nature of fractures in bedrock. Water can flow rapidly through large, interconnected fractures.

The water table rises and falls with the seasons. During a wet season, such as spring in a temperate climate, rain seeps into the ground and the water table rises. The processes by which ground water is replenished are collectively called **recharge**. During a dry season, the water table falls. It is common for the water levels in wells to rise and fall through the year and in wet years and years of drought.

In a temperate climate, ground water seeps into streams. Figure 9–18 shows a stream and water table in

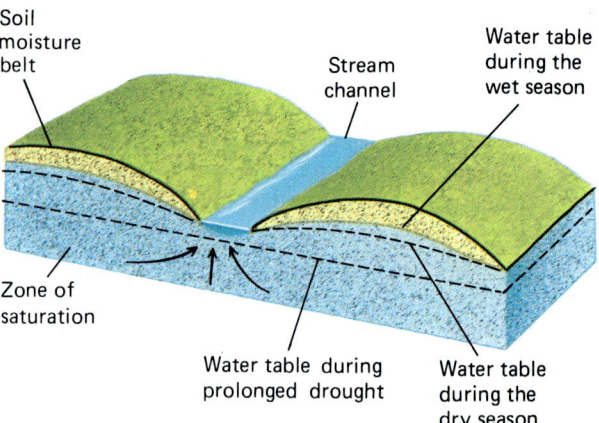

FIGURE 9–18
The water table follows topography, rising beneath hills and sinking beneath valleys. It also rises during the rainy season and falls during dry times.

hilly country. In general, the water table follows the contours of the land, rising and falling with the topography. Just as streams always flow downhill, ground water always seeps from areas where the water table is highest toward areas where it is lowest. Streams follow gulleys and valleys. Therefore, ground water normally flows toward streams and seeps into stream beds, which is why streams continue to flow even when rain has not fallen for weeks or months.

In contrast, in a desert or semiarid region, the water table commonly lies below stream beds, and water seeps downward from a stream to the water table. Most of the time, desert stream channels are dry. When streams do flow, the water usually comes from wetter environments in nearby mountains, although sometimes desert storms fill the channels for short periods of time. Thus, a desert stream feeds the ground-water reservoir, but in temperate climates, ground water feeds the stream (Fig. 9–19).

Springs

A **spring** is a place where ground water flows or seeps onto the surface. A spring forms wherever the water table intersects the land surface (Fig. 9–20). In some places,

FIGURE 9–19
A desert stream lies above the water table. Courthouse Wash, Arches National Park, Utah. See Figure 14–7 for the same view during the dry season.

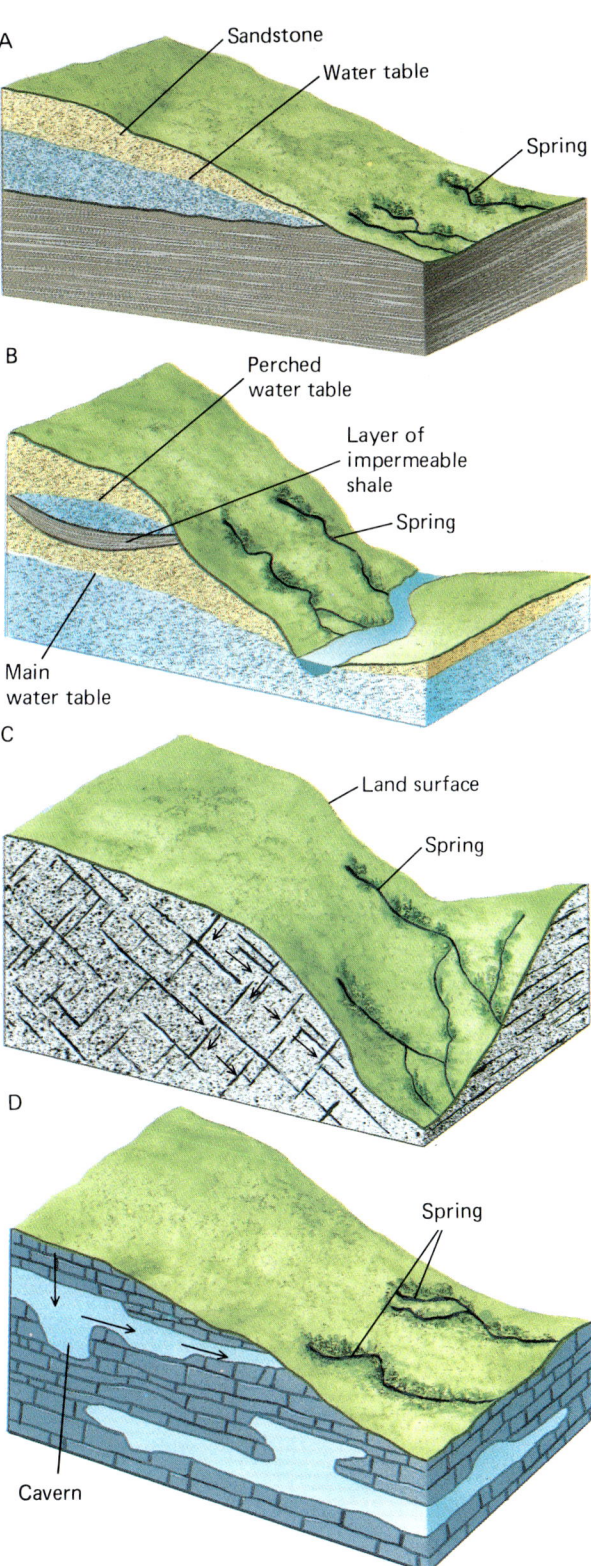

FIGURE 9–20
Springs form where the water table intersects the land surface. This situation can occur where (A) the land surface intersects a contact between permeable and impermeable rock layers; (B) a layer of impermeable rock or clay lies "perched" above the main water table; (C) water flows from fractures in otherwise impermeable bedrock; and (D) water flows from caverns onto the surface.

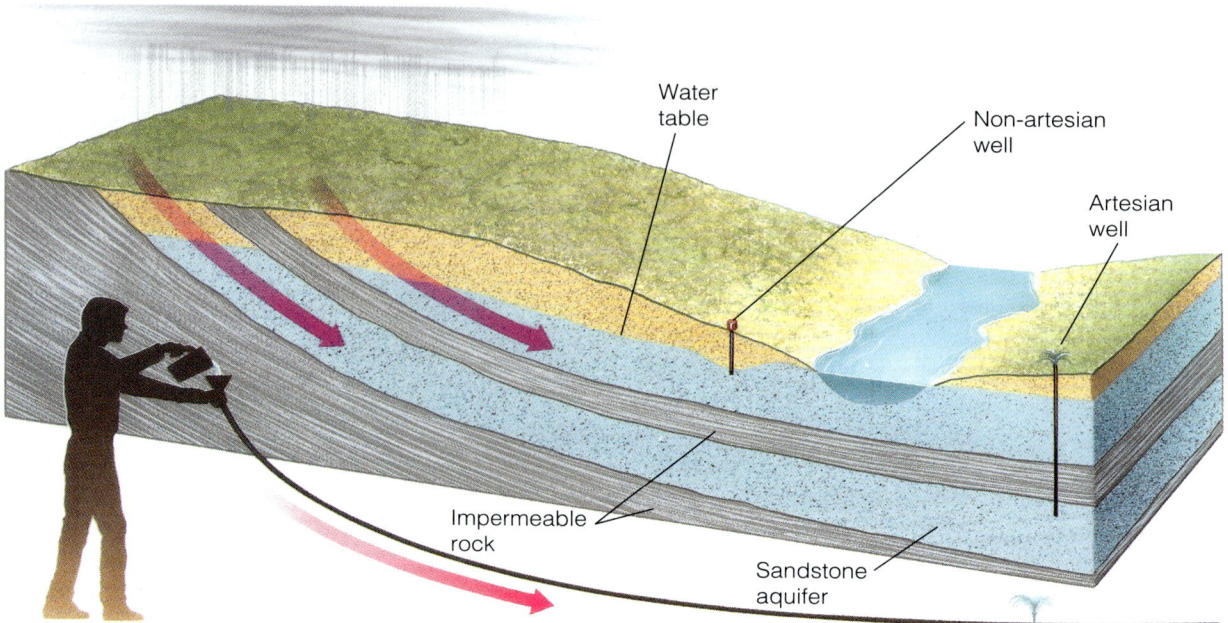

FIGURE 9–21
An artesian aquifer forms where a tilted layer of permeable rock, such as sandstone, lies sandwiched between layers of impermeable rock, such as shale. Water rises in an artesian well without being pumped. A hose with a hole (inset) shows why an artesian well flows spontaneously.

a layer of impermeable rock or clay lies above the main water table, creating a local saturated zone, the top of which is called a **perched water table**. A perched water table can also intersect the land surface to form a spring (Fig. 9–20B).

Artesian Wells

Figure 9–21 shows a layer of permeable sandstone sandwiched between two layers of impermeable shale. On the left, the strata are tilted. An inclined aquifer bounded on the top and bottom by impermeable rock is called an **artesian aquifer**. Water in the lower part of the aquifer is under pressure from the weight of water above. Therefore, if a well is drilled into the sandstone, water rises without being pumped. If pressure is sufficient, the water spurts out onto the surface. A well of this kind is called an **artesian well**. As an analogy, think of a water-filled hose. If one end is held high with the lower end sealed, and you puncture the hose below the high point, water squirts out.

Caverns, Sinkholes, and Karst Topography

Just as streams erode valleys and form flood plains, ground water also creates landforms. Rainwater reacts with atmospheric carbon dioxide to produce a slightly acidic solution capable of dissolving limestone. A **cavern** forms when acidic water seeps into a crack in limestone, dissolving the rock and enlarging the crack. Mammoth Cave in Kentucky and Carlsbad Caverns in New Mexico are two famous caverns formed in this way.

Sinkholes

If the roof of a cavern collapses, a **sinkhole** forms on the Earth's surface. A sinkhole can also form as limestone dissolves from the surface downward (Fig. 9–22). A well-documented sinkhole formed in May 1981 in Winter Park, Florida. During the initial collapse, a three-bedroom house, half a swimming pool, and six Porsches in a dealer's lot all fell into the underground cavern. Within a few days, the sinkhole was 200 meters wide and 50 meters deep, and it had devoured additional buildings and roads (Fig. 9–23).

Although sinkholes form naturally, human activities can accelerate the process. The Winter Park sinkhole formed when the water table dropped, removing support for the ceiling of the cavern. The water table fell as a result of a severe drought augmented by excessive removal of ground water by humans.

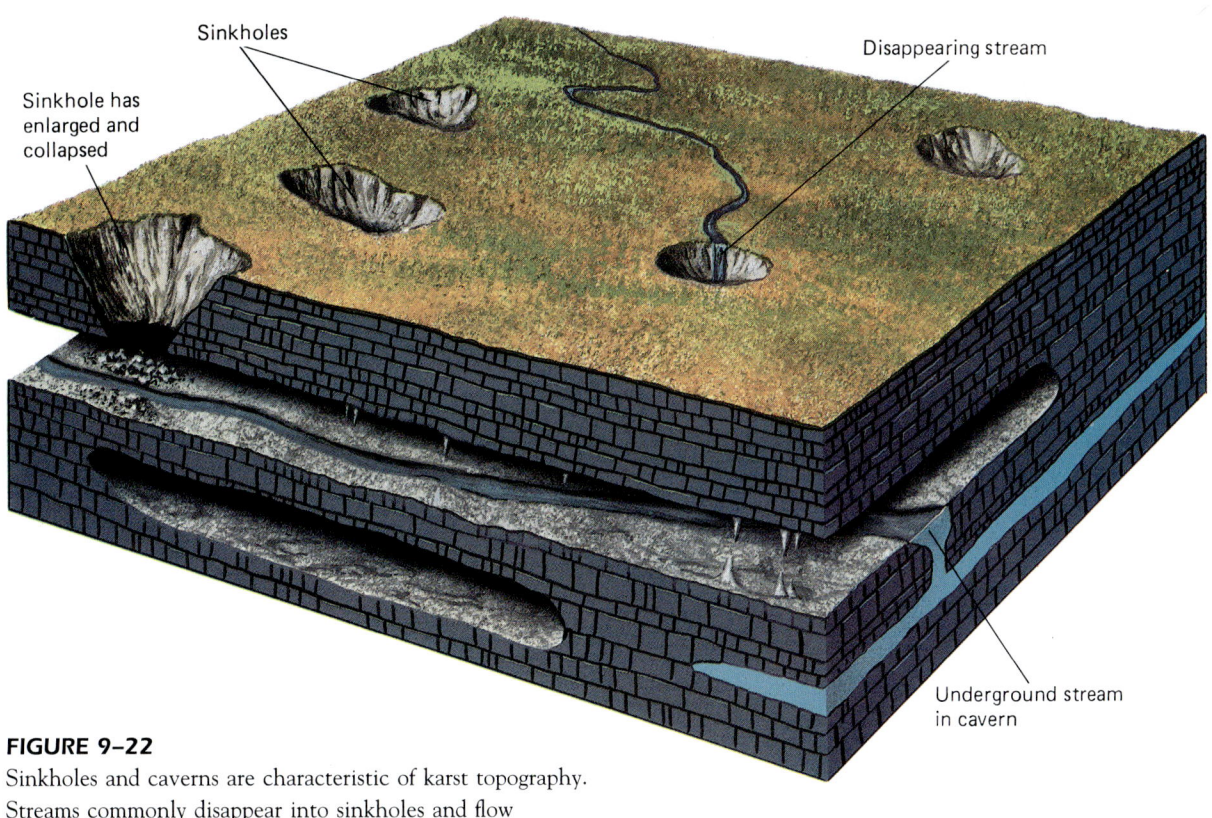

FIGURE 9–22
Sinkholes and caverns are characteristic of karst topography. Streams commonly disappear into sinkholes and flow through the caverns to emerge elsewhere.

FIGURE 9–23
This sinkhole in Winter Park, Florida, collapsed suddenly in May 1981, swallowing several houses and a Porsche agency.
(Wide World Photos/Associated Press)

Karst Topography

Karst topography forms in broad regions underlain by limestone and other readily soluble rocks. Caverns and sinkholes are common features of karst topography (Fig. 9–24). Surface streams often pour into sinkholes and disappear into caverns. In the area around Mammoth Caves in Kentucky, streams are given names such as Sinking Creek, an indication of their fate.

The word *karst* is derived from a region in Croatia where this type of topography is well developed. Karst landscapes are found in Alabama, Tennessee, Kentucky, southern Indiana, and northern and central Florida. Extensive karst landscapes also occur in China.

9.5 Lakes

Lakes and lake shores make up some of the most attractive recreational and living environments on Earth. Clean, sparkling water, abundant wildlife, beautiful scenery, many kinds of aquatic recreation, and fresh breezes all come to mind when we think of going to the lake. Despite

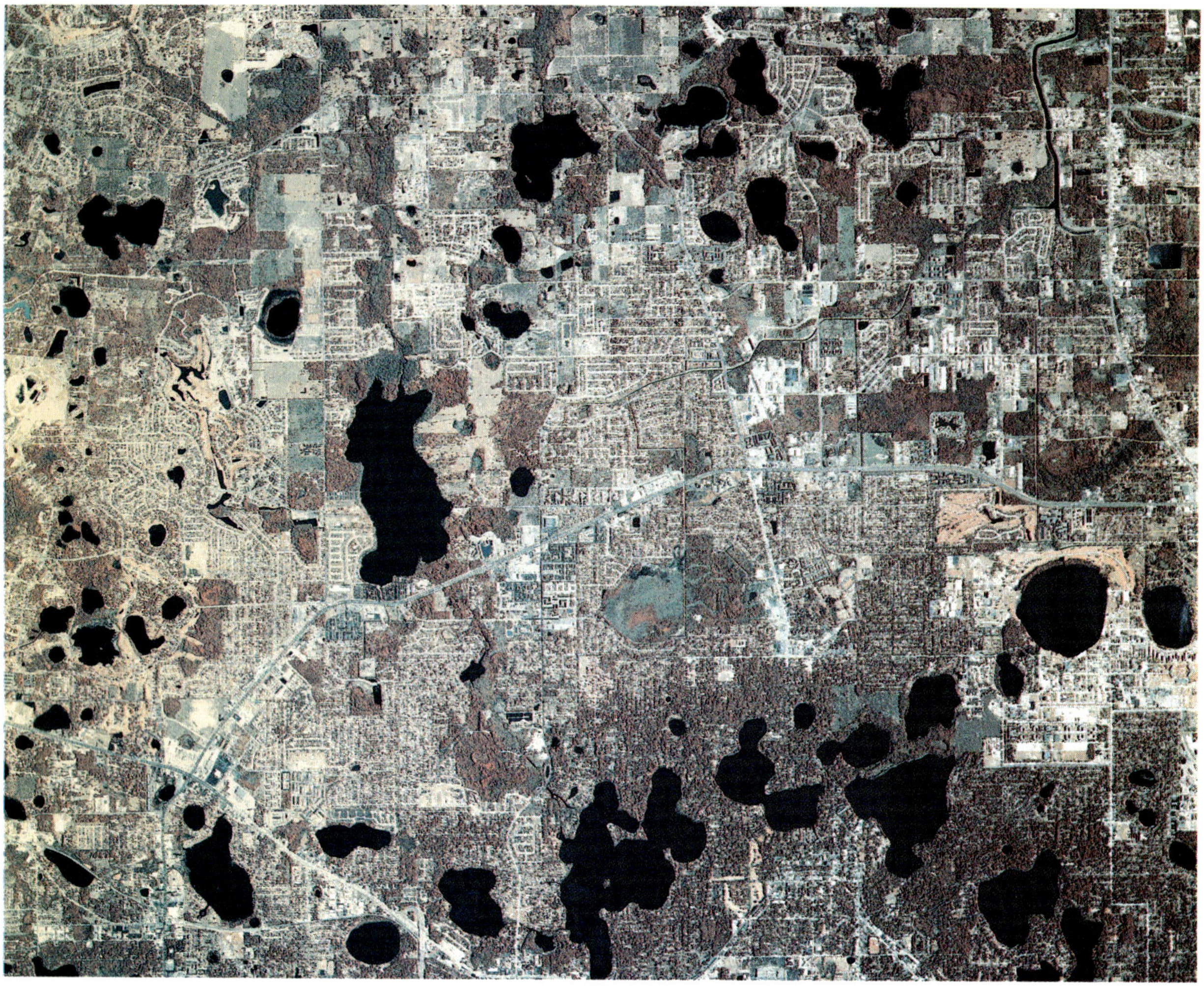

FIGURE 9–24
Aerial view of karst topography, Winter Park, Florida. The lakes are sinkholes.

the great value that we place on them, lakes are among the most fragile and ephemeral landforms. Modern, post–ice age humans live in a special time in Earth history when the Earth's surface is dotted with numerous beautiful lakes.

The Life Cycle of a Lake

A **lake** is a large inland body of standing water that occupies a depression in the land surface (Fig. 9–25). A lake forms where the water table intersects the surface in such a depression. But streams flowing into the lake carry sediment, which fills the depression in a relatively short time, geologically speaking. Soon the lake becomes a swamp, and with time the swamp fills further to become a meadow or forest with a stream flowing through it.

FIGURE 9–25
Lago Nube in Bolivia's Cordillera Apolobamba.

If most lakes fill quickly with sediment, why are lakes so abundant today? Most, although certainly not all, lakes exist in places that were covered by glaciers during the most recent ice age. As described in Chapter 13, less than 20,000 years ago, great continental ice sheets extended well south of the Canadian border, and mountain glaciers flowed into nearby lowlands as far south as New Mexico and Arizona. Similar ice sheets and alpine glaciers scoured higher latitudes of the Southern Hemisphere. We are just now emerging from that glacial episode.

The glaciers created lakes in several different ways. Flowing ice eroded numerous depressions in the land surface, which then filled with water. The Great Lakes and the Finger Lakes of upper New York State are examples of large lakes occupying glacially scoured depressions.

The glaciers also deposited huge amounts of sediment as they melted and retreated. Because mountain glaciers flow down stream valleys, some of these great piles of glacial debris formed dams across the valleys. When the glaciers melted, streams flowed down the valleys again but were blocked by the dams. Many modern lakes, such as Montana's Flathead Lake, occupy such glacially dammed valleys (Fig. 9–26). In addition, the melting glaciers left huge blocks of ice buried within the sediment that they deposited. Over time, the ice blocks also melted, forming depressions that filled with water. Many thousands of small lakes and ponds, called **potholes**, formed in this way. Pothole lakes are common in Montana, the Dakotas, Minnesota, Wisconsin, and the southern Canadian prairie (Fig. 9–27).

Most of these glacial lakes formed within the past 10,000 years, and sediment is rapidly filling many of

FIGURE 9–27
Pothole lakes in the Flathead Valley, Montana.

them. In the near future, geologically speaking, they will fill with mud and cease to exist. Many smaller lakes have already filled and are now swamps; others will continue to exist for hundreds or thousands of years longer. The largest, such as the Great Lakes, may continue to exist for tens of thousands of years. But the life spans of lakes such as these are limited, and it will take another glacial episode to replace them.

Lakes also form by nonglacial means. As described in Chapter 6, a volcanic eruption can create a crater that fills with water to form a lake, such as Crater Lake in Oregon. Other lakes form in abandoned river channels, such as the oxbow lakes on the Mississippi River flood plain, or in flat land with a shallow water table, such as Lake Okeechobee of the Florida Everglades. These types of lakes, too, fill with sediment and, as a result, have limited lifetimes.

A few lakes, however, form in ways that extend their lives far beyond that of a normal lake. For example, Russia's Lake Baikal is a large, deep lake lying in a depression created by an active fault. Although rivers pour sediment into the lake, movement of the fault repeatedly deepens the basin as it fills. As a result, the lake has continued to exist for more than a million years—so long that indigenous species of seals and other animals and fish have evolved in its ecosystem.

Oligotrophic and Eutrophic Lakes

When plants and animals die in a lake, their bodies settle slowly to the bottom, carrying essential nutrients with them. If the water is deep enough, the nutrients accumu-

FIGURE 9–26
A moraine (a pile of glacial debris) near one end of Flathead Lake, Montana. A series of such moraines dammed the valley and formed the lake.

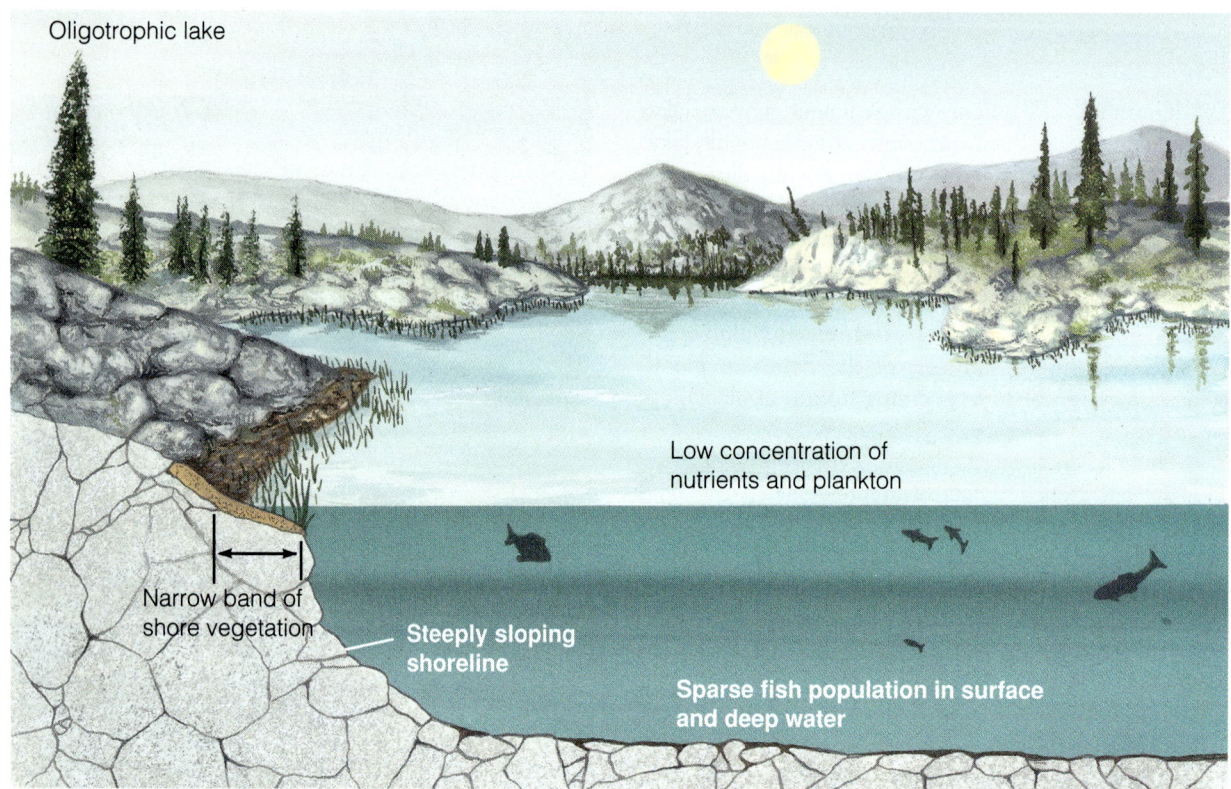

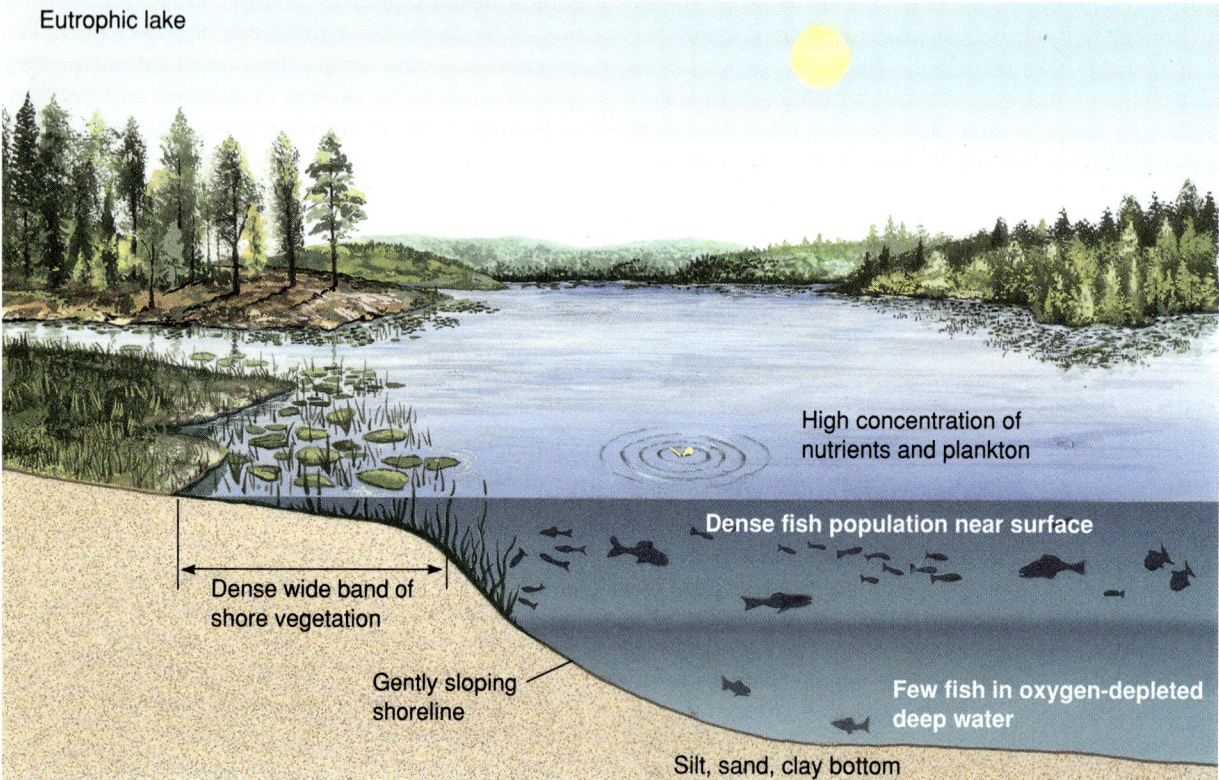

FIGURE 9-28
(A) An oligotrophic, or nutrient-poor, lake typically contains pure water and few organisms.
(B) A eutrophic, or nutrient-rich, lake commonly contains many fish and other organisms.

late below the zone where adequate sunlight for plant growth penetrates the water. Thus, in a deep lake, sunlight is available near the surface, but nutrients are available only on the bottom. Plankton (small, free-floating organisms) grow poorly on the surface due to the lack of nutrients, and bottom-rooted plants cannot grow due to lack of sunlight. Thus, the lakes contain nearly pure water with low concentrations of nitrates, phosphates, and other critical nutrients that sustain aquatic food webs. The purity of the water gives these lakes a deep blue color that we associate with a clean, healthy lake. Such a lake is called **oligotrophic**, meaning "poorly nourished" (Fig. 9–28A). Oligotrophic lakes have low productivities, meaning that they sustain relatively few living organisms, although a lake of this type typically contains a few huge trout or similar game fish.

As a lake fills with sediment, however, it becomes shallower and sunlight reaches more and more of the lake bottom. The sunlight allows bottom-rooted plants to proliferate. As they die and rot, their litter adds nutrients to the lake water. As a result, plankton increase in numbers, as do fish and other organisms. The lake becomes so productive that its surface may become covered with a green scum of plankton or a dense mat of rooted plants. As the lake becomes more productive, the fallen litter contributes to the sediment filling the lake, and eventually the lake becomes a swamp. A lake of this kind with a high nutrient supply is called a **eutrophic** lake (Fig. 9–29B). Eutrophication occurs naturally as part of the life cycle of a lake. However, addition of nutrients in the forms of sewage and other types of pollution has greatly accelerated the eutrophication of many lakes. This subject is discussed in Chapter 11.

A

B

FIGURE 9–29
(A) Crater Lake, Oregon, has a low nutrient content and therefore the plant population is low, leaving clear, blue water. *(Rich Buzzelli/Tom Stack and Associates)* (B) A eutrophic lake in western New York, covered with a slimy mat of algae and bacteria. *(W. A. Banaszewski/Visuals Unlimited)*

Fresh Water and Salty Lakes

Most lakes contain fresh water, but a few, such as Utah's Great Salt Lake, are salty, some more so than the oceans (Fig. 9–30). Most lakes are fresh-water lakes because streams flow both into and out of them. The constant flow of water through the lakes keeps their water fresh. In contrast, a salty lake forms when streams flow into the lake but no streams flow out. Water leaves the lake only by evaporation and small amounts of seepage into the ground. All streams, even the purest mountain brooks, contain small amounts of dissolved salts. The streams carry the salt to the lake, but evaporation removes only pure water. Thus, over time the small amounts of dissolved salts carried in by the streams concentrate in the lake water. Such salty lakes usually occur in desert and semiarid basins, where dry air and sunshine evaporate water effectively.

FIGURE 9–30
Mining salt deposits near the shore of Great Salt Lake, Utah.

Temperature Layering and Turnover in Lakes

If you have ever dived into a deep lake on a summer day, you probably discovered that the top meter or so of lake water can be much warmer than deeper water. Thermal stratification occurs because sunshine warms the upper layer of water, making it less dense than the cooler, deeper water. The warm, less dense water floats on the cooler, denser water. The boundary between the warm and cool layers is called the **thermocline** (Fig. 9–31).

In temperate climates, colder autumn weather cools the upper layer to a temperature below that of deeper water. Now the surface water is more dense than the deeper water. Consequently it sinks, mixing the surface and deep waters and equalizing the water temperature throughout the lake. This process is called **fall turnover**.

As winter comes and the lake continues to cool, the surface water eventually reaches the critical temperature of 4°C. At this temperature water attains its maximum density: It is less dense at both warmer and cooler temperatures. Therefore, when the surface water temperature falls below 4°C, the coldest water floats to the surface. Fall turnover stops and the lake becomes layered again. Eventually, the surface water cools to 0°C and freezes. Because ice is less dense than water, it remains on the lake surface and the deeper water is warmer than the frozen lake surface. When spring comes, the ice melts and the surface water temperature rises to 4°C. Because

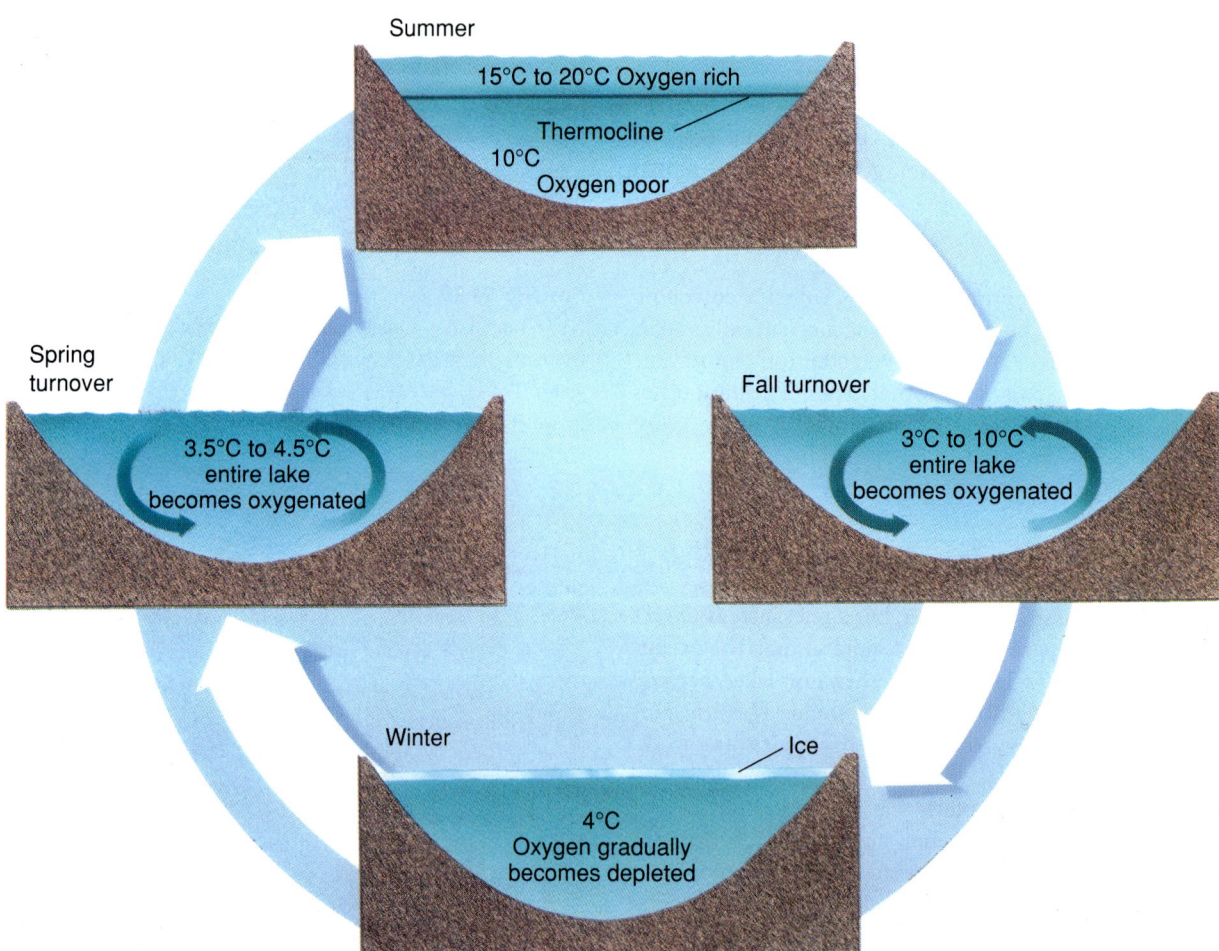

FIGURE 9–31

Lakes in temperate climates develop temperature layering in both summer and winter. As a result, the oxygen in bottom waters becomes depleted. In fall and spring, water temperature becomes constant throughout the lake and turnover brings new supplies of oxygen to the deep waters.

of its density, the surface water sinks and mixing occurs again in **spring turnover**. As summer comes, the lake again develops thermal layering.

Turnover in a temperate lake is important to the aquatic ecosystem. During summer and winter when the lake water is layered, bottom-dwelling organisms may use up most or all of the oxygen in deep waters. At the same time, surface organisms may deplete the dissolved nutrients in surface waters. However, surface water is rich in oxygen because it is in contact with the atmosphere, and deep water may be rich in nutrients because it is in contact with bottom sediment. Turnover enriches the deep water with oxygen and, at the same time, supplies nutrients to the surface water. The latter effect often becomes evident in the form of an algal bloom—a sudden and obvious increase in the amount of floating green algae on a lake's surface—in spring and fall.

Turnover normally occurs only in lakes in temperate climates, where air temperature varies with the seasons. In contrast, many tropical lakes experience no turnover and keep their thermal layering intact for years. As a result, their ecosystems can differ greatly from those of temperate lakes. Lack of lake turnover caused a disaster in Lower Nyos, a small village in Cameroon, Africa, in 1986. Lake Nyos occupies a volcanic crater. Carbon dioxide gas from the dormant volcano had slowly seeped into the bottom of the lake for many years. If the lake had turned over seasonally as a temperate lake does, the gas would have escaped to the atmosphere safely in small, seasonal doses. However, because of the lack of turnover, the gas accumulated in the bottom waters and mud of the lake bed. Then in 1986, for an unknown reason a large amount of the gas suddenly escaped from the lake, forming a cloud that enveloped the village for a brief time. Carbon dioxide is not poisonous—it is a constituent of our atmosphere. But the cloud displaced the normal oxygen-rich air, and 1700 people died of asphyxiation.

The Lake Nyos disaster is not a unique geological oddity. Geologists monitor the active volcanoes in Ecuador in an attempt to predict life- and property-threatening eruptions. They monitor the many volcanic lakes of the region for gas accumulations similar to that of Lake Nyos.

9.6 Wetlands

Known across North America as swamps, bogs, marshes, sloughs, mud flats, and many other names, **wetlands** are the boundaries between land and water. Some are water soaked or flooded throughout the entire year; others are dry for much of the year and wet only during times of high water. Some are wet only during exceptionally wet years and may be dry for several years at a time. North American wetlands include frozen Arctic tundra, warm Louisiana swamps, coastal Florida mangrove swamps, boggy mountain meadows of the Rockies, and the immense swamps of interior Alaska (Fig. 9–32). Wetland

A

B

FIGURE 9–32
North American wetlands include (A) the Florida Everglades *(Everglades Natural Historical Association)* and (B) boggy mountain meadows of Montana.

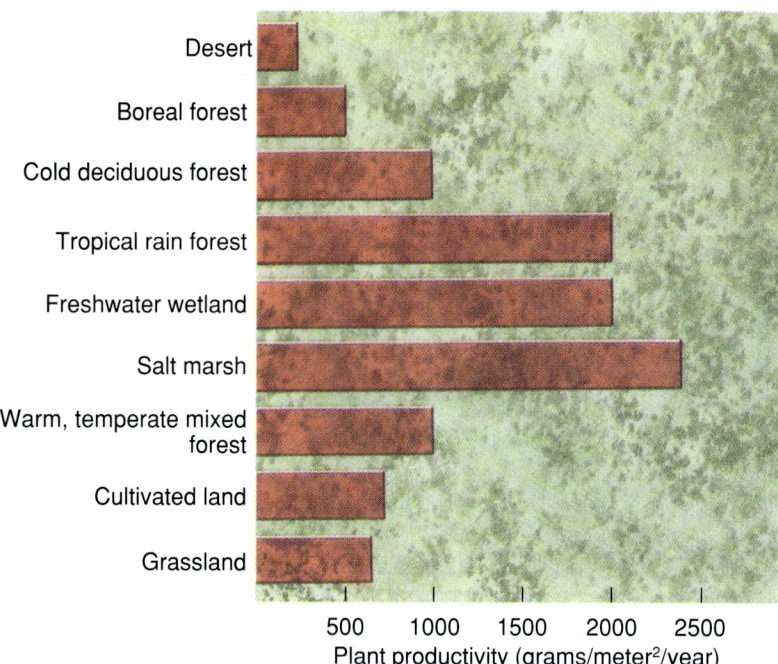

FIGURE 9–33
Wetlands are among the most productive environments on Earth.

ecosystems vary so greatly that the concept of a wetland defies a simple definition. Wetlands share certain properties in common, however: The ground is wet for at least part of the time; the soils reflect anaerobic (lacking oxygen) conditions; and the vegetation consists of plants such as red maple, cattails, bullrushes, mangroves, and other species adapted to periodic flooding or water saturation.

Wetlands are among the most biologically productive environments on Earth (Fig. 9–33). Two thirds of the Atlantic fish and shellfish consumed by humans rely on coastal wetlands for at least part of their life cycles. Between one third and one half of the endangered species of both plants and animals in the United States also depend on wetlands for survival. More than 400 of the 800 species of protected migratory birds and one third of all resident bird species feed, breed, and rest in wetlands.

Wetlands and Humans

When European settlers first arrived in North America, about 87 million hectares of wetlands existed outside of Alaska. Over the past century, farmers, ranchers, and developers (commonly assisted and subsidized by the U.S. government) drained or filled more than half of the original wetlands in what are now the lower 48 states. California and several upper midwestern states have lost more than 80 percent of their wetlands (Fig. 9–34). Wetlands now make up between 6 and 9 percent of the area of the lower 48 states and as much as 60 percent, or about 80 million hectares, of Alaska (Fig. 9–35). Of the 42 million hectares of wetlands that still exist in the lower 48 states, 75 percent are privately owned. About 95 percent are fresh-water wetlands, and the other 5 percent are salty tidewater marshes and estuarian swampland. Currently, between 120,000 and 200,000 hectares of wetlands are destroyed each year.

Wild ducks and geese are among the migratory birds that depend on wetlands for breeding, food, and cover. Many wildlife biologists think that their numbers may reflect the health of America's wetlands. Between 1975 and 1990, the number of breeding ducks and geese declined from 45 million to 31 million, a drop of 31 percent. Some species have been affected more than others. According to the U.S. Fish and Wildlife waterfowl census, the population of northern pintails declined from 10 million in 1956 to 2.5 million in 1988. Black ducks and mallards are half as numerous now as they were in the 1950s.

Why have we destroyed more than half of our wetlands, and why are we continuing to destroy them? Americans have long viewed wetlands as mosquito-infested, malarial swamps occupying land that can be farmed or otherwise developed if drained or filled. In

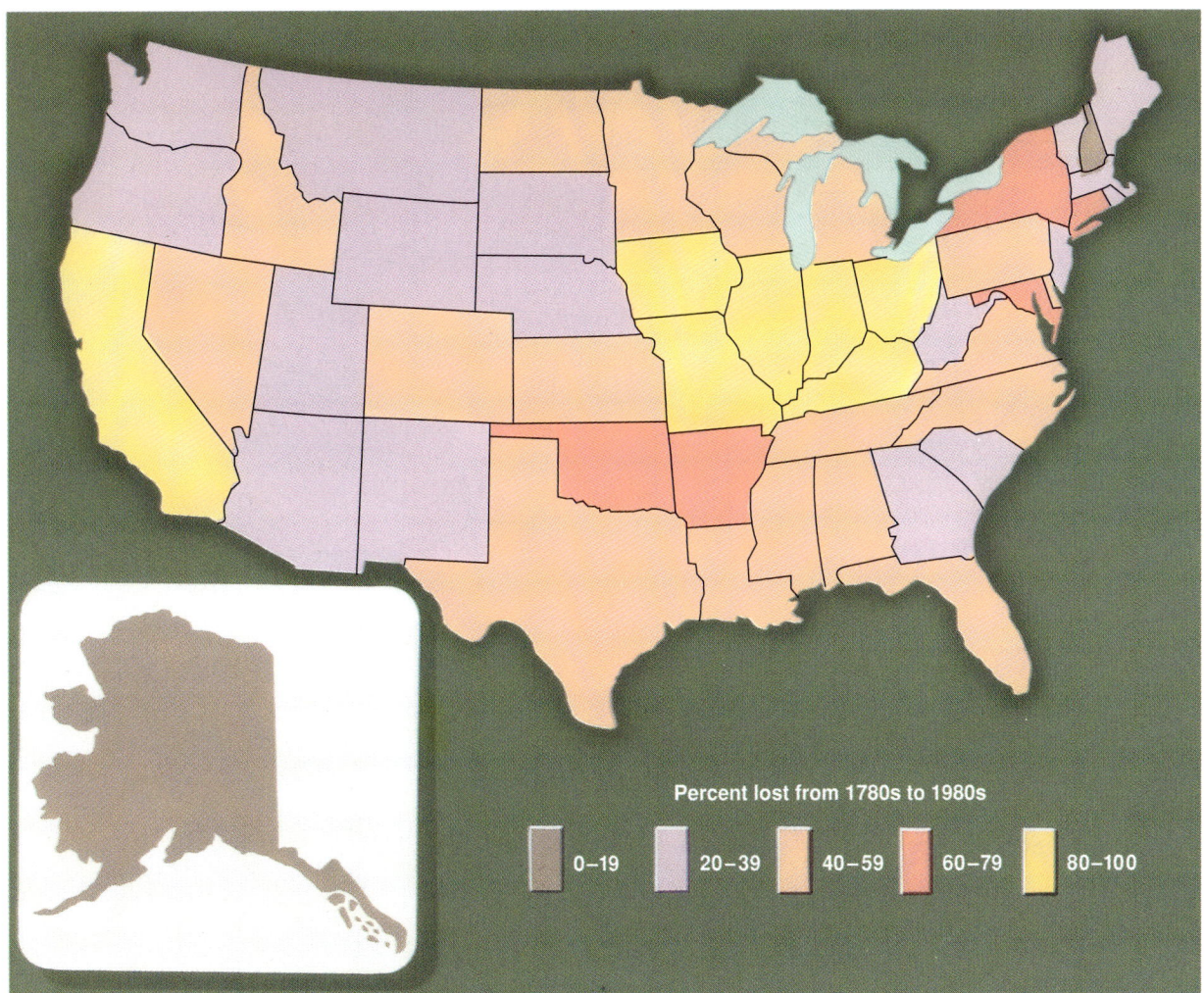

FIGURE 9–34
Wetland losses in the United States in the past 200 years.

the mid-1800s the federal government passed legislation known as the Swamp Land Acts, which established an official policy to fill and drain wetlands wherever possible to convert them to agricultural uses. Since then, farmers and ranchers, supported by both federal and state programs, have worked industriously to convert swampland to fields and pastures. Residential and commercial development have also contributed to the losses.

These government policies and private practices have ignored the beneficial qualities of wetlands. Bacteria, algae, and other aquatic organisms consume many pollutants and degrade them to harmless byproducts. Because such organisms abound in wetlands, these ecosystems are natural sewage treatment systems. Wetlands also mitigate flooding by absorbing excess water that might otherwise overrun towns and farms. One reason that the 1993 Mississippi River floods caused so much damage is that levees prevent flooding of the natural wetlands of the flood plain. (See case history, pp. 214–215.)

The importance of wetlands in water purification, flood control, and wildlife habitat did not become widely recognized until the 1960s. Only since then have large-scale attempts been made to halt the draining and filling of wetlands. At present, the focus of federal and state laws has changed from destruction of wetlands to their protection and preservation. But it has proved more difficult to reverse practice than to reverse policy, and wetland losses continue today.

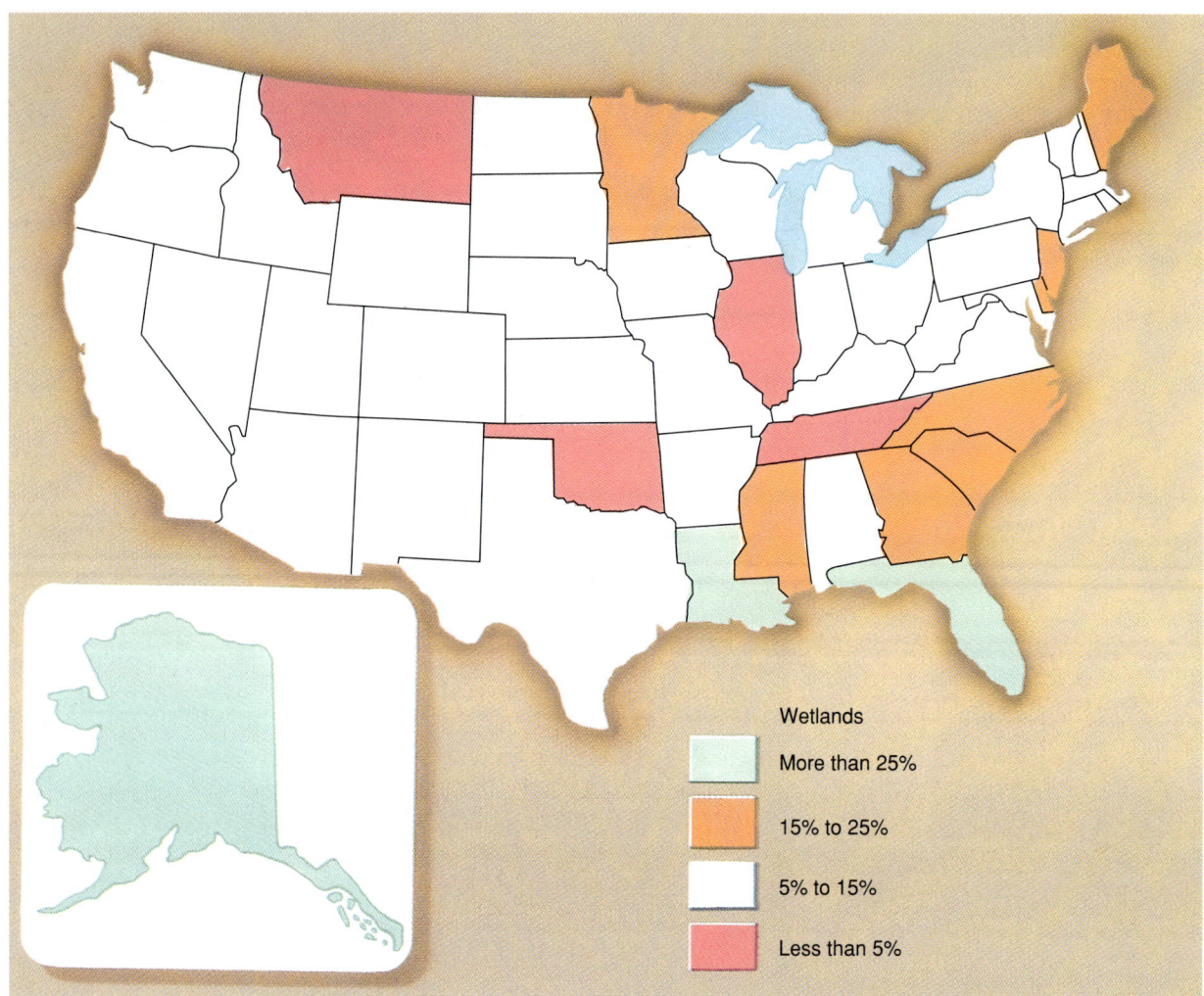

FIGURE 9–35
Abundance of wetlands in the United States.

CASE HISTORY

The Florida Everglades

Southern Florida is a vast, low-lying peninsula of porous limestone. Water saturates the bedrock as if it were a sponge, and the water table is so high that much of the peninsula is wetlands. In its natural state prior to 100 years ago, the peninsula had a fresh-water ecosystem that was unique on Earth. Rain falling on the lakes south of Orlando flowed into the Kissimmee River (Fig. 9–36).

The Kissimmee carried the water southward through palmetto groves and grasslands to Lake Okeechobee, an 1800-square-kilometer lake with an average depth of only 3.6 meters.

During the rainy summers, the lake spilled over its southern rim, forming an 80-kilometer-wide sheet of water that flowed southward at a rate of about 30 meters per day. This process repeatedly renewed the fresh-water ecosystem of the Everglades, keeping the swamps wet and the uncountable small lakes and ponds full of water during the wet season. The water took about a year to flow from Lake Okeechobee into Florida Bay at the southern tip of the peninsula, or into the Gulf of Mexico via Shark River Slough. In the process, some of it flowed through what is now suburban Miami.

Animals and plants were uniquely adapted to the Everglades. The swamp consisted of a sawgrass plain dotted with tree islands. The tree islands provided nesting grounds and shelter for many species of birds, mammals, and reptiles. During the wet season, the sawgrass grew rapidly. If this growth had continued year round, the Everglades marsh would have filled in with plant matter and turned into a forest. Instead, in the dry winter the water level fell and the grasses dried up. Almost every year, fires started by lightning or people raced across the plain. The timing of these fires was crucial to the ecosystem. The sawgrass dried up during the start of the drought, even when a few centimeters of water remained on the ground. If the grasses burned at this time, the fire did little permanent damage to the marsh. Water protected the peaty soils and the roots of the plants. The grass fires swept rapidly across the plain and burned around the tree islands but did not consume them.

As the yearly drought continued, the water level dropped still further. If not for another peculiar ecological adaptation, the swamp would have dried up and most of the animals in it would have died. But

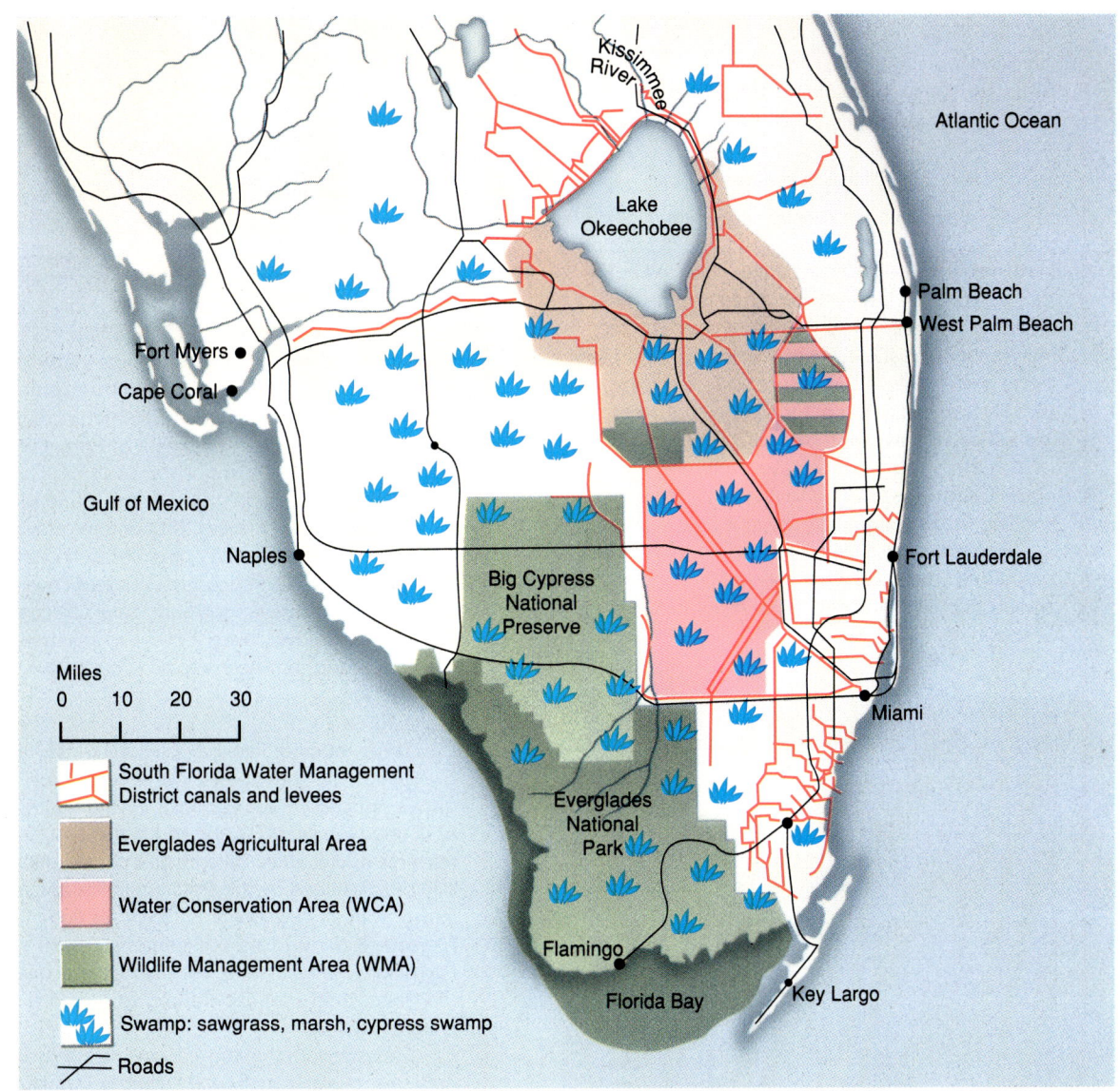

FIGURE 9–36
The Everglades ecosystem of southern Florida has been changed by agriculture and the construction of canals and levees.

FIGURE 9–37
Alligators scoop out depressions to collect water during the dry season in the Everglades. *(C. Walker, Everglades Natural Historical Association)*

the alligators saved the swamp ecosystem (Fig. 9–37). These animals scooped out large depressions in the marsh with their tails. Water collected in the "gator holes," and as the plain dried up, fish migrated to the holes to survive the drought. In fact, much of the aquatic life of the region concentrated in the gator holes to survive until the next rainy season. This concentration of food enabled many species of animals to feed, raise their young, and survive. In these complex ways the plants and animals of the Everglades adapted to a cycle of seasonal flooding, growth, fire, and drought (Fig. 9–38).

However, the growing human population in southern Florida has diverted ever-increasing amounts of the water needed to keep the Ever-

FIGURE 9–38
The animals and plants of the Everglades have adapted to a complex cycle of seasonal flooding, drought, and fire. *(Lynn M. Stone)*

glades wet. Today, more than 5 million people live in south Florida. Each one uses 750 liters of fresh water per day, most of it extracted from the Everglades ecosystem. Intensive cultivation of sugarcane and vegetables on 180,000 hectares of drained swampland in the Everglades Agricultural Area extracts even more water from the ecosystem. In addition, human sewage and agricultural pollutants have changed the composition of the water that flows through the system. As a result of these and other changes, 60 percent of south Florida's wetlands have vanished, and the Everglades is considered one of the most endangered large ecosystems on Earth.

The changes began just after the Civil War, when a boat canal was built to connect Lake Okeechobee to the Gulf Coast. The canal lowered the level of the lake and interrupted the southward flow of fresh water to the Everglades. Then, beginning early in the 1900s, canals, dams, and locks were built to drain the swamps and increase the amount of land available for agriculture. As a result, the level of Lake Okeechobee dropped further, and the supply of fresh water to the Everglades diminished again. Disastrous hurricanes in 1926 and 1928 burst levees surrounding the lake, causing floods that killed 2000 people. Another hurricane in 1947 did much additional damage. In an effort to control nature's fury and at the same time further increase the amount of agricultural land, the U.S. Army Corps of Engineers began the Central and Southern Florida Flood Control Project in 1948. Today, 3060 kilometers of canals and levees, 30 pumping stations, hundreds of culverts and control gates, and three human-made water impoundment areas covering 3300 square kilometers surround Lake Okeechobee. The Kissimmee River, once a meandering stream 160 kilometers long, has been replaced by a 100-meter-wide, 80-kilometer-long straight canal and has been renamed Canal 38.

These modifications have separated the modern Everglades into three parts. In the north, the Everglades Agricultural Area consists of 280,000 hectares of drained wetlands. To the south and east, 530,000 hectares of wetlands, called water conservation areas, are managed by the State of Florida and are protected from all development except for flood control projects. The third part, at Florida's southern tip, is Everglades National Park, the first national park in the United States set aside purely to protect a threatened ecosystem.

Intensive farming in the Everglades Agricultural Area diverts water that once flowed southward through the Everglades. At least as importantly, however, the farms add nutrients to the water that does flow into the surrounding wetlands. For example, to grow sugarcane on land that once was under water, the land is drained with pumps. The newly exposed soil is rich in organic matter. When this soil is exposed to air, it oxidizes and releases both nitrates and phosphates. In addition, the growers use fertilizers, which supply more nutrients to the cultivated ground. North of the Agricultural Area, manure from 45,000 dairy cows adds to the phosphates entering the aquatic system. Rains flush the nutrients into surrounding canals, which carry them to nearby wetlands. These nutrient-rich waters reenter the southward-flowing water system and pass through the remainder of the wild Everglades.

The nutrient-contaminated runoff from the Agricultural Area contains 10 to 20 times normal concentrations of phosphates and nitrates. Some of it flows into Lake Okeechobee, hastening the natural aging and eutrophication of that shallow lake. In August 1986, nutrient-fed algal blooms covered more than 15 percent of the lake surface. In an effort to preserve the lake, state agencies stopped pumping agricultural runoff to the lake and instead pumped it into the water conservation areas.

Two consequences of these hydrologic modifications threaten the Everglades: (1) The reduced water flow from north to south is drying up much of the remaining wetlands, and (2) pollutants and nutrients from the Everglades Agricultural Area are increasing plant growth, which threatens to fill the marsh with litter. The chemicals also alter the composition of Everglades water, which in turn affects populations of some of the Everglades' famous birds, animals, and plants. In the past 25 years, the number of nesting pairs of breeding wood storks, an endangered species, has dropped by 80 percent. The number of nesting wading birds has declined from 300,000 sixty years ago to 15,000 today. Cattails are replacing native sawgrass at rates up to 1.6 hectares per day in some parts of the water conservation areas. More than 6000 hectares of turtle grass, an abundant sea grass in the Florida Bay estuary, have disappeared since 1987.

Although federal and state agencies and private conservation groups are now working to preserve the Everglades ecosystem, more than half of it no longer exists; it has been irreversibly converted to very different environments. The future of what remains is in doubt.

SUMMARY

Only about 0.65 percent of the Earth's water is fresh. The rest is salty seawater and glacial ice. **Evaporation, transpiration, precipitation**, and **runoff** continuously recycle water among land, sea, and the atmosphere in the **hydrologic cycle**.

A **stream** is any body of water flowing in a **channel**. The velocity of a stream is determined by its **gradient** and **discharge**. A **flood** occurs when a stream overflows its banks and flows over its **flood plain**. Floods are the most common and costly of all geologic hazards.

Ultimate base level is the lowest elevation to which a stream can erode its bed. It is usually sea level. A lake or resistant rock can form a **local**, or **temporary base level**. A **graded stream** has a smooth, concave profile. **Downcutting**, lateral erosion, and mass wasting combine to form a stream valley. Mountain streams downcut rapidly and form V-shaped valleys, whereas lower-gradient streams form wider valleys by lateral erosion and mass wasting. A **drainage basin** is the region drained by a single stream. Streams erode the land surface to flatten rugged topography, but tectonic activity uplifts the land at the same time that erosion levels it.

Much of the rain that falls on land seeps into soil and bedrock to become **ground water**. Ground water saturates the upper few kilometers of soil and bedrock to a level called the **water table**, which provides drinking water for more than half of the population of the United States and is a major source of irrigation and industrial water. **Porosity** is the proportion of rock or soil that consists of open space. **Permeability** is the speed with which fluid can move through pores in soil or bedrock. An **aquifer** is a body of rock that can yield economically significant quantities of water. Aquifers are porous and permeable.

Most ground water moves slowly, about 4 centimeters per day. **Springs** occur where the water table intersects the surface of the land. Dipping layers of permeable and impermeable rock can produce an **artesian aquifer**.

Caverns form where ground water dissolves limestone. A **sinkhole** forms when the roof of a limestone cavern collapses. **Karst topography**, with numerous caves, sinkholes, and subterranean streams, is characteristic of limestone regions.

Lakes are short-lived landforms because streams fill them with sediment. Many modern lakes were created by recent glaciers; as a result, we live in an unusual time of abundant lakes. The life history of a lake commonly involves a progression from an **oligotrophic** lake to a **eutrophic** lake to a swamp and, finally, to a flat meadow or forest. Temperate lakes normally develop temperature layering but experience **turnover** twice a year, which mixes the lake water and supplies oxygen to deep waters and nutrients to surface waters.

Wetlands are among the most biologically productive environments on Earth. In addition, they are natural water purification systems, and they mitigate flood effects by absorbing flood waters that might otherwise destroy property. Despite these facts, public and private efforts have destroyed more than half of the wetlands that existed in the lower 48 states when European settlers first arrived. The rate of wetland loss has diminished since the 1960s, when the focus of federal and state laws changed from draining wetlands to preserving them. However, we continue to lose wetland ecosystems at a rate of 120,000 to 200,000 hectares per year.

KEY TERMS

Hydrologic cycle 206
Transpiration 206
Precipitation 206
Runoff 206
Surface water 206
Ground water 206
Residence time 206
Stream 207
River 207
Tributary 207
Channel 207
Bed 207
Bank 207
Flood 207
Flood plain 207
Gradient 207
Discharge 208
Competence 208

Capacity 208
Downcutting 209
Base level 209
Ultimate base level 209
Local base level 209
Temporary base level 209
Graded stream 209
Meander 210
Oxbow lake 210
Drainage divide 211
Drainage basin 211
Natural levee 213
Artificial levee 214
Porosity 216
Permeability 216
Zone of saturation 217
Water table 217
Zone of aeration 217

Soil moisture belt 217
Aquifer 218
Recharge 218
Spring 219
Perched water table 220
Artesian aquifer 220
Artesian well 220
Cavern 220
Sinkhole 220
Karst topography 221
Lake 222
Oligotrophic 225
Eutrophic 225
Thermocline 226
Fall turnover 226
Spring turnover 227
Wetlands 227

REVIEW QUESTIONS

1. In which physical state (solid, liquid, or vapor) does most of the Earth's free water exist? Which physical state accounts for the least?

2. Describe the movement of water through the hydrologic cycle.

3. Describe the residence times of water in the various parts of the hydrologic cycle.

4. Describe the factors that determine the velocity of stream flow and how those factors interact.

5. What is meant by the term *100-year flood*? By the term *10-year flood*?

6. Give two examples of natural features that create temporary base levels. Why are they temporary?

7. Draw a profile of a graded stream and one of an ungraded stream.

8. Explain how a stream forms and shapes a valley.

9. In what type of terrain would you be likely to find a V-shaped valley? Where would you be likely to find a meandering stream with a broad flood plain?

10. How can a stream become rejuvenated? Give an example of a landform created by a rejuvenated stream.

11. (a) Draw a cross section of soil and shallow bedrock, showing the zone of saturation, water table, and zone of aeration. (b) Explain each of the preceding terms.

12. What is an aquifer, and how does water reach it?

13. Explain why bedrock or soil must be both porous and permeable to be an aquifer.

14. Compare the movement of ground water in an aquifer with that of water in a stream.

15. How does an artesian aquifer differ from a normal one? Why does water from an artesian well rise without being pumped?

16. What is karst topography? How can it be recognized? How does it form?

17. Why are most lakes short-lived landforms?

18. What geological conditions create a long-lived lake?

19. Describe the differences between an oligotrophic lake and a eutrophic lake.

20. What is a thermocline? Under what conditions does a thermocline form in a lake?

21. What conditions cause mixing, or turnover, of surface and deep waters in a temperate lake?

22. Describe a wetland. What types of environments are included in the wetlands category?

23. What proportion of the wetlands that existed in what are now the lower 48 states when the early European settlers arrived no longer exist?

24. Describe how wetlands mitigate flooding.

DISCUSSION QUESTIONS

1. Discuss why the residence times of water vary greatly among the different parts of the hydrologic cycle.

2. Describe the ways in which (a) a rise and (b) a fall in the average global temperature could affect the Earth's hydrologic cycle.

3. A stream is 50 meters wide at a certain point. A bridge is built across the stream, and the abutments extend into the channel, narrowing it to 40 meters. Discuss the changes that might occur as a result of this constriction.

4. Describe and discuss effects of flood control structures, such as artificial levees, on the damage caused by the 1993 Mississippi River floods.

5. The National Flood Insurance Program (NFIP) is a federally sponsored insurance for people who live on flood plains. In exchange for low-cost insurance policies, the NFIP establishes rigorous building codes for new construction in flood plains. However, older buildings are not subject to these standards. As long as damage from one flood is less than 50 percent of the value of the structure, it is eligible for subsidized insurance against the next flood. Outline potential benefits and drawbacks of this program.

6. Imagine that a 100-year flood has just occurred on a river near your home. You want to open a small business in the area. Your accountant advises that your business and building have an economic life expectancy of 50 years. Would it be safe to build on the flood plain? Why or why not?

7. Imagine that you live on a hill 25 meters above a nearby stream. You drill a well 40 meters deep and do not reach water. Explain.

8. Discuss and explain how a stream can supply water to an underground aquifer in one environment, but an aquifer can supply water to a stream in another environment.

9. Why are caverns, sinkholes, and karst topography most commonly found in limestone terrain?

10. Most substances become denser when they freeze. Water is anomalous in that ice is less dense than water. Outline the seasonal temperature profiles of a lake if ice were denser than water. How would aquatic ecosystems be affected?

11. Discuss why wetlands support such large populations of wildlife.

12. Discuss the mechanisms by which wetland ecosystems purify contaminated water.

13. Discuss how changing public perceptions may have played a role in reversing the governmental policy toward wetlands.

CHAPTER 10

Human Use of Water

10.1 Water in the United States: Supply and Demand
10.2 Water Diversion and Environmental Impacts
10.3 Salinization
10.4 Water Use in the Western United States
10.5 Conservation
10.6 Water and International Politics

More than 2 centuries ago, when fresh water seemed as inexhaustible as the buffalo that roamed the high plains, Benjamin Franklin warned, "When the well's dry, we know the worth of water." Today, as we reach, and in some cases exceed, the limits of our renewable water resources, the well is running dry and we are discovering the real value and cost of water.
  The problem with our water supply is not that the amount of water has run low. As pointed out in the preceding chapter, the hydrologic cycle continuously replenishes fresh water on land. The amount of available fresh water is about the same as it was 2 centuries ago. The problem is that the demand has risen dramatically until it has approached or even exceeded the daily supply in many parts of the world. Two centuries ago, less than 4 million people inhabited the United States; today we number more than 250 million. At the same time, global population has increased by a factor of 5. The growing population, both of the United States and globally, has created demands that stretch our fresh-water supplies to, and in some cases beyond, their renewable limits (Fig. 10–1). In addition, technological advances have made it possible to both consume and pollute our water supplies at rates that would have been inconceivable to Ben Franklin.

Modern agriculture is highly dependent on irrigation.

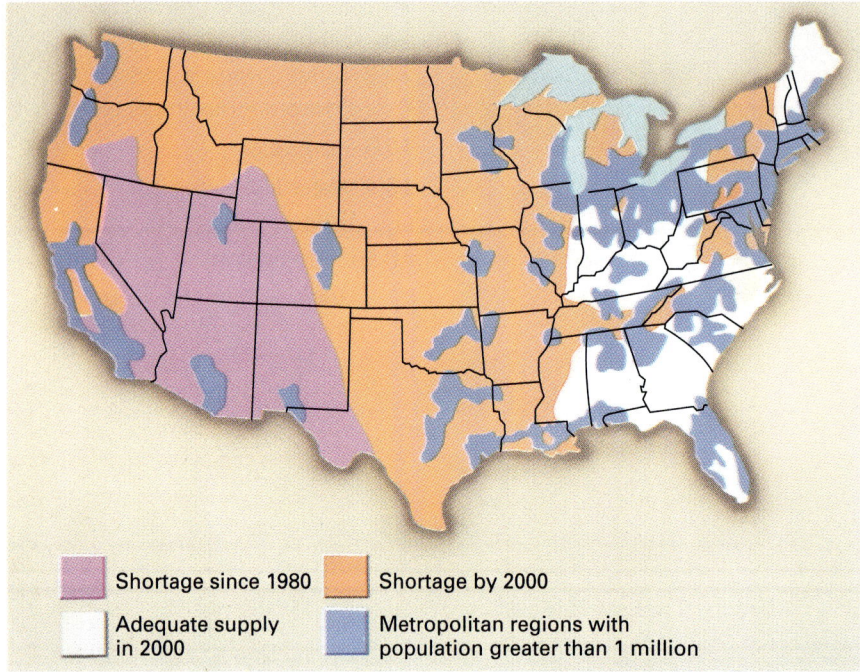

FIGURE 10–1
Water shortages have affected the American Southwest since 1980 and will affect much of the remaining United States, including many metropolitan areas, by 2000. *(Compiled from USGS and U.S. Water Resources Council data)*

10.1 Water in the United States: Supply and Demand

About 17 trillion liters of precipitation fall on the United States each day. Of that amount, 11 trillion liters evaporate and return directly to the atmosphere, leaving 6 trillion liters to flow into streams and to recharge ground water (Fig. 10–2). Americans use about 2 trillion liters per day for all purposes. Thus, the United States receives about three times as much water as it uses. As mentioned in Chapter 9, 30 times more water is stored as ground water than as surface water. However, because surface water flows much more rapidly than ground water, it is replenished more rapidly.

Precipitation is not distributed evenly. Some of the driest regions of the United States are those that use the greatest amounts of water (Fig. 10–3). It is both ironic and inefficient that some of our most productive farmland lies in the deserts and near-deserts of the American West—California, Texas, Arizona, and nearby states. These farms must be watered by extremely expensive, usually publicly funded irrigation systems. At the same

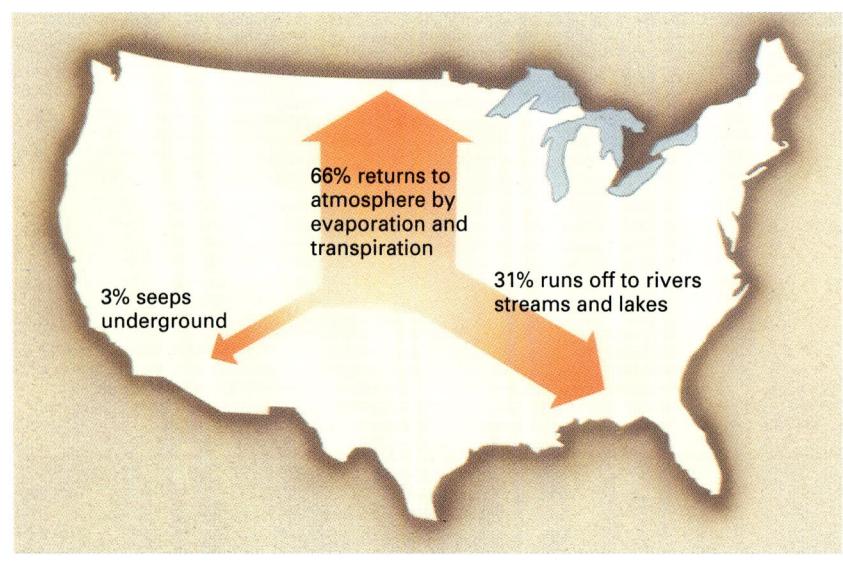

FIGURE 10–2
The United States water budget.

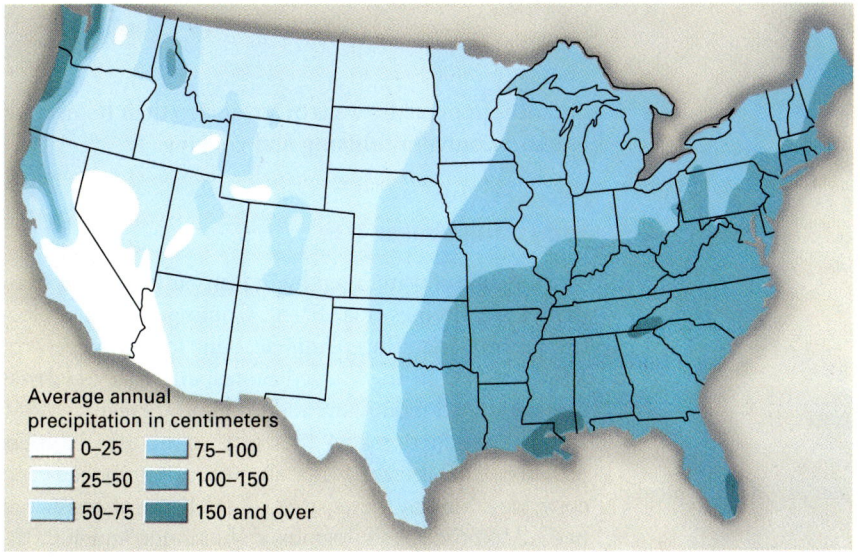

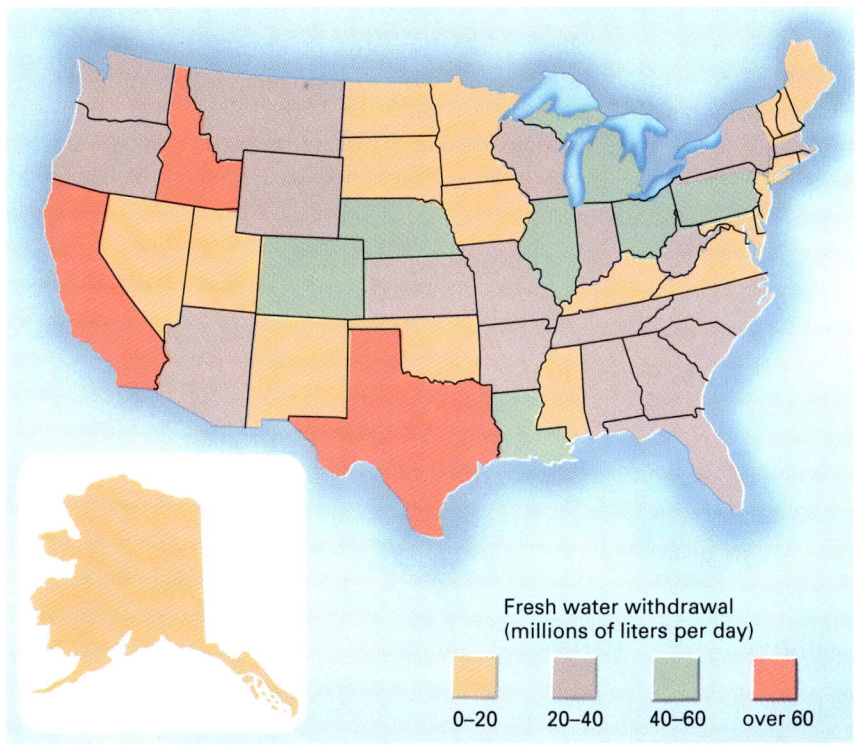

FIGURE 10-3
(A) Average annual precipitation varies greatly throughout the United States. (B) Average annual water use also varies greatly. But regions using the greatest amounts of water often receive little rain.

time, farmers in southeastern states, where rainfall is abundant and irrigation often unnecessary, are paid subsidies to refrain from growing the same crops that are raised and watered by federal irrigation systems in the West.

Demands on water resources fall into three categories: **domestic, industrial**, and **agricultural**. Combining all three categories, the United States uses three times as much water per day—8000 liters per person—as the average European country and hundreds of times more than many developing and less-developed nations.

Domestic Water Use

Domestic use accounts for only 10 percent of the water used in the United States (Fig. 10–4). Each adult drinks about 4 liters of water per day. In the average home, each American uses an additional 4 liters of water a day for cooking. In comparison, flushing a toilet once uses 20 liters, enough to satisfy the drinking and cooking requirements of one person for 2.5 days (Fig. 10–5). The water used to machine-wash one load of laundry or to fill a bathtub is enough for 3 weeks of cooking and drinking.

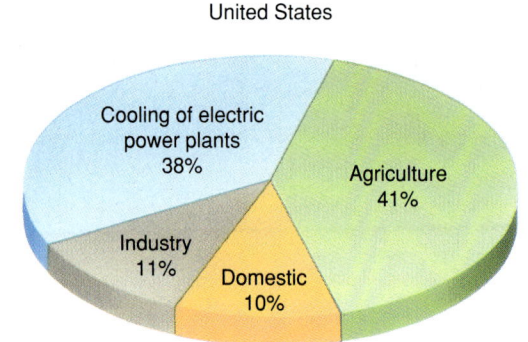

FIGURE 10-4
Industrial, agricultural, and domestic water use in the United States. *(Compiled from Worldwatch Institute data)*

Watering ornamental lawns and gardens uses much greater amounts. All domestic uses of water combined account for an average of 800 liters per day for each person in the United States. In contrast, more than half of the Earth's people get by on less than 100 liters per day. Many of these people must carry their water long distances from public wells or streams to their homes and use water only for drinking and cooking.

Industrial Water Use

The amount of water used by industry is far greater than that used in homes (Fig. 10–6). Cooling systems of electric power generating plants account for 38 percent of all water used in the United States; all other industrial uses combined require an additional 11 percent.

The production of 1 liter of gasoline uses 10 liters of water; 1000 liters of water are needed to produce each copy of a Sunday newspaper, 9000 liters of water are needed to produce a kilogram of aluminum, and 400,000 liters of water are needed to make a single automobile.

Agricultural Water Use

Irrigation and other agricultural uses account for 41 percent of water use in the United States (Fig. 10–7). For example, 1800 liters of water are used to produce 1 dozen eggs, and the production of 1/4 pound of hamburger

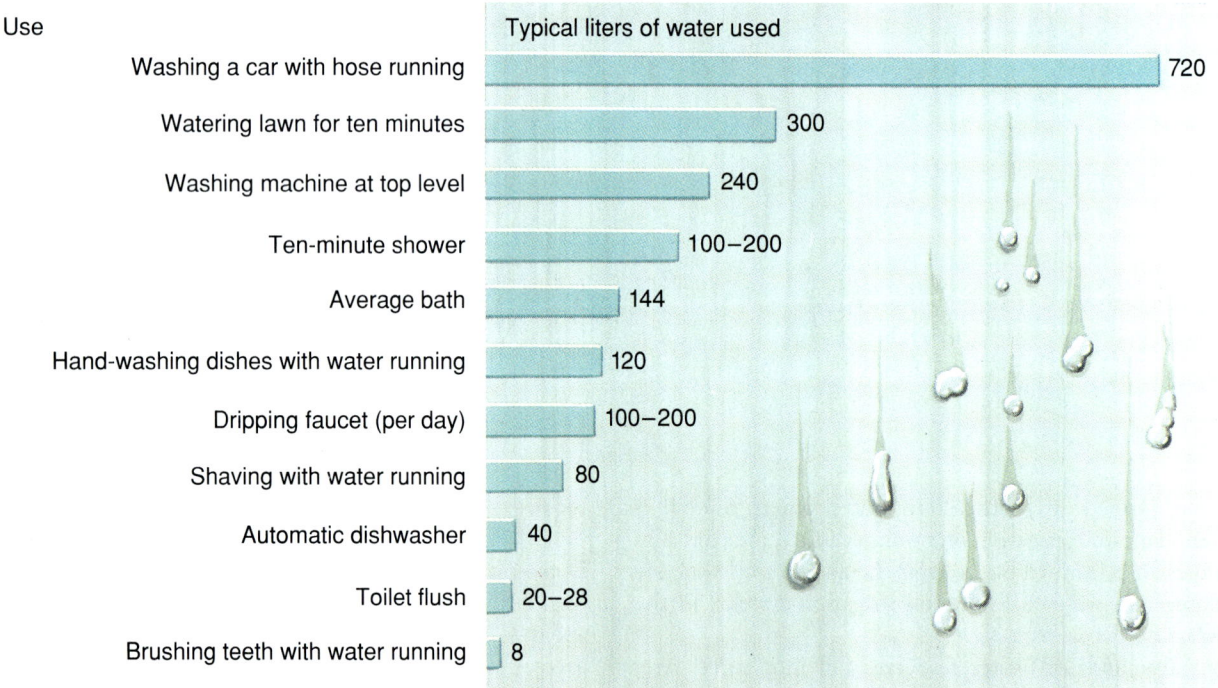

FIGURE 10-5
Domestic water use in the United States varies greatly with the activity. *(Compiled from American Water Works Association data)*

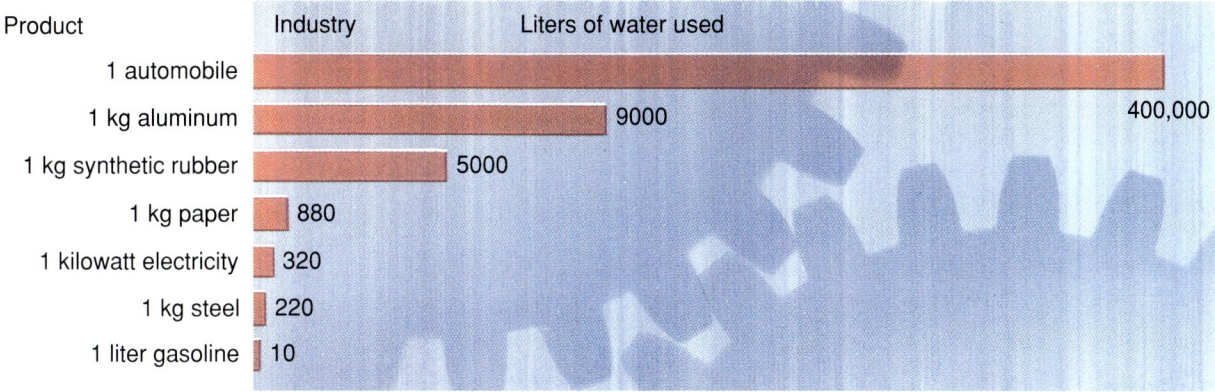

FIGURE 10–6
Amounts of water used to produce common industrial products in the United States.
(Compiled from USGS data)

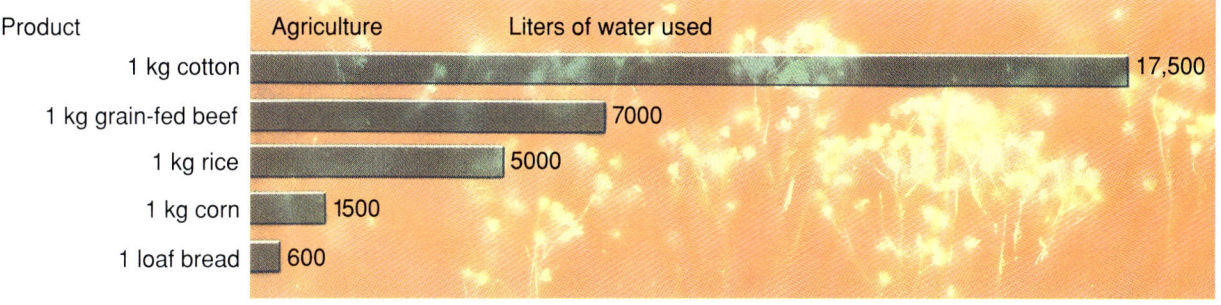

FIGURE 10–7
Amounts of water used to produce common agricultural products in the United States.
(Compiled from USGS data)

FIGURE 10–8
An irrigated cotton field in the Central Valley, California, stands out against the desert hills in the background. *(Craig Aurness/Westlight)*

requires 750 liters of water. Seventeen thousand liters of water are used to grow a single kilogram of cotton (Fig. 10–8).

Croplands are watered by rain and irrigation. Today, rainfall waters 84 percent of the world's farmland; irrigation waters the remaining 16 percent. However, irrigation turns some of the Earth's driest but most fertile land into highly productive farmland: 36 percent of the world's food grows on the 16 percent of farmland that is irrigated. The immensely productive fruit- and vegetable-growing region of California's Central Valley relies entirely on irrigation for its water supply (Fig. 10–9). The productivity of the Great Plains in the western United States would decline by one third to one half if irrigation were to cease. Globally, many nations rely on irrigation for more than half of their food production, and in dry regions an even higher proportion of farmland must be irrigated (Table 10–1).

FIGURE 10-9
Farmers withdraw 41 percent of the water used in the United States. (Bob Waterman/Westlight)

TABLE 10-1 Net Irrigated Area, Top 20 Countries and World, 1989		
Country	Net Irrigated Area[1] (thousand hectares)	Share of Cropland That Is Irrigated (percent)
China	45,349	47
India	43,039	25
Soviet Union	21,064	9
United States	20,162	11
Pakistan	16,220	78
Indonesia	7,550	36
Iran	5,750	39
Mexico	5,150	21
Thailand	4,230	19
Romania	3,450	33
Spain	3,360	17
Italy	3,100	26
Japan	2,868	62
Bangladesh	2,738	29
Brazil	2,700	3
Afghanistan	2,660	33
Egypt	2,585	100
Iraq	2,550	47
Turkey	2,220	8
Sudan	1,890	15
Other	36,664	7
World	235,299	16

[1] Area actually irrigated; does not take into account double cropping.
Sources: Intergovernmental Panel on Climate Change, *Policymakers' Summary of the Potential Impacts of Climate Change: Report from Working Group II to IPCC* (Geneva: World Meteorological Organization/U.N. Environment Programme, 1990); Paul E. Waggoner, "U.S. Water Resources Versus an Announced But Uncertain Climate Change," *Science*, March 1, 1991.

Water Withdrawal and Water Consumption

Most of the water used by industry and used domestically returns to the Earth near the place from which it was taken. For example, river water pumped through an electric generating station to cool the exhaust is returned almost immediately to the river. Little is lost, although all of it is heated, which causes thermal pollution, discussed in Chapter 11. Water used to flush a toilet in a city is pumped to a sewage treatment plant, purified, and discharged into a nearby stream. In these examples, the water is **withdrawn** and then returned to its place of origin.

In contrast, most irrigation water evaporates and disperses with the wind. Thus, although industry accounts for most of the water withdrawn in the United States, agriculture accounts for most of the water that is **consumed** and not returned to its point of origin (Fig. 10–10). Globally, about two thirds of the water that is withdrawn is consumed.

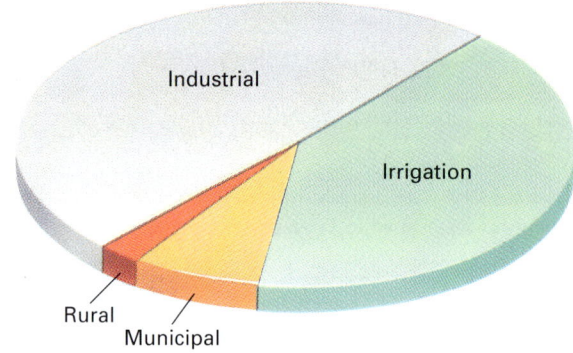

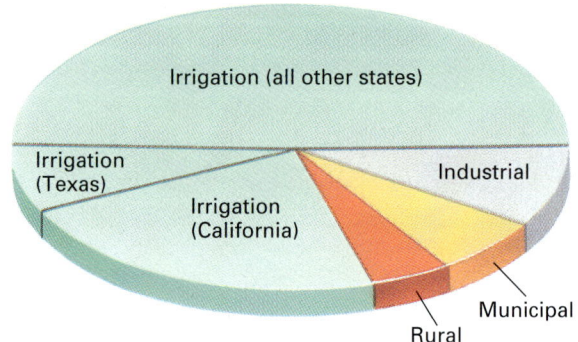

FIGURE 10-10
Although industry accounts for about half of all water *withdrawn* in the United States (top figure), agriculture accounts for more than three quarters of the water that is *consumed* (bottom figure).

10.2 Water Diversion and Environmental Impacts

It seems reasonable to withdraw surface water from a river, lake, or other abundant source and transport it to dry regions where it is needed. In some parts of our country, surface water is scarce but huge reservoirs of ground water lie only a few tens of meters beneath the surface. This ground water can be pumped to the surface for human use. In both cases, water is **diverted** from its natural place and path in the hydrologic cycle and transported to a new place and path to serve human needs.

Water **diversion projects** are nearly as old as agriculture. Archeologists have uncovered extensive irrigation systems in ancient Babylon, and Roman aqueducts still supply water to some European cities. However, the rapidly improving technology of the past 5 or 6 decades has made it possible to move unprecedented amounts of water. Diversion projects have radically changed the distribution of water in many parts of the United States. As mentioned earlier, some of the most productive agricultural regions of the United States lie in the deserts of California, Texas, and Arizona. In these areas, the productivity is entirely supported by diversion of surface water from distant sources or by pumping of ground water. Desert cities such as Phoenix, Los Angeles, Albuquerque, and Tucson support large populations that use hundreds of times more water than is locally available and replenished by the hydrologic cycle. Although water diversion projects support the agricultural productivity of many farming regions of the United States and supply water to most cities, large diversion projects commonly create environmental problems.

Surface-Water Diversion Projects: Dams and Delivery Systems and Environmental Impacts

Most surface-water diversion projects require construction of a **dam** on a stream or lake outlet to collect water in a **reservoir** (Fig. 10–11). Canals, aqueducts, or pipes then transport the water to the consumer. Because dams and delivery systems are extremely expensive to build and maintain, all large dams and aqueduct systems in the United States have been built by the federal government—the U.S. Army Corps of Engineers and/or the Bureau of Reclamation. In some cases, state governments have contributed to the construction and maintenance. Although farmers, ranchers, and cities pay for the water, the amount that they pay is only a small fraction of the actual cost of the water delivery system. As a result, the American taxpayer directly subsidizes all users of this water.

FIGURE 10–11

The Glen Canyon Dam forms Lake Powell on the Colorado River, just upstream from Grand Canyon.

Since antiquity, people have understood that they can create a reservoir by building a dam across a stream. In a region where water is abundant in the spring but scarce in the summer, a reservoir provides flood control and stores water for irrigation and domestic uses. In addition, the potential energy of the water in a dammed reservoir can be harnessed to generate electricity. Energy produced in this manner is called hydroelectric energy. Today, about 5 percent of the total world consumption of energy is supplied by hydroelectric generators. Dams are beneficial to humans, but they also create undesired side effects.

Loss of Water

The reservoir formed by a dam provides more surface area for evaporation and more bottom area for seepage into bedrock than did the stream that preceded it. For example, about 270,000 cubic meters of water per year evaporate from Lake Powell, the lake behind the Glen Canyon Dam across the Colorado River. As a result of these losses, even if no water were diverted from the reservoir, less water would be available downstream. In addition, many irrigation canals are simply ditches excavated in soil or bedrock, and they leak profusely.

Silting

A stream deposits its sediment load when it slows down upon entering a reservoir. Rates of sediment accumulation in reservoirs vary. In a few instances, engineers and politicians have made expensive miscalculations of sedimentation rates. The reservoir behind the Sanmenxia Dam on the Yellow River in China filled 4 years after the dam was finished, making both the dam and reservoir useless. The Tarbela Dam in Pakistan took 9 years to build, and the reservoir is expected to fill with sediment in 20 years. Other reservoirs have longer life expectancies. Lake Mead, behind Hoover Dam in Arizona and Nevada, lost 6 percent of its capacity in its first 35 years as a result of sediment accumulation. At rates such as this, the lakes behind high dams can last hundreds of years, but not forever.

Erosion

A stream has a certain amount of energy that it uses to carry sediment and overcome resistance to flow. Stream sediment settles out in a reservoir, and as a result, the water that flows through or over a dam is clear. Downstream from the dam, the stream has more energy than it needs to overcome resistance to flow because it is not carrying any sediment. Therefore, it flows faster and erodes its bed and banks to increase its sediment load again. This increased erosion may cause both recreational and economic losses. The beautiful beaches on the Colorado River in Grand Canyon are disappearing rapidly as a result of the water management practices of the Glen Canyon Dam, just upstream from the canyon (Fig. 10–12).

If erosion deepens the bed of a large stream, erosion of the tributary streams increases because their temporary base level is lowered. In some cases, this effect causes soil loss on agricultural land.

FIGURE 10–12
Water management practices of the Glen Canyon Dam have eroded the sandy beaches of Grand Canyon to small remnants.

Risk of Disaster

Flood plains are attractive for farming, industry, and commerce, especially when dams promise flood control, irrigation water, and electric power. But unusually heavy rainfall can fill a reservoir and overflow a dam. Furthermore, dams have been known to break. Under such circumstances, the population in the flood plain is vulnerable to disaster.

CASE HISTORY

The Teton Dam Disaster

The U.S. Bureau of Reclamation built the Teton Dam in the potato-farming region of eastern Idaho during the early 1970s, despite warnings by geologists from the U.S. Geological Survey and the University of Montana that the bedrock on which it was built was leaky, weak, and unsafe. The Bureau closed the gates and began filling the reservoir in October 1975. By the following spring, the reservoir was nearly full. In mid-May of 1976, the Teton River was swollen with spring snowmelt, and the water level in the reservoir was rising at a rate of more than a meter per day, much faster than planned.

On June 3, a heavy-equipment operator arriving for work early in the morning noticed water seeping from two sites in the right-hand canyon wall below the dam. The following day, another leak appeared. Apparently, water was seeping through the bedrock and around the dam, exactly as geologists had predicted. By Saturday morning, June 5, a muddy torrent was pouring from a hole in the bedrock near the right abutment of the dam. The amount of water leaking around the dam was visibly increasing, minute by minute. At 9:30 A.M., the dam manager called the sheriffs of counties downstream, telling them to prepare to evacuate 12,000 people. Two bulldozer operators frantically attempted to fill the hole, but the growing gushers washed away the rocks and dirt faster than they could dump it in. At 11:30 the sides of the hole collapsed and both bulldozers fell into the chasm. The operators escaped by jumping from their machines and clawing up the collapsing canyon wall. At 11:55 A.M., the top of the dam fell into the reservoir. Water poured through the gap, and one third of the dam disappeared in less than a minute (Fig. 10–13).

One of the greatest floods since the last ice age then roared down the Teton River canyon. Six miles below the dam, the canyon empties onto a broad flood plain. The town of Wilford, near the junction of the canyon and the flood plain, no longer appears on Idaho maps. The flood waters swept away 154 of the 155 buildings in the town. The waters destroyed the lower half of Rexburg, a larger town farther downstream (Fig. 10–14).

FIGURE 10–13
(A) The Teton Dam before its failure. (B) The Teton Dam as the lake water poured through the breach.
(Bureau of Reclamation)

FIGURE 10–14
After the Teton Dam failed, the city of Rexburg, Idaho, was inundated by silty water. *(Bureau of Reclamation)*

Only 11 people died in the flood because of the effective warnings and evacuations carried out by civil agencies. But if the flood had occurred in the middle of the night, when warnings and evacuations would have been more difficult, thousands might have been killed. The flood damaged or destroyed 4000 homes and 350 businesses. Damage estimates were $2 billion, and the raging waters stripped the topsoil from tens of thousands of acres of farmland. Ironically, many of the area residents are calling for the dam to be rebuilt.

FIGURE 10–15
Glen Canyon before construction of the Glen Canyon Dam.
(Fred S. Finch/U.S. Bureau of Reclamation)

Recreational and Aesthetic Losses

Dams are often built across narrow canyons to minimize engineering and construction costs. But when the canyons are flooded, rare and sometimes unique ecosystems are destroyed. Glen Canyon Dam on the Colorado River just above Grand Canyon floods one of the most spectacular of all desert canyons in the West. Today the canyon is submerged for more than 300 kilometers above the dam and can be seen only in photos (Fig. 10–15). Both the Bureau of Reclamation and the U.S. Army Corps of Engineers, supported by several western state governments, have made repeated attempts to place additional dams on the Colorado River within Grand Canyon. One of the Bureau's arguments for putting dams in Grand Canyon was that a lake in the canyon would allow tourists to see the canyon more easily from motorboats. The Sierra Club responded with newspaper ads asking if we should also flood the Sistine Chapel so that tourists can get nearer to the ceiling to see Michelangelo's paintings. In addition to losses of beautiful and remote places, dams destroy white-water recreation areas. The flooding of canyons, however, provides lakes that can be used for fishing and other types of water sports.

Ecological Disruptions

A river is an integral part of the ecosystem through which it flows, and when the river is altered, the ecosystem changes in response. Many of the changes are detrimental. For example, before the Aswan Dam was built, the Nile River flooded every spring, depositing nutrient-rich sediments over the Nile flood plain. When the dam was built and the flood waters were controlled, this source of free fertilizer was eliminated. Although the energy produced by the dam is used to manufacture commercial fertilizers, many of the poorer farmers in the valley cannot afford to purchase what they once received free from the river. In addition, increased erosion below the dam has destroyed many farms.

Deterioration of Water Quality at the Source

When water is diverted from a drainage basin, the quality of the remaining water can be affected. Fresh water from a stream or river becomes progressively saltier as it enters and mixes with ocean water. A fast-flowing river carries its fresh water well out to sea. A slow-flowing tidal basin—such as the Hudson River valley in New York—is saline well inland. Therefore, if diversion of water reduces river flow sufficiently, salty ocean water migrates upstream. This phenomenon, called **reverse flow**, can reduce agricultural productivity on the lower flood plain. Changes in water flow can also disrupt the life cycles of fish and wildlife.

Energy Consumption

The great aqueducts of ancient Rome run downhill; no pumps were needed. Farmers in ancient Egypt pumped the waters of the Nile a few feet up to their farms by human or animal power; many do the same today. In contrast, modern water diversion projects use electrically driven pumps to carry water uphill. Diversion projects move hundreds of cubic meters per second and therefore need much energy. One estimate for an expanded California State Water Project predicts the use of 10 billion kilowatt-hours of electricity in the year 2000—about as much power as is used in 2 million homes today.

Ground-Water Diversion and Environmental Impacts

Ground water provides drinking water for more than half of the population of North America and is a major source of water for irrigation and industry. It is a valuable resource because:

1. It is abundant. As mentioned previously, 30 times more fresh water exists underground than in surface reservoirs.
2. Because ground water moves so slowly, it is stored below the Earth's surface and remains available during dry periods.
3. In some regions, ground water flows from wet environments to arid ones, making water available in dry areas.

Most ground water is extracted from wells, which are dug or drilled in valleys close to the levels of streams or lakes. If ground water is pumped continuously and rapidly, the water table can drop significantly. The first disturbance occurs near the well. If water is withdrawn faster than it can flow into the well from the aquifer, a **cone of depression** forms (Fig. 10–16). When the pump is turned off, ground water flows back toward the well

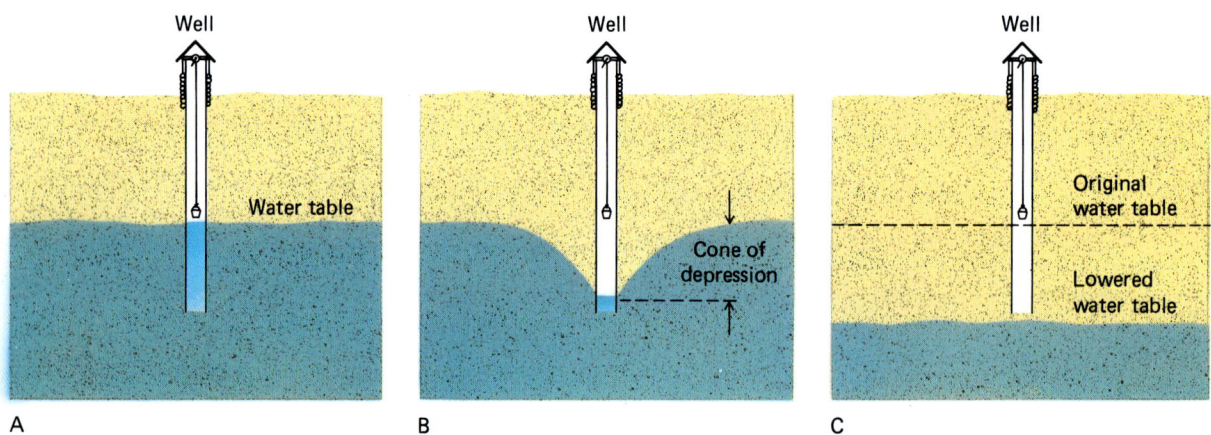

FIGURE 10–16
(A) A well is drilled into an aquifer. (B) A pump draws water from the well faster than it can flow into the well through the aquifer, and a cone of depression forms. (C) If the pump continues to extract water more rapidly than it flows to the well, the water table falls.

in a matter of days or weeks if the aquifer has good permeability, and the cone of depression disappears. On the other hand, if water is continuously removed more rapidly than it can flow to the well through the aquifer, then the water table drops.

Before the development of advanced drilling and pumping technologies, human impact on ground water was minimal. Today, however, deep wells and high-speed pumps can extract ground water more rapidly than the hydrologic cycle replaces it. Where such excessive pumping is practiced, the water table falls as the ground-water reservoir becomes **depleted**. In some cases the aquifer is no longer able to supply enough water to continue to support the farms or cities that have overexploited it. This situation is common in the central, western, and southwestern United States (Fig. 10–17).

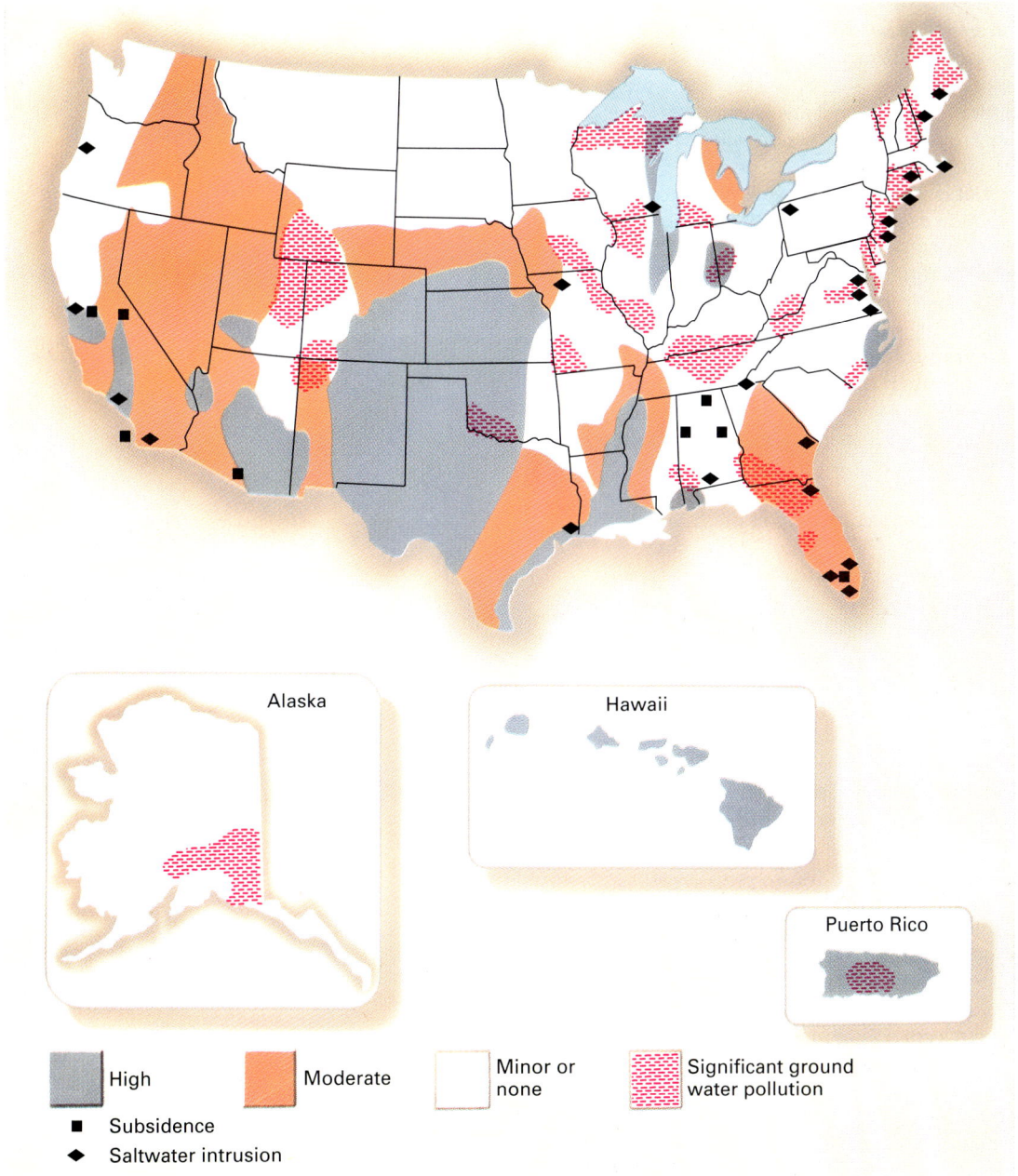

FIGURE 10–17
Areas of subsidence and ground-water pollution in the United States. (*Data compiled from USGS and U.S. Water Resources Council*)

CASE HISTORY

The High Plains and the Ogallala Aquifer

Most of the high plains in western and midwestern North America receive scant rainfall. Early settlers prospered in rainy years and suffered during drought. In the 1930s, two sequences of events combined to change agriculture in this region. One was a great drought that destroyed crops and exposed the soil to erosion. Dry winds blew across the land, eroding the parched soil and carrying it for hundreds and even thousands of kilometers. Thousands of families lost their farms, and the region was dubbed the Dust Bowl. The second event was the arrival of inexpensive technology. Electric lines were built to service rural regions, and relatively cheap pumps and irrigation systems were developed. With the specter of drought fresh in people's memories and the tools to avert future calamities available, the age of modern irrigation began.

Figure 10–18 shows a cross section and map of the Ogallala aquifer in the central high plains. The aquifer extends almost 900 kilometers from the Rocky Mountains eastward across the prairie, and from Texas into South Dakota. It consists of porous sandstone and conglomerate within 350 meters of the surface. The aquifer averages about 65 meters in thickness and contains a vast amount of water. Between 1930 and 1980, about 170,000 wells were drilled into the Ogallala aquifer, and extensive irrigation systems were installed throughout Kansas, Nebraska, Oklahoma, the Texas panhandle, and parts of neighboring states.

Today, farmers and hydrologists are concerned that the Ogallala aquifer is being depleted. They estimate that half of the water has already been removed from parts of the aquifer, and pumping rates are increasing. Water moves through the Ogallala aquifer at an average rate of about 15 meters per year. Most of the water accumulated when the last Pleistocene ice sheet melted. But because the high plains receive little rain, the aquifer is mostly recharged by rain and snowmelt in the Rocky Mountains, hundreds of kilometers to the west. At a flow rate of 15 meters per year, ground water takes 60,000 years to travel from the mountains to the eastern edge of the aquifer. Under such conditions, deep ground water is, for all practical purposes, nonrenewable. Just as coal and petroleum are called fossil fuels, deep ground water is sometimes called **fossil water**. The removal of deep ground water is therefore analogous to mining.

If the present pattern of water use continues, wells in the Ogallala aquifer beneath the high plains will dry up early in the next century. In the mid-1980s, about 5 million hectares of land were irrigated from this aquifer. (This is an area about the size of the states of Massachusetts, Vermont, and

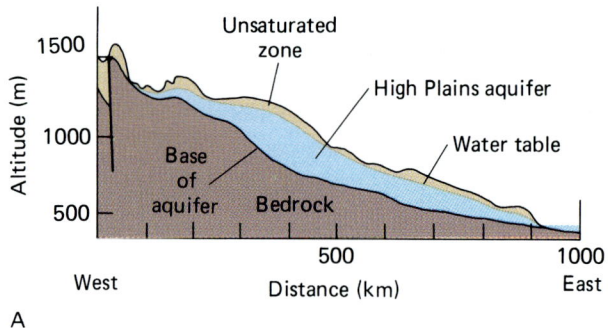

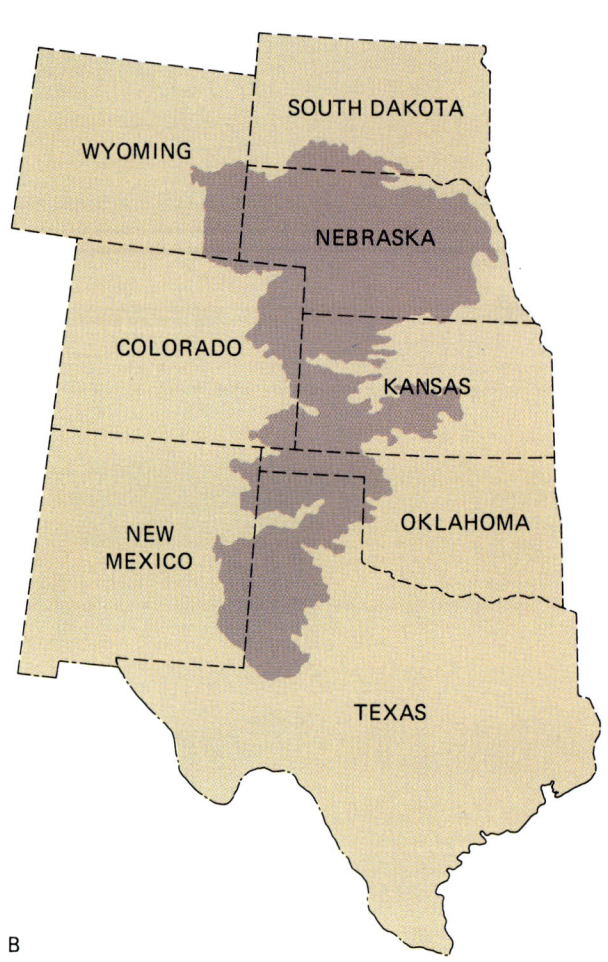

FIGURE 10–18
The Ogallala aquifer. (A) Cross-sectional view of the high-plains aquifer. The deep ground water in the Ogallala system originates in the Rocky Mountains. However, because ground water moves slowly and the distances are great, the reserve is considered to be nonrenewable on a human time scale. (B) A map of the Ogallala aquifer. (After Gutentag et al., 1984, *U.S. Geological Professional Paper 1400-B*)

> Connecticut combined.) About 40 percent of the cattle in the United States are fed with corn and sorghum raised in this region, and large quantities of grain and cotton are grown there as well. If the aquifer is depleted and another source of irrigation water is not found, productivity in the central high plains is expected to decline by 80 percent. Farmers will go bankrupt, and food prices throughout the nation will rise.

Subsidence

Excessive removal of ground water can cause **subsidence**, the sinking or settling of the Earth's surface. When water is withdrawn from an aquifer, rock or soil particles may shift closer to each other, filling some of the space left by the lost water. As a result, the volume of the aquifer decreases and the overlying ground subsides. Removal of oil from petroleum reservoirs has the same effect.

Subsidence rates can reach 5 to 10 centimeters per year, depending on the rate of ground-water removal and the nature of the aquifer. Some areas in the San Joaquin Valley of California have sunk nearly 10 meters. The land surface has subsided by as much as 3 meters in the Houston–Galveston area of Texas. The problem is particularly severe when it occurs beneath a city. For example, Mexico City is built on an old marsh. Over the years, as the weight of buildings and roadways has increased and much of the ground water has been removed, parts of the city have settled as much as 8.5 meters. Many millions of dollars have been spent to maintain this complex city on its unstable base. Similar problems are occurring in Phoenix, Arizona, and other U.S. cities (Fig. 10–19).

Unfortunately, subsidence is not a reversible process. When rock and soil contract, their porosity is permanently reduced so that ground-water reserves cannot be completely recharged even if water becomes abundant again.

Saltwater Intrusion

Two types of ground water occur in coastal areas: fresh ground water and salty ground water that seeps in from the sea. Fresh water floats on top of salty water because it is less dense. If too much fresh water is removed from the aquifer, the zone of fresh-water saturation shrinks and salty ground water rises to replace the fresh water (Fig. 10–20). As a result, salt water intrudes into the shallower portion of the aquifer and commonly rises to the level of wells. It is unfit for drinking, irrigation, or industrial use. **Saltwater intrusion** has affected much of south Florida's coastal ground-water reservoirs.

FIGURE 10–19
Ground-water removal contributed to subsidence and structural damage in Jacksonville, Florida. (*Wendell Metzen/Bruce Coleman, Inc.*)

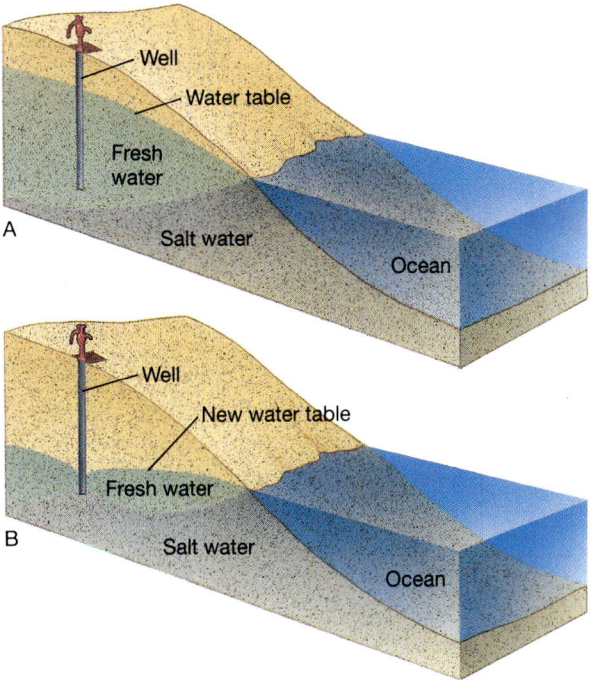

FIGURE 10–20
Saltwater intrusion. (A) Fresh water lies above salt water, and the water in the well is fit to drink. (B) If too much water is removed, the water table declines. The level of salt water rises and contaminates the well.

10.3 Salinization

Most of the world's large rivers contain about 100 parts per million (ppm) dissolved salts. Thus, 1 liter of average river water contains one tenth of a gram of dissolved salts. Although most rivers do not taste salty, they carry tremendous amounts of material to the oceans in this manner. For example, the Mississippi River transports 120 million tons of dissolved salts to the ocean each year. When river water is used to irrigate crops, much of it evaporates, leaving the salts in the soil. Over a period of years, the salts accumulate in a process called **salinization**. If desert or semidesert soils are irrigated for long periods of time, they become too salty for crops.

The great civilizations of Mesopotamia, where much of Western art, science, and literature originated, arose in a desert between the Tigris and Euphrates rivers of western Asia. The agriculture of this region depended on an extensive irrigation system, one of the great achievements of early civilization. Archaeologists dig up ancient irrigation canals, farming tools, and grinding stones in the desert. Today, much of this once-fertile region is barren, eroded, and desolate. This ecological catastrophe was the outcome of salinization caused by irrigation.

In the United States, salinization lowers crop yields on 25 to 30 percent of irrigated farmland, more than 5 million hectares (Fig. 10–21). Globally, the problem affects 10 percent of irrigated land, some 25 million hectares, and it is increasing at a rate of 1 million to 1.5 million hectares per year. Salt buildup is most severe in regions such as the San Joaquin Valley in California, where intensive irrigation is used to convert desert to rich farmland. In some soils, enough calcium carbonate has precipitated from irrigation water to form a rock-hard layer called caliche. To continue growing crops, the farmers must break up and remove the caliche with heavy machinery at great expense. Caliche development can be slowed by removing excess salty water with a drainage system, but it is expensive to install tiles and pipes to collect and dispose of salty water. One technique for preventing salinization is to use less irrigation water by moistening the roots of individual plants with perforated pipes rather than by sprinkling an entire field. However, this technique, called drip irrigation, is expensive to maintain and therefore is applicable only to cash-intensive vegetables.

10.4 Water Use in the Western United States

CASE HISTORY

The Great American Desert

If you started an airplane trip from Seattle, Washington, and flew eastward, at first you would pass over numerous cities, towns, and farms watered by the abundant moisture that flows eastward from the Pacific Ocean. Then, after 100 kilometers, you would pass over the Cascade Range, part of the great western mountain chain that extends from the Mexican border north through California, Oregon, Washington, and into British Columbia. These high western mountains wring out most of the remaining moisture from the clouds. Here, precipitation drops almost immediately, from as much as 400 centimeters per year on the western slope to as little as 10 centimeters on the eastern side of the mountains. Precipitation remains low—except at high elevations—between the mountains and the 100th meridian, the line of longitude running through the Dakotas, Nebraska, Kansas, and Abilene, Texas. The 100th meridian, in turn, separates the dry western half of our nation from the much wetter eastern half.

A desert is any region that receives less than 20 centimeters of rain annually, and land with 50 centimeters or less of rain annually is hostile to most kinds of farming. John Wesley Powell, one of the founders of American geology, explored the area between the Sierra Nevada–Cascade crest and the 100th meridian in the years following the Civil War. He recognized that most of this region is arid or semiarid, and he called it the **Great American Desert** (Fig. 10–22). Powell pointed out that if we evenly distributed all of the surface water between the Columbia River and the Gulf of Mexico, this part of North America would still be a desert. The region simply does not receive enough precipitation or stream flow from the mountains to alter its essentially arid and semiarid environments.

FIGURE 10–21
Salinization has reduced the productivity of millions of hectares of agricultural land globally. Central Nevada.

10.4 Water Use in the Western United States

FIGURE 10–22
The Great American Desert extends from the 100th meridian to the west coast of our continent. Eastern Arizona.

Powell was unaware of the vast amounts of ground water underlying much of the region, nearly all of it a gift from the melting glaciers of the last ice age. However, this legacy is being used so rapidly that it will last no more than 100 years.

Despite Powell's warning about inadequate water, Americans flocked to settle the Great American Desert. As mentioned earlier, several parts of this region now use huge amounts of water for cities, industry, and agriculture, although they receive little precipitation. For example, arid and semiarid California, Nevada, Arizona, Texas, and Idaho use the greatest quantities of water in the United States. Some cities, including Phoenix, El Paso, Reno, and Las Vegas, are located in real deserts and receive less than 20 centimeters of precipitation annually. Such places rely on surface-water diversion projects and/or overpumping of ground water for nearly all of their water (Fig. 10–23).

Billions of dollars have been spent on heroic projects to divert rivers and pump ground water to serve the Great American Desert. For example, 1200 major dams and the two largest irrigation projects in the world have been built in California. Yet all this effort may, like the great irrigation canals of Mesopotamia, hold the desert at bay for only a short time. Marc Reisner's *Cadillac Desert* chronicles the water

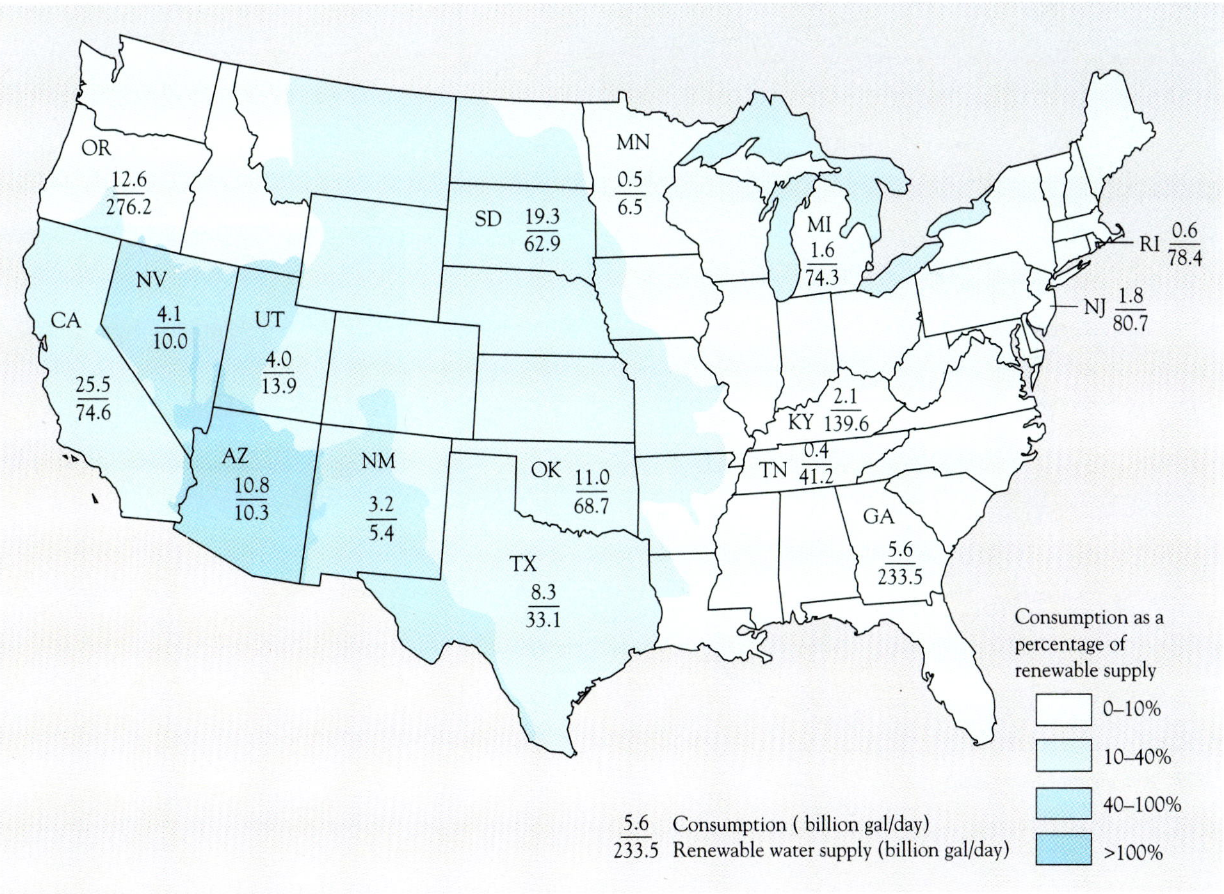

FIGURE 10–23
Consumption of ground water as a percentage of renewable supply. In Arizona consumption is greater than supply, whereas in Georgia it is much less.

252 Chapter 10 Human Use of Water

history of the Great American Desert and forecasts a bleak agricultural future for the region.[1] Reisner predicts that, within the coming 5 decades, millions of hectares of the region's farmland will fall out of production as a result of exhaustion of ground-water reserves. Many reservoirs will silt up, further diminishing the supplies of irrigation water. Salinization will render more of the region's cropland sterile. If this scenario is correct, the irrigation-fueled agricultural productivity of the Great American Desert will occupy but a brief moment in history, less than that of any other large agricultural system in human records.

[1] Marc Reisner, *Cadillac Desert*, New York. Viking, 1986.

CASE HISTORY

The Colorado River

The Colorado River drains the southwestern portion of the Great American Desert. Starting from the snowy mountains of Colorado, Wyoming, Utah, and New Mexico, it flows across the arid Colorado Plateau and southward into Mexico, where it empties into the Gulf of California (Fig. 10–24A).

In the 1920s, the Colorado's discharge was 18 billion cubic meters of water per year. Because the river flows through a desert, farmers, ranchers, cities, and industrial users along its entire length competed for rights to use the water. In 1922 the government apportioned 9 billion cubic meters for the Upper Basin (Colorado, Utah, Wyoming, and New Mexico) and the remaining 9 billion cubic meters for Lower Basin users in Arizona, California, and Nevada. No

A B

FIGURE 10–24
(A) The Colorado River drains much of the American Southwest. (B) The river often dries up completely near the Mexican border. *(Dan Lamont/Matrix)*

water was set aside for maintenance of the river ecosystems or for Mexican users south of the border. Twenty years later, an international treaty awarded Mexico 1.8 billion cubic meters to be taken equally from Upper and Lower Basin users. However, the treaty made no stipulations concerning water quality.

The Colorado and many of its tributaries flow across sedimentary rocks, many of which contain soluble salt deposits. As a result, the Colorado is naturally a salty river. In addition, the United States government built ten large dams on the Colorado, and evaporation from the reservoirs has further concentrated the salty water. In 1961, the water flowing into Mexico contained 27,000 ppm salts, compared with an average salinity of about 100 ppm for the world's large rivers. For comparison, 27,000 ppm is 77 percent as salty as seawater. Mexican farmers used this water for irrigation, and their crops died. In 1973, the United States government built a desalinization plant to reduce the salinity of Mexico's share of the water.

Measurements of the total discharge of 18 billion cubic meters were made during a time of abundant rainfall. Drought from 1930 to 1968 reduced the Colorado's average discharge to 15.6 billion cubic meters, less than the amount of water that had been allocated. As a result, the river completely dried up several times (Fig. 10–24B).

Future demands on Colorado River water may exceed the discharge by even greater amounts. Large reserves of oil shale lie in the Upper Colorado Basin. Abundant water is needed to process the shale, and energy companies have purchased rights to more than 1.2 billion cubic meters of water per year. When shale mining and processing become economical, more people will move to the region, creating additional demand on the scarce water resources.

CASE HISTORY

The Los Angeles Water Project

In *Cadillac Desert*, Marc Reisner described Los Angeles in the late 1800s as a "torpid, suppurating, stunted little slum." Too far from the California gold fields to attract the miners or their money, it sat neglected in its arid coastal basin with neither a seaport nor a railroad. The average annual precipitation of the San Fernando Valley, where Los Angeles is located, is 65 centimeters, most of which falls during a few winter weeks. The Los Angeles River, which flows through the valley, frequently flooded in winter and diminished to a trickle in summer. The city's main attraction was as a haven for the persecuted, both virtuous and criminal.

Among the virtuous who came to Los Angeles were Mormons, who had more experience with irrigation than anyone else. Others learned from the Mormons, and by the late 1800s, irrigated farms in the Los Angeles basin were producing a wealth of fruits and vegetables: corn, cabbages, oranges, avocados, dates. It seemed that almost anything grew bigger and better with the combination of sun, warmth, rich soil, and irrigation. Suddenly, Los Angeles was an attractive, growing town. In 1848 the town had a population of 1600. The population passed 100,000 by 1900 and 200,000 by 1904. Then the city ran out of water. Wells dried up, and the river was inadequate to meet demand. To supply human needs, the city government prohibited lawn watering and shut off irrigation wells in the nearby San Fernando Valley. The town was surrounded on three sides by deserts and on the fourth by the Pacific. There was no nearby source of fresh water.

However, 250 miles to the north, the Owens River flowed out of the Sierra Nevada and watered the beautiful, green Owens Valley. The city of Los Angeles bought water rights, bit by bit, from farmers and ranchers in the Owens Valley and at the headwaters of the river. The city then spent millions of dollars building the Los Angeles Aqueduct across some of the most difficult, earthquake-prone terrain in North America. When finished, the aqueduct was 357 kilometers long; 85 of those kilometers were tunnel (Fig. 10–25). Siphons and pumps carried the water over hills and mountains too treacherous for tunneling. On November 5, 1913, the first water poured from the aqueduct into the San Fernando Valley (Fig. 10–26).

The following 10 to 15 years were unusually rainy in the Los Angeles area and in the Owens Valley. The rain replenished the ground water underlying the Los Angeles basin, and the Los Angeles River flowed freely again. As a result, little or none of the Owens River water coming through the aqueduct was needed by the city. Instead, most of the water was diverted to farms in the San Fernando Valley. Irrigated farmland in that valley increased from 1200 hectares in 1913, when the aqueduct opened, to more than 30,000 hectares in 1918. In addition, some of the excess water now available to LA was used to water lawns again. Almost every house had a green lawn, and Santa Monica Boulevard changed from a dusty rut to a palm-lined oasis, creating the impression that Los Angeles wasn't really so dry after all.

In the 1920s, normal dryness and drought returned. Los Angeles and the farms of the San Fernando Valley demanded, and received, more water from the Owens River, and the Owens Valley became drier and drier. The mood of the ranchers and farmers in the Owens Valley grew steadily worse,

FIGURE 10–25

The Los Angeles Aqueduct and related water diversion systems in California and Arizona.

and in 1924 they threatened the aqueduct with sabotage. William Mulholland, superintendent of the Los Angeles Department of Water and Power, remarked that it was too bad that so many of the orchard trees in the Owens Valley had died of thirst because now there weren't enough left to hang all the Owens Valley troublemakers.

By the mid-1930s, Los Angeles owned 95 percent of the farmland and 85 percent of the residential and commercial property in the Owens Valley. Although the city leased some of the farmland back to the farmers and ranchers, the water supply became so unpredictable that agriculture slowly dried up with the land. As LA continued to grow, the water from the Owens River was not sufficient, so the Department of Water and Power drilled wells in the Owens Valley and began pumping its ground-water aquifer dry.

FIGURE 10–26

The Los Angeles Aqueduct transports water across the Owens Valley desert to the San Fernando Valley, near Los Angeles. *(Charles O'Rear/Westlight)*

Today, the Owens Valley is a desert (Fig. 10–27). Its remaining citizens pump gas, sell beer, and make up motel beds for tourists driving through on their way to somewhere else. Ironically, the water that once irrigated local farms and ranches in the green, beautiful Owens Valley was diverted through the LA Aqueduct, at great cost to taxpayers, to irrigate farms in the naturally arid San Fernando Valley. The aqueduct converted a farmer's paradise to a desert and a desert to a farmer's paradise.

Today additional diversion projects and aqueducts divert water from California rivers nearly as far north as San Francisco and from the Colorado River to LA, San Diego, and other areas of southern California. The related $3.9 billion Central Arizona Project pumps water uphill from the Colorado River to Phoenix and Tucson. In the process, it removes water from the Colorado that was used by San Diego until 1985. But as a result, Arizona has reduced its dependency on ground water from more than 90 percent to less than 65 percent.

More than 80 percent of the water diverted to southern California is used to irrigate desert and near-desert cropland. Only 17 percent goes to the cities. During the past 4 decades, Central Valley farmers have paid 5 percent of the actual cost of water delivery for their irrigation water; the rest is subsidized by taxpayers. Most of the farmland is, or until recently was, owned by agribusinesses. For example, in the San Joaquin Valley the major landowners include Chevron USA, a subsidiary of Standard Oil Company of California; the Tejon Ranch, one of the largest California land companies; Getty Oil Company; Shell Oil Company; Prudential Insurance Corporation; Mitsubishi; and Tenneco.

The water supplied to these farms commonly irrigates crops such as cotton, rice, and alfalfa. Recall from Section 10.1 that cotton requires 17,000 liters per kilogram; rice and alfalfa also use great amounts of water. At the same time, the U.S. government pays large subsidies to farmers in naturally wet southeastern and south central states to *not* grow the same crops. The irrigation water used to support cattle ranching and other livestock in California is enough to supply the needs of the entire human population of the state. The livestock could be raised equally well in areas not requiring irrigation.

In the process of supplying southern California cities and agribusinesses with heavily subsidized water, federal and state governments have dammed and drained large rivers, lakes, and wetlands; eliminated thousands of hectares of fish and wildlife habitat; destroyed hundreds of kilometers of salmon spawning beds; and flooded hundreds of kilometers of whitewater recreational rivers. The farming and ranching have contaminated waterways with pesticides and fertilizers and have increased the salinity of downstream rivers. To put these environmental consequences of California's water diversion projects in perspective, agriculture contributes 2.5 percent to California's total economic productivity.

10.5 Conservation

Dams, diversion projects, and wells alleviate water shortages by increasing the supply. An alternative approach is to decrease demand through conservation. Frequently conservation is both cheaper and less disruptive to the environment than either ground-water extraction or surface-water diversion.

Individuals can practice conservation in the home by using low-flow shower heads to reduce the amount of water for washing and by using low-volume toilets to reduce the amount of water used for flushing. Although these methods are both effective and inexpensive, they don't make a large impact on water consumption, simply because in-home domestic use represents only a small percentage of total water consumption. A far more effective domestic conservation measure would be to eliminate lawn watering. **Xeriscaping** is a type of landscaping using indigenous plants that can survive on semiarid and desert amounts of rainfall. If you live in the desert, plant cactus in the front yard and cover the remaining soil with decorative stones.

Agriculture, particularly irrigation, is by far the largest water consumer in most places with inadequate local sources. Immense amounts of water can be conserved by raising crops that require large amounts of water in regions with high rainfall and growing crops that require less water in dry regions. In addition, many irrigation methods are grossly inefficient. Flooding fields

FIGURE 10–27
Without the Owens River, the Owens Valley is a desert.

FIGURE 10–28
Flood irrigation of a rice paddy in western China (A) and sprinkler irrigation in western Montana (B) both lose large amounts of water to evaporation.

or spraying water into the air and then allowing it to fall onto the entire cultivated surface are the most common irrigation techniques and cause the greatest amount of evaporation (Fig. 10–28). Drip irrigation reduces salinization and conserves water (Fig. 10–29). Drip irrigation is not used extensively because it is expensive. However, if farmers were charged the real price of the water that they use, drip irrigation would become economical.

FIGURE 10–29
Drip irrigation conserves great amounts of water by watering the crop plants directly and leaving the intervening soil dry. *(USDA)*

10.6 Water and International Politics

Forty percent of the Earth's people use water from rivers that flow through two or more nations. The United States and Mexico share the Colorado River; India and Bangladesh share the Ganges; India and Pakistan share the Indus; Czechoslovakia and Hungary share the Danube; the Jordan River flows through Israel, Jordan, Syria, and Lebanon. In Africa alone, 57 different lakes and river basins are shared by two or more countries. Frequently, political problems arise when nations must share limited water resources. The problem becomes severe when the nations are already unfriendly toward one another for other reasons. Jordan's King Hussain said in 1990 that water was the only issue that could take him to war with Israel. Egyptian president Anwar Sadat stated, after signing the peace accords with Israel, that "the only issue that could take Egypt to war again is water."

Avoidance of conflict over inadequate water resources requires agreements (a) to share the available water and (b) to share the consequences of shortages. However, current international water law offers little help in resolving these issues. For example, the Colorado River originates in the mountains of the southwestern United States but flows into Mexico. So does the water belong to the United States or to Mexico? In general, upstream nations have been unwilling to agree to the principle that common water resources should be cooperatively governed, managed, and shared. Some nations maintain that they have absolute control over the fate of water within their borders and have no responsibility to downstream neighbors.

Any international code of water use must be based on three fundamental principles:

1. Water users in one country must not cause major harm to water users in other countries downstream.
2. Water users in one country must inform neighbors of actions that may affect them before the actions are taken. (For example, if a dam is built, engineers must inform downstream users that they plan to stop or reduce river flow to fill a reservoir.)
3. People must distribute water equitably from a shared river basin.

Unfortunately, these principles, especially the last one, are open to such subjective and self-interested interpretations that their intent is easily and often subverted. The sharing of the Jordan River by Israel, Jordan, Syria, and Lebanon presents an unusually difficult problem because the region is desert and the nations are already hostile. In other regions, international cooperation has been more successful. In 1960, India and Pakistan signed the Indus Waters Act, mediated and funded by the World Bank. The act partitioned water use equitably and established a permanent commission to oversee continued cooperation between the two nations. In sub-Saharan Africa, the nations of Mali, Mauritania, and Senegal cooperate in the management of the Senegal River. The large dams that they have jointly built have created some environmental nightmares, but at least the nations are not warring over the water. The eight nations in the African Zambezi River basin cooperate under the United Nations Environment Program to ensure fair distribution of water in that basin.

International sharing of water and the consequences of its shortage also necessitate water conservation to maximize the amount of water available to all. Unfortunately, although conservation is the easiest and cheapest way to increase water availability, most water-starved regions of the world still practice it only when shortages are acute.

No nation or region can be economically, socially, or politically stable unless it has an adequate water supply. More political leaders are realizing that national stability depends on water and that an adequate water supply depends, in turn, on international cooperation. In 1994, Jordan and Israel signed a peace accord that included agreements over water allocations. Any future international conflicts over water could lead to unprecedented numbers of small wars among water-starved nations.

SUMMARY

Overpopulation has created demands for water that stretch, and in some cases exceed, the amount of water that is available, both globally and in the United States. The United States receives about three times as much water, in the form of precipitation, as it uses. But some of the driest regions use the greatest amounts of water. As a result, those regions must import water from wetter regions or pump it from shrinking ground-water reservoirs.

Water use falls into three categories. **Domestic** use accounts for 10 percent of U.S. water consumption, **industrial** use accounts for 49 percent, and **agricultural** uses require 41 percent. Most of the water used by homes and industry is **withdrawn** and then returned to streams or ground-water reservoirs near the site of withdrawal. But most of the water used by agriculture is **consumed**, because it evaporates.

Water **diversion projects** collect and transport surface and ground water from places where water is available to places where it is needed. In a surface diversion project, a **dam** is built on a stream or lake outlet to create a **reservoir**. Canals, aqueducts, or pipes then carry the water to consumers. Dams supply large amounts of water to places where it is needed, and some generate hydroelectric energy. But they are so costly that their construction is heavily subsidized, and they create undesirable environmental effects, including water loss, siltation, erosion downstream, destruction when a dam fails, recreational and aesthetic losses, ecological disruptions, and deterioration of water quality. Water delivery systems require large quantities of energy to pump the water uphill, and **salinization** of farmland often accompanies irrigation in dry regions.

Ground-water projects pump ground water to the surface for human use but, because ground water flows so slowly, the extraction often creates a **cone of depression** that grows as pumping **depletes** the aquifer and causes **subsidence**. Depletion of coastal aquifers causes **saltwater intrusion**.

The **Great American Desert** is a mostly arid and semi-arid region of the United States that reaches from the Sierra Nevada–Cascade crest to the 100th meridian. Americans have built great cities and extensive farms and ranches in the region, all supplied by irrigation systems. Diminishing water reserves and increasing costs of water diversion projects suggest that their future is uncertain.

KEY TERMS

Domestic water use 239
Industrial water use 240
Agricultural water use 240
Withdrawal 242
Consumption 242
Diversion projects 243
Dam 243
Reservoir 243
Cone of depression 246
Depletion 247
Fossil water 248
Subsidence 249
Saltwater intrusion 249
Salinization 250
Great American Desert 250
Xeriscaping 255

REVIEW QUESTIONS

1. The United States receives three times more water as precipitation than it uses. Why do water shortages exist in many parts of the country?

2. Describe the three main categories of water use in the United States. What proportion of total U.S. water use falls into each category?

3. Explain the differences among water use, water withdrawal, and water consumption.

4. Why is agriculture responsible for the greatest proportion of water consumption, whereas industry uses the greatest amount of water?

5. What are the two main sources of water exploited by water diversion projects?

6. Describe the beneficial effects of dams and their associated water delivery systems.

7. Describe the negative environmental and other effects of dams and their associated water delivery systems.

8. Why does salinization commonly result from desert irrigation?

9. Describe the factors that make ground water a valuable resource.

10. Describe three problems caused by excessive pumping of ground water.

11. Why does ground-water depletion cause longer-term problems than might result from the draining of a surface reservoir?

12. How and why does land subside when ground water is depleted?

13. If land subsides when an underlying aquifer is depleted, will it rise to its original level when pumping stops and the aquifer is recharged? Explain your answer.

14. Why does saltwater intrusion affect only coastal areas?

15. Describe the geographic region that John Wesley Powell called the Great American Desert.

16. Why is the Great American Desert such a dry region?

17. Describe efficient and inefficient irrigation systems. Why are inefficient systems so common?

DISCUSSION QUESTIONS

1. Consider how (a) a rise or (b) a fall in average global temperature might affect surface- and ground-water reserves in the United States.

2. Discuss the relative merits and disadvantages of surface- and ground-water diversion projects.

3. No major dam has been built in the United States without federal support, and the price charged to irrigators and other users for the water from a major dam has never paid for the cost of building the dam. Discuss why we build dams.

4. Develop scenarios for the future of farming and ranching in regions currently obtaining water from the Ogallala aquifer.

5. Discuss how water conservation in each of the three main categories of water use in the United States can affect the total amount of water available for all types of uses.

6. Discuss why cotton and rice are grown in semiarid California at the same time that farmers in naturally wet southeastern and south-central states are paid federal subsidies to not grow the same crops.

7. Compare and contrast the causes and effects of reverse flow and saltwater intrusion.

8. In the book *Cadillac Desert*, Marc Reisner describes the transformation of the American West from its natural desert and semidesert environment to a region with great cities and a vast agricultural economy. He argues that the West

cannot sustain the cities or the agriculture. Discuss how this transformation occurred, and discuss the validity of his hypothesis that the current system cannot continue indefinitely.

9. Consider reasons why desert and semidesert cities such as Phoenix, Las Vegas, and Los Angeles have become so heavily populated despite the lack of adequate water resources.

10. Abundant water exists in the Columbia Basin and in the Fraser River basin in Canada, whereas the Colorado Basin to the south is water-deficient. Argue for or against a proposal to divert water from the Pacific Northwest to the arid Southwest.

11. Discuss the role that water plays in the relations between the United States and its neighboring countries.

CHAPTER 11

Waste Disposal and Fresh-Water Pollution

11.1 Public Awareness of Water Pollution
11.2 Water Pollutants
11.3 Pollution by Nutrients
11.4 Industrial Organic Wastes
11.5 Surface Water in the United States 20 Years After the Clean Water Act
11.6 Ground-Water Pollution
11.7 Deep Injection Wells and Ground-Water Pollution
11.8 Ground Water and Nuclear Waste Disposal
11.9 Municipal Waste Disposal
11.10 Alternatives to Waste Disposal

All organisms produce wastes, but most natural ecosystems don't become polluted because a variety of organisms consume the wastes and recycle the nutrients. Cities don't operate in this manner. Millions of people crowd into an area that cannot possibly support them or dispose of their wastes. To survive, city dwellers import food and export wastes.

During the years immediately following World War II, urban populations increased rapidly and industries flourished. As a consequence, both the quantity and variety of wastes increased. Although some of these wastes were disposed of in environmentally sound ways, billions of tons of refuse were dumped into rivers and buried in shallow landfills. In New York City today, about 5 billion liters of sewage flow into treatment plants, and garbage collectors handle about 35,000 tons of refuse every day.

Water pollution is common throughout the world. Tunis, Tunisia. (SuperStock)

11.1 Public Awareness of Water Pollution

CASE HISTORY

The Cuyohoga River Fire and the Clean Water Act

In the early days of the Industrial Revolution, factories and sewage lines dumped untreated wastes into rivers. The first sewage treatment plant in the United States was built in Washington, D.C., in 1889, more than 100 years after the Revolutionary War. Soon other cities followed suit, but few laws regulated industrial waste discharge.

In November 1952, an oily film on the Cuyohoga River near Cleveland, Ohio, caught fire, spreading flame and smoke across the water (Fig. 11-1). The image of a burning river triggered public awareness of water pollution, and people called for action. Nearly 20 years later, in 1970, President Nixon declared that "the 1970s absolutely must be the years when America pays its debt to the past by reclaiming the purity of its air, its waters, and our living environment. It is literally now or never." The **Clean Water Act**, passed in 1972 over Nixon's veto, stated that

> The objective of this Act is to restore and maintain the chemical, physical, and biological integrity of the Nation's waters. In order to achieve this objective it is hereby declared that. . . . (1) it is the national goal that the discharge of pollutants into the navigable waters be eliminated by 1985; (2) it is the national goal that wherever attainable, an interim goal of water quality which provides for the protection and propagation of fish, shellfish, and wildlife and provides for recreation in and on the water be achieved by July 1, 1983; (3) it is the national policy that the discharge of toxic pollutants in toxic amounts be prohibited; . . ."

Thus, the Clean Water Act set an ambitious agenda for cleaning the nation's rivers, lakes, and wetlands. Later in this chapter, we will describe the successes and failures of this program.

FIGURE 11-1
The Cuyohoga River fire.

CASE HISTORY

Love Canal and the Superfund

Hooker Chemical Company's main plant was located in Niagara Falls, New York. Early in the 1940s, Hooker purchased an abandoned canal called Love Canal. During the following years, the company disposed of approximately 19,000 tons of chemical wastes by loading them into 55-gallon steel drums and storing them in the canal. In 1953, the company covered one of the dump sites with dirt and sold the land for $1 to the Board of Education of Niagara Falls, which built a school and a playground on the site.

The steel drums eventually leaked, and the chemicals seeped into ground water. In the spring of 1977, heavy rains raised the water table and turned the area around Love Canal into a muddy swamp. But it was no ordinary swamp; poisonous compounds from the leaking drums mingled with the water and soil. The toxic fluid soaked the playground, seeped into basements, and saturated gardens and lawns. Children who attended the school and adults who lived nearby developed epilepsy, liver malfunctions, miscarriages, skin sores, rectal bleeding, and severe headaches. An abnormal number of newborn babies in the area had birth defects (Fig. 11-2).

The Love Canal incident demonstrated that wastes buried in landfills are removed from sight but not from the ecosystems that sustain us. Any wastes spilled onto the ground or buried beneath the surface are potential hazards that can pollute the soil and ground water. In December 1979, the U.S. Congress passed the Comprehensive Environmental Response, Compensation, and Liability Act, commonly known as **Superfund** and abbreviated **CERCLA**. This law provides an emergency fund to clean up chemical hazards and imposes stiff fines on any company maintaining a dump site that pollutes the environment.

FIGURE 11-2
Illness, birth defects, and death caused people in the Love Canal area to abandon their homes. *(Joe Traver/Gamma Liaison)*

FIGURE 11-3
Industry and agriculture are among the many sources of water pollution. (A) The Exxon oil refinery in Linden, New Jersey. (B) A feedlot in Idaho.

11.2 Water Pollutants

The terminology and epidemiology of chemical pollutants were discussed in Chapter 1. Seven categories of pollutants degrade surface and ground waters (Fig. 11–3).

1. **Sewage** is wastewater from toilets, sinks, washing machines, and other household drains. It includes biodegradable organic material such as human and food wastes, soaps, and detergents. Sewage also includes some industrial chemicals because people flush paints, solvents, pesticides, and other household chemicals down the toilet. Industrial wastes may end up in the sewer systems because of illegal dumping.
2. **Disease organisms**, such as typhoid and cholera, are carried into waterways in the sewage of infected people.
3. **Plant nutrients**, such as phosphates and nitrates, flow into surface and ground waters from nonpoint sources such as croplands. Phosphate detergents and phosphates and nitrates from feedlots also fall into this category.
4. Many **industrial organic compounds** are similar enough to natural materials that decay organisms consume them. Thus they are biodegradable. Many others are so foreign to natural food chains that they are not decomposed by environmental chemicals or consumed by decay organisms. Non-biodegradable pollutants such as dioxin and DDT are especially troublesome because they persist in the environment. Many are toxic or carcinogenic.
5. **Toxic inorganic compounds** include mine wastes, road salt, and industrial metals such as cadmium, arsenic, mercury, and lead. This topic is discussed in Chapter 15, "Minerals and Mining."
6. When **sediment** enters surface waters, the soil particles neither fertilize nor poison the aquatic system. However, as explained in Chapter 7, the sediment muddies streams and buries aquatic habitats and thus degrades the quality of an ecosystem.
7. **Radioactive materials** include wastes from the mining of radioactive ores, nuclear power plants, nuclear weapons, and medical and scientific applications.

Pollution control strategies differ for different pollutants. Sewage treatment plants can't purify radioactive wastes. In turn, radioactive waste repositories are much too expensive for sewage. Despite the wide range of options available, all pollution control strategies fall into the following four categories.

Dispersal: An old saying goes, "Dilution is the solution to pollution." If you dump a small amount of a chemical into a river, the compound may become so diluted that it becomes harmless. However, some chemicals, such as dioxin and plutonium, are toxic even in minuscule concentrations. In addition, the tragedy of the commons comes into play here because, if everyone dumps a small amount of waste into the river, then the concentrations of pollutants may rise to harmful levels.

- **Destruction**: The world is no longer littered with dinosaur feces because animal waste degrades in natural environments. The cheapest method of handling many biodegradable wastes, such as sewage, is to provide an environment favorable for rapid decay. Even nonbiodegradable compounds can be destroyed by strong acids, bases, oxidizing agents, or high temperatures. When a chemical compound is destroyed, the molecule breaks apart and the atoms recombine to form simpler compounds such as carbon dioxide and water. Radioactive substances and certain mine wastes cannot be destroyed because the pollutants are atoms or ions. Atoms and ions cannot be degraded by ordinary chemical or physical processes.
- **Isolation**: The owners of Hooker Chemical Company tried to isolate their wastes and store them where they would not disperse into the environment. Isolated wastes must be protected from weathering, accidents, catastrophic events such as earthquakes and volcanoes, terrorism, and war.
- **Recycling** is the reuse of waste. When a waste product is recycled, it is converted to a useful material.

11.3 Pollution by Nutrients

Cultural Eutrophication

Recall from Chapter 9 that a lake naturally becomes **eutrophic** when a large amount of organic matter accumulates in the lake and is consumed by decay organisms. As long as oxygen is available, **aerobic** decay organisms predominate. (An aerobic organism uses oxygen as it consumes organic matter.) The decay organisms release nutrients that nourish a population bloom of plants and animals. In turn, these organisms consume most or all of the dissolved oxygen, and fish begin to die. Thus an increased supply of nutrients causes three changes in the lake ecosystem: (1) dissolved oxygen is depleted; (2) species of fish and other aquatic animals that require abundant oxygen die; (3) other organisms that thrive with a lower oxygen concentration proliferate in the nutrient-rich water. In this way, a clear, sparkling stream or lake that supports trout and salmon is replaced by an algae-choked waterway with carp and water worms. Note that organic matter doesn't poison aquatic systems; it nourishes them. However, this nourishment alters species distribution and transforms the aquatic ecosystem into one that is less pleasant for humans.

The **biological oxygen demand**, or **BOD**, is the amount of dissolved oxygen required by aerobic decay organisms to decompose the organic matter in a given amount of water. Sewage is rich in organic nutrients and creates a high BOD. Thus, it leads to oxygen depletion. If a small amount of sewage is dumped into a waterway, decay organisms consume some but not all of the oxygen. In the process, they release nutrients, leading to aquatic blooms and alteration of the species distribution in a lake or stream. However, if a large amount of sewage is dumped into a waterway, the decay organisms consume all the oxygen. As a result, most aquatic life dies. Then, **anaerobic** bacteria (bacteria that live without oxygen) take over, releasing noxious hydrogen sulfide gas as they feast on the organic matter (Fig. 11–4). Even carp and worms cannot survive in such an environment. As they die and rot, the hydrogen sulfide that bubbles to the surface smells like rotten eggs.

Fertilizers and detergents can also cause eutrophication of lakes and streams. Fertilizers contain several essential plant nutrients. Many detergents contain phosphates, a nutrient that is frequently in short supply in natural waters. Eutrophication caused by human pollutants of all kinds is called **cultural eutrophication** (Fig. 11–5).

FIGURE 11–4

This stream flowing through La Paz, Bolivia, is an open sewer. Decay organisms have consumed all of the oxygen in the water, and anaerobic bacteria have taken over.

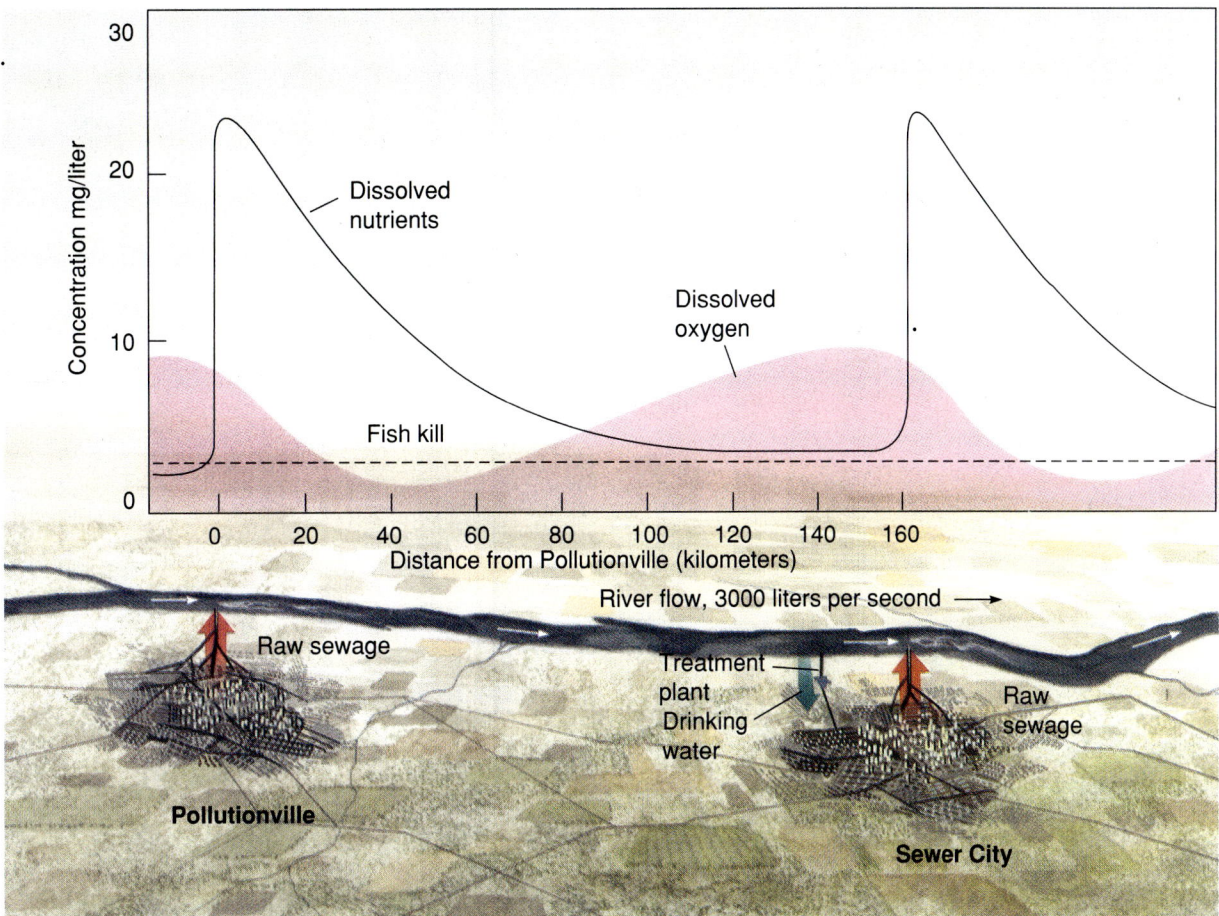

FIGURE 11–5
If raw sewage is discharged into a river (left side of the figure), the concentration of dissolved nutrients increases immediately. Decay organisms use up the dissolved oxygen as they consume the nutrients. Therefore both the nutrients and oxygen concentration decrease. When the dissolved oxygen concentration falls below a threshold level, fish die. When the nutrient concentration falls sufficiently, decomposition slows down, the river becomes aerated again, and fish can live in the water. However, if a second discharge adds a new load of nutrients, the process is repeated.

Municipal Sewage Treatment

If you live in a city or town, the wastewater from toilets, sinks, and showers flows through a pipe to a sewage treatment plant (Fig. 11–6). At the plant the water first flows through a series of screens that remove large, solid objects. Then it is pumped into a chamber and left for an hour or two to allow time for suspended solids to settle out. The screening and settling comprise **primary sewage treatment**. After primary treatment, the water is almost clear, but it still contains abundant disease organisms, a high concentration of organic matter, and dissolved material. **Secondary sewage treatment** purifies the water further by pumping it into an aerated pond or chamber where bacteria and other decay organisms feast on the organic compounds, breaking them down to simple molecules. Once the decay process is complete, the bacteria are killed with a disinfectant. Frequently the wastewater is then discharged into a nearby river. However, although most of the biodegradable wastes are removed by this process, other harmful compounds are not. **Tertiary sewage treatment** is a series of chemical processes designed to remove specific pollutants, such as industrial organic compounds, phosphates, and nitrates,

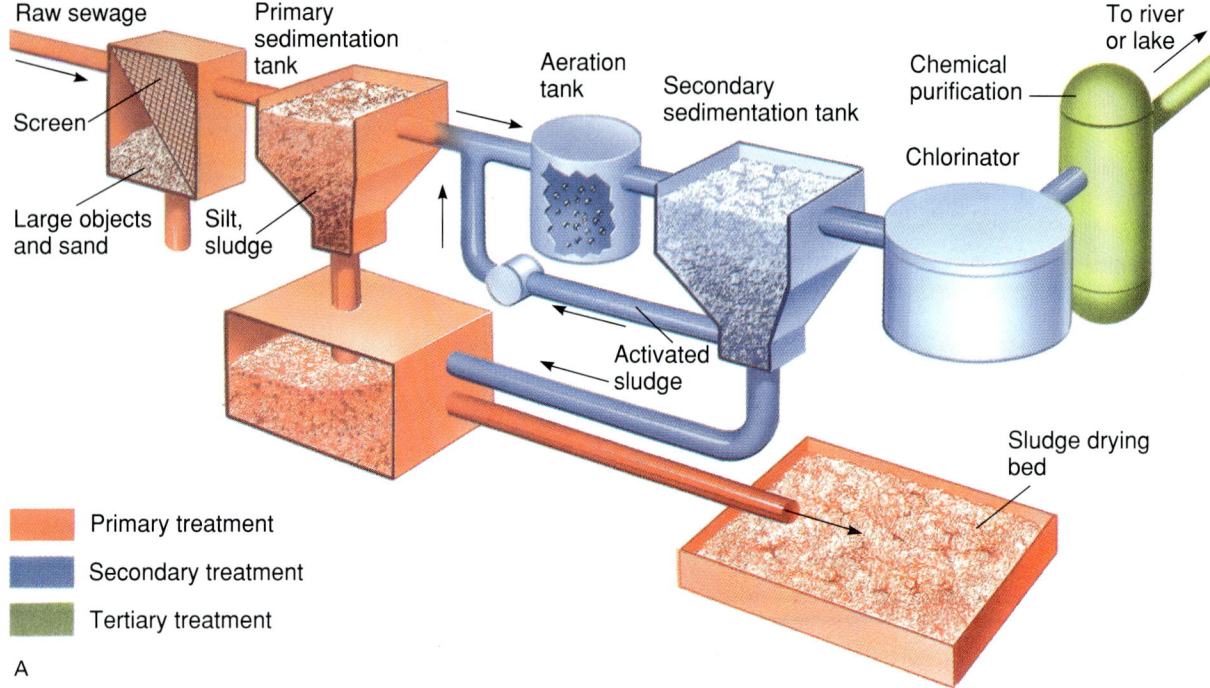

FIGURE 11-6
(A) Schematic of a sewage treatment plant. Primary treatment removes heavy solids by screening and settling. Secondary treatment purifies the wastewater by exposing it to air and bacteria. A portion of the bacteria-laden sludge from the secondary sedimentation tank is returned to the aeration tank. This return maintains an active bacterial colony in the aeration tank. Tertiary treatment is a series of chemical processes to remove specific pollutants. (B) Aerial view of a Denver sewage treatment plant.

that are not extracted by the first two treatments. After the purified sewage is discharged into the river, the next town downstream takes river water, disinfects it, and uses it for drinking.

Sewage treatment plants are legally allowed to dump their effluent into natural waterways. If treatment is complete, the discharged water will not pollute aquatic ecosystems, but many sewage plants are designed poorly or operated ineffectively. In 1991, the Environmental Protection Agency (EPA) reported that 527 cities failed to meet federal standards for their sewage treatment plants.

Thirty-four cities on the East Coast were simply screening out large solids and discharging the water directly into rivers or the ocean.

11.4 Industrial Organic Wastes

Although some industrial organic wastes are biodegradable and nontoxic, many are persistent and poisonous. The EPA defines a **hazardous waste** as any material that is (1) flammable (such as petroleum products or organic solvents), (2) explosive or volatile enough to release toxic fumes, (3) corrosive (such as strong acids and bases), or (4) toxic, carcinogenic, or mutagenic. No one knows for certain how much hazardous waste is produced in the United States, but estimates range between 240 million and 2.4 billion tons per year.

Dispersal of Industrial Wastes

Despite the strict wording of the Clean Water Act, the EPA issued 43,457 permits to allow factories to discharge wastes into waterways in the United States between 1972 and 1991 (Fig. 11–7). The government issued those permits because EPA scientists considered the wastes nontoxic once they were diluted. Environmentalists disagree and argue that this government policy legalizes pollution and endangers public health. The arguments between these two factions center around two issues: the health effects of low doses of pollutants and the biomagnification of pollutants.

If you drink a glass of water polluted with a few thousandths of a percent of benzene, the water will taste foul, but it will not harm you. But if you drink this water every day for 20 years, then you may get cancer. As explained in Chapter 1 and discussed further in Section 11.5, it is difficult to assess accurately the health effects of low concentrations of pollutants.

Scientists have also learned that the biological impact of a pollutant may be much greater than one might expect from its concentration alone. Organochloride pesticides such as DDT, DDD, and related compounds are only slightly soluble in water.[1] Therefore, the concentra-

[1]An organochloride is an organic compound that contains chlorine.

FIGURE 11–8

A crop duster sprays pesticide on crops. Winds carry some of the pesticide into surface waters. *(D. Kirkland/Sygma)*

tions of these pesticides in lakes and streams are frequently low. However, organochloride compounds accumulate in bottom sediment and concentrate in plant and animal tissue. In the 1950s, government workers sprayed Clear Lake in California with DDD to control insect pests (Fig. 11–8). After the project was completed, the water contained 0.02 parts per million (ppm) of DDD. Plankton that lived in Clear Lake concentrated and stored some of the pesticide until there were 5 ppm in their tissues. Plant-eating fish ate the plankton and further concentrated the poison to 40 to 300 ppm of DDD. Predatory fish and birds had as much as 2000 ppm of DDD in their tissues (Fig. 11–9). **Biomagnification** is the increase in concentration of a pollutant as it moves through a food chain. In Clear Lake, bird populations declined because birds with high organochloride concentrations lay thin-shelled eggs that commonly break before they hatch. The fish were unfit for human consumption.

Destroying Industrial Wastes

Nonbiodegradable organic wastes can be destroyed either by treating them with chemicals such as strong acids or by burning them in a process called **incineration** (Fig. 11–10). Incineration is also used for municipal garbage (see Section 11.9). An incinerator operating at high temperature destroys nearly all organic wastes. Modern air pollution control equipment can capture the harmful gases and particulates that are not destroyed, so that the process is environmentally clean. However, if combustion is incomplete and control measures inadequate, the decomposition products escape as air pollutants. In some cases, air pollutants from incinerators are more hazardous than the original wastes. As a result, few people want a hazardous waste incinerator in their back yards.

FIGURE 11–7

Soap foam from the ILWD treatment plant.

FIGURE 11–9
Biomagnification of the pesticide DDD in Clear Lake, California.

FIGURE 11–10
A plasma torch produces a 10,000°C flame that is hot enough to destroy toxic molecules. This type of technology is used in modern hazardous waste incinerators. (*Westinghouse Electric Company*)

Isolation of Industrial Wastes

Isolation is often the cheapest option for managing industrial organic waste. Bernard Reilly, a corporate lawyer for DuPont Chemical Company, argued that "something like dioxin isn't going anywhere. . . . You can keep it in place for a couple of million dollars or you can remove it for $100 million." The problem with this technique and argument is that toxic chemicals often do escape, as discussed in Section 11.6.

11.5 Surface Water in the United States 20 Years After the Clean Water Act

Analyzing the result of the Clean Water Act is like trying to decide whether a glass is half empty or half full. Between 1972 and 1994, the proportion of the U.S. population served by sewage treatment plants jumped from 32 to 74 percent. In the same period, aquatic emissions of industrial toxic pollutants declined by 99 percent, and discharge of metal ions from mining and metal refining declined by 98 percent. But at the same time, the number of pristine waterways declined and the number of polluted waterways increased. Figure 11–11 shows trends in water quality for rivers and lakes in the United States between 1982 and 1989.

How could the number of polluted waterways increase at the same time that the quantity of pollutants decreased? One possibility is that pollution may not be getting worse, but our ability to detect it is improving.

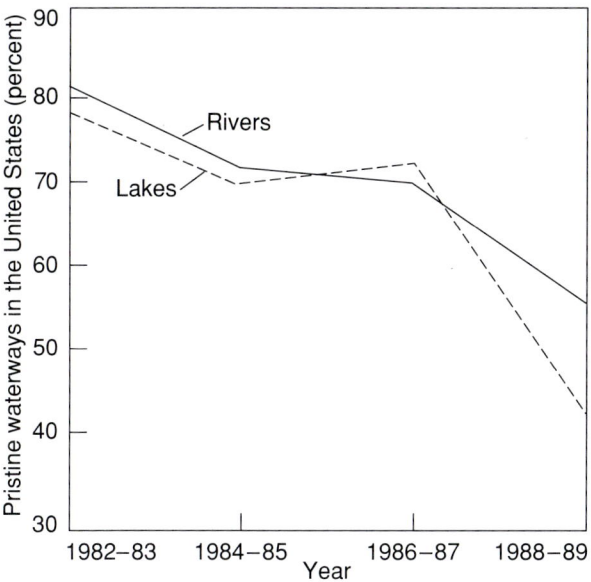

FIGURE 11–11
The percentage of pristine rivers and lakes in the United States between 1982 and 1989. A pristine waterway is defined by the Clean Water Act as one that fully supports the uses and aquatic ecosystems that it supported in its natural state. (From *The Clean Water Act 20 Years Later* by Adler, Landman, and Cameron)

FIGURE 11–12
The Ganges River is a sewer, a public bathing place, and a source of drinking water. Waterborne diseases are a common source of sickness and death in India and other less-developed countries. (Mike Barlow/Dembinsky Photo Associates)

Over the past 2 decades, the criteria for defining a waterway "fully supporting original aquatic ecosystems"[2] have become stricter, so a level of pollution deemed insignificant in 1970 is judged more serious today. In addition, the number of testing sites has increased and thus we have improved detection of water pollution.

A second possibility is that aquatic systems are being degraded by **synergistic effects**, interactions between several independent factors that combine to produce a larger effect. For example, fish may survive in mildly polluted waterways. They may also survive habitat destruction caused by siltation or by waterfront development. But if both effects occur simultaneously, the combined pressures kill the fish.

Water Pollution and Human Health

Worldwide, water pollution is responsible for more human illness than any other environmental factor. Approximately one fourth of the human population does not have access to safe drinking water. In the poorest regions of Africa and Asia, 80 percent of the rural population drinks unhealthful water. Waterborne infectious diseases are the major cause of infant mortality in the less developed countries and kill tens of millions of children every year. For the millions of people threatened by these diseases, even the most rudimentary water purification systems are unaffordable, and in many cases sewers and local drinking water supplies are identical (Fig. 11–12).

In the United States and other developed countries, the picture is very different. In the United States, we use purified water to irrigate lawns; in India it is often not available for drinking. Shortly after Congress enacted the Clean Water Act, it approved a related bill, the Safe Drinking Water Act, designed to regulate contamination of the nation's water by disease organisms and chemicals.

If a person becomes sick from bacterially or virally contaminated water, it is relatively easy to trace the source of the infection. Although the number of infections due to drinking contaminated water in the United States is low, it increased in the decade following the passage of the Safe Drinking Water Act because bacterial and viral contamination of drinking water have increased.

---
[2]Criteria as defined in the Clean Water Act.

FOCUS ON

Thermal Pollution

Water pollution is usually thought of as the addition of foreign matter to an aquatic ecosystem. But heat can also pollute water in a process called **thermal pollution**.

In a fossil fuel or nuclear electric generator, an energy source (coal, oil, gas, or nuclear fuel) boils water to form steam. The steam expands against the blades of a turbine, and the spinning turbine runs the generator to produce electricity. The exhaust steam must then be cooled to maintain efficient operation. The cheapest cooling agent is often river or ocean water. A 1000-megawatt power plant heats 10 million liters of water by 35°C every hour.

Fish are cold-blooded animals. This means that their body temperatures increase or decrease with the temperature of the water. When water temperature rises, all the body processes of a fish (its metabolism) speed up. As a result, the animal needs more oxygen, just as you need to breathe harder when you speed up your metabolism by running. But hot water holds less dissolved oxygen than cold water. Therefore, fish accustomed to cold water may suffocate in warm water. In addition, warm water can cause outright death through failure of the nervous system. Many aquatic animals lay their eggs in the springtime when the water naturally becomes warm. If a power plant heats the water in mid-winter, some organisms may start laying, but if the eggs hatch at this time, the young may not find the food they need to survive.

The relationship between health and chemical contamination is more difficult to establish. Hundreds of thousands of different chemicals are released into U.S. waterways every year. The Safe Drinking Water Act ruled that the EPA must set maximum contaminant standards for the 700 pollutants most likely to cause health problems. Thus, the law allows measurable quantities of chemical pollutants in drinking water but states that the pollutants cannot exceed specific guidelines. By 1990, the EPA had established standards for only 65 of the 700 pollutants. In that year, government scientists reported 100,000 violations of the existing standards. However, the EPA issued only 200 fines or injunctions to correct infractions.

Recall from Chapter 1 that scientists study the health effects of a suspected carcinogen by feeding large doses of the compound to rats. However, no one knows whether or not it is valid to apply the resultant data to low-dose impacts on humans. Another technique for assessing the health effects of chemical contamination is to map the geographical incidence of a disease. This approach is called epidemiology, the study of epidemics. For example, people living in the northeastern portion of the United States have a higher risk of dying from intestinal cancer than do persons living in the Southwest. These data suggest but do not prove that regional pollutants may cause intestinal cancer. Medical statisticians point out that many other factors contribute to geographical incidence of disease. These factors include different genetic or economic backgrounds, diets, levels of exercise, and a variety of other factors. Despite limitations, geographic studies can sometimes identify regions in which chemical water pollutants cause disease. However, epidemiological studies provide information only after harmful effects have occurred. Thus, an epidemiologist will be able to demonstrate that a pollutant causes cancer only after many people have contracted the disease.

11.6 Ground-Water Pollution

Wastes that are buried or spread over the land surface may migrate downward to pollute ground water (Fig. 11–13). More than 50 percent of the people in the United States drink ground water. Recent studies have shown that 45 percent of municipal ground-water supplies in the United States are contaminated with synthetic organic chemicals; wells in 38 states contain pesticide levels high enough to pose a threat to health (Fig. 11–14); in New Jersey, every major aquifer is contaminated (Fig. 11–15); in Florida, where the water table is only 3 meters below the surface in places and 92 percent of the population drinks ground water, more than 1000 wells have been closed due to contamination and over 90 percent of the remaining wells have detectable levels of industrial organic compounds.

11.6 Ground-Water Pollution 271

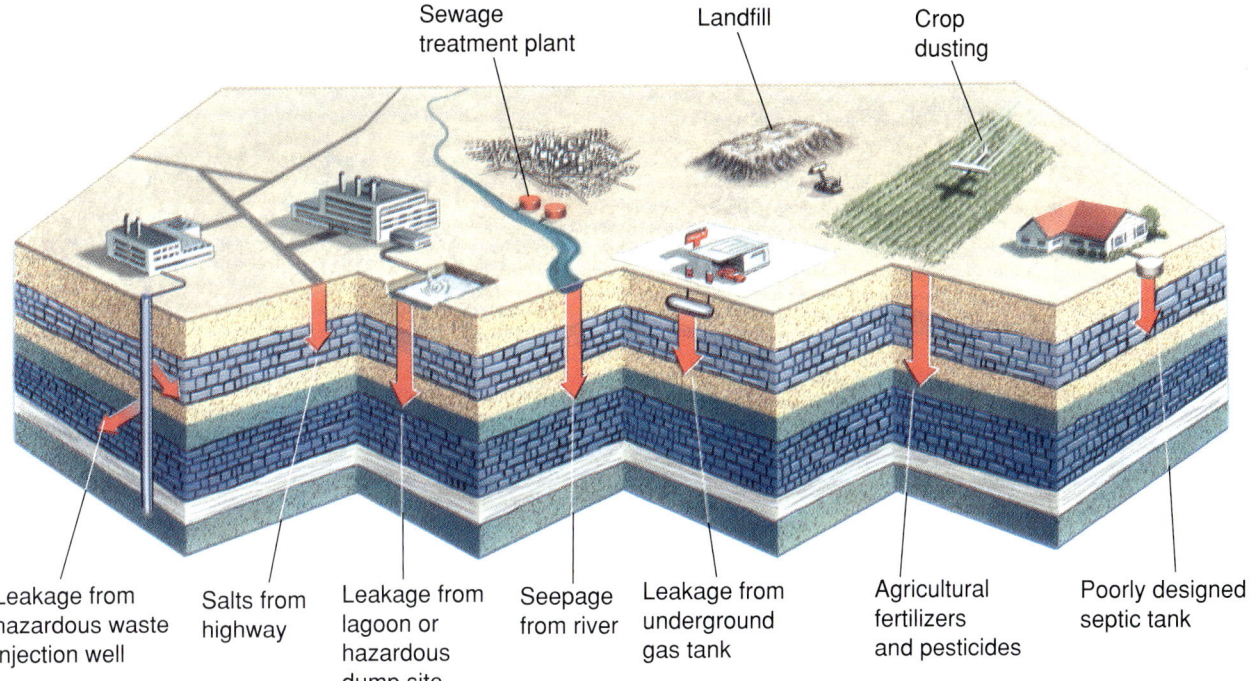

FIGURE 11-13
Sources of ground-water pollution.

FIGURE 11-14
Agricultural chemicals are a major source of ground-water pollution.

FIGURE 11-15
An oil refinery in New Jersey. New Jersey suffers from some of the worst ground-water pollution in North America as a result of heavy industry.

Sewage in Ground Water

In rural areas, it is too expensive to pump sewage to a central treatment facility, so domestic wastes are discarded underground, through a **septic system**. Sewage is flushed into a large tank, typically with a 4000-liter storage capacity. The solids settle to the bottom of the tank and are pumped out every 5 years or so. The wastewater flows out of the tank and into a series of underground trenches filled with gravel. As wastewater trickles through the gravel and into the soil, bacteria decompose organic matter and thus purify the water.

For purification to be effective, wastewater must move slowly through well-aerated soil or rock. Sandy soils and sandstone have the proper permeability for effective purification. If movement is slow and oxygen is available, decay organisms feed on the sewage, decomposing it to harmless byproducts before the polluted water travels very far (Fig. 11–16). On the other hand, if a septic tank is installed in coarse gravel, fractured granite, or cavernous limestone, the sewage may flow rapidly through the large openings and may pollute a well or a stream before the decay organisms decompose it.

Industrial Wastes in Ground Water

Industrial wastes are often dumped onto the ground or buried in shallow landfills like that at Love Canal. All of these sites are potential sources of ground-water pollution. When the Superfund was established in 1979, the EPA identified 20,766 hazardous waste sites in the United States. By 1989 the General Accounting Office estimated that there may be as many as 400,000 hazardous waste sites. Many are small, involving a few rusting drums in a backlot, but others contaminate large areas (Fig. 11–17). By 1993, 1300 sites had been targeted for cleanup, but only 217 of these projects had been completed, at an average cost of $27 million per site.

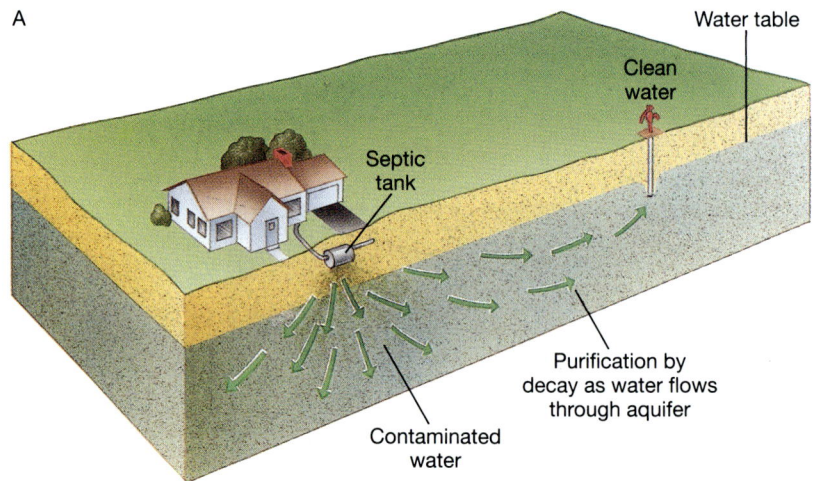

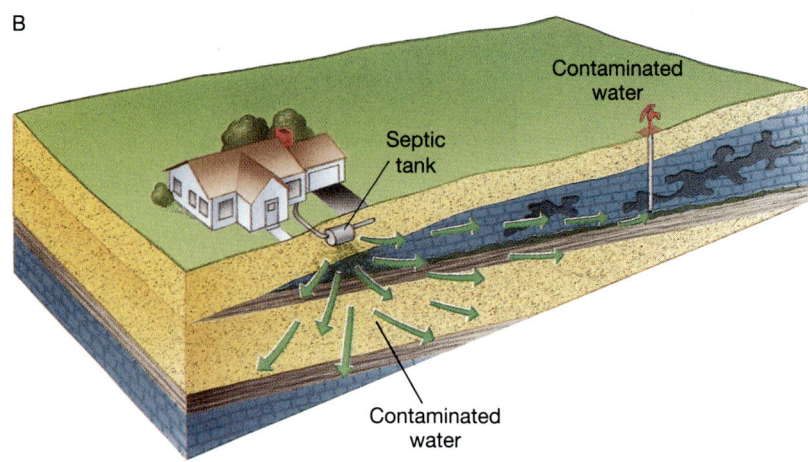

FIGURE 11–16
(A) Water moves slowly through permeable sandstone, giving natural processes enough time to purify it before it reaches a well. (B) But water flows through cavernous limestone too rapidly to be purified and pollutes a nearby well.

FIGURE 11–17
A hazardous waste dump site photographed in 1989 shows violations of environmental protection laws. *(Jeff Amberg/Gamma Liaison)*

Approximately 2 million underground storage tanks exist in the United States. Half hold gasoline at service stations, and the remainder contain chemicals. Many of the tanks are untreated steel. Although most are not hazardous waste sites at present, if they rust and start to leak, their contents will spill into the soil.

Once a nonbiodegradable compound contaminates ground water, it may persist for a long time. Recall that ground water flows slowly, so pollutants are not quickly diluted or dispersed. In addition, ground water does not have access to as much air as surface water does. Therefore, oxidation of persistent chemicals is less effective and less complete underground than it is in surface water.

Because natural processes are not effective in removing chemical contaminants from ground water, clean-up can be difficult and expensive. Contaminated ground water can be contained by building impermeable barriers to isolate it from the aquifer, or it can be pumped to the surface for further treatment and purification. Frequently, containment and pumping are used simultaneously. Another control measure, called **bioremediation**, uses microorganisms to decompose contaminants. Recall that a sewage treatment plant and a septic tank operate by exposing the sewage to air and a healthy population of bacteria. In a similar manner, chemical pollutants in ground water will decompose if air and bacteria are pumped into the soil. Bioremediation is effective only if the rate of bacterial decay is rapid compared with the rate of ground-water flow.

All of these measures are expensive, and none is completely effective. To avoid creating new problems for future generations, it is important to prevent leakage from existing hazardous waste sites. Thus, underground storage containers should be lined and leak detectors installed. Landfills and waste treatment ponds should be lined with impermeable clay or plastic so the contaminants are isolated from aquifers. Factories should be monitored closely to ensure that hazardous wastes are disposed of responsibly.

CASE HISTORY

The Lipari Landfill, Pitman, New Jersey

The Lipari Landfill opened in 1958, before modern environmental laws were enacted. It is situated in an abandoned sand and gravel pit near Pitman, New Jersey. Three million gallons of chemical wastes were dumped legally into the site before it was closed and covered in 1971. (Fig. 11–18) But engineers only sealed the surface, and wastes seeped into the ground water. By the mid-1970s, nearby Alcyon Lake turned orange and purple. Local residents said that fumes rising from the soil brought tears to their eyes and sometimes even left a bittersweet taste on their tongues. In 1983, four years after the Superfund legislation was enacted, the EPA called Lipari the worst hazardous waste site in the United States. New Jersey Senator Frank Lautenberg proclaimed that "Lipari is a symbol" and promised rapid action.

The dump site lies in permeable sand and gravel. EPA engineers learned that an impermeable clay layer lies beneath the dump. They reasoned that the toxic wastes were seeping downward to the clay layer and then flowing laterally on top of the clay to contaminate ground water and the lake. Therefore, they attempted to isolate the landfill by digging a trench around its perimeter to the clay and then fill-

FIGURE 11–18
The Lipari Landfill, Pitman, New Jersey.

ing the trench with concrete. However, the year after the wall was finished, approximately 2600 gallons of polluted water leaked outside the perimeter. Perhaps the wall had cracked, the bond between the wall and the clay was permeable, or the clay layer itself was fractured.

Next, the EPA sunk numerous wells into the landfill and into the ground outside the wall. The engineers pumped ocean water into some of the wells and removed polluted water from others. Then they pumped the polluted water to a purification plant (Fig. 11–19). This process, begun in 1989, is expected to continue for 7 to 10 years.

Local citizens are not happy. They contend that the plan is analogous to pouring water into a leaky bathtub and that the process will increase ground-water contamination. In addition, some of the wastes in the landfill were sealed in metal drums. What will happen if the drums rust through after the flushing project is completed? These critics argue that the only solution is to dig up the landfill, remove the polluted material, and recover the sealed drums. In addition, they want the EPA to dredge Alcyon Lake and remove the contaminated sediment. But it is expensive to purify and isolate thousands of cubic yards of contaminated sediment.

CASE HISTORY

Ground-Water Pollution in the San Joaquin Valley

Ground-water pollution is not restricted to industrial areas. Some of the worst problems arise in agricultural areas where nonpoint source pollution by pesticides and fertilizers is common. The San Joaquin Valley in California is a rich farming district. In the early 1980s, scientists discovered high concentrations of the pesticide DBCP (1,2-dibromo-3-chloropropane) in the drinking water in the region. The source of the DBCP was no mystery: It was sprayed heavily on crops to kill soil pests.

It was a relatively simple matter for chemists to prove that the DBCP had contaminated local drinking water. Ground-water geologists easily traced the route traveled by these materials from the farms to the wells. Once scientists had identified the source of drinking-water pollution, the problem became an issue of public policy. Environmentalists held that pesticides should be regulated so that they do not contaminate drinking water. But farmers claimed that they needed DBCP to raise crops economically and feed the population. They also asked, "We understand that DBCP is present in the drinking water, but how do we know that it is harmful in the concentration observed?"

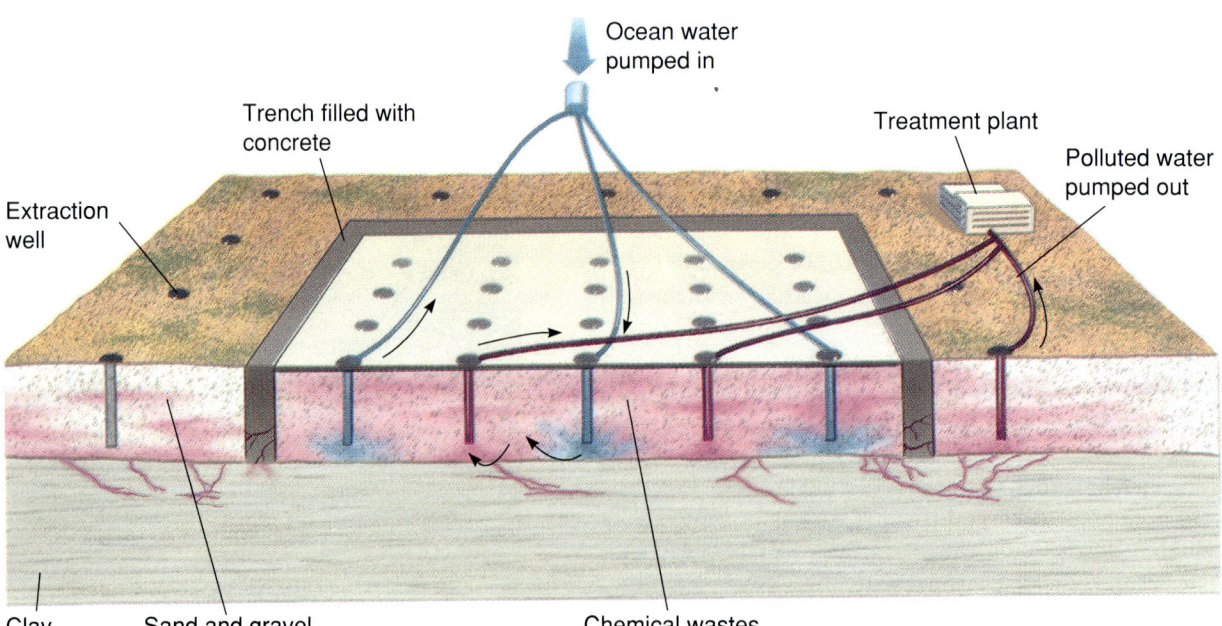

FIGURE 11–19
Schematic of the Lipari Landfill showing techniques used to contain and remove pollutants.

> McFarland and Fowler are two small towns in the San Joaquin Valley. McFarland has a population of 6000, and Fowler has a population of 3000. Based on national averages, one would expect 0.02 cases of childhood leukemia per 1000 residents per year. Therefore, the two towns considered together should have had about one case of childhood leukemia between 1980 and 1985 (0.02 cases/1000 residents/year × 9000 residents × 5 years = 0.9, or about one case). In the first half of the 1980s, six cases of childhood leukemia and eight other childhood cancers occurred in the two towns. Officials began to suspect DBCP. Yet farmers and pesticide manufacturers raised the following objections:
>
> 1. Despite the small probability of so many cases of leukemia and cancer occurring by chance, the residents of McFarland and Fowler might just have been unlucky.
> 2. It is possible that an environmental contaminant other than DBCP caused the cancers. After all, no one has studied the toxicology of all the chemicals in the local ground water.
>
> The arguments, emotions, and legal questions involved in such issues are complex. We tell this story to emphasize that science is not isolated from the broader questions that confront society. If you were a ground-water geologist involved in this case, your professional testimony would be directed specifically toward questions about the source and path of the pollution. However, your involvement would expose you to legal, economic, and social aspects of the problem.

11.7 Deep Injection Wells and Ground-Water Pollution

Liquid chemical wastes are sometimes disposed of by injecting them into permeable rock layers below an aquifer. **Injection wells** are drilled into porous rock—commonly sandstone or fractured limestone—so that the liquid wastes can migrate from the well and flow into deep zones beneath an aquifer that is used for water supplies. However, this permeable layer must be isolated from shallower ground-water reserves by impermeable rock such as shale. A geologic cross section of an injection well is shown in Figure 11–20A. To penetrate below drinking-water supplies, the injection wells are commonly hundreds to thousands of meters deep. A 3000-meter well may cost $1 million, but injection wells are often the cheapest technique for disposing of hazardous wastes. If designed and operated properly, an injection well removes hazardous materials from the biosphere. However, three potential problems exist.

An injection well passes through surface soil and aquifers before reaching the deeper, isolated disposal layer. If the wastes are corrosive or if the pressure is high, the injection pipe may leak and the wastes may then pollute soil or an aquifer used for water supplies.

A second problem with this method of disposal is that the contaminants may migrate from the disposal zone into an aquifer. Even though deep ground water moves slowly, an injection well should isolate contaminants for thousands of years. Therefore, geologists must study the thickness and orientation of rock layers far from the well. If the strata are tilted and the wastes are less dense than the water in the deep aquifer, then the wastes may rise and eventually escape into a drinking or irrigation water aquifer (Fig. 11–20B). Alternatively, a dome in the rock may trap the wastes forever (Fig. 11–20C).

Catastrophic geologic events such as earthquakes and volcanoes can fracture subterranean rock and release the hazardous wastes. Therefore, geologists must also study geological history before recommending a site for an injection well. In some cases, the high fluid pressure in an injection well may trigger earthquakes, which could fracture overlying rock and release the wastes.

11.8 Ground Water and Nuclear Waste Disposal

In a nuclear reactor, radioactive uranium nuclei split into smaller atoms and subatomic particles called **fission products**. Many fission products are themselves radioactive. Most are also useless and must be disposed of despite their dangerous levels of radioactivity.

In the United States, military processing plants, 111 commercial nuclear reactors, and numerous laboratories and hospitals generate approximately 3000 tons of radioactive wastes every year. Consider plutonium-239. Plutonium does not occur naturally on Earth; all of this element that exists on our planet has been manufactured in nuclear reactors. "A thousandth of a gram of plutonium taken into the lungs as an invisible speck of dust will kill anyone.... Even a millionth of a gram is likely, eventually, to cause lung or bone cancer."[3] Plutonium and other radioactive compounds are elements and cannot be destroyed by any biological or chemical process. However, each radioactive isotope decays naturally at a rate measured in terms of its half-life. One **half-life** is

---

[3] John McPhee, *The Curve of Binding Energy*. New York, Ballantine Books, 1975.

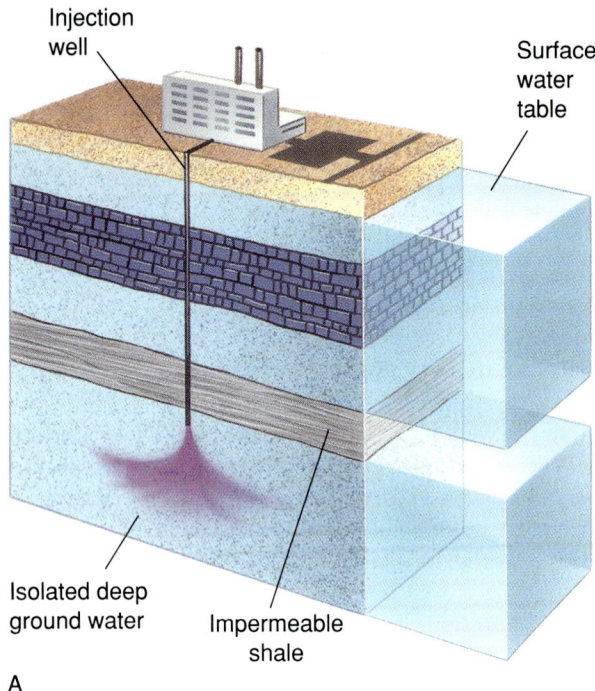
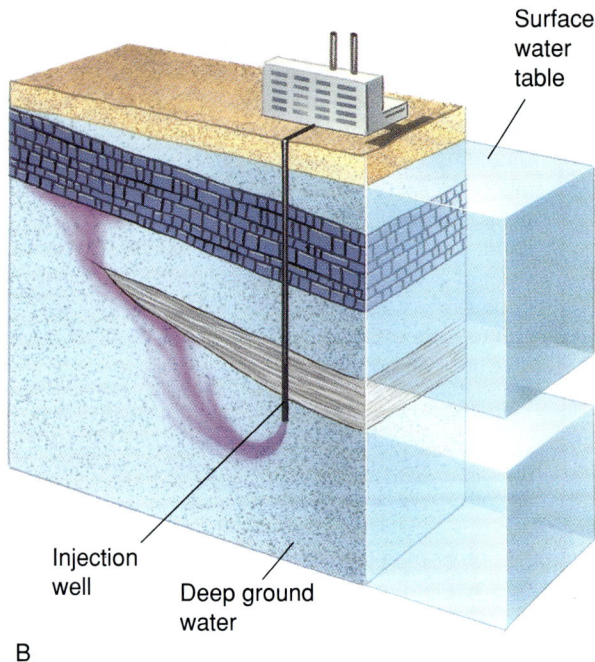
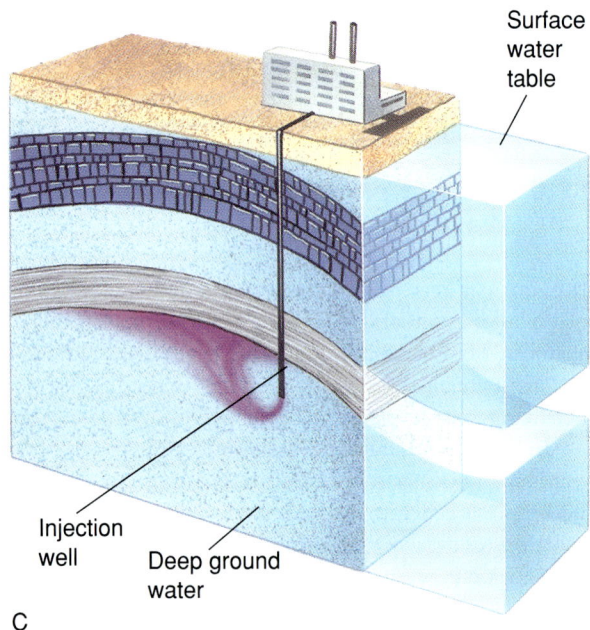

FIGURE 11-20

Injection wells can isolate hazardous wastes. (A) In some localities, shallow ground water is isolated from deep ground water by a layer of impermeable rock, such as shale. If hazardous wastes are injected into permeable rock below the impermeable barrier, the wastes are isolated. Before drilling an injection well, geologists must study the orientation of rock layers at the well site. (B) In this example, rock layers are tilted and the shale is not continuous, so injected wastes migrate into shallow ground water and contaminate an aquifer. (C) However, if the rock is folded into a dome, the contaminants are trapped and thus stored effectively.

the time required for half of a sample to decompose. Half-lives range from fractions of a second to billions of years. The half-life of plutonium-239 is 24,000 years. That means that if 1 gram of plutonium exists today, 1/2 gram will be left after 24,000 years. After another 24,000 years, 1/4 gram will be left. The plutonium produced today will remain a problem for thousands of future generations.

The only feasible disposal method for radioactive wastes is to store them in a place safe from geological hazards and human intervention and to allow them to decay naturally. The U.S. Department of Energy defines a permanent repository as one that will isolate radioactive wastes for 10,000 years. To keep these dangerous materials safely isolated for such a long time, a repository must meet at least three geological criteria:

1. A radioactive waste repository must be completely safe from earthquakes and volcanic eruptions.
2. A repository must be safe from landslides, soil creep, and other forms of mass wasting.
3. A repository must be free from floods and seeping ground water, which might corrode containers and carry wastes into aquifers.

Let us examine three nuclear waste repositories.

CASE HISTORY

Hanford, Washington

Hanford, Washington, was chosen in 1942 as the site of the first nuclear reactor in the United States. The reactor's main purpose was to produce plutonium for atomic bombs to be used against Japan during World War II. In retrospect, the primary consideration clearly was to build the bombs as quickly as possible, whereas the safe disposal of nuclear wastes was given low priority. As a result, radioactive wastes were dumped in ditches and ponds, stored in wooden cribs and steel tanks, and injected into wells and underground drains.

The Hanford site consists of nine nuclear reactors, seven processing plants, and 1377 hazardous waste sites all located on 1500 square kilometers of land on the Columbia River Plateau (Fig. 11–21). Hanford lies on a basalt plateau formed by a sequence of volcanic eruptions about 15 million years ago. Much of the basalt bedrock is fractured and interlaced with fissures and tunnels that formed as the lava cooled. Thus, the bedrock is extremely permeable to ground water.

Military contractors at Hanford stopped producing radioactive material for bombs in 1986. Previously classified documents revealed that 4100 tons of uranium, 20 tons of plutonium, 2000 tons of other radioactive materials, and 11 million cubic meters of hazardous chemical wastes were buried or otherwise stored on the site. In the late 1940s and early 1950s, some of the acids, organic chemicals, and nuclear wastes were dumped together in wooden cribs. The acids dissolved the uranium compounds and the corrosive, radioactive solution leaked from the cribs into the fractured basalt. The solution percolated downward until it encountered an impermeable layer of basalt, and then it flowed laterally until it reached an old well drilled through the basalt layer. It then flowed downward through the well, contaminating a local aquifer. Today, the radioactivity of water pumped from wells in the region exceeds federal drinking-water standards by a factor of 1300.

Engineers at Hanford also stored some of the wastes in 149 single-walled tanks that hold from 225,000 to 4 million liters each (Fig. 11–22). Almost half of these tanks have corroded and leaked, spill-

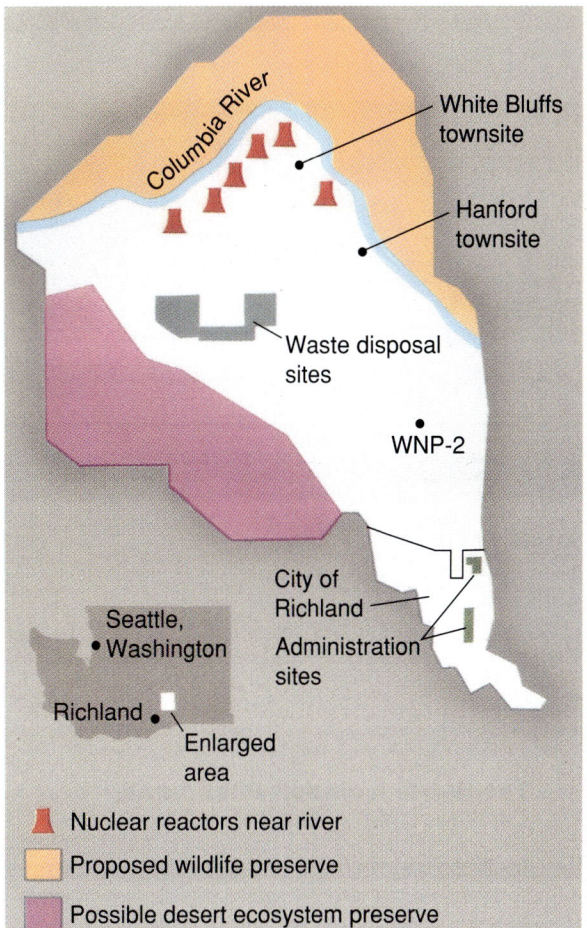

FIGURE 11–21

Site map of the Hanford facility. (Redrawn from *Chemical and Engineering News*)

ing their contents into the ground water. Twenty-eight newer double-walled tanks have not leaked. However, even the stable tanks pose an environmental threat. Viscous chemicals containing radioactive materials have settled to the bottoms of the tanks. The liquid solutions overlying this sludge are, in turn, topped by semisolid crusts. Chemical reactions within the tanks generate hydrogen, an explosive gas. In one tank, called the "burp tank," every 3 months the hydrogen builds up enough pressure to dislodge the sludge, break through the crust, and escape with a giant burp.

Clean-up of the Hanford Reserve is expected to take 30 years and cost from $50 billion to $100 billion. But even then, safety is not guaranteed. Scientists plan to separate radioactive wastes from chemical ones and then encase the radioactive wastes in glass or concrete and bury them in the middle of the Hanford site. Earthquakes or moving ground water could disrupt this storage and release radioactivity.

FIGURE 11–22
Clean-up of leaking radioactive waste storage facilities at Hanford. The leakage problem is so great that clean-up is expected to cost $50 billion to $100 billion and to continue for 30 years. *(Gary Payne/Gamma Liaison)*

The Waste Isolation Pilot Plant, Carlsbad, New Mexico

The Department of Energy (DOE) chose a 200-million-year-old salt deposit near Carlsbad, New Mexico, as a repository for nuclear wastes from weapons production. Salt (sodium chloride) is extremely soluble in water. Therefore, natural salt deposits are usually found in dry environments. If ground water were present, the salt would have dissolved. Thus, the DOE reasoned that the Carlsbad site must be free from dangers associated with seeping ground water.

DOE contractors dug 56 caverns 650 meters below ground to store the waste (Fig. 11–23). If this site is approved, wastes will be sealed in drums, deposited in the salt cavern, and then abandoned. Because the walls of the cavern are unstable, the salt is expected to collapse, enveloping the drums and isolating them from the environment. According to DOE scientists, the isolation will last for tens of thousands of years.

However, geologists have recently learned that small amounts of water seep into the cavern. Over the centuries, even a tiny seepage will dissolve enough salt to create a salt brine that will corrode the drums and carry the radioactive materials into the ground water. In January 1992, an earthquake that measured 4.8 on the Richter scale occurred near the repository site. The repository was undamaged because it is designed to withstand a magnitude 5.5 quake directly under the site, but the quake was an indication that the site may not be as stable as geologists originally assumed. Finally, scientists studying the burp tank in Hanford reasoned that similar reactions could occur in the drums to be buried in Carlsbad. Escaping gases could cause an explosion that would jeopardize the integrity of the site. Due to all these concerns, in October 1993 the DOE agreed to conduct more studies before burying wastes at the Carlsbad site.

The Yucca Mountain Repository

In December 1987, the U.S. Congress chose a site near Yucca Mountain, Nevada, about 175 kilometers from Las Vegas, as the national burial ground for all spent reactor fuel unless sound environmental objections were raised.

Bedrock at the Yucca Mountain site is welded tuff, a hard volcanic rock. The tuffs erupted from several large volcanoes that were active from 16 million to 6 million years ago. Later volcanism created the Lathrop Wells cinder cone 24 kilometers from the proposed repository. The last eruption near Lathrop Wells occurred 15,000 to 25,000 years ago. Geologists have mapped 32 faults that have moved during the past 2 million years adjacent to the Yucca Mountain site. The site itself is located within a structural block bounded by parallel faults (Fig. 11–24). Critics of the Yucca Mountain site argue that recent earthquakes and volcanoes prove that the area is geologically active.

The environment is desert dry, and the water table lies 550 meters beneath the surface. The repository is a series of tunnels and caverns dug into the tuff 300 meters beneath the surface and 250 meters above the water table. Critics have suggested that an earthquake could drive deep ground water upward, where it would contact the radioactive wastes. Because radioactive decay produces heat, the wastes may be hot enough to boil water and create steam. Steam trapped underground could build up enough pressure to rupture containment vessels and cavern walls. Even in the absence of an earthquake, slow seepage of waste from the surface will percolate through the repository site to the water table sometime between 9000 and 80,000 years from now. The lower end of this estimate is within the 10,000-year mandate for isolation. If the climate becomes appreciably wetter, which is possible over thousands of years, ground-water flow may accelerate and the water table may rise. If rocks beneath the site were fractured by an earthquake, then contaminated ground water might disperse more rapidly than predicted. Furthermore, critics point out that construction of the repository will involve blasting and drilling, and these activities could fracture underlying rock, opening conduits for flowing water.

To stop development of the Yucca Mountain site, the state of Nevada refused to issue air quality permits to operate drilling rigs at the repository. As this book is being written, the legal battle continues. Supporters of the repository argue that we need nuclear power and, therefore, as a society we must accept a certain level of risk. Furthermore, the Yucca Mountain Repository is safer than the temporary storage sites now being used. At present, 24,000 tons of radioactive waste lie in temporary storage at nuclear power plants across the United States. Supporters and critics of the Yucca Mountain Repository agree that these temporary sites are not secure. Yet the requirements of permanent storage are so stringent and the danger is so great that the debate continues.

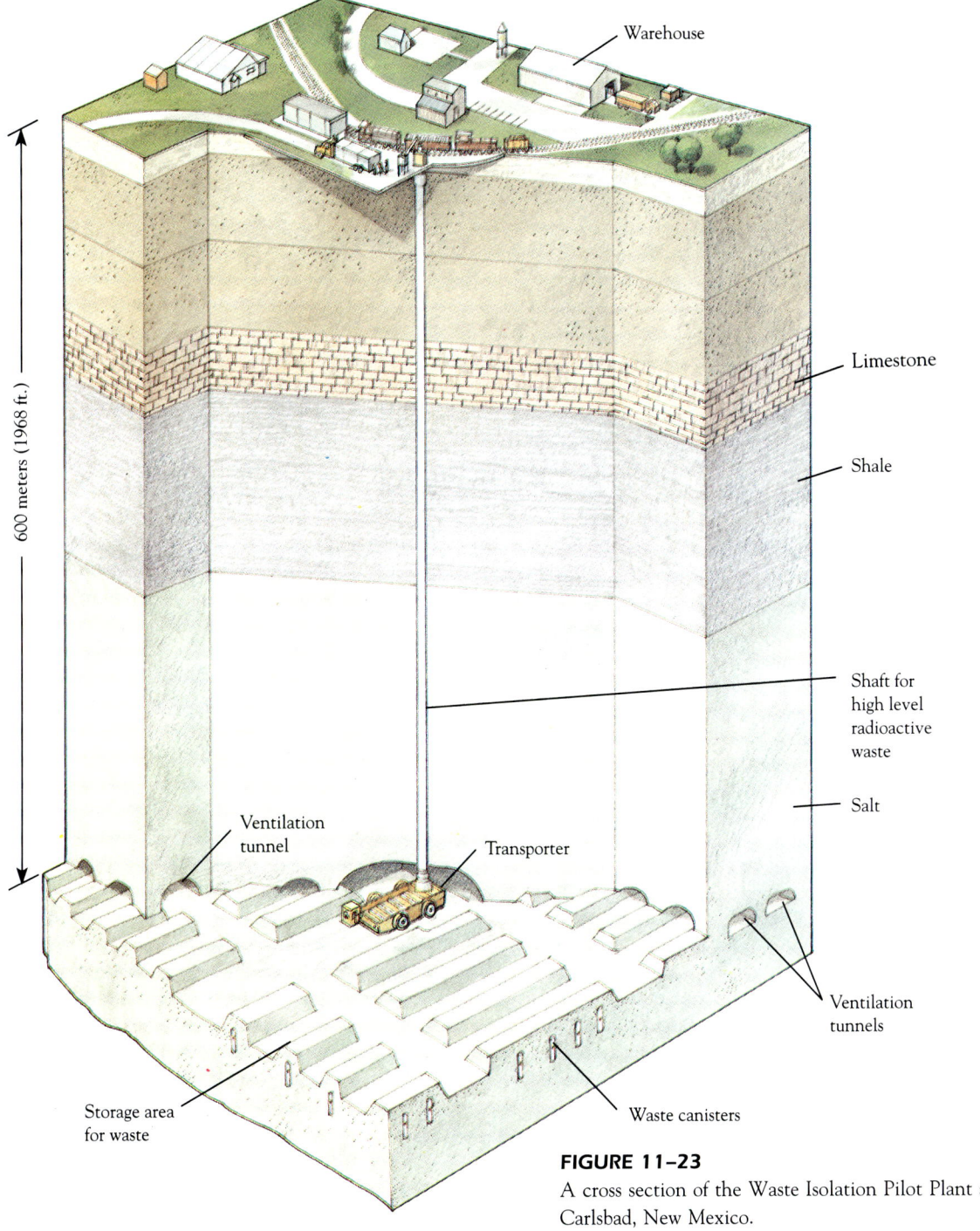

FIGURE 11–23
A cross section of the Waste Isolation Pilot Plant at Carlsbad, New Mexico.

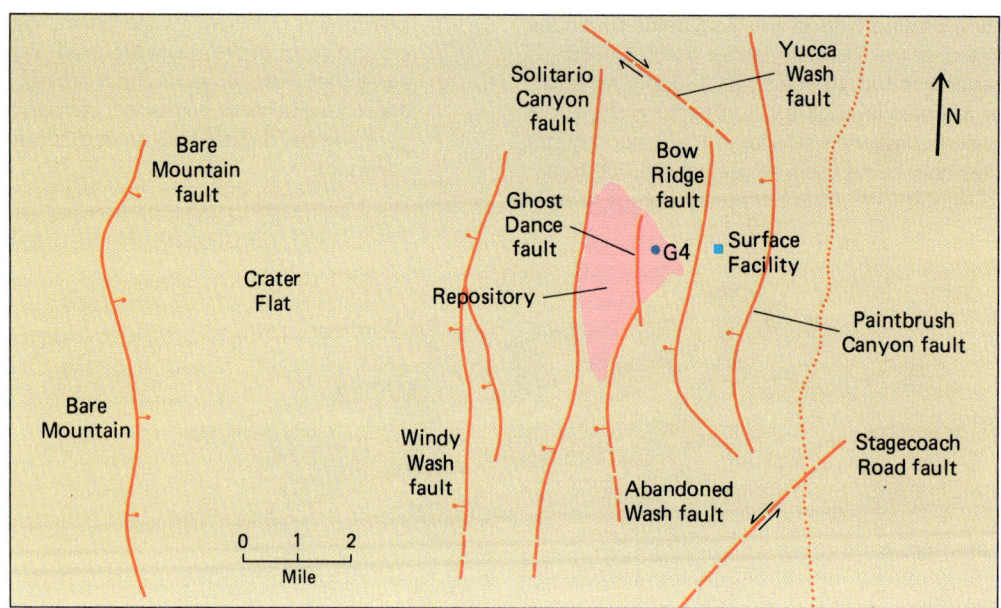

FIGURE 11–24
A map of faults near the proposed Yucca Mountain site. The red lines are faults, places where rock has fractured and moved in the past. (Redrawn from *Geotimes,* January 1989)

11.9 Municipal Waste Disposal

Sanitary Landfills

In the United States, the average person produces 1.8 kilograms of household garbage every day, 660 kilograms per year (Fig. 11–25). Despite active recycling programs, both the per-capita and overall trash production continue to increase (Fig. 11–26). About 75 percent of municipal trash is dumped in **sanitary landfills**. After waste is brought to a sanitary landfill, it is compacted with bulldozers or other heavy machinery. Each day, 15 to 30 centimeters of soil is pushed over the trash to exclude air and rodents (Fig. 11–27). When a site is full, the trash is buried under a thick layer of soil.

Most household trash consists of biodegradable materials such as paper, cardboard, and food scraps (Fig. 11–28). However, oxygen must be available for biological decay to be effective. Once trash is buried and compacted, air is excluded and microbial decomposition slows down. As a result, many materials that are normally biodegradable persist for decades.

Landfills, like hazardous waste dumps, can be sources of ground-water pollution. Most older sites were chosen without concern for local geology or ground-water pollution. Some household garbage, such as paints, solvents, and pesticide residues, is toxic and contributes to groundwater pollution. Modern landfills are built over an impermeable layer of clay or bedrock to isolate them from ground water. However, if the basin fills up, contaminated water may spill over the sides and pollute an aquifer (Fig. 11–29).

Several other environmental problems are associated with sanitary landfills. When trash decomposes in the presence of air, the main pollutants are carbon dioxide, a greenhouse gas, and sulfur dioxide, a contributor to acid rain. When trash decomposes in the absence of oxygen, the main pollutants are methane, another greenhouse gas, and hydrogen sulfide, a toxic air pollutant that smells like rotten eggs. Thus landfills pollute the air as well as ground water. In addition, habitats are destroyed when undeveloped land is converted to a landfill. Many large metropolitan areas have used up their available sites for landfills and now transport their trash farther into the countryside. In these instances, transportation costs rise. Finally, disposal in a landfill represents a depletion of resources. Food wastes and sewage sludge that could be used as fertilizers are buried. Paper and wood scraps that could be recycled are lost, and nonrenewable metals become hard to recover.

11.9 Municipal Waste Disposal 281

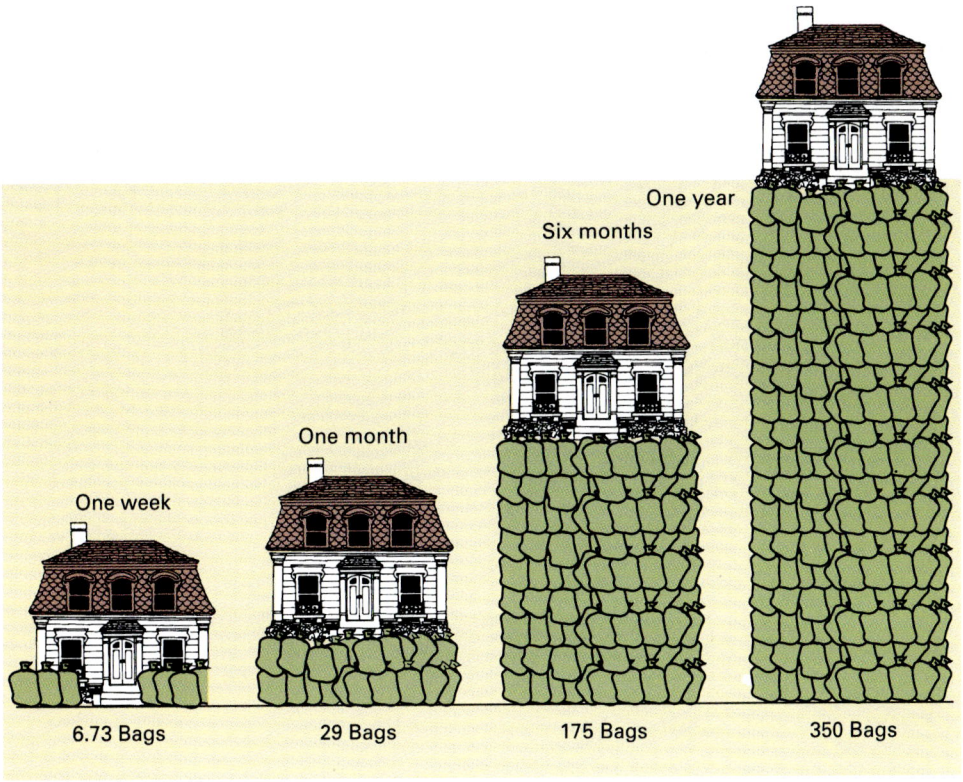

FIGURE 11–25
The volume of municipal solid waste generated by an average American household compared with the volume of an average American house. (From *Environment* by Raven, Berg, and Johnson)

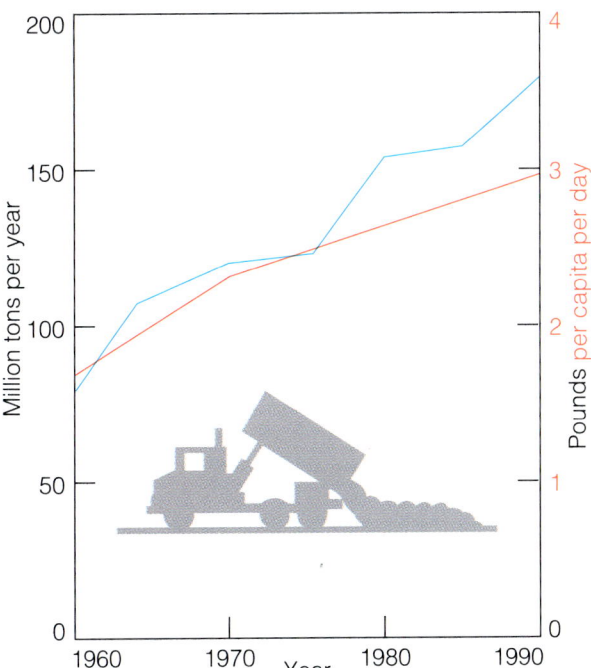

FIGURE 11–26
Both total municipal solid waste (blue line) and solid waste per person (red line) in the United States increased between 1960 and 1990. (Data from EPA)

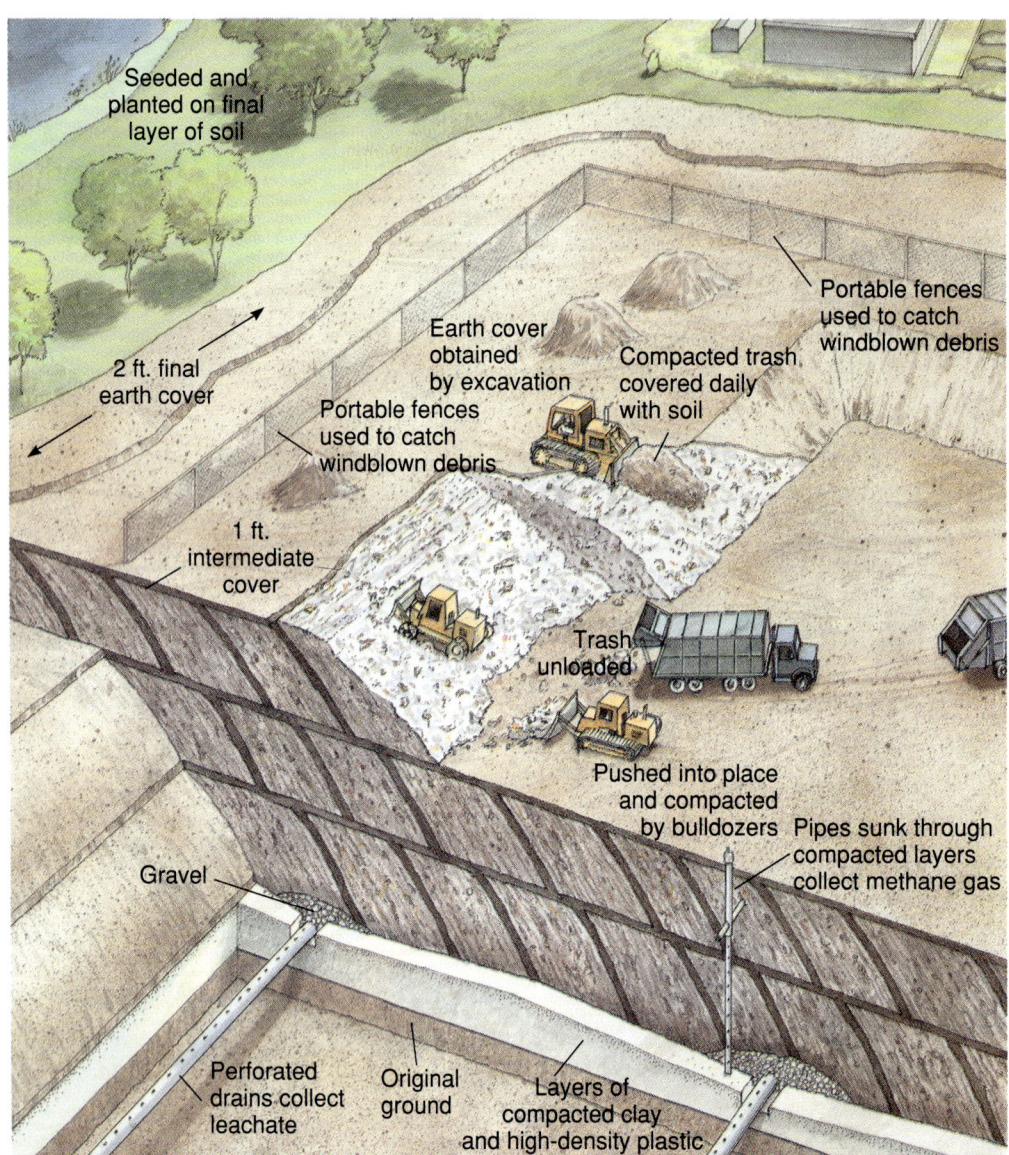

FIGURE 11–27
A modern sanitary landfill.

Fresh Kills Landfill

The Fresh Kills Landfill in New York City is one of the oldest in the United States and the largest in the world (Fig. 11–30). It was started in 1948 over a tidal wetland. Today it covers 1200 hectares (12 sq km) and contains 67 million cubic meters of refuse; 17,000 tons are added every day. In 1948, many people didn't appreciate the value of wetlands, and the site was considered useless land. Today, people realize that Fresh Kills has destroyed a vital haven for waterfowl migrating along the eastern seaboard.

By 1998 the growing pile is expected to rise 165 meters above sea level and thus become the highest hill in New York City. When refuse is piled on top of refuse, air is excluded, so decay is slow. Core drilling in the landfill has brought newspapers from the early 1950s to the surface, and the text is still readable.

For the first few decades of operation, trash was dumped directly into the swamp. Although new areas of the landfill are lined with an impermeable barrier to prevent leaching, approximately 3 million liters of contaminated water leach into the ground every day through the unlined portion of the landfill.

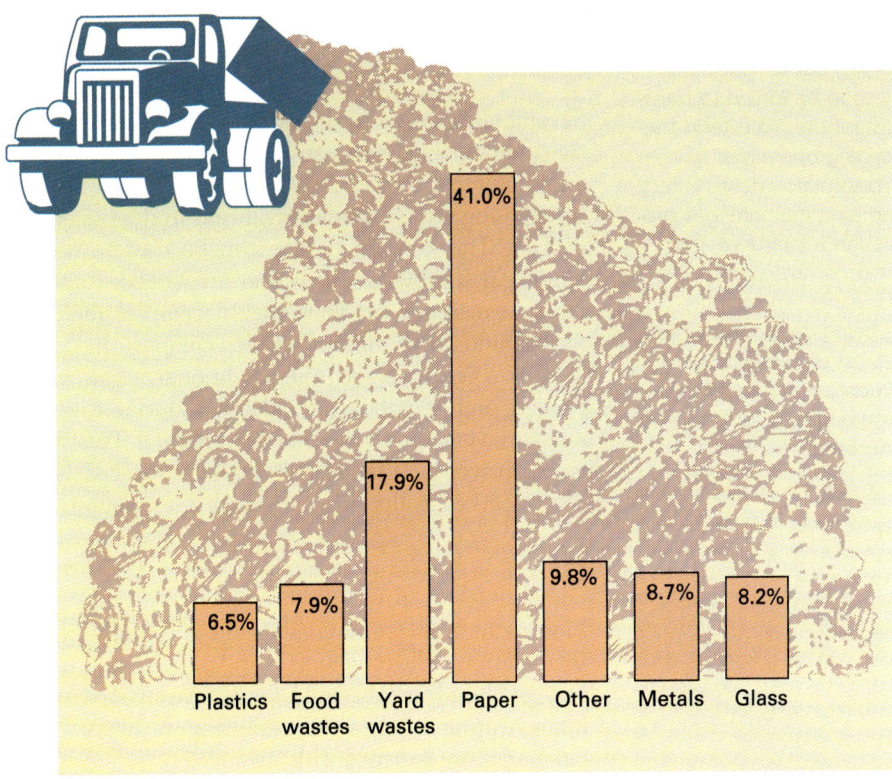

FIGURE 11–28
Composition of household trash.

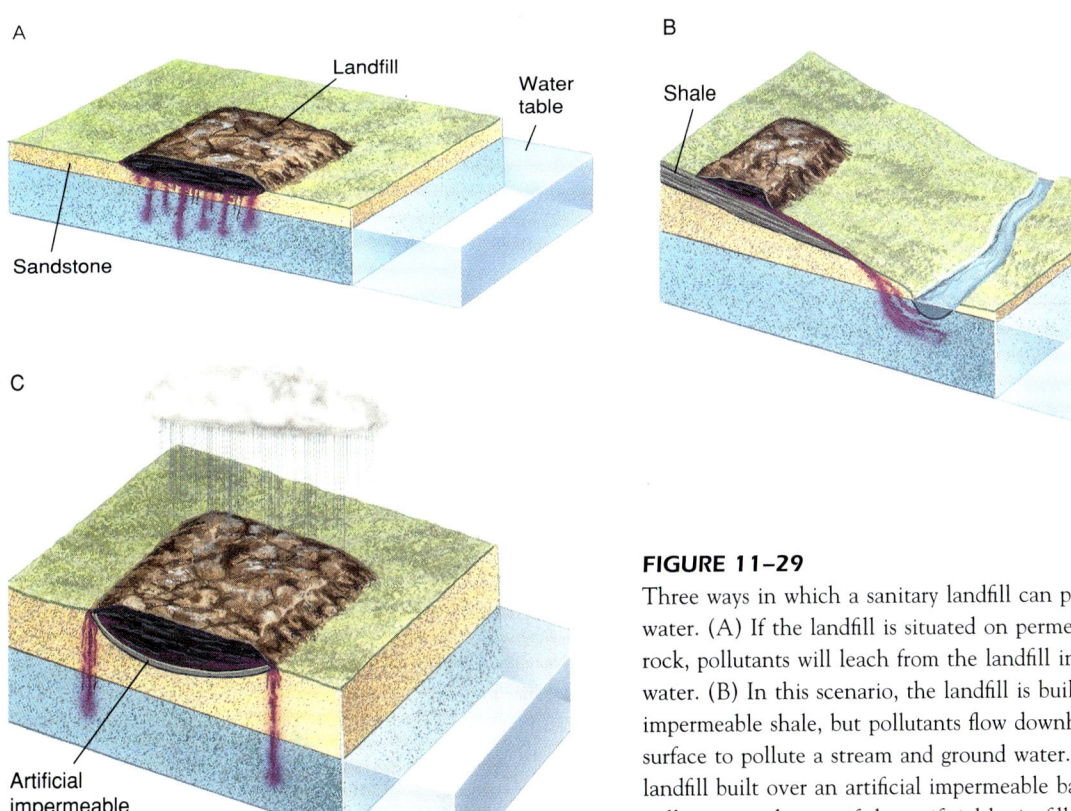

FIGURE 11–29
Three ways in which a sanitary landfill can pollute ground water. (A) If the landfill is situated on permeable soil and rock, pollutants will leach from the landfill into the ground water. (B) In this scenario, the landfill is built on a layer of impermeable shale, but pollutants flow downhill over the surface to pollute a stream and ground water. (C) A modern landfill built over an artificial impermeable barrier may pollute ground water if the artificial basin fills up and overflows.

Today the landfill is running out of space, and engineers fear that they will soon be forced to transport city refuse to a more distant site. To extend the life of Fresh Kills, several recycling operations have been initiated or planned. Construction debris is recycled, yard trash is composted, and methane gas produced by organic decomposition is captured and sold. A plant is currently under construction to sort trash and recycle glass, cans, paper, and plastic.

Recycling technology is being developed simultaneously at many locations across the United States. For example, engineers at Puente Hills Landfill near Los Angeles are retrofitting garbage trucks to run on methane produced by rotting trash.

Composting

Several alternatives to the sanitary landfill exist. Recall that municipal sewage treatment facilities, septic systems, and bioremediation of ground-water pollution all work by exposing waste to decay organisms in the presence of oxygen. The same technique can be used to degrade municipal solid wastes. If wastes are shredded and agitated to maintain air flow, they decompose rapidly to form a humus-like substance called **compost** that can be used as a fertilizer and soil conditioner. Glass, metal, plastics, and household hazardous wastes must be separated from organic refuse before composting. This process is expensive, but the cost is offset by reduced landfill costs, reduced pollution, and sale of the compost.

Incineration

Most municipal trash is combustible and can be burned. Incineration reduces volume by 60 percent, and the energy produced can be used for industrial processes or home heating. Modern trash incinerators, like waste incinerators, are designed with advanced pollution control (Fig. 11–31). However, burning always produces carbon dioxide. Small concentrations of organic toxins and toxic metals such as lead, cadmium, and mercury that are present in the trash usually escape as well. Although 128 trash-to-energy incinerators were operating in the United States in 1990, many new projects were blocked by citizen opposition or were canceled due to high cost.

FIGURE 11–30
Municipal refuse is barged from Manhattan to the Fresh Kills Landfill on Staten Island. *(Louie Psihoyos/Matrix International Inc.)*

FIGURE 11–31
This mobile toxic incinerator is designed to destroy hazardous industrial wastes, but the plume from the stack contains steam and small amounts of hazardous wastes. *(Tim Lynch/Gamma Liaison)*

11.10 Alternatives to Waste Disposal

Reducing Consumption

Dispersal, destruction, and isolation of wastes are all unsatisfactory for one reason or another. An alternative to trash disposal is to reduce the quantity of material we use and then throw away. If people consume less, then there is less manufacturing waste and fewer products to throw into the landfill.

Consider the daily newspaper. Logging disrupts soil, pollutes streams, and destroys natural ecosystems. Manufacture of paper from the logs produces air and water pollution. Transportation of the newspapers consumes fossil fuels, contributes to the greenhouse effect, and adds to urban air pollution. Finally, the newspapers become a solid-waste problem. Yet the news is available through television, radio, and computer networks. Thus, you can conserve resources and reduce pollution by obtaining the news electronically and not buying the newspaper.

You can also reduce consumption by using manufactured items as long or as efficiently as possible. If you enjoy newspapers more than electronic media, you can read the paper in the library, where it is accessible to many people. Similarly, environmentally conscious shoppers refill glass bottles and take their groceries home in durable cloth bags. Durable goods such as automobiles and refrigerators can be repaired when they are old rather than being thrown away and replaced.

Recycling

When an item is recycled, it is first destroyed and then treated to extract its useful raw materials (Fig. 11–32). Discarded metal can be melted and recast into new products. Used tires can be shredded and converted to raw rubber. Old newspapers can be pulped and converted to new paper (Fig. 11–33). Spoiled meat and meat scraps from slaughterhouses can be rendered and converted to tallow and animal feed.

Recycling conserves material resources. In addition, more energy is needed to extract and refine virgin resources than to process recycled materials. For example, nearly 20 times more energy is consumed to produce aluminum from virgin ore as from scrap aluminum. More than twice as much energy is needed to manufacture steel from ore as from scrap, or to manufacture paper

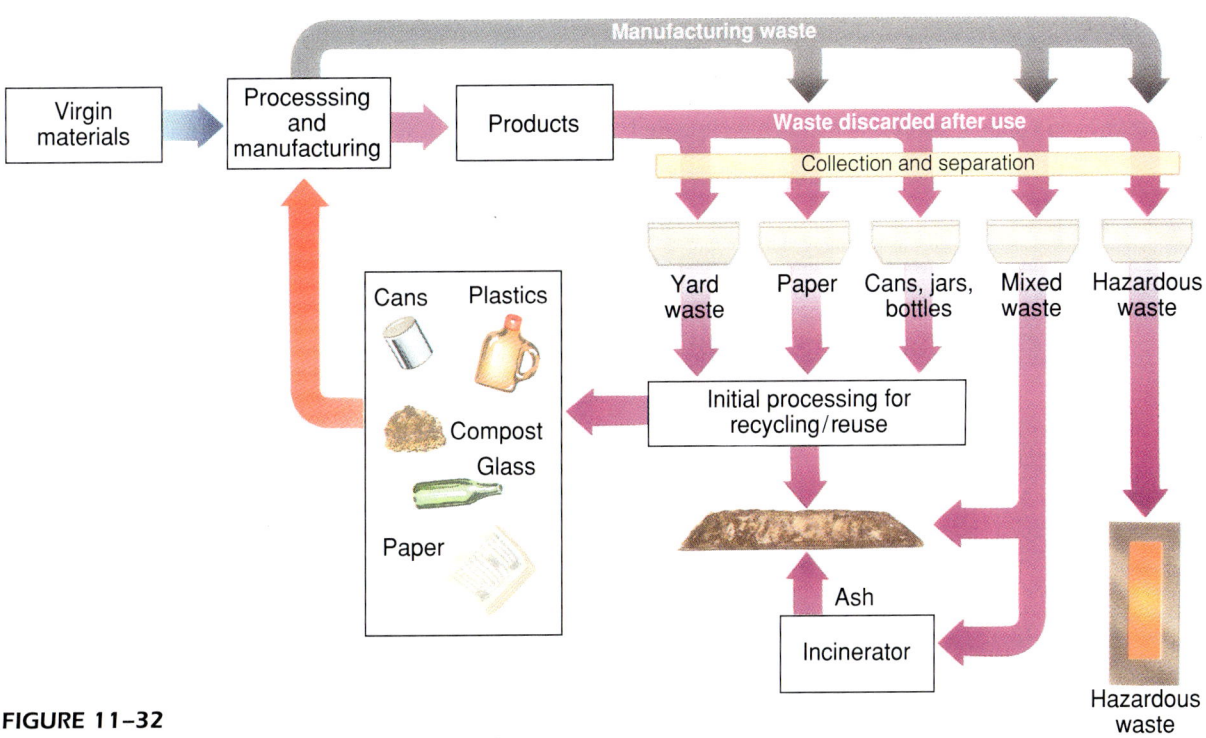

FIGURE 11–32
Waste management in a society that recycles its refuse.

FIGURE 11-33
A conveyor belt brings waste paper to a pulper for recycling. Recycling conserves resources and reduces waste. *(Warren Faubel/Bruce Coleman, Inc.)*

from trees as from recycled paper (Table 11–1). On the other hand, recyclable scrap is often widely dispersed, and fuel is used to transport it to processing plants.

In most cases, recycling also emits less pollution than production from virgin resources. The EPA has estimated that recycling all the metal and paper in municipal trash in the United States would prevent the release of more than 2000 tons of air pollutants and 700 tons of water pollutants every year.

In the United States, recycling has increased in recent years but rates still lag behind those in other industrial countries (Fig. 11–34). Recycling becomes attractive when three conditions are met:

1. The cost of disposal is high.
2. If a large and concentrated supply of scrap is available, investors are likely to build factories to recycle the scrap. Thus, if a large proportion of the population recycles, then recycling becomes economical. Voluntary recycling programs have become increasingly popular during the past few decades. In addition, many cities, faced with increasing costs and adverse environmental consequences of landfills or incinerators, require citizens to separate their trash for recycling. For example, laws in Seattle have reduced solid waste volume by 40 percent compared with that of a decade ago. In other cities, mixed trash is collected and sorted in central processing facilities.
3. For recycling to become economical, people must buy goods made from recycled materials. In recent years, many recycling operations have closed down because there is no market for their products. Environmentalists argue that if people were taxed for the environmental consequences of resource consumption and solid waste disposal, then the market for recycled goods would increase.

	Paper (%)	Aluminum (%)	Iron and Steel (%)	Glass (%)
TABLE 11–1 Environmental Benefits of Recycling*				
Energy use reduction	30–55	90–95	60–70	5–25
Spoil and solid waste reduction	130**	100	95	80
Air pollution reduction	75	95	85	20

* Reduction of environmental insults during manufacturing when recycled materials are substituted for materials made from virgin resources.
**More than 100 percent reduction is possible because 1.3 kg of waste paper is required to produce 1 kg of recycled paper. If all paper were recycled, the waste reduction, of course, would equal only 100 percent.

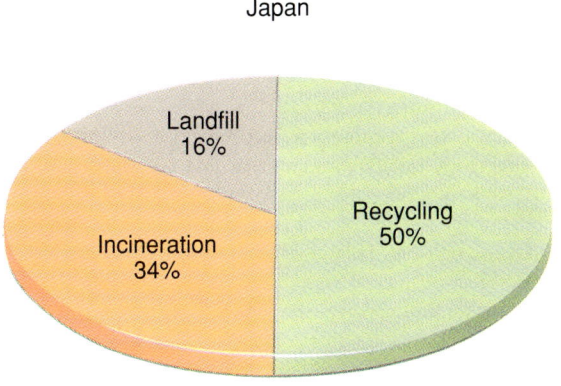
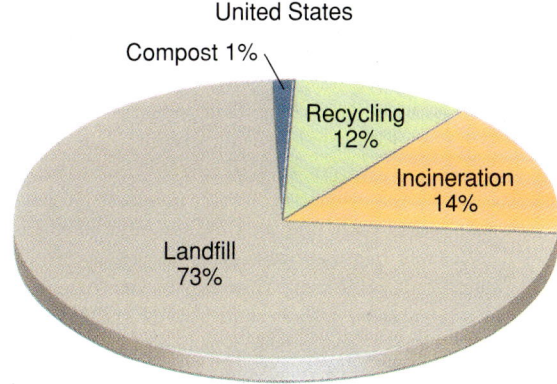

FIGURE 11–34
Solid waste disposal in Japan and the United States. Notice that the Japanese recycle four times as much municipal trash as the Americans do. (Data from INFORM and EPA)

SUMMARY

The Clean Water Act regulates discharge of pollutants into surface waters, and the Comprehensive Environmental Response, Compensation, and Liability Act (**CERCLA**), commonly known as Superfund, provides funding and guidelines to clean up hazardous waste sites. The seven categories of water pollutants are sewage, disease organisms, plant nutrients, industrial organic compounds, toxic inorganic compounds, sediment, and radioactive materials. Wastes can be **dispersed, destroyed, isolated** and **stored**, or **recycled**.

Eutrophication occurs when added nutrients fertilize microorganisms and aquatic plants, causing their populations to increase rapidly. These organisms consume so much dissolved oxygen that fish die.

Primary sewage treatment screens out large objects and allows others to settle. In **secondary treatment**, microorganisms consume the nutrients in the presence of oxygen. **Tertiary sewage treatment** removes specific pollutants such as industrial organic compounds, phosphates, and nitrates.

The federal government issues permits to discharge chemical wastes into surface waters. It is difficult to determine the health effects of these pollutants. Certain chemicals **biomagnify** in the food chain, so the concentration in the water is not a true measure of their effect on the ecosystem. Industrial organic chemicals can also be incinerated or stored in permanent repositories.

In a properly designed septic system, sewage is decomposed by bacteria as it percolates slowly through underground trenches filled with gravel. Industrial wastes in ground water decompose and migrate slowly. Polluted ground water can be purified by pumping it to the surface or by **bioremediation**, but both of these processes are expensive and only partially effective. Deep injection wells isolate wastes if they are designed and maintained properly.

Nuclear wastes pose an unresolved threat to ground water. **Sanitary landfills** remove wastes from sight but may also pollute ground water. Decomposition is slow in landfills. Municipal waste is also removed by **composting** and **incineration**. Waste volume can also be reduced by reducing consumption and by recycling.

KEY TERMS

Clean Water Act 262
Superfund 262
CERCLA 262
Sewage 263
Dispersal 263
Isolation 264
Recycling 264
Aerobic 264

Biological oxygen demand 264
Anaerobic 264
Cultural eutrophication 264
Primary sewage treatment 265
Secondary sewage treatment 265
Tertiary sewage treatment 265
Hazardous waste 266
Biomagnification 267

Incineration 267
Synergistic effects 269
Bioremediation 273
Injection well 275
Fission products 275
Half-life 275
Sanitary landfill 280
Compost 284

REVIEW QUESTIONS

1. List the three major goals of the Clean Water Act.
2. What is the purpose of the Comprehensive Environmental Response, Compensation, and Liability Act? What is another name for this legislation?
3. List the seven categories of water pollutants. What is/are the source(s) of each, and what are the harmful effects?
4. Discuss four treatment strategies for wastes.
5. What is eutrophication? How could a nontoxic substance such as cannery waste become a water pollutant?
6. Explain how a sewage treatment plant operates. What pollutants are removed by primary, secondary, and tertiary treatment?
7. What is a hazardous waste?
8. Explain how a pollutant can become magnified in an aquatic food chain.
9. Discuss the progress between 1972 and 1994 under the Clean Water Act.
10. Explain how water in a septic tank is purified. Why is purification more effective in some types of rock or soil than in others?
11. Explain why pollutants persist in ground water longer than they do in surface water.
12. Draw a cross section of a deep injection well that would effectively isolate contaminants and one that would pollute public ground-water reservoirs.
13. List three sources of radioactive wastes.
14. Define half-life. How does the half-life concept relate to the problem of nuclear waste disposal?
15. Compare and contrast the isolation strategies used at the Waste Isolation Pilot Plant and the Yucca Mountain Repository.
16. Explain how a sanitary landfill operates. How can landfills pollute the air and the ground water?
17. List the advantages and disadvantages of composting and incineration as alternatives to sanitary landfills.
18. Explain why recycling is an attractive alternative to disposal.
19. List the three conditions that encourage recycling.

DISCUSSION QUESTIONS

1. Discuss potential sources of surface- and ground-water pollution in your area.
2. Disposal of chemical wastes in Love Canal was unsatisfactory because toxic wastes escaped after a few decades. If you were in charge of establishing policy, how would you define permanent isolation?
3. Discuss the similarities and differences among a sewage treatment plant, a septic system, and bioremediation.

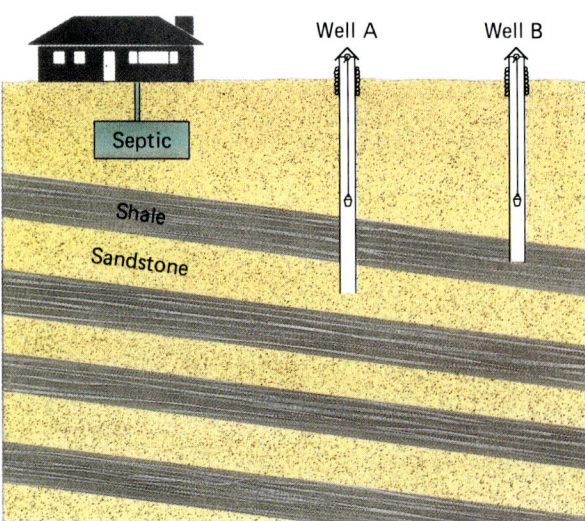 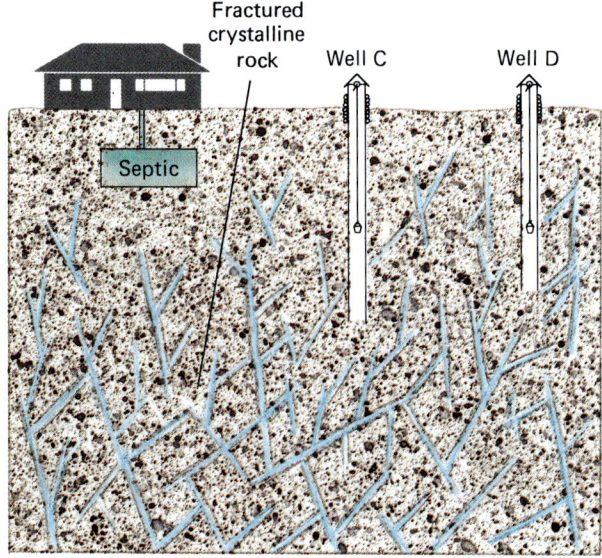

FIGURE 11–35

4. Categorize each of the following as dispersal, destruction, or isolation: secondary sewage treatment; a septic tank; a deep injection well; a sanitary landfill; a hazardous waste incinerator; dumping the effluent from a sewage treatment plant into the river.

5. Would DDT contamination raise the BOD of a lake appreciably? Why or why not?

6. Discuss difficulties in comparing surface water quality today with that of 20 years ago. What historical data sources would you look for to make your comparison? Discuss limitations of these data.

7. Give examples of synergistic effects from activities besides water pollution (examples include human disease, auto mechanics, etc.).

8. Discuss the problems of relying on epidemiological data to determine the health effects of environmental contamination.

9. Compare and contrast a septic system with a municipal sewage treatment plant.

10. Explain why a hazardous organic chemical like DDT can be destroyed, whereas mine wastes such as arsenic cannot.

11. Which of the wells in Figure 11–35 would you expect to contain water? Which would you expect to be polluted? Explain your answers.

12. Argue for or against the following: If old newspapers are burned or composted, the carbon is released into the atmosphere, enhancing the greenhouse effect. However, decomposition is slow in landfills, so the carbon is stored. Therefore, landfills represent an environmentally attractive disposal method for newspapers.

13. If old cars were repaired and used for a long time, then pollution from the steel and auto industries would be reduced. However, new cars emit significantly less pollution than old ones. With this background, discuss the statement "If people consume less, then there is less manufacturing waste and fewer products to throw into the landfill."

14. Discuss the probability of the potential failure modes for the Carlsbad and Yucca Mountain Repositories. How does this assessment relate to the precautionary principle discussed in Chapter 1?

CHAPTER 12

Coastlines and Coastal Pollution

12.1 Emergent and Submergent Coastlines
12.2 Erosion and Transport Along Coastlines
12.3 Erosion of the Atlantic Coast
12.4 Erosion of the Gulf Coast
12.5 Erosion of the Pacific Coast
12.6 Coastal Development and Public Policy
12.7 Ocean Pollution
12.8 Reefs
12.9 Global Warming and Sea-Level Rise

The seashore has always been an attractive place to live. Coastal regions are cooler in summer and warmer in winter than continental interiors. The sea provides food and transportation. In addition, we enjoy the salt air and find the rhythmic pounding of surf soothing and relaxing. Vacationers sail, swim, and fish along the shore. For all these reasons, coastlines have become heavily urbanized and industrialized. In the United States, 75 percent of the population, 40 percent of the manufacturing plants, and 65 percent of the electrical power generators are located within 80 kilometers of the oceans or the Great Lakes.

The seashore also provides habitats for many plants and animals. Coastal plains and the continental shelves comprise 10 percent of the Earth's surface but support 25 percent of its life. Seventy-five percent of the fish caught every year are harvested from coastal oceans. In recent decades, cities have grown to the water's edge as developers girdle river mouths and ocean bays with concrete. Habitat loss and pollution have created an environmental crisis along many of our coastlines.

A barrier island consists of sand that is continuously deposited and eroded by ocean currents that flow parallel to a coast. The south shore of Long Island.

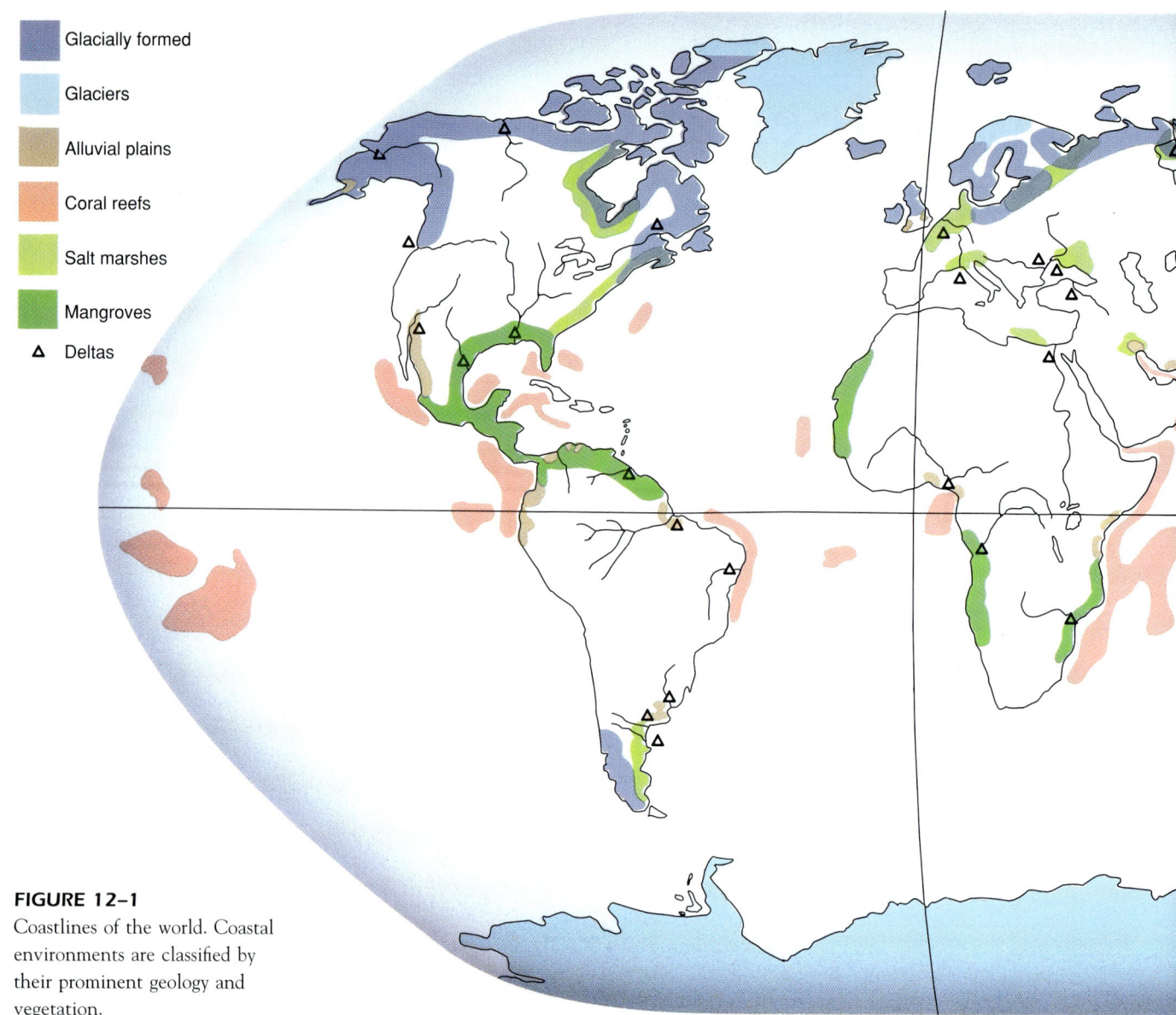

FIGURE 12–1
Coastlines of the world. Coastal environments are classified by their prominent geology and vegetation.

Coastlines are among the most geologically active zones on Earth (Fig. 12–1). Waves and currents erode the shore and transport sediment. Along many coasts, converging tectonic plates buckle the edge of the continent and create mountain ranges adjacent to the sea. In other regions, volcanoes pour lava into the ocean, armoring the beaches with wave-resistant rock. Starting about 30,000 years ago, massive quantities of seawater evaporated from the oceans and accumulated as glacial ice on the continents. By 18,000 years ago, continental glaciers had grown so massive that sea level had fallen by 100 meters. When the glaciers melted, sea level rose again.

12.1 Emergent and Submergent Coastlines

Geologists have found drowned river valleys and fossils of terrestrial animals beneath the sea (Fig. 12–2). In other regions they have found fossil fish and other marine organisms far inland. Thus we learn that coastlines change, sometimes dramatically. An **emergent coastline** forms when land rises or sea level falls. Thus a portion of the continental shelf that was previously under water becomes exposed as dry land. In contrast, a **submergent coastline** develops when coastal land sinks or sea level rises. The sea then floods low-lying areas, and the shore-

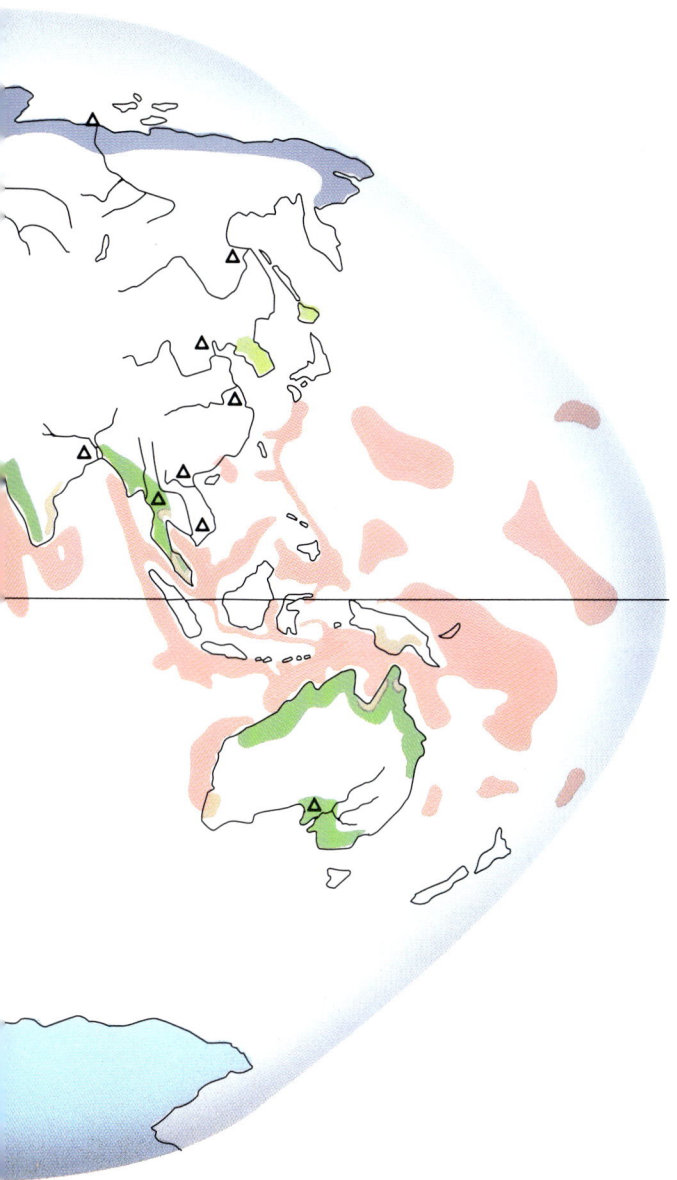

FIGURE 12–2

Geologists have found fossils of land mammals far offshore from the Siberian coast near the Bering Straits.

line moves inland (Fig. 12–3). Emergence or submergence can affect a portion of a coastline, or they can occur simultaneously on all continents.

Factors That Cause Emergent and Submergent Coastlines

Mountain building and related tectonic processes can cause a coastline in one region to rise or sink while those in other regions remain unaffected. Other processes can also depress or elevate a portion of a coastline. For example, about 18,000 years ago, a huge continental glacier covered most of Scandinavia. The glacier was so heavy that Scandinavia sank isostatically (Fig. 12–4). As the crust settled, the displaced asthenosphere flowed southward, causing Holland to rise. When the ice melted, the process reversed. Today, Scandinavia is rebounding. Rock in the asthenosphere is flowing back northward and Holland is sinking. As the land sinks, the Dutch are spending billions of dollars to build dikes to protect their country from the encroaching sea.

In addition to these regional phenomena, sea level can change globally. Global changes in sea level, called **eustatic** changes, occur by three mechanisms: growth and melting of glaciers, changes in water temperature, and changes in the volume of underwater mountain ranges.

During an ice age, vast amounts of water move from the sea to continental glaciers, and sea level falls. When the glaciers melt, sea level rises again. During the maximum extent of the Pleistocene glaciation, sea level fell to about 100 meters below its current level. If the remaining continental glaciers on Greenland and Antarctica were to melt, sea level would rise another 60 meters. Thus, glaciers cause emergence and submergence simultaneously by two different mechanisms. The weight of a glacier can locally depress a coastline, as it did in Scandinavia. At the same time, the transfer of water from the seas to glacier ice lowers sea level globally, thereby causing emergence of all coastlines. The net effect of these two opposing trends on any particular place may be difficult to predict.

Water expands and contracts when its temperature changes. If you take a glass of room-temperature water

294   Chapter 12   Coastlines and Coastal Pollution

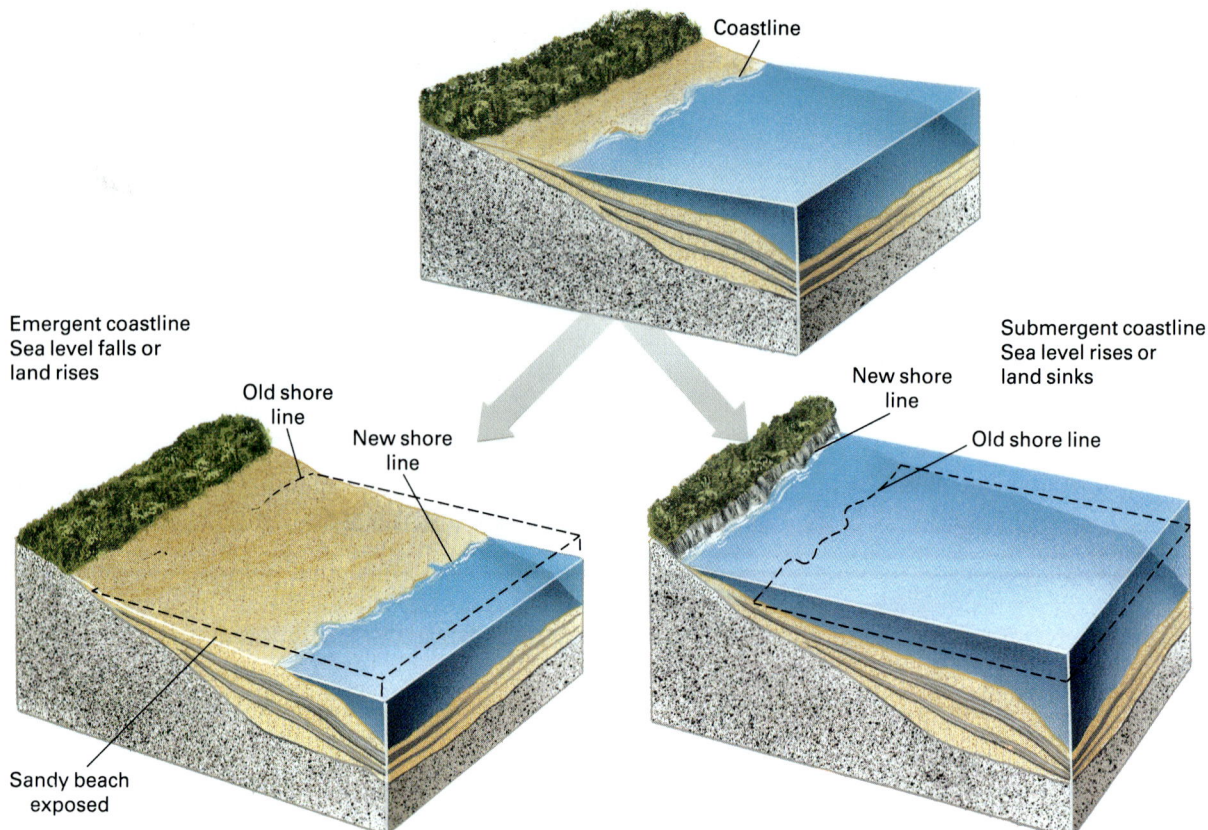

FIGURE 12–3
Emergent and submergent coastlines. If sea level falls or if the land rises, the sediment-laden shallow sea floor is exposed, forming sandy beaches. If coastal land sinks or sea level rises, areas that were once land are flooded. Irregular shorelines develop and beaches are commonly sediment-poor.

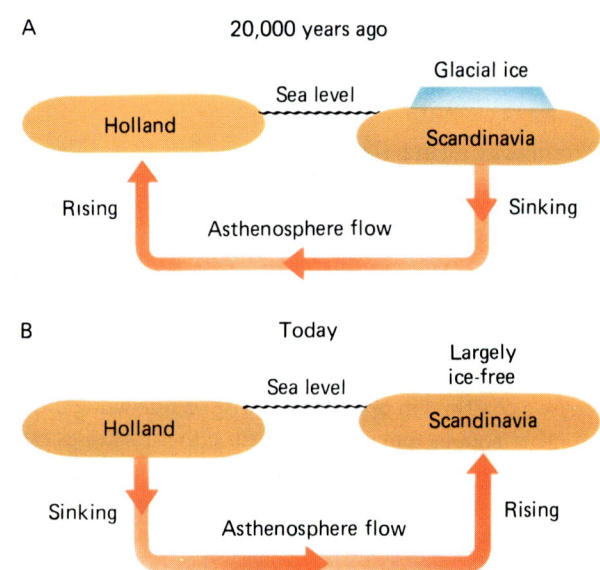

FIGURE 12–4
About 18,000 years ago, the weight of a continental glacier depressed Scandinavia. The displaced asthenosphere flowed laterally and raised Holland. Today the ice has melted, Scandinavia is rebounding, and Holland is sinking into the sea.

and place it in the refrigerator, the volume will decrease and the water level will drop. Although this change is barely noticeable in the glass, the oceans are so deep that if the volume changes by even a small amount, sea level changes appreciably. For every 1°C change in water temperature, sea level changes by 2 meters. Temperature changes and glaciation are linked. When global temperature rises, seawater expands and glaciers melt; when temperature falls, seawater contracts and glaciers grow.

Several times in the past, shallow seas flooded low-lying portions of all continents simultaneously. As a result, these continental regions are covered by marine limestone, shale, and sandstone. Most of the central United States, from the Rockies to the Appalachians, is covered by a veneer of these sedimentary rocks.

Because the ages of these sedimentary rocks do not coincide with times of glacial melting, we must look for a different mechanism to explain these episodes of global sea-level rise and continental flooding. Recall that a long, narrow submarine mountain range called the mid-oceanic ridge circles the globe beneath the world's major oceans. This submarine mountain chain displaces a huge volume of seawater. If the mid-oceanic ridge were smaller, it would displace less seawater and sea level would fall. If it were larger, sea level would rise. Therefore, one way in which global sea level could change is by an increase or decrease in the volume of the mid-oceanic ridge.

Geologists have deduced that the mid-oceanic ridge was unusually large during Late Cretaceous time, about 110 million to 85 million years ago. This high-volume ridge displaced great quantities of seawater, causing a global rise in sea level. Major flooding of low-lying portions of continents created the Cretaceous sedimentary rocks that we observe on several continents today. Then, about 20 million years before the demise of the dinosaurs, the mid-oceanic ridges shrank and sea level fell again.

Sediment-Rich and Sediment-Poor Coastlines

Rivers transport a huge amount of sediment and deposit it on deltas where the rivers enter the sea. The delta of a large river like the Mississippi may cover thousands of square kilometers. In some regions, glaciers deposited large quantities of sediment along the coast during the last ice age. Much of this sediment accumulated just below sea level in the shallow water near the coast. If a coastline rises or sea level falls, the vast supply of sediment from deltas and glacial deposits is exposed. Thus, **emergent coastlines are commonly rich in sediment** and sandy beaches are abundant.

In contrast, **submergent coastlines are commonly sediment-poor** and are characterized by steep, rocky shores (Fig. 12–5). In many areas on land, bedrock is exposed or covered by a thin layer of soil. If this type of sediment-poor terrain is submerged and if rivers do not supply large amounts of sediment, the coastline is rocky. Submergent coasts are commonly irregular, with many bays and headlands. The coast of Maine, with its numerous inlets and rocky bluffs, is a submergent coastline (Fig. 12–6). Small sandy beaches form in protected coves, but most of the beaches are rocky and bordered by cliffs.

A

B

FIGURE 12–5
(A) A sandy beach on the southern shore of Long Island, New York. (B) A rocky beach on the northern coast of California.

FIGURE 12-6
The Maine coast is a rocky submergent coastline.

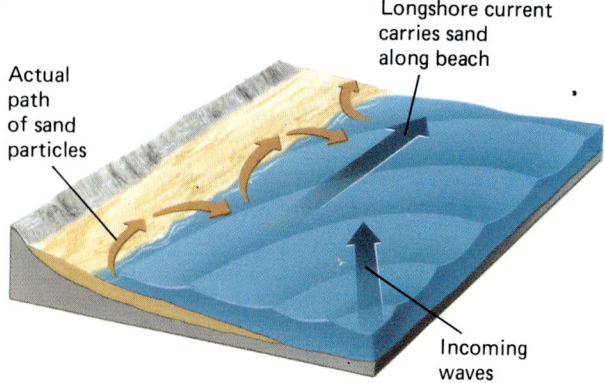

FIGURE 12-7
A longshore current develops where waves strike shore at an angle.

12.2 Erosion and Transport Along Coastlines

Currents and waves erode and transport coastal sediment. A **current** is a continuous flow of water in a particular direction. In contrast, a **wave** is an oscillation of the sea surface. Although a sea wave may travel for hundreds of kilometers across the open ocean, the water in a wave travels in small ellipses, ending where it began. When a wave approaches shore and enters shallow water, its base begins to touch bottom. Friction between the lower part of the wave and the sea floor slows the bottom of the wave. But the sea floor does not slow the upper part of the wave, which rushes forward until the crest rides over the trough. At this point it collapses forward, or **breaks,** as it rolls into shallower water. Once a wave breaks, its water flows toward the beach as a chaotic, turbulent mass called **surf**. A breaking wave transports sediment shoreward; as the wave recedes, it carries the sediment back out to sea.

If waves strike the shore at an angle, water moves along the beach a short distance with each wave. Thus the water zigzags, as shown in Figure 12–7. The net result is a **longshore current** that flows parallel to the coast. Longshore currents flow in the surf zone and a little farther out to sea and may travel for tens or even hundreds of kilometers. They transport sand and other sediment parallel to the coast (Fig. 12–8). For example, much of the sand on the beaches of the Carolinas and Georgia eroded from New York's Hudson River delta and from glacial deposits on Long Island and southern New England. Longshore currents have carried it hundreds of kilometers southward along the coast.

Waves and longshore currents deposit sand and gravel in characteristic landforms. A long ridge of sand or gravel extending out from a beach is called a **spit** (Fig. 12–9). Longshore currents and waves continuously deposit and erode the sand on a spit.

A **barrier island** is a long, narrow, low-lying island that parallels the shoreline (Fig. 12–10). It looks like a beach or spit separated from the mainland by a sheltered body of water called a **lagoon**. A chain of barrier islands extends along the east coast of the United States from New York all the way to Florida. The islands are so nearly continuous that a small boat can travel the entire coast inside the barrier island system and remain protected from the open ocean most of the time.

Barrier islands form in several ways. The two essential ingredients are a large supply of sand and waves or currents to transport it along the coast. If a coast is shallow for several kilometers outward from shore, breaking storm waves may carry sand toward shore and deposit it as a low-lying barrier island. If a longshore current veers out to sea, it slows down and deposits its sediment where it reaches deeper water. If this sediment is further piled up by waves, a barrier island may form. Yet another mechanism involves sea-level change. Underwater sandbars may be exposed as coastlines emerge. Alternatively, sand dunes or beaches may form barrier islands if a coastline sinks. Many seaside resorts are built on spits and barrier islands, and developers often ignore the fact that

12.2 Erosion and Transport Along Coastlines 297

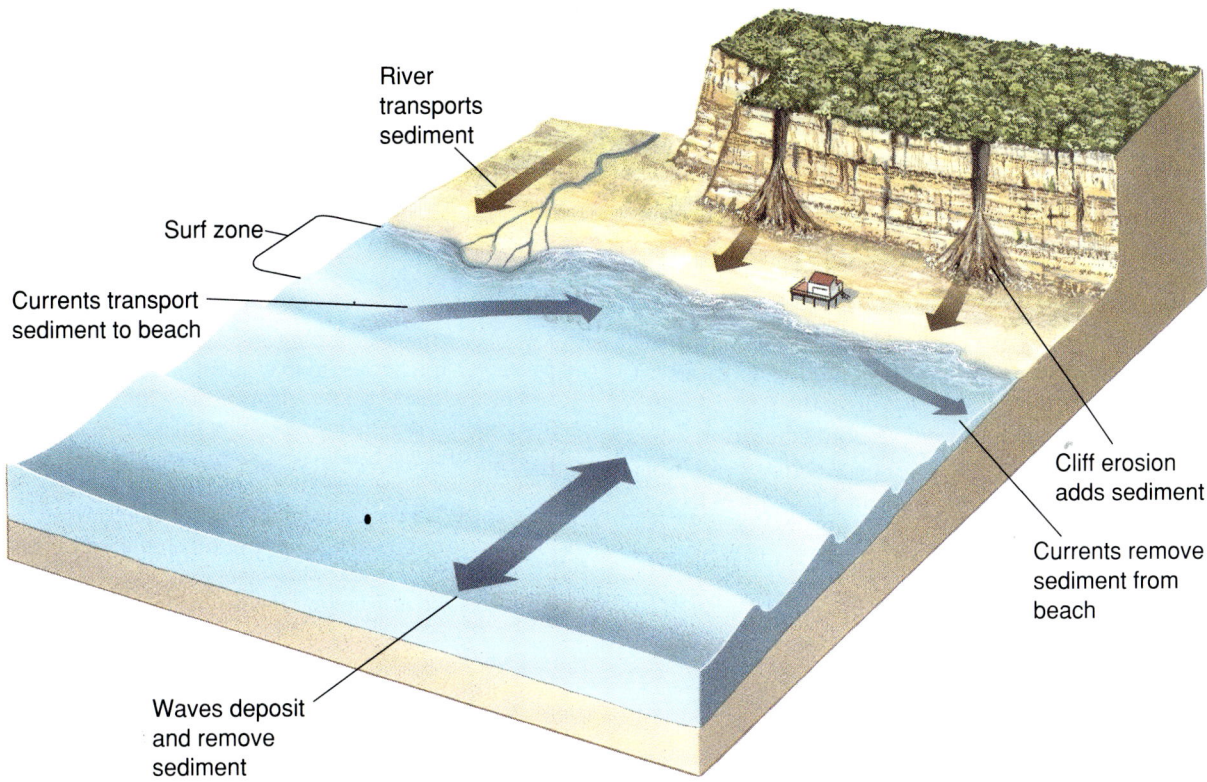

FIGURE 12-8
The growth and shrinkage of a beach depend on the sum total of erosion and deposition.

FIGURE 12-9
Aerial view of a spit on the south coast of Long Island, New York.

FIGURE 12-10
Aerial view of a barrier island along the south coast of Long Island, New York.

they are transient and changing landforms. If the rate of erosion exceeds that of deposition for a few years in a row, a spit or a barrier island can shrink or disappear completely, leading to destruction of beach homes and resorts.

12.3 Erosion of the Atlantic Coast

The Gulf Stream is a marine current that warms the Atlantic coast from the tip of Florida to northeastern Maine. But the same winds that drive the tropical water of the Gulf Stream northward also bring hurricanes that periodically batter the coast. The Atlantic coast from Massachusetts to Florida is sediment-rich, with extensive barrier islands. In contrast, the Maine coast is rocky and sediment-poor. Recently, development has triggered increased erosion along the Atlantic coast, as shown in Table 12–1.

TABLE 12–1 Rates of Yearly Seaward (+) and Landward (−) Migration of the High Tide Line Along the Atlantic Coast

Region	Change (m/yr)
Maine	−0.4
New Hampshire	−0.5
Massachusetts	−0.9
Rhode Island	−0.5
New York	+0.1
New Jersey	−1.0
Delaware	+0.1
Maryland	−1.5
Virginia	−4.2
North Carolina	−0.6
South Carolina	−2.0
Georgia	+0.7
Florida	−0.1
Average	−0.8

CASE HISTORY

Long Island

Long Island extends eastward from New York City and is separated from Connecticut by Long Island Sound (Fig. 12–11). A series of narrow, low barrier islands lies along the southern coast of Long Island. Longshore currents flow westward, eroding sand from the beaches and barrier islands and carrying it toward New Jersey. At the same time, they carry in new sand eroded from glacial deposits at the eastern end of the island.

Are the beaches of Long Island stable or unstable? The answer to that question depends on our time perspective. Over geologic time, the beach is unstable. The glacial deposits at the eastern end of the island will become exhausted and the flow of sand will cease. Then the entire coastline will erode and the barrier islands and beaches will disappear. However, this change will not occur in the near future because of the vast amount of sand still available at the east end of the island. Thus, the beaches are stable over a period of many years or decades. Over these time spans, longshore currents replace sand at the same rate at which they erode it and carry it westward.

If we narrow our time perspective and look at a Long Island beach over a season or during a single storm, it may shrink or expand. Over such short times, the rates of erosion and deposition are not equal. Violent winter waves and currents erode beaches, whereas sand accumulates on the beaches during the calmer summer months. In an effort to prevent these seasonal fluctuations and protect their personal beaches, Long Island property owners have built stone barriers called **groins**. If a groin is built from shore out into the water, it intercepts the steady flow of sand moving from the east and keeps that particular part of the beach from eroding (Fig. 12–12). But the groin impedes the overall flow of sand. West of the groin the beach erodes as usual, but the sand is not replenished because the upstream groin traps it. As a result, beaches down current from the groin erode away (Fig. 12–13).

The landowners living downcurrent from a groin may decide to build another groin and pass the problem farther downstream (Fig. 12–12C). The situation has a domino effect, with the net result that millions of dollars must be spent in attempts to stabilize a system that was naturally stable in its own dynamic manner.

Storms pose another dilemma. Hurricanes strike Long Island in the late summer or fall, generating storm waves that completely overrun the barrier islands, flattening dunes and eroding beaches. When the storms are over, gentler waves and longshore currents carry sediment back to the beach and rebuild it. As the sand accumulates again, salt marshes rejuvenate and the dune grasses grow back within a few months.

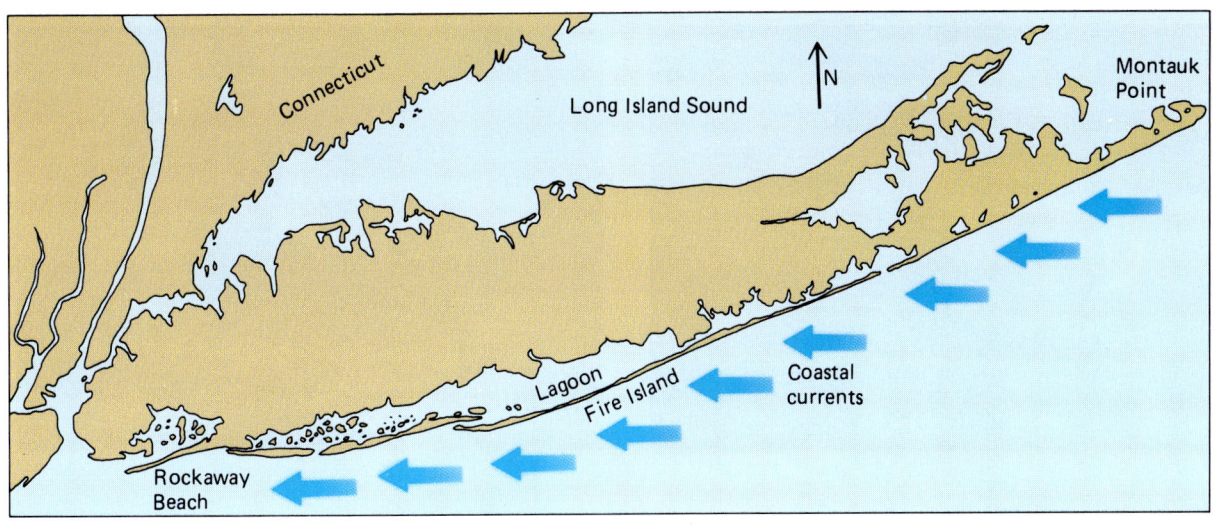

FIGURE 12-11
A map of Long Island showing the movement of longshore currents.

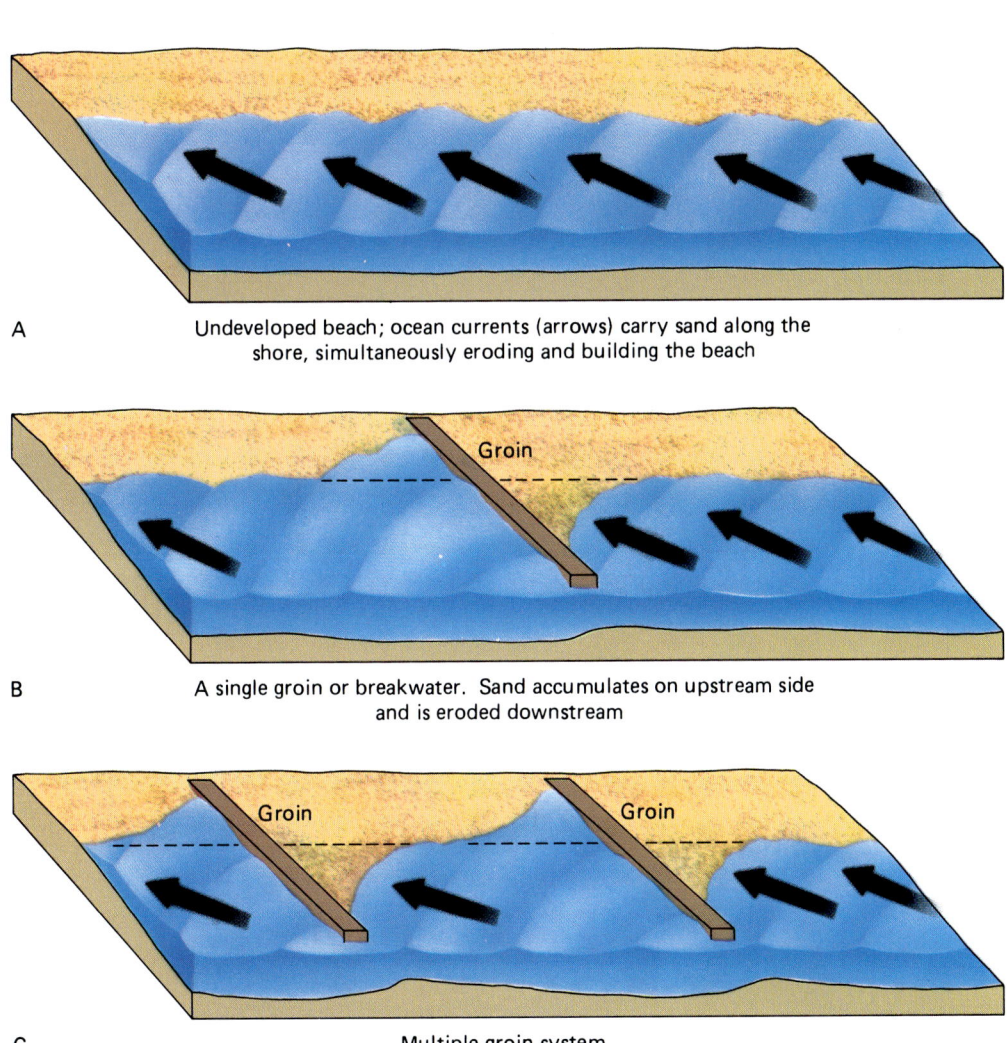

FIGURE 12-12
The effect of groins on a beach. (A) An undeveloped beach. No net change occurs as sand is simultaneously eroded and deposited by longshore currents. (B) A single groin, or breakwater. Sand accumulates on the upstream side and is eroded downstream. (C) A multiple-groin system.

FIGURE 12–13
(A) Longshore currents on Long Island move from east to west. As a result, sand accumulates on the upstream (east) side of a groin, and the west side is eroded. Here waves lap against the foundation of the house along the eroded portion of the beach. (B) A close-up of the house in part (A).

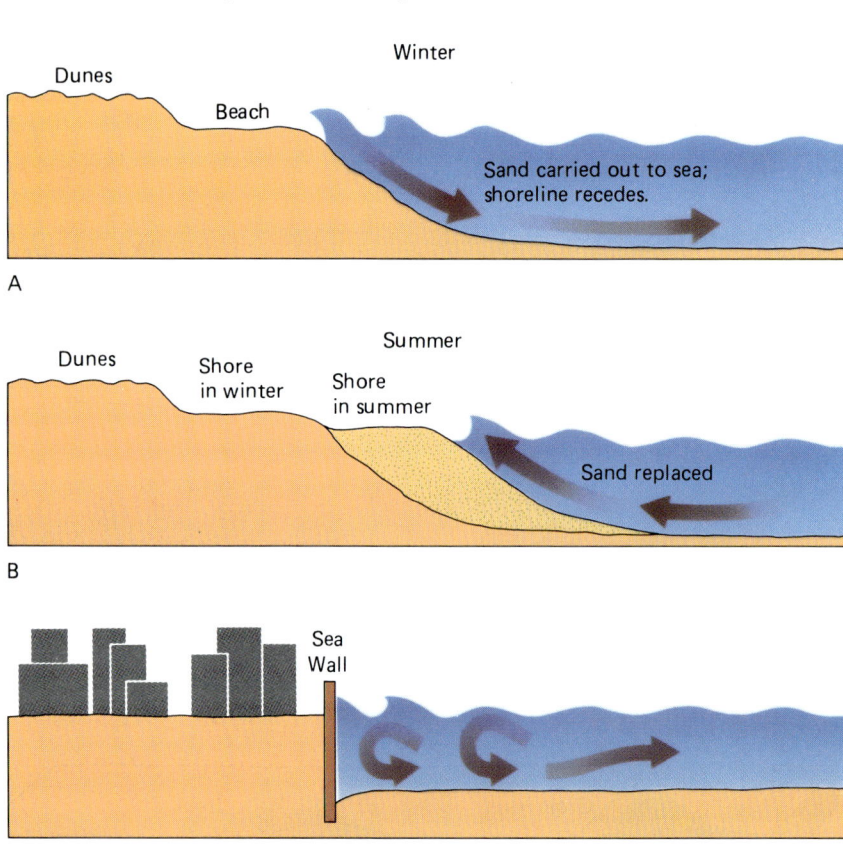

FIGURE 12–14
(A) In a natural system, the violent winter waves often move sand out to sea. (B) The gentler waves that occur in summer push sand toward shore and thereby rebuild the beach. (C) If a sea wall is erected, waves strike the wall and may eventually destroy it.
(D) The beach is severely eroded.

These short-term fluctuations are not compatible with human activity, however. People build houses, resorts, and hotels on or near the shifting sands. The owner of a home or resort hotel cannot allow the buildings to be flooded or washed away. Therefore, property owners construct large sea walls along the beach. When a storm wave rolls across a low-lying beach, it dissipates its energy gradually as it flows over the dunes and pushes the sand inland. The beach is like a judo master who defeats an opponent by yielding with the attack, not countering it head on. A sea wall interrupts this gradual absorption of wave energy. The waves crash violently against the barrier and erode the base of the sea wall. When the sand erodes beneath the wall, the wall collapses. It may seem surprising that a reinforced concrete sea wall is more likely to be permanently destroyed than a beach of grasses and sand dunes, yet this is often the case (Fig. 12–14)

12.4 Erosion of the Gulf Coast

The Gulf of Mexico is protected from the open ocean, so under normal conditions wave energy in the gulf is less than along the Atlantic or Pacific coast. Sea-level change and erosion rates in this region are summarized in Table 12–2. Most of the Gulf Coast is dominated by an extensive series of barrier islands that extend from Florida all the way to Mexico's Yucatán Peninsula. Lagoons lie between the barrier islands and the mainland. The largest of these lagoons is Laguna Madre, which is formed by Padre Island and neighboring islands. This barrier island system is broken by the extensive Mississippi Delta that dominates most of the Louisiana coast.

TABLE 12–2 Rates of Yearly Seaward (+) and Landward (−) Migration of the High Tide Line Along the Gulf Coast

Region	Change (m/yr)
Florida	−0.4
Alabama	−1.1
Mississippi	−0.6
Louisiana	−4.2
Texas	−1.2
Average	−1.8

CASE HISTORY

The Mississippi Delta

A stream deposits most of its sediment where it flows into the still water of a lake or the sea, to form a **delta**. Part of a delta lies above water level, and the remainder consists of a shallow underwater plain. A delta grows as its river deposits sediment. Flooding is an important part of this process, because during a flood a river overflows its banks and deposits sediment over the delta. However, three natural processes reduce the amount of land exposed above sea level on a marine delta:

1. *Sea-level change:* Because a marine delta barely rises above sea level, a small rise in sea level can flood large areas. Most evidence indicates that global sea level is now rising.
2. *Subsidence:* The weight of accumulated sediment in a large delta, such as that of the Mississippi, causes the delta to subside isostatically. Subsidence leads to flooding of coastal beaches, marshes, and plains.
3. *Erosion:* Coastal currents, waves, and hurricanes erode a delta.

Although continued sediment addition by the Mississippi River has enlarged the delta for nearly 200 million years, the amount of land exposed on the delta has decreased over the past 50 years. Three factors have combined to cause this trend:

1. *Decreased sedimentation:* Flood control projects have interrupted the natural deposition of sediment on the portion of the delta that lies above sea level. The river has been channeled to reduce floods, and as a result most of the sediment accumulates on the submarine portion of the delta.
2. *Increased subsidence:* The two most common liquids in the Earth's crust are water and petroleum. Large amounts of both lie below the Mississippi delta, and both have been extracted in huge quantities in recent years. When water is removed from near-surface wells and oil is pumped from deep reservoirs, the loss of fluid causes the ground to settle (Fig. 12–15). This subsidence, combined with the natural isostatic subsidence mentioned above, allows the ocean to flood parts of the delta.
3. *Increased erosion:* In its natural state, erosion along the southern coast of Louisiana did not cause rapid loss of land area. When the delta coast was developed, houses and roadways were built and sea walls were erected to protect them. First, the natural vegetation was removed. Second, as explained in the preceding case history, when waves strike a solid barrier, erosion increases. In addition, during the industrialization of the delta, many canals were dredged for navigation or for the development

FIGURE 12–15
Aerial view of an oil drilling platform on the Mississippi delta. Oil drilling leads to subsidence of the low-lying delta. *(Cameramann, International)*

of oil fields. The canals allowed salt water to flow inland. In turn, the salt water killed the fresh-water plants that protected the soil.

Today, more than 150 square kilometers of land are being lost every year from the Mississippi delta. This amounts to 1 hectare every 35 minutes (1 acre every 14 minutes). If current trends continue, the sea will flood the shaded area in Figure 12–16 by the middle of the next century and the beach will recede into the suburbs of New Orleans. The flooding will cause severe economic losses and destroy wildlife habitats.

12.5 Erosion of the Pacific Coast

Whereas the Gulf Stream warms the Atlantic coast, the California current flowing southward from the North Pacific cools the Pacific coast. As a result, surfers in southern California wear wet suits even in summer, whereas bathers on Long Island enjoy relatively warm water, even though Long Island is farther north. Much of the Pacific coast is sediment-poor and is bounded by cliffs and rocky headlands. Erosion rates are generally lower than along the Atlantic and Gulf coasts. However, as illustrated by the following case history, local erosion causes serious problems (Table 12–3).

TABLE 12–3 Rates of Yearly Seaward (+) and Landward (−) Migration of the High Tide Line Along the Pacific Coast	
Region	Change (m/yr)
California	−0.1
Oregon	−0.1
Washington	+0.5
Alaska	−2.4
Average	0.0

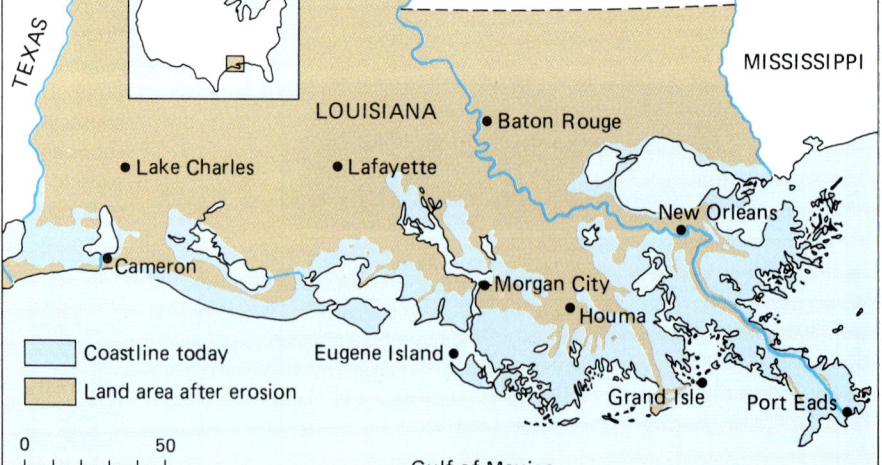

FIGURE 12–16
Coastal erosion in southern Louisiana. The outer boundary shows the existing coastline. If erosion continues unchecked, vast areas of land will be flooded.

CASE HISTORY

San Diego County

In the early 1970s, real estate developers planned a series of condominiums and homes atop a seaside bluff near Encinitas and Solana Beach. Before starting construction, they hired consultants to evaluate the geological hazards. One consultant studied aerial photographs of the coast taken between 1928 and 1973. He saw no significant erosion during that time and concluded that the cliff was stable. Consequently, developers built condominiums as close as 3.5 meters from the cliff face. Between 1975 and 1990, many portions of the cliffs collapsed. Some buildings were lost, and other residents spent millions of dollars to save their homes (Fig. 12–17).

In 1880, surveyors for the California Southern Railroad made accurate maps of the southern California coast between Los Angeles and San Diego. Four years later, severe storms lashed the region. Water saturated the ground and triggered massive landslides both inland and along the coast. The slides destroyed all but two of the railroad trestles between San Bernardino and San Diego. Several other storms in the late 1800s and early 1900s also caused sea-cliff collapse and beach erosion. Because all of these disasters occurred before 1928, the geologist working for the real estate developers was unaware of them.

Over geologic time, all cliffs are unstable, and any structure built on a cliff edge is temporary. Therefore, before construction, developers must balance the geologic stability against the expected life span of the structure. In this example, the geologist evaluated stability over only a 45-year period.

Urbanization had increased coastal erosion along the San Diego coast. Small boat harbors and jetties built at Camp Pendleton and Oceanside to the north trapped sediment carried southward by longshore currents. Once the sediment flow was interrupted, beaches to the south eroded rapidly. After the beaches disappeared, ocean waves battered directly against cliffs.

Coastal cliffs are not only subject to erosion from ocean waves but are also vulnerable to forces from the land. In the San Diego County development, people planted lawns around their new homes and condominiums. Irrigation added five to ten times as much water to the soil as comes from rain. Excess water adds weight and lubricates rock and soil, promoting landslides.

Yet another problem occurred because engineers channeled local streams into drainage pipes and diverted them toward the beaches. Water runs more rapidly through a straight pipe than it does along a meandering stream, and therefore water flowing from the pipes erodes the coast.

12.6 Coastal Development and Public Policy

When Hurricane Hugo struck the North Carolina coast in 1989 with 220-kilometer-per-hour winds, waves flooded coastal areas, destroying 1900 homes and killing 40 people. Property damage totaled $3 billion. Hurricane Andrew hit Florida and Louisiana in 1992, killing 50 people, destroying 100,000 homes, and causing $30 billion in property damage (Fig. 12–18). In September 1992, Hurricane Iniki struck Hawaii, causing millions of dollars in damage. Three months later, a winter storm ravaged the Atlantic coast from New Jersey to Massachusetts, toppling homes and reshaping barrier islands. The blizzard of 1993 pounded the Atlantic coast from Canada to Florida, leaving ravaged buildings and 100 people dead.

Between 1947 and 1969, an average of 3.3 severe hurricanes battered the Atlantic coast every year.[1] This average dropped to 1.5 hurricanes per year from 1970 to 1987, but it has increased since then. Scientists from the National Hurricane Center predict that hurricane

FIGURE 12-17
Coastal erosion has undermined this house.

[1] A severe hurricane is defined as one with winds greater than 111 kilometers per hour (kph).

FIGURE 12–18
High winds from Hurricane Andrew destroyed these houses in Florida City. *(AP/Wide World Photos)*

frequency will be high throughout the 1990s. Faced with this sobering prediction, people must establish policy on personal, local, and national levels.

One question is, "Who should pay for flood damage from severe storms?" Private insurance companies refuse to offer flood insurance because they argue that only those who are vulnerable to floods will purchase policies. To fill the gap, the federal government enacted the National Flood Insurance Program (NFIP) in 1968. The program provides inexpensive flood insurance for existing structures, but it prohibits insurance coverage for new development on the most vulnerable flood plains and beaches. It does not, however, prevent people from rebuilding destroyed property on the same site and insuring the new structure as inexpensively as they did the old one. One criticism of this law points out that floods occur over and over again in the same places. Thus, people who have flaunted geological change in the past can continue to do so in the future, at the taxpayers' expense (Fig. 12–19). The NFIP was designed to be self-sufficient financially, but it has operated at a deficit since 1978. With the recent hurricanes and the Mississippi River flood in 1993, the red ink continues to flow.

FIGURE 12–19
These expensive hotels in Miami have been built on geologically transient barrier islands. Much of the rest of the city lies on reclaimed swamps surrounded by lagoons. *(Comstock)*

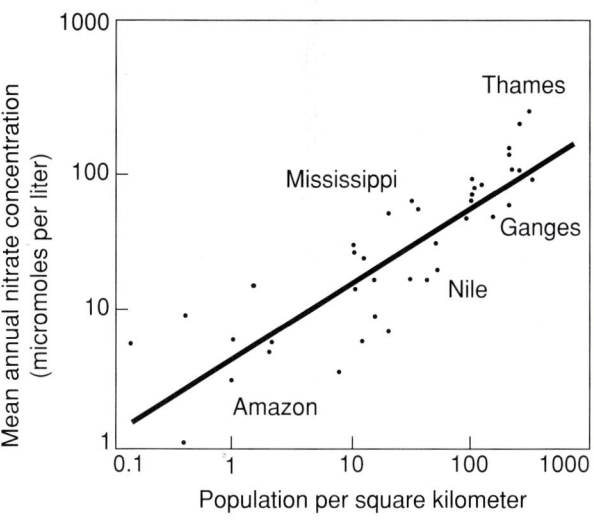

FIGURE 12–20
The mean annual nitrate concentration of rivers is proportional to the population densities in the drainage basins.

FIGURE 12–21
An Aquasphere Project volunteer collects debris washed up on a beach. Some beaches have been contaminated with medical and other hazardous wastes. *(Mark Elias/Wide World Photos)*

12.7 Ocean Pollution

Nearly all rivers and most ground water carry their pollutants into the sea. In addition, people dump sewage, industrial waste, and municipal trash directly into the ocean. The ocean is large enough to dilute these pollutants, but because dispersal is slow, pollutants concentrate along shallow coastlines, especially in protected bays and estuaries with limited access to the deep sea. Worldwide, coastal pollution is roughly proportional to the population density of the drainage basins that feed water into each coastal area (Fig. 12–20).

Sewage and Municipal Waste in Coastal Waters

In the United States, 35 percent of municipal sewage from coastal cities is dumped into the ocean without treatment. Prior to 1990, many cities also dumped municipal trash offshore. During the summer of 1988, used hypodermic syringes, intravenous tubing, and blood sample vials washed up on beaches along the Eastern Seaboard, and health inspectors recorded high levels of disease-causing bacteria in coastal waters (Fig. 12–21). Nearly 500 beaches from Maine to Florida were closed for health reasons. In response to public outcry, Congress passed the Ocean Dumping Ban Act, banning ocean disposal of municipal trash. However, the act has had little effect on sewage treatment plants, and in 1992 the number of beach closings increased to 2600. The fivefold increase in beach closures reflects both continued pollution and increased public awareness and monitoring.

Sewage also contains nutrients that cause ocean water to become eutrophic. Microscopic organisms called **phytoplankton** conduct most photosynthesis in the oceans. During the short days and low light of winter, phytoplankton grow slowly, but the populations bloom in early spring with the returning heat and sunlight. In recent years, both the number and intensity of phytoplankton blooms have increased dramatically, upsetting marine ecosystems. Researchers suspect that these blooms are caused by increased nutrient concentrations. Some phytoplankton are toxic to people and fish, causing devastating outbreaks named after the color of the plankton: "red tides," "yellow foams," and "emerald slimes."

Toxic Chemicals in Coastal Waters

When commercial whaling was banned along the mouth of the St. Lawrence River in the 1950s, biologists estimated the beluga whale population at about 1200. By 1988 the population had declined to 450. In that year, marine biologists from the University of Western Ontario performed autopsies on 72 dead whales that had washed up on the beaches. They detected 30 chemical pollutants, including DDT, PCBs (polychlorinated biphenyls),

FIGURE 12–22
In recent years, unusual numbers of whales have become beached and died. Many marine biologists suspect that the beachings are related to oceanic pollution. *(Gamma Liaison)*

FIGURE 12–23
Plastic litter commonly entangles and kills wildlife. *(Nicholas Conte/Bruce Coleman, Inc.)*

Mirex (a pesticide), mercury, and cadmium (Fig. 12–22). A majority of the whales had died of septicemia, a form of blood poisoning that attacks the immune system. Other diseases included pneumonia, hepatitis, perforated gastric ulcers, and bladder cancer. Dr. Joseph Cummings, a researcher on the project, believes that a strong link exists between the toxic substances and the fatal diseases in whales.

Toxins also affect people who eat contaminated seafood. According to the EPA, fish from 4000 test sites in the United States were contaminated with high enough concentrations of toxins to pose a threat to human health.

Plastic Litter in Coastal Water

Plastic bottles, fishing nets and line, six-pack rings, gloves, lids, bags, egg cartons, and a host of other products float around in the ocean and wash up on beaches. Although most plastic products are chemically inert, they trap and kill sea animals by encircling their mouths, gills, necks, or other body parts (Fig. 12–23). In the Pribilof Islands in western Alaska, 40,000 seals die every year from becoming entangled in plastic. These animals are naturally curious and play with fragments of fishing net or other detritus, often with fatal results.

Habitat Destruction and Coastal Ecosystems

Although coastal seas account for only 10 percent of the ocean surface area and 0.5 percent of its volume, approximately 90 percent of marine animals rely on estuaries, river mouths, and coastal wetlands for some portion of their life cycles. But worldwide, over half of the coastal wetlands have been destroyed by commercial development. The world fish catch rose between 1950 and 1989 and then leveled off between 1989 and 1992 (Fig. 12–24). The combined effects of overfishing, pollution, and habitat destruction caused the leveling.

> ### Oil in the Ocean—The Wreck of the *Exxon Valdez*
>
> Petroleum from Alaska's north coast is pumped through the Alaska pipeline to Valdez on the south coast of Alaska (Fig. 12–25). From there it is loaded into tankers that carry it to refineries. In March 1989, as the supertanker *Exxon Valdez* steamed out of port, the captain, who had been drinking, left the bridge in the hands of an inexperienced third mate. As the mate steered around floating ice, he ran the ship aground on a rock, spilling 42 million liters (11 million gallons) of petroleum into Prince William Sound (Fig. 12–26).

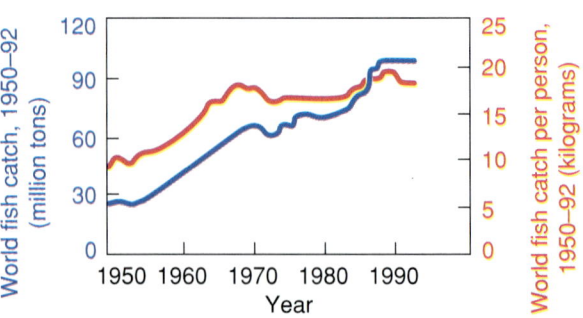

FIGURE 12–24
The world fish harvest rose from 1950 to 1989, and then fell from 1989 to 1992.

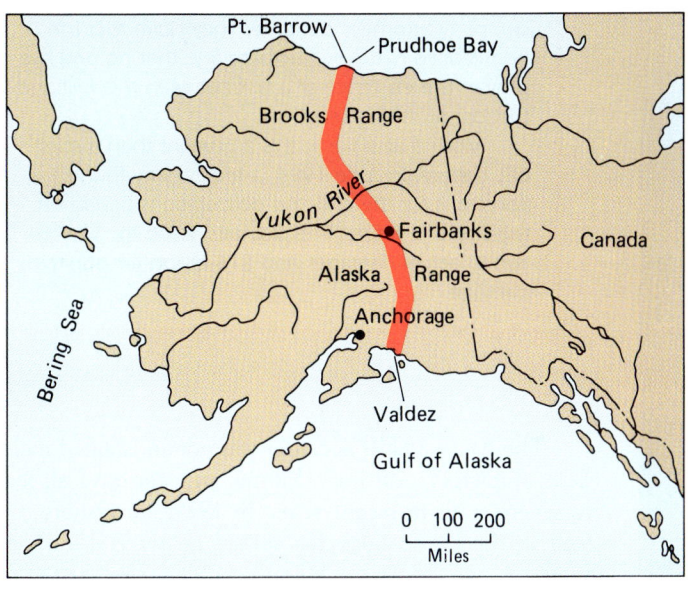

FIGURE 12–25

(A) The route of the Alaska pipeline. (B) The Alaska pipeline is built in a zig-zag fashion so that it will not rupture when the permafrost soil expands and contracts. *(Alyeska Pipeline Company)*

FIGURE 12–26

The *Exxon Valdez* after it ran aground and began to spill oil. The slick appears off the ship's bow. *(Wide World Photos)*

In preparation for such an emergency, shipping companies had stored booms and pumps at the dock in Valdez. If they had been deployed quickly, they might have contained and retrieved much of the spilled oil, but days passed before the clean-up effort was coordinated. During the following few weeks, oil fouled more than 5000 kilometers of coastline, killing 30,000 birds, 3500 to 5500 sea otters, 30 seals, 22 whales, and an unknown number of fish. The Exxon Oil Company spent $2 billion to doctor sea mammals, wash oil off birds, and scour beaches (Fig. 12–27). Some scientists argued that the clean-up work did more harm than good in many cases. The detergents used to scour the rocks killed many intertidal organisms. The crews trampled shoreline organisms such as barnacles, clams, mussels, eelgrass, and rockweed, killing them and destroying their habitats. As a result, many untreated beaches regained productivity faster than those that had been cleaned.

Although the short-term effects of the oil spill were devastating, the long-term ecosystem damage is difficult to evaluate. Petroleum is a complex mixture of many different compounds. The lightest ones are volatile enough to evaporate within days or weeks. The heaviest ones glob together into tar balls that persist for decades but are relatively harmless. Compounds of medium molecular weight remain in the environment for years until bacteria slowly degrade them. In 1993, four years after the accident,

FIGURE 12–27
A cleanup crew uses hot water and steam in an effort to wash oil from a rocky beach on the outer Kenai Peninsula, Alaska, in August 1989. *(Doug Loshbaugh)*

scientists from Exxon and the National Oceanic and Atmospheric Administration (NOAA) debated the long-term effects of the spill. Both groups of scientists agreed that a wide variety of plants and animals in Prince William Sound contained high concentrations of hydrocarbons in their tissue. Government scientists claimed that the tissue analysis proved that contamination from the spill lingered in the ecosystem. However, Exxon scientists disagreed. They pointed out that every sample of crude oil has a specific ratio of hydrocarbons that distinguishes it from all other samples, much as a fingerprint identifies an individual. They then pointed out that the hydrocarbon ratios from tissue samples did not match those of the oil spilled by the *Exxon Valdez*. Therefore, the Exxon scientists argued, the tissue contamination originated from other sources such as natural seeps, discharges from fishing boats, and lingering contamination from older spills. Government scientists argued back that hydrocarbon ratios change when the oil is ingested by living organisms, so the fingerprint changes.

An additional argument concerns fish populations. Salmon fishing is one of the main industries in Prince William Sound. Fish harvests were abnormally low from 1990 through 1993. Fishermen and women believe that the decline in fish populations was caused by the oil spill. However, again the oil company attorneys disagree. They claim that fish populations naturally fluctuate and that no one has proven the existence of a link between the spill and the loss of fishing revenues.

More is at stake in this argument than a scientific debate. In June 1994, a federal jury found Exxon liable for the spill and deliberations began on the size of the fine. The plaintiffs asked for $1.5 billion in actual damages and $15 billion for punitive damage.

The *Exxon Valdez* accident was not an isolated incident; shipwrecks, offshore drilling, and war have all led to severe spills in recent years. In 1979, an offshore oil well in the Gulf of Mexico spilled nearly 700 million liters; in 1991, Sadaam Hussein deliberately poured 1 billion liters of crude oil into the Persian Gulf; in 1993, the tanker *Braer* lost power off the Shetland Islands and crashed into the rocks, spilling most of its 100 million liters of light crude oil. Each event was caused by a sequence of seemingly avoidable circumstances—a negligent captain, a broken drill pipe, an angry dictator, or an engine malfunction and a delayed rescue tug. But accidents also occur; between 1980 and 1988, tankers in U.S. waters were involved in 468 groundings, 371 collisions, 97 rammings, and 55 fires or explosions. In 1990, in the wake of the *Exxon Valdez* disaster, Congress passed a bill designed to reduce tanker accidents. The law increases ship-owner liability, establishes Coast Guard spill response teams, and requires double hulls for all oil transport vessels. (If the *Exxon Valdez* had been equipped with a double hull, it would not have spilled oil after it went aground.) This legislation should reduce oil pollution, but ships have sunk as long as people have sailed the seas, so even the strictest precautions cannot guarantee an end to the problem.

Estuaries and the Pollution of Chesapeake Bay

An **estuary** is a coastal inlet formed where the sea floods a river mouth. Fresh and salt water mix in the estuary, and tides interact with the river flow. Estuaries are extremely rich marine environments. Streams transport nutrients to the coast, and the shallow water provides habitats for marine organisms. These areas also make excellent harbors and therefore are prime sites for industrial activity. As a result, many estuaries have become seriously polluted in recent years.

Chesapeake Bay is an estuary formed by submergence of the Susquehanna River valley (Fig. 12–28). It is approximately 100 kilometers long, averages 10 to 15 kilometers wide, and contains numerous bays and inlets. Despite its great size, Chesapeake Bay is only 7 to 10 meters deep near its mouth, and its greatest depth is 50 meters.

Three major cities—Washington, Baltimore, and Harrisburg—lie along Chesapeake Bay and its tributaries. Between 1950 and 1985, the population in the watershed increased by 50 percent to 15 million, and the land area devoted to urban uses tripled (Fig. 12–29). As the Chesapeake watershed became heavily urbanized, the bay became polluted. Eroding soils carry nonpoint pollution such as silt, fertilizers, and pesticides into Chesapeake Bay. The pesticides are toxic to marine organisms. Silt destroys habitats by clogging spaces between rocks and covering plant roots, thereby reducing the flow of oxygen. Fertilizers lead to eutrophication, which disrupts aquatic ecosystems. Numerous factories and sewage treatment plants discharge toxic compounds into the bay and increase the biological oxygen demand.

Pollution and overfishing have combined to devastate animal and plant populations in Chesapeake Bay. In 1960, 2.7 million kilograms of striped bass were harvested in the bay; by 1985, the catch had dwindled by 90 percent to 270,000 kilograms (Fig. 12–30). The bay once supported the richest oyster beds in the world; today the oyster population has dropped by 99 percent.

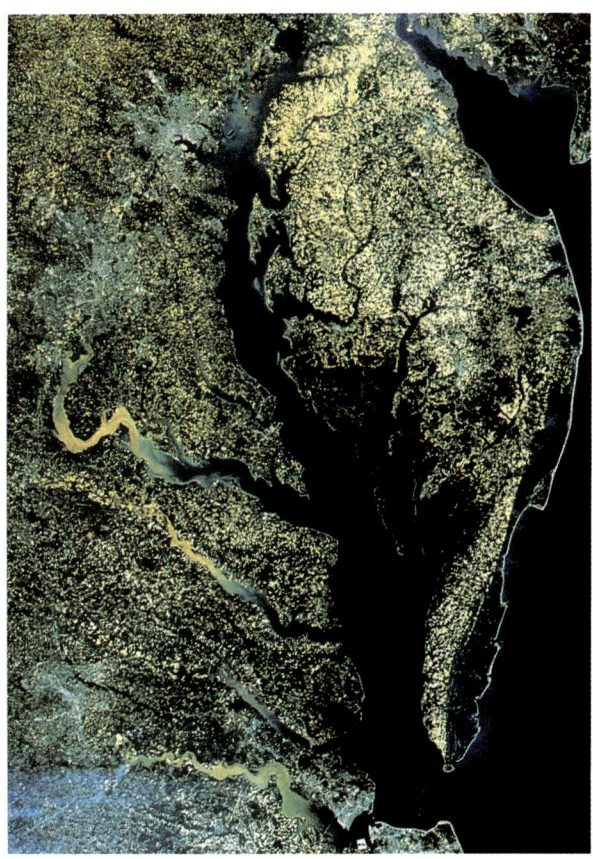

FIGURE 12–28
A satellite view of Chesapeake Bay. *(Chesapeake Bay Foundation)*

FIGURE 12–29
Intensive development has permanently altered the shoreline of Chesapeake Bay. *(Chesapeake Bay Foundation)*

FIGURE 12–30
A fisherman pulls in nets on Chesapeake Bay.
(Chesapeake Bay Foundation)

> By the early 1980s, people realized that the goals of the Clean Water Act were not being met and the EPA was not enforcing the law. Consequently, in 1984, two citizens' groups—the Chesapeake Bay Foundation and the Natural Resources Defense Council—filed a lawsuit against the Bethlehem Steel Corporation for illegally dumping wastes into the estuary. According to the suit, Bethlehem Steel had discharged excessive quantities of pollutants in the bay on more than 300 different occasions, and the EPA had failed to enforce compliance with the law. The environmental groups won their suit, and Bethlehem Steel was fined over a million dollars. Other industrial polluters began clean-up to avoid similar suits.
>
> In 1984, in an unusual example of interstate cooperation, the governments of Maryland, Virginia, Pennsylvania, and the District of Columbia agreed to work together to save the estuary. As a first step, all the states tightened regulations on easily identified pollution point sources such as municipal sewage treatment facilities. In 1985, Maryland banned phosphate from detergents, and in 1987 Virginia followed suit. State-run agricultural education programs have shown farmers that they can maintain high yields with fewer fertilizers and pesticides than they have been accustomed to using.
>
> All of these efforts have improved the water quality of Chesapeake Bay from 1980 to 1993. Fish populations are making a slow recovery. But as long as cities line its coasts, the bay can never return to its pristine state. Developers have built concrete wharves over saltwater marshes. Runoff from roadways, parking lots, and storm drains carries oil into the estuary. Air pollutants drift into the water; runoff from farmers' fields can be reduced but not eliminated. Chemicals that seeped into the ground water decades ago slowly percolate into the estuary's waters. The environmental challenge ahead is to identify each pollution source and balance the cost of pollution control against the economic and aesthetic value of a cleaner bay.

Pollution of the Central Oceans

To date, the central oceans remain relatively pristine. Sewage and other biodegradable pollutants decompose, and nonbiodegradable pollutants are sufficiently diluted in the 1.4 billion cubic kilometers of seawater. For these reasons, some people argue that deep-sea disposal of hazardous chemical waste is environmentally safer than disposal on land or by incineration. Therefore, according to this argument, municipal and industrial wastes should be barged to deep offshore waters and dumped on the central ocean floor. However, not everyone agrees. Some scientists contend that if we continue to pollute the oceans, we will eventually overwhelm their ability to dilute toxins. They argue further that ocean dumping merely delays hard decisions on waste disposal that should be addressed now, before new problems arise.

12.8 Reefs

A **reef** is a wave-resistant ridge or mound built by corals, algae, and other organisms. Because these organisms need sunlight and warm, clear water to thrive, reefs develop in shallow tropical seas where no suspended clay or silt muddies the water (Fig. 12–31). As reef-building organisms die, their offspring grow on their remains. Thus, only the outer and topmost portions of a reef contain living organisms.

The South Pacific and portions of the Indian Ocean are dotted with numerous islands called atolls. An **atoll** is a circular reef that forms a ring of islands around a lagoon. Atolls vary from 1 to 130 kilometers in diameter. They are surrounded by deep water of the open sea. If corals and other reef-building organisms cannot live in deep water, how did atolls form? Charles Darwin studied this question during his famous voyage on the *Beagle* from 1831 to 1836. He reasoned that, to make an atoll, a reef must have formed in shallow water on the flanks of a volcanic island. Eventually the island sank, but the reef continued to grow upward so that the living portion of the reef always remained in shallow water (Fig. 12–32). This proposal was not well received at first because scientists could not accept the idea that volcanic islands sank. However, when scientists drilled into a Pacific atoll shortly after World War II and found volcanic rock hundreds of meters beneath the reef, Darwin's original hypothesis was reconsidered. Today we know that, as the rock beneath a volcanic island cools after the volcano becomes extinct, the rock becomes denser and sinks isostatically, carrying the island down with it.

Reefs around the world have suffered severe epidemics of disease and predation within the last decade. Studies of fossils show that epidemics and even extinctions have affected reefs periodically for millions of years. However, the past few decades are different from any other interval in the Earth's history because of the phenomenal growth of human population and industry. Human activ-

FIGURE 12-31
(A) Aerial view of a reef adjacent to the island of Palau in the southwestern Pacific. Note that the waves break on the reef, leaving calmer, shallow water between the reef and the island. (B) Underwater photograph of castle coral, a reef-building organism. *(Larry Davis)*

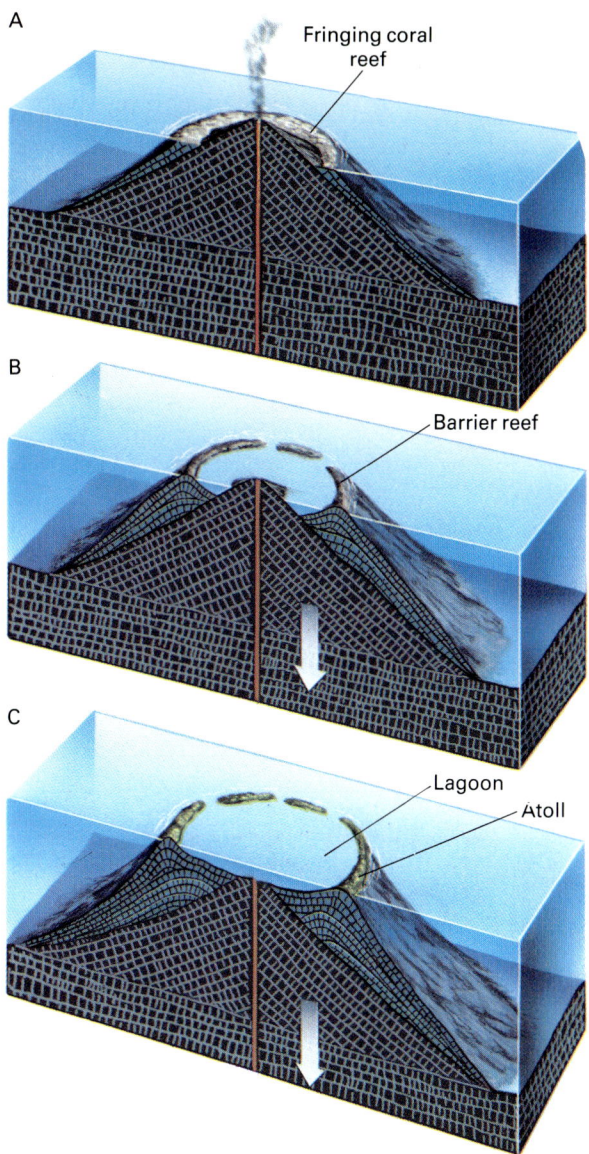

FIGURE 12-32
Formation of an atoll.

ity has caused the extinction of uncounted species and has decimated the populations of many more. Are the recent epidemics among reef organisms part of a natural process or a result of human activity? One possible explanation for the epidemics is that sewage provides nutrients for algae and other organisms that smother reef-building organisms. In addition, many scientists are concerned that air pollutants are increasing global temperature. Reef-building organisms could be adversely affected by a rise in seawater temperature. Both of these explanations are difficult to test. It is often difficult to locate the sources of chemical compounds in seawater. Some may originate as industrial pollutants, whereas others occur naturally. In addition, global warming remains controversial. Thus, we do not know whether the reefs are dying in response to natural causes or human impact.

12.9 Global Warming and Sea-Level Rise

Sea level has risen and fallen repeatedly in the geologic past, and coastlines have emerged and submerged throughout Earth's history. It is sobering to realize that coastlines will continue to change in the future. Can sea level rise sufficiently to flood coastal cities and towns?

During the past 40,000 years, sea level has fluctuated by 150 meters, primarily in response to growth and melting of glaciers (Fig. 12–33). The rapid sea-level rise that started about 18,000 years ago began to level off about 7000 years ago. By coincidence, humans began to build cities about 7000 years ago. Thus, civilization has developed during a period when sea level has been relatively constant.

Sea level started rising again about 75 years ago, at a rate of about 2.0 to 2.5 millimeters per year (Fig. 12–34). The change in a single year is small, but it is half as fast as the dramatic postglacial sea-level rise. Many climatologists predict that the greenhouse effect or other causes

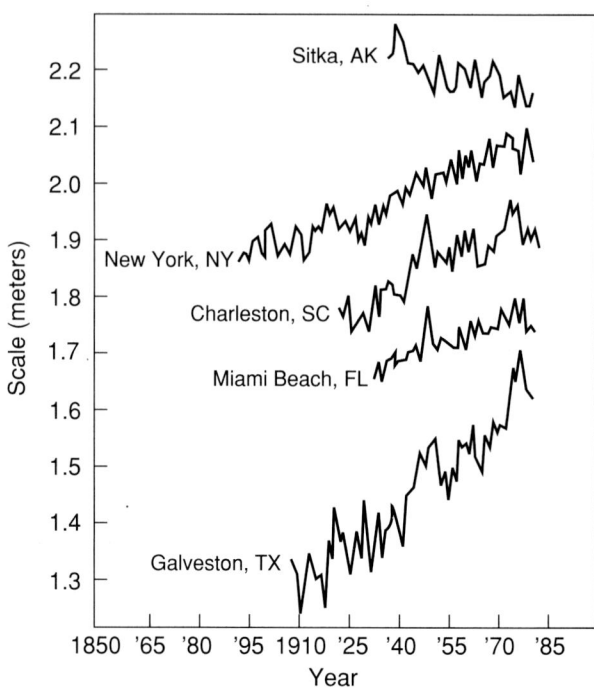

FIGURE 12–34
Coastal emergence and submergence at several locations in the United States. Land subsidence in Galveston has led to rapid submergence, and tectonic uplift along the Alaskan coast has led to local emergence in Sitka. (Stephen H. Schneider, *Global Warming*, p. 164)

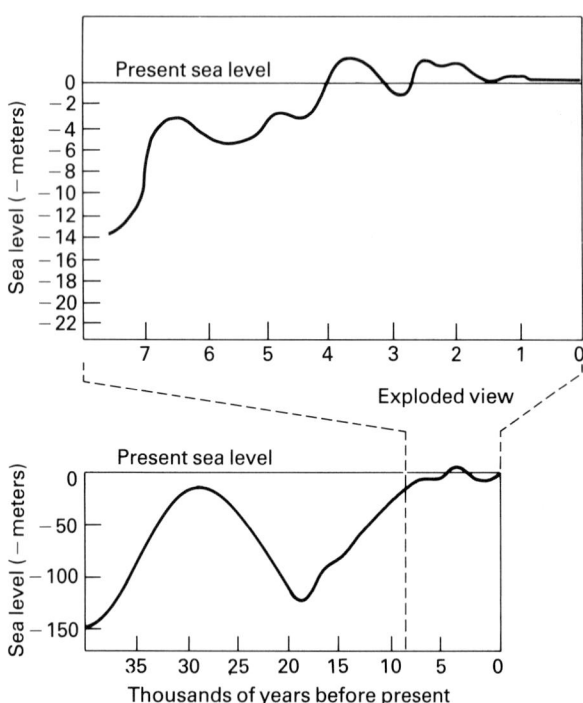

FIGURE 12–33
Sea-level change during the past 40,000 years. The lower graph shows an exploded view of sea-level change over the past 7000 years. (J. D. Hansom, *Coasts* [Cambridge: Cambridge University Press, 1988])

will raise global temperature during the next century. If global warming occurs, it will increase the volume of ocean water and simultaneously melt portions of the huge Greenland and Antarctic ice caps, thus raising sea level on a global scale.

Some scientists disagree with this hypothesis. Climatological data and modeling suggest but do not prove greenhouse warming. In addition, some scientists argue that, even if global warming occurs, it will lead paradoxically to growth of glaciers and a falling sea level. They suggest that if global temperature increases, more water will evaporate from the oceans. More evaporation leads to more precipitation. Antarctica and Greenland are so cold that they would remain below freezing for most of the year even if the average temperature rose a few degrees. Thus, precipitation would fall as snow at higher latitudes and glaciers would grow. Water would be removed from the oceans.

12.9 Global Warming and Sea-Level Rise 313

Consequences of this change vary with location. The wealthy, developed nations would build massive barriers to protect cities and harbors. In regions where global sea-level rise is compounded by local tectonic sinking, dikes are already in place or planned. Portions of Holland lie below sea level. In London, where the high-tide level has risen by 1 meter in the past century, multimillion-dollar storm gates have been built on the Thames River. A similar system is now planned to protect the city of Venice from further flooding.

If sea level rises as predicted, people in the United States will spend about $10 billion per year to protect developed coastlines (Fig. 12–35). The cost would exceed that of any construction project in history. Wetlands, farms, and houses that are not valuable enough to be protected will be lost. If sea level rises by 1 meter, 20,000 square kilometers of land in the United States will be lost, about the area of the state of Massachusetts. An additional 17,000 square kilometers of coastal wetlands will become flooded. In addition, storm damage and coastal erosion will increase.

Problems will be more severe in poorer countries that do not have the capital to finance coastal protection. A 1-meter rise in sea level would flood portions of the Nile delta, displacing 10 million people and decreasing Egypt's agricultural productivity by 15 percent. Seventeen percent of the land area of Bangladesh would be flooded, displacing 38 million inhabitants (Fig. 12–36).

FIGURE 12–35
Expensive condominiums along the Delaware coast will be flooded if sea level rises.

Although we cannot disprove this argument, the majority of researchers working in the field believe that sea-level rise is likely to continue. Although estimates vary, many scientists predict a 1-meter rise in sea level by the year 2100.

FIGURE 12–36
A 1-meter sea-level rise would flood 17 percent of the land area of Bangladesh and displace 38 million people. *(Wide World Photos)*

SUMMARY

If land rises or sea level falls, the coastline migrates seaward and abandons old beaches above sea level, forming an **emergent coastline**. In contrast, a **submergent coastline** forms when land sinks or sea level rises. **Eustatic** sea-level changes occur by growth and melting of glaciers, changes in water temperature, and changes in the volume of submarine mountain ranges. An emergent coastline is commonly **sediment-rich** because the sediment-rich continental shelf is exposed. A submergent coastline is usually **sediment-poor** because the sediment-rich part of the shelf is submerged below the reach of erosion processes.

Most coastal sediment is eroded from deltas and glacial deposits and transported by **longshore currents**. Longshore currents form **spits**, which are long ridges of sand or gravel extending out from a beach. A **barrier island** is a long, narrow, low-lying island that parallels the shoreline.

Human intervention such as the building of **groins** may upset the natural movement of coastal sediment and alter patterns of erosion and deposition on beaches. Sewage introduces pathogenic organisms and nutrients to coastal aquatic ecosystems, often leading to toxic phytoplankton blooms. Toxic chemicals poison fish and marine mammals, and plastic litter kills them by entangling mouths, gills, or fins. Oil spills routinely pollute coastal ecosystems.

An **estuary** is a coastal inlet formed when a river mouth is flooded by the ocean. Chesapeake Bay is cleaner than it was a decade ago, but it seems unlikely that it will return to its pristine state in the foreseeable future.

A **reef** is a wave-resistant ridge or mound built by corals, algae, and other organisms. Reefs abound in tropical seas. An **atoll** is a circular reef that forms a ring of islands around a lagoon. Atolls form around sinking volcanic islands.

Sea level has fluctuated by 150 meters over the past 40,000 years. Within the past 75 years, sea level has risen 2.0 to 2.5 millimeters per year. A continued or accelerated sea-level rise would flood coastal cities.

KEY TERMS

Emergent coastline 292
Submergent coastline 292
Eustatic 293
Sediment-rich 295
Sediment-poor 295
Current 296

Wave 296
Surf 296
Longshore current 296
Spit 296
Barrier island 296
Lagoon 296

Groin 298
Delta 301
Phytoplankton 305
Estuary 308
Reef 310
Atoll 310

REVIEW QUESTIONS

1. Explain the difference between an emergent coastline and a submergent coastline. How does each type of coast form?
2. Explain the difference between eustatic and local sea-level changes.
3. List and explain three mechanisms for eustatic sea-level change.
4. What is isostatic adjustment? How does it affect sea level?
5. Explain why emergent and submergent coastlines are characterized by different quantities of sediment.
6. List the two major sources of coastal sediment.
7. Compare and contrast a barrier island and a spit.
8. How is a longshore current generated, and how do longshore currents affect coastlines?
9. How does a barrier island form? Are barrier islands more likely to be found on emergent or submergent coastlines?
10. What is a groin? How does it affect the beach in its immediate vicinity? How does it affect the entire shoreline?
11. Explain how industrialization has led to erosion of the Louisiana coastline.
12. Discuss both natural and human-caused factors in the erosion of bluffs in San Diego County.
13. What are phytoplankton? What is their significance in ocean ecosystems?
14. Discuss the sources and effects of three categories of coastal pollution.
15. Explain how an atoll forms.
16. Explain how warming of the atmosphere could lead to a rise in sea level.
17. What is an estuary, and why are many estuaries severely polluted?

DISCUSSION QUESTIONS

1. Explain why coastal erosion is commonly episodic.
2. Imagine that you find fossils of marine-dwelling organisms in 250-million-year-old rock in the interior of a continent. How can you tell whether these rocks were submerged by eustatic or local factors?
3. We stated in the text that submergent coastlines are commonly sediment-poor. However, the reverse is not necessarily true; sediment-poor coastlines are not necessarily submergent. Outline a sequence of events that would produce a rocky emergent coastline. Outline a sequence of events that would produce a sandy submergent coastline.
4. Is Chesapeake Bay an emergent or submergent coastline? Briefly outline a plausible sequence of events that could have led to the formation of this coastline.
5. Prepare a three-way debate. Have one student argue that the government should support the construction of groins. Have the second student argue that the government should prohibit the construction of groins. Have the third student defend the position that groins should be permitted but not supported.
6. Prepare a debate on whether government funding should be used to repair storm damage to property on barrier islands.
7. Solutions to environmental problems can be divided into two general categories: social solutions and technical solutions. Social solutions involve changes in attitudes and life styles but generally do not require expensive industrial or technological processes. Technical solutions do not mandate social adjustments but require advanced and often expensive engineering. Describe a social solution and a technical solution to the problem of coastline changes on Long Island. Which approach do you believe would be more effective?
8. Imagine that sea level rises by 1 meter in the next century. How far inland would the shoreline advance if the beach (a) consisted of a vertical cliff? (b) Sloped steeply at a 45° angle? (c) Sloped gently at a 5° angle? (d) Were almost flat and rose only 1°?
9. From Figure 12–23, calculate the average annual rate of sea-level change between 15,000 and 10,000 years ago. How does this compare with the current rate of change, 2.0 to 2.5 millimeters per year?

CHAPTER 13

Glaciers and Ice Ages

13.1 Types of Glaciers
13.2 Glacial Movement
13.3 Glacial Erosion
13.4 Glacial Deposits
13.5 The Ice Ages

Most people have never seen a glacier or walked on its surface. We often think of these ice rivers as features of high mountains and the frozen polar regions. Yet anyone who lives in the northern third of the United States is familiar with landscapes carved and shaped by glaciers.

A flowing glacier picks up huge amounts of rock and soil. When it melts, it deposits that sediment. The smooth, low, rounded hills of upper New York State, Wisconsin, and Minnesota are piles of gravel deposited by great ice sheets as they melted. The barren sediment deposited by the melting ice was unprotected by plants. Strong winds blowing from the dying glaciers eroded the sand and dust and then deposited them again. As a result, great fields of sand dunes and windblown silt formed in the wakes of the glaciers.

One hundred thousand years ago, the world was free of ice except for the polar ice caps of Antarctica and Greenland. Then, in a relatively short period of time—perhaps only a few thousand years—the Earth's climate cooled by a few degrees. As winter snow failed to melt in summer, the polar ice caps grew and spread into lower latitudes. At the same time, glaciers formed near the summits of high mountains, even near the equator. They flowed down mountain valleys into nearby lowlands. When the glaciers reached their maximum size 18,000 years ago, they covered one third of the Earth's continents. About 15,000 years ago, Earth's climate warmed again by a few degrees. As a result, the glaciers melted rapidly.

Geologists often speak of events that occurred millions or billions of years ago. To put the time of this most recent major glacial episode in perspective, Cro-Magnon artists were painting on cave walls in central Europe 35,000 years ago, and agricultural societies developed about 10,000 years ago.

A tidewater glacier flows into the sea. Ellesmere Island, Canada.

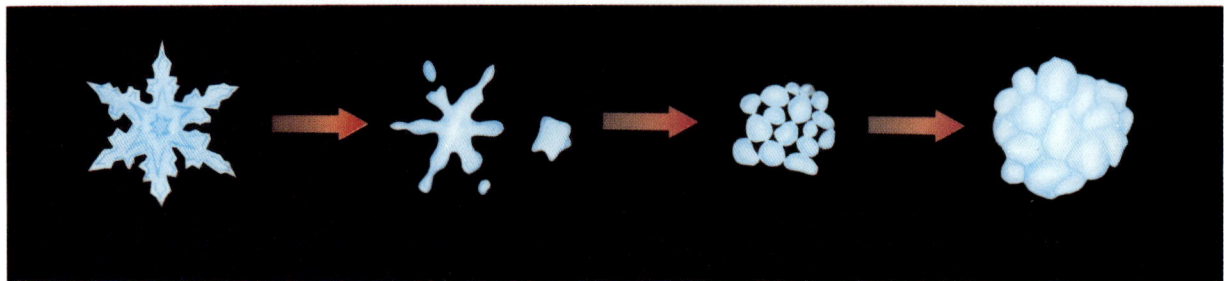

FIGURE 13–1
Newly fallen snow changes through several stages into glacial ice.

13.1 Types of Glaciers

In most temperature regions, winter snow melts entirely during the summer. However, in certain cold, wet environments, only part of each winter's snow melts. The remainder accumulates year after year and becomes denser as new snow buries it to ever greater depths. If snow survives through one summer, it converts to rounded ice grains called **firn** (Fig. 13–1). Mountaineers like firn because the sharp points of their ice axes and crampons sink into it easily and hold firmly. If firn is buried deeper in the snowpack and is subjected to thawing and freezing, the crystals pack together so tightly that water cannot percolate through the ice.

A **glacier** is a massive, long-lasting accumulation of compacted snow and ice. Glaciers form only on land, wherever the amount of snow that falls in winter exceeds the amount that melts in summer. Glaciers in mountain regions flow slowly downhill. Glaciers on level land flow outward under their own weight, just as cold honey poured onto a tabletop spreads outward slowly.

Glaciers form in two environments. Alpine glaciers form at all latitudes on high, snowy mountains. Continental glaciers form at all elevations in the cold, polar regions.

Alpine Glaciers

Mountains are generally colder and wetter than adjacent lowlands. Near the mountain summits, winter snowfall is deep and summers are short and cool. These conditions create **alpine glaciers** (Fig. 13–2). Alpine glaciers exist on every continent—in the Arctic and Antarctic, in temperate regions, and in the tropics. Glaciers cover the summits of Mount Kenya in Africa and Mount Cayambe in South America, even though both peaks are near the equator. Some alpine glaciers at high latitudes flow great distances from the peaks onto adjacent lowlands. For example, the Kahiltna Glacier flows down the southwest side of Denali (Mount McKinley) in Alaska. It is about 65 kilometers long, 12 kilometers across at its widest point, and about 700 meters thick. Most alpine glaciers are smaller; some are larger.

The development of an alpine glacier depends on both temperature and precipitation. The average annual temperature in the state of Washington is warmer than that in Montana. But alpine glaciers in Washington are larger and flow to lower elevations than those in Montana because more snow falls in Washington. Washington's mountains receive such heavy winter snowfall that even

FIGURE 13–2
This alpine glacier flows around granite peaks in British Columbia, Canada.

though summer melting is rapid, large quantities of snow accumulate every year. In Montana, snowfall is light enough that most of it melts in the summer, and thus most of Montana's mountains have no glaciers.

Continental Glaciers

In Greenland and Antarctica, winters are so long and cold and summers so short and cool that glaciers are not confined to the mountains but cover most of the land regardless of elevation. An **ice sheet**, or **continental glacier**, covers an area of 50,000 square kilometers or more (Fig. 13–3). The ice spreads outward in all directions under its own weight. Together, the ice sheets of Greenland and Antarctica make up 99 percent of the world's ice. About three-fourths of the Earth's fresh water is frozen into these continental glaciers. The Greenland sheet is more than 2.7 kilometers thick in places and covers 1.8 million square kilometers. Yet it is small compared with the Antarctic ice sheet, which covers about 13 million square kilometers, almost 1.5 times the size of the United States. If the Antarctic ice sheet melted, the meltwater would create a river the size of the Mississippi that would flow for 50,000 years. In contrast, much of the Arctic is ocean, and because glaciers cannot form on oceans, no ice sheet exists at the North Pole. Instead, much of the Arctic Ocean is commonly covered by sea ice.

Continental glaciers preserve many geological records. Scientists study air bubbles trapped in old ice layers to learn about atmospheric composition at times when the ice formed. They study subtle changes in the ratio of heavy oxygen to the more abundant light oxygen ($^{18}O/^{16}O$) to determine prehistoric climate change. Some search for meteorites on the surface of the Antarctic ice cap, and others drill cores into the ice to find layers of volcanic dust.

At certain times in the past, average global temperature was lower than at present. Polar and temperate climates were colder, and possibly wetter, than now. Consequently, vast continental glaciers covered much of North America, Europe, Asia, and parts of the southern continents. These times, called the **ice ages**, are discussed later in this chapter.

13.2 Glacial Movement

Imagine that you set two poles in dry ground on opposite sides of a glacier, and a third pole in the glacier in a straight line with the other two. After a few months, the center pole would have moved downslope, forming a triangle with the other two (Fig. 13–4). This simple experiment tells us that the glacier moved downhill. Rates of glacial movement vary with steepness, precipita-

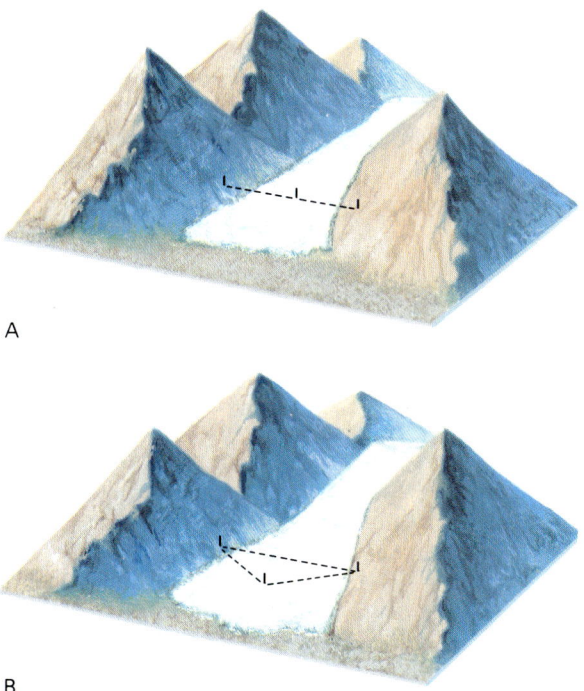

FIGURE 13–4

A glacier flows downslope. If three stakes are set in a straight line with two on land and one on a glacier (A), the stake on the ice will move (B).

FIGURE 13–3

The Antarctic ice sheet, Victoria Land. *(Mugs Stump)*

tion, and temperature. In the coastal ranges of Alaska, where annual precipitation is high and average temperature is relatively high (for glaciers), some glaciers move several meters a day. In contrast, in the interior of Alaska, where conditions are generally cold and dry, glaciers move only a few centimeters a day. At these rates, ice flows the length of an alpine glacier in hundreds to a few thousand years. In some instances, a glacier may **surge** at a speed of 10 to 50 meters per day. A glacier moves by two mechanisms: basal slip and plastic flow.

In **basal slip**, the entire glacier slides over bedrock in the same way that a bar of soap slides down a tilted board. Just as wet soap slides more easily than dry soap, an accumulation of water between bedrock and the base of a glacier accelerates basal slip. Several factors cause water to accumulate near the base of a glacier. Earth heat rises to melt ice near bedrock. Friction from glacial movement also generates heat. In addition, when ice melts and becomes water, it shrinks. Thus, pressure favors melting and the weight of overlying ice can melt the base of a glacier. Additionally, during summer, water flowing over the surface of a glacier may seep downward to the glacier's base.

Glacial ice also flows like a very viscous fluid. Near the surface of a glacier, pressure is low and the ice is brittle, like an ice cube or ice on a frozen lake. If you hit it with a hammer, it shatters. However, at depths greater than 40 to 50 meters, pressure is high enough that the ice flows plastically. As a result, this portion of the glacier moves by plastic flow in addition to basal slip. In **plastic flow**, glacial ice deforms and flows as a very viscous fluid rather than fracturing.

When a glacier flows over uneven bedrock, the deeper plastic ice bends and flows over bedrock bumps, but the brittle upper ice cracks, forming **crevasses** (Fig. 13–5). Crevasses form only in the brittle upper 40 to 50 meters of a glacier, but do not continue into the lower plastic zone. Crevasses open and close slowly as a glacier moves.

An ice fall is a section of a glacier where many crevasses alternate with towering ice pinnacles. The pinnacles form where ice blocks broken from crevasse walls rotate as the glacier moves. If you were a skilled mountaineer, you might climb down into a crevasse. The surrounding walls are a pastel blue, and sunlight filters through the narrow opening above. Almost certainly, you would hear and feel the ice shift and crack. With crampons and ice axes, you could scale one of the pinnacles, but it would be dangerous to do so because the glacier moves continuously and such pinnacles often topple without warning. Many mountaineers have been crushed by falling ice while traveling through ice falls.

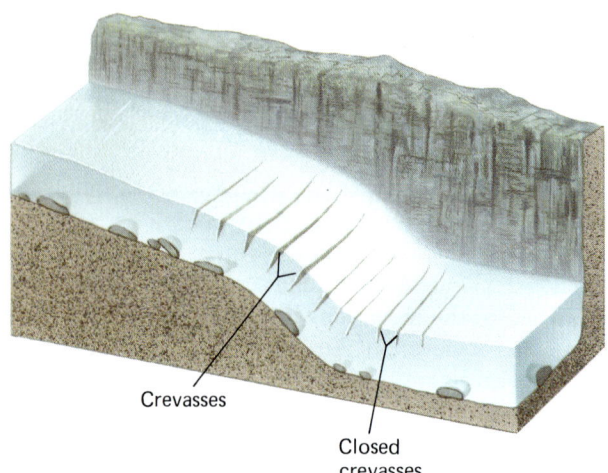

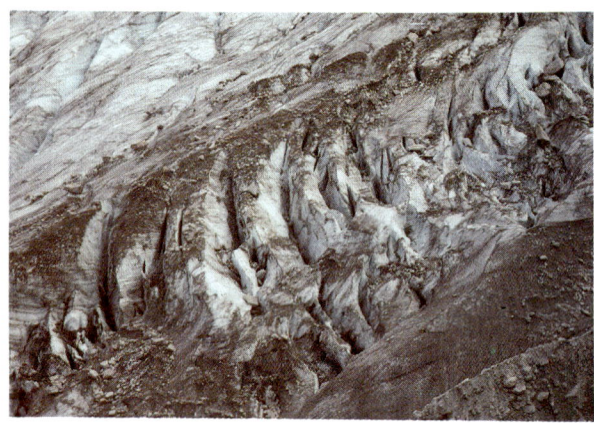

FIGURE 13–5

(A) Crevasses form in the upper, brittle zone of a glacier where the ice flows over uneven bedrock. (B) Crevasses in the Bugaboo Mountains of British Columbia.

The 1986 Hubbard Glacier Surge

CASE HISTORY

The motion of a glacier is normally masked by its apparent stillness. It is easy to find evidence of glacial erosion, but a glacier takes hundreds or thousands of years to scour mountains and deposit moraines. If you walk on a glacier, you are likely to hear it crack and rumble as the ice shifts and fractures, but you rarely see the ice move. In an unusual event in the mid-1980s, however, the Hubbard Glacier in Alaska surged and changed significantly in a matter of a few months.

As shown in Figure 13–6, the Hubbard Glacier flows out of the St. Elias Range along the Canada–

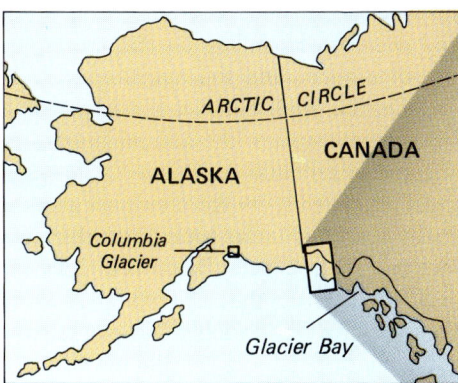
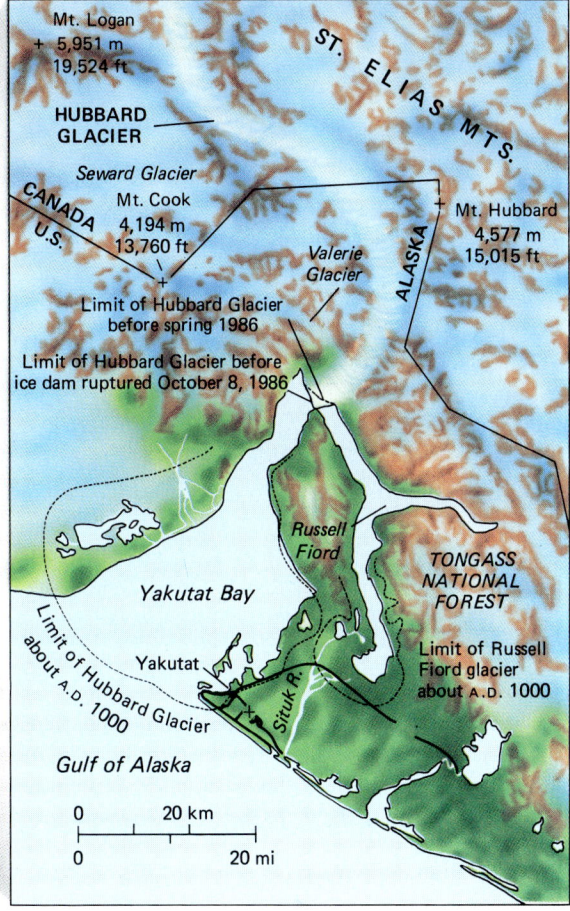

FIGURE 13–6
The Hubbard Glacier flows into Yakutat Bay from the St. Elias Mountains near the Canada–Alaska border.

Alaska border and meets the sea at Russell Fjord. The Hubbard is joined by the Valerie Glacier just before the Hubbard flows into Russell Fjord. Under normal conditions, both glaciers flow at a rate of about 15 centimeters per day. In the spring of 1986, however, the Valerie accelerated radically. It started to move at 35 meters per day (about the length of a football field every 3 days). As a result, the lower Hubbard Glacier sped up to 15 meters per day. The Hubbard then advanced across Russell Fjord, forming a dam that isolated the fjord from the sea.

The ice dam rose 30 meters above sea level, extended to the floor of the fjord, and was nearly half a kilometer wide. By sealing off the strait, it converted the fjord to a salty inland lake. Fresh water flowed into the lake from nearby streams and glaciers, and as a result the lake started to rise about 0.5 meter per day. The fresh water diluted the seawater, and as the salinity declined, many saltwater species trapped in the fjord began to die. Seals and porpoises, deprived of their food supply, starved. Residents in the nearby town of Yakutat feared that continued flooding would threaten the village and the rich salmon spawning grounds nearby. Then in October, approximately 3 months after the ice dam formed, it broke and the fjord was once again connected to the ocean.

Glacial surges are fairly common. Normally, water collects at the base of an alpine glacier, promoting a constant rate of basal slip. The water flows from the snout of the glacier. Just prior to a glacial surge, the amount of meltwater flowing from the glacier decreases greatly, indicating that the drainage beneath the ice is blocked. As a result of the blockage, excess water accumulates under the ice. The glacier then surges forward on this layer of water. Eventually, drainage becomes reestablished beneath the ice, excess water escapes, and the glacier resumes its normal, slow movement.

FIGURE 13–7
The terminus of an alpine glacier on Baffin Island, Canada, in midsummer. Dirty, old ice lies in the lower part of the glacier below the firn line, whereas clean, white snow lies higher up on the ice above the firn line. *(Steve Sheriff)*

The Mass Balance of a Glacier

Consider an alpine glacier flowing from high mountains into a lowland valley (Fig. 13–7). At the upper end of the glacier, snowfall is heavy, temperatures are low for much of the year, and avalanches carry large quantities of snow from the surrounding peaks onto the ice. Thus, more snow accumulates in winter than melts in summer, and snow piles up from year to year. This higher-elevation part of the glacier is called the **accumulation area**. There the glacier's surface is covered by snow year round; the snow is often powdery in winter and slushy in summer.

Lower in the valley, the temperature is higher throughout the year, and less snow falls. This lower part of a glacier, where more snow melts in summer than accumulates in winter, is called the **ablation area**. When the snow melts, a surface of old, hard glacial ice is left behind. The **firn line**, or **snowline**, is the boundary between permanent snow and seasonal snow. The firn line shifts up and down the glacier from year to year, depending on weather. Why is there any ice at all in the ablation area? Glacial ice flows downward from the accumulation area to the ablation area and continuously replenishes it. Even farther down valley, the rate of glacial flow cannot keep pace with melting, so the glacier ends at its **terminus** (Fig. 13–8).

Glaciers grow and shrink. If annual snowfall increases or average temperature drops, more snow accumulates; then the firn line of an alpine glacier descends to a lower elevation, and the glacier grows thicker. At first the terminus may remain stable, but eventually it advances down the valley. The lag time between a change in climate and a glacial advance may range from a few years to several decades, depending on the size of the glacier, its rate of motion, and the magnitude of the climate change. On the other hand, if annual snowfall decreases or the climate warms, the accumulation area shrinks and the glacier retreats.

When a glacier retreats, its ice continues to flow downhill, but the terminus melts back faster than the glacier flows downslope. In Glacier Bay, Alaska, glaciers have retreated 60 kilometers in the past 125 years, leaving a barren landscape of rock and rubble. No plants grow there, and only seabirds perch on the otherwise lifeless rock. Over the centuries, their droppings will mix with windblown silt and weathered rock to form thin soil. At first, lichens will grow on the bare rock, and then mosses will take hold in sheltered niches that hold soil. The mosses will be followed by grasses, bushes, and finally trees as vegetation reclaims the landscape.

13.3 Glacial Erosion

A flowing glacier erodes soil and scours bedrock. It does not only scrape the Earth's surface like a bulldozer, however. In addition, meltwater seeps into the bedrock beneath a glacier and then freezes, fracturing and loosening the rocks.

Once a glacier picks up rocks and other sediment, it carries them along. Ice is too soft to wear away bedrock, but rocks embedded in the base of a glacier scrape across bedrock like sandpaper pushed by a giant's hand. As the ice flows, the embedded rocks gouge parallel grooves and scratches, called **glacial striations**, into bedrock (Fig. 13–9). Striations show the direction of ice movement. Geologists use them to map the movements that took place during the most recent ice age.

Erosional Landforms Created by Alpine Glaciers

Let's take an imaginary journey through a mountain range that was glaciated in the past but is now mostly ice-free. We start with a helicopter ride to the summit

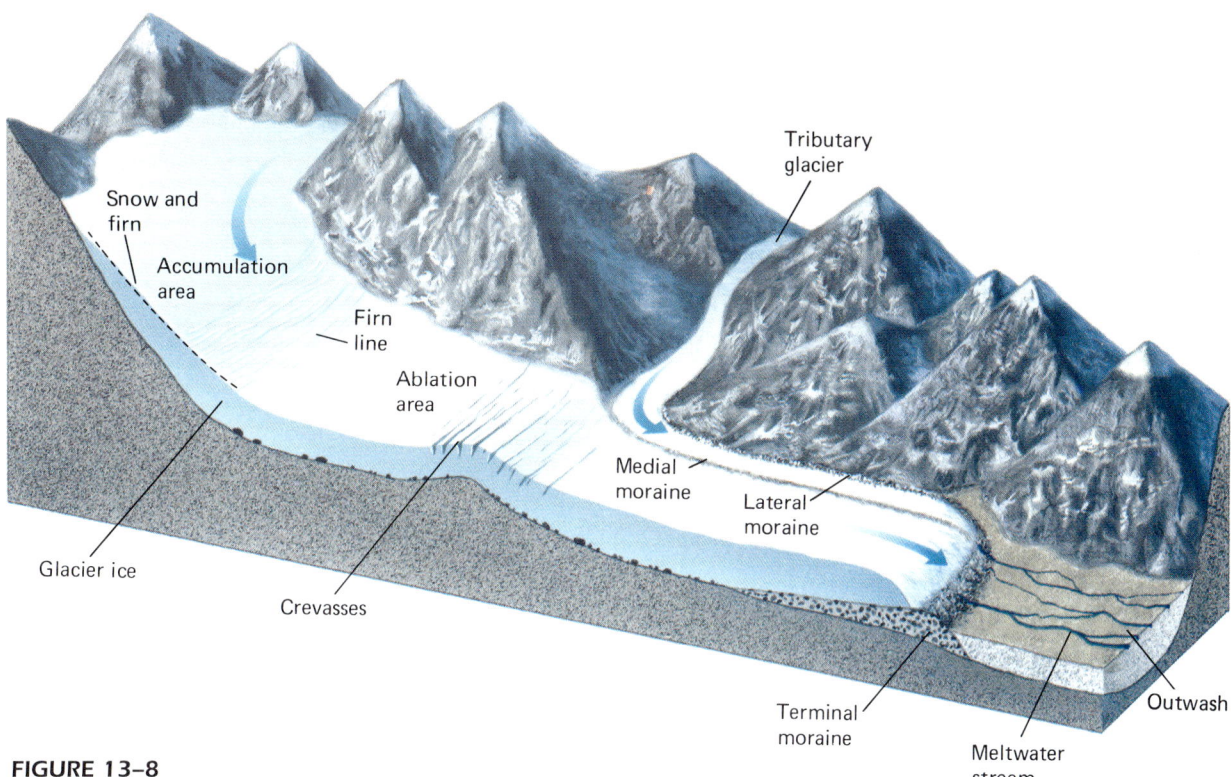

FIGURE 13–8

Prominent features of an alpine glacier. The firn line is the boundary between permanent snow and seasonal snow.

of a high, rocky peak. Our first view from the helicopter is of sharp, jagged mountains rising steeply above smooth, rounded valleys (Fig. 13–10A).

A mountain stream commonly erodes downward into its bed, cutting a steep-sided, V-shaped valley. A glacier, however, is not confined to a narrow stream bed but instead fills its entire valley. As a result, it erodes outward against the valley walls as well as downward into its bed, scouring the sides as well as the bottom of its valley and forming broad, rounded, **U-shaped valleys** (Fig. 13–11).

We land on one of the peaks and step out of the helicopter. Beneath us, a steep cliff drops off into a horseshoe-shaped depression gouged out of the mountainside. This depression is called a **cirque**. A small glacier at the head of the cirque reminds us of the larger mass of ice that existed there in a colder, wetter time.

To understand how a glacier creates a cirque, imagine a gently rounded mountain. As snow accumulates and a glacier forms, the ice begins to flow down the mountainside (Fig. 13–10B). The flowing ice erodes a small depression that grows slowly as the glacier grows larger (Fig. 13–10C). As erosion enlarges the cirque, its walls become steeper and higher. Rocks tumble from the cirque walls onto the glacier, and the flowing ice carries this debris from the cirque to lower parts of the valley (Fig. 13–10D). When the glacier finally melts, it leaves a steep-walled, rounded cirque.

If glaciers erode three or more cirques into different sides of a peak, they may create a steep, pyramid-shaped rock summit called a **horn**. The Matterhorn in the Swiss

FIGURE 13–9

Rocks embedded in the base of a glacier gouge glacial striations into bedrock.

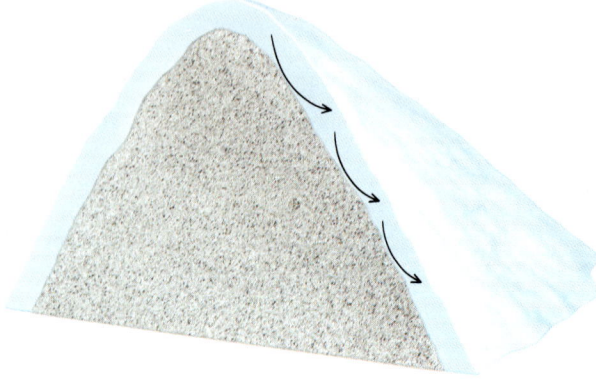
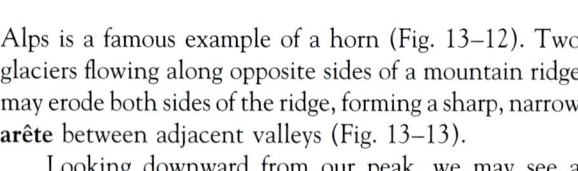

FIGURE 13–10
(A) A cirque rises above a mountain meadow in the Canadian Rockies. (B) Snow accumulates, and a glacier begins to flow downslope from the summit of a peak. (C) Glacial plucking erodes a small depression in the mountainside. (D) Continued glacial erosion and weathering enlarge the depression. When the glacier melts, it leaves a cirque carved in the side of the peak, as in the photograph.

Alps is a famous example of a horn (Fig. 13–12). Two glaciers flowing along opposite sides of a mountain ridge may erode both sides of the ridge, forming a sharp, narrow **arête** between adjacent valleys (Fig. 13–13).

Looking downward from our peak, we may see a waterfall where a small, high valley empties into a larger, deeper one. A small glacial valley lying high above the floor of the main valley is called a **hanging valley** (Fig. 13–14). To understand how a hanging valley forms, imagine these mountain valleys filled with glaciers, as they were several millennia ago (Fig. 13–15). The main glacier, flowing through the lower valley, gouged a deep trough. In contrast, the smaller tributary glacier did not scour the rock as deeply. As a result, when the ice melted,

FIGURE 13–11
A U-shaped glacial valley, Purcell Mountains, British Columbia.

324

13.4 Glacial Deposits 325

FIGURE 13–12
The Matterhorn formed as three glaciers eroded cirques into the peak from three different sides. *(Swiss Tourist Board)*

the floor of the tributary was considerably higher than that of the main valley, forming an abrupt drop where it enters the main valley.

In many coastal regions, deep, narrow inlets called **fjords** extend far inland. Most fjords are glacially carved valleys that were later flooded by rising seas as the glaciers melted (Fig. 13–16).

Continental glaciers erode the landscape just as alpine glaciers do. The main difference is that continental glaciers are considerably thicker and not confined to valleys. Therefore, they scour the entire landscape and sometimes cover whole mountain ranges. An alpine glacier is like an engraver's knife that scratches thin, deep lines into the landscape, whereas a continental glacier is more closely analogous to a bulldozer. The most spectacular landforms created by continental glaciers are those formed by the deposition of sediment carried by the ice, which is described in the next section.

13.4 Glacial Deposits

In the 1800s, before geologists understood that continental glaciers covered vast parts of the land only 10,000 to 20,000 years ago, they recognized that large deposits of

FIGURE 13–13
An arête in the Bugaboo Mountains in British Columbia.

FIGURE 13–14
Two hanging valleys in Yosemite National Park. *(Science Graphics/Ward's Natural Science Establishment, Inc.)*

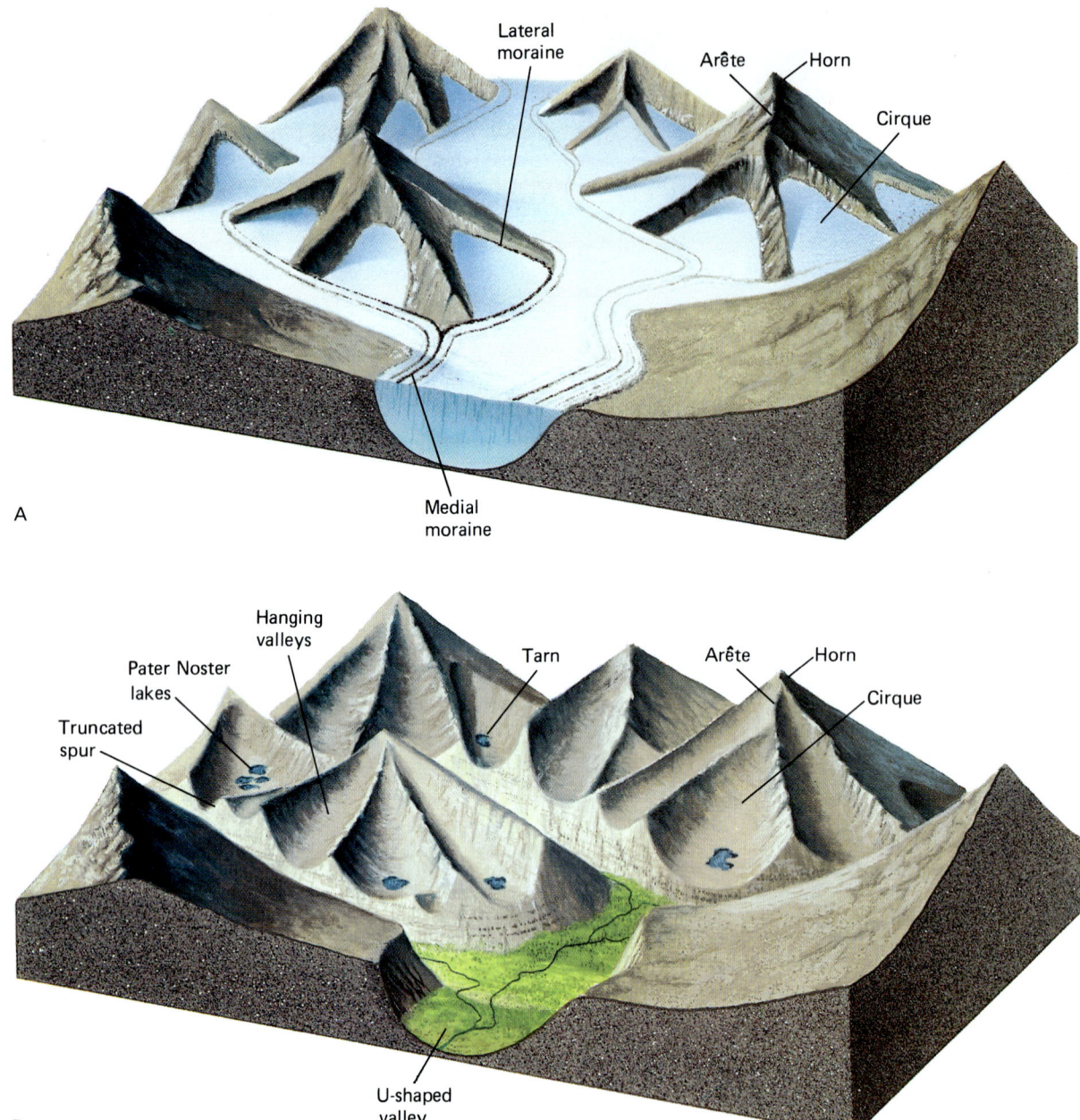

FIGURE 13–15
Alpine glaciers create several different landforms. (A) Mountains covered by alpine glaciers. (B) The same mountains after the glaciers have melted.

sand and gravel found in some places had been transported from distant sources. A popular theory suggested that this material had been carried by icebergs during catastrophic floods. The deposits were called "drift" after this inferred mode of transport.

Today we know that glaciers carried and deposited drift. Although the term *drift* is a misnomer, it remains in common use. Now geologists define **drift** as all rock or sediment transported and deposited by a glacier. Glacial drift averages 6 meters thick over the rocky hills and pastures of New England and 30 meters deep over the plains of Illinois.

Drift is divided into two categories. **Till** is deposited directly by glacial ice. **Stratified drift** is sediment that

FIGURE 13–17
An end moraine is a ridge of till piled up at a glacier's terminus. Baffin Island, Canada.

FIGURE 13–16
Thousand-meter-high cliffs surround a steep-sided fjord on Baffin Island, Canada.

was first carried by a glacier and then transported and deposited by a stream.

Landforms Composed of Till

Ice is so much more viscous than water that it carries all sizes of sediment together. As a result, till, a mixture of clay, sand, and gravel, is deposited in an unsorted, unstratified, jumbled mass.

Moraines

A **moraine** is a mound or ridge of till deposited by a glacier. Think of a glacier as a giant conveyor belt. If you place a number of suitcases on a conveyor belt, it carries them to its end and dumps them in a heap. Similarly, a glacier carries sediment and then drops it at its terminus. If a glacier is neither advancing nor retreating, its terminus may remain in the same place for years, and the sediment piles up at the terminus in a ridge called an **end moraine** (Fig. 13–17). An end moraine that forms when a glacier is at its greatest advance, before beginning to retreat, is called a **terminal moraine**. Other types of moraines are defined briefly in Table 13-1.

An end moraine deposited by a large alpine glacier may be so high that it would take an hour to climb to its top. A steep moraine can be difficult and dangerous to climb. Till is commonly loose, and large boulders are mixed randomly with rocks, cobbles, sand, and clay. A careless hiker can dislodge boulders and cause a dangerous rockfall.

Terminal moraines show the extent of the most recent continental glaciers. In North America the moraines form a broad, undulating line across the northern United States from Montana to New York. Enough time has passed that soil and vegetation cover these moraines (Fig. 13–18).

TABLE 13–1 Types of Moraines	
Recessional moraine	A moraine that forms if a glacier starts to retreat and then stabilizes long enough to form a new end moraine.
Ground moraine	A relatively thin layer of till that forms when a glacier retreats at an even rate.
Lateral moraine	Till that collects on the outside boundaries of an alpine glacier.
Medial moraine	A moraine in the middle of a glacier, formed by coalescing lateral moraines of two tributary glaciers.

328   Chapter 13   Glaciers and Ice Ages

FIGURE 13-18
This moraine in New York State marks the farthest advance of the latest continental ice sheet.

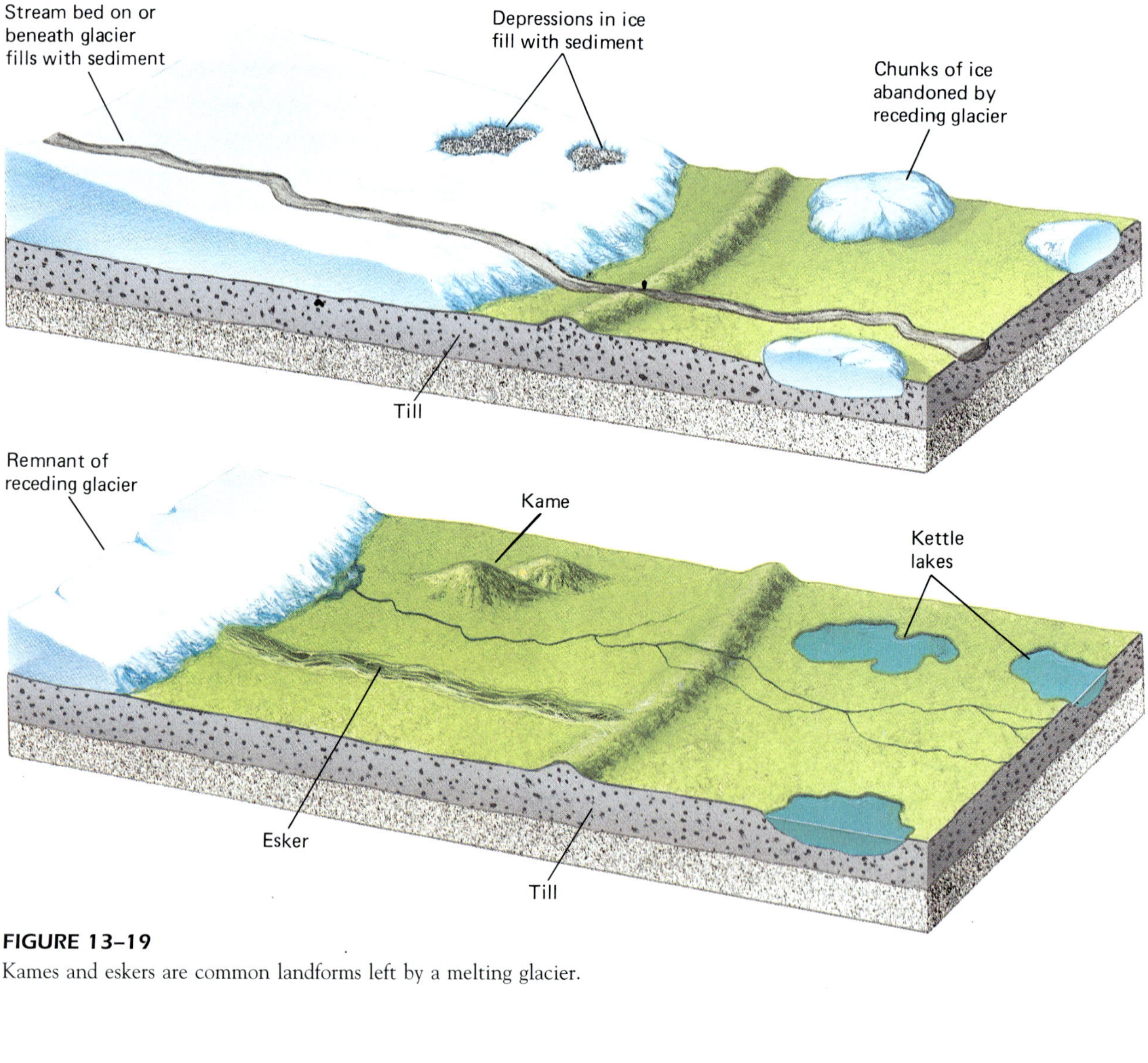

FIGURE 13-19
Kames and eskers are common landforms left by a melting glacier.

Landforms Composed of Stratified Drift

During summer, when snow and ice melt rapidly, streams flow over the surface of a glacier. Many are so deep and wide that hikers cannot jump across them easily. A stream flowing on a glacier commonly runs into a crevasse and plunges into the interior of the glacier. Some water finds its way to the bottom of the ice and flows over bedrock or drift beneath the glacier. Eventually, all of this water flows from the terminus.

Because a glacier erodes so much sediment, a stream flowing from a glacier commonly carries large amounts of silt, sand, and gravel. The stream eventually deposits this sediment downstream from the glacier as **outwash**.

As a glacier melts, streams flowing on top of, within, and beneath the ice commonly deposit small mounds of sand and gravel called **kames**. A kame can form as a fan or delta at the margin of a melting glacier, or where sediment collects in a crevasse or other depression in the ice. An **esker** is a long, snakelike ridge that forms as the channel deposit of a stream that flowed within or beneath a glacier (Fig. 13–19).

Because kames, eskers, and outwash are stream deposits, they are sorted and show sedimentary bedding. Thus, they are easily distinguished from the unsorted and unstratified till deposited directly by glacial ice.

13.5 The Ice Ages

Geologists have found moraines and bedrock striations far from mountain glaciers. Once they understood that glaciers had formed these features, they realized that, sometime in the past, glaciers covered much of the Earth's continents at high latitudes. A time when alpine glaciers descend into lowland valleys and continental glaciers spread over higher latitudes is called an **ice age** or a **glacial epoch**.

Geologic evidence shows that the Earth has been warm and relatively ice-free for at least 90 percent of the past 2.5 billion years. However, at least five major ice ages have occurred during that time (Fig. 13–20).

FIGURE 13–20

Average global temperature has varied during the past 1 billion years. The times of lowest temperature coincide with ice ages. The right-hand scale is an expanded view to show temperature fluctuations during the Pleistocene Ice Age.

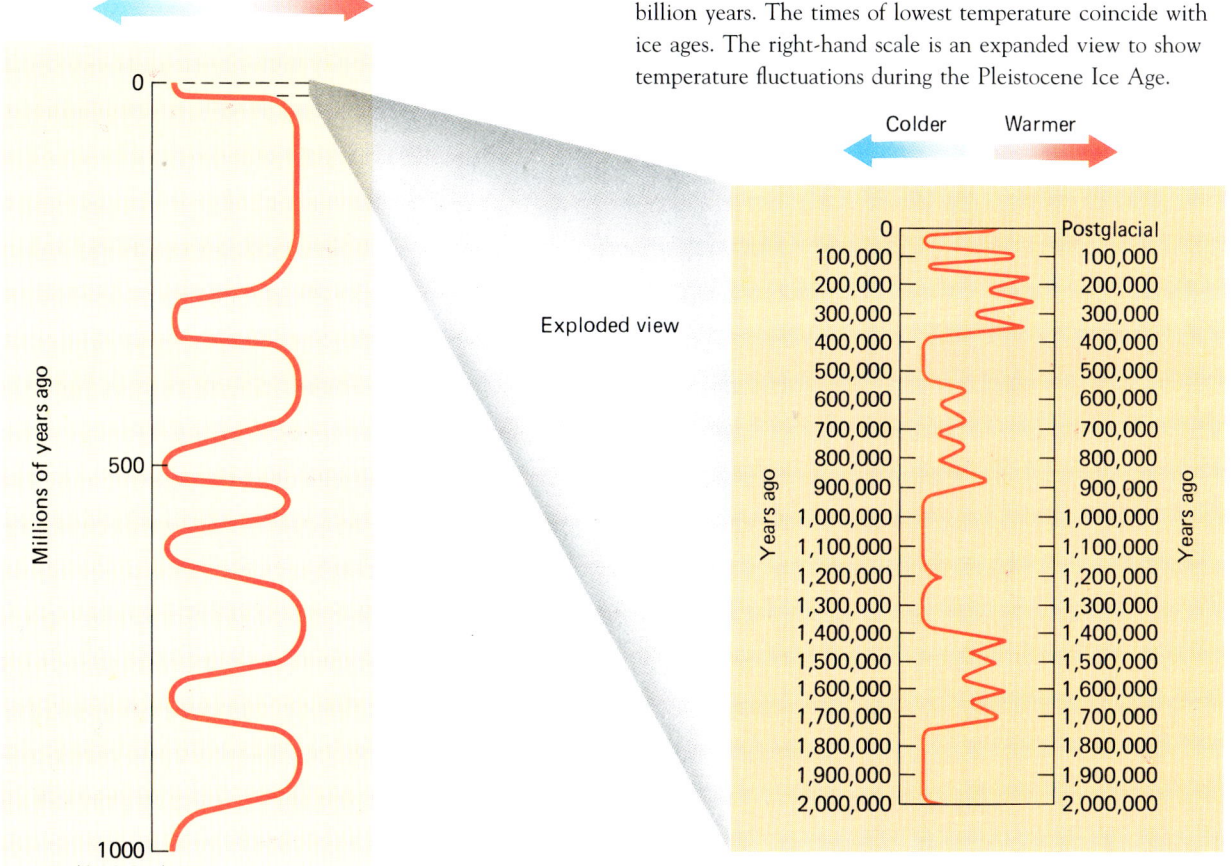

The most recent ice age occurred mainly during the Pleistocene Epoch and is called the **Pleistocene Ice Age**. It began about 2 million years ago, and many geologists think we are still in this ice age. However, the Earth has not been glaciated continuously during the Pleistocene Ice Age; instead, continental ice sheets and alpine glaciers have grown and then melted away several times. Figure 13–20 also shows fluctuations in average global temperature during the Pleistocene Epoch.

Effects of Pleistocene Continental Glaciers

In Chapter 1, we defined environmental geoscience as the study of geology as it applies to living organisms and the quality of life on our planet. Many of the environmental issues discussed in this book relate to changes caused by human activities, such as depletion of ground water, the damming of rivers, or the urbanization of coastal wetlands. But 18,000 years ago, glaciers covered one third of the Earth's land area (Fig. 13–21). Those continental glaciers altered conditions on Earth far more dramatically than any human activity.

The thickest portion of the North American ice sheet was in Canada, southeast of Hudson Bay. The ice flowed outward from that region, eroding soil and scouring bedrock (Fig. 13–22). At the same time, alpine glaciers flowed from the mountains into the lowland valleys. The ice widened and deepened old stream valleys and reshaped mountains. When the ice stopped, it deposited huge quantities of till. Today, terminal moraines lie south of the Great Lakes and extend westward into Montana and eastward into southern New York and Long Island. Moraines and outwash cover much of the northern Great Plains. These deposits form the fertile soil of North America's "breadbasket." When the glaciers melted, (see Chapter 10), water seeped into the ground, recharging the Ogallala Aquifer, which is now used to irrigate the central high plains.

Recall from Chapter 12 that, when glaciers grow, they take water from the oceans and sea level falls. When glaciers melt, sea level rises again. When the Pleistocene glaciers reached their maximum 18,000 years ago, global sea level fell about 100 meters below its present elevation (Fig. 13–23). As submerged continental shelves became exposed, the land area increased by 8 percent (although about one third of the land was ice-covered).

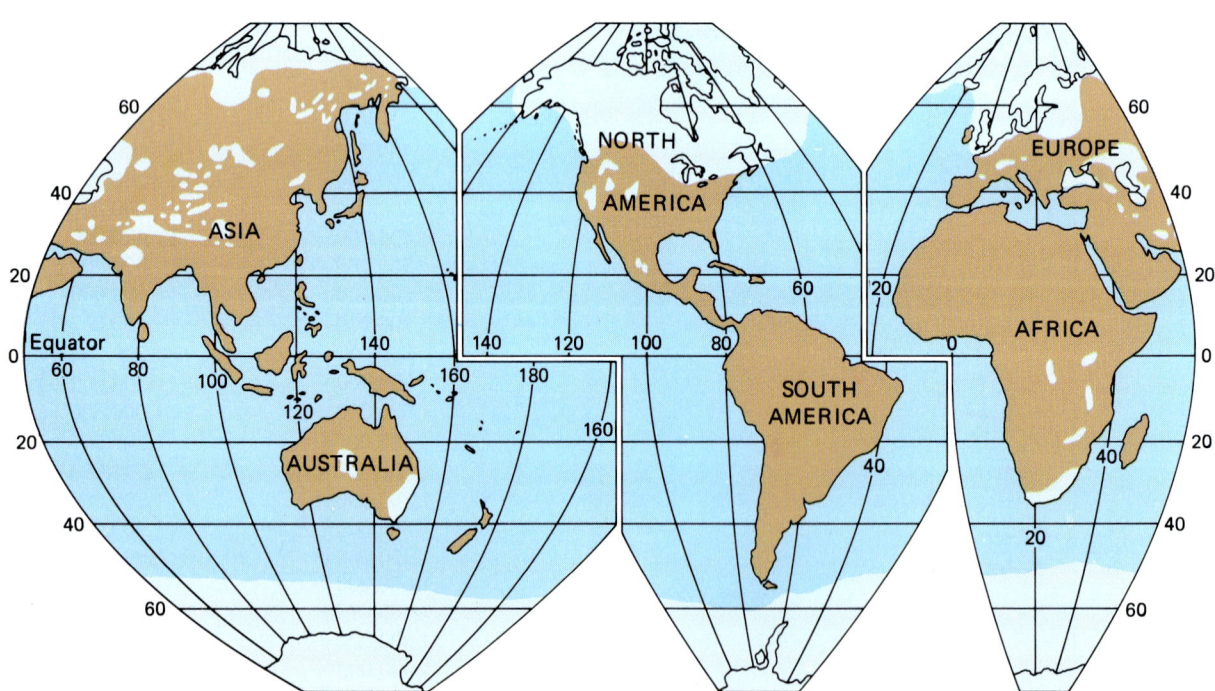

FIGURE 13–21

Glaciers covered about one third of the Earth's continents 18,000 years ago, during the most recent glacial maximum.

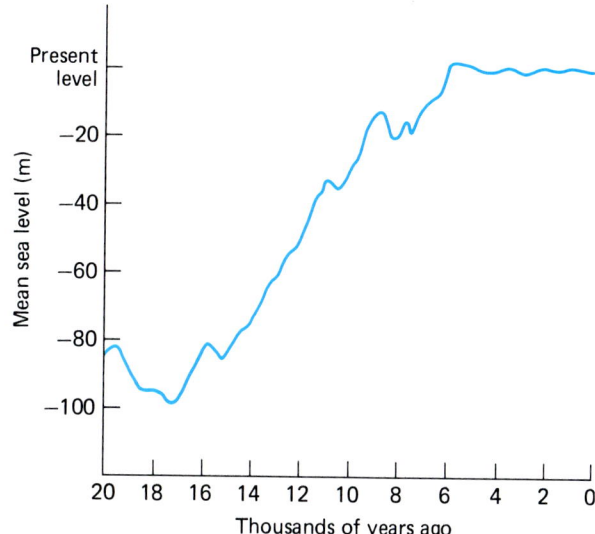

FIGURE 13-22
An ice sheet covered most of northern North America 18,000 years ago. It was thickest near Hudson Bay and from there flowed outward in all directions, as shown by arrows.

FIGURE 13-23
Sea level has risen by about 100 meters during the past 18,000 years. It was lowest about 18,000 years ago when glaciers were largest.

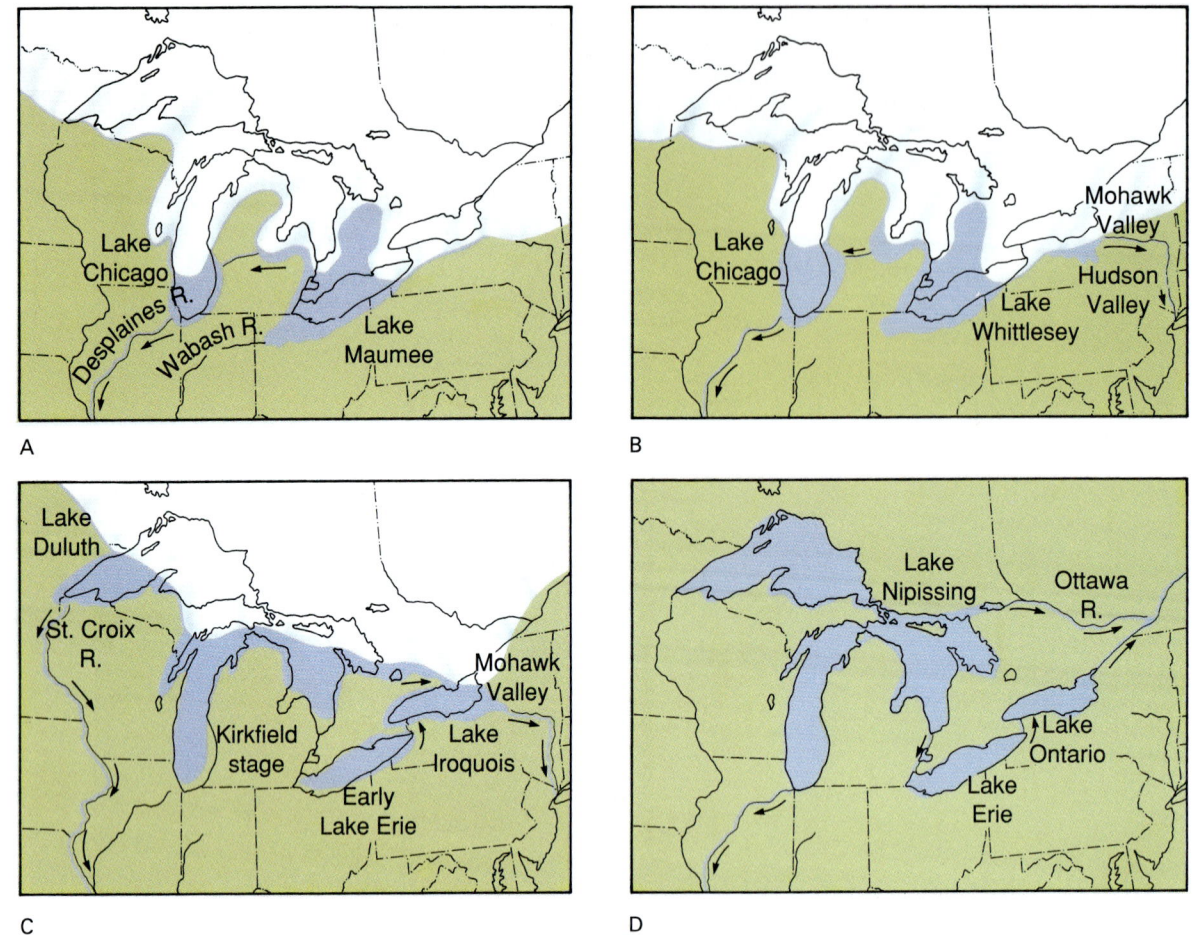

FIGURE 13-24
Continental glaciers scoured the Great Lakes Basin and altered its drainage pattern several times.

Pleistocene Glaciers and the Great Lakes

Before the onset of the Pleistocene glaciation, several major river valleys flowed through what is now the Great Lakes Basin. Ice flowed into these valleys as the glaciers expanded. The glaciers scoured and deepened the valleys and deposited moraines at their southern edges. Ice dammed the St. Lawrence outlet to the Atlantic Ocean, so most of the water from central North America flowed southward into the Mississippi drainage (Fig. 13-24A). The Mississippi had a steeper gradient than it does today because the great accumulation of ice on land had lowered sea level. As a result, rapid downcutting formed a V-shaped valley with a narrow flood plain in the Mississippi Valley.

About 18,000 years ago, the ice began to melt and the glaciers started to retreat. The meltwater carried huge quantities of silt, sand, and gravel into the Mississippi Valley. More sediment flowed into the Mississippi than the current could transport, so the river spread out into many small channels, forming a **braided stream**.

As the glacier continued to melt, lakes formed behind the moraine dams all along the receding ice front. When enough water accumulated in the eastern basin (where Lake Ontario is today), water breached the dam and flowed into the Hudson River valley (Fig. 13-24B).

Over the next five thousand years, the ice receded, sea level rose, and the part of the continent that had been depressed by the massive glacier rebounded isostatically. These three trends caused the drainage patterns to change several times (Figs. 13-24C and D). About 9000 years ago, the lakes drained eastward and southward through the St. Lawrence and Mississippi Rivers. A few thousand years later, the Mississippi River link was abandoned and all the water flowed through the St. Lawrence, as it does today.

> When its supply of water was reduced, the Mississippi slowed down and carried much less sediment. Gradually, the braided channels coalesced, forming the modern meandering river with its broad flood plain.

CASE HISTORY

Glacial Lake Missoula and the Greatest Flood in North America

About 18,000 years ago, a giant tongue of ice pushed southward from Canada into the United States along the Idaho–Montana border. The ice dammed the Clark Fork River, forming Glacial Lake Missoula, which was 600 meters deep and contained 2000 cubic kilometers of water—as much as modern Lake Ontario. Ice is a disastrously poor material for a dam because it floats on water. When the lake became deep enough, it floated the ice dam and the water poured across northern Idaho and Washington. But this was no ordinary flood like the 1993 Mississippi River flood. It was even orders of magnitude more catastrophic than modern disasters such as the flood that occurred after the Teton Dam collapsed. Geologists estimate that a wall of water 600 meters high raced down valley. The water spread out as the valley opened up, and continued down the Columbia River valley. The raging torrent transported rocks from western Montana and deposited some of them 300 meters above river level near the Columbia River Gorge, 450 km from the failed dam. The rushing water eroded deep stream valleys in eastern and central Washington, which now contain tiny streams or no water at all. The area is called the **channeled scablands**.

Causes of Ice Ages

At least five major ice ages have occurred in the past 2.5 billion years, and the most recent, the Pleistocene Ice Age, was characterized by several glacial advances and retreats. The causes of both the ice ages and the Pleistocene glacial fluctuations were related to global climate changes.

In considering the causes of ice ages, two questions arise:

1. Why have ice ages occurred at several times in the Earth's history?
2. Why did continental glaciers advance and then retreat several times during the Pleistocene Ice Age?

Scientists have suggested a wide range of explanations both for the ice ages and for Pleistocene advances and retreats. They include changes in the energy output of the Sun, variations in the Earth's orbit around the Sun, interactions between the Earth's magnetic field and climate, changes in the positions of the continents due to tectonic motion, changes in deep ocean currents, surges in Antarctic glaciers, and movements of the pack ice in the Arctic Ocean. Several research teams have suggested a relationship between volcanic activity and climate change. As explained in Chapter 6, volcanic eruptions blast dust into the upper atmosphere, which leads to global cooling.

Two of the most widely discussed theories involve movement of the continents and variations in the Earth's orbit.

Major Glacial Epochs and Plate Tectonics

One explanation for long-term climatic change and the onsets of ice ages comes from plate tectonics. According to this hypothesis, ice ages occurred at times when the continents were positioned close to the poles. For example, during the Permian Period (about 250 million years ago), all the continents were gathered into one giant landmass called Pangaea, part of which lay at the South Pole. Glaciation was extensive at this time. As Pangaea broke apart and the continents moved toward the equator, the Permian Ice Age ended.

Cycles of the Pleistocene Ice Age and Orbital Variations

The movements of continents can be used to explain climate changes over periods of tens or hundreds of millions of years. However, these processes are too slow to explain the more rapid onset and retreat of the Pleistocene Ice Age and the glacial fluctuations during that time.

One explanation for Pleistocene glacial advances and retreats is that the Earth's temperature may change in response to variations in the Earth's orbit. Three types of orbital variations occur (Fig. 13–25):

1. The shape of the Earth's orbit around the Sun changes on about a 100,000-year cycle. This is known as **eccentricity**.
2. The Earth's axis is currently tilted at about 23.5° with respect to a line perpendicular to the plane of its orbit around the Sun. The **tilt** changes by 1.5° on about a 41,000-year cycle.
3. The Earth's axis, which now points directly toward the North Star, circles like a spinning top. This circling, called **precession**, completes a full cycle every 23,000 years.

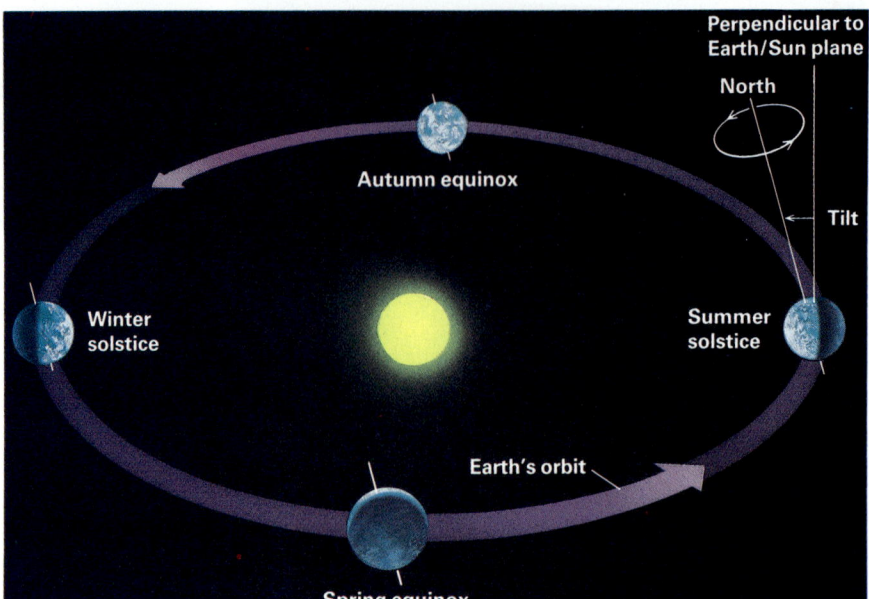

FIGURE 13-25
Earth orbital variations may explain the temperature oscillations and glacial advances and retreats during the Pleistocene Epoch. Orbital variations occur over time spans of tens of thousands of years.

Orbital changes do not greatly affect the total *amount* of solar energy that strikes the Earth, but they do affect the *distribution* of sunlight with respect to latitude and season. Seasonal changes in the amount of sunlight reaching the Earth can cause cooler summers, which lead to glacier growth. One important factor is the summer temperature. In high latitudes, snow falls during the winter, even during interglacial periods. If summers are hot, the snow melts, but if summers are cool, the snow accumulates, leading to growth and spreading of glaciers.

Early in the twentieth century, a Yugoslavian astronomer, Milutin Milankovitch, plotted the cycles of the three different orbital variations. He then calculated their combined effects on the Earth's climate. The calculations showed that the three orbital variations combine to generate alternating periods of warm and cool summers in higher latitudes. Moreover, the timing of the calculated warming and cooling episodes coincide with the known timing of Pleistocene glacial advances and retreats. Some scientists agree with the Milankovitch theory, but others criticize it. Scientists continue to debate the causes of climate change and glaciation.

SUMMARY

If snow survives through one summer, it converts to granular ice crystals called **firn**. A **glacier** is a thick mass of ice that forms on land and flows plastically. A glacier forms wherever winter snowfall exceeds summer melting. **Alpine glaciers** occur in mountainous regions; a **continental glacier** covers a large part of a continent. A glacier moves by both **basal slip** and **plastic flow**. The upper 40 to 50 meters of a glacier are too brittle to flow, and large cracks called **crevasses** develop in this layer.

In the **accumulation area** of a glacier, more snow falls than melts, whereas in the **ablation area**, melting exceeds accumulation. The **snowline**, or **firn line**, is the boundary between permanent and seasonal snow. The end of a glacier is called the **terminus**.

Glaciers scour both the bottoms and the sides of their valleys, giving the valleys a characteristic U shape. Alpine glaciers create **cirques**, steep-sided depressions eroded into peaks; **arêtes**, thin, sharp ridges separating two glaciated valleys; **horns**, pyramid-shaped peaks formed by the intersection of three or more cirques; and **hanging valleys**, formed by tributary glaciers that "hang" high above the floor of a larger valley.

Drift is any rock or sediment transported and deposited by a glacier. The unsorted drift deposited directly by a glacier is **till**. Most glacial terrain is characterized by large mounds of till known as **moraines**. **Outwash** is **stratified drift** deposited by streams flowing from the terminus of a glacier.

At least five major **ice ages** occurred during the past 2.5 billion years. The most recent happened during the Pleistocene Epoch, when both alpine and continental glaciers created many topographic features that are prominent today.

KEY TERMS

Firn 318
Glacier 318
Alpine glacier 318
Ice sheet 318
Continental glacier 319
Ice age 319
Basal slip 320
Plastic flow 320
Crevasse 320
Accumulation area 322
Ablation area 322

Firn line 322
Terminus 322
Glacial striation 322
U-shaped valley 323
Cirque 323
Horn 323
Arête 324
Hanging valley 324
Fjord 325
Drift 326
Till 326

Stratified drift 326
Moraine 327
End moraine 327
Terminal moraine 327
Outwash 329
Kame 329
Esker 329
Glacial epoch 329
Pleistocene Ice Age 330
Braided stream 332
Channeled scablands 333

REVIEW QUESTIONS

1. Differentiate between an alpine glacier and a continental glacier. Where are alpine glaciers found today? Where are continental glaciers found today?

2. Distinguish between basal slip and plastic flow.

3. Why are crevasses only about 40 to 50 meters deep, even though many glaciers are much thicker?

4. Describe the surface of a glacier in the summer and in the winter in (a) the accumulation area and (b) the ablation area.

5. How do icebergs form?

6. Describe how glacial erosion can create (a) a cirque, (b) striated bedrock, and (c) an arête.

7. Describe the formation of arêtes, horns, and hanging valleys.

8. Why are kames and eskers features of receding glaciers?

9. Describe four types of topographic features left behind by the continental ice sheets.

DISCUSSION QUESTIONS

1. Outline the changes that would occur in a glacier if (a) the average annual temperature rose and precipitation decreased; (b) the temperature remained constant but precipitation increased; and (c) the temperature decreased and precipitation remained constant.

2. In some regions of northern Canada, both summer and winter temperatures are cool enough for glaciers to form. Speculate on why continental glaciers are not forming in the region.

3. In the 1980s and early 1990s, global temperature was rising. Some climatologists have argued that global warming could lead to increased glaciation. Give a plausible mechanism for this scenario.

4. In some mountain ranges, the tops of mountain peaks are jagged and covered by rubble, whereas the lower elevations of the mountains are rubble-free. Give a plausible explanation for these observations.

5. Imagine that you encountered some gravelly sediment. How would you determine whether it was deposited by a stream or a glacier?

6. Compare and contrast erosion, transport, and deposition of sediment by wind, streams, mass wasting, and glaciers.

CHAPTER 14
Deserts and Desertification

14.1 Deserts of the World
14.2 Water in Deserts
14.3 Wind Erosion
14.4 Dunes
14.5 Expansion and Contraction of Deserts
14.6 Land Degradation in Desert and Semiarid Lands

The word *desert* commonly evokes an image of thirsty travelers crawling across an infinity of lifeless sand dunes. This image accurately depicts some deserts, but not others. Many deserts are rocky and even mountainous, with colorful cliffs or peaks towering over plateaus and narrow canyons. Cactus, sage, grasses, and other plants may dot the landscape. After a rainstorm, millions of multicolored flowers bloom and color the desert. In winter, a thin layer of snow may cover the ground. A **desert** is defined as any region that receives less than 20 centimeters (8 inches) of rain per year and consequently supports little or no vegetation. Most deserts are surrounded by **semiarid** zones that receive more moisture than a true desert but less than surrounding regions.

Although some particularly dry deserts are nearly lifeless, most deserts receive enough rainfall to support sparse vegetation. Desert plants employ two different types of strategies. Annual plants grow and produce seeds quickly in response to sporadic rainfall. Then the plants die and their seeds lie dormant until the soil becomes moist again. In contrast, perennial desert plants grow slowly and conserve moisture. For example, cacti have neither leaves nor stems, thus exposing minimal surface area to the dry desert air. The stalks have a thick, nearly waterproof skin that further conserves water. Desert animals have also adapted to their dry environment. For example, the African oryx, a large antelope, varies its metabolism, storing heat during the day and releasing it at night. The North American kangaroo rat feeds on dry seeds; it conserves water by producing highly concentrated urine.

Traditionally, many desert societies were nomadic. Small groups traveled across the land between water holes with flocks of camels, sheep, and goats, moving on before the animals overgrazed the vegetation. Other desert cultures developed irrigation systems to raise crops close to rivers

Courthouse Wash in Arches National Park, Utah, is a dry stream bed for much of the year.

338   Chapter 14   Deserts and Desertification

A

B

FIGURE 14–1
(A) Some deserts are rocky. Arches National Park, Utah. (B) Others are sandy. Lago Poopó, Bolivia.

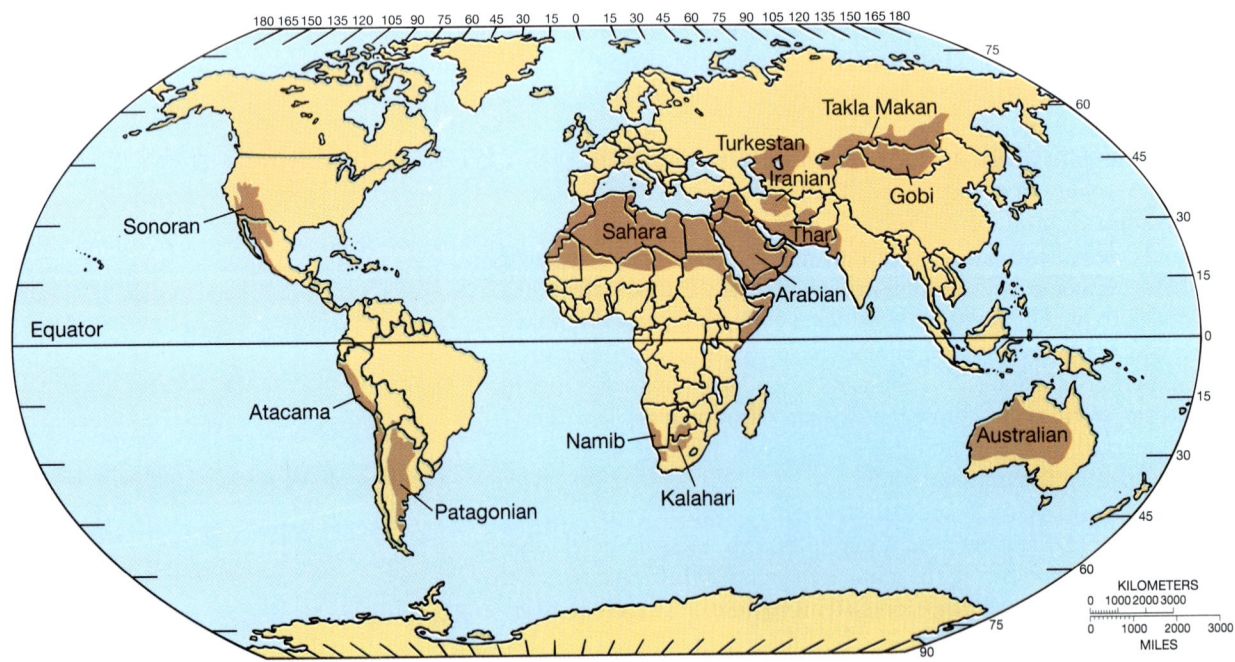

FIGURE 14–2
The major deserts of the world highlighted in brown concentrate near 30° north and south latitudes. Most of the deserts are surrounded by semiarid lands.

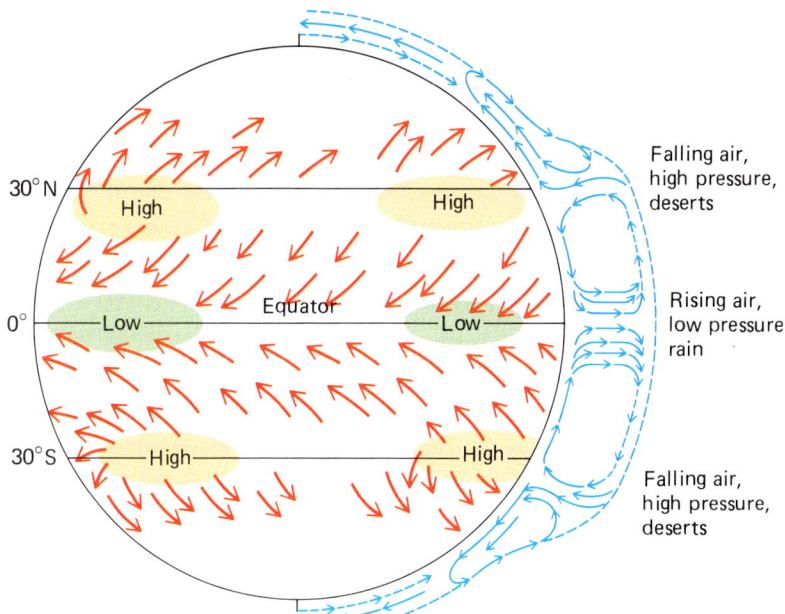

FIGURE 14-3
Global wind patterns favor development of deserts at 30° north and south latitudes. The arrows drawn inside the globe indicate surface winds. The arrows to the right show both vertical and horizontal movement of air on the surface and at higher elevations.

and wells. Today, advanced irrigation systems have improved human adaptation to dry environments and enabled 13 percent of the world's population to live in deserts. Two thirds of the world's crude oil lies beneath the deserts of the Middle East, transforming some of the poorest nations of the world into the richest. Perhaps in the future, vast arrays of solar cells will convert desert sunlight to electricity.

14.1 Deserts of the World

If you were to visit the great deserts of the Earth, you would see coastal deserts along the beaches of Chile, shifting dunes in the Sahara, deep red sandstone canyons in southern Utah, and stark granite mountains in Arizona (Fig. 14–1). The world's deserts are similar to one another only in that they all receive scant rainfall.

The Effect of Latitude

About 25 percent of the Earth's land surface outside the polar regions is desert (Fig. 14–2). Most of these arid regions lie at about 30° both south and north of the equator. Why do deserts concentrate at these latitudes?

The Sun shines most directly near the equator, warming the air near the Earth's surface. The warm air absorbs moisture from the equatorial oceans. This warm, moist air rises and cools. The cooling condenses the water vapor, which falls as rain. For this reason, vast tropical rainforests grow near the equator. The air, which is now drier because of the loss of moisture, flows northward and southward at high altitudes. This air falls back toward the Earth's surface at about 30° north and south latitudes (Fig. 14–3). As air falls, it is compressed. The compression heats the air and enables it to hold more water vapor. As a result, water evaporates from the surface into the air. Because the falling air absorbs water, rainfall is infrequent and deserts are common at about 30° north and south latitudes.

Effect of Topography: Rain-Shadow Deserts

When moisture-laden air flows over a mountain range, it rises. As the air rises, it cools and the water vapor frequently condenses into droplets that fall as rain or snow. These conditions cause abundant precipitation on the windward side and the crest of the range. After the air has passed the crest, it flows down the leeward (or downwind) side of the mountains (Fig. 14–4). This air has already lost much of its moisture. As in the case of sinking air at 30° latitude, the air is compressed as it falls, creating a dry, high-pressure zone called a **rain-shadow desert** on the leeward side of the range. Figure 14–5 shows that leeward valleys in California are much drier than the mountains to the west.

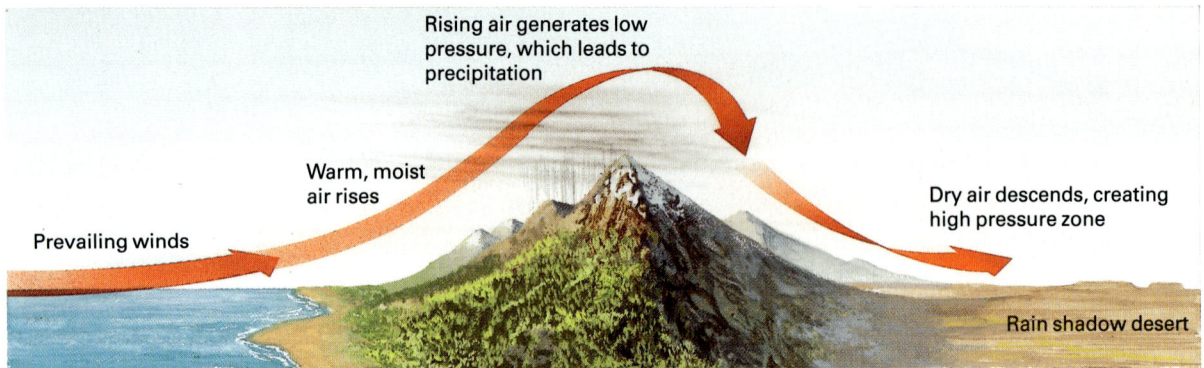

FIGURE 14–4
Warm, moist air from the ocean rises as it flows over a mountain range. As it rises, it cools, and water vapor condenses to form rain on the windward side and crest of the range. A rain-shadow desert develops where the dry, descending air on the lee side of the range absorbs moisture.

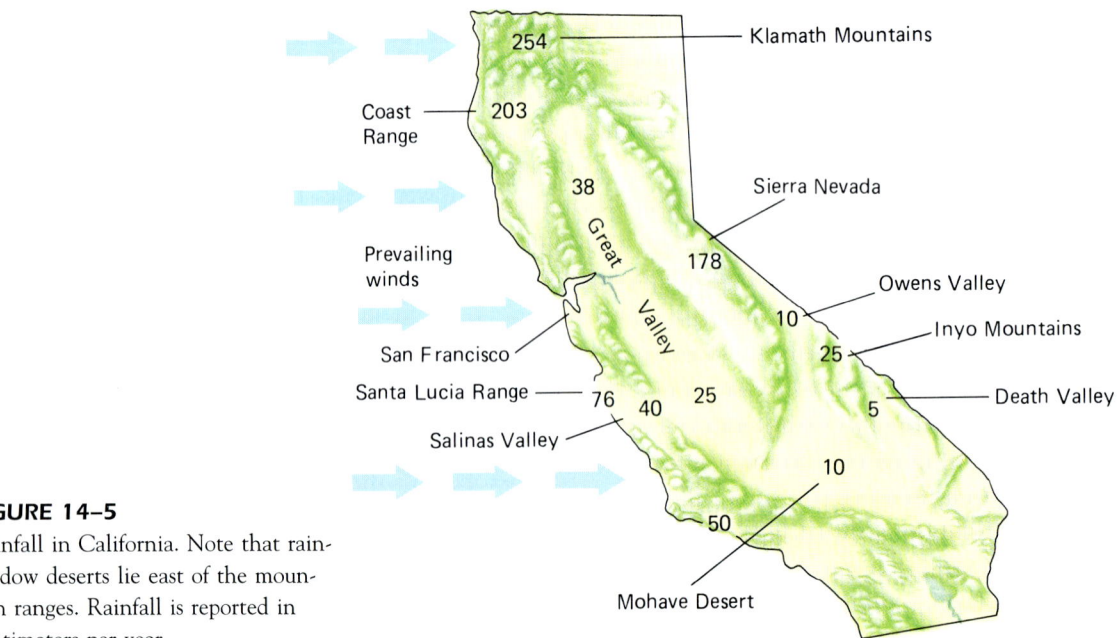

FIGURE 14–5
Rainfall in California. Note that rain-shadow deserts lie east of the mountain ranges. Rainfall is reported in centimeters per year.

Continental Interiors

Because most evaporation occurs over the oceans, one might expect that coastal areas would be moist and climates would become drier with increasing distance from the sea. This generalization is often true, but notable exceptions exist. Winds carry moisture to many continental interiors, and some ocean currents produce coastal deserts.

The Takla Makan Desert in western China is the driest continental interior desert in the world. In the Turlan Depression, 154 meters below sea level, rainfall is less than 2.5 centimeters (1 inch) per year. The Takla Makan is not only in the interior of Asia, the world's largest continent, but it is also almost completely surrounded by high mountains. As a result, the rain-shadow effect adds to the dryness of the region.

Deserts Formed by Cold Ocean Currents

The Atacama Desert on the west coast of South America is so dry that portions of Peru and Chile have received no rainfall for a decade or more. The desert exists because cool ocean currents flow along the west coast of South America, chilling the sea air. When the cool marine air encounters warm land, the air becomes warmer. The warm air absorbs moisture, creating a coastal desert. The Namib Desert on the west coast of Africa is similarly formed by cold ocean currents. Coastal deserts such as the Atacama and the Namib receive more moisture from fog than they do from rain.

14.2 Water in Deserts

Desert Streams

Large rivers flow through some deserts. For example, the Colorado River crosses the arid southwestern United States (Fig. 14–6), and the Nile flows through the North African desert. Desert rivers receive most of their water from wet, mountainous areas bordering the arid lands. Thus, they flow continuously all year long.

FIGURE 14–7
Courthouse Wash, Utah. Although the wash is dry, the vegetation along its banks is lush compared with the cactus-covered hillside beneath the rocks in the background.

In contrast, most smaller desert streams flow for only part of the year. Recall from Chapter 9 that in a moist environment, ground water continuously feeds streams. In a desert, however, the water table is often so low that water seeps out of the stream bed into the ground below. As a result, many desert streams flow for only a short time after a rainstorm, or during spring when winter snow is melting. A stream bed that is dry for most of the year is called a **wash** (Fig. 14–7).

Because ground water can flow for hundreds of kilometers from a wet mountain range into a desert, wells and springs are also important sources of water in deserts. Although natural processes recharge some desert ground-water reservoirs rapidly, other ground-water reservoirs are called "fossil water" because they are recharged extremely slowly or were originally filled when the climate was wetter. When modern pumps extract large amounts of fossil water rapidly to irrigate farmland or supply a city, the aquifer may dry up in a few decades. This phenomenon is occurring today in Phoenix, Tucson, Las Vegas, and other large cities of the Great American Desert.

FIGURE 14–6
The Colorado River has cut Grand Canyon through the Arizona desert.

Today geologists study satellite data to locate ground water in remote areas. Geologists from a World Bank development team used Earth-penetrating radar to reveal images of old river beds under the shifting sands of the Sahara. They located an aquifer in the desert in northern Egypt, and proposed drilling wells and expanding irrigation to enrich the region. Village elders asked the geologists whether the underground water was like the Nile, which flows continuously, or like an oil well, which will someday dry up. When the geologists explained that the water was more like an oil reservoir, the village council said that they did not want the well. They argued that if their children and grandchildren became accustomed to a plentiful water supply, they would lose the skills needed for survival when the wells went dry. The villagers were proponents of **sustainable development**, development that lasts indefinitely. On the other hand, the World Bank in this instance advocated development of the aquifer, even though it is not sustainable. The argument between these opposing viewpoints applies to many environmental issues.

Flash Floods and Debris Flows

Consider what happens when rain falls on a desert. Little vegetation is present to absorb moisture or to protect the soil from erosion. Much of the land surface may be covered by bare rock. Therefore, the rain does not soak in as fast as it falls, and the excess runs over the surface into gullies and washes. During a heavy rainstorm, a dry stream bed may fill with water so rapidly that a **flash flood** occurs. A flash flood is one in which flooding occurs suddenly and the water then recedes rapidly. Occasionally, novices to desert camping pitch their tents in washes, where they find soft sand to sleep on and shelter from the wind. However, if a thunderstorm occurs upstream during the night, a flash flood may fill the wash with a wall of water mixed with rocks and boulders. People have been killed by such floods. By mid-morning of the next day, the wash may contain only a tiny trickle, and within 24 hours it may be completely dry again. In contrast, in a large river such as the Mississippi, flood waters rise and fall over a period of days or weeks, not hours or minutes.

Playas

During the wet season, streams, ground water, and rain fill a desert lake. However, during the dry season, little water flows into the lake while water evaporates and seeps into the ground. If the water loss is great enough, the lake dries up completely. A desert lake that dries up periodically is called a **playa lake**, and the dry lake bed is called a **playa** (Fig. 14–8). Recall from Chapters 9 and 10 that water dissolves ions from rock and soil. When this slightly salty water accumulates in a playa lake and then evaporates, the ions precipitate to deposit minerals on the playa. Over years of repetition of this process, economically valuable **evaporite deposits**, such as those of Death Valley, may accumulate. Examples include salt, borax (a boron mineral used in glass, ceramics, agricultural chemicals, and pharmaceuticals), and gypsum (hydrated calcium sulfate used in cement and wallboard) (Fig. 14–9).

FIGURE 14–8

Mudcracks pattern the floor of a playa in Utah.

FIGURE 14–9

Borax and other valuable minerals are abundant in the evaporite deposits of Death Valley. Mule teams hauled the ore from the valley in the 1800s. *(U.S. Borax)*

FIGURE 14–10
(A) Wind erodes silt and sand, but leaves larger rocks behind to form desert pavement. (B) Desert pavement is a continuous cover of stones left behind when wind blows silt and sand away.

One can hardly imagine strong desert winds without thinking about a hide-behind-your-camel type of sandstorm. Wind erosion, called **deflation**, is a selective process. Because air is much less dense than water, wind can move only small particles such as sand and silt. Thus, wind erosion leaves pebbles and rocks behind to form a continuous cover of stones called **desert pavement** (Fig. 14–10). Once desert pavement forms, the supply of sediment is cut off. Thus, most deserts are rocky and covered with gravel, and sand dunes are relatively rare.

Wind can erode and carry only the smallest and lightest particles. Sand grains, which are relatively large and heavy, are usually lifted less than 1 meter in the air and are transported only a short distance. In contrast, wind carries fine silt in suspension. Skiers in the Alps commonly encounter a silty surface on the snow, blown from the Sahara Desert and carried across the Mediterranean Sea. The even smaller clay particles stick together, and consequently wind does not move them.

Wind by itself is not abrasive enough to erode rocks, but windblown sand and silt are effective agents of erosion. Because wind carries sand only a meter or less above the ground, abrasion concentrates close to the ground. If you see a tall desert pinnacle topped by a delicately perched cap, it was probably *not* created by wind erosion. The pinnacle is too high. However, if the base of a pinnacle is sculpted, windblown sand is probably the responsible agent (Fig. 14–11).

FIGURE 14–11
Wind abrasion selectively eroded the base of this rock because windblown sand moves mostly near the ground surface. Lago Poopó, Bolivia.

14.3 Wind Erosion

When wind blows through a forest or across a grassy prairie, the trees and grass protect the soil from erosion. In addition, in moist environments, rain commonly accompanies a windstorm. The moisture binds soil particles together, so little wind erosion occurs. In contrast, seacoasts, recently glaciated areas, and deserts commonly have little or no vegetation to protect the soil. Therefore, wind erodes soil in these environments.

FIGURE 14–12
Sand dunes are abundant on the Oregon coast.

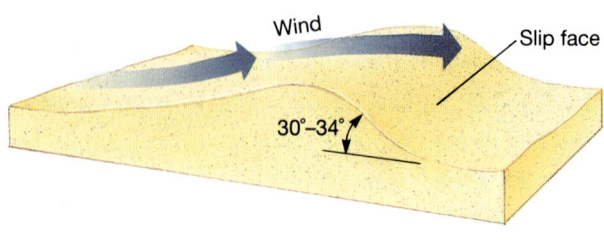

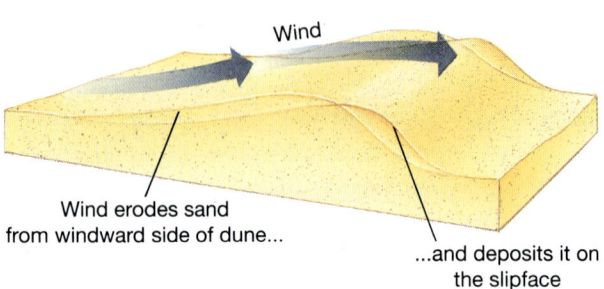

FIGURE 14–14
A dune migrates in a downwind direction as wind erodes sand from the dune's windward side and deposits it on the slip face.

14.4 Dunes

Mounds or ridges of windblown sand called **dunes** cover approximately 20 percent of the world's desert land. The largest dune field is the Rub Al Khali (Empty Quarter) in Arabia, which covers 560,000 square kilometers, larger than the state of California. Dunes also form in environments where vegetation is scarce, sand is plentiful, and wind is common—such as regions where glaciers have melted recently and along sandy coastlines (Fig. 14–12).

If wind blows over a rock or a small clump of vegetation, the downwind side of the obstacle provides a small, sheltered area where the wind slows down. Grains of sand settle out in this protected zone. The growing mound of sand creates a larger windbreak, and more sand accumulates, forming a dune. Dunes commonly grow to heights of 30 to 100 meters, and some giants exceed 500 meters (Fig. 14–13). In places they are tens or even hundreds of kilometers long.

Most dunes are asymmetrical. Wind erodes sand from the windward side of a dune, and then the sand slides down the lee side, where it accumulates. In this way, a dune migrates in the downwind direction (Fig. 14–14). The leeward face of a dune is called the **slip face**. Typically the slip face is about twice as steep as the windward face. Wind speed, sand supply, and vegetation all influence the shape and orientation of a dune.

Migrating dunes can overrun buildings and highways. For example, U.S. Highway 95 runs across the Nevada desert. Dunes advance across the highway near the town of Winnemucca several times a year (Fig. 14–15). Highway crews must remove as much as 4000 cubic meters of sand to reopen the road.

FIGURE 14–13
Desert dunes. Coral Pinks State Park, Utah.

FIGURE 14–15
Windblown sand frequently blocks Highway 95 about 3 miles north of Winnemucca, Nevada.

Attempts are often made to stabilize dunes in inhabited areas. One solution is to plant vegetation to reduce deflation and stop dune migration. The main problem with this approach is that desert dunes commonly form in regions that are too dry to support vegetation. Another approach to dune control is to build artificial windbreaks to create dunes in places where they do the least harm. For example, a fence traps blowing sand and forms a dune, thereby protecting areas downwind. Fencing is a temporary solution, however, because eventually the dune covers the fence and migrates. In Saudi Arabia, engineers sometimes stabilize dunes by covering them with tarry wastes from petroleum refining.

14.5 Expansion and Contraction of Deserts

Rocks and landforms contain many clues to ancient climates. By studying these clues, geologists have learned that deserts form, grow, shrink, and disappear over time. When other sediment buries dunes and the sand becomes lithified, the resulting sandstone retains the original sedimentary structures of the dunes. For example, Figure 14–16 shows a rock face in Zion National Park in Utah. Notice the steeply dipping sedimentary layers. This rock has not been tilted by tectonic forces. It is a lithified dune, and the dipping beds are the layering of the dune's slip face. The bedding dips in the direction in which the wind was blowing when it deposited the sand. The layers dip in various directions because wind directions changed. Because dunes can form on beaches or near melting glaciers, fossil dunes do not necessarily indicate that the region was once a desert.

FIGURE 14–17
Vertical rills, formed by running water, indicate that the Sphinx was built in a wet climate. The Chephren Pyramid, in the background, only shows evidence of wind erosion. Thus geologists conclude that the pyramid was built at a later date, in a drier climate. *(Cameramann International)*

The Sphinx lies on the banks of the Nile River adjacent to other monuments built by the ancient Egyptians (Fig. 14–17). Many Egyptologists assumed that all of these structures were built at the same time. However, a geologist noticed that small vertical rills covered the top and sides of the sphinx, whereas adjacent structures showed only horizontal grooves close to the desert surface. Running water eroded the rills, but windblown sand formed the horizontal grooves near the desert floor. Thus, by studying weathering patterns, scientists learned that the Sphinx was built before adjacent monuments, at a time when the climate was wetter. This conclusion that North Africa was once wetter than it is today is supported by radar images showing ancient river channels beneath the dunes of the Sahara.

Major regional climatic changes have occurred as a result of tectonic plate movements. Two hundred million years ago, Great Britain lay at 30° north latitude, the latitude of the modern Sahara Desert. As a result, the 200-million-year-old New Red Sandstone of the Midlands is composed of lithified ancient dunes. Deserts also shrink and grow because of more rapid climate changes resulting from variations in Earth's orbit, solar output, greenhouse effects, changes in ocean currents, and other meteorological conditions.

FIGURE 14–16
Cross-bedded sandstone in Zion National Park preserves the sedimentary bedding of ancient sand dunes.

14.6 Land Degradation in Desert and Semiarid Lands

Six thousand years ago, advanced civilizations thrived along desert rivers such as the Tigris-Euphrates and the Nile. The Babylonians and Egyptians used buckets, wheels, canals, and pumps to irrigate fields near the rivers. Desert nomads herded their flocks in the semiarid grasslands that lay between the rivers and deserts (Fig. 14–18). Prior to the twentieth century, nomads moved across deserts and semiarid lands with little regard for national boundaries, traveling with the seasons and abandoning an area after it had been grazed for a short period of time. This constant movement prevented overgrazing. In addition, population levels of the nomadic tribes were stable and low.

The Sahara is the largest desert on the planet. South of the Sahara lies the semiarid Sahel. During the 1960s, unusually heavy rains caused the Sahel to bloom. People expanded their flocks to take advantage of the additional forage. Rich countries contributed foreign aid. As a result, medical attention and sanitation improved, and the human population grew dramatically. Many people predicted a new era of prosperity for the Sahel, but the favorable rains were an anomaly. In the late 1960s and early 1970s, drought destroyed the range. During this period, governments in North Africa began to enforce

FIGURE 14–19
Cattle in the Sahel have eaten all ground vegetation, leaving the soil bare and vulnerable to erosion. Famine often follows large-scale range degradation in Africa. *(R. E. Ford/Terraphotographics/Biological Photo Service)*

FIGURE 14–18
A tent camp of desert nomads in Niger, North Africa. In earlier times, the herders moved with the seasons, alternately grazing a portion of the range and then leaving it idle so the grasses could regenerate. When nomads become settled, their cattle frequently overgraze the fragile semiarid ecosystems. *(Jason Laure)*

national borders more strictly, curtailing nomadism. Civil and international war brought instability to the region, and famine struck (Fig. 14–19). Reports issued in the 1970s and 1980s claimed that the Sahara was advancing steadily southward into the Sahel at a rate of 5 kilometers per year. Scientists argued that overgrazing, farming, and firewood gathering had caused the desert advance. This growth of the desert caused by human mismanagement has been called **desertification**.

More recent research has shown that overgrazing a semiarid region causes land degradation but does not cause a desert to grow. The Sahel–Sahara desert boundary is clearly visible on satellite photographs as a line of vegetation. Researchers plotted changes in the size of the desert by studying satellite photographs taken be-

14.6 Land Degradation in Desert and Semiarid Lands 347

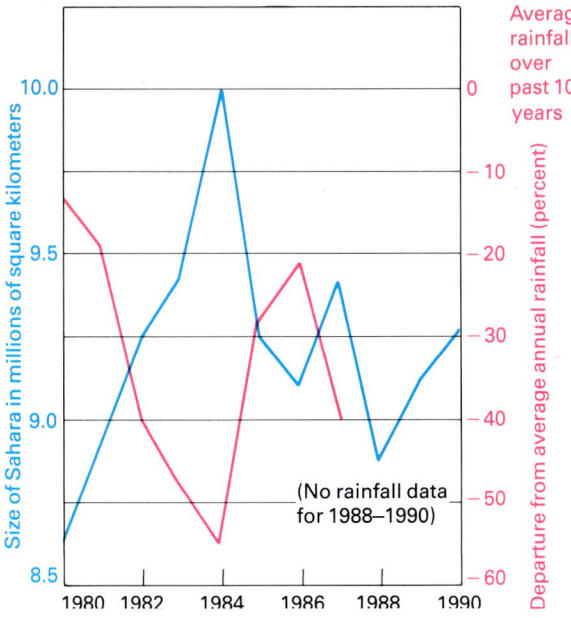

FIGURE 14–20
Expansion and contraction of the Sahara Desert from 1980 to 1990. Note that the Sahara has expanded (blue line) when rainfall has decreased (red line). Thus, rainfall, not land management, is the prime factor governing the size of the Sahara. (Compton J. Tucker, Harold E. Dregne, and Wilbur W. Newcomb, "Expansion and Contraction of the Sahara Desert from 1980 to 1990," *Science* 253 [July 19, 1991], 299ff)

tween 1980 and 1990. As shown in Figure 14–20, the desert expands (blue line) when rainfall declines (red line). Thus, decreasing rainfall, not overgrazing, was responsible for the expansion of the desert.

In another project, researchers studied natural and overgrazed range in semiarid grasslands in southern New Mexico.[1] The natural range consisted mainly of a homogeneous blanket of grass. In regions where cattle had overgrazed the land, woody creosote and mesquite had replaced the grass. The net primary productivities (the total plant growth) in the natural and overgrazed systems were nearly identical; thus, overgrazing did not reduce plant growth. It did, however, change the types of plants and their spatial distribution. The shrub invasion caused by excess grazing reduced the economic value of the range.

Grasses have shallow roots that absorb falling rain. Excessive grazing destroys the grasses as the cattle eat the grass down to the roots. In addition, the heavy animals pack the soil with their hooves, blocking the natural seepage of air and water. When soil is devoid of vegetation and baked in the sunlight, it becomes so impermeable that water evaporates or runs off before it soaks in. Increased runoff erodes the soil and carries off nutrients. The water, soil, and nutrients collect in shallow depressions and desert washes. In some regions, shrubs accumulate in these moist, nutrient-rich low points. Thus, the original range had a homogeneous distribution of plants, water, and nutrients, whereas the overgrazed range was heterogeneous, alternating thick shrubbery with bare ground (Fig. 14–21). This redistribution of soil and nutrients alters conditions so that the original grassland ecosystem cannot become reestablished quickly or easily. Thus, range deterioration is long lasting, and the limited productivity of the land declines.

[1] William H. Schlesinger et al., "Biological Feedbacks in Global Desertification," *Science* 247 (March 1990), 1043.

FIGURE 14–21
Sagebrush with bare ground between the bushes indicates that rangeland has been overgrazed. *(Dave Stoecklein)*

SUMMARY

Deserts have an annual precipitation of less than 20 centimeters (8 inches) and are usually surrounded by **semiarid** regions. The world's largest deserts occur near 30° north and south latitudes, where warm, dry, descending air absorbs moisture from the land. Deserts also occur in the rain shadows of mountains, in continental interiors, and in coastal regions adjacent to cold ocean currents.

Desert streams are often dry for much of the year but may develop **flash floods** when rainfall occurs. **Playa lakes** are desert lakes that dry up periodically, leaving abandoned lake beds called **playas**.

Wind is an important agent of erosion, transport, and deposition of sediment in environments where little vegetation is present to protect soil. Sparse vegetation is common in deserts, on sandy coastlines, and in regions recently abandoned by glaciers. Wind carries sand grains short distances at a meter or less above the ground, but it can transport silt great distances at higher elevations. Windblown particles are abrasive, but because the heaviest grains travel close to the surface, abrasion occurs mainly near ground level.

A **dune** is a mound or ridge of wind-deposited sand. Most dunes are asymmetrical, with gently sloping windward sides and steeper **slip faces** on the lee sides. Dunes migrate in a downwind direction. Deserts have expanded, contracted, and migrated throughout geological time. Overgrazing does not alter the total vegetation in a desert or semiarid range, but it does degrade the economic value of the landscape.

KEY TERMS

Desert 337
Semiarid 337
Rain-shadow desert 339
Wash 341
Sustainable development 342

Flash flood 342
Playa lake 342
Playa 342
Evaporite deposit 342
Deflation 343

Desert pavement 343
Dune 344
Slip face 344
Desertification 346

REVIEW QUESTIONS

1. Why are many deserts concentrated along zones at 30° latitude in both the Northern and Southern Hemispheres?
2. List and discuss four conditions that produce deserts.
3. Why do flash floods and debris flows occur in deserts?
4. Compare and contrast floods in the desert with those in more humid environments.
5. Why is wind erosion more prominent in deserts and on sandy coastlines than in humid regions?
6. Describe the formation of desert pavement.
7. What is meant by the term *desertification*? Why is land degradation a better term for the changes that occur when a semiarid range is overgrazed?
8. Discuss the changes that occur in a semiarid range when it is overgrazed.

DISCUSSION QUESTIONS

1. Coastal regions include some of the wettest and some of the driest environments on Earth. Briefly outline the climatological conditions that produce coastal rainforests versus coastal deserts.
2. Explain why soil moisture content might be more useful than total rainfall in defining a desert. How could one region have a higher soil moisture content and lower total rainfall than another region?
3. Imagine that you live on a planet in a distant solar system. With no prior information on the topography or climate of the Earth, you design a robotic spacecraft to land on Earth. The spacecraft has arms that can reach out a few me-

ters from the landing site to collect material for chemical analysis. It also has instruments to measure the immediate meteorological conditions and cameras that can focus on anything within a range of 100 meters. The batteries on your radio transmitter have a life expectancy of 2 weeks. The spacecraft lands and you begin to receive data. What information would convince you that the spacecraft has landed in a desert?

4. Deserts are defined as areas with low rainfall, yet water is an active agent of erosion in desert landscapes. Explain this apparent contradiction.

5. Compare and contrast erosion, transport, and deposition by wind with erosion, transport, and deposition by streams and glaciers.

6. Imagine that you are looking at a satellite photograph of a distant planet. What deductions can you make if you see numerous sand dunes?

7. What type of environment would produce fossilized seashells embedded in lithified sand dunes?

8. Discuss two types of tectonic change that could produce deserts in previously humid environments.

9. Explain why the original vegetation may be slow to become reestablished once a range is degraded by overgrazing.

10. Explain how overgrazing could affect soil moisture content.

PART 4

Mineral and Energy Resources

15 Mineral Resources and Mining
16 Fossil and Uranium Fuels
17 Alternative Energy Resources

The setting Sun illuminates a field of oil drilling towers. (SuperStock)

CHAPTER 15
Mineral Resources and Mining

15.1 Geologic Resources
15.2 How Ore Forms
15.3 Mining and Refining of Metals
15.4 Future Availability of Minerals

Many animals use tools. Apes use sticks as back scratchers and occasionally as clubs. Woodpecker finches on Ecuador's Galapagos Islands use small twigs to prod insects from trees. But no other species uses tools to the extent that humans do, and no other animal controls fire. Tools and fire are essential to civilization. Prehistoric people made wooden tools, and wood was the major fuel of antiquity. Prehistoric people also used geologic resources such as flint and obsidian to make weapons and hide scrapers. About 7000 B.C., people learned to bake clay in a fire to make pottery. Archaeologists have found copper trinkets in Turkey that date to 6500 B.C.; 1500 years later, copper farm implements and weapons were widely used in Mesopotamia. With time, human use of geological resources became increasingly sophisticated. Today, the silicon chip that operates your computer, the titanium valves in a space probe, and the gasoline that powers your car are all derived from the Earth.

Heavy equipment working in an open pit mine. (SuperStock)

15.1 Geologic Resources

In Chapter 1 we defined a resource as any source of raw materials. Geologic resources fall into two major categories.

1. *Energy resources.* Petroleum, coal, and natural gas are called **fossil fuels** because they formed from the remains of plants and animals that lived in the geologic past. Uranium is the basic fuel for nuclear reactors. Energy resources are discussed in Chapter 16.
2. *Mineral resources.* Mineral resources can be further subdivided into two categories, **metals** and **nonmetallic resources.** A metal is any chemical element characterized by a metallic luster, ductility, and the ability to conduct electricity and heat.

About 40 metals are commercially important. Some, such as iron, lead, copper, aluminum, silver, and gold, are familiar to all of us (Fig. 15–1). Others, such as vanadium, titanium, and tellurium, are less well known but are vital to industry.

A nonmetallic resource is any useful material that does not have metallic properties. Sand and gravel are mined for road building and for the manufacture of concrete. Limestone is used to produce cement. Granite, limestone, marble, and other rocks are quarried for use as building stone. Phosphorus and potassium are essential fertilizers extracted from certain types of rocks. Sulfur is used widely in the chemical industry. Clay minerals are baked to produce ceramics (Table 15–1).

FIGURE 15–1

In the early 1900s, miners extracted gold, copper, and other metals from underground mines such as this one, 600 meters below the surface in Butte, Montana. *(Montana Historical Society)*

TABLE 15-1 Some Important Elements and Their Uses

Mineral	Type	Some Uses
Aluminum (Al)	Metal	Structural materials (airplanes, automobiles), packaging (beverage cans, toothpaste tubes), fireworks
Borax ($Na_2B_4O_7$)	Nonmetal	Diverse manufacturing uses—glass, enamel, artificial gems, soaps, antiseptics
Chromium (Cr)	Metal	Chrome plate, pigments, steel alloys (tools, jet engines, bearings)
Cobalt (Co)	Metal	Pigments, alloys (jet engines, tool bits), medicine, varnishes
Copper (Cu)	Metal	Alloy ingredient in gold jewelry, silverware, brass, and bronze; electrical wiring, pipes, cooking utensils
Gold (Au)	Metal	Jewelry, money, dentistry, alloys, specialty electronics
Gravel	Nonmetal	Concrete (buildings, roads)
Gypsum ($CaSO_4 \cdot 2H_2O$)	Nonmetal	Plaster of Paris, wallboard, soil treatments
Iron (Fe)	Metal	Basic ingredient of steel (buildings, machinery)
Lead (Pb)	Metal	Pipes, solder, battery electrodes, pigments
Magnesium (Mg)	Metal	Alloys (aircraft), firecrackers, bombs, flashbulbs
Manganese (Mn)	Metal	Steel, alloys (steamship propellers, gears), batteries, chemicals
Mercury (Hg)	Liquid metal	Thermometers, barometers, dental inlays, electric switches, streetlamps, medicine
Molybdenum (Mo)	Metal	Steel alloys, lamp filaments, boiler plates, rifle barrels
Nickel (Ni)	Metal	Money, alloys, metal plating
Phosphorus (P)	Nonmetal	Medicine, fertilizers, detergents
Platinum (Pt)	Metal	Jewelry, delicate instruments, electrical equipment, cancer chemotherapy, industrial catalyst
Potassium (K)*	Nonmetal	Salts used in fertilizers, soaps, glass, photography, medicine, explosives, matches, gunpowder
Common salt (NaCl)	Nonmetal	Food additive
Sand (largely SiO_2)	Nonmetal	Glass, concrete (buildings, roads)
Silicon (Si)	Nonmetal	Electronics, solar batteries, ceramics, silicones
Silver (Ag)	Metal	Jewelry, silverware, photography, alloys
Sulfur (S)	Nonmetal	Insecticides, rubber tires, paint, matches, papermaking, photography, rayon, medicine, explosives
Tin (Sn)	Metal	Cans and containers, alloys, solder, utensils
Titanium (Ti)	Metal	Paints; manufacture of aircraft, satellites, and chemical equipment
Tungsten (W)	Metal	High-temperature applications, light bulb filaments, dentistry
Zinc (Zn)	Metal	Brass, metal coatings, electrodes in batteries, medicine (zinc salts)

*Potassium, which is very reactive chemically, is never found free in nature; it is always combined with other elements.

Nonmetallic Resources

When we think about becoming rich from mining, we usually think of finding gold. As mundane as it seems, however, more money has been made mining sand and gravel than mining gold. For example, in the United States in 1992, $5 billion worth of sand and gravel were mined and sold, but only $3.5 billion worth of gold was extracted from the Earth.

Concrete is an essential and versatile building material. Reinforced with steel, it is used to build roads, bridges, and buildings. Concrete is a mixture of cement, sand, and gravel. Sand and gravel are mined from stream and glacial deposits, sand dunes, and beaches (Fig. 15–2). Cement is made by heating a mixture of crushed limestone and clay.

FIGURE 15–2

More money has been made mining sand and gravel than mining gold. Bitterroot Valley, Montana.

FIGURE 15–3

(A) A Chinese quarryman splitting a large granite block. (B) After the rock is split, circular saws cut it into thin slabs for floors and walls.

Many buildings are faced with stone—usually granite or limestone, although marble, slate, sandstone, and other rocks are also used. Stone is mined from **quarries** cut into bedrock (Fig. 15–3). Phosphorus and potassium are valuable fertilizers extracted from sedimentary rocks, and common table salt is also mined from sedimentary deposits.

Look at the room around you and think about how many utensils and building materials are manufactured from mineral resources. The "lead" in your pencil is a mixture of graphite and clay. Your coffee cup may be ceramic and therefore made of clay; if not ceramic, it is probably plastic, manufactured from coal or petroleum. The inside walls of your building are probably lined with wallboard or plaster, both of which are made of gypsum, which is taken from sedimentary rocks. The exterior of the building may be brick, which is baked clay, or faced with granite or limestone. You may be wearing jewelry that contains a semiprecious stone, such as turquoise or topaz, or perhaps even a precious gem, such as a ruby or diamond.

Metal Deposits

If you walked outside, picked up any rock at random, and sent it to a laboratory for analysis, the report would probably show that the rock contains measurable amounts of iron, gold, silver, aluminum, and other valuable metals. In most rocks, the concentrations of these metals are so low that the extraction cost would be much greater than the income gained by selling the metals. In certain locations, however, geologic processes have enriched the rocks with metals many times above their normal concentrations (Table 15–2).

A **mineral deposit** is a local enrichment of one or more minerals. **Ore** is any natural material sufficiently enriched in one or more minerals to be mined profitably. Geologists and mining engineers usually use *ore* to refer to metal deposits, and the term is commonly accompanied by the name of the valuable metal—for example, *iron ore* or *silver ore*. Table 15–2 shows that the concentrations of elements in ore may exceed those in ordinary rock by factors of more than 100,000.

TABLE 15–2 Comparison of Concentrations of Specific Elements in Earth's Crust with Concentrations Needed to Operate a Commercial Mine

Element	Natural Concentration in Crust (% by weight)	Concentration Required to Operate a Commercial Mine (% by weight)	Enrichment Factor
Aluminum	8	24–32	3–4
Iron	5.8	40	6–7
Copper	0.0058	0.46–0.58	80–100
Nickel	0.0072	1.08	150
Zinc	0.0082	2.46	300
Uranium	0.00016	0.19	1,200
Lead	0.00010	0.2	2,000
Gold	0.0000002	0.0008	4,000
Mercury	0.000002	0.2	100,000

Mineral reserves are the known supplies of an ore in the ground. The concept can apply on a global or national scale, or it can apply to the amount of ore remaining in a particular mine.

15.2 How Ore Forms

One of the primary professional objectives of many geologists is to find new ore deposits. Successful exploration requires an understanding of the processes that concentrate elements to form ore. For example, platinum concentrates in certain types of igneous rocks. Therefore, if you were exploring for platinum, you would not look in shale or limestone.

Magmatic Processes

Crystal Settling

Cooling magma does not solidify and crystallize all at once. Instead, high-temperature minerals crystallize first. Lower-temperature minerals crystallize later, when the temperature drops.

Solid minerals are denser than liquid magma. Consequently, early-formed crystals sink to the bottom of a magma chamber in a process called **crystal settling** (Fig. 15–4). These crystals accumulate as a layer that commonly consists of a single mineral or a mixture of minerals that form at the same temperature. If the minerals contain valuable metals, an ore deposit may form. The largest ore deposit formed in this way is found in the Bushveldt intrusion of South Africa. It is about 375 by 300 kilometers in area—roughly the size of the state of Maine—and about 7 kilometers thick. Large quantities of chromium and platinum are mined in the Bushveldt.

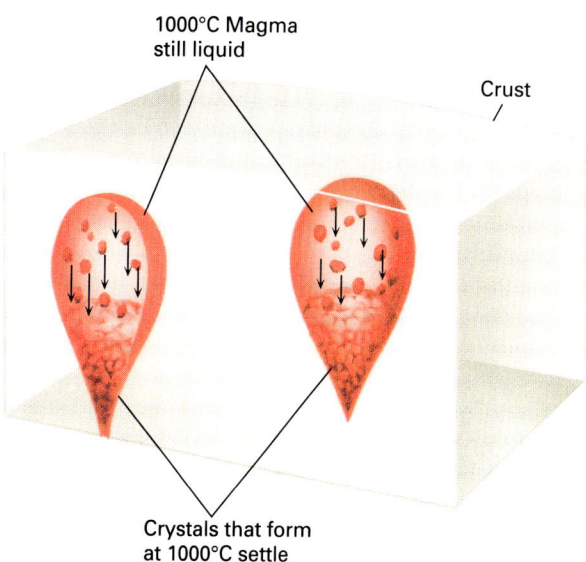

FIGURE 15–4
Early-formed crystals settle and concentrate near the bottom of a magma chamber.

FIGURE 15–5
Large crystals of lithium minerals are mined from this pegmatite in the Black Hills of South Dakota. The woman is sitting on a single crystal.

Pegmatites

Most granitic magma contains up to 10 percent water and crystallizes deep within the crust—at depths of 10 or 15 kilometers. However, under unusual circumstances water-rich granitic magma may rise to a few kilometers or less below the Earth's surface. The water then causes large crystals to grow as the magma solidifies. The extremely coarse-grained rock that results from this process is called **pegmatite** (Fig. 15–5). Crystals in common granite are a few millimeters to a few centimeters across, but crystals of feldspar and other minerals in pegmatite may be as large as 10 meters long and 3 meters in diameter. Feldspar is often mined from pegmatite for the manufacture of ceramics. Certain rare elements also concentrate in pegmatites. Beryllium occurs in the mineral beryl; emeralds are gem-quality specimens of beryl. When beryl is not of gem quality, it is mined for its beryllium content. Some pegmatites are rich in lithium and are mined for that element. Pegmatites are also sources of many semiprecious gems, including tourmaline, aquamarine, and many attractive forms of quartz.

Kimberlites

Rare igneous rocks called **kimberlites** are the world's main source of diamonds. In a few locations in South Africa, Russia, Arkansas, Wyoming, and Canada's Northwest Territories, cylindrical **pipes** of kimberlite extend from the asthenosphere to the Earth's surface. They are conduits that once carried magma on its way to erupt at a volcano, but they are now filled with the last bit of magma that solidified within the conduit. They are both scientifically fascinating and economically important. Most known pipes formed between 70 million and 140 million years ago, and for an unknown reason, most intruded into continental crust older than 2.5 billion years. Pipes are interesting because the kimberlite rock that composes them originated in the asthenosphere. This rock is among the few direct samples of the upper mantle available to geologists.

Under the intense pressures of the upper mantle, small amounts of carbon in the kimberlite magma crystallized as diamond before the magma rose toward the Earth's surface. The best evidence indicates that, after the diamonds formed, the kimberlite magma shot upward through the Earth's crust at very high, perhaps even supersonic, speed. Thus, the diamonds crystallized more than 200 kilometers beneath the surface and then rose with the magma to depths of a few kilometers or less, where they can be accessed by modern mining technology (Fig. 15–6).

Volcanic Vent Deposits

Sulfur, used primarily for sulfuric acid in industrial applications, forms as a pure, yellow deposit precipitated from the gases that escape from some volcanic vents (Fig. 15–7). Such deposits are sometimes mined even as the sulfur-rich fumes continue to escape from the volcano.

FIGURE 15–6
A diamond in kimberlite. *(Smithsonian Institution)*

FIGURE 15–7
Yellow sulfur coats the vent of Ollagüe volcano in southern Bolivia.

Hydrothermal Processes

Magma and underground rock contain large amounts of hot water. The water is typically salty with the same dissolved ions found in seawater. This mixture of hot water and dissolved ions is a **hydrothermal solution**. (The word *hydrothermal* combines the roots *hydro* for water and *thermal* for hot.)

Hydrothermal solutions are corrosive: They dissolve metals such as copper, gold, lead, zinc, and silver from hot rock or magma. Recall that most rocks contain low concentrations of many metals. For example, average crustal rock contains 0.0000002 percent gold, 0.0058 percent copper, and 0.0001 percent lead. Although the metals are present in extremely low concentrations, hydrothermal solutions percolate slowly through vast volumes of rock, dissolving and accumulating large amounts of the metals. The solutions then deposit the metals when they encounter changes in temperature, pressure, or chemical environment (Fig. 15–8). In this way, hydrothermal solutions scavenge metals from large volumes of average crustal rocks and then deposit them locally to form ore.

If the metals precipitate in fractures in rock, a hydrothermal **vein** deposit forms. Ore veins range from less than a millimeter to several meters in width. They can be incredibly rich. Single gold or silver veins have yielded several million dollars' worth of ore.

The same hydrothermal solutions that flow rapidly through open fractures to form veins may also soak into large volumes of country rock around the fractures. If metals precipitate in the rock, they may create a large but much less concentrated **disseminated ore deposit**. Because they commonly form from the same solutions, vein deposits and disseminated deposits are often found together. In many early mining districts, miners dug shafts and tunnels to follow the rich veins. After the veins were exhausted, later miners used huge power shovels to extract low-grade ore from disseminated deposits surrounding the veins.

Disseminated copper deposits, with accompanying veins, are abundant in a zone extending along the entire western margin of North and South America (Fig. 15–9). They are most commonly associated with large granitic plutons and are called **porphyry copper deposits**. (*Porphyry* is a term for an igneous rock in which large crystals, usually feldspar, are set in a finer-grained matrix. Most porphyry copper deposits are found in granitic porphyries.) Both the plutons and the copper deposits formed as a result of subduction that occurred as North and

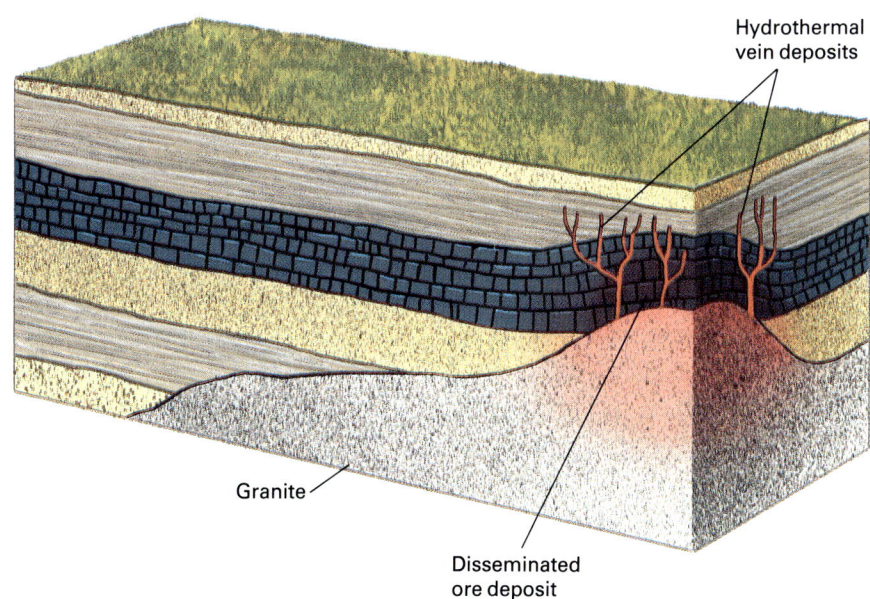

FIGURE 15–8
Hydrothermal ore deposits form when hot water deposits metals in bedrock.

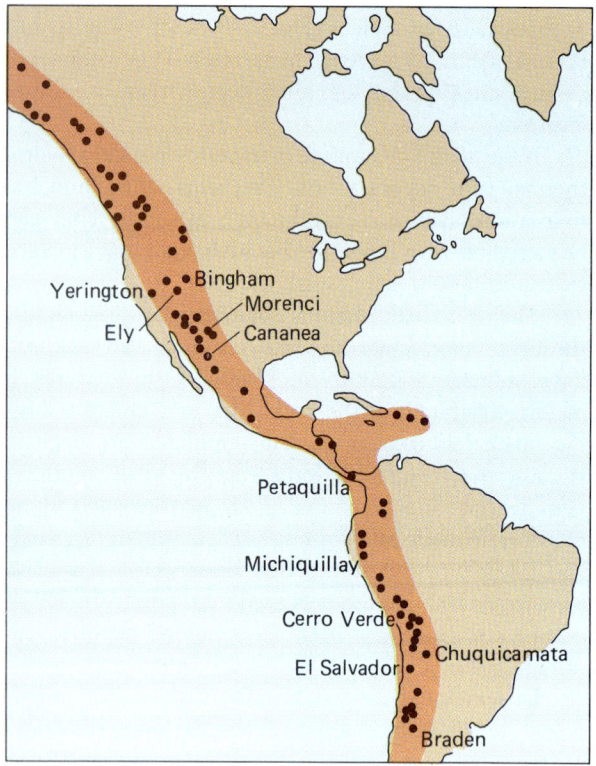

FIGURE 15–9
Porphyry copper deposits in the Western Hemisphere lie along modern and ancient tectonic plate boundaries. (Courtesy of Brian Skinner and Stephen Porter, *The Dynamic Earth: An Introduction to Physical Geology*, first edition.)

South America migrated westward during the breakup of Pangaea described in Chapter 4. Other metals, including lead, zinc, molybdenum, gold, and silver, are found with porphyry coppers. Examples of such deposits occur at Butte, Montana; Bingham, Utah; Bisbee, Arizona; and Ely, Nevada.

Hydrothermal ore deposits also form in regions of the sea floor where magmatic activity occurs—near the mid-oceanic ridge and submarine volcanoes. Here, seawater seeps through vast volumes of young, hot, fractured basalt and dissolves metals from the rocks. Later, as the metal-bearing solutions cool and chemical conditions change, huge, highly concentrated deposits of iron, copper, lead, and zinc precipitate within the sea-floor sediment and basalt (Fig. 15–10). Tectonic activity may eventually carry submarine hydrothermal deposits of this type to the Earth's surface. The lead–zinc ore of the Sullivan Mine in British Columbia, the copper deposits of New Brunswick and Cyprus, and the gold deposits of Kirkland Lake, Ontario, and Kalgoorlie, Australia, all formed in this manner.

Hydrothermal solutions can be seen as jets of black water, called **black smokers**, spouting from fractures in the East Pacific rise (Fig. 15–11). The black color is caused by precipitation of very fine-grained metal sulfide minerals as the hydrothermal solutions cool by contact with seawater.

Sedimentary Processes

Two types of sedimentary processes form ore deposits: sedimentary sorting and precipitation.

Sedimentary Sorting: Placer Deposits

Gold occurs naturally as a pure metal and is denser than any other mineral. Therefore, if a mixture of gold dust, sand, and gravel is swirled in a glass of water or in a gold pan, the gold falls to the bottom first (Fig. 15–12). Differential settling also occurs in nature. Many streams carry silt, sand, and gravel with an occasional small grain of gold. The gold settles first when the current slows down. Thus, grains of gold concentrate near bedrock or in coarse gravel, forming a **placer deposit** (Fig. 15–13).

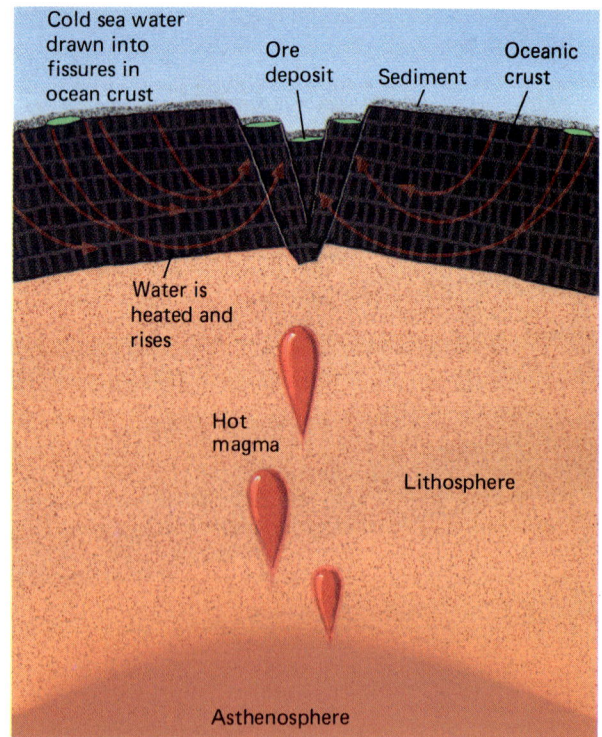

FIGURE 15–10
Submarine hydrothermal ore deposits precipitate from circulating seawater near the mid-oceanic ridge.

FIGURE 15–11
A black smoker spouting from the East Pacific rise. The "smoke" consists of tiny mineral grains that precipitate where hot solutions meet cold ocean water. The hot, nutrient-rich water sustains thriving plant and animal communities. (*Dudley Foster, Woods Hole Oceanographic Institution*)

FIGURE 15–12
Panning for gold in Montana. (*Montana Historical Society*)

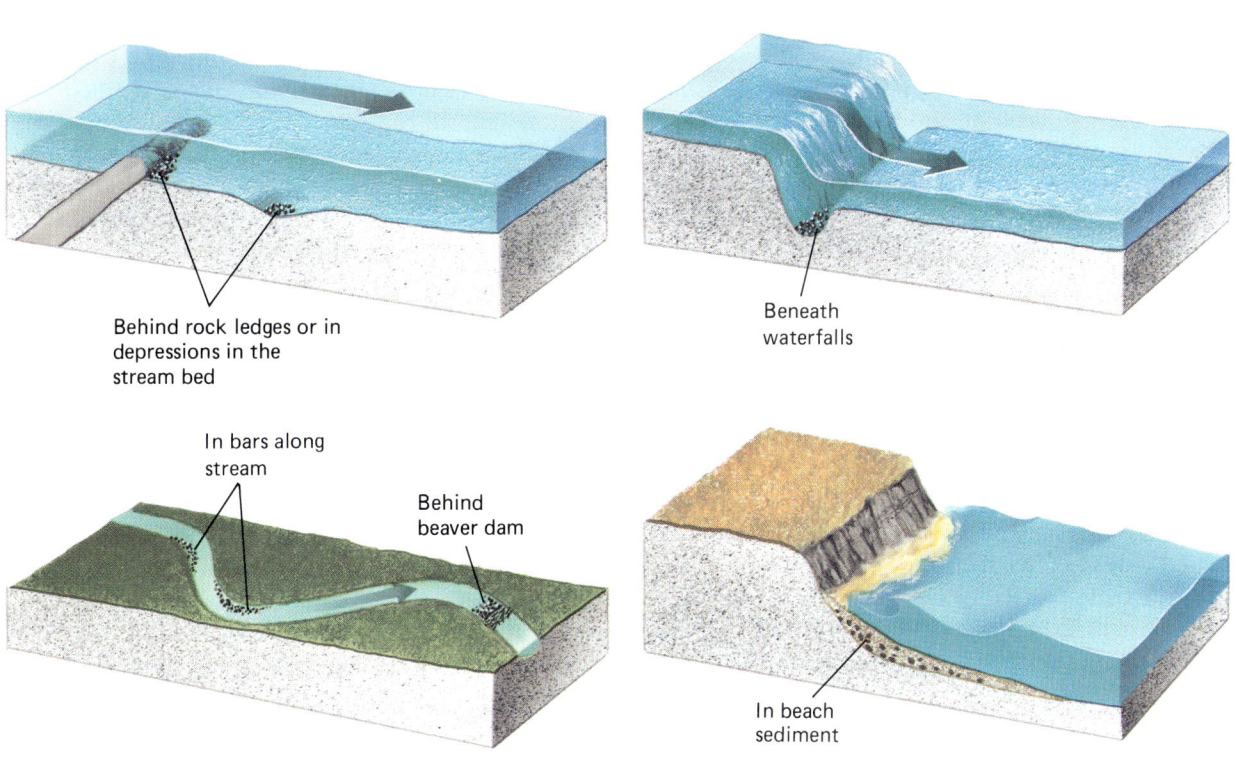

FIGURE 15–13
Placer deposits form where water currents slow down and deposit heavy minerals.

Precipitation

As ground water percolates through rock, it dissolves minerals and carries off dissolved ions. In most environments, this water eventually flows into streams and then to the sea. Some of the dissolved ions, such as sodium and chloride, make seawater salty. In deserts, however, lakes develop with no outlet to the ocean. Water flows into the lakes but can escape only by evaporation. As the water evaporates, the dissolved ions concentrate until they precipitate.

You can perform a simple demonstration of evaporation and precipitation. Fill a bowl with warm water and add a few teaspoons of table salt. The salt dissolves and you see only a clear liquid. Set the bowl aside for a few days until the water evaporates. The salt precipitates and encrusts the sides and bottom of the bowl.

Salts that form by evaporation are called **evaporite deposits**. Evaporite deposits formed in desert lakes include table salt, borax, sodium sulfate, and sodium carbonate. These salts are used in the production of paper, soap, and medicines, and for the tanning of leather.

Several times during the past 500 million years, shallow seas covered large regions of North America and all other continents. At times, those seas were so poorly connected to the open oceans that water did not circulate freely between them and the oceans. Consequently, evaporation concentrated the dissolved ions until salts precipitated as **marine evaporites**. Periodically, storms flushed new seawater from the open ocean into the shallow seas, providing a new supply of salt. In this way, thick marine evaporite beds formed. Evaporites underlie nearly 30 percent of North America. Table salt, gypsum (used to manufacture plaster and sheetrock), and potassium salts (used in fertilizer) are mined extensively from marine evaporites.

About 1 billion tons of iron are mined every year, 90 percent from **banded iron formations**, which are sedimentary layers of iron-rich minerals sandwiched between beds of silicate minerals. The alternating layers are a few centimeters thick and give the rocks their banded appearance (Fig. 15–14).

The most abundant and economically important banded iron formations developed between 2.6 and 1.9 billion years ago. What happened during that interval to create these peculiar rocks? To answer this question, we must understand how iron interacts with oxygen. In an oxygen-poor environment, iron dissolves in water. If the oxygen concentration rises, iron reacts with the oxygen to form iron oxide minerals that precipitate rapidly. During the early history of the Earth, very little molecular oxygen was present in the atmosphere, and as a result the seas contained large quantities of dissolved iron.

FIGURE 15–14
A banded iron formation. The red bands are iron minerals; the dark layers are chert. (*Ward's Natural Science Establishment*)

Then plants evolved. As they became abundant, the concentration of oxygen in the atmosphere increased because plants release oxygen into the atmosphere during photosynthesis. About 2.6 billion years ago, enough oxygen had accumulated in the atmosphere to react with the dissolved iron. As a result, vast quantities of iron precipitated, producing the banded iron formations.

Weathering Processes

In environments with high rainfall, the abundant water dissolves and removes most of the soluble ions from soil and rock near the Earth's surface. The relatively insoluble ions left behind form **residual deposits**. Both aluminum and iron have very low solubilities in water. **Bauxite**, our principal source of aluminum, forms in this manner, and in some instances iron also concentrates enough to become ore. Most bauxite deposits occur in warm, rainy tropical or subtropical environments such as Jamaica, Cuba, Guinea, and Australia, where chemical weathering occurs rapidly (Fig. 15–15).

Weathering processes also form immense amounts of clay, as described in Chapter 5. One clay mineral, called kaolin, is the primary constituent of porcelain and other fine ceramics. Another, smectite, is mixed with water and other minerals to make drilling mud, a clay-rich slurry used to cool and lubricate drill bits in the drilling of deep oil wells. Several types of clay are used to line irrigation ditches and ponds because they form impermeable barriers.

Metamorphic Processes

Recall from Chapter 3 that a metamorphic rock forms when heat and pressure alter the mineralogy and texture

FIGURE 15–15
Bauxite ore forms by intense weathering of aluminum-rich rocks. (*J. Bridge/USGS*)

of any preexisting rock. Metamorphic processes can also cause migration of hydrothermal fluids, which, in turn, deposit metal ores. Thus, some hydrothermal ore deposits are also metamorphic in origin. Graphite, the main component of the "lead" in pencils, forms when metamorphism affects the carbon in some organic-rich rocks. Asbestos is a commercial name for two different minerals that form under similar metamorphic conditions. As mentioned earlier, marble, a valuable building stone and sculptor's material, forms by metamorphism of limestone.

FOCUS ON

Manganese Nodules on the Sea Floor

As our domestic reserves of some metals dwindle and political and economic instability threaten foreign sources, many geologists are looking to metal deposits on the sea floor.

About 25 to 50 percent of the Pacific Ocean floor is covered with golf-ball- to bowling-ball-size **manganese nodules** (Fig. 1). The average nodule contains 20 to 30 percent manganese, 6 percent iron, about 1 percent each of copper and nickel, and lesser amounts of other metals. (Much of the remaining 60 to 70 percent consists of anions such as oxygen chemically bonded to the metals.) The metals may be introduced into seawater by volcanic activity at the mid-oceanic ridge. The dissolved metals then precipitate to form the nodules by chemical reaction between seawater and sea-floor sediment.

The nodules grow by about ten layers of atoms per year, which amounts to 3 millimeters per million years. Curiously, they are found only on the surface of, but never within, the mud on the ocean floor. Because sediment accumulates much faster than the nodules grow, why doesn't it bury the nodules as they form? Worms burrow into the mud, and other animals pile sediment against the nodules to build protective shelters. Some geologists suggest that these activities constantly lift the nodules onto the surface.

A trillion or more tons of manganese nodules lie on the sea floor. They contain several valuable metals that could be harvested without drilling or blasting.

One can imagine a scenario in which exploration is carried out with undersea television cameras and the nodules are collected by a vacuum diver, a scoop, or a robot. But despite its tremendous wealth, the ocean floor is not the easiest environment in which to operate complex machinery, so at present exploitation of manganese nodules is not profitable.

FIGURE 1
Manganese nodules cover the sea floor on the Blake Plateau east of Florida. The nodules range from a few centimeters to about 30 centimeters in diameter, so this photograph covers a few hundred square meters. (Frank Manheim, Woods Hole Oceanographic Institution)

15.3 Mining and Refining of Metals

Mining

Ore is commonly covered and surrounded by soil and rock that is not of ore quality. As a result, miners must dig up and discard large amounts of waste rock, called **mine spoil**, as they expose and extract the ore.

Mineral resources are mined either in underground tunnels or in surface mines. A **tunnel mine** does not directly disturb the surface of the land. Nevertheless, most tunnel mines have large spoil dumps at their entrances, because the bedrock dug out to get at the ore and make the tunnel must go somewhere.

In many situations, a tunnel mine is more expensive and less efficient to work than a **surface** or **open pit mine**, in which bedrock is drilled, blasted, and excavated by heavy machinery to extract ore. The process creates a gigantic open pit in the Earth. The largest human-created hole on Earth is the Bingham Canyon, Utah, open pit copper mine. It is 4 kilometers in diameter and 0.8 kilometer deep. The mine produces 230,000 tons of copper a year and smaller amounts of gold, silver, and molybdenum (Fig. 15–16).

FIGURE 15–16
An aerial view of the Bingham Canyon, Utah, open pit copper mine. *(Agricultural Stabilization and Conservation Service/USDA)*

Mine spoil from open pit mines is piled up on the surface. In the United States, the Surface Mining Control and Reclamation Act requires that mining companies restore mined land so that it can be used for the same purposes for which it was used before the mining operation was started. In addition, a tax is levied to reclaim

FOCUS ON

The 1872 Mining Law

The 1872 Mining Law still governs the mining of metal ores on public land in the United States. In the mid-1800s, when the country was growing rapidly and much of the continent had not been explored, the intent of the law was to encourage the exploration and development of America's natural resources and to foster economic growth. The law states that any individual or corporation can obtain mineral rights to metal ores on public land. The miner need only prove that he or she can make a profit by mining the deposit and do a small amount of development work on the claim each year. The miner or corporation pays no royalties or other fees to the government or public for the ore extracted. In some cases, companies or private citizens can buy land for as little as $5 per acre.

In recent years, conservationists and other public interest groups have attacked the 1872 law, claiming that it is one of the biggest giveaway programs in the United States. For example, in 1994 the federal government sold 2000 acres near Elko, Nevada, to American Barrick Resources Inc. for $9765. Geologists estimate that the property contains $10 billion worth of gold. (Profits will be much less because of mining expenses.) In another instance, a mining company bought 160 acres near Keystone, Colorado. It operated mines on a portion of the land and sold the rest at $11,000 per acre for ski condominiums.

In response to the public interest groups, defenders of the 1872 law argue that mineral exploration entails a great deal of risk and expense because the probability of finding new ore is low. The potential rewards must be great to make exploration attractive to investors. Defenders also argue that the law has worked well (at least for the mining industry) for more than a century and has produced one of the world's most productive and efficient mining industries.

land that was mined and destroyed before the law was enacted. However, environmentalists claim that the government has been lax in enforcing the act and that 6000 surface mine sites have not been restored in compliance with the law.

In the United States, mining of all mineral resources (not including coal) produces at least six times more solid waste than do all towns and cities combined. Unrestored surface mines cover an area of about 90,000 square kilometers, slightly smaller than the state of Virginia. Although this figure includes coal strip mines, it does not include abandoned sand and gravel mines and rock quarries, which probably account for an even larger area.

Mine spoil is prone to erosion, and the muddy runoff pours into nearby streams and destroys aquatic habitats. In addition to siltation problems, many chemicals found in mine spoils poison streams and ground water. Sulfur is present in large quantities in many metal ores as well as in coal. The sulfur reacts with water in the presence of air to produce sulfuric acid (H_2SO_4). If pollution control is inadequate, the sulfuric acid then runs off into streams and ground water below the mine or mill. This type of pollution is called **acid mine drainage**. Heavy metals, such as lead, cadmium, zinc, and arsenic, are common in many metal ores. The case history of the Butte Superfund site describes some of the effects of heavy metal poisoning on streams and ground water.

Refining

Gold, silver, and a few other elements exist in their pure forms in the crust, but most elements combine naturally with others. Thus, pure iron is rarely found in the Earth's crust. The iron in iron ore is chemically bonded to oxygen, forming iron oxides. Rust is one example of an iron oxide. After iron ore is dug up, it is crushed and separated from waste rock. The iron must then be separated from oxygen and converted to the pure metal. In the process, the oxygen escapes into the atmosphere or combines with hydrogen to form water. Because oxygen is a natural component of air, this process is harmless.

However, many metal ores consist of the valuable metal combined with sulfur. When sulfide minerals are refined, the sulfur commonly escapes as gaseous air pollutants such as hydrogen sulfide and sulfuric acid. These gases contribute to acid precipitation and other air pollutants. Air pollution has created "dead zones" around many older smelters. Sulfur, of course, is not the only polluting chemical from mining operations. Many other elements and compounds, particularly heavy metals such as lead, cadmium, zinc, and arsenic, are released by ore refining as well as mining. Many of these metals and their compounds are toxic and pollute both air and water.

CASE HISTORY

The Largest Superfund Site—Butte, Montana

The largest complex of toxic waste sites in the United States was produced not by a chemical factory or petroleum refinery but by the mines and smelters in Butte and Anaconda in Montana (Fig. 15–17). Remember that most ore is not pure metal but rock containing metal-rich minerals. Some of the ore from the mines in Butte contained as little as 0.3 percent copper, with even smaller concentrations of manganese, zinc, arsenic, lead, cadmium, and a variety of other metals. Mining and refining began in Butte in the early 1800s, before pollution control laws existed. When the ore was extracted, the waste rock was simply piled up near the mine mouths. The smelters in Butte and nearby Anaconda disposed of their mill wastes in the same way. Rain and streams eroded the mine and mill spoils into the Clark Fork River, silting the stream bed and destroying aquatic habitats. But that was only the beginning of the problem, because the smelting process generated additional wastes.

In the boom days in Butte, during the late 1800s and early 1900s, metal smelting was inefficient, so large quantities of arsenic, cadmium, zinc, lead, copper, and other metals were discarded in the waste heaps. The Butte ores also contained large amounts of sulfur, which was discarded with the other wastes. Furthermore, Butte's copper ore contains gold, which the early miners attempted to extract by dissolving it in mercury, and the waste mercury was also dumped on the piles. Today in the Butte–Anaconda area there are about 25 square kilometers (about 10 square miles) of mine spoils and mill wastes averaging 15 meters (about 50 feet) high, or about as tall as a five-story building.

These wastes have poisoned both the ground water and streams flowing from the Butte–Anaconda

FIGURE 15–17
Mill wastes surround the now-abandoned smelter at Anaconda, Montana.

FIGURE 15–18
Copper leached from the Anaconda mill tailings colors this soil green. Copper, arsenic, cadmium, and other heavy metals have poisoned the soil so that nothing can grow. *(Johnnie Moore)*

area. The sulfur compounds have reacted with water to form sulfuric acid, which has spread into streams and ground-water reservoirs. Metal contamination has stunted and killed plants, the animals that eat those plants, and the microorganisms in the soil in which those plants grow (Fig. 15–18). People are also poisoned through metal contamination. One study showed that workers in a cadmium recovery plant suffered a higher incidence of lung cancer than the public at large. In another study, skin cancer was prevalent among a group of people who drank well water contaminated with arsenic.

The sulfur and metals have migrated up to 220 kilometers downstream from Butte and Anaconda along the Clark Fork River (Fig. 15–19). Windblown dust from the spoils has contaminated entire counties. In the 1980s, arsenic levels had risen so high in wells of Milltown—a small town on the Clark Fork River about 200 kilometers downstream from Butte—that government officials ordered the residents to stop drinking well water. An alternative drinking water supply was installed at great public cost. No one knows how to clean up such a mess or how much it will cost. The problem is that thousands of square kilometers of land surface are contaminated. The Clark Fork Coalition, a local environmental organization, has estimated that it will cost more than $1 billion to clean up the Butte mine, the Anaconda smelter, and the Clark Fork drainage. Even a partial clean-up, under way now, will cost tens to hundreds of millions of dollars.

FIGURE 15–19
In a healthy ecosystem, lush vegetation grows on riverbanks. Heavy metals and silt from the Anaconda smelter have poisoned the banks of the upper Clark Fork River in Montana. *(Johnnie Moore)*

Aluminum, Power, and Dams

Aluminum is one of the most important metals in the modern world. Second only to iron in dollar value and total tons produced annually, it is light, strong, corrosion-resistant, easily worked, a good conductor of electricity and heat, and, perhaps best of all, cheap—66¢ a pound in June 1994. Those properties make modern airplanes possible; similar aircraft built of steel would be so heavy that they couldn't fly. Other lightweight metals such as titanium would be too expensive.

The United States is the largest per-capita consumer of aluminum. The average U.S. citizen uses 42 pounds per year. Thirty-one percent goes to pack-aging, most of which is used for beer and soda pop cans; about 15 percent is used in automobiles; 19 percent is used in building materials such as siding, doors, and windows; and 10 percent is used for electric wiring.

Environmental Impacts of Aluminum Mining and Refining

As described in Section 15.2, bauxite (the only ore of aluminum) forms by intensive weathering of aluminum-rich rocks such as granite. The aluminum concentrates in the soil while rainwater dissolves and carries off more soluble elements. Because of this mode of formation, aluminum ore occurs as a widespread thin layer at the Earth's surface. As a result, almost all commercial bauxite deposits are mined from open pits. Worldwatch Institute's John Young states that "bauxite mining probably destroys more surface area than the mining of any other metal" (Fig. 15–20). Two thirds of the world's supply of bauxite is mined in Australia and Guinea.

FIGURE 15-20
A bauxite strip mine in Georgia. (*J. Bridge/USGS*)

FIGURE 15-21
The bare bedrock and scrub brush of Teakettle Mountain stand out in stark contrast to the heavily forested peaks of Glacier Park in the background.

Most other metals are extracted from their respective ores by heating the ore in furnaces with a chemical reagent to release the pure metal. Aluminum, however, is so tightly bonded to oxygen and hydrogen in bauxite that separation of metallic aluminum requires a complex and energy-intensive process. First, the bauxite is treated to produce alumina, which is aluminum oxide. The process creates about a ton of "red mud" wastes—consisting of clay, iron oxides, other metals, and alkaline solutions (strong bases, such as caustic soda)—for each ton of alumina produced. Then the alumina is dissolved in molten aluminum fluoride and a strong electrical current is passed through the solution. The current breaks the bonds between the aluminum and oxygen and allows extraction of the pure metal.

A poorly managed aluminum smelter can create severe environmental damage. The aluminum refining process releases large amounts of fluorine, which may commonly escape into the atmosphere. Fluorine and fluoride compounds are toxic to plants. They also accumulate in the teeth and bones of humans and other animals and make them so brittle that the teeth sustain tiny fractures and wear away rapidly, and the bones fracture spontaneously. The fluorides also damage soft tissue.

The Columbia Falls aluminum smelter lies just outside the west entrance to Glacier National Park, Montana, and beneath the once heavily forested Teakettle Mountain. It is the most important source of income for residents of Columbia Falls. Small dairy farms once operated in the area, and local ranchers raised beef cattle. During its early years of operation, the smelter had ineffective pollution control systems, and much of the fluorine escaped directly into the air above the plant. The air pollution killed all the trees on Teakettle Mountain (Fig. 15-21). Many of the dairy and beef cattle died. Some starved because their teeth wore away; others died from the results of bone fractures. Abnormally high fluoride concentrations were found in the animals, and many ranchers think that the animals died as a result of fluoride pollution from the plant. As modern air pollution laws became effective, fluorine emissions were reduced. Brush and small trees started to reclaim the land around the smelter, and ranchers were once again able to raise cattle nearby. However, abnormally high fluoride concentrations have been found in soils in Glacier Park, and park officials still express concern about the long-term effects of fluorine pollution in the park.

Aluminum Production and Electrical Consumption

Aluminum production uses so much electrical power that the metal has been called "congealed electricity." The production of a dozen aluminum beer cans uses the energy equivalent of 1 liter of gasoline. Although many cans are recycled, the number thrown away in the United States represents enough electricity to power Baltimore.

An average smelter in the United States produces 100,000 tons of aluminum per year and consumes the same amount of electricity as a small city. Aluminum smelters must run 24 hours a day. If one is shut down for more than 6 hours, molten aluminum can solidify in the smelting pots, causing expensive damage to the plant. Thus, not only do aluminum smelters use huge amounts of electricity, they also require an uninterrupted supply. As a result of the demand for electricity, aluminum smelters are located near cheap, reliable power sources rather than near bauxite deposits. For example, the United States is the world's leading aluminum producer, but little bauxite is mined in the United States. Canada is the third leading producer but has no bauxite mines.

We all know that our own electric bills are high and have been rising for more than 2 decades. How can the aluminum industry continue to supply aluminum so cheaply? How can we make throw-away beer and soda pop cans of the metal? Almost all aluminum smelters pay substantially lower prices for power than do other users in the same area; thus, other power consumers subsidize the aluminum industry.

Aluminum and Dams

The rise and current status of the U.S. aluminum industry are closely connected with World War II, when, for the first time, air power became critical to winning a war. Bombers require aluminum, aluminum requires electricity, and during World War II the cheapest source of power was hydroelectric energy obtained by damming rivers. By 1939, Germany had passed the United States in aluminum production and was manufacturing warplanes at an unprecedented rate. The United States responded by completing the Bonneville and Grand Coulee dams in the Pacific Northwest and turning much of their power to aluminum and aircraft production. By 1943, the United States was producing four times more aluminum than Germany, and ultimately half of U.S. wartime aluminum production came from the Pacific Northwest.

Before the war, Alcoa held a monopoly on U.S. aluminum production. After the Japanese attacked Pearl Harbor, the U.S. government constructed its own aluminum plants and hired Alcoa to run them. At the war's end, the government created competition in the aluminum industry by selling its wartime smelters to Reynolds and Kaiser, thus creating two new aluminum producers. In the 1950s, rapidly increasing domestic aluminum use and continued military requirements escalated the demand for aluminum and cheap power. As a result, the U.S. Army Corps of Engineers and the Bureau of Reclamation built 36 large dams on the Columbia River, and the competing aluminum companies constructed smelters nearby.

Aluminum production in the Pacific Northwest has caused many environmental changes. The great dams have converted the Columbia River to a series of huge lakes and have decimated the salmon and steelhead populations that once migrated upstream in huge numbers to spawn in the Rocky Mountain headwaters. Grizzly bears and other predators that relied on the fish for food are now gone from many of the headwater streams. In 1991, the U.S. National Marine Fisheries Service declared the Snake River sockeye salmon an endangered species because its numbers had dwindled to the point where it was in danger of extinction. Several other migratory fish are similarly endangered and under consideration for formal designation as endangered species.

At present, no one knows how to solve the problem of the threatened survival of these fish species. Engineers built fish ladders around the dams so that returning adults could migrate upstream. Some fish manage to navigate the ladders, but many fail. In addition, mortality rates are extremely high for juvenile fish as they swim downstream to return to the sea. The fingerlings are disoriented by the absence of current in the reservoirs above the dams, and their metabolisms are disrupted because the reservoirs are warmer than a free-flowing stream. Finally, most of the young fish must pass through whirling turbine blades near the bottom of each dam. The blades either kill them outright or so disorient them that they are vulnerable to predators that wait in the pools below. Trucking the migrating fish around the dams has failed. Yet the Endangered Species Act demands that the species not be allowed to become extinct. Removal of the giant, tremendously expensive dams seems unlikely.

Aluminum and Conservation

Although conservation and recycling of aluminum products will not solve the environmental problems that result from existing dams, they do offer ways to reduce electricity consumption by the aluminum industry. In many applications, aluminum actually saves energy. Use of the light metal in airplanes, autos, and trains reduces their weight so much that the amount of energy saved by burning less fuel is greater than the energy used to produce the aluminum. The use of aluminum in cans and other forms of packaging—31 percent of all U.S. aluminum use—is wasteful of energy, however. If we were paying the real price of aluminum (including unsubsidized energy costs and environmental costs), aluminum would be too expensive to use in throw-away containers. Thus, the use of aluminum in durable goods is energy-efficient, but its use in throw-away applications is wasteful.

The greatest potential for energy conservation lies in recycling. Only 6 percent as much energy is required to produce aluminum products from scrap metal as from ore. In 1990, nearly one fourth of the aluminum produced in the world—more than 5 million tons—came from recycling of scrap. Japan currently obtains about 40 percent of its aluminum from recycling, the United States about 20 percent, and Europe between 20 and 30 percent.

CASE HISTORY

The Gold Bug

Rare and beautiful, gold is the most highly prized metal in the world, although a few others command higher prices. Humans have committed almost every known crime in pursuit of the soft, malleable, yet durable metal. Today, we are in the middle of a modern gold rush, and widespread and intensive environmental destruction is among the newest of transgressions caused by the search for gold.

The U.S. Bureau of Mines estimates that 95,000 tons of gold are currently in world circulation—59,000 tons owned by corporations and individuals, and 36,000 tons held by governments. About 15 percent of the gold mined each year is used by industry; one third of the industrial use goes to specialized electronics applications, because gold is the best electrical conductor known. Jewelry accounts for the other 85 percent.

A drastic price increase is primarily responsible for the modern gold rush. From 1970 to 1994, the price almost quadrupled from about $100 per ounce to $380, although it briefly soared to $850 in January 1980.[1] As a result of the higher prices, subeconomic deposits suddenly became profitable. Many of the newly minable deposits are in the United States, particularly in Nevada, which produces 177 tons of gold per year.

Gold Mining in the United States: Cyanide Heap Leaching

Most of these new mines extract gold from extremely low-grade deposits, with gold concentrations as low as 0.00006 percent, or .02 ounce per ton. At the August 1994 price of $380 per ounce, that concentration amounts to $7.60 of gold per ton of ore. A technique called cyanide heap leaching was developed to extract the tiny amounts of gold from such huge volumes of rock. The process takes advantage of the fact that, although gold is soluble in almost nothing else, it dissolves in a cyanide solution. Miners use heavy equipment to dig the ore from open pit mines. Then they spread great piles of finely crushed ore on asphalt or plastic pads and spray the heaps with a cyanide solution. The cyanide dissolves the gold as it percolates down through the crushed rock and then runs off on the impermeable asphalt

---

[1]The weight of gold is always expressed in troy ounces. One pound contains 12 troy ounces, whereas a pound contains 16 of the more commonly used avoirdupois ounces.

FIGURE 15–22
Cyanide solutions leach gold from ore in this heap leach operation at Candelaria, Nevada. *(John G. Cleary)*

or plastic to collection vats. The gold-bearing cyanide solutions are then treated to extract the gold (Fig. 15–22).

The mining and heap leaching of such low-grade ore causes at least two kinds of environmental problems. First, the process creates immense piles of waste. At such low ore grades, 3000 to 5000 tons of raw, crushed, leached rock must be disposed of for each kilogram of gold recovered. Thus, annual recovery of 177 tons of gold from Nevada mines creates more than 500 million tons of waste per year.

Second, cyanide is highly toxic. In a well-constructed and well-managed heap leach operation, the cyanide can be looped through a closed system so that none is lost; this is why mining companies can obtain permits to use the poisonous compound. But in practice, the cyanide solutions commonly escape and enter surface and ground water. Cyanide solutions are chemically unstable: The cyanide quickly decomposes in surface water, where oxygen is plentiful and acidic conditions prevail. Nevertheless, escaping cyanide killed at least 1000 birds during the first year of operation of the McCoy/Cove mine near Battle Mountain, Nevada. In October 1990, a dam in the leaching pond at the Brewer Gold Mine near Jefferson, South Carolina, broke. Ten million gallons of cyanide solution poured into the Lynches River and killed approximately 10,000 fish. Repeated cyanide leaks from the Summitville mine near Del Norte, Colorado, have exterminated all life for 17 miles in the Alamosa River. The owner of this mine has gone bankrupt. The Environmental Protection Agency is spending $800,000 a month to prevent further cyanide leaks, and the clean-up will eventually cost $20 million in public funds.

Cyanide can persist at toxic levels for much longer periods in ground water. As a result, it poses a long-term threat to reservoirs used for human consumption, livestock, and irrigation.

In June 1993, an executive geologist with one of North America's leading gold producers spoke to us about one of Nevada's largest open pit–heap leach gold operations. "We're not making any profit on it," he said. "It's a break-even operation. The only reason we keep it open is that it's a lot cheaper to keep it running than it is to shut it down and start paying for the clean-up. It's a Superfund site waiting to happen."

Gold Mining in South America: Mercury Contamination

The current gold rush is also creating environmental havoc of a different kind in South America. Concentrations of placer gold have accumulated in many streams in Brazil, Guyana, Ecuador, Bolivia, Venezuela, and other nations. Thousands of independent miners dig out and wash immense amounts of gravel, sand, and mud to extract the gold (Fig. 15–23), and corporations operate huge dredges on some of the larger deposits (Fig. 15–23). With ineffective environmental regulations, dredging has added so much silt to the Upper Mazaruni River in Guyana that the water is undrinkable for 65 kilometers downstream from the dredges. Similar siltation is occurring near most other placer sites.

Far more serious is the contamination of the Amazon ecosystem and other South American rivers with mercury. Mercury is another of the very few liquids that dissolve gold. Nearly all South American placer miners, both the independents and the corporations, use mercury to extract the gold from their sluices, dredges, and gold pans. They then use a torch to evaporate the mercury, leaving pure gold behind for recovery. Most of the mercury escapes into the atmosphere as vapor or into the streams in liquid form. Gold mining introduces about 100 tons of mercury into the Amazon ecosystem each year. It is not known how much escapes into other river systems.

Mercury is toxic. It accumulates in plants and animals and biomagnifies as it rises through the food chain. It causes severe neurological diseases and birth defects in both animals and humans. Mercury poisoning is particularly insidious because it often occurs years after a person is exposed to the metal. Mercury levels in fish in several Amazon tributaries and other South American streams now exceed safe levels for human consumption. In addition, mercury poisoning has begun to appear among people living in Amazon riverside villages, where fish are a major food source.

Gold mining in the Amazon basin has also upset the lives of indigenous people in the region. Many have abandoned their traditional life style to join the gold rush. Environmental destruction and disease have displaced others. The Yanomami Indians live in the northern Brazil jungle and, until recently, avoided all contact with the outside world. Recent gold discoveries have brought swarms of miners into their region. During this time, the Yanomami have lost 15 percent of their population to malaria. In response to their plight, Brazil banned mining in a 10-million-hectare area reserved for the Yanomami in 1991. However, recent reports indicate that miners are illegally intruding into the reservation in their quest for gold.

FIGURE 15–23
Independent miners sort gold ore by hand in the Amazon Basin of Brazil. *(Errol Sehnke/Bruce Coleman, Inc.)*

15.4 Future Availability of Minerals

Mineral Reserves

Mining depletes mineral reserves by decreasing the amount of ore remaining in the ground; but reserves may increase in two ways. First, geologists may discover new mineral deposits, thereby adding to the known amount of ore. Second, **subeconomic mineral deposits**—those in which the concentrated minerals cannot be mined at a profit—can become profitable if the price of that type of metal increases or if improvements in mining or refining technology allow cheaper extraction.

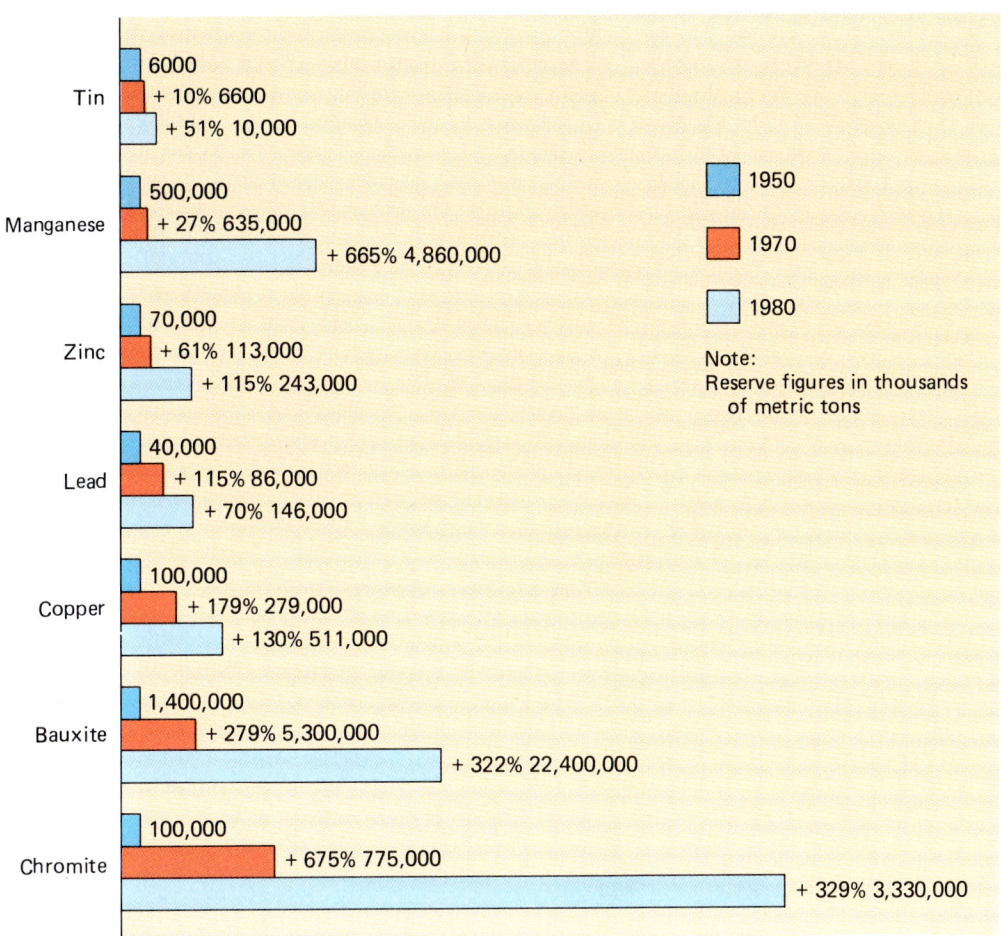

FIGURE 15–24

Changes in the reserves of some types of ore from 1950 to 1980.

Many known mineral deposits are subeconomic. For example, a deposit containing 30 percent iron is not ore because it would not be profitable to extract the metal from the rock. But if technology improved so that the iron could be refined more cheaply or if the price of iron increased, then the deposit would suddenly become ore.

Consider an example of the changing nature of reserves. In 1966 geologists estimated that global reserves of iron were about 5 billion tons.[2] At that time, world consumption of iron was about 280 million tons per year. Assuming that consumption continued at the 1966 rate, the global iron reserves identified in 1966 would have been exhausted in 18 years (5 billion tons/280 million tons per year = 18 years), and we would have run out of iron ore in 1984. Something must have changed, because iron ore is still available and cheap. The critical change resulted from development of inexpensive methods for processing iron ores of lower grade. Thus, deposits that were subeconomic in 1966, and therefore not counted as reserves, are now ore. Figure 15–24 shows that the reserves of many mineral resources have increased dramatically in recent years. Additionally, many regions of the Earth, especially some in less developed countries, have not been explored thoroughly for ore, so new deposits probably will be discovered.

Depletion of Mineral Reserves

The Romans discovered veins of nearly pure copper on the island of Cyprus. Those deposits have been mined and are now gone. In modern times, many rich and easily accessible ores, such as the high-grade iron deposits of

---

[2]B. Mason, *Principles of Geochemistry*, 3rd ed. (New York: John Wiley, 1966), Appendix III.

the Mesabi Range in the north-central United States, are being used rapidly or have already been exhausted. Ore deposits are nonrenewable. But our technological life will not end with the exhaustion of rich deposits, because less concentrated ore is still available.

The history of mining has followed a sequence of steps as reserves have been discovered, used, and depleted. The first miners exploited high-grade, easily accessible deposits. For example, shallow deposits of tin were discovered in prehistoric times. In the Bronze Age (3500 to 1000 B.C.), people learned to blend tin with copper to make bronze vessels, jewelry, and weapons that were stronger and harder than those made from any previously known material. Thus, tin became a much-sought-after resource.

When local surface deposits were exhausted, the second step in mining history commenced as miners explored farther from home for other rich and easily mined deposits. Thus, both the Phoenicians and the Romans sailed from the Mediterranean to Britain for new sources of tin. Exploration for mineral resources was the impetus behind much early geographic exploration of the world. The Spanish search for silver and gold led to the European exploration and eventual conquest of both North and South America. The quest for gold also caused the California gold rush of 1849 and the Alaska rush of 1898 (Fig. 15–25). Much of the initial settlement of both states occurred because of the gold bug. Thus, the quest for mineral resources has been the foundation of much world trade. It has also been, and continues to be, the cause of colonization, exploitation, environmental destruction, international rivalries, and war.

Finally, as rich, easily accessible reserves all over the world were depleted, miners dug deeper and extracted lower-grade ore that is chemically more difficult to refine.

The Modern Geopolitics of Metal Reserves

The Earth's mineral resources are unevenly distributed, and no single nation is self-sufficient in all mineral re-

FIGURE 15–25
Digging for gold in Klondike gravel during the Alaskan gold rush. *(The Bettmann Archive)*

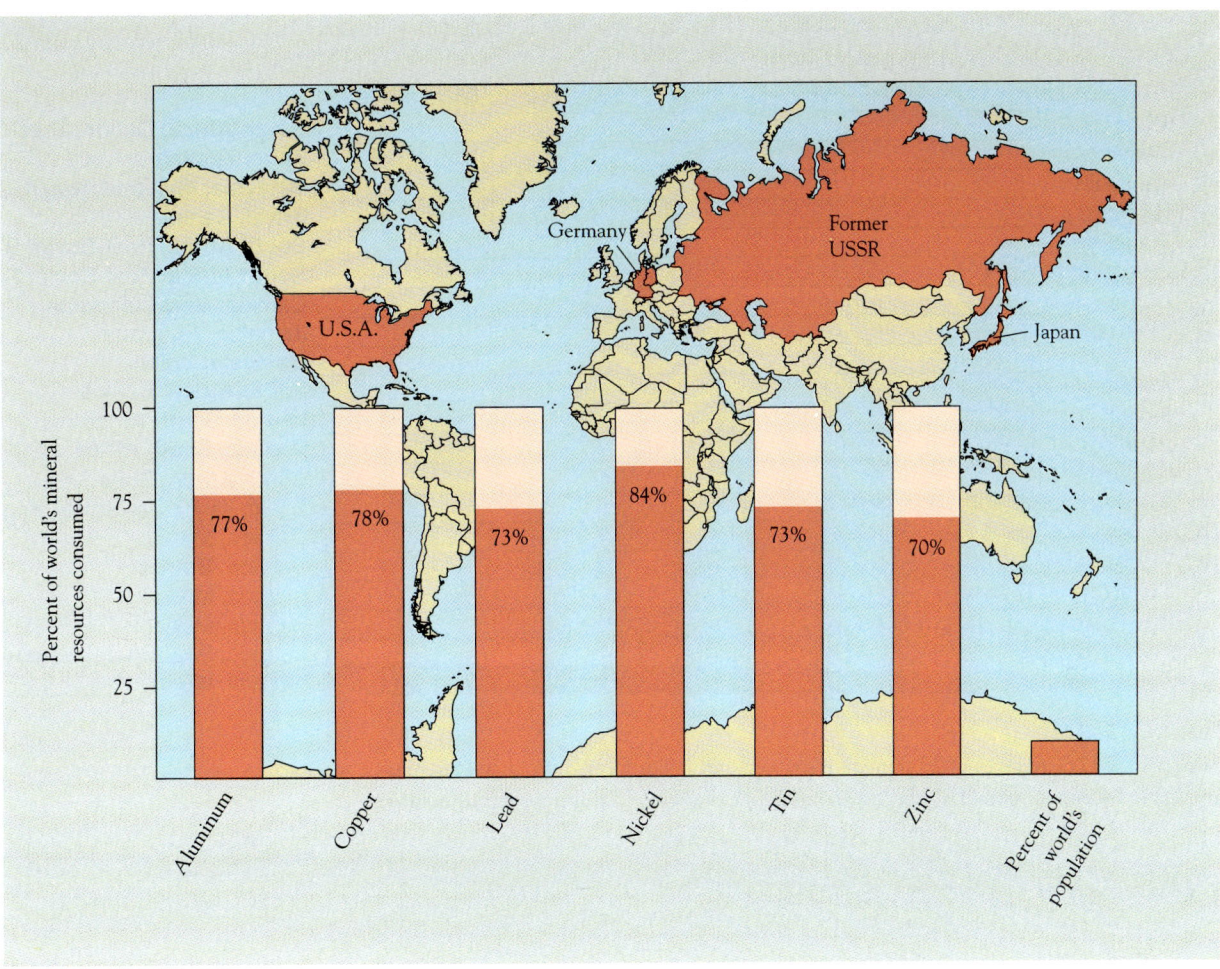

FIGURE 15–26

Four nations—the United States, Germany, the former Soviet Union, and Japan—consume about 75 percent of the most intensively used metals, although they account for only 14 percent of world population. (Raven, Berg, and Johnson, *Environment*)

sources. For example, almost two thirds of the world's known molybdenum deposits, and more than one third of the lead deposits, are in the United States. Two thirds of the aluminum deposits are found in two nations, Australia and Guinea. The United States uses 40 percent of all aluminum produced in the world, yet it contains no large economic aluminum deposits. Zambia and Zaire supply half of the world's cobalt, although neither nation uses the metal for its own industry. China contains about 75 percent of the world's tungsten reserves. Five nations—the United States, the former Soviet Union, South Africa, Canada, and Australia—supply most of the mineral resources used by modern societies. In contrast, other nations have few mineral resources. Japan has almost no metal reserves; despite its thriving economy and high productivity, it relies entirely on imports for mineral and fuel resources.

As is true for most natural resources, developed nations consume an extremely high proportion of the Earth's mineral resources. Figure 15–26 shows that four nations—the United States, Japan, Germany, and the former Soviet Union—consume about 75 percent of the most intensively used metals, although they account for only 14 percent of world population. However, as mentioned earlier, three of these nations are also among the top five mineral producers.

A **critical mineral** is one that is essential to the national economy. A **strategic mineral** is necessary for

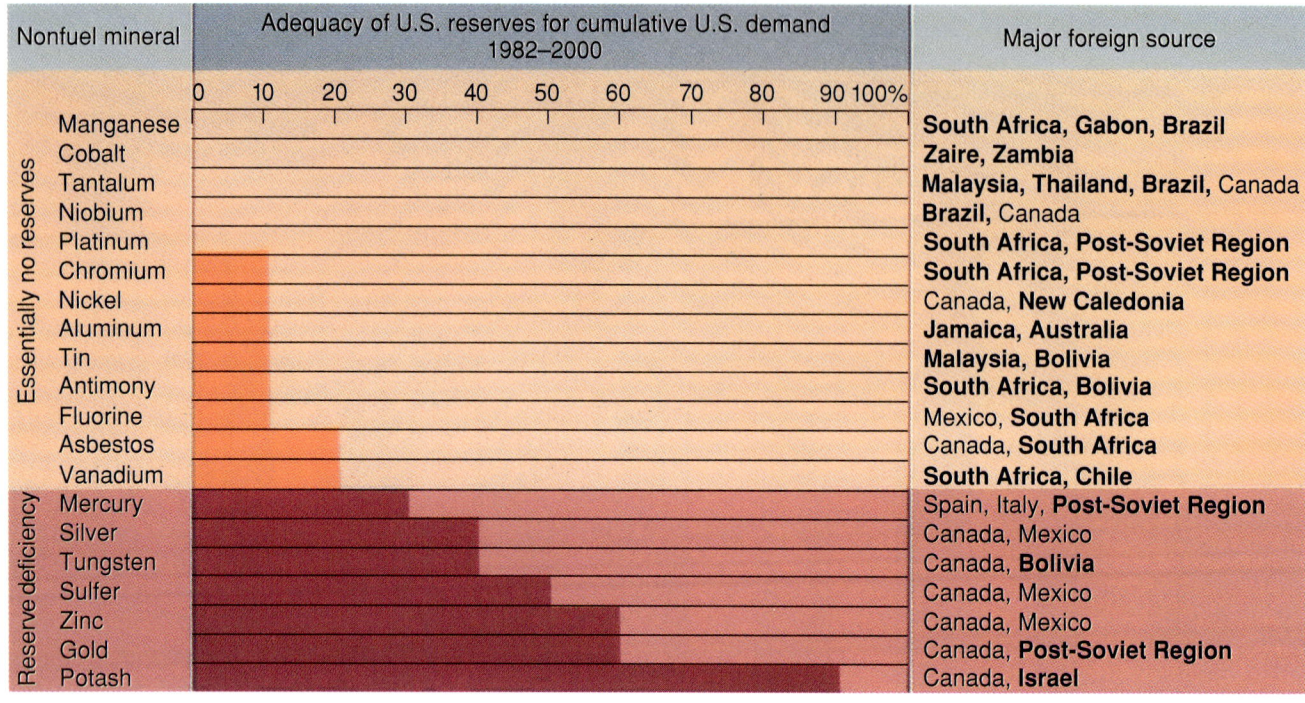

FIGURE 15-27
Adequacy of U.S. reserves of 20 mineral resources, and major foreign suppliers. Suppliers considered unreliable because of potential military, political, or economic disruptions are in boldface.

national defense. Currently, the United States depends on 25 other countries for more than half of the approximately 40 critical and strategic minerals. Some must be imported because we have no reserves of our own. We have reserves of others but we consume them more rapidly than we mine them. We import others because it is cheaper to buy them abroad than to mine and refine our own reserves. Figure 15-27 shows that the United States has insufficient reserves of 20 important mineral resources to meet demands through the year 2000, and it has almost no reserves of more than half of those resources.

Several foreign sources of critical minerals are unreliable because of potential military, political, or economic disruptions (Table 15-3). As mentioned earlier, the African nations of Zaire and Zambia supply half of the world's cobalt. Cobalt is an essential component of alloys used in jet engines and other specialized machinery. In January 1978, the government of Zaire reduced cobalt production in an effort to force the price to rise. Later that year, a civil war caused another disruption in the cobalt supply from Zaire. Prices rose rapidly from $11 to $25 per kilogram and then briefly skyrocketed to $120 per kilogram.

Forecasting the Future

Table 15-3 shows that 19 metals, including iron, aluminum, chromium, and nickel will still be abundant by the year 2100. However, shortages of other strategic and critical minerals—including gold, silver, copper, lead, and tin—are expected to develop.

Metal atoms are never destroyed; they are only dispersed. Thus, when we speak of a shortage of a metal, we don't mean that the total quantity of that element on Earth is diminished; we mean that it is spread around and hard to collect again. We can remedy mineral shortages by exploiting low-grade resources or by conserving, substituting, and recycling metals.

Exploitation of Low-Grade Resources

Figure 15-28 shows the energy required to extract and refine 1 ton of copper and 1 ton of aluminum from different concentrations of ore. Notice that relatively little energy is needed for rich ores. But as concentrations decrease, energy consumption rises rapidly. As we will

TABLE 15-3 Predictions of Future Supplies of Elements Used in Industry

Availability	Metals	Nonmetals
Limitless supply	Sodium, magnesium, calcium	Nitrogen, oxygen, neon, argon, xenon, krypton, chlorine, bromine, silicon
Abundant supply	Aluminum, gallium, iron, potassium	Sulfur, hydrogen, carbon
Abundant only if research continues to improve extraction efficiency	Lithium, strontium, rubidium, chromium, nickel, cobalt, platinum, palladium, rhodium, ruthenium, iridium, osmium	Boron, iodine
Might be in short supply by the year 2100	Gold, silver, mercury, copper, zinc, lead, arsenic, tin, molybdenum, plus 22 other metals	Helium, phosphorus, fluorine

Source: Goeller, H. E., and A. Zucker, "Infinite Resources: The Ultimate Strategy," *Science* 223 (1984), p. 456.

see in Chapter 16, energy resources are also limited, so depletion of fuels and depletion of minerals are interconnected.

Mines disturb the Earth's surface; mineral processing creates air and water pollution; and solid waste dumps cover valuable land. Just as more energy is required to refine low-grade ore than high-grade ore, more pollution generally results from processing lower-grade ore. The pollution can be controlled, but such measures are expensive and add to the total cost of mining and refining.

Conservation, Substitution, and Recycling

The economic well-being of a nation is measured in terms of its gross national product (GNP), which is the value of the total quantity of goods and services produced. The GNP rises if goods are manufactured, sold, used, and quickly discarded so that consumers purchase new products and maintain the cycle. This kind of value system encourages consumption of raw materials such as timber, fuel, and minerals. Although consumption of resources enriches us today, it may impoverish future generations. Conservation—using less—is one way to extend the life spans of our nonrenewable mineral resources.

Many metals can be replaced by cheaper, lighter, and more readily available materials in certain applications. Today, many products that were once made of metal are now constructed of plastic. Bridges are often built of reinforced concrete rather than pure steel. Aluminum wire is frequently used instead of copper.

As the prices of ores rise, recycling becomes economical. Recycling is also encouraged by higher costs of trash disposal and by programs focused on creating a cleaner environment.

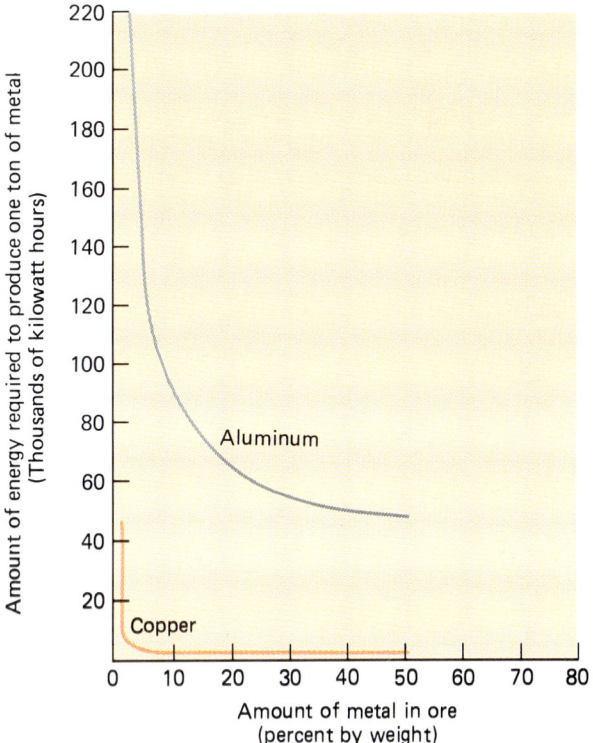

FIGURE 15-28

The energy required to mine and refine aluminum and copper ores as a function of the metal concentration of the ore. Note that there is little change in energy requirements for very concentrated ores, but then, as the metal concentration drops off, the energy requirements rise very rapidly. However, almost no mines contain 70 percent aluminum or 20 percent copper. Most mining is already being done in the steep section of the curve, so considerably more energy will be needed as lower-grade ores are exploited.

SUMMARY

Mineral resources include **metallic** and **nonmetallic** rocks and minerals. A **mineral deposit** is a local enrichment of one or more minerals. **Ore** is any deposit of metal-bearing minerals that can be mined at a profit. **Mineral reserves** are the known supplies of an ore remaining in the ground. Ore forms by magmatic, hydrothermal, sedimentary, weathering, and metamorphic processes.

Mineral resources are mined either in **tunnel mines** or in **open pit** mines. Both mining processes produce **mine spoil**, waste rock that erodes easily and weathers to form **acid mine drainage**. Mine spoil also liberates toxic heavy metals that poison streams and ground water. Smelters refine ore to produce pure metals. Smelters produce air pollution, and the waste materials from smelting erode and weather to pollute surface and ground water.

Mining depletes mineral reserves, but reserves can increase in two ways. First, geologists can discover new ore bodies. Second, **subeconomic** deposits can become profitable if the price of ore increases or if cheaper technologies of mining and refining are developed. Reserves of many metals have increased in recent years as a result of these two kinds of changes.

A **critical mineral** is one that is essential to the economy of a nation. A **strategic mineral** is one that is necessary for national defense. No nation is self-sufficient in all critical and strategic minerals because the Earth's mineral resources are unevenly distributed. The United States depends on imports for more than half of its critical and strategic mineral resources. Exploitation of low-grade mineral resources can supplement dwindling reserves, but mining and refining of low-grade ore use large amounts of energy and produce more pollution than processing of high-grade ore. In addition to finding new ore bodies, mineral reserves can be increased by conservation, recycling, and substitution of other materials for nonrenewable mineral resources.

KEY TERMS

Fossil fuel 354
Metal 354
Nonmetallic resource 354
Quarry 350
Mineral deposit 356
Ore 356
Mineral reserve 357
Crystal settling 357
Pegmatite 358
Kimberlite 358

Pipe 358
Hydrothermal solution 359
Disseminated ore deposit 359
Porphyry copper deposit 359
Black smoker 360
Placer deposit 360
Evaporite deposit 362
Marine evaporite 362
Banded iron formation 362
Residual deposit 362

Bauxite 362
Manganese nodule 363
Mine spoil 364
Tunnel mine 364
Surface mine 364
Open pit mine 364
Acid mine drainage 365
Subeconomic mineral deposit 370
Critical mineral 373
Strategic mineral 373

REVIEW QUESTIONS

1. Describe the two main categories of mineral resources. What characteristics differentiate them?
2. List several examples of metal resources.
3. List several examples of nonmetallic mineral resources.
4. Look around you and list all of the objects you see that are made from mineral resources. Look at Table 15–1 and add to your list the particular mineral resource used to manufacture each object.
5. What is ore? What are mineral reserves?
6. Explain crystal settling.
7. What types of mineral resources occur in pegmatites?
8. Explain how kimberlites contain diamonds that form in the mantle about 200 kilometers beneath the Earth's surface.
9. How do pure sulfur deposits form?
10. Describe the formation of hydrothermal ore deposits.
11. Why are vein deposits commonly associated with disseminated ore deposits of the same metals?
12. Describe how hydrothermal ore deposits form beneath the sea. How are they eventually exposed on continents?
13. Why does gold concentrate in placer deposits? What other mineral resources might concentrate in similar environments?
14. Describe the formation of marine evaporites.
15. Describe the formation of banded iron formations.
16. How and why do bauxite deposits form?
17. Compare the environmental impact of underground mines with that of surface mines.
18. Describe the types of environmental problems associated with mine spoils.

19. What kinds of environmental problems are created by the refining of metal ores?

20. What special kinds of problems characterize mining and smelting of aluminum ore?

21. What special kinds of problems characterize gold mining and refining?

22. Describe factors that can change estimates of mineral reserves.

23. Describe the differences between strategic mineral resources and critical mineral resources.

24. If most elements are widely distributed in ordinary rocks, why should we worry about someday running short?

25. Explain how the availability of mineral resources depends on the availability of energy, on other environmental issues, and on political considerations.

DISCUSSION QUESTIONS

1. Discuss the factors that determine whether or not a mineral deposit is ore.

2. Why are nonmetallic mineral resources, such as sand, gravel, limestone, and sedimentary deposits of phosphorus and potassium minerals, so important to our national economy?

3. Prepare a class debate. Have one side argue that we are quickly approaching a national crisis resulting from dwindling reserves of critical and strategic mineral resources, and have the other side argue that society will adjust to changes in reserves without disruption.

4. What factors can make our metal reserves last longer? What factors can deplete them rapidly?

5. It is common for a single mine to contain ores of two or more metals. Discuss how geological processes might favor concentration of two metals in a single deposit.

6. Discuss the changes that might occur in the ways in which we use aluminum if the aluminum industry were to pay the same energy rates that we pay for electricity supplied to our homes.

7. Compare the benefits resulting from extraction of gold from the Earth with the environmental and human costs of gold mining and refining.

8. Imagine that you plan to write a novel about a society that lives on the Earth in the year 2100. In your story, a nuclear war has occurred and major industries have been destroyed. Write a brief description of the types of materials that would be available to your characters, and explain how they would extract and process them.

9. List ten objects that you own. Of what resources are they made? How long will each of the objects be used before it is discarded? Will the materials eventually be recycled or deposited in the trash? Discuss ways of conserving resources in your own life.

10. Explain why the environmental effects and energy consumption of the mining and refining of mineral resources are likely to increase in the future. How can the environmental damage and energy consumption be mitigated?

11. Who pays the cost of reclaiming surface mines? What costs arise from surface mines if they are not reclaimed, and who pays these expenses? Which costs are most immediately apparent to the average citizen?

CHAPTER 16

Fossil and Uranium Fuels

16.1 Fossil Fuels
16.2 Coal
16.3 Petroleum and Natural Gas
16.4 Secondary Recovery and Synfuels
16.5 Uranium and Atomic Energy
16.6 The Future of Our Energy Supply
16.7 The Geopolitics of Fossil Fuels
16.8 Energy Strategies for the United States

Most people in developed countries live in luxury that would have been unimaginable to royalty a few centuries ago. A twist of the thermostat warms or cools your house. Hot water flows out of a shower head at the flick of a wrist. Push a few buttons and press gently on the gas pedal, and you can speed across the country, listening to music in air-conditioned comfort, traveling farther in an hour than a person on horseback could ride in a day. The vegetables, spices, and wines in a corner grocery store are richer and more varied than the larders of medieval kings.

Behind these luxuries lie environmental problems that threaten our future. The air conditioning, the hot water, your automobile, and the fresh kiwi fruit in the middle of winter all depend on plentiful, cheap energy. Fossil fuels provide most of the energy used in industrial societies. Mining, refining, and burning of fossil fuels deface the land, pollute air and water, and may cause greenhouse warming. Nuclear power produces radiation that can cause cancer and birth defects. In 1991, the world went to war over oil. As the supply of readily available fossil fuel becomes exhausted, we face an economic and physical disruption of our society unprecedented in modern history.

Roughnecks work to change drilling pipe in an oil field. (SuperStock)

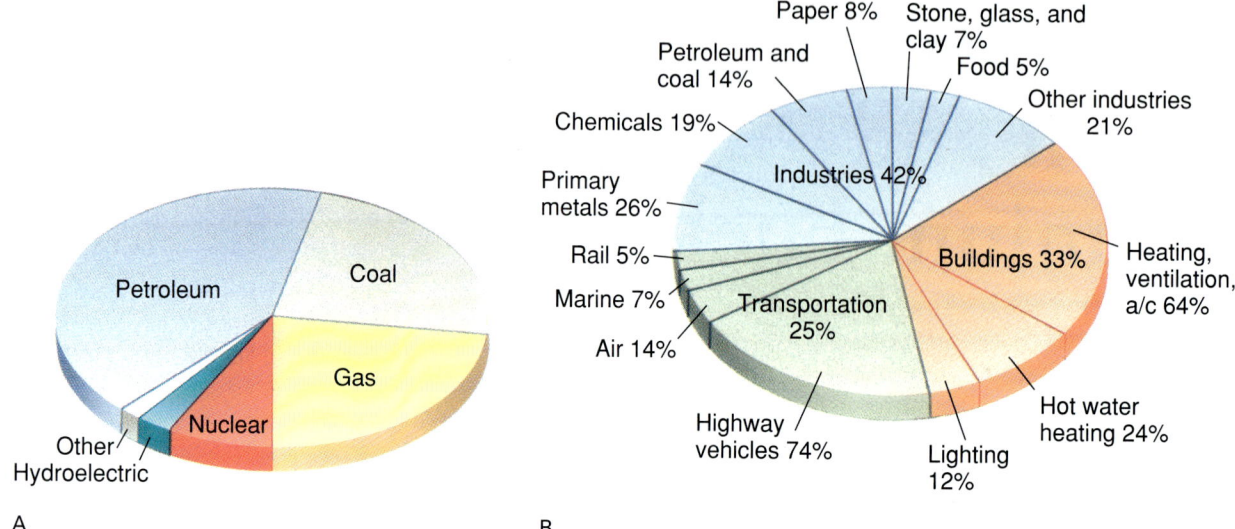

FIGURE 16-1
(A) Sources of energy production and (B) use of energy in the United States. The numbers on the outside of the circle represent percentages of each use sector. (*Annual Energy Review 1992*, EIA)

16.1 Fossil Fuels

In prehistoric times, fuel was used to provide heat. Wood, the major fuel of antiquity, was burned to warm living quarters and to cook food. During the late 1700s and early 1800s, at the start of the Industrial Revolution, people invented engines to convert the heat of burning fuels to work. As the forests of Europe became depleted, people turned to fossil fuels for their energy needs. The three major fossil fuels are coal, petroleum, and natural gas. All formed from the partially decayed remains of living organisms. Coal was used first because it is easily mined and because it can be burned without refining. The first commercial petroleum well was drilled in the United States in 1859, ushering in a new energy age. Today, energy from fossil fuels is used to produce work and heat and to transmit and manage information in the radios, televisions, stereos, telephones, and computers that have become ubiquitous in our society. Energy use in the United States is summarized in Figure 16–1.

In the early days of the Industrial Revolution, coal powered steam locomotives and even a few automobiles, but today coal is used almost exclusively in stationary facilities. About 60 percent of the coal is used in electric generating plants. The remainder is used to make steel or to produce steam in factories. Coal is abundant, but it is dirty and difficult to handle, burns slowly relative to gasoline, and emits noxious air pollutants that must be removed by expensive and bulky control devices.

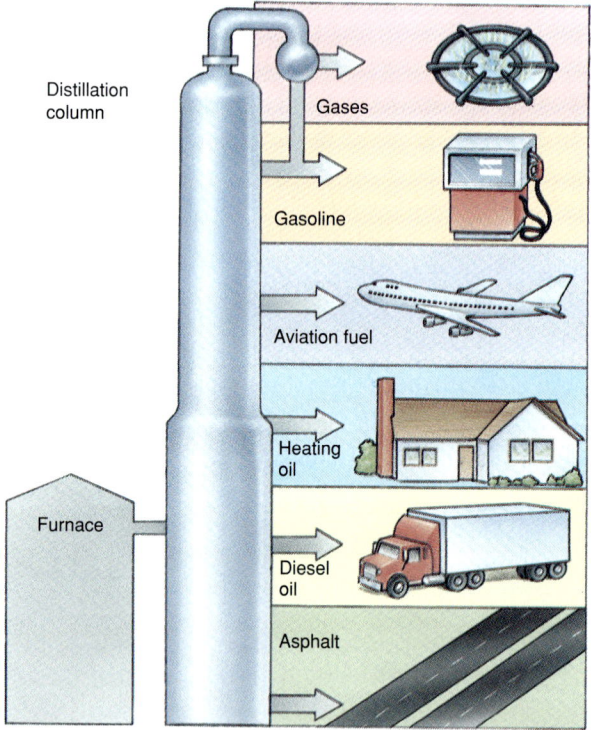

FIGURE 16-2
An oil refinery treats crude oil with heat and chemicals to produce useful products. The products are then separated by distillation. The molecules that boil at the lowest temperature rise to the top of the column, while the heavy tars are extracted from the bottom.

Petroleum is the most versatile of the fossil fuels. Crude oil, as it is pumped from the ground, is a gooey, viscous, dark liquid made up of thousands of different chemical compounds. It is then refined to produce propane, gasoline, jet fuel, heating oil, motor oil, road tar, and other useful materials. During refining, the crude oil is heated under pressure to break apart its large molecules. The mixture is then separated in multistory distillation columns. Some of the chemicals in the oil are used for the manufacture of plastics, medicines, and many other products (Fig. 16–2).

Every step in the extraction, transport, refining, and use of coal and petroleum is a potential source of environmental degradation (Fig. 16–3). On the other hand, natural gas is extracted as a nearly pure compound—methane (CH_4), and it is burned without refining. Also, it contains few impurities, and it releases no sulfur or other pollutants when it burns. As a result, gas is becoming increasingly popular and will be even more desirable if pollution laws become stricter.

16.2 Coal

Formation of Coal

In forests and grasslands, dead plants fall to the ground where they mix with oxygen and water. The oxygen supports organisms that decompose the litter. The water flows downward through the soil and removes some decay products; others are recycled back to surface plants. As a result, little organic matter accumulates except in the topsoil.

In some swamps, however, plants grow and die so rapidly that plant litter accumulates faster than it decomposes. Newly fallen vegetation covers the older, partially rotted material, so atmospheric oxygen cannot penetrate into the deeper layers. Therefore, decomposition stops before it is complete, leaving brown, partially decayed plant matter called **peat**. Commonly, peat is then buried by mud deposited in the swamp. Increasing temperature

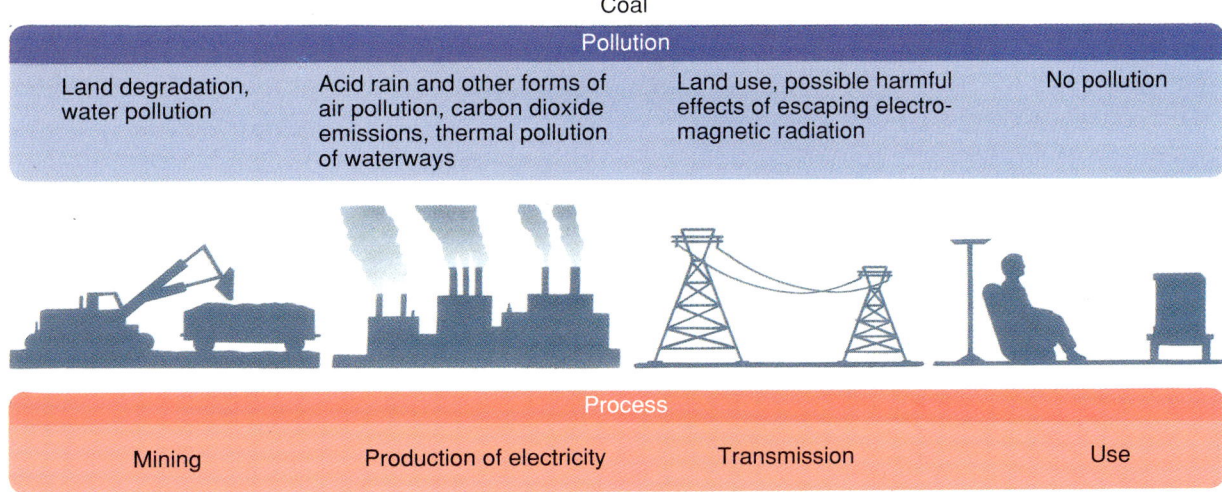

FIGURE 16–3
Every step in the extraction, transportation, refining, and use of oil and coal can pollute the environment.

FIGURE 16–4
Peat and coal form as sediment buries organic litter in a swamp.

and pressure force out water and other volatiles. Plant matter is composed mainly of carbon, hydrogen, and oxygen. During burial, most of the hydrogen and oxygen escape as gases, and the proportion of carbon increases. The result is **coal**, a combustible rock composed mainly of carbon (Fig. 16–4). Types of coal are distinguished by hardness, carbon content, and heat content.

Coal Mining

Coal is mined in both underground tunnels and surface mines (Fig. 16–5). Although tunnel mines do not directly disturb the land surface, many abandoned mines collapse, and occasionally houses fall into the holes (Fig. 16–6). Over 800,000 hectares (2 million acres) of land in

FIGURE 16–5
An underground coal mine.

FIGURE 16–6
The land surface has subsided and collapsed over an underground coal mine in Wyoming. *(C. R. Dunrud/USGS)*

central Appalachia have settled into underground mine shafts. In addition, the mine spoil is often deposited as mounds of loose, unvegetated rock. Mine spoil erodes easily, and the muddy runoff silts nearby streams and destroys aquatic habitats. Most coal contains sulfur. When the coal is exposed, some of this sulfur reacts with water and air to produce sulfuric acid (H_2SO_4). If pollution control is inadequate, the sulfuric acid oozes out of mines and tailings piles to pollute streams and ground water below the mine or mill, creating acid mine drainage similar to that produced when metal ores are extracted. Tunnel mines are also unhealthful for the miners. Coal dust collects in miners' lungs to cause **black lung disease**, a lethal form of emphysema.

Tunnel mines are often more expensive and less efficient than surface mines (Fig. 16–7). For this reason, most modern coal mining is done by large power shovels

FIGURE 16–7

An aerial view of the Decker strip coal mine in Montana. *(E. N. Hinrichs/USGS)*

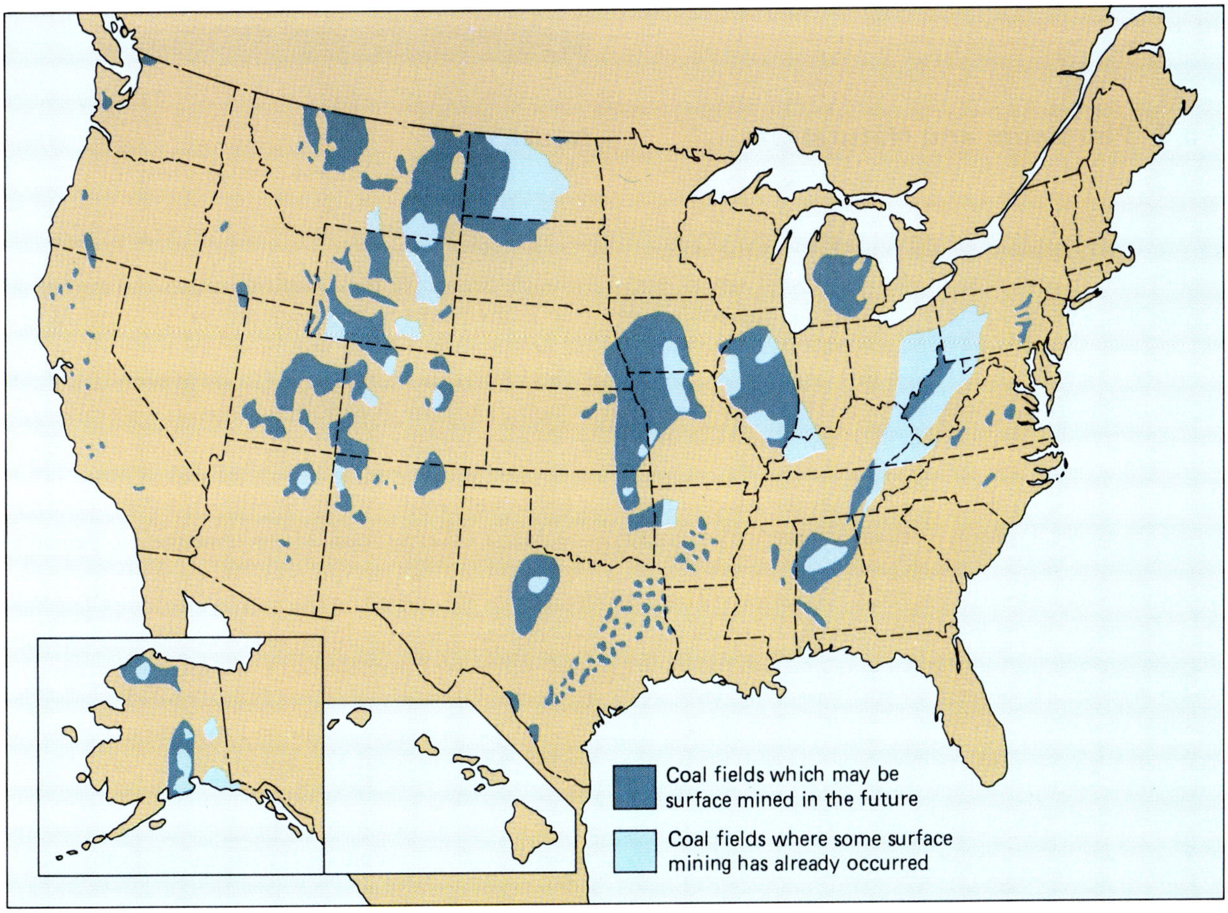

excavating coal from huge open pit mines. Figure 16–8 shows coal fields that may be surface mined in the United States. Many of these deposits lie under farmland in the

FIGURE 16–8

Areas in the United States where economically attractive coal deposits are found. Lightly shaded areas indicate coal fields where some surface mining has already occurred. Dark shading indicates coal fields that may be surface mined in the future.

383

West and Midwest. Thus, a potential conflict of resource use arises. Should land be used for fuel or food? Recall from Chapter 15 that surface mines can be restored by replacing the overburden and topsoil. Recall, also, that many mining companies have failed to adhere to the strict wording of the Surface Mining Control and Reclamation Act.

Burning Coal

Of all the fossil fuels, coal produces the most air pollution when burned. Mineral matter in coal escapes as fly ash when the coal is burned. Before strict pollution control laws were passed, fly ash covered the cities around coal-burning facilities with a black, gritty film.

Sulfur present in coal escapes into the atmosphere as sulfur dioxide, which reacts to form sulfuric acid. The sulfuric acid then falls to the Earth as acid precipitation. When any fuels are burned, the carbon in the fuel reacts to form carbon dioxide, which is the major contributor to the greenhouse effect.

16.3 Petroleum and Natural Gas

Formation of Petroleum

Streams carry organic matter from land to the oceans and deposit it with mud in shallow coastal waters. As plants and animals living in the sea die and settle to the bottom, they add more organic matter to the mud. Younger sediment then buries this organic-rich mud. Rising temperature and pressure caused by burial convert the mud to shale or mudstone. At the same time, the rising temperature and pressure also convert the organic matter to liquid **petroleum** that is finely dispersed in the rock (Fig. 16–9). Typically, petroleum forms in the range from 50° to 100°C. As a comparison, lukewarm tea is about 50°C, and water boils at 100°C. But dispersed oil in shale cannot be pumped from an oil well. A commercial petroleum **reservoir** forms when oil flows from the shale and concentrates in other rock. Such a reservoir develops in three phases.

The shale or other sedimentary rock in which oil originally forms is called the **source rock**. Oil cannot be recovered from shale because shale is relatively **impermeable**; that is, liquids do not flow through it rapidly.

If conditions are favorable, petroleum is forced out of the source rock and migrates to a nearby layer of porous sandstone or limestone. If liquids can flow readily through porous rock, the rock is said to be **permeable**.

Because petroleum is less dense than water and much less dense than rock, it rises or migrates horizontally

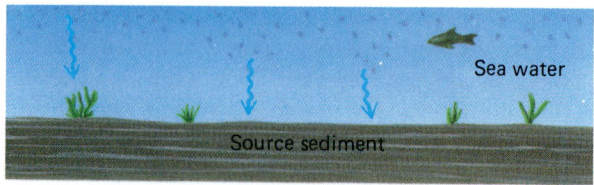

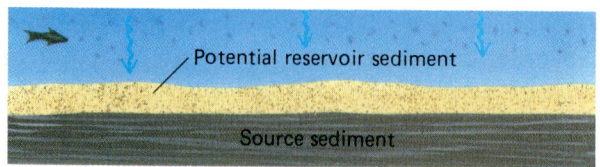

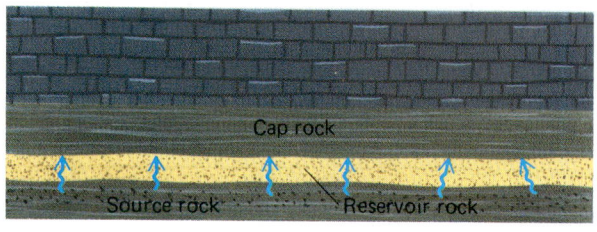

FIGURE 16–9
The stages in the development of an oil reservoir.

through permeable rock until it reaches a trap. An **oil trap** is any barrier to the migration of oil or gas. When oil or gas seeps into a trap, it accumulates there and forms a reservoir. Most reservoirs form in permeable sandstone or limestone. A petroleum reservoir is not an underground pool or lake of oil. It consists of permeable, porous rock filled with oil, more like an oil-soaked sponge than a bottle full of oil.

Many oil traps form where impermeable **cap rock** covers permeable reservoir rock and prevents the petroleum from escaping. The cap rock is often another shale bed because shale has low permeability. Other oil traps form where rock is distorted or fractured. Four different types of oil traps are shown in Figure 16–10. If alternate layers of permeable sandstone and impermeable shale are folded, petroleum may concentrate in the sandstone at the high point of the fold (Fig. 16–10A). Alternatively, a trap can form by faulting or tilting of layered rock (Figs. 16–10B and C). In some regions, layers of salt rise to form salt domes because salt is less dense than the surrounding rocks. The salt dome may distort the surrounding rock to form an oil trap (Fig. 16–10D).

Geological activity can destroy an oil trap as well as create one. A fault may fracture the trap, or the trap may be uplifted and eroded. In either case, the petroleum

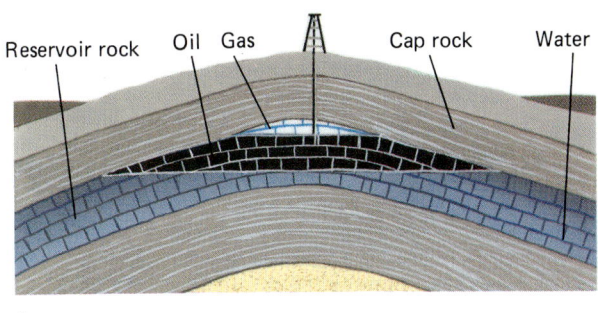
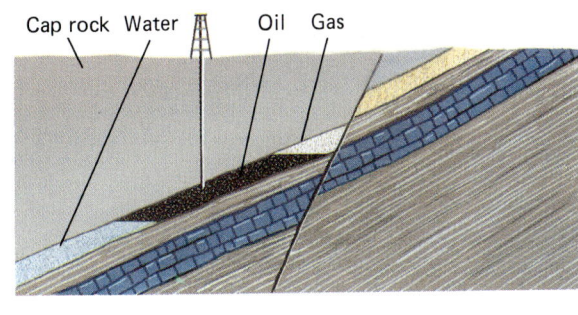
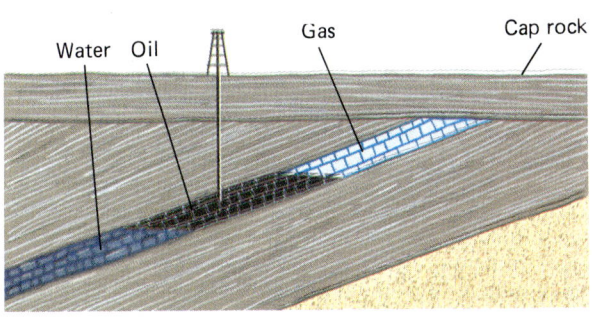
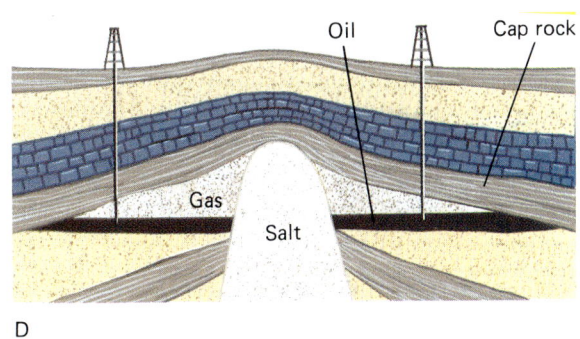

FIGURE 16–10

Four different types of oil traps. (A) Petroleum rises into permeable limestone and is trapped by a dome of impermeable shale. (B) A trap forms where a fault has moved impermeable shale against permeable sandstone. (C) Horizontally bedded shale overlies inclined, petroleum-bearing limestone. (D) In a salt dome, a sedimentary salt deposit rises and deforms overlying strata to create a dome trap.

escapes once the trap is destroyed. Sixty percent of all oil wells are found in relatively young rocks that formed during the Cenozoic Era. Undoubtedly, much petroleum that had formed in older Mesozoic and Paleozoic rocks escaped long ago and decomposed at the Earth's surface.

Petroleum Extraction, Transport, and Refining

Relatively little environmental damage occurs when an oil well is drilled in a temperate or arid locality. An oil well occupies only a few hundred square meters of land. Wells are even drilled in people's backyards with minimal environmental problems (Fig. 16–11). As these relatively accessible reserves are being depleted, however, oil companies have begun to extract petroleum from more fragile environments, such as the ocean floor and the Arctic tundra.

Rich offshore oil reserves exist on continental shelves in many parts of the world, including the coast of southern California, the Gulf of Mexico, and the North Sea in Europe. To obtain the oil, engineers first build platforms on pilings driven into the ocean floor.

FIGURE 16–11

After an oil well has been drilled, the expensive drill platform is removed and replaced by a pumper, such as this one on the Wyoming prairie.

FIGURE 16–12
An offshore oil drilling platform. (Schlumberger Inc.)

FIGURE 16–13
An oil refinery in Linden, New Jersey.

Drilling rigs are then mounted on these steel islands (Fig. 16–12). Despite great care, accidents occur during drilling and extraction of oil. Broken pipes, excess pressure in wells, and difficulty in capping a new well have all led to blowouts, spills, and oil fires. When these accidents occur on land, they can destroy small areas of farmland but do not affect entire ecosystems. When accidents occur at sea, however, millions of barrels of oil can spread throughout the waters, poisoning marine life and disrupting marine ecosystems. Significant oil spills have occurred in virtually all offshore drilling areas.

Petroleum is commonly shipped long distances by huge oil tankers. As discussed in Chapter 12, oil spills from both offshore drilling and tanker accidents pollute the world's coastal oceans.

Oil refineries are also sources of pollution. A large refinery covers hundreds of acres and processes thousands of barrels of petroleum every day (Fig. 16–13). Although all refineries use expensive pollution control equipment, these devices are never completely effective. As a result, toxic and carcinogenic chemicals escape into the atmosphere.

Natural Gas

Natural gas, or methane (CH_4), forms naturally when crude oil is heated to between 100° and 150°C during burial. Many oil wells contain natural gas that floats above the heavier liquid petroleum. In other instances, the lighter, more mobile gas escaped and was trapped elsewhere in a separate reservoir. Over geologic time, large quantities of natural gas have escaped into the atmosphere.

16.4 Secondary Recovery and Synfuels

Today, coal is forming in some swamps. Oil and gas are forming in organic-rich muddy sediments in many marine basins and coastal areas. However, because these processes are much slower than current rates of consumption, fossil fuels are **nonrenewable resources**. As discussed in Section 16.6, fossil fuel availability will diminish during the early part of the twenty-first century and the price will certainly rise. The problem will become acute for oil and gas within a generation, although coal reserves are more plentiful. Will people start riding horses to town again when the supply of petroleum runs out (Fig. 16–14)? The answer is almost certainly no because many alternative energy sources, described in the next chapter, are available. In addition, the amount of available oil in known fields can be extended by secondary recovery methods. Another approach to increasing energy resources is to synthesize liquid fuels from nonliquid fuels. Synthetic fuels derived from coal and other sources are called **synfuels**.

Secondary Recovery from Oil Wells

On the average, about half of the oil in a reservoir is too viscous to be pumped to the surface by conventional techniques. This oil is left behind after a well has "gone dry" but can be extracted by **secondary recovery** techniques. One method is to force superheated steam into old wells. The steam heats the oil so that it can flow through the rock and into the well. Because energy is needed to heat the steam, this type of extraction is not always cost-effective or energy-efficient. Another process involves pumping detergent into the reservoir. The detergent dissolves the oil and carries it to the well. The petroleum is then recovered and the detergent recycled.

In the United States, more than 300 billion barrels of secondary oil remain in oil fields. In 1992, people in the United States consumed 6.2 billion barrels of petroleum. Thus, if all the secondary petroleum could be recovered, it would supply the United States for nearly 50 years. However, not all of the 300 billion barrels can be recovered, and the energy yield from secondary extraction will be reduced by the energy required to extract it. Nevertheless, when petroleum reserves are depleted and the price rises, secondary recovery will become significant.

Tar Sands

In some regions, large sand deposits are permeated with heavy oil and an oil-like substance called **bitumen** that is too thick to be pumped. The richest tar sands exist in Alberta (Canada), Utah, and Venezuela. In Alberta alone, tar sands contain an estimated 1 trillion barrels of petroleum. About 10 percent of this fuel lies close enough to the surface to be mined in open pits, like coal (Fig. 16–15). The oil-rich sand is dug up and heated with steam to make the bitumen fluid enough to separate from the sand. The bitumen is treated chemically to convert it to crude oil, which is then refined. At present, several companies are operating tar sands profitably, producing 11 percent of Canada's petroleum. Deeper deposits, constituting the remaining 90 percent of the reserve, can be extracted using subsurface techniques similar to those discussed for secondary recovery. A portion of these larger deposits will be economical to mine when the price of oil stabilizes above $20 per barrel (in August 1994, petroleum cost $18 a barrel).

FIGURE 16–14

Will people start riding horses to town again when the petroleum supply runs out? *(David Stoecklein)*

FIGURE 16–15
Mining tar sands in Alberta, Canada. *(Syncrude Canada, Limited)*

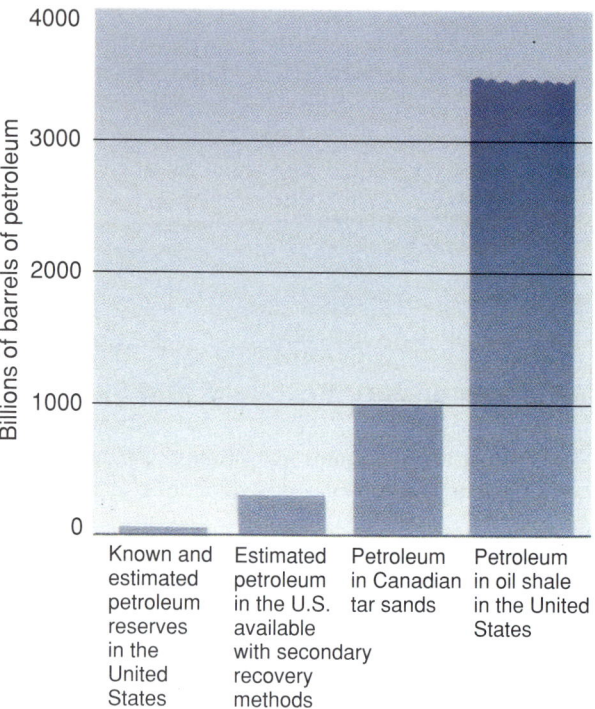

FIGURE 16–16
Secondary recovery, tar sands, and oil shale increase our petroleum reserves significantly.

Oil Shale

Some shales and other sedimentary rocks contain a waxy, solid organic substance called **kerogen**, the precursor of liquid petroleum. Kerogen-bearing rock is called **oil shale**. If oil shale is mined and heated in the presence of water, the kerogen converts to petroleum. In the United States, oil shales contain the energy equivalent of 2 to 5 trillion barrels of petroleum, enough to fuel the nation for 325 to 800 years at the 1992 consumption rate (Fig. 16–16). However, many oil shales are of such low grade that they require more energy to mine and convert the kerogen to petroleum than is generated by burning the oil, so they will probably never be used for fuel. Oil from higher-grade oil shales in the United States would supply this country for about 75 years if consumption rates remained at 1992 levels. Worldwide deposits are not as rich, so the global situation is not as promising.

Oil shale can be mined either in surface mines or underground. In either case, the mining creates environmental disruptions similar to those created by coal mining. But additional problems specific to shale development also occur. When oil shale is broken into small pieces and is exposed to the atmosphere, it absorbs water and swells. Thus, the volume of rock increases even though oil is removed from it. As an analogy, take a potato, slice it into thin slices, and fry the slices to make potato chips. Now place the potato chips in a pile. Which occupies more volume, the potato chips or the potato? After the kerogen is removed, the expanded shale must be dumped somewhere, and the piles of rock are unsightly and easily eroded.

Another problem inherent in oil shale development is water consumption. Approximately two barrels of water are needed to produce each barrel of oil from shale. Oil shale occurs most abundantly in the semiarid western United States (Fig. 16–17). In this region, water is also needed for agriculture, domestic uses, and industry (Fig. 16–18). Proponents of oil shale development claim that when oil prices rise and oil shale becomes economical, new dams can be built to retain water, and industries can move elsewhere. Moreover, most of the ranchers in the region use water inefficiently, and much of the land is marginally productive. Therefore, according to this argument, it might be in the national interest to use the water for oil shale development. But the law allocates water rights according to history of use. The original ranchers and farmers passed down the water rights to later generations or sold them with the land. It would be unfair and perhaps unconstitutional to change water laws and harm individuals, even if the nation as a whole might benefit.

When oil prices skyrocketed from $22 per barrel in 1978 to $45 per barrel in 1981, major oil companies explored oil shale reserves and built experimental recovery plants. However, when prices plummeted a few years later, most of this activity came to a halt. We can expect renewed interest in oil shale when the wholesale price of oil doubles to about $35 per barrel.

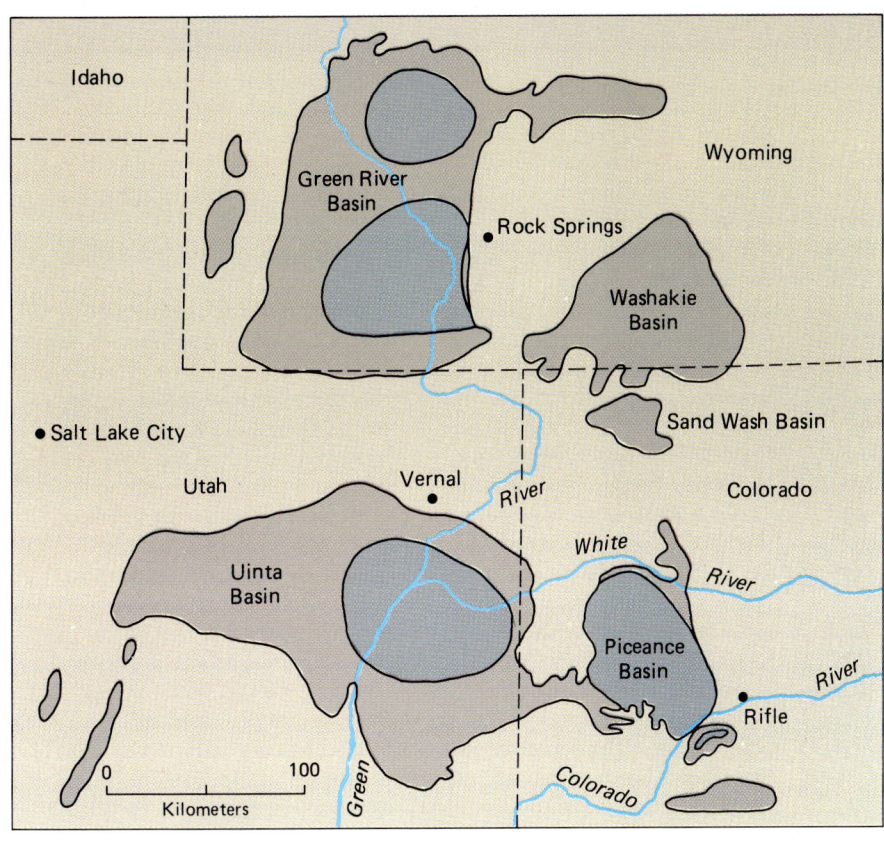

FIGURE 16–17
Locations of major oil shale deposits in the United States. (*Adapted from U.S. Geological Survey Circular 2223, 1965*)

FIGURE 16–18
The area near Rifle, Colorado, will experience conflict over water use if the great oil shale deposits are exploited.

Liquefaction of Coal

During World War II, Allied warships and bombers cut off Germany's oil supply. Germany is endowed with plentiful coal deposits, but coal cannot be used directly in conventional automobiles, airplanes, and tanks. So the Germans developed chemical procedures to convert coal to liquid fuels. Today, this 50-year-old technology has been improved and is practiced both in the laboratory and in pilot plants. However, coal conversion is about as costly as oil shale extraction and thus is presently too expensive to compete with direct petroleum extraction.

Gas Hydrates

In the 1930s, when the first high-pressure natural gas pipelines were built, engineers were troubled by solids that formed in the pipe and blocked the flow of gas. They learned that natural gas (methane) reacts with water under the high pressure and low temperatures found in some pipelines to form an icelike solid called **methyl**

FOCUS ON

Formation of Fossil Fuels: A Novel Idea

Conventional geological theory states that fossil fuels form from the remains of ancient organisms, as explained in the text. However, Dr. Thomas Gold of Cornell University has proposed an alternative hypothesis to explain the origins of oil and gas.

Petroleum is composed primarily of hydrocarbons, energy-rich molecules made from carbon and hydrogen. Natural gas is pure methane, which is the simplest hydrocarbon, CH_4. Hydrocarbons also exist throughout the Solar System and have been detected on Jupiter, Saturn, Uranus, Neptune, Pluto, many planetary moons, asteroids, meteorites, and comets. Abiological (nonliving) processes created these extraterrestrial hydrocarbons 5 or more billion years ago, when the incipient Solar System was a cloud of dust and gas rotating slowly in space.

Dr. Gold proposes that large quantities of these abiological hydrocarbons accumulated in the Earth as it formed. Because hydrocarbons are less dense than rock, they have been rising slowly toward the surface throughout geological time. Oil and gas reservoirs are places where these hydrocarbons have risen from the lower crust and upper mantle and have concentrated in traps.

Critics point out that oil and gas deposits contain complex molecules, such as DNA fragments, that are synthesized only in living organisms and are not found elsewhere in the Solar System. Thus, oil and gas could only have formed from living organisms, as conventional hypotheses propose. Dr. Gold counters that these biological molecules are not evidence for the origins of the fuel deposits, but have been added to the oil and gas by a secondary process. Deep-drilling experiments have uncovered bacteria and other microbes 5 to 10 kilometers beneath the Earth's surface. According to Dr. Gold, these deep-crust microbes eat the hydrocarbons and contaminate them with biogenic molecules.

Several observations support Gold's novel and controversial hypothesis. First, Russian geologists have dug deep wells beneath conventional oil and gas deposits and in every case have found additional hydrocarbon deposits down to and into basement rock. If oil and gas reserves formed from hydrocarbons rising from the deep crust and mantle, one would expect this observed vertical profile. However, if fossil fuels formed by biological processes, it would be unlikely for conditions to remain favorable for concentration of organic material in the same place over the hundreds of millions of years represented by the entire vertical sequence of rocks containing the hydrocarbons.

Second, every deep hole drilled into igneous or metamorphic basement rock has revealed hydrocarbons. In one well drilled through granitic rock in Sweden, geologists found a thick hydrocarbon sludge 6.7 kilometers beneath the surface. They collected 15 tons of this thick, petroleum-like liquid. In addition, enough methane and hydrogen rose from the hole to burn in a steady flame at the top of the well. The closest sedimentary rocks were 14 kilometers from these hydrocarbon deposits.

Dr. Gold also points out that the methyl hydrate deposits in marine sediments on the sea floor and beneath the Arctic permafrost are too abundant to have formed by concentration of gases escaping from decaying organisms, as most geologists have postulated.

If Dr. Gold's hypothesis is correct, our fossil fuel reserves may be many times greater than geologists now estimate. Yet the hypothesis remains speculative, and the reserves, if they exist, may be very deep or widely dispersed. Dr. Gold recently wrote us a letter and explained, "I have had over 100 individual reprint requests for the article[1] but only two from petroleum geologists. Evidently they continue to be skeptical."

[1] Thomas Gold, "The Deep, Hot Biosphere," *Proceedings of the National Academy of Science*, 89 (July 1992), 6045–6049.

---

hydrate. The problem was solved by carefully drying the gas before it was fed into the pipe.

In the 1970s, oceanographers built specialized drilling ships to study the sediments of the continental shelves. When certain sediment cores were brought to the surface and stored on deck, sudden gas releases blew the sediment out of the drill pipes. Researchers learned that below a depth of about 500 meters, seawater temperature is low enough and the pressure is high enough to convert methane trapped in sea-floor mud to methyl hydrate. When brought to the surface, the hydrate decomposed to release methane gas. Scientists have found similar methyl hydrate deposits in Arctic permafrost. Many geologists contend that the sea-floor methane was originally produced by microbes, but not everyone agrees (see "Focus on Formation of Fossil Fuels: A Novel Idea").

Using both drill samples and studies of seismic waves, scientists have located sea-floor methyl hydrates worldwide. Geochemist Keith Kvenvolden of the United States Geological Survey estimates that methyl hydrate deposits hold twice as much carbon as all conventional coal, oil, and gas reserves combined.

At present, commercial extraction of methyl hydrates is impractical. It is expensive to drill in deep water, and before such ventures are attempted, energy companies must develop more reliable estimates of the hydrate concentrations. A thin layer spread throughout the continental shelves would be prohibitively expensive to exploit, whereas concentrated reservoirs would be more attractive.

In addition to their possible commercial applications, methyl hydrates have considerable scientific interest. Tectonic activity at subduction zones or landslides on continental slopes could reduce pressure and release methane from hydrate deposits. In turn, increased atmospheric methane could trigger greenhouse warming. Ice core studies show that global atmospheric methane concentration has altered rapidly in the past, perhaps by sudden releases of oceanic methyl hydrates.

16.5 Uranium and Atomic Energy

How a Nuclear Reactor Works

In 1896, the French scientist Henri Becquerel was studying the interaction of sunlight with mineral crystals. For 2 days storm clouds obscured the sun, and Becquerel placed the crystals in a drawer while he waited for better weather. By chance, some of the crystals were composed of uranium salts; and also by chance, Becquerel stored photographic plates in the same drawer. A few days later, he found, to his surprise, that the photographic plates had been exposed, as if they had been left out in the light. After careful experimentation, Becquerel learned that uranium spontaneously emits energy that exposes photographic film. The spontaneous emission of radiation from an element is called **radioactivity**. Radioactivity emanates from atomic nuclei and is fundamentally different from chemical reactions that occur as a result of changes in the electron configuration of atoms and molecules (Fig. 16–19). The natural radioactivity of uranium and other radioactive elements is partially responsible for heating the interior of the Earth.

Modern nuclear power plants use **nuclear fission** to produce heat and generate electricity (Fig. 16–20). One isotope of uranium, U-235, is the major fuel. When a U-235 nucleus is bombarded with a neutron, it breaks apart, forming decay products (hence the word *fission*,

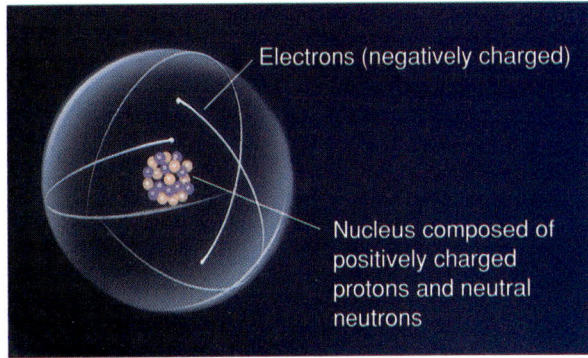

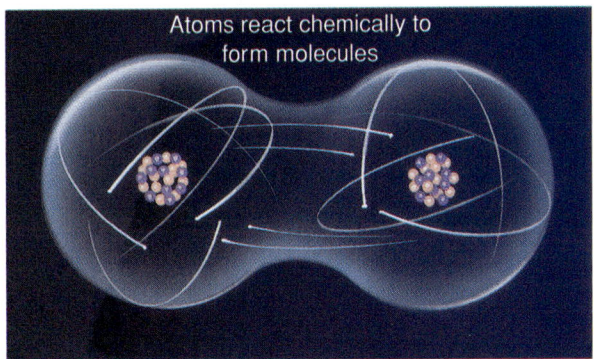

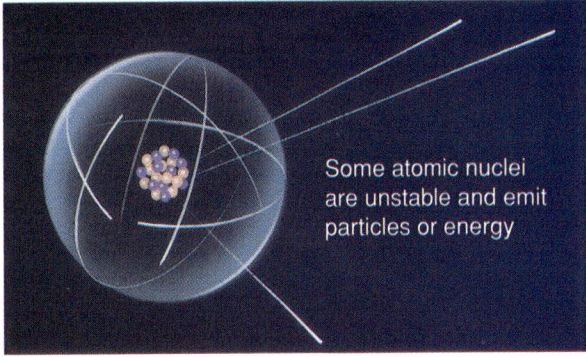

FIGURE 16–19
(A) An atom consists of a small, dense, positively charged nucleus surrounded by a diffuse cloud of negatively charged electrons. (B) A chemical reaction occurs when the electrons interact. (C) Radioactivity emanates from atomic nuclei and is therefore fundamentally different from a chemical reaction.

which means "splitting"). Because this fission is initiated by neutron bombardment, it is not a spontaneous process and is different from natural radioactivity.

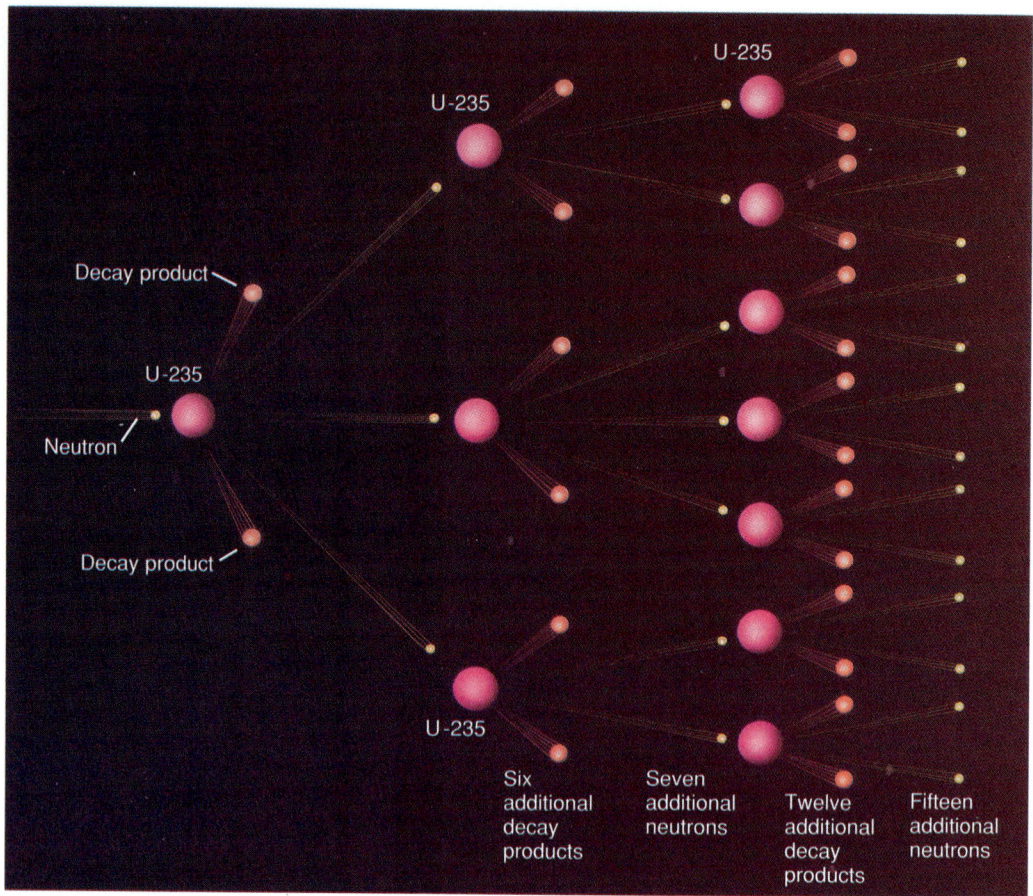

FIGURE 16–20

A branching chain reaction. When a neutron strikes a uranium-235 nucleus, the nucleus splits into two roughly equal decay products and emits two or three neutrons. These neutrons can then initiate additional reactions, which produce more neutrons. A branching chain reaction accelerates rapidly through a mass of concentrated uranium-235.

A U-235 nucleus emits two or three neutrons when it splits (depending on which reaction pathway it follows). Because one neutron starts the reaction but two or three neutrons are produced, the reaction can spread through a sample quickly. The situation is analogous to a forest fire. Imagine that one tree is struck by lightning and set on fire. The heat from the burning tree ignites two nearby trees, those trees ignite four more, and so on. Soon the whole forest is burning. A process that accelerates in this manner is called a **branching chain reaction** (Fig. 16–20).

A neutron within a mass of uranium is like a bullet flying through a forest. If the trees are far apart and the forest is small, a bullet may escape without hitting a tree. However, if the trees are close together and the forest is large, a bullet will eventually hit a tree. If a U-235 sample is small, most of the neutrons will escape and the reaction will not branch. If the sample is large and concentrated, most of the neutrons will hit nuclei, causing a branching chain reaction.

To make an atomic bomb, scientists slam several small pieces of concentrated uranium-235 together rapidly to create a large, concentrated mass of uranium-235. They then supply a few neutrons, and the bomb explodes. The trick to operating a nuclear power plant is to control the neutrons so that less than half of them hit other U-235 nuclei. In this way, each fission reaction triggers only one other fission reaction. Thus, the chain reaction doesn't branch and the fission proceeds at a controlled rate.

Natural uranium ore contains only 0.7 percent U-235. Because there are so few fissionable nuclei, most neutrons hit other atoms and do not cause a branching

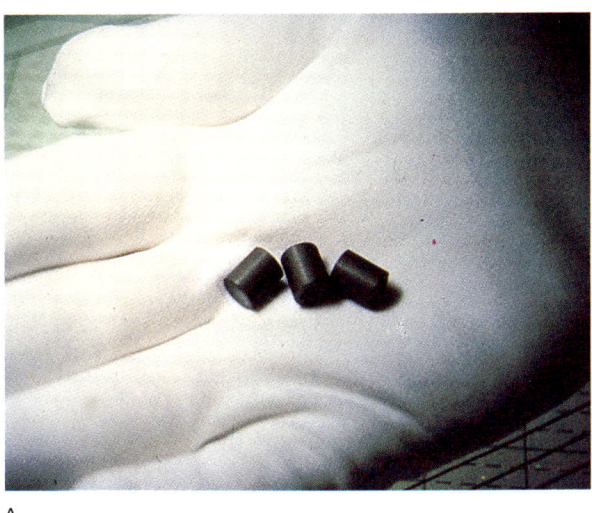

FIGURE 16–21
(A) Fuel pellets containing enriched uranium-235. Each pellet contains the energy equivalent of 1 ton of coal. (B) Fuel pellets are encased in narrow rods that are bundled together and lowered into the reactor core. *(Courtesy Westinghouse Electric Corp.)*

reaction. Thus, natural uranium ore cannot be used as nuclear fuel. However, if uranium ore is enriched to about 3 percent U-235, the concentration is just right for fission to proceed at a controlled rate. If a neutron strikes a U-235 atom, it splits, emitting two or three neutrons. But at this low concentration only one of these strikes another U-235 atom. The others, by chance, strike other nuclei and do not cause a reaction.

Because of the low grade of the uranium fuel used in nuclear power plants, the reaction cannot possibly branch to create an explosion. There are many problems with nuclear reactors, but they cannot explode like an atomic bomb.

The fuel-grade uranium is compressed into small pellets. Each pellet could easily fit into your hand but contains the energy equivalent of 1 ton of coal. A column of pellets is encased in a 2-meter-long pipe called a **fuel rod** (Fig. 16–21). A typical nuclear power plant contains about 50,000 fuel rods bundled into assemblies of 200 rods each. **Control rods** made of special neutron-absorbing alloys are spaced among the fuel rods. The control rods fine-tune the reactor. If the reaction speeds up because too many neutrons are striking other uranium atoms, then the power plant operator lowers the control rods to absorb more neutrons and slow down the reaction. If fission slows down because too many neutrons are absorbed, the operator raises the control rods. If an accident occurs and all internal power systems fail, the control rods fall into the reactor core and quench the fission.

The reactor core produces tremendous amounts of heat. Some fluid, usually water, is pumped through the reactor core to cool it. The cooling water (which is now radioactive due to exposure to the reactor core) is then passed through a radiator, where it heats another source of water to produce steam. The steam drives a turbine, which in turn generates electricity (Fig. 16–22).

Environmental Problems from Nuclear Reactors

Health Effects of Routine Radioactive Emissions

Experiments with animals and humans, observations of atomic bomb victims in Japan, and studies of accidents in the nuclear power and weapons industries show that a high dose of radioactivity damages human tissue, causes cancer and mutations, and, in extremes, is lethal. On

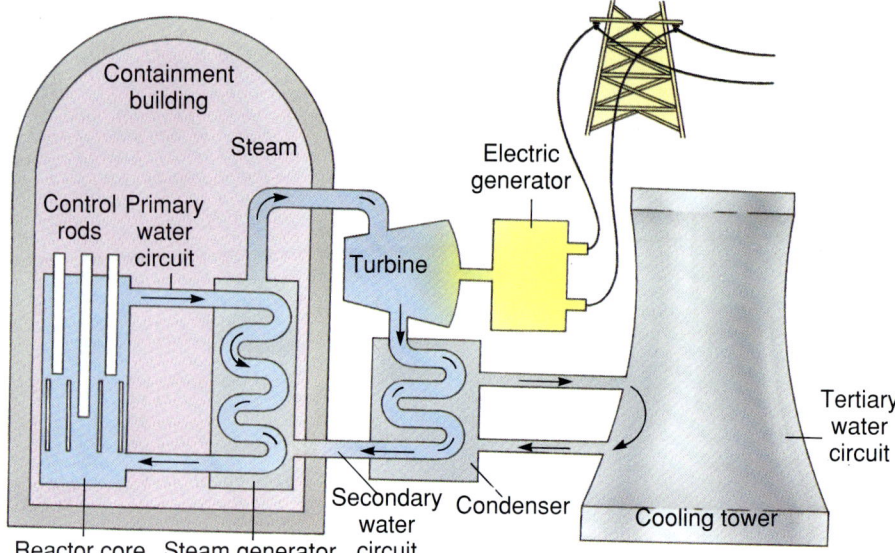

FIGURE 16–22
Schematic diagram of a nuclear power plant.

the other hand, scientists debate the effects of the low doses of radioactivity that escape during routine operations in a nuclear power plant. All living organisms on Earth are exposed to low levels of background radiation from cosmic rays and radioactive elements in the Earth's crust. In addition, most people in developed countries are exposed to medical X-rays. A nuclear power plant is carefully shielded with lead linings and a concrete housing, so little radiation escapes from the plant. Thus, routine operation of nuclear power plants contributes only 0.1 percent of the average human radiation dose (Fig. 16–23). Proponents of nuclear power point out that coal is contaminated with radioactive minerals. Because coal-fired power plants burn many tons of fuel, routine radioactive emissions from a coal-fired plant are actually greater than those from a nuclear power plant.

Radioactive Wastes

Every step in the mining, processing, and use of nuclear fuel produces radioactive waste products that create environmental problems (Fig. 16–24). When uranium ore is mined, the mine spoil is radioactive. Wastes are produced during enrichment. Finally, the fission process itself produces radioactive waste. When a U-235 nucleus breaks apart, it splits into two radioactive decay products. These new nuclei are more radioactive than the original uranium. The new nuclei cannot be used in a reactor, either because they are not fissionable or because they are too expensive to purify. Thus, a nuclear power plant creates radioactive wastes. After several months in a reactor, the U-235 concentration in the fuel rods drops until the fuel pellets are no longer useful. The fuel rods, with their radioactive wastes, are then removed. In some countries, these pellets are reprocessed to recover U-235, but in

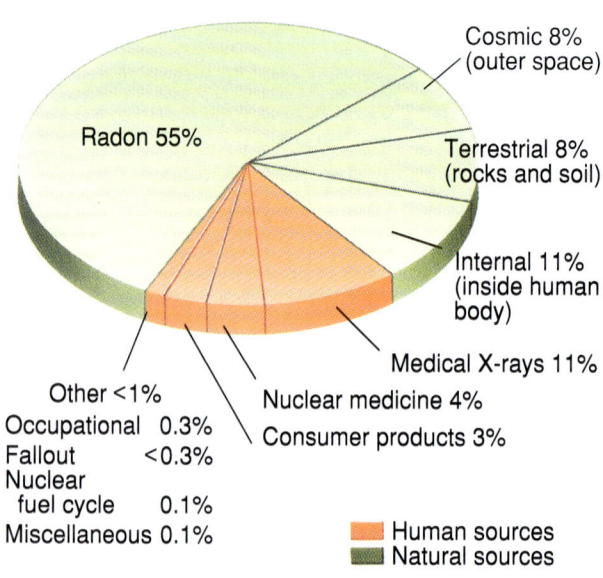

FIGURE 16–23
An average person in the United States is exposed to several sources of radiation. Note that 82 percent of the radiation comes from natural and unavoidable sources and only 0.1 percent comes from the nuclear fuel cycle.

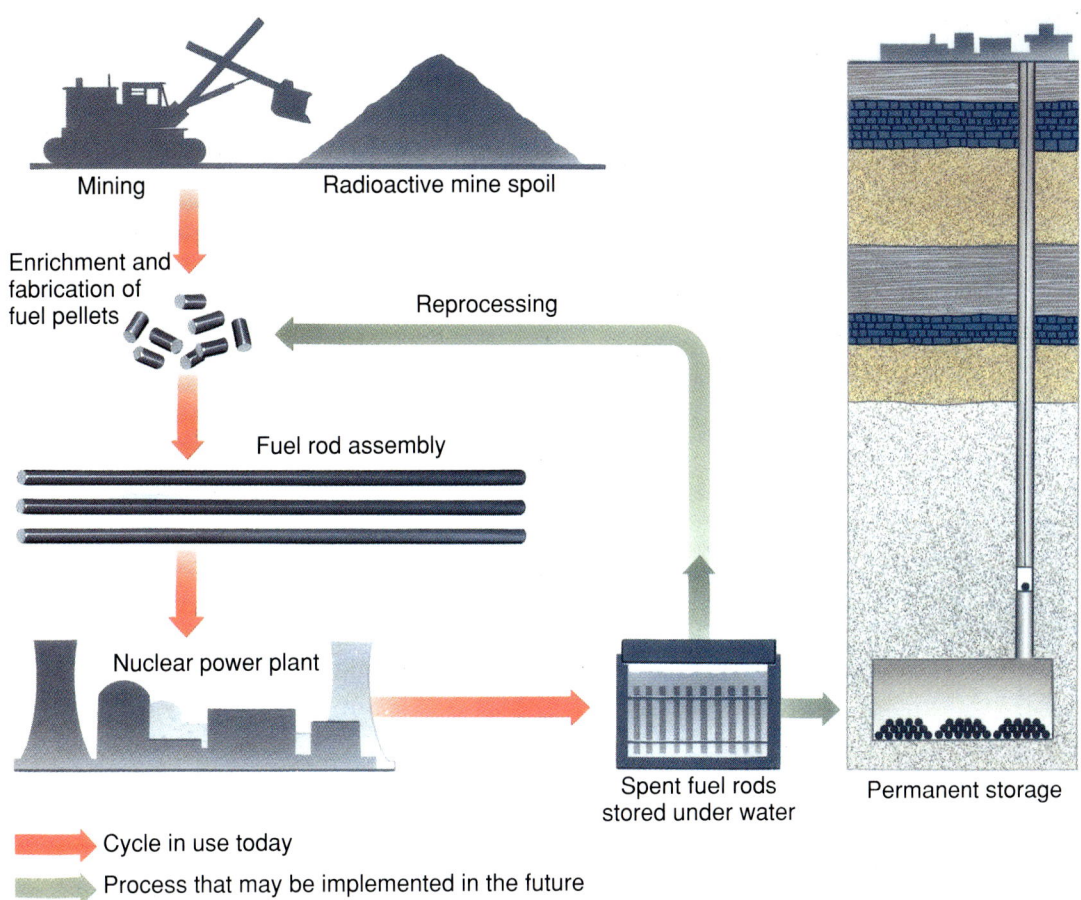

FIGURE 16–24

The nuclear fuel cycle. Although reprocessing is practiced in some countries, it is not economical in the United States; the result is more nuclear wastes to be buried.

the United States this is not economical and the pellets are discarded. In Chapter 11 we discussed proposals for storing radioactive wastes. To date, no satisfactory solution has been found.

Decommissioning Nuclear Power Plants

Another problem arises from **decommissioning** nuclear power plants. Gradually, the metal walls of a nuclear reactor become brittle from the constant neutron bombardment. At the same time, repeated pressure and temperature fluctuations weaken the pipes carrying steam to the reactor's turbines. If the fatigued metals crack, dangerous radioactivity can escape. Therefore, a nuclear power plant has a useful life span of only 30 to 40 years. Because the nuclear energy age began about 30 years ago, decommissioning has recently been added to the woes of the nuclear power industry. At the end of 1992, two reactors in the United States were retired, and 67 are scheduled for decommissioning by 2015.

Two options have been proposed for decommissioning. **Dismantlement** involves removing spent fuel, taking the reactor apart, and shipping all radioactive waste and contaminated metal to a permanent underground storage facility. Utility officials estimate that the cost will be from $170 million to $1 billion for each power plant. Due to the cost, the danger to workers, and the fact that at present no permanent underground storage site exists, no reactors have been dismantled. A second approach is to entomb the reactor with concrete and dirt, build a fence around it, and post guards to keep people away. This procedure, called **isolation**, creates radioactive waste storage sites throughout the countryside that are vulnerable to natural disasters and terrorism.

FOCUS ON

Plutonium: One of the Deadliest Environmental Contaminants

Plutonium-239 is one of the deadliest poisons known. If a person inadvertently ingested even a millionth of a gram, he or she would be likely to die prematurely of cancer. Plutonium-239 is a radioactive isotope with a half-life of 24,000 years. This isotope existed during the earliest history of the Earth, but because the half-life is so short compared with geologic time, all of the natural plutonium decayed billions of years ago.

Uranium-235 is the major fuel in conventional fission reactors. However, uranium-235 makes up less than 1 percent of natural uranium. The more abundant isotope, uranium-238, does not split as readily. If uranium-238 is bombarded with a neutron, it reacts to form plutonium-239. Plutonium-239 undergoes fission readily and can be used in reactors and bombs. One way to stretch the reserves of nuclear fuels is to produce plutonium from nonfissionable uranium-238. A reactor that simultaneously creates plutonium and generates energy is called a **breeder** reactor, because the fissionable fuel—plutonium-239—is "bred."

Although breeders were seriously considered a few decades ago, they are more complicated to operate and hence more expensive than the nonbreeders now in use. In addition, they generate large quantities of plutonium, which then becomes an environmental hazard. Today, world uranium reserves are high, and even the less expensive nonbreeders are more costly than other energy sources. As a result, no one is talking seriously about building commercial breeder reactors for power production. However, some plutonium inadvertently forms in nonbreeders, and additional supplies are produced during the development of nuclear warheads. As a result, the disposal of waste plutonium remains a significant environmental issue.

Accidents at Nuclear Power Plants

Although a nuclear reactor cannot explode like an atomic bomb, serious accidents have occurred. The greatest danger involves failure of the cooling system. If the flow of coolant is interrupted, then the core will quickly heat up until it melts into a white-hot mixture of radioactive fuel, radioactive wastes, and molten metal. This scenario has been called the **China syndrome** because the glowing mass could melt through the floor of the reactor and into the ground (although, of course, it would never melt all the way through the Earth to China, as the name implies).

On March 29, 1979, one of the reactors at the Three Mile Island nuclear power plant near Harrisburg, Pennsylvania, lost its cooling water due to a series of human and mechanical failures (Fig. 16–25). About half of the reactor core melted, but the situation was brought under control before the molten liquids breached the floor of the reactor and entered the environment. Although a small amount of radiation escaped into the air, a study 10 years after the accident concluded that cancer rates in the vicinity were not statistically higher than in surrounding areas. Clean-up and confinement of the damaged reactor cost about $1 billion.

FIGURE 16–25
The cooling towers of the Three Mile Island nuclear power plant loom over Harrisburg, Pennsylvania. (*Neal Palumbo/Gamma Liaison*)

16.5 Uranium and Atomic Energy 397

On April 26, 1986, an explosion tore through the roof of the Chernobyl nuclear power plant in the former Soviet Union (Fig. 16–26). The cooling system had failed and the intense heat had shattered the fuel rods into red-hot fragments. These fragments reacted chemically with the metal fuel casings and a small amount of cooling water left in the core to produce hydrogen gas. The hydrogen gas mixed with air and exploded, ejecting the radioactive fuel into the atmosphere as a fine dust. Thus, although the reactor exploded, it was a chemical explosion and not an atomic bomb–type explosion.

The explosion and intense radiation killed 36 plant workers and fire fighters, and an additional 237 people received such high doses of radioactivity that they died soon thereafter or are expected to die prematurely. Another 500,000 nearby residents are being monitored for

FIGURE 16–26
The Chernobyl nuclear reactor after the accident. (A) The arrow in the upper photograph shows the roof region destroyed by the explosion. (B) Workers then entombed the damaged reactor in concrete. *(Bettmann Archive)*

potential long-term effects, and 4 million Europeans far from the plant received radiation doses high enough to concern health officials (Fig. 16–27).

Poor design, inadequate safety devices, and a poorly trained staff caused the Chernobyl accident. Proponents of nuclear power contend that such accidents are avoidable. They point out that the Three Mile Island accident, although expensive, did not release dangerous levels of radiation. No one was killed; thus, safety systems do work. However, not everyone agrees. The Nuclear Regulatory Commission estimated that the probability of a complete core meltdown in the United States in the next 20 years is 15 to 45 percent. Scientists from the Sandia National Laboratory calculate that if everything went wrong, a reactor accident could lead to 100,000 immediate deaths, 40,000 later deaths from cancer, and $150 billion in damages.

The Present and Future Status of the Nuclear Power Industry

Over the past 40 years, optimism about nuclear power has waxed and waned. Initially, proponents of nuclear power overlooked many hazards and economic problems and wrote statements such as "We can look forward to universal comfort, practically free transportation, and unlimited supplies of materials." By the mid-1970s, this rosy forecast had changed. Construction of new reactors had become so costly that electricity generated by nuclear power was more expensive than that generated by coal-

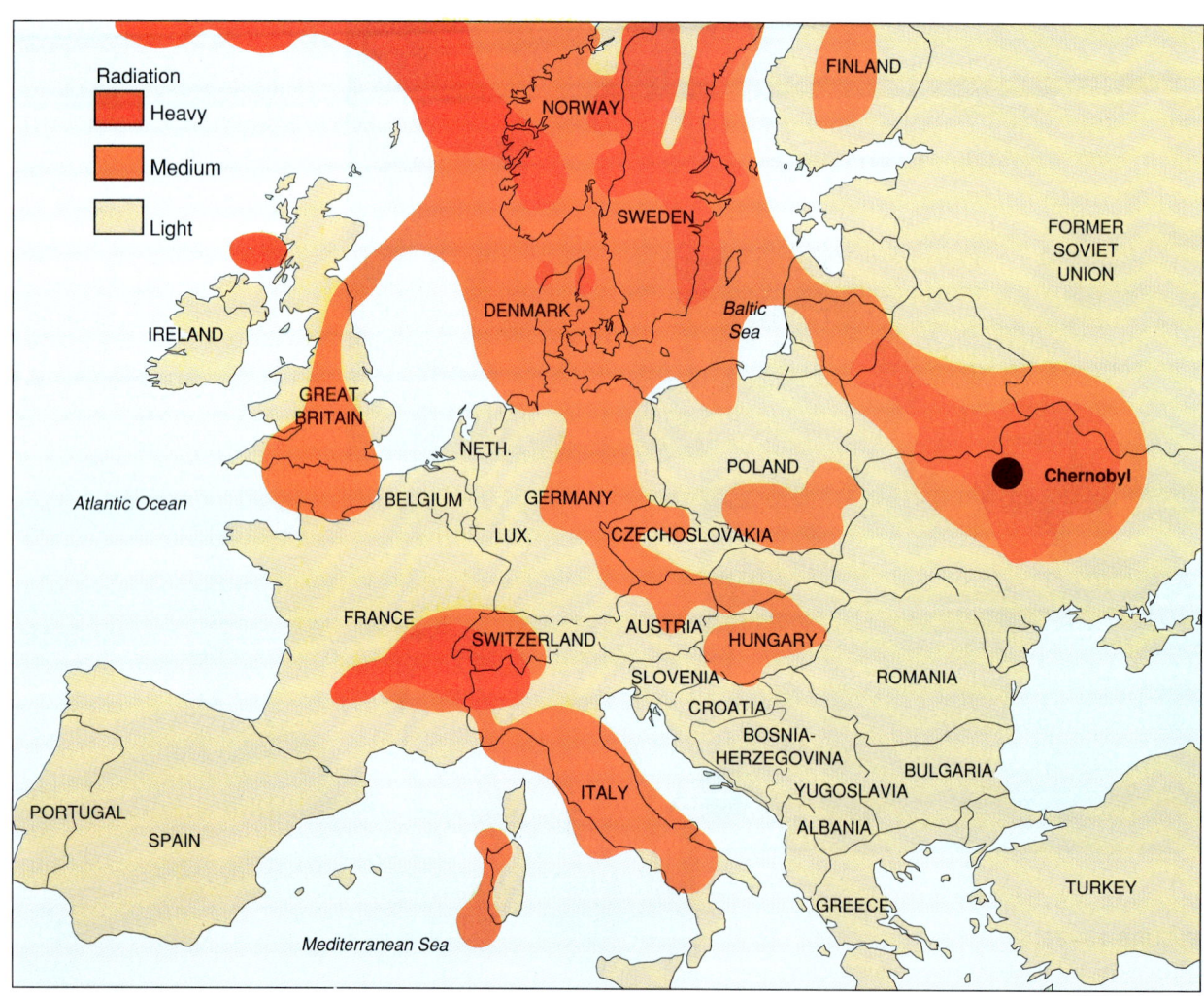

FIGURE 16–27

Radioactive fallout from the Chernobyl disaster. Variable winds and rainfall led to an erratic fallout pattern.

nuclear power program "the largest managerial disaster in U.S. business history, involving $1 trillion in wasted investment and $10 billion in direct losses to stock holders" (Fig. 16–29).

In 1993, 109 commercial reactors were operating in the United States. These generators produced 23 percent of the total electricity that year. As explained earlier, many will become obsolete and will be decommissioned in the near future.

Worldwide, a similar scenario has occurred and the amount of energy produced by nuclear power plants has begun to drop. However, nuclear power may regain importance if fossil fuels become expensive and alternative energy sources are not developed.

16.6 The Future of Our Energy Supply

Humans and their immediate ancestors have been on Earth for roughly 2 million years. Coal has been mined in quantity for a few hundred years, and petroleum has been extracted for a century. Most forecasts estimate that fossil fuel reserves will become depleted in a few hundred years. Thus, the fossil fuel age is destined to be a tiny sliver of time in the history of the human race.

In the United States, petroleum consumption doubled from 1948 to 1970. If oil consumption were to continue to double every 22 years, 660 years from now we would consume as much petroleum annually as there is water in the world's oceans. However, the rapid rise in energy consumption after World War II was not sustained. Petroleum consumption in the United States increased steadily until 1973. Then oil-producing countries realized how valuable and limited their fuels were, so

FIGURE 16–28
A nuclear power plant in Washington between Portland, Oregon, and Seattle. The domed structure on the left is the reactor; the larger structure on the right is the cooling tower.

fired power plants (Fig. 16–28). In addition, public concern about accidents and radioactive waste disposal became acute. Finally, the demand for electricity rose less than expected, so the nation didn't need new generating capacity. As a result, growth of the nuclear power industry came to an abrupt halt. After 1974, many planned nuclear power plants were canceled; after 1977, no new orders were placed for nuclear power plants in the United States. *Forbes* business magazine called the United States

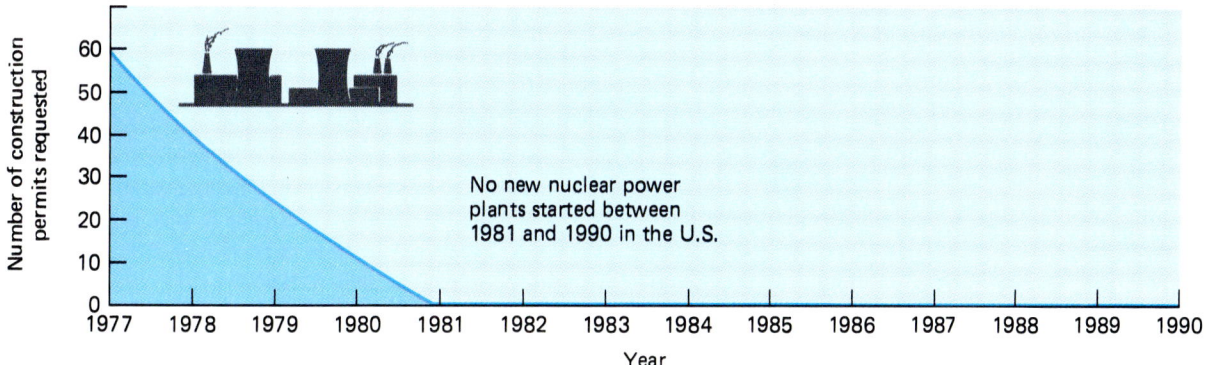

FIGURE 16–29
The number of construction permits requested by the nuclear power industry between 1977 and 1990. The rapid decline reflects a smaller-than-predicted growth in electric consumption and the fact that nuclear power currently is not economical. (*Annual Energy Review 1991, Energy Information Administration*)

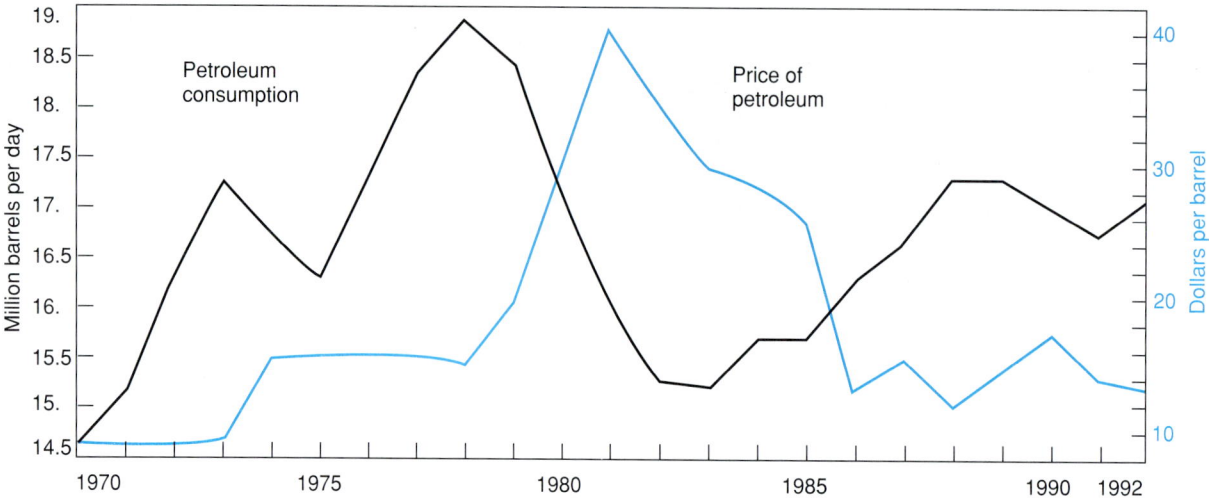

FIGURE 16–30

Curves showing oil consumption (black) and the price of oil (blue) in the United States from 1970 to 1992. Oil consumption increased when the price decreased or was stable. Oil consumption decreased after the price increased in 1973 and 1978.

FIGURE 16–31

Motorists waited in long lines during the gasoline shortages of 1973. *(Tony Korody/Sygma)*

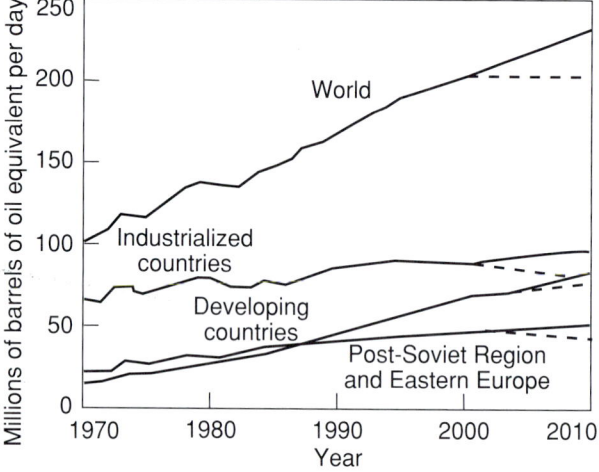

FIGURE 16–32

Past and predicted global energy demand. The demand in industrialized countries is now high, but will stabilize or decrease as a result of population stability and conservation. However, growing population and economic development will cause energy use to increase in the developing countries. (Dotted lines show energy demand if conservation becomes effective.) (William Fulkerson, Roddie R. Judkins, and Manoj K. Sanghvi, "Energy from Fossil Fuels," *Scientific American*, September 1990, p. 129)

they raised prices (Figs. 16–30 and 16–31). As prices went up, consumers used less fuel. People drove fewer miles and purchased smaller, more fuel-efficient automobiles. They added insulation to buildings and saved energy in many other ways. Conservation resulted in declines in petroleum use both in the United States and in the world after the price increases in 1973 and 1975. Consumption generally increased after the price declined dramatically in 1982.

Future rates of petroleum and total energy consumption in the developed nations are expected to stabilize or rise slowly (Fig. 16–32). Population in these nations is stable or increasing gradually, and although people's standard of living continues to increase, more efficient energy use is holding consumption down.

In contrast, energy consumption in developing and less developed countries is expected to rise. In 1993,

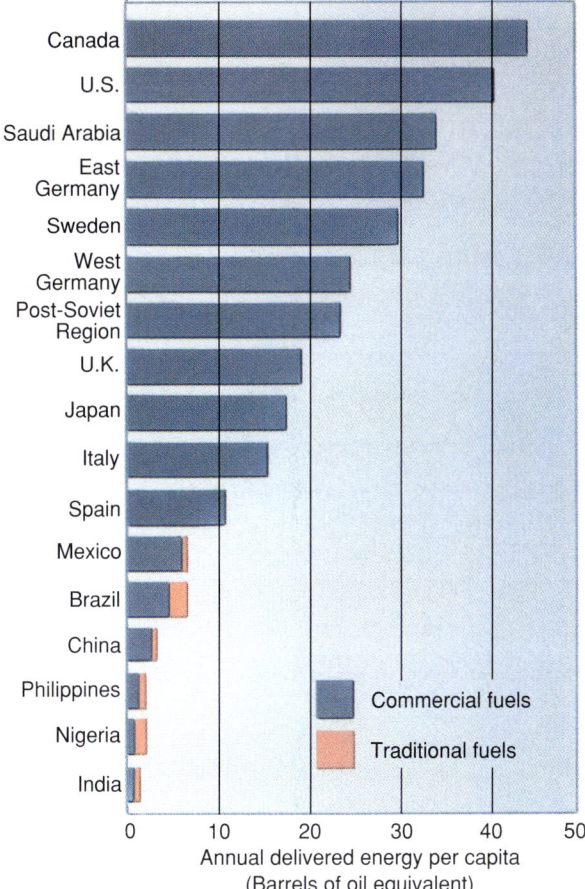

FIGURE 16-33

Global per-capita energy use in 1990. The average Canadian used more than 20 times as much energy as the average person in India. (William Fulkerson, Roddie R. Judkins, and Manoj K. Sanghvi, "Energy From Fossil Fuels," *Scientific American*, September 1990, p. 129)

FIGURE 16-34

Bicycle commuters on the streets of Chengdu, China.

78 percent of the world's population lived in the less developed and developing nations, but these people accounted for only 26 percent of the total global energy consumption (Fig. 16-33). Two trends are likely to increase energy consumption in less developed and developing nations. First, populations in developing and less developed countries are growing faster than those in the industrial nations. Second, economic growth in some poor countries, especially those in Asia, is rapid. The gross national product of China is growing at about 15 percent a year. With a population of more than 1 billion, that growth translates into a tremendous demand for goods and services, many of which depend on energy. This growth in the less developed and developing countries creates an ever-increasing demand for fossil fuels. For example, bicycles are now the most common form of human transportation in China (Fig. 16-34). As the economy grows, people will be able to afford automobiles. The increased demand for petroleum will affect world energy supplies.

Many predictions have been made regarding the future availability of petroleum. Although they differ as to the exact schedule, they all share the conclusion that, sometime in the next century, petroleum supply will be insufficient to meet demand. In a recent article in *Scientific American*, the authors estimated that global petroleum reserves are large enough to supply the world for 60 years at 1988 consumption rates.[2] People in their twenties in the early part of the 1990s will see, within their lifetimes, a restructuring of society resulting from a petroleum shortage. When liquid fuels become scarce, the price will rise and synfuels will become economical. Alternative energy resources will be developed. It is difficult to predict whether or not this transition will occur smoothly.

The authors of the *Scientific American* article estimate that global natural gas reserves will last 120 years if consumption remains constant. However, consumption is increasing. If natural gas consumption increases, reserves will be depleted in about 75 to 80 years (Fig. 16-35).

Large reserves of coal exist in many parts of the world. Near-surface deposits that can be mined at current prices will supply the world until about the year 2200 if consumption rates remain constant. Deeper deposits, up to 1.5 kilometers below the surface, contain even more coal—enough to supply the world for 1500 years at current consumption rates—but these reserves will be expensive to exploit. These estimates will change if prices and demand for coal-based synfuels rise or if environmental problems resulting from mining slow production.

[2] William Fulkerson, Roddie R. Judkins, and Manoj K. Sanghvi, "Energy from Fossil Fuels," *Scientific American*, September 1990, 129.

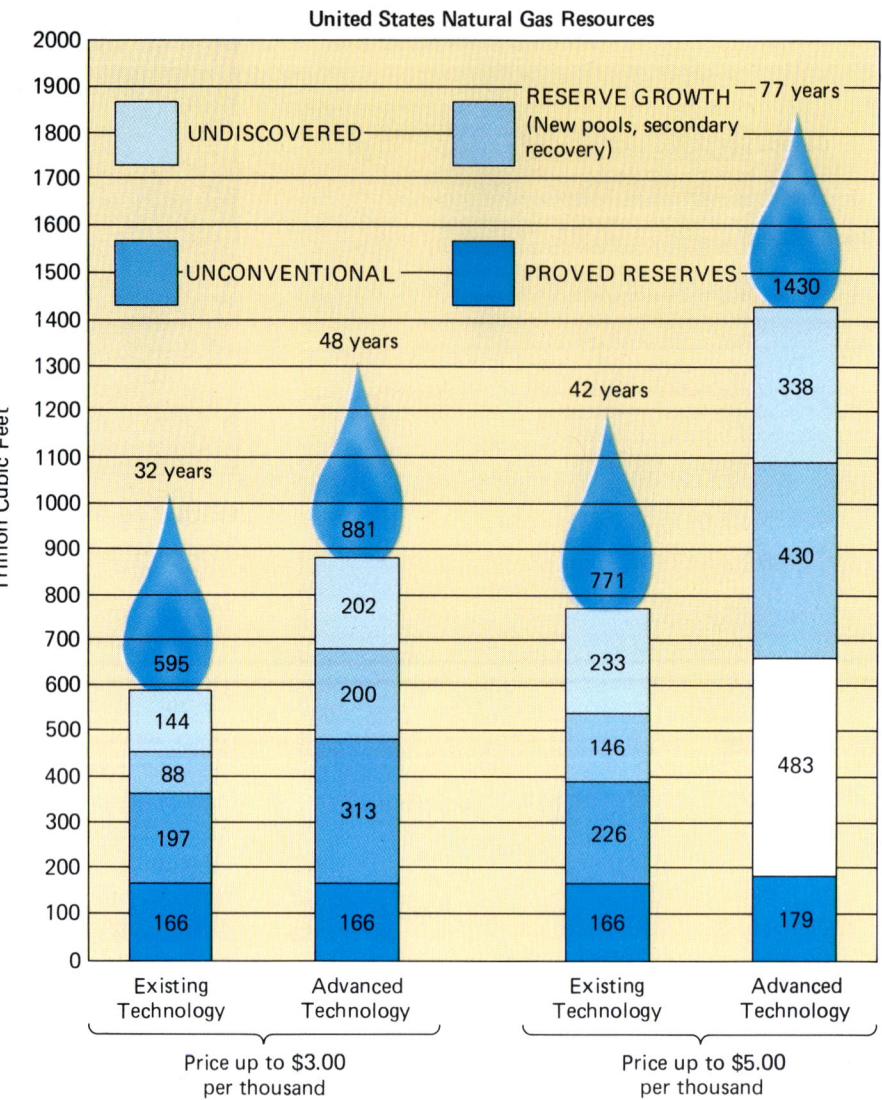

FIGURE 16–35
Future availability of natural gas, given four different possibilities of price and extraction technology. The numbers above each column indicate the years that our gas reserves will last, given the conditions outlined in the column. All predictions assume that consumption rates remain constant. (American Association of Petroleum as reported in *Geotimes*, November 1989)

The world's current uranium reserves are about 6 million tons. With these reserves, the present rate of nuclear power generation could be maintained for about 100 years without the mining of low-grade uranium ore. Low-grade ores are much more abundant, but mining them could result in large-scale environmental degradation.

16.7 The Geopolitics of Fossil Fuels

Fossil fuel reserves are not distributed evenly; some countries are richly endowed, and others have virtually no resources. Large quantities of coal formed worldwide in the Carboniferous Period, between 360 and 285 million years ago, and again during Cretaceous time, between 144 and 66 million years ago, when warm, humid swamps covered broad areas of low-lying land. During the Carboniferous Period, the continents were joined together to form Pangaea. The landmasses now located in the Southern Hemisphere were clustered close to the South Pole and were too cold to support extensive coal swamps. As a result, most of the coal reserves are now concentrated in the Northern Hemisphere.

Petroleum reserves are also unevenly distributed. The largest petroleum reserves are in the Middle East, but the greatest consumption is in North America (Fig. 16–36). In the absence of political disruptions, this unequal distribution of energy resources stimulates world trade, which enriches many nations. However, twice in the 1970s, the major oil-producing nations reduced oil production and raised the price, causing economic problems in oil-importing nations. In 1990 Iraq invaded

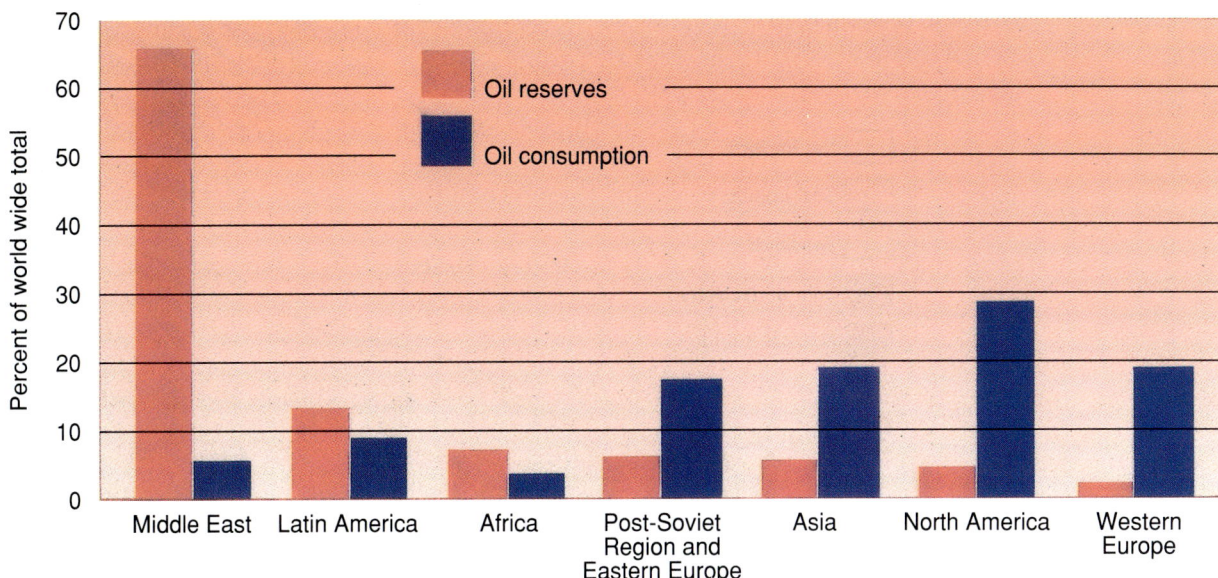

FIGURE 16–36
Global oil reserves and oil consumption.

Kuwait and threw global petroleum markets into turmoil (Fig. 16–37). Within a week, the price of petroleum skyrocketed from below $20 per barrel to $40 per barrel; it then fell again after the brief Gulf War.

A major problem for many nations is how to maintain or improve people's standard of living in the face of a limited and uncertain fossil fuel supply. We will look at this problem in the remainder of this chapter and in all of the next.

FIGURE 16–37
American bombs falling on Baghdad during the Gulf War, 1992. The war reminded us that petroleum reserves are distributed unevenly and are subject to disruption.
(Gamma Liaison)

16.8 Energy Strategies for the United States

The United States has abundant coal and oil reserves. In 1992, we produced more coal than we needed and exported $3.4 billion worth of coal. Before 1957, the United States was also an oil-exporting nation. However, growing consumption in the 1950s and 1960s reversed the balance and we became an oil-importing nation. Domestic oil production peaked in 1970 and has declined since then. For every meter of exploratory drilling, we now find half as much oil as we did in the 1950s. These trends have led to a steady increase in oil imports (Fig. 16–38). In 1992, the United States was the third largest oil-producing nation (behind Saudi Arabia and Russia), but our consumption was so high that we imported 45 percent of our petroleum, at an annual cost of $25 billion. This cash outflow has had a negative impact on our national economy.

In 1989, before the Gulf War, the United States spent between $15 billion and $54 billion to safeguard petroleum supplies and shipping lanes in the Middle East. This wide range in estimates reflects the difficulties of assigning costs to military operations. The Gulf War in 1990–1991 added another $30 billion to the cost of protection. Using the lowest estimates, these figures add about $23.50 to the cost of each barrel of petroleum. Thus, the cost of defending oil imports more than doubled the cost of imported oil. Such military expenditures are not added to the cost of gasoline at the pump, but they must be paid for nevertheless.

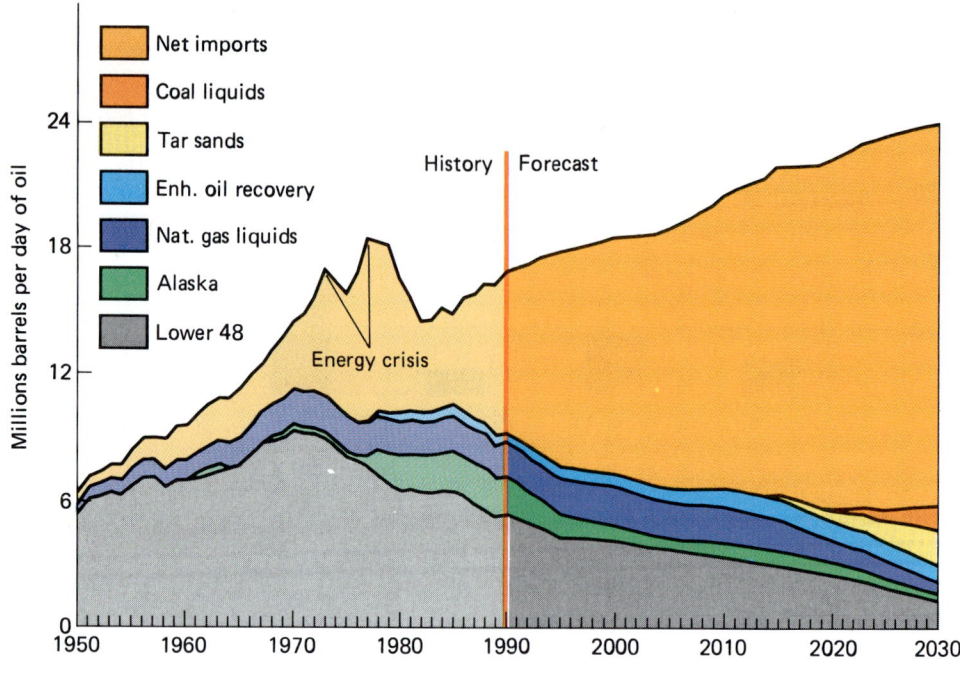

FIGURE 16–38

Sources of oil used in the United States. The left side of the graph shows what has happened in the past. The right side shows what will happen in the future if current trends continue. Imports grew from near zero in 1950 to about 45 percent of consumption in 1990 because of increased consumption and decreased internal production. If these trends continue, our dependence on imports will rise steadily. (From *Improving Technology: Modeling Energy Futures for the National Energy Strategy*, Energy Information Administration paper SR/NES/90-01)

In short, the United States no longer has an abundant, low-cost energy resource.... Indeed, the U.S. energy picture is growing closer to Europe and Japan, which lack cheap domestic energy sources and have to work hard to function amidst high energy costs. The U.S. has a long way to go considering that it uses around twice as much oil per person as European nations do.[3]

According to a 1989 USGS report, approximately 82 billion barrels of known and estimated undiscovered oil reserves still exist in the United States. If imports are not impeded by political instability or war, they will about double the oil supply, extending our supply for 30 years, or to about the year 2020. However, these numbers do not address the environmental impact of extracting the remaining domestic oil.

Existing U.S. oil fields are being depleted, production is declining, and discovery rates per foot of drilling have decreased by half since 1950. What about finding new oil fields? Let us examine one case in particular. In 1980 Congress set aside an 18-million-acre wildlife refuge in northeastern Alaska called the Arctic National Wildlife Refuge (ANWR) (Fig. 16–39). After the conclusion of the Gulf War in 1991, President Bush advocated drilling for oil in the refuge. The argument for this proposal is that our life style, economy, and national security depend on petroleum, and therefore it is imperative to drill for domestic oil wherever it exists. The argument against drilling in ANWR is that the refuge is one of the finest wildlife sanctuaries in the world and that the government must protect what little is left of such places. The U.S. Department of the Interior estimates that 3.2 billion barrels of oil lie beneath ANWR. Under normal operat-

---

[3]Christopher Flavin, "Conquering U.S. Oil Dependence," *Worldwatch*, January–February 1991.

FIGURE 16–39
Caribou on the coast in the Alaskan National Wildlife Refuge.

ing conditions, the oil reserves would be extracted over 30 years. Production would peak at 500,000 barrels per day and average about 270,000 barrels per day. Today we import about 8 million barrels per day. Therefore, the average contribution from ANWR would result in a 3 percent reduction of imports. Calculations show that production of other domestic reserves would result in similarly small reductions of imports. Thus, we cannot obtain petroleum independence simply by exploiting fragile environments (Fig. 16–40). In 1991 Congress voted to maintain protection of ANWR and prohibit drilling.

Another approach to reducing our dependence on imported oil is to use more abundant fuels for transportation and other applications that currently consume petroleum. For example, coal and nuclear energy can be used to create electricity for electric transportation systems. All the major automobile manufacturers have designed prototype electric cars, and several models are expected to be in limited production soon. Electric trains are commonly used in most cities, and an expansion of these mass transit systems would extend the life of our limited petroleum resources. Although advances in battery technology would encourage the use of electric cars, a transition to an electric transportation system depends mostly on other factors. Commuters aren't accustomed to electric cars, and people change their habits slowly. A 1000-megawatt coal-fired electric generating plant takes a minimum of 10 years from the planning stage to completion and costs $1.5 billion, and a nuclear power plant is more expensive and requires more time to plan and build. Yet we would need 3000 new electric generating plants to provide the energy equivalent of our current automobile petroleum consumption.

Similarly, great expense and time are required to develop oil shale mines and synfuel plants. Corporations aren't willing to build these facilities as long as the price of petroleum remains low. Unfortunately, if the price or availability of petroleum changes rapidly, we will have to deal with the decade or more lag time and tremendous capital investment involved in conversion to alternative fuels.

Renewable energy resources—wind, solar, geothermal, biomass, hydroelectric, and tidal—may be cheaper, easier, and quicker to implement than synfuels and nuclear power sources. This important topic is discussed in Chapter 17. Finally, the fastest and cheapest way to extend our fossil fuel reserves and build a sustainable energy future is to conserve energy. Chapter 17 also discusses advances in heating, transportation, and communication technologies that allow us to use less energy to perform common tasks.

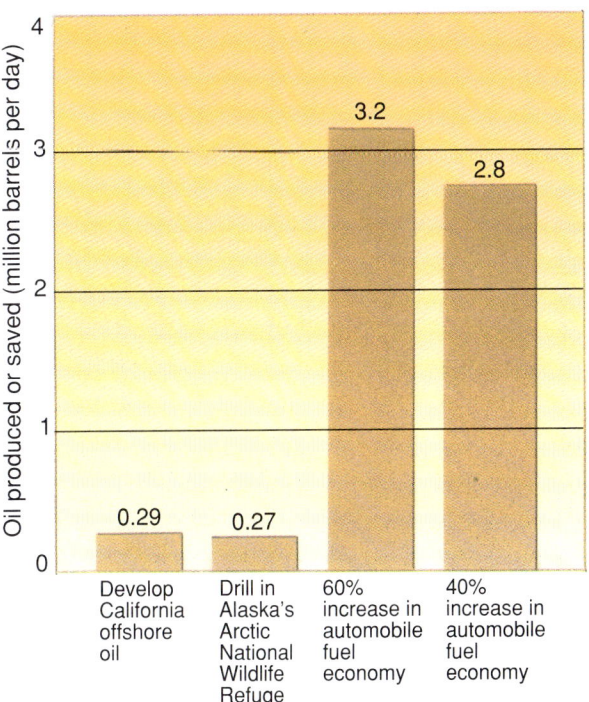

FIGURE 16–40
Petroleum savings from conservation versus drilling in California offshore fields and in ANWR. (From Union of Concerned Scientists, *Nucleus* 12, No. 3 [1990])

SUMMARY

Fossil fuels include oil, gas, and coal. If oxygen and flowing water are excluded, plant matter decays partially to form **peat**. Peat converts to **coal** when it is buried and subjected to elevated temperature and pressure. Coal is abundant, but it is dirty, burns slowly relative to gasoline, and emits noxious air pollutants. Coal is mined either in tunnels or in surface mines. Tunnel mines are unhealthful for the miners and often subside. Surface mines can be reclaimed; but, if operated carelessly, they deface the land and create acid mine drainage.

Petroleum forms from the remains of organisms that settle to the ocean floor and are incorporated into **source rock**. The organic matter converts to liquid oil when it is buried and heated. The petroleum then migrates to a **reservoir**, where it is retained by an **oil trap**. Crude oil is a dark liquid made up of a mixture of thousands of different chemicals. It is refined to produce propane, gasoline, jet fuel, heating oil, motor oil, road tar, plastics, medicines, and other useful materials. Relatively little environmental damage occurs when oil wells are drilled in temperate and arid regions. However, oil spills have occurred in virtually all offshore drilling operations. Oil refineries are also sources of pollution.

Natural gas, or methane (CH_4), forms when crude oil is heated to between 100° and 150°C during burial. Natural gas is extracted as a nearly pure compound—methane, CH_4. It is a popular fuel because it burns cleanly.

Fossil fuels are **nonrenewable resources** and will eventually be depleted. Additional supplies of petroleum can be recovered by secondary extraction from old wells, tar sands, **oil shale**, liquefaction of coal, and perhaps marine methyl hydrates. Pollution and water depletion problems arise from the mining and refining of oil shale.

When a neutron hits a uranium-235 nucleus, the nucleus undergoes **fission**, releasing two or three neutrons. If uranium-235 is concentrated, each released neutron initiates two or three additional reactions, creating a **branching chain reaction**. In a nuclear power reactor, the concentration of uranium-235 is kept low and neutrons are absorbed by **control rods**, so the reaction proceeds at a controlled rate.

Mining, processing, and disposal of nuclear fuels and decommissioning of old plants are all potential sources of radioactivity that can endanger health or threaten ecosystems and ground water. Accidents at nuclear power plants have occurred in the past and may occur again.

Nuclear power is expensive, and questions about safety and disposal of nuclear wastes have not been answered to everyone's satisfaction. As a result, no new nuclear power plants have been ordered since 1977. Inexpensive uranium ore will be available for a century or more.

Petroleum and natural gas reserves will be depleted within the next century, and near-surface coal deposits will last until the year 2200. When easily accessible fossil fuel deposits are exhausted, deep coal reserves and synfuels will be available at an increased cost. Energy supplies are subject to political disruption. Oil reserves in the United States are being exhausted, although coal is plentiful. Today, we import about half of our petroleum, which leaves us vulnerable to political upheaval.

KEY TERMS

Peat 381
Coal 382
Black lung disease 383
Petroleum 384
Reservoir 384
Source rock 384
Permeable 384
Oil trap 384

Cap rock 384
Nonrenewable resource 387
Synfuels 387
Secondary recovery 387
Bitumen 387
Kerogen 388
Oil shale 388
Methyl hydrate 389

Radioactivity 391
Nuclear fission 391
Branching chain reaction 392
Fuel rod 393
Control rod 393
Decommissioning 395
Dismantlement 395
China syndrome 396

REVIEW QUESTIONS

1. Explain how coal forms. Why does it form in some environments but not in others?

2. Explain the importance of source rock, reservoir rock, cap rock, and oil traps in the formation of petroleum reserves.

3. Discuss three sources of petroleum that will be available after conventional wells go dry. Explain the advantages and disadvantages of each.

4. What are gas hydrates? Where are they found? What is the potential for exploiting gas hydrates as fuels?

5. List the relative advantages and disadvantages of coal, petroleum, and natural gas.

6. Compare the environmental impacts of underground coal mines with those of surface mines.

7. Review the environmental impacts of drilling, transporting, and refining petroleum.

8. Why does extraction of fuel from oil shale create more environmental problems than drilling for oil in conventional wells?

9. Explain why branching chain reactions can produce explosions and how they are harnessed to obtain a controlled power source.

10. Explain how a nuclear reactor works. Discuss the behavior of neutrons, the importance of control rods, and how the heat from the reaction is harnessed to produce useful energy.

11. Discuss the status of the nuclear power industry in the United States.

12. Explain why most coal reserves are in the Northern Hemisphere.

13. Discuss the prospects for the availability of petroleum and coal in the next 20, 50, and 100 years. What uncertainties are inherent in these predictions?

DISCUSSION QUESTIONS

1. If you were searching for petroleum, would you search primarily in sedimentary rock, metamorphic rock, or igneous rock? Explain.

2. Which of the following materials would you expect to find on the Moon: iron, gold, aluminum, coal, or petroleum? Defend your answer.

3. What fuels supply the energy to provide the light you study by, to heat your house, to cook your food, and to transport you to school? Supply a flow chart showing the major processes and resultant environmental problems for each fuel, from source to end use.

4. What sources of energy do you think your children will use for the functions in question 3? What about your grandchildren? Defend your answers.

5. Imagine that you are a space traveler and are abandoned on an unknown planet in a distant solar system. What clues would you look for if you were searching for fossil fuels?

6. Is an impermeable cap rock necessary to preserve coal deposits? Why or why not?

7. Discuss problems in predicting the future availability of fossil fuel reserves. What is the value of the predictions?

8. Prepare a class debate. Have one side argue that we are quickly approaching an energy crisis that will debilitate our society, and have the other side argue that society will adjust to changes in energy supply without disruption.

9. Compare the depletion of mineral reserves with the depletion of fossil fuels. How are the two problems similar, and how are they different?

10. Prepare a class debate. Have one side argue that energy independence is as important to national security as a strong military and should demand a larger share of the national budget. The other side should counter that the current allocation of federal money is satisfactory.

11. Explain why the environmental effects of extracting fossil fuels are likely to increase as fuel reserves are gradually exhausted.

CHAPTER 17

Alternative Energy Resources

17.1 Energy Conservation
17.2 Alternative Energy Resources
17.3 Solar Energy
17.4 Hydroelectric Energy
17.5 Geothermal Energy
17.6 Wind Energy
17.7 Biomass Energy
17.8 Energy from the Seas
17.9 Hydrogen Gas as a Fuel
17.10 Nuclear Fusion

Several energy resources are available when we exhaust our reserves of oil, gas, and coal. They include solar energy; hydroelectric power; geothermal energy; wind energy; biomass energy; tidal, wave, and heat energy from the seas; and nuclear fission (discussed in Chapter 16). Each of these energy resources, with the exception of fission fuels, is renewable and available in nearly continuous supply. Furthermore, the technologies to exploit these resources are available, and most are in current use in the United States or elsewhere in the world. In addition to these proven resources, nuclear fusion, although now technologically impracticable, may provide nearly unlimited energy in the future.

Fossil fuels and fission-driven nuclear reactors produce such a high proportion of energy in the industrialized world that they are generally considered to be the normal, or standard, sources of energy. Consequently, all other energy resources are called **alternative energy resources**.

Alternative energy resources will provide us with energy when fossil fuel reserves run low or become too expensive. However, oil, gas, and coal are uniquely suited to certain important applications. These special applications include the use of oil for production of light, liquid, portable fuels for automobiles and trucks and the use of oil and coal as raw material for plastics, nylon, and other synthetics. At present, no technologically and economically viable substitutes for these uses of fossil fuels are known. Thus, although alternative energy resources reassure us that we will not run out of energy when we run out of fossil fuels, it is also important to begin using alternative resources now to preserve our remaining fossil fuels for their most important applications.

Despite the availability of alternative energy resources, the single most effective way to increase the amount of energy available to us now, and to prolong the availability of cheap fossil fuels, is to conserve energy by changing to efficient equipment, design, and human habits.

Wind provides renewable energy in many parts of the world.

410 Chapter 17 Alternative Energy Resources

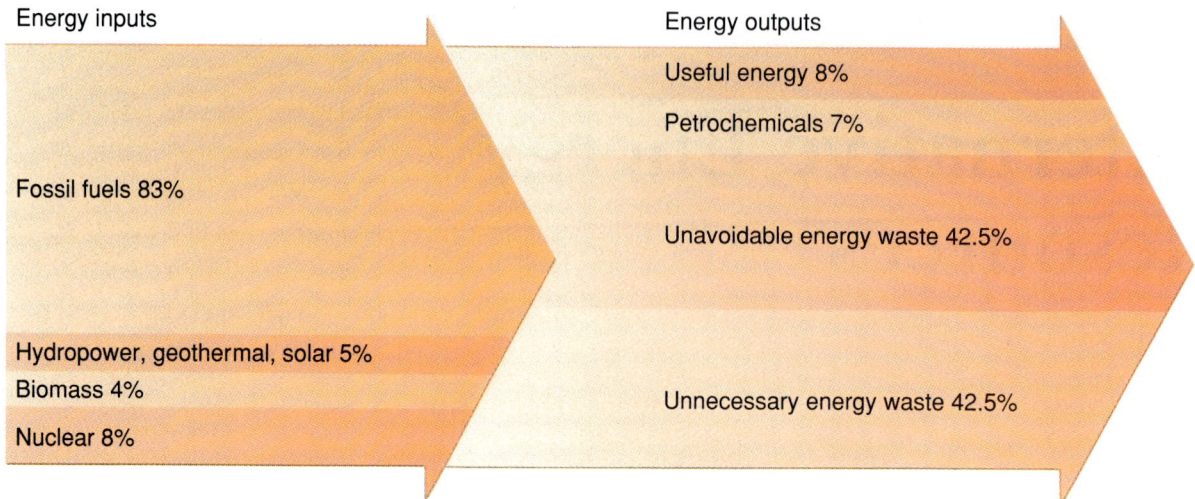

FIGURE 17–1
Only 15 percent of the energy generated in the United States performs useful tasks; 85 percent is wasted. Forty-two percent of the waste is unavoidable but the remainder could be avoided with conservation.

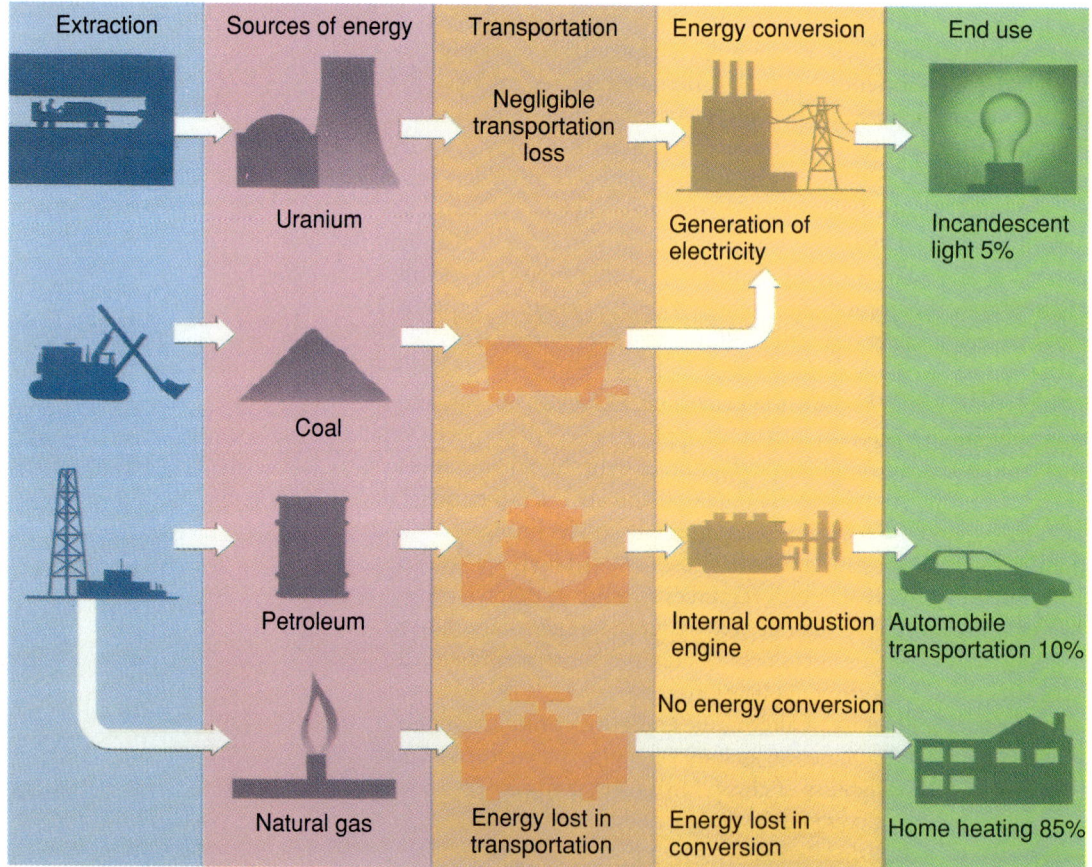

FIGURE 17–2
The end-use efficiencies of important energy-consumption systems.

17.1 Energy Conservation

Eighty-five percent of all commercially produced energy is wasted (Fig. 17–1). Thus, only about 15 percent of provided energy results in useful work or useful products (Fig. 17–2). Of the 85 percent that is wasted, about half is unavoidably lost because the Second Law of Thermodynamics requires that energy loss invariably accompanies the conversion of heat energy to work. The other half is wasted because of inefficient machinery, building design, and human practices. Figure 17–3 summarizes the energy efficiencies of common devices.

Let's follow the energy path from source to end use in lighting a room with an incandescent light bulb in a city where a coal-fired generating plant provides the electricity. First, energy is used to operate the mining equipment to mine the coal. More energy is used to ship the coal to the electrical generating plant. Then energy is lost in the conversion of heat to work in the turbine. More energy is lost as electricity passes through transmission lines. Finally, much of the energy that reaches the light bulb is converted to heat rather than light. As a result of these energy expenditures and losses, only 5 percent of the original energy in the coal actually provides light; the remaining 95 percent is used to extract and ship the coal and is dissipated as waste heat along the way.

Obviously, the combination of a coal-fired generating plant and an incandescent light bulb is not an energy-efficient source of light. Many generating systems and common energy-consuming devices are similarly inefficient and wasteful. Why do we waste energy when most of our energy comes from nonrenewable fossil fuels? In most cases, the generating systems and devices were invented and designed when technology was less highly developed than it is now, energy was cheap and plentiful, and few perceived the limited nature of fossil fuels. In addition, the initial costs of most inefficient devices are less than those of more efficient devices that perform the same tasks, even though the efficient devices cost less to operate in the long run. As a result, we continue to use incandescent bulbs with an energy efficiency of 5 percent rather than fluorescent bulbs, which cost more but last longer and generate light with an efficiency of 22 percent.

Conservation as an Alternative Energy Resource

Construction of a new hydroelectric dam and power plant, coal-fired generating plant, or nuclear reactor is much more expensive than conserving energy by switching to more efficient technologies and by adopting new habits. We also do not have to tolerate the environmental degradation that accompanies construction and operation of these energy sources. Thus, conservation is the cheapest and most immediately available alternative energy resource. Conservation also reduces our dependence on foreign oil and extends the life of our limited petroleum reserves. Two kinds of conservation strategies exist.

Social solutions involve decisions to use current energy systems more efficiently. Thus, they entail personal choices and sacrifices. For example, if you choose to carpool rather than drive your own car, you save fuel but inconvenience yourself by coordinating your schedule with your neighbor, driving or walking to a meeting place, and waiting if your friend is late. However, if everyone carpooled or used mass transit, highways would be less congested and there would be fewer traffic jams and more parking places (Fig. 17–4).

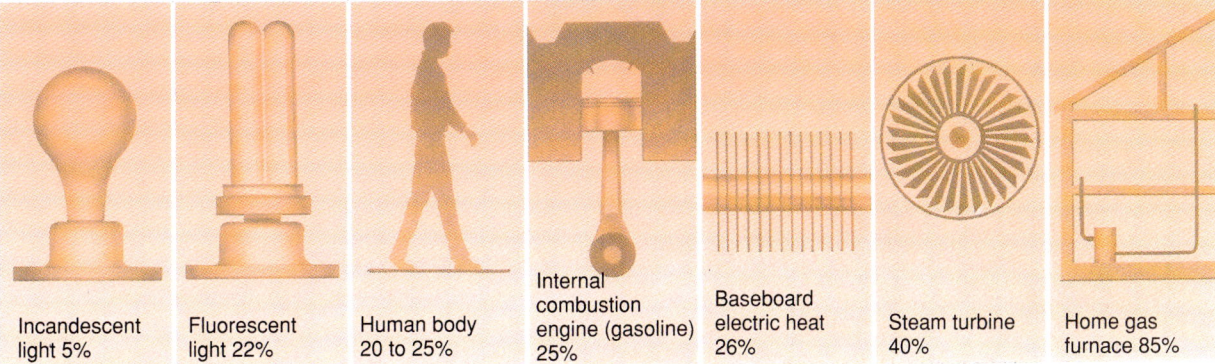

FIGURE 17–3
The efficiencies of important energy-consuming devices.

FOCUS ON

Conversion of Heat from Fuels to Work and Electricity

Prehistoric people used fuels to heat their homes and cook food. In contrast, humans, animals, wind, or falling water performed the work. For centuries, scientists thought that heat and work were different phenomena, but in 1849 James Joule proved that heat and work are interchangeable forms of energy. Heat can be converted to work, and work can be converted to heat. **The First Law of Thermodynamics states that energy is always conserved.** Thus, no energy is ever lost or gained during any process.

James Watt ushered in the Industrial Age in the late 1700s by building the first commercial steam engine, a device that converts heat to work. Early steam locomotives used the energy from burning wood or coal to pull strings of freight cars. Scientists soon learned that, in every heat engine, some of the energy always escapes as hot exhaust. Energy is always conserved because the energy from the fuel equals the work produced plus the heat that escapes through the exhaust. However, because some of the heat escapes in exhaust, not all the energy from the fuel is available to perform work.

Energy Budget of a Heat Engine

Heat input = work output + heat in exhaust

Thus, although energy is conserved, the exhaust heat is lost in the sense that it is not available to perform work. This observation is so consistent that it has become a law of physics. **The Second Law of Thermodynamics states that no heat engine is 100 percent efficient.**

One consequence of the second law is that energy cannot be recycled indefinitely. The waste heat that escapes as exhaust dissipates into the environment, warming the air or coolant water slightly. That energy isn't lost, but it is no longer available to produce useful work.

To understand the Second Law, consider a steam-powered engine used to drive a generator and produce electricity. A fuel, usually coal, oil, or uranium, produces heat to boil water. The water expands as it converts to steam and, in turn, the steam flows past the blades of a fanlike device called a **turbine,** causing the turbine to spin (Fig. 1). The turbine is connected to an electric generator.

Now think about this process on a molecular level. Steam is hot. When a substance is hot, the molecules in it move quickly. Thus, hot, fast-moving molecules collide with the turbine blades

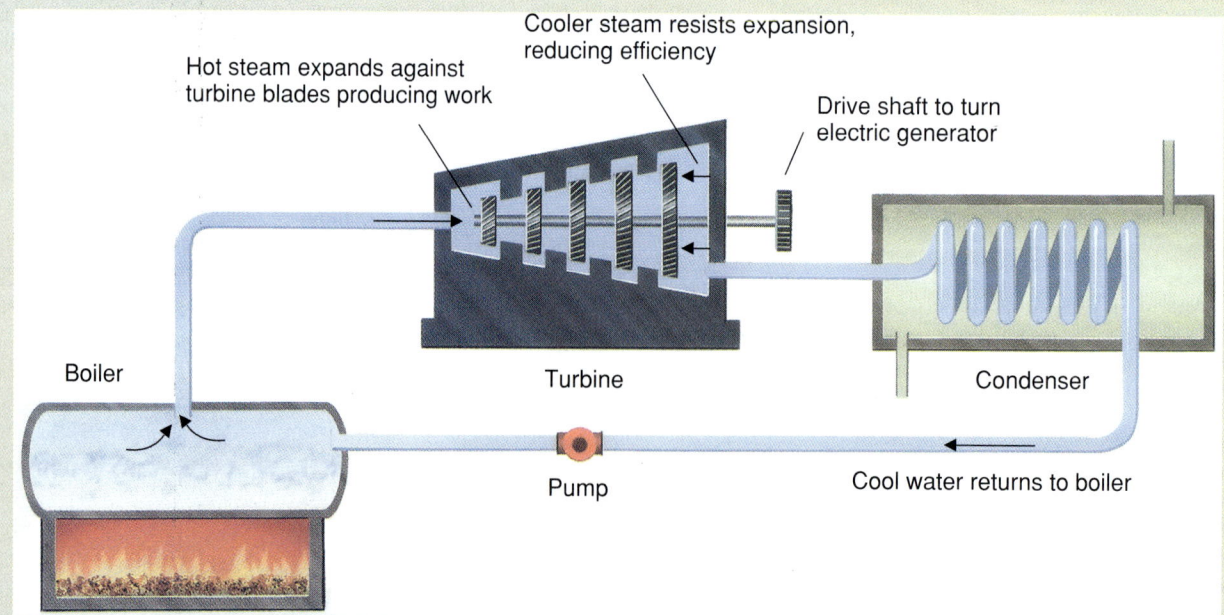

FIGURE 1

To produce electricity, a boiler generates steam that spins a turbine. The steam is then cooled in a condenser and the water returned to the boiler. The turbine drives an electrical generator.

and force them to spin. The steam cools as it passes through the turbine and expands. Although the exhaust steam is cooler than the steam leaving the boiler, the exhaust is still hot. Just as you would burn your fingers if you touched the tailpipe of a car, you would not want to put your hand in the exhaust of an electrical generating plant. The hot exhaust steam pushes back against the onrushing hotter steam. This back pressure slows the turbine and causes the efficiency loss expressed in the Second Law. To maximize efficiency, the initial steam must be as hot as possible and the exhaust must be as cool as possible to reduce the back pressure. In practice, the highest steam temperatures are limited by the metallurgy of turbine blades. Above 550°C, the metals bend, fatigue, and crack. The optimal exhaust temperature is limited by the cheapest coolant—usually river or ocean water, which ranges from about 10° to 15°C. The theoretical efficiency of a heat engine operating between these two temperatures is 65 percent. However, unavoidable inefficiencies in the cooling cycle, fluctuations in steam temperature, friction, and other losses further reduce the actual efficiency to 40 percent. Thus, 60 percent of the energy content of the fuel is dissipated and unavailable to perform useful work. This efficiency varies by a few percent, depending on the age and design of the plant. An additional 5 percent is dissipated in operating the plant, and 9 percent is lost from electricity transmission wires, so that only 26 percent of the original energy from a fuel is available as electricity in the outlet at your home (Fig. 2).

In our society, electricity is used to produce work, provide heat, and power electronic devices. Even with the turbine losses, an electric motor is more efficient than a small gasoline engine because temperatures are more carefully controlled in stationary steam turbines than they are in portable gasoline engines. Thus, an electric drill is more efficient than a gas-powered drill. It is also lighter, quieter, easier to maintain, and less polluting.

Telephones, stereos, and computers can only operate with electricity. No other energy source will do. However, electric heat is inefficient. Electric baseboard heaters, water heaters, and stoves are all 100 percent efficient at converting electricity to heat. The inefficiency arises from energy losses at the power plant and in the transmission wires.

One way to conserve energy resources is to use electricity where it is needed or efficient, but not use it to provide heat.

FIGURE 2
Only 26 percent of the original energy from a fuel is available as electricity from one outlet in your home.

FIGURE 17–4
Modern highways are congested by commuters driving alone rather than carpooling.

FIGURE 17–5
Managers of modern office buildings often leave the lights on at night, when the buildings are mostly unoccupied. This practice wastes large quantities of energy. The Los Angeles skyline at night. *(James Blank/Bruce Coleman, Inc.)*

Technical solutions involve switching to more efficient implements. For example, a fluorescent light bulb uses one fourth as much electricity as an incandescent bulb. A car that runs 40 miles on a gallon of gasoline can carry you as quickly and comfortably as an old gas hog that gets 12 miles per gallon. In 1989, people in the United States used 14.5 million barrels of oil per day *less* than they would have if they had retained the technology of 1973. Further improvements are possible. An additional 40 percent increase in average automobile fuel economy would save 2.8 million barrels of oil a day, about one third of the current import level.

Conservation Strategies

Several kinds of energy conservation, involving both social and technical solutions, can be put into practice in the United States at once. Many involve little or no cost; others have initial costs but pay for themselves within a few years. Consider as examples three categories of energy use: energy for buildings and homes, energy for industry, and energy use in transportation.

Conservation in Buildings and Homes

Residential and commercial buildings consume about 36 percent of all the energy produced in the United States. Most of that energy is used for heating, cooling, lighting, cooking, and water heating. Improvements in design and insulation of homes and other buildings to minimize the use of furnaces and air conditioners costs money initially, but the changes almost always pay for themselves within a few years by reducing power bills.

For example, a super-insulated home or commercial building costs about 5 percent more to construct than a conventional one, but the energy savings usually compensate for the extra cost within about 5 years. After that time, the super-insulated construction saves money. However, developers wish to keep purchase prices low to attract prospective buyers. As a result, they commonly build initially cheaper, energy-inefficient structures and furnish them with similarly cheap, inefficient appliances. This penny-wise but pound-foolish approach costs homeowners about 50 percent more money to buy and operate the house over a lifetime. The extra cost is nearly all in the form of higher energy bills.

About 25 percent of the electricity generated in the United States is used for lighting. As mentioned previously, a fluorescent bulb uses one fourth of the energy consumed by a comparable incandescent bulb. Fluorescent bulbs cost about $15 each, but they last ten times longer and save three times as much money as they cost. As a result, switching to more efficient light sources would conserve the electricity generated by about 120 large generating plants and would save $30 billion in fuel and maintenance expenses. Just turning off the lights when no one is in a room can save huge amounts of electricity and power costs (Figs. 17–5 and 17–6). Similar changes in use patterns would result in large energy savings for other appliances, such as water heaters, refrigerators, and air conditioning and heating systems.

FIGURE 17-6
This satellite photo shows that cities light up the nighttime sky over North America. *(Defense Meteorological Satellite Program)*

Conservation in Industry

Industry consumes another 36 percent of the energy used in the United States. About 70 percent of the electricity consumed by industry—half of all electricity produced in the United States—drives electric motors. Most such motors are inefficient because they run only at full speed and are slowed by brakes to operate at the proper speeds to perform their tasks. This approach is like driving your car with the gas pedal on the floor and controlling your speed with the brakes. Replacing older electric motors with variable-speed motors would save an amount of electricity generated by 150 large power plants. The reduced rate of electricity consumption would pay for the new motors in about a year. After that time, continued use of the new motors would save money in the form of lower power bills.

Recognizing that money can be saved by conserving energy, U.S. industry reduced its energy consumption by about 70 percent per unit of production between 1973 and 1983. Since 1983, however, industrial efficiency has been relatively static. As a result, industry still wastes great amounts of energy.

Conservation in Transportation

About 28 percent of all energy consumed in the United States and two thirds of all the oil consumed are used to transport people and goods. Americans own one third of all automobiles in the world and drive as much as the rest of the world's population combined. Ten percent of all oil consumed in the world is used by Americans on their way to and from work, about two thirds of them driving alone. The efficiency of auto and truck engines is about 10 percent. Thus, much energy can be saved by changing to more efficient cars and trucks and by using them more efficiently. Carpooling, mass transit, walking, and using bicycles save immense amounts of fossil fuel.

Several auto companies, including Volvo, Toyota, and Volkswagen, have produced prototypes of full-sized cars with fuel efficiencies of 30 to 60 kilometers per liter (70 to 140 miles per gallon). These cars, manufactured with strong and light materials, meet current safety and pollution standards, carry four or five passengers, and accelerate as rapidly as many current models. If mass produced, they would cost somewhat more than conventional autos, but the fuel savings would more than pay for the higher initial cost.

Electric cars reduce oil consumption and are essentially pollution-free on the highway. Coal- or oil-fueled power plants used to generate electricity to recharge the batteries would still produce air pollution. However, because emission controls at power plants are more efficient than those on cars, the total amount of pollution produced would be less. But if used in large numbers, electric cars would increase demands on our electrical generating capacity and require the construction of many new power plants. In addition, because they run for only about 200 kilometers before the batteries need recharging, they are useful only for urban commuting and short trips. Batteries must be replaced after about 40,000 kilometers at a cost of about $1500. Thus, an electric car costs about twice as much to operate as a conventional auto. Nevertheless, electric cars are being considered seriously in California, where smog is a big problem.

17.2 Alternative Energy Resources

In 1992, fossil fuels provided about 88 percent of the energy consumed in the United States; fission-type nuclear reactors provided another 7.6 percent. Three and a half percent was hydroelectric power. Solar, geothermal, wind, biomass, and ocean wave and current sources each provided less than 1 percent.

Although the energy produced by alternative sources is small compared to energy provided by fossil and nuclear fuels, alternative energy resources have much greater potential. The state of California has established a program to exploit alternative and renewable energy resources, including wind, solar, hydroelectric, and geothermal power, for more than 40 percent of its electricity.

At present, a serious drawback characterizes all alternative energy sources: They cannot be used directly in commercially available automobiles. Eventually, electric cars, trains, and trolleys may dominate our transportation systems, but the changes will not occur quickly or cheaply.

17.3 Solar Energy

The solar energy received by Earth every 30 seconds is equal to the 1992 daily energy consumption of all nations combined. In the least sunny parts of the United States (except for Alaska), enough solar energy falls on a 9- by 9-meter area—smaller than the average house roof—to meet the energy needs of an average American family (Fig. 17–7). Some, although not all, of this energy can be harnessed and converted to heat and electricity for human use. Current technologies enable us to use solar energy in three ways: passive solar heating, active solar heating, and electricity production by solar cells.

Passive Solar Heating

The simplest way to use solar energy is to design and orient a building so that the Sun heats it directly. This use of the Sun's heat is called **passive solar heating**. The use of passive solar energy is older than civilization. On a cold day, animals lie in the sun rather than shiver in the shade. Prehistoric cave-dwellers probably did too. Anyone who spends time outside instinctively does the same thing today. Anasazi Indians living in the American Southwest chose south-facing caves and ledges for their cliff dwellings (Fig. 17–8). A well-chosen cliff-dwelling site has an overhanging ledge above it, so that the low winter Sun shines directly onto the houses but the ledge shades the houses from the higher summer Sun, keeping the dwellings cool in hot weather. In contrast, in the

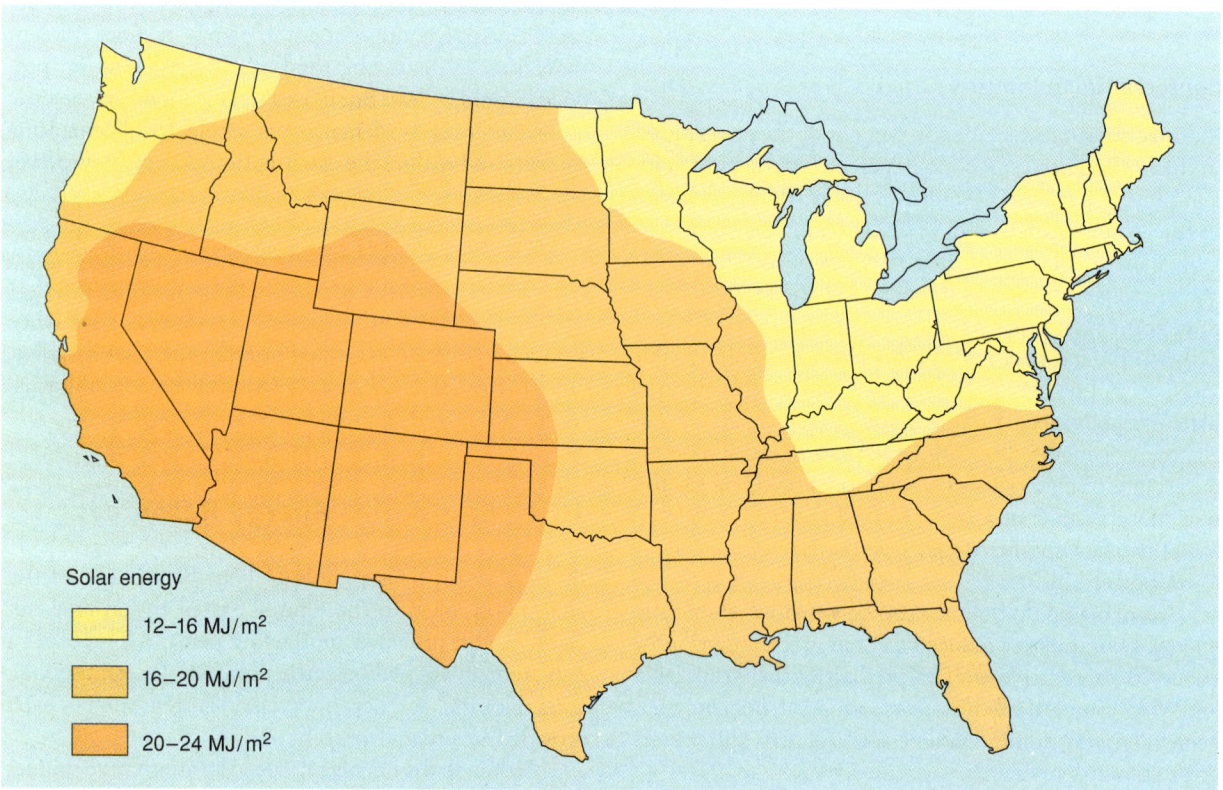

FIGURE 17–7
In the United States, the greatest amount of solar energy falls on the Southwest; the smallest amounts fall on the Northeast and Northwest. The units are megajoules per square meter. Even in the least sunny parts of the country, rooftop collectors can capture enough solar energy to supply a family with most of its domestic needs.

FIGURE 17–8
The Spruce Tree House in Mesa Verde National Park was built on a south-facing cliff. The overhanging rock ledge shades the homes from the high-angle summer Sun, but the lower winter Sun shines in to warm the buildings. *(Martin G. Kleinsorge)*

FIGURE 17–9
This passive solar home, designed by Jim Logan of Boulder, Colorado, costs less than $20 per month for all energy bills.

modern Southwest most people live in poorly insulated and poorly oriented houses and trailers. As a result, large amounts of electricity must be used to heat them in winter and cool them in summer.

A passive solar house is built with large, double-pane, south-facing windows and only small windows on the north side (Fig. 17–9). A massive stone, adobe, or concrete wall is situated opposite the windows so that the Sun shines directly onto it. Solar energy passes freely through glass and is absorbed by masonry (Fig. 17–10). During the day, the masonry absorbs and stores heat, thereby keeping the house from becoming too hot. At night, the masonry radiates the heat, keeping the house warm. Drapes or insulated shutters are closed at night to prevent conductive heat loss from the windows. Thick insulation in the walls and roof retains additional heat. Finally, an efficient solar house is well sealed so that warm air cannot escape and cold air does not seep in.

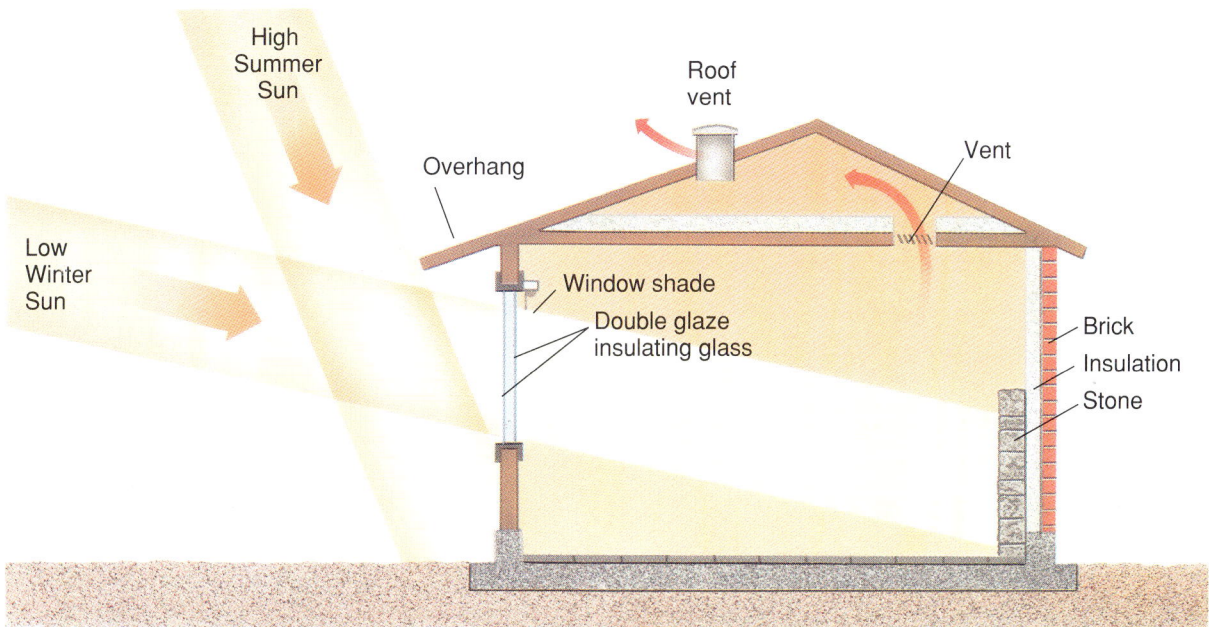

FIGURE 17–10
A passive solar house absorbs the Sun's energy during the day and releases the heat at night.

Even in the cold climate of the northern Rocky Mountains, passive solar energy provides well-designed homes with more than 50 percent of the heat needed in winter.

Active Solar Heating

In **active solar heating** systems, **solar collectors** trap the Sun's energy and use it to heat water. Pumps then circulate the hot water through radiators to heat a building, or the inhabitants use the hot water directly for washing and bathing. One type of collector consists of a coil of copper pipe fastened to a blackened metal base (Fig. 17–11). The assembly is enclosed in a box covered with clear plastic or glass. Water circulates through the piping and into a tank. Solar energy passes freely through the clear glass or plastic and is absorbed by the blackened metal base and copper piping. The glass prevents the heat from radiating back out of the box, so the pipes and water become hot. On a 25°C day, a good collector can heat water to 75°C. The heated water then passes into the home. This type of system is called active because pumps circulate the water and thermostats regulate the flow.

Active solar heating is commonly used in conjunction with a conventional hot water heater or furnace because on cloudy or winter days it cannot heat water to a high enough temperature for domestic use. Nevertheless, active hot water systems can save large amounts of fuel and usually pay for themselves in 5 to 10 years.

Electricity Production by Solar Cells

Certain types of materials generate electricity when light shines on them. A device that produces electricity directly from sunlight is called a **solar cell**. Small solar cells are familiar power sources in hand-held calculators and small battery rechargers. Larger versions produce domestic and industrial energy in remote places away from power lines. Solar cells also produce electricity for some streetlights and traffic signals. Modern solar cells are still a relatively expensive way to produce electricity. Although they cannot compete with other commercial systems at present, their potential for producing cheap, pollution-free energy is so great that a few research breakthroughs could alter the way in which the world produces its electricity.

A modern solar cell is a wafer or film of silicon, the second most abundant element in the Earth's crust. Trace amounts of rare elements, such as gallium, arsenic, and cadmium, are added to the silicon to produce a semiconductor. Sunlight energizes electrons in the semiconductor so that they travel through the crystal, producing an electric current. Because each cell produces only a small amount of electricity, many must be connected to generate a usable amount of power (Fig. 17–12).

Several problems must be overcome before solar cells become economically viable sources of electricity:

1. They are expensive. In 1992, Photocomm, Inc., the largest supplier in the United States, quoted a price of $8 to $10 per watt to supply and install a solar-cell generating system in a home. If fuels, passive solar design, and active solar energy are used to provide heat, and if energy-efficient appliances reduce electricity consumption, solar cells can now supply an average-size house with electricity for an initial investment of about $17,000. New production technologies are expected to

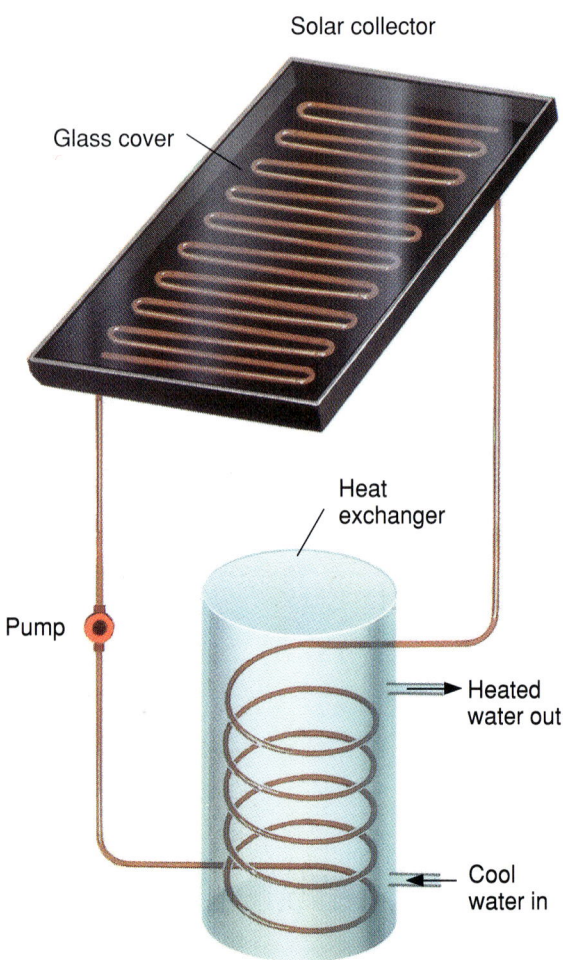

FIGURE 17–11
A solar collector uses the Sun's energy to heat water.

FIGURE 17-12
This bank of solar cells generates electricity for the visitor center at Natural Bridges National Monument, Utah.

FIGURE 17-13
Rooftop solar cells can generate enough energy to supply all the electricity for an energy-efficient home.

lower the price of solar cells to $1.50 to $2 per watt. If this lower price is combined with energy conservation and the use of energy-efficient appliances, the initial cost of installing a solar-cell generating system in a home falls to about $4500.

2. Solar cells generate electricity only during sunlight hours. Thus, large and expensive banks of batteries must be used to store energy for use at night and on dark days.
3. Some critics argue that solar cells take up too much space. Twenty-seven square meters of solar panels are required to power an energy-efficient house with a 2000-watt power supply. Proponents of solar energy point out that most roofs cover more than 27 square meters, so the space problem is easily overcome by simply mounting the solar cells on the roof (Fig. 17-13).

Critics also argue that the space problem is overwhelming in the case of large-capacity solar generating plants. For example, a 100-megawatt plant (one generating 100 million watts) would supply 50,000 people with electricity but would cover about 2 square kilometers of land with solar panels. Defenders of solar cells point out, however, that this is about the same area occupied by a strip coal mine and its waste disposal areas for the generation of the same level of power. They also observe that the environmental degradation created by 2 kilometers of solar panels is far less than that created by a strip coal mine.

Many scientists feel that solar cells have the potential to supply between 50 and 100 percent of all electricity used in the United States. The Union of Concerned Scientists, an organization of 100,000 American scientists, stated in 1990 that

> of all the forms of energy we have—renewable or nonrenewable—solar has the greatest potential for providing clean, safe, reliable power. The supply is inexhaustible, and the cost, which has fallen dramatically in the past few years, should be competitive with fossil fuels within the decade. Unlike fossil fuels, however, solar energy does not increase global warming, nor does it leave us dependent on foreign suppliers. And unlike nuclear power, it presents no problems of safety, disposal of radioactive wastes, or danger that nuclear materials will fall into the wrong hands.

Solar technologies can provide both electricity and heat and thus can be used for virtually all energy needs in the United States.

17.4 Hydroelectric Energy

If a river is dammed, the energy of water dropping downward through the dam can be harnessed to produce electricity. Energy generated in this way is called **hydroelectric energy** (Fig. 17-14). Hydroelectric generators supply between 15 and 20 percent of the world's electricity. They provide about 3.5 percent of all energy consumed in

FIGURE 17-14
The Flaming Gorge Dam in Utah generates hydroelectric power.

the United States, but about 15 percent of our electricity. There is much potential for further development of hydroelectric sites. In addition, many small flood control dams now exist without turbines. It is possible to modify or rebuild these dams and add turbines to them. The Federal Power Commission estimates that if dams were built to harness undeveloped hydropower potential in the United States, the hydroelectric energy output would triple. As a result, nearly half of our electricity would come from dams.

Hydropower is a clean and renewable energy source. A hydroelectric generating plant does not pollute the water that powers it, and the hydrologic cycle continually replenishes the water flowing down the stream. However, dams and hydropower systems are so expensive to build that the construction costs must be borne by taxpayers, and, as discussed in Chapter 10, those costs are rarely included in the price of electricity paid by consumers. In addition, dams and hydropower systems radically alter stream ecosystems.

The high monetary and environmental costs of dam building, low oil prices, potential of alternative energy resources, lack of political support, and criticisms from conservation and other public interest groups make it unlikely that major new dam projects will start up in the United States in the near future. Other countries, however, are planning large-scale development of hydropower resources. Huge hydroelectric dam projects are scheduled in China, Brazil, India, Nepal, and Canada. China has about 10 percent of the Earth's undeveloped hydropower potential. By the year 2000, it is expected to become the world's largest hydroelectric producer. Canada is planning a huge dam project at James Bay, near the Ontario–Quebec border. The project will cause massive environmental changes in the area and displace and disrupt the life styles of many native people. Canada doesn't need the power and plans to sell it to New York at prices lower than production costs. Many Canadians are opposed to the project for a wide variety of reasons, and New York residents may not need the power. As this book was being written, the fate of the project was uncertain.

17.5 Geothermal Energy

Three different processes heat ground water:

1. The Earth's temperature increases by about 25°C per kilometer in the upper portion of the crust. Therefore, if ground water descends through cracks to depths of 2 to 3 kilometers, Earth heat warms it by 50° to 75°C.
2. In regions of recent volcanism, magma or hot igneous rock may remain near the surface and can heat ground water at relatively shallow depths.
3. Many hot springs have the odor of rotten eggs from small amounts of hydrogen sulfide (H_2S) dissolved in the hot water. These springs are heated by chemical reaction between sulfide minerals, such as pyrite (FeS_2), and water. The reactions release heat. Hydrogen sulfide, produced by the reaction, rises with the heated ground water and gives it the strong odor.

Hot ground water can be used to generate electricity, or it can be used directly to heat homes and other buildings. Energy extracted from the Earth's heat is called **geothermal energy**. The United States is the largest producer of geothermal electricity in the world, with a production capacity of 2200 megawatts. This is equivalent to the power output of two large nuclear reactors and provides about 1 million people with all of their electrical needs. However, this amount is minuscule compared to the amount of available geothermal energy. The United States Geological Survey estimates that the upper 4 kilometers of rock beneath the United States contains heat energy equivalent to 3000 trillion barrels of oil. This is enough energy to run the country for the next 200,000 years at current consumption rates.

The major problem with modern methods of extracting geothermal energy from the Earth is that they work only where ground water is naturally heated. Only a limited number of such "wet" sites exist, where abundant ground water *and* hot rock or magma occur at the same place (Fig. 17–15). However, many "dry" sites exist where rising magma has heated rocks close to the surface but little ground water is available. Wells can be

17.5 Geothermal Energy 421

FIGURE 17-15
Hot water from these California hot springs is piped to a generating plant to produce electricity. *(Pacific Gas & Electric)*

However, problems arise with attempts to extract heat from dry sites. Hot rock usually lies more than 3 or 4 kilometers below the surface. It is expensive and technically difficult to drill a deep well into hot rock. In addition, the drilling and heat extraction process uses large amounts of water. Most shallow, hot rock in North America is found in the West, where tectonism and igneous activity have recently taken place. Much of this region is semiarid.

Scientists and engineers are developing methods for extracting energy from dry Earth heat at a pilot project at Fenton Hill, New Mexico. They drilled two separate wells side by side (Fig. 17–16). Pumps force water down the injection well to a depth of about 4 kilometers. The water circulates through hot, fractured granite at the bottom of the well and then flows to the extraction well, where it returns to the surface. The scientists have succeeded in pumping water into the well at 20°C and extracting steam 12 hours later at 190°C.

drilled in these places. Pumps then circulate water through the hot rock and back to the surface, where it produces steam that drives turbines to generate electricity.

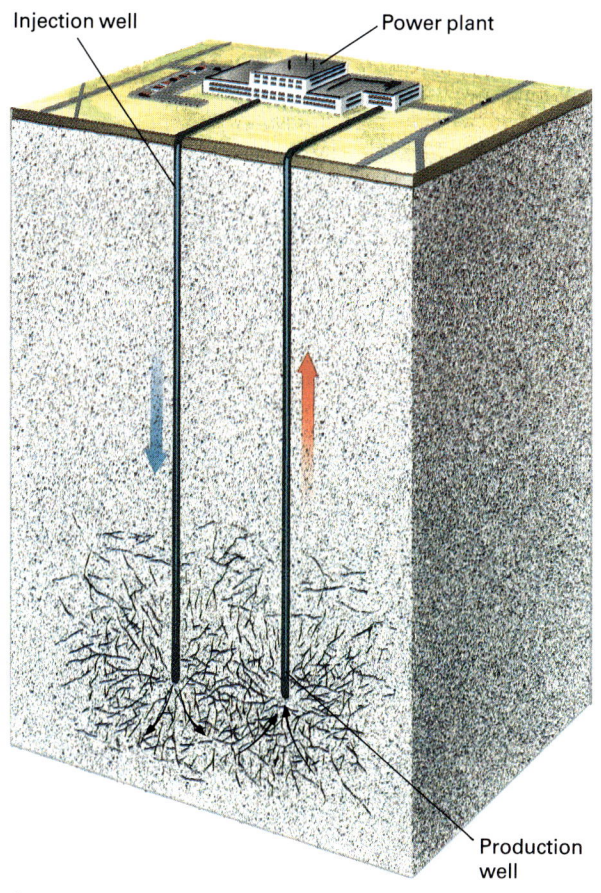
A

B

FIGURE 17-16
(A) A schematic view of the Fenton Hill, New Mexico, dry geothermal energy plant. (B) Steam escapes from hot water heated by deep, hot rock at Fenton Hill. *(Los Alamos National Laboratory)*

In an era when petroleum prices are low, there is little incentive to develop expensive geothermal projects. The economic picture is expected to change within the next few decades as petroleum resources become limited and prices rise.

17.6 Wind Energy

The Sun warms the Earth's atmosphere unevenly. Masses of warmer air expand and rise. Cooler air flows along the Earth's surface to replace the rising warm air. We feel the flow of air as wind. Because the Sun's heat creates wind, wind energy is a form of solar energy. Windmills can harness the wind's energy in areas where wind is strong and steady.

Wind power is a nonpolluting and renewable energy resource. From the 1930s to the 1950s, before rural electrification projects brought commercial power to many parts of the United States, isolated ranchers and farmers used windmills to pump water and generate electricity for their homes. On some old ranches the towers and wind blades still stand, reminding us of life in those times (Fig. 17–17). A few individuals now use modern wind generators to supply power to homes, but most modern wind generation now occurs on "wind farms," clusters of efficient wind turbines concentrated in areas of strong, steady wind (Fig. 17–18).

FIGURE 17–18
A wind farm in Oahu, Hawaii. *(Douglas Peebles)*

In 1992 in the United States, 16,000 wind generators produced 2 billion kilowatt-hours of electricity at a cost of 8 cents per kilowatt-hour. The average price of electricity from conventional sources in 1992 was 6.7 cents per kilowatt-hour. The total capacity of commercial wind generators in the United States is 1600 megawatts, as much as one and one half nuclear power plants. Most of our wind generators are in California, but the largest single wind farm in the United States is on Hawaii's island of Oahu. A huge, untapped potential for wind generation exists in several midwestern and western states, where winds blow strongly and almost continuously. Some energy experts predict that technological advances will reduce the cost of electricity generated by wind turbines to about 3.5 cents per kilowatt-hour, even at nonideal sites.

17.7 Biomass Energy

In 1990, plant fuels produced 2.7 quads (quadrillion BTUs) of energy. For comparison, nuclear power plants generated 6.16 quads. Wood was the most productive of all biofuels, generating 2.2 quads; controlled garbage incineration produced 0.4 quads; and alcohol fuels produced 0.1 quad. Wood-burning stoves produce so much smoke and particulate pollution, however, that their use is banned in some cities and restricted in others to reduce winter air pollution. Recently, wood stoves have been developed that are 65 percent efficient and produce only 10 percent as much pollution as older wood stoves. But they are expensive, and wood will probably be significant only in rural areas (Fig. 17–19).

FIGURE 17–17
In the first half of this century, windmills like this one were used to pump water in remote regions. Other windmills generated electricity for isolated ranchers.

FIGURE 17-19
Wood is the most common biofuel in rural regions.

Some biofuel experts suggest that trees grow too slowly to be used as a renewable energy source. Instead, they suggest that crops such as sugarcane, pineapples, corn, soybeans, peanuts, sunflowers, or even wheat should be grown for the energy content in their sugars, oils, or starches. In this case, the plant products are fermented to make ethanol, a useful fuel. Ethanol burns more cleanly in automobile engines than conventional gasoline and is used as an additive to reduce urban air pollution. In the early summer of 1994, President Clinton initiated a program to increase the bio-ethanol production in the United States. The plan sounds attractive—free fuel from the ground. The reality is that, although bio-ethanol reduces air pollution, it will not reduce fossil fuel consumption. Farmers use fossil fuels to manufacture fertilizers and to power tractors. More fuel is needed to transport the crops and operate the fermentation vats. According to the European Farm Commission, the fuel consumption needed to produce ethanol from wheat or corn equals the energy produced. Thus there is no net energy gain.

More than half of the household trash in the United States and Canada is paper, which could be collected and used as fuel. If this material is burned in controlled incinerators to produce useful energy, two problems are solved simultaneously. The volume of solid waste is reduced and fossil fuel resources are conserved. However, toxic materials in household waste can produce air pollution if pollution control devices are not maintained adequately. In the United States, construction of trash incinerators boomed in the 1980s as public landfills neared capacity. However, public concern about air pollution from these incinerators has led to a recent decline in construction. The largest electrical generator that operates solely on trash is located in Rotterdam, Holland. This facility generates 550 megawatts, enough to continuously supply 250,000 people with their domestic needs.

17.8 Energy from the Seas

Tidal Energy

A tide is the cyclical rise and fall of ocean water caused by the gravitational force of the Moon and, to a lesser extent, of the Sun. If islands or a narrow-mouthed bay constricts tidal flow, the moving water funnels into a tidal current. Twice a day, seawater flows into the bays with the rising tide, and twice a day it rushes outward. The energy of these currents can be harnessed by a tidal dam and turbine (Fig. 17-20).

During the 1800s, tidal power was popular in the United States, especially along the coast of Maine, with its narrow bays and inlets. At that time, it was an attractive alternative to bulky steam engines, which consumed large quantities of wood. However, when fossil fuels became inexpensive in the 1900s, almost all tidal power plants were abandoned. Today the potential still exists, but even in the most favorable sites tidal dams are not always economical.

FIGURE 17-20
Tidal currents can be harnessed to produce electricity.

In addition to being costly, tidal dams create environmental problems. Tidal bays are commonly productive estuaries where fish breed, and they are also popular places for recreation. If these areas are dammed, some of the natural qualities and resources are lost. Nevertheless, proponents of tidal projects point out that when fossil fuels become scarce, tidal energy may become attractive.

Wave Energy

Waves strike every coastline in the world, and their combined energy is enormous. In 1990, engineers designed and built a wave generator in Northern Ireland, shown in Figure 17–21. When water from an incoming wave rises in a concrete chamber, it forces air through a narrow valve. The air spins a rotor that drives an electrical generator. The generator produces electricity for a small coastal village.

To date, the major barrier to widespread use of wave power has been economic. Although the wave energy along an entire coastline is enormous, the potential in any one place is small. Therefore, many wave generators are needed to harness large quantities of energy. At present, the generators are too expensive to be practical. However, improvements in design and rising oil prices may reverse the economic balance sheet.

Ocean Thermal Power

It is also possible to use the heat of seawater to produce electricity. In tropical regions, the sea surface is approximately 20°C warmer than subsurface water. The warm water is hot enough to vaporize a pressurized liquid such as ammonia. The gaseous ammonia can drive a turbine, just as hot steam drives a turbine in a coal-fired plant. In an ocean thermal generator, the gaseous ammonia is cooled by the subsurface water and recycles indefinitely, as shown in Figure 17–22. Although the efficiency of such an engine is low, so much heat energy is available in the oceans that a large amount of electricity could be produced. Today the costs of ocean thermal power plants are so high that they are uneconomical, even though they produce electricity without consuming any fuel.

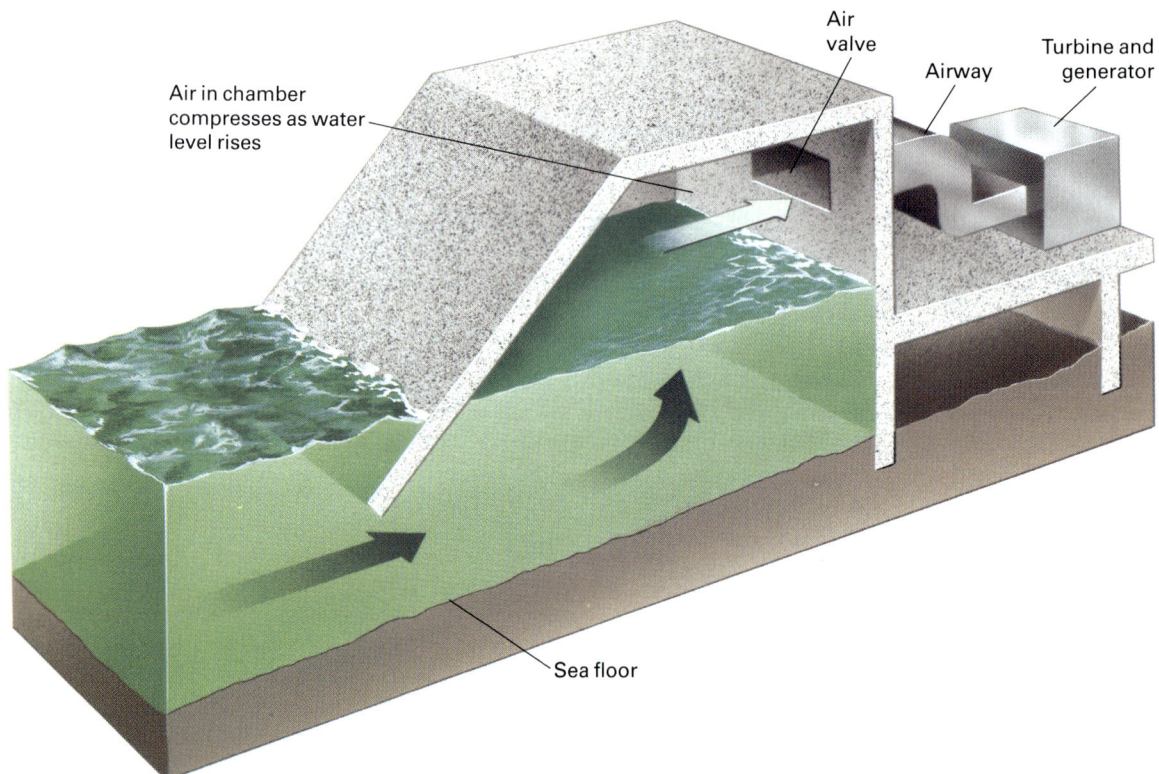

FIGURE 17–21
A schematic view of one design for harnessing wave energy. An incoming wave raises the water level inside a concrete chamber. Air inside the chamber is compressed and spins a turbine that powers a generator.

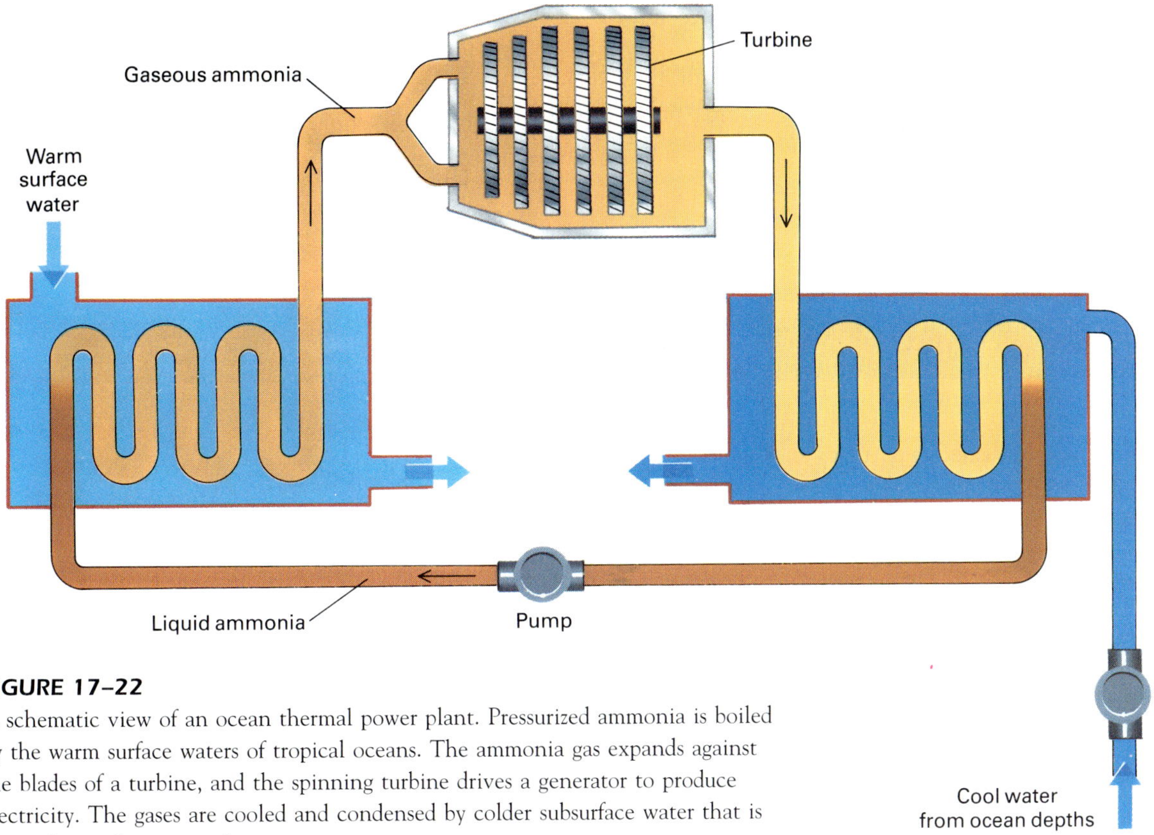

FIGURE 17–22
A schematic view of an ocean thermal power plant. Pressurized ammonia is boiled by the warm surface waters of tropical oceans. The ammonia gas expands against the blades of a turbine, and the spinning turbine drives a generator to produce electricity. The gases are cooled and condensed by colder subsurface water that is pumped into the power plant.

17.9 Hydrogen Gas as a Fuel

When hydrogen gas burns, it combines with atmospheric oxygen to produce water vapor. Although water vapor is a greenhouse gas, it is nonpolluting. Thus, hydrogen is a clean fuel. In addition, it is energy-rich: Combustion of 1 kilogram of hydrogen produces two and one half times as much energy as does a kilogram of gasoline.

Hydrogen gas can be liberated from water by passing an electric current through the water. Because water is so abundant, the hydrogen supply is nearly limitless. After extraction, hydrogen can be stored under high pressure as a gas or combined with metals to form solid compounds called **hydrides**. In its gaseous form, the hydrogen can be shipped like natural gas or propane to homes and other consumers. The hydrides can be heated to release the hydrogen gas to fuel autos, furnaces, or an electricity-generating fuel cell. Gaseous hydrogen explodes easily, as do natural gas and propane, but the hydrides neither explode nor burn. Thus they are safe and portable energy sources.

The extraction of hydrogen from water uses more energy than can be gotten by burning the hydrogen. It seems that such an energy budget would preclude hydrogen's use as a fuel. However, hydrogen is energy-rich, portable, and clean and thus is suited as a fuel for automobiles and other forms of transportation. Most other alternative energy resources described in this chapter are not light and portable. A plausible scenario for the future of transportation is one in which heavy, land-based power plants produce electricity from wind, solar, geothermal, and other renewable energy resources. Some of that electricity is used to extract hydrogen from seawater and convert it to hydrides. The hydrides, in turn, fuel autos, trucks, trains, and even airplanes, providing a nearly limitless and pollution-free energy resource for transportation. At current prices, it costs about $1.50 to extract an amount of hydrogen with the energy equivalent of a gallon of gasoline. Thus, hydrogen fuel already closely approaches the cost of gasoline and diesel fuel in many parts of the United States and is much cheaper than these fuels in many other nations.

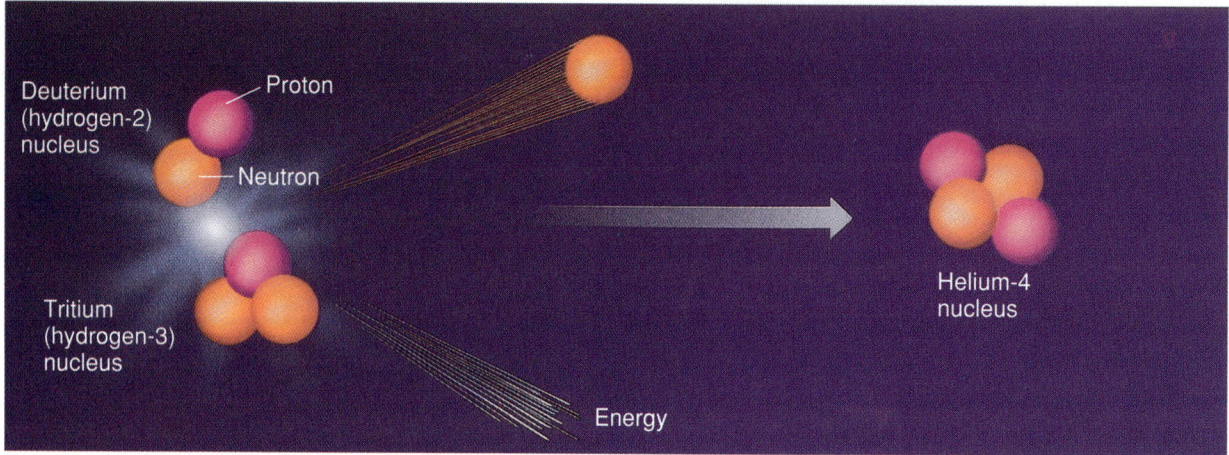

FIGURE 17–23
Fusion releases great amounts of energy as two nuclei combine to form a single, heavier nucleus.

17.10 Nuclear Fusion

Nuclear fusion occurs when nuclei of light elements combine to form heavier nuclei (Fig. 17–23). The energy of stars such as the Sun is produced when hydrogen nuclei combine to form helium nuclei. Nuclear fusion liberates a tremendous amount of energy relative to the amount of fuel consumed. A single gram of fusion fuel produces an amount of energy equivalent to 9400 liters of gasoline.

Two objects with the same electrical charge repel each other. Because all atomic nuclei are positively charged, the electrical force holds them apart. However, *at very close distances*, a strong attractive nuclear force overwhelms the repulsive electrical force and draws the nuclei together. In a star the size of our Sun, hydrogen nuclei are energized by immense gravitational forces and by high temperature. They collide with each other so rapidly that they overcome the electrical repulsion and nuclear attraction takes over. In a series of reactions, hydrogen nuclei combine to form helium nuclei, releasing large quantities of energy.

Scientists have long understood that fusion is a potential source of energy for both weapons and peaceful uses. One technological challenge in harnessing fusion for energy production is to achieve the tremendously high temperatures needed to start the reaction. Three isotopes of hydrogen exist. Protium, or light hydrogen, is most abundant but fuses at the highest temperature. Deuterium and tritium are much less abundant but fuse at lower temperatures. Fusion between deuterium and tritium starts when the temperature reaches 40 million degrees Celsius, much hotter than any conventional flame.

To control a fusion reaction, deuterium and tritium must be fed into a reaction chamber at a slow but steady rate. However, no substance on Earth can remain solid at 40 million degrees, and therefore scientists cannot build a conventional container. Instead they envision a "magnetic bottle," which consists of a magnetic field that confines the charged nuclear particles.

In 1972, the Department of Energy built the Tomak fusion test reactor at Princeton University. In 1993, after 21 years of research, scientists produced 6 megawatts of power in a controlled reaction that lasted about 1 second. However, they used 30 megawatts to heat the chamber to fusion temperature. Other experiments using intense laser pulses have also produced encouraging results but no self-sustaining reaction. Thus, at present, although development is progressing, we are far from even experimental fusion energy production, and no one expects commercial fusion to be available for decades.

In 1992, two scientists from the University of Utah in Salt Lake City announced that they had produced a fusion reaction at room temperature by creating a chemical environment in an electrochemical cell that allowed hydrogen nuclei to approach one another closely enough to fuse. Because cold fusion would mean an endless supply of cheap, clean energy and an end to the world's energy problems, their announcement created immediate excitement around the globe. Many scientists were skeptical, however, and questioned both the experimental methods and the results. In the months that followed, some of the world's best laboratories attempted to duplicate the Salt Lake City experiments, but without success. Today, most scientists discount cold fusion, but continued research occasionally produces results suggesting that fusion at low temperatures is possible, although no one really knows what is going on.

SUMMARY

When fossil fuel reserves become depleted or too costly to use for energy production, we will turn to **alternative energy resources** to produce electricity and other forms of power for homes, transportation, and industry. However, it is important to start using alternative energy resources now to save some fossil fuels for their unique applications.

A large proportion of the energy consumed by humans is wasted by inefficient implements, machinery, building design, and human habits. Consequently, of all the alternative energy resources available to us, **conservation** can increase the amount of available energy in the quickest and least expensive manner.

Renewable alternative energy resources in use today include **solar energy**; **hydroelectric power**; **geothermal energy**; **wind energy**; **biomass energy**; energy from **tidal currents**, **waves**, and **heat**; and **hydrogen fuel**. These proven energy resources will probably become steadily more popular as the availability of fossil fuels declines and prices increase. **Nuclear fusion** is currently impracticable, but if scientists can overcome technological barriers, it may provide the world with cheap, clean, and unlimited power in the future.

KEY TERMS

Alternative energy resource 409
Social solution 411
Technical solution 414
Passive solar heating 416
Active solar heating 418
Solar collector 418
Solar cell 418
Hydroelectric energy 419
Geothermal energy 420
Hydrides 425
Nuclear fusion 426

REVIEW QUESTIONS

1. List the alternative energy resources described in this chapter. Can you think of others?
2. Describe the differences between nonrenewable and renewable energy resources.
3. Why is it important to begin using alternative energy resources before we exhaust fossil fuel reserves?
4. Describe political problems associated with America's heavy reliance on petroleum.
5. How does conservation act as an alternative energy resource?
6. What proportion of commercially produced energy is wasted?
7. What proportion of commercially produced energy could be saved by using efficient equipment and habits?
8. List the amount of energy produced in the United States by each of the conventional and alternative energy resources.
9. Why are alternative energy resources incompatible with current transportation modes in the United States?
10. Describe each of the ways in which solar energy can be used.
11. What design and construction characteristics distinguish a passive solar home from a conventional house?
12. Describe the differences between passive and active solar heating systems.
13. What problems must be overcome before solar cell electricity generation can become widely available?
14. Why does the Union of Concerned Scientists feel that solar energy has the greatest potential of all alternative energy resources?
15. How is the energy of a flowing stream harnessed to produce electricity?
16. Describe environmental problems associated with hydroelectric power generation.
17. How is usable energy extracted from the Earth's heat?
18. What role does water play in the exploitation of geothermal energy?
19. Describe the conditions that create an ideal site for harnessing wind energy.
20. Describe several different ways in which biomass energy is used.
21. In what forms can energy be obtained from the seas?
22. Describe the unique advantages of hydrogen fuel.
23. Describe a nuclear fusion reaction.
24. What problems must be overcome before nuclear fusion can be used to generate usable energy?

DISCUSSION QUESTIONS

1. In the 1950s, power companies encouraged increased electricity consumption by advertising that people should "live better electrically." Today many power companies across the United States offer customers cash rebates for adding extra insulation or purchasing more efficient appliances so consumers will use less electricity. Why is it good business practice for a company that sells electricity to encourage conservation?

2. Imagine that you are planning to build a new home near your school. What alternative energy sources are economically attractive? How much would they add to the initial price of a home? How much would you save in energy costs every year?

3. In Europe, many countries tax petroleum fuels heavily to reduce dependence on foreign imports. In the United States, significant increases in fuel taxes have been bitterly opposed. Prepare a class debate. Have one side argue for a $1-per-gallon gasoline tax while the other side opposes the idea.

4. Discuss why energy-inefficient light sources and machinery are still used heavily, despite the availability of efficient implements.

5. Discuss ways in which alternative energy resources can become compatible with transportation in the United States.

6. Discuss the environmental, economic, and political implications of the development of a solar cell that produces electricity more cheaply than conventional sources.

7. Sketch a design for a passive solar home. Include instructions regarding orientation and placement of the home.

8. Wind energy, hydroelectric power, and energy from sea waves are sometimes called "other forms of solar energy." Discuss how the Sun is the source of those alternative energy resources.

9. Discuss the ultimate sources of geothermal energy.

10. Why is the concept of cold fusion such a fascinating one to scientists and others involved in energy production?

11. Because controlled fusion has been so elusive, scientists have suggested that useful energy could be extracted first by exploding an underground hydrogen bomb. Then water could be pumped into the chamber. The hot rock would heat the water, and the hot water could be used to generate electricity. Discuss the pros and cons of this idea.

12. Discuss ways in which the government could encourage energy efficiency. Discuss the pros and cons of various government programs.

APPENDIX A

The Elements

Key:
26
Fe
55.847

26 — Atomic number (Z)
Fe — Element symbol
55.847 — Atomic mass of naturally occurring isotopic mixture; for radioactive elements, numbers in parentheses are mass numbers of most stable isotopes

IA	IIA	IIIB	IVB	VB	VIB	VIIB	VIII			IB	IIB	IIIA	IVA	VA	VIA	VIIA	O
1 H 1.0079																1 H 1.0079	2 He 4.00260
3 Li 6.941	4 Be 9.01218											5 B 10.81	6 C 12.011	7 N 14.0067	8 O 15.9994	9 F 18.998403	10 Ne 20.179
11 Na 22.98977	12 Mg 24.305											13 Al 26.98154	14 Si 28.0855	15 P 30.97376	16 S 32.06	17 Cl 35.453	18 Ar 39.948
19 K 39.0983	20 Ca 40.08	21 Sc 44.9559	22 Ti 47.90	23 V 50.9415	24 Cr 51.996	25 Mn 54.9380	26 Fe 55.847	27 Co 58.9332	28 Ni 58.70	29 Cu 63.546	30 Zn 65.38	31 Ga 69.72	32 Ge 72.59	33 As 74.9216	34 Se 78.96	35 Br 79.904	36 Kr 83.80
37 Rb 85.4678	38 Sr 87.62	39 Y 88.9059	40 Zr 91.22	41 Nb 92.9064	42 Mo 95.94	43 Tc (98)	44 Ru 101.07	45 Rh 102.9055	46 Pd 106.4	47 Ag 107.868	48 Cd 112.41	49 In 114.82	50 Sn 118.69	51 Sb 121.75	52 Te 127.60	53 I 126.9045	54 Xe 131.30
55 Cs 132.9054	56 Ba 137.33	57 *La 138.9055	72 Hf 178.49	73 Ta 180.9479	74 W 183.85	75 Re 186.207	76 Os 190.2	77 Ir 192.22	78 Pt 195.09	79 Au 196.9665	80 Hg 200.59	81 Tl 204.37	82 Pb 207.2	83 Bi 208.9804	84 Po (209)	85 At (210)	86 Rn (222)
87 Fr (223)	88 Ra 226.0254	89 †Ac 227.0278	104 Unq (261)	105 Unp (262)	106 Unh (263)	107 Uns	109										

*Lanthanide Series

58 Ce 140.12	59 Pr 140.9077	60 Nd 144.24	61 Pm (145)	62 Sm 150.4	63 Eu 151.96	64 Gd 157.25	65 Tb 158.9254	66 Dy 162.50	67 Ho 164.9304	68 Er 167.26	69 Tm 168.9342	70 Yb 173.04	71 Lu 174.967

†Actinide Series

90 Th 232.0381	91 Pa 231.0359	92 U 238.029	93 Np 237.0482	94 Pu (244)	95 Am (243)	96 Cm (247)	97 Bk (247)	98 Cf (251)	99 Es (252)	100 Fm (257)	101 Md (258)	102 No (259)	103 Lr (260)

Note: Atomic masses shown here are 1977 IUPAC values.

International Table of Atomic Weights

Based on the assigned relative atomic mass of $^{12}C = 12$.

The following values apply to elements as they exist in materials of terrestrial origin and to certain artificial elements. When used with the footnotes, they are reliable to ±1 in the last digit, or ±3 if that digit is in small type.

	Symbol	Atomic Number	Atomic Weight		Symbol	Atomic Number	Atomic Weight		Symbol	Atomic Number	Atomic Weight
Actinium	Ac	89		Hafnium	Hf	72	178.4_9	Promethium	Pm	61	
Aluminum	Al	13	26.98154^a	Helium	He	2	$4.00260^{b,c}$	Protactinium	Pa	91	$231.0359^{a,f}$
Americium	Am	95		Holmium	Ho	67	164.9304^a	Radium	Ra	88	$226.0254^{f,g}$
Antimony	Sb	51	121.7_5	Hydrogen	H	1	$1.0079^{b,d}$	Radon	Rn	86	
Argon	Ar	18	$39.94_8^{b,c,d,g}$	Indium	In	49	114.82	Rhenium	Re	75	186.207
Arsenic	As	33	74.9216^a	Iodine	I	53	126.9045^a	Rhodium	Rh	45	102.9055^a
Astatine	At	85		Iridium	Ir	77	192.2_2	Rubidium	Rb	37	85.467_8^c
Barium	Ba	56	137.3_4	Iron	Fe	26	55.84_7	Ruthenium	Ru	44	101.0_7
Berkelum	Bk	97		Krypton	Kr	36	83.80	Samarium	Sm	62	150.4
Beryllium	Be	4	9.01218^a	Lanthanum	La	57	138.905_5^b	Scandium	Sc	21	44.9559^a
Bismuth	Bi	83	208.9804^a	Lawrencium	Lr	103		Selenium	Se	34	78.9_6
Boron	B	5	$10.81^{c,d,e}$	Lead	Pb	82	$207.2^{d,g}$	Silicon	Si	14	28.08_6^d
Bromine	Br	35	79.904^c	Lithium	Li	3	$6.94_1^{c,d,e,g}$	Silver	Ag	47	107.868^c
Cadmium	Cd	48	112.40	Lutetium	Lu	71	174.97	Sodium	Na	11	22.98977^a
Calcium	Ca	20	40.08^g	Magnesium	Mg	12	$24.305^{c,g}$	Strontium	Sr	38	87.62^g
Californium	Cf	98		Manganese	Mn	25	54.9380^a	Sulfur	S	16	32.06^d
Carbon	C	6	$12.011^{b,d}$	Mendelevium	Md	101		Tantalum	Ta	73	180.947_9^b
Cerium	Ce	58	140.12	Mercury	Hg	80	200.5_9	Technetium	Tc	43	98.9062^f
Cesium	Cs	55	132.9054^a	Molybdenum	Mo	42	95.9_4	Tellurium	Te	52	127.6_0
Chlorine	Cl	17	35.453^c	Neodymium	Nd	60	144.2_4	Terbium	Tb	65	158.9254^a
Chromium	Cr	24	51.996^c	Neon	Ne	10	20.17_9^c	Thallium	Tl	81	204.3_7
Cobalt	Co	27	58.9332^a	Neptunium	Np	93	237.0482^f	Thorium	Th	90	232.0381^f
Copper	Cu	29	$63.54^{c,d}$	Nickel	Ni	28	58.70	Thulium	Tm	69	168.9342^a
Curium	Cm	96		Niobium	Nb	41	92.9064^a	Tin	Sn	50	118.6_9
Dysprosium	Dy	66	162.5_0	Nitrogen	N	7	$14.0067^{b,c}$	Titanium	Ti	22	47.9_0
Einsteinium	Es	99		Nobelium	No	102		Tungsten	W	74	183.8_5
Erbium	Er	68	167.2_6	Osmium	Os	76	190.2	Uranium	U	92	$238.029^{b,c,e}$
Europium	Eu	63	151.96	Oxygen	O	8	$15.999_4^{b,c,d}$	Vanadium	V	23	$50.941_4^{b,c}$
Fermium	Fm	100		Palladium	Pd	46	106.4	Wolfram			(see Tungsten)
Fluorine	F	9	18.99840^a	Phosphorus	P	15	30.97376^a	Xenon	Xe	54	131.30
Francium	Fr	87		Platinum	Pt	78	195.0_9	Ytterbium	Yb	70	173.04_4
Gadolinium	Gd	64	157.2_5	Plutonium	Pu	94		Yttrium	Y	39	88.9059^a
Gallium	Ga	31	69.72	Polonium	Po	84		Zinc	Zn	30	65.38
Germanium	Ge	32	72.5_9	Potassium	K	19	39.09_8	Zirconium	Zr	40	91.22
Gold	Au	79	196.9665^a	Praseodynium	Pr	59	140.9077^a				

[a] Mononuclidic element.
[b] Element with one predominant isotope (about 99 to 100 percent abundance).
[c] Element for which the atomic weight is based on calibrated measurements by comparison with synthetic mixtures of known isotopic compositions.
[d] Element for which known variation in isotopic abundance in terrestrial samples limits the precision of the atomic weight given.
[e] Element for which users are cautioned against the possibility of large variations in atomic weight due to inadvertent or undisclosed artificial isotopic separation in commercially available materials.
[f] Most commonly available long-lived isotope.
[g] In some geological specimens this element has a highly anomalous isotopic composition, corresponding to an atomic weight significantly different from that given.

APPENDIX B
Systems of Measurement

1 THE SI SYSTEM

In the past, scientists from different parts of the world have used different systems of measurement. However, global cooperation and communication make it essential to adopt a standard system. The International System of Units (SI) defines units of measurement as well as prefixes for multiplying or dividing the units by decimal factors. Some primary and derived units important to geologists follow.

Time

The SI unit is the **second** (s or sec), which used to be based on the rotation of the Earth but is now related to the vibration of atoms of cesium-133. SI prefixes are used for fractions of a second (such as milliseconds or microseconds), but the common words **minutes, hours,** and **days** are still used to express multiples of seconds.

Length

The SI unit is the **meter** (m), which used to be based on a standard platinum bar but is now defined in terms of wavelengths of light. The closest English equivalent is the **yard** (0.914 m). A **mile** is 1.61 kilometers (km). An **inch** is exactly 2.54 centimeters (cm).

Area

Area is length squared, as in **square meter, square foot,** and so on. The SI unit of area is the **are** (a), which is 100 sq m. More commonly used is the **hectare** (ha), which is 100 ares, or a square that is 100 m on each side. (The length of a U.S. football field plus one end zone is just about 100 m.) A hectare is 2.47 acres. An **acre** is 43,560 sq ft, which is a plot of (for example) 220 ft by 198 ft.

Volume

Volume is length cubed, as a **cubic centimeter** (cm^3), **cubic foot** (ft^3), and so on. The SI unit is the **liter** (L), which is 1000 cm^3. A **quart** is 0.946 L; a U.S. liquid **gallon** (gal) is 3.785 L. A **barrel** of petroleum (U.S.) is 42 gal, or 159 L.

Mass

Mass is the amount of matter in an object. **Weight** is the force of gravity on an object. To illustrate the difference, an astronaut in space has no weight but still has mass. On Earth, the two properties are directly proportional and the terms are often used interchangeably. The SI unit of mass is the **kilogram** (kg), which is based on a standard platinum mass. An avoirdupois (avdp) **pound** (lb), is a unit of weight. On the surface of the Earth, 1 lb is equal to 0.454 kg. A **metric ton,** also written as **tonne,** is 1000 kg, or about 2205 lb.

Temperature

The Celsius scale is used in most laboratories to measure temperature. On the Celsius scale the freezing point of water is 0°C and the boiling point of water is 100°C.

The SI unit of temperature is the **kelvin.** The coldest possible temperature, −273°C, is zero on the Kelvin scale. The size of 1 degree Kelvin is equal to 1 degree Celsius.

$$\text{Celsius temperature (°C)} = \text{Kelvin temperature (K)} - 273 \text{ K}$$

Fahrenheit temperature (°F) is not used in scientific writing, although it is still popular in English-speaking countries. An illustration of conversion between Fahrenheit and Celsius follows.

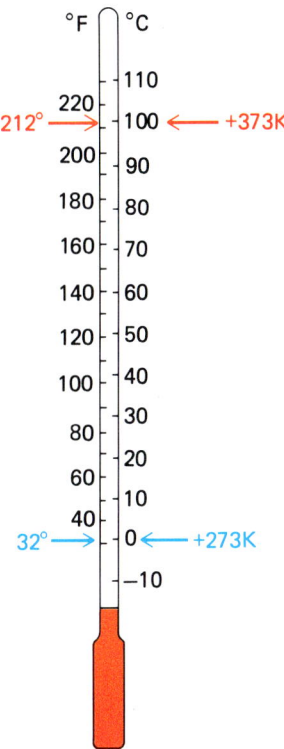

Energy

Energy is a measure of work or heat. The two were once thought to be different quantities. Hence, two different sets of units were adopted and still persist, although we now know that work and heat are both forms of energy.

The SI unit of energy is the **joule** (J), the work required to exert a force of 1 newton through a distance of 1 m. In turn, a newton is the force that gives a mass of 1 kg an acceleration of 1 m/sec^2. In human terms, a joule is not much—it is about the amount of work required to lift a 100-g weight to a height of 1 m. Therefore, joule units are too small for discussions of machines, power plants, or energy policy. Larger units are

megajoule, MJ = 10^6 J (a day's work by one person)
gigajoule, GJ = 10^9 J (the energy in half a tank of gasoline)

The energy unit used for heat is the **calorie** (cal), which is exactly 4.184 J. One calorie is just enough energy to warm 1 g of water 1°C. The more common unit used in measuring food energy is the **kilocalorie,** (kcal), which is 1000 cal. When **Calorie** is spelled with a capital C, it means kcal. If a cookbook says that a jelly doughnut has 185 calories, that is an error—it should say 185 Calories (capital C) or 185 kcal. A value of 185 calories (small c) would be the energy in about one quarter of a thin slice of cucumber.

The unit of energy in the British system is the **British thermal unit** (Btu), which is the energy needed to warm 1 lb of water 1°F.

$$1 \text{ Btu} = 1054 \text{ J} = 1.054 \text{ kJ} = 252 \text{ cal}$$

The unit often referred to in discussions of national energy policies is the **quad,** which is 1 quadrillion Btu, or 10^{15} Btu.

Some approximate energy values are

1 barrel (42 gal) of petroleum = 5900 MJ
1 ton of coal = 29,000 MJ
1 quad = 170 million barrels of oil, or 34 million tons of coal

II PREFIXES FOR USE WITH BASIC UNITS OF THE METRIC SYSTEM

Prefix	Symbol†	Power		Equivalent
geo*		10^{20}		
tera	T	10^{12}	= 1,000,000,000,000	Trillion
giga	G	10^9	= 1,000,000,000	Billion
mega	M	10^6	= 1,000,000	Million
kilo	k	10^3	= 1,000	Thousand
hecto	h	10^2	= 100	Hundred
deca	da	10^1	= 10	Ten
———	—	10^0	= 1	One
deci	d	10^{-1}	= .1	Tenth
centi	c	10^{-2}	= .01	Hundredth
milli	m	10^{-3}	= .001	Thousandth
micro	μ	10^{-6}	= .000001	Millionth
nano	n	10^{-9}	= .000000001	Billionth
pico	p	10^{-12}	= .000000000001	Trillionth

* Not an official SI prefix but commonly used to describe very large quantities such as the mass of water in the oceans.

† The SI rules specify that SI symbols are not followed by periods, nor are they changed in the plural. Thus, it is correct to write "The tree is 10 m high," not "10 m. high" or "10 ms high."

III HANDY CONVERSION FACTORS

To Convert From	To	Multiply By	
Centimeters	Feet	0.0328 ft/cm	
	Inches	0.394 in/cm	
	Meters	0.01 m/cm	(exactly)
	Micrometers (microns)	1000 µm/cm	(")
	Miles (statute)	6.214×10^{-6} mi/cm	
	Millimeters	10 mm/cm	(exactly)
Feet	Centimeters	30.48 cm/ft	(exactly)
	Inches	12 in/ft	(")
	Meters	0.3048 m/ft	(")
	Micrometers (microns)	304,800 µm/ft	(")
	Miles (statute)	0.000189 mi/ft	
Grams	Kilograms	0.001 kg/g	(exactly)
	Micrograms	1×10^6 µg/g	(")
	Ounces (avdp.)	0.03527 oz/g	
	Pounds (avdp.)	0.002205 lb/g	
Hectares	Acres	2.47 acres/ha	
Inches	Centimeters	2.54 cm/in	(exactly)
	Feet	0.0833 ft/in	
	Meters	0.0254 m/in	(exactly)
	Yards	0.0278 yd/in	
Kilograms	Ounces (avdp.)	35.27 oz/kg	
	Pounds (avdp.)	2.205 lb/kg	
Kilometers	Miles	0.6214 mi/km	
Meters	Centimeters	100 cm/m	(exactly)
	Feet	3.2808 ft/m	
	Inches	39.37 in/m	
	Kilometers	0.001 km/m	(exactly)
	Miles (statute)	0.0006214 mi/m	
	Millimeters	1000 mm/m	(exactly)
	Yards	1.0936 yd/m	
Miles (statute)	Centimeters	160,934 cm/mi	
	Feet	5280 ft/mi	(exactly)
	Inches	63,360 in/mi	(exactly)
	Kilometers	1.609 km/mi	
	Meters	1609 m/mi	
	Yards	1760 yd/mi	(exactly)
Ounces (avdp.)	Grams	28.35 g/oz	
	Pounds (avdp.)	0.0625 lb/oz	(exactly)
Pounds (avdp.)	Grams	453.6 g/lb	
	Kilograms	0.454 kg/lb	
	Ounces (avdp.)	16 oz/lb	(exactly)

IV EXPONENTIAL OR SCIENTIFIC NOTATION

Exponential or scientific notation is used by scientists all over the world. This system is based on exponents of 10, which are shorthand notations for repeated multiplications or divisions.

A positive exponent next to a number specifies how many times that number is to be multiplied by itself. Thus, the number 10^2 (read "ten squared" or "ten to the second power") is exponential notation for $10 \cdot 10 = 100$. Similarly, $3^4 = 3 \cdot 3 \cdot 3 \cdot 3 = 81$. The reciprocals of these numbers are expressed by negative exponents. Thus, $10^{-2} = 1/10^2 = 1/(10 \cdot 10) = 1/100 = 0.01$.

To write 10^4 in longhand form, you simply start with the number 1 and move the decimal four places to the right: 10000. Similarly, to write 10^{-4} you start with the number 1 and move the decimal four places to the left: 0.0001 .

It is just as easy to go the other way—that is, to convert a number written in longhand form to an exponential expression. Thus, the decimal place of the number 1,000,000 is six places to the right of 1:

$$1\underbrace{000\,000}_{6 \text{ places}} = 10^6$$

Similarly, the decimal place of the number 0.000001 is six places to the left of 1, and

$$0.000001 = 10^{-6}$$
$$6 \text{ places}$$

What about a number like 3,000,000? If you write it $3 \cdot 1,000,000$, the exponential expression is simply $3 \cdot 10^6$. Thus, the mass of the Earth, which, expressed in long numerical form, is 3,120,000,000,000,000,000,000,000 kg, can be written more conveniently as $3.12 \cdot 10^{24}$ kg.

APPENDIX C
Rock Symbols

The symbols used in this book for types of rocks are shown below:

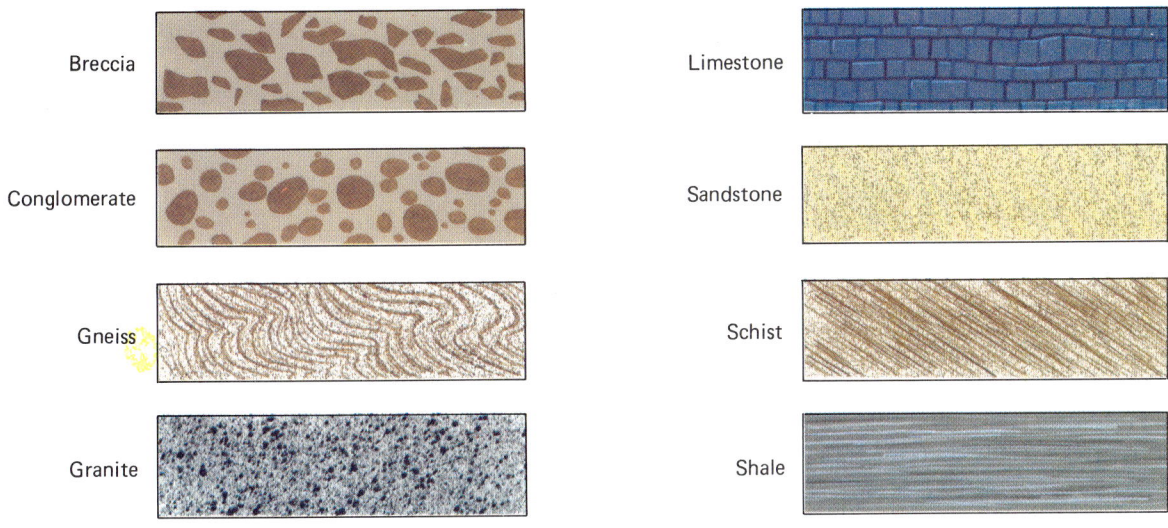

In this book we have adopted consistent colors and style for depicting magma and layers in the upper mantle and crust.

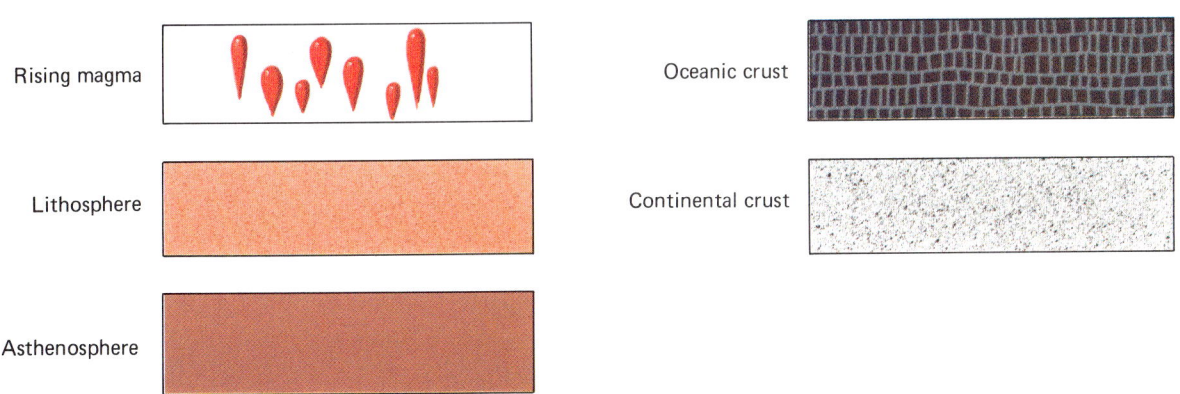

GLOSSARY

A horizon The uppermost layer of soil, composed of a mixture of organic matter and leached and weathered minerals. (syn: topsoil)

aa A lava flow that has a jagged, rubbly, broken surface.

ablation area The lower portion of a glacier where more snow melts in summer than accumulates in winter, so there is a net loss of glacial ice. (syn: zone of wastage)

abrasion The mechanical wearing and grinding of rock surfaces by friction and impact.

absolute humidity See *humidity*.

absolute time Time measured in years.

abyssal fan A large, fan-shaped accumulation of sediment deposited at the bases of many submarine canyons adjacent to the deep-sea floor. (syn: submarine fan)

abyssal plain A flat, level, largely featureless part of the ocean floor between the mid-oceanic ridge and the continental rise.

accessory mineral A mineral that is seen often, but usually only in small amounts. Although common, accessory minerals are not abundant enough to classify as rock-forming minerals.

accreted terrain A landmass that originated as an island arc or a microcontinent that was later added onto a continent.

accumulation area The upper part of a glacier where accumulation of snow during the winter exceeds melting during the summer, causing a net gain of glacial ice.

acid precipitation (often called acid rain) A condition in which natural precipitation becomes acidic after reacting with air pollutants.

acid mine drainage Solutions with a pH lower than 4.5 that flow from mines and mine wastes. The acidity results from oxidation of sulfide minerals exposed during mining, which produces sulfuric acid. The acid dissolves minerals from surrounding rocks, which, in turn, further degrade water quality.

active continental margin A continental margin characterized by subduction of an oceanic lithospheric plate beneath a continental plate. (syn: Andean margin)

active solar heating The use of solar collectors to trap the Sun's energy and heat water with it. Pumps then circulate the hot water through radiators to heat a building, or the inhabitants use the hot water directly for washing and bathing.

active volcano A volcano that is erupting or is expected to erupt.

adiabatic temperature changes Temperature changes that occur without gain or loss of heat.

advection In meteorology, the horizontal component of a convection current in air (i.e., the surface movement that is commonly called wind).

aerobic decay organism An organism that uses oxygen as it consumes organic matter.

air mass A large body of air that has approximately the same temperature and humidity throughout.

albedo The reflectivity of a surface. A mirror or bright, snowy surface reflects most of the incoming light and has a high albedo, whereas a rough, flat road surface has a low albedo.

alfisol A soil with low humus content and a clay-rich B horizon, indicating extensive leaching but less than in ultisols; found in the Mississippi Valley.

alluvial fan A fanlike accumulation of sediment created where a steep stream slows down rapidly as it reaches a relatively flat valley floor.

alpine glacier A glacier that forms in mountainous terrain.

alternative energy resources All energy resources other than fossil fuels and nuclear fission, including solar energy; hydroelectric power; geothermal energy; wind energy; biomass energy; tidal, wave, and heat energy from the seas; and nuclear fusion.

ammonoid An extinct subclass of swimming cephalopods that lived from Devonian through Cretaceous time.

amphibole A group of double-chain silicate minerals. Hornblende is a common amphibole.

anaerobic bacteria Bacteria that live without oxygen.

Andean margin A continental margin characterized by subduction of an oceanic lithospheric plate beneath a continental plate. (syn: active continental margin)

andesite A fine-grained gray or green volcanic rock intermediate in composition between basalt and granite, consisting of about equal amounts of plagioclase feldspar and mafic minerals.

angle of repose The maximum slope or angle at which loose material remains stable.

angular unconformity An unconformity in which younger sediment or sedimentary rocks rest on the eroded surface of tilted or folded older rocks.

anion An ion that has a negative charge.

antecedent stream A stream that was established before local uplift started and cut its channel at the same rate that the land was rising.

anticline A fold in rock that resembles an arch; the fold is convex upward and the oldest rocks are in the middle.

apparent polar wandering The apparent migration of the Earth's magnetic poles through time. In fact, the magnetic poles remain nearly stationary. The appearance of polar migration is caused by movement of continents through time.

aquifer A porous and permeable body of rock that can yield economically significant quantities of ground water.

Archean Eon A division of geologic time 3.8 to 2.5 billion years ago. The oldest known rocks formed at the beginning of or just prior to the start of the Archean Eon.

arête A sharp narrow ridge between adjacent valleys formed by glacial erosion.

aridosol An alkaline soil frequently containing caliche layers (pedocals).

arkose A feldspar-rich sandstone formed when granite disintegrates rapidly.

artesian aquifer An inclined aquifer that is bounded top and bottom by layers of impermeable rock so the water is under pressure.

artesian well A well drilled into an artesian aquifer in which the water rises without pumping and in some cases spurts to the surface.

artificial levee An artificial embankment along a watercourse or an arm of the sea, built to protect land from flooding.

asbestos An industrial name for a group of carcinogenic minerals that crystallize as long, thin fibers. The two most common types are fibrous varieties of the minerals chrysotile and amphibole.

asbestosis An often lethal lung disease caused by prolonged exposure to asbestos fibers.

aseismic ridge A submarine mountain chain with little or no earthquake activity.

ash (volcanic) Fine pyroclastic material less than 2 millimeters in diameter.

ash flow A mixture of volcanic ash, larger pyroclastic particles, and gas that flows rapidly along the Earth's surface as a result of an explosive volcanic eruption. (syn: nuée ardente)

asteroid One of the many small celestial bodies in orbit around the Sun. Most asteroids orbit between Mars and Jupiter.

asthenosphere The portion of the upper mantle beneath the lithosphere. It consists of weak, plastic rock and extends from a depth of about 100 kilometers to about 350 kilometers below the surface of the Earth.

atmosphere The gaseous fluid surrounding Earth or any other planet or celestial object. Also a unit of pressure approximately equal to air pressure at sea level; about 14.7 pounds per square inch or 1033.3 grams per square centimeter.

atoll A circular reef that surrounds a lagoon and is bounded on the outside by deep water of the open sea.

atom The fundamental unit of elements, consisting of a small, dense, positively charged center called a nucleus surrounded by a diffuse cloud of negatively charged electrons.

aulacogen A tectonic trough on a craton bounded by normal faults and commonly filled with sediment. An aulacogen forms when one limb of a continental rift becomes inactive shortly after it forms.

axial plane An imaginary plane that runs through the axis of a fold, dividing it as symmetrically as possible into two halves.

B horizon The soil layer just below the A horizon where ions leached from the A horizon accumulate.

back arc basin A sedimentary basin on the opposite side of the magmatic arc from the trench, either in an island arc or in an Andean continental margin.

backshore The upper zone of a beach that is usually dry but is washed by waves during storms.

bajada A broad depositional surface extending outward from a mountain front formed by the merging of alluvial fans.

banded iron formation A sedimentary rock composed of thin layers of iron-rich minerals sandwiched between beds of silicate minerals.

bank The side of a stream channel.

banks The rising slopes bordering the two sides of a stream channel.

bar An elongate mound of sediment, usually composed of sand or gravel, in a stream channel or along a coastline.

barchan dune A crescent-shaped dune, highest in the center, with the tips facing downwind.

barometer A device used to measure barometric pressure.

barometric pressure The pressure exerted by the Earth's atmosphere.

barrier island A long, narrow, low-lying island that extends parallel to the shoreline.

barrier reef A reef separated from the coast by a deep, wide lagoon.

basal slip Movement of the entire mass of a glacier along the bedrock.

basalt A dark-colored, very fine-grained, mafic, volcanic rock composed of about half calcium-rich plagioclase feldspar and half pyroxene.

base level The deepest level to which a stream can erode its bed. The ultimate base level is usually sea level, but this is seldom attained.

basement rocks The older granitic and related metamorphic rocks of the Earth's crust that make up the foundations of continents.

basin A low area of the Earth's crust, of tectonic origin, commonly filled with sediment.

batholith A large plutonic mass of intrusive rock with more than 100 square kilometers of surface exposed.

bauxite A gray, yellow, or reddish brown rock composed of a mixture of aluminum oxides and hydroxides. It is the principal ore of aluminum.

baymouth bar A bar that extends partially or completely across the entrance to a bay.

beach Any strip of shoreline washed by waves or tides.

beach terrace A level portion of old beach elevated above the modern beach by uplift of the shoreline or fall of sea level.

bed The floor of a stream channel. Also the thinnest layer in sedimentary rocks, commonly ranging in thickness from a centimeter to a meter or two.

bed load That portion of a stream's load that is transported on or immediately above the stream bed.

bedding Layering that develops as sediments are deposited.

bedrock The solid rock that underlies soil or regolith.

Benioff zone An inclined zone of earthquake activity that traces the upper portion of a subducting plate in a subduction zone.

big bang An event thought to mark the beginning of our Universe. The big bang theory postulates that 10 to 20 billion years ago, all matter exploded from an infinitely compressed state.

biodegradable Capable of being consumed or otherwise decomposed by decay organisms or other natural processes.

biodiversity The number and variety of organisms; includes genetic diversity, ecological diversity, and species diversity.

biogeochemical cycle The movement of a nutrient through the atmosphere, the hydrosphere, the biosphere, and the solid Earth.

biological oxygen demand (BOD) The amount of dissolved oxygen used by decay organisms to decompose the organic matter in a given amount of water.

biomagnification The increase in concentration of a pollutant as it moves through a food chain.

biomass energy Electricity or other forms of energy produced by combustion of plant fuels.

biome A community of plants living in a large geographic area characterized by a particular climate.

bioremediation The use of microorganisms to decompose contaminants.

biosphere The thin zone near the Earth's surface occupied by living organisms; it includes the uppermost solid Earth, the hydrosphere, and lower parts of the atmosphere.

biotite A black, rock-forming mineral of the mica group.

bitumen A general name for solid and semisolid hydrocarbons that are fusible and soluble in carbon bisulfide. Petroleum, asphalt, natural mineral wax, and asphaltites are all bitumens.

black lung disease A lethal form of emphysema common among coal miners.

black smoker A jet of black water spouting from a fracture in the sea floor, commonly near the mid-oceanic ridge. The black color is caused by precipitation of very fine-grained metal sulfide minerals as hydrothermal solutions cool by contact with seawater.

black hole A small region of space that contains matter packed so densely that light cannot escape from its intense gravitational field.

blowout A small depression created by wind erosion.

body waves Seismic waves that travel through the interior of the Earth.

boulder A rounded rock fragment larger than a cobble (diameter greater than 256 centimeters).

Bowen's reaction series A series of minerals in which any early-formed mineral crystallizing from a cooling magma reacts with the magma to form minerals lower in the series.

braided stream A stream that divides into a network of branching and reuniting shallow channels separated by mid-channel bars.

breccia A coarse-grained sedimentary rock composed of angular, broken fragments cemented together.

breeder reactor A reactor that simultaneously creates plutonium and generates energy.

brittle fracture The rupture that occurs when a rock breaks sharply.

butte A flat-topped mountain with several steep cliff faces. A butte is smaller and more tower-like than a mesa.

C horizon The lowest soil layer, composed of partly weathered bedrock grading downward into unweathered parent rock.

calcite A common rock-forming mineral, $CaCO_3$.

caldera A large circular depression caused by an explosive volcanic eruption.

caliche A hard soil layer formed when calcium carbonate precipitates and cements the soil.

calving A process in which large chunks of ice break off from tidewater glaciers to form icebergs.

canyon A stream-cut gorge enclosed by steep cliffs. Canyons are typical of arid and semiarid regions.

cap rock An impermeable rock, usually shale, that prevents oil or gas from escaping upward from a reservoir.

capacity The maximum quantity of sediment that a stream can carry.

capillary action The action by which water is pulled upward through small pores by electrical attraction to the pore walls.

capillary fringe A zone above the water table in which the pores are filled with water due to capillary action.

carbon cycle The movement of carbon through the atmosphere, the hydrosphere, the biosphere, and the solid Earth.

carbonate One of a group of minerals whose major anion is the $(CO_3)^{2-}$ complex anion.

carbonate rock A rock, such as limestone or dolomite, that is made up primarily of carbonate minerals.

carbonization A process in which a fossil forms when the volatile components of the soft tissues are driven off, leaving behind a thin film of carbon.

carcinogen A cancer-causing substance or process.

carrying capacity The maximum population of a given species that can live in a region.

cast A fossil formed when sedimentary rock or mineral matter fills a natural mold.

cation A positively charged ion.

cavern An underground cavity or series of chambers created when ground water dissolves large amounts of rock, usually limestone. (syn: cave)

cementation The process by which clastic sediment is lithified by precipitation of a mineral cement among the grains of the sediment.

Cenozoic Era The latest of the four eras into which geologic time is subdivided; 65 million years ago to the present.

chalk A very fine-grained, soft, earthy, white to gray bioclastic limestone made of the shells and skeletons of marine microorganisms.

channel An elongate depression in which a stream flows.

channelization An attempt to solve the problem of stream channel sedimentation by dredging artificial channels across meanders.

chemical weathering The chemical decomposition of rocks and minerals by exposure to air, water, and other chemicals in the environment.

chert A hard, dense sedimentary rock composed of microcrystalline quartz. (syn: flint)

chromosphere A turbulent, diffuse, gaseous layer of the Sun that lies above the photosphere.

cinder cone A small volcano, as high as 300 meters, made up of loose pyroclastic fragments blasted out of a central vent.

cinders (volcanic) Glassy pyroclastic volcanic fragments 4 to 32 millimeters in size.

cirque A steep-walled, semicircular depression eroded into a mountain peak by a glacier.

cirrus cloud A wispy, high-level cloud.

clastic sediment Sediment composed of fragments of weathered rock that have been transported some distance from their place of origin.

clastic sedimentary rocks Rocks composed of lithified clastic sediment.

clay Any clastic mineral particle less than 1/256 millimeter in diameter. Also a group of layered silicate minerals.

claystone A fine-grained clastic sedimentary rock composed predominantly of clay minerals and small amounts of quartz and other minerals of clay size.

clearcutting A logging method in which all the trees are removed.

cleavage The tendency of some minerals to break along certain crystallographic planes.

climate The composite pattern of long-term weather conditions that can be expected in a given region. Climate refers to yearly cycles of temperature, wind, rainfall, and so forth, and not to daily variations (see *weather*).

closed universe The hypothesis that if the mass and gravitational force of the Universe are sufficient, all the galaxies will eventually slow down, reverse direction, and fall back to a common center.

cloud A collection of minute water droplets or ice crystals in air.

coal A flammable organic sedimentary rock formed from partially decomposed plant material and composed mainly of carbon.

cobbles Rounded rock fragments in the 64- to 256-millimeter size range, larger than pebbles and smaller than boulders.

column A dripstone or speleothem formed when a stalactite and a stalagmite meet and fuse.

columnar joints The regularly spaced cracks that commonly develop in lava flows, forming five- or six-sided columns.

coma The bright outer sheath of a comet.

comet An interplanetary body composed of loosely bound rock and ice that forms a bright head and extended fuzzy-appearing tail when it enters the inner portion of the Solar System.

compaction Tighter packing of sedimentary grains, causing weak lithification and a decrease in porosity, usually resulting from the weight of overlying sediment.

competence A measure of the largest particles that a stream can transport.

composite volcano A volcano that consists of alternate layers of unconsolidated pyroclastic material and lava flows. (syn: stratovolcano)

compressive stress Stress that acts to shorten an object by squeezing it.

concordant Pertaining to an igneous intrusion that is parallel to the layering of country rock.

condensation The conversion of water vapor to liquid.

cone of depression A conelike depression in the water table formed when water is pumped out of a well more rapidly than it can flow through the aquifer.

conformable The condition in which sedimentary layers were deposited continuously without interruption.

conglomerate A coarse-grained clastic sedimentary rock, composed of rounded fragments larger than 2 millimeters in diameter, cemented in a fine-grained matrix of sand or silt.

consumed water Water that is extracted from a surface or groundwater source and transported to a distant place. Most irrigation water evaporates and is carried by wind away from its source. Globally, about two thirds of the water that is withdrawn is consumed (see *withdrawn water*).

contact A boundary between two different rock types or between rocks of different ages.

contact metamorphic ore deposit An ore deposit formed by contact metamorphism.

contact metamorphism Metamorphism caused by heating of country rock and/or addition of fluids from a nearby igneous intrusion.

continental crust The predominantly granitic portion of the crust, 20 to 70 kilometers thick, that makes up the continents.

continental drift The theory, proposed by Alfred Wegener, that continents were once joined together and later split and drifted apart. The continental drift theory has been replaced by the more complete plate tectonics theory.

continental glacier A glacier that forms a continuous cover of ice over areas of 50,000 square kilometers or more and spreads outward in all directions under the influence of its own weight. (syn: ice sheet)

continental margin The region between the shoreline of a continent and the deep ocean basins, including the continental shelf, continental slope, and continental rise. Also the region where thick, granitic continental crust joins thinner, basaltic oceanic crust.

continental margin basin A sediment-filled depression or other thick accumulation of sediment and sedimentary rocks near the margin of a continent.

continental rifting The process by which a continent is pulled apart at a divergent plate boundary.

continental rise An apron of sediment between the continental slope and the deep-sea floor.

continental shelf A shallow, nearly level area of continental crust, covered by sediment and sedimentary rocks, that is submerged below sea level at the edge of a continent between the shoreline and the continental slope.

continental slope The relatively steep (3° to 6°) underwater slope between the continental shelf and the continental rise.

continental suture The junction created where two continents collide and weld into a single mass of continental crust.

contour A line on a map connecting points of equal value to show the elevation of the land surface.

contour interval The difference in value between two adjacent contours on a map.

control rod A column of neutron-absorbing alloys that are spaced among fuel rods to fine-tune nuclear fission in a reactor.

convection current A current in a fluid or plastic material, formed when heated materials rise and cooler materials sink.

convergent plate boundary A boundary where two lithospheric plates collide head-on.

coquina A bioclastic limestone consisting of coarse shell fragments cemented together.

core The innermost region of the Earth, probably consisting of iron and nickel.

Coriolis effect The deflection of air or water currents caused by the rotation of the Earth.

corona The luminous, irregular envelope of highly ionized gas outside the chromosphere of the Sun.

correlation Demonstration of the age equivalence of rocks or geologic features that come from different locations.

cost–benefit analysis A system of analysis that attempts to weigh the cost of an act or policy, such as pollution control, directly against the economic benefits.

country rock The older rock intruded by a younger igneous rock or mineral deposit.

crater A bowl-like depression at the summit of a volcano.

craton A segment of continental crust, usually in the interior of a continent, that has been tectonically stable for a long time (commonly a billion years or longer).

creep The slow downslope movement of unconsolidated material under the influence of gravity.

crest (of a wave) The highest part of a wave.

crevasse A fracture or crack in the upper 40 to 50 meters of a glacier.

critical mineral A mineral that is essential to the national economy.

crop rotation A practice in which the type of crop grown on a field is changed from year to year.

cross bedding An arrangement of small beds at an angle to the main sedimentary layering.

cross-cutting relationship Any relationship in which younger rocks or geological structures interrupt or cut across older rocks or structures.

cross section An illustration that shows an imaginary vertical slice of the Earth to depict topography or geologic features, such as rock type or ground-water reserves.

crust The Earth's outermost layer, about 7 to 70 kilometers thick, composed of relatively low-density silicate rocks.

crystal A solid element or compound whose atoms are arranged in a regular, orderly, periodically repeated array.

crystal habit The shape in which individual crystals grow and the manner in which crystals grow together in aggregates.

crystal settling A process in which the crystals that solidify first from a cooling magma settle to the bottom of a magma chamber because the solid minerals are denser than liquid magma.

cultural eutrophication Eutrophication caused by human pollutants of all kinds.

cumulus cloud A column-like cloud with a flat bottom and a billowy top.

Curie point The temperature below which rocks can retain magnetism.

current A continuous flow of water in a concerted direction.

cyclone A low-pressure region with its accompanying surface wind. Also a synonym for a tropical cyclone or hurricane.

daughter isotope An isotope formed by radioactive decay of another isotope.

debris flow A type of mass wasting in which particles move as a fluid and more than half of the particles are larger than sand.

decay organism An organism that consumes dead plants, animals, and their waste products in the environment. Most are microscopic in size.

deflation Erosion by wind.

deformation Folding, faulting, and other changes in the shapes of rocks or minerals in response to mechanical forces, such as those that occur in tectonically active regions.

degeneracy pressure The strength of atomic particles that holds a white dwarf star together, preventing further collapse.

delta The nearly flat, alluvial, fan-shaped tract of land at the mouth of a stream.

demographic rate A measure of how fast the size of a population, or a segment of a population, is changing. It is expressed as the ratio of a particular change to the population as a whole.

demographic transition A series of demographic changes from high birth and death rates to low birth and death rates.

demography The statistical study of human populations.

dendritic drainage pattern A pattern of stream tributaries which branches like the veins in a leaf. It often indicates uniform underlying bedrock.

denitrifying bacteria Bacteria that break up nitrogen-containing organic molecules for their own food. In the process, they convert the nitrogen back to N_2 gas, which returns to the atmosphere.

deposition The laying down of rock-forming materials by any natural agent.

depositional environment Any setting in which sediment is deposited.

depositional remanent magnetism Remanent magnetism resulting from mechanical orientation of magnetic mineral grains during sedimentation.

desert A region with less than 25 centimeters of rainfall per year. Also defined as a region that supports only a sparse plant cover.

desert pavement A continuous cover of stones that is created as wind erodes fine sediment, leaving larger rocks behind.

desertification A process by which semiarid land is converted to desert, often by improper farming or by climate change.

developed country An industrialized nation characterized by a low fertility rate, a low infant mortality rate, and high per-capita income. Examples are the United States, Japan, Canada, and Germany.

developing country A nation that is not highly industrialized and is characterized by a high fertility rate, a high infant mortality rate, and low per-capita income. Examples include many countries in Africa, Asia, and Latin America.

dew Moisture condensed onto objects from the atmosphere, usually during the night, when the ground and leaf surfaces become cooler than the surrounding air.

differential weathering The process by which certain rocks weather more rapidly than adjacent rocks, usually resulting in an uneven surface.

dike A sheetlike igneous rock that cuts across the structure of country rock.

dike swarm A group of dikes that form in parallel or radial sets.

dinosaur An extinct reptile of the subclass *Archosauria*, which lived from Triassic to Cretaceous time.

diorite A rock that is the medium- to coarse-grained plutonic equivalent of andesite.

dip The angle of inclination of a bedding plane, measured from the horizontal.

discharge The volume of water flowing downstream per unit of time. It is measured in cubic meters per second.

disconformity A type of unconformity in which the sedimentary layers above and below the unconformity are parallel.

discordant Pertaining to a dike or other feature that cuts across sedimentary layers or other kinds of layering in country rock.

disseminated ore deposit A large, low-grade ore deposit in which generally fine-grained, metal-bearing minerals are widely scattered throughout a rock body in sufficient concentration to make the deposit economical to mine.

dissolved load The portion of a stream's sediment load that is carried in solution.

distributary A channel that flows outward from the main stream channel, such as is commonly found in deltas.

divergent plate boundary The boundary or zone where lithospheric plates separate from each other. (syn: spreading center)

docking The accretion of island arcs or microcontinents onto a continental margin.

doldrums A region of the Earth near the equator in which hot, humid air moves vertically upward, forming a vast low-pressure region. Local squalls and rainstorms are common, and steady winds are rare.

dolomite A common rock-forming mineral, $CaMg(CO_3)_2$.

dome A circular or elliptical anticlinal structure.

Doppler effect The observed change in frequency of light or sound that occurs when the source of the wave is moving relative to the observer.

dormant volcano A volcano that is not now erupting but has erupted in the past and will probably do so again.

doubling time The time required for the Earth's population (or anything else) to double.

downcutting Downward erosion by a stream.

drainage basin The region that is ultimately drained by a single river.

drainage divide A ridge or other topographically high region that separates adjacent drainage basins.

drift (glacial) Any rock or sediment transported and deposited by a glacier or by glacial meltwater.

dripstone A deposit formed in a cavern when calcite precipitates from dripping water.

drumlin An elongate hill formed when a glacier flows over and reshapes a mound of till or stratified drift.

dry adiabatic lapse rate The rate of cooling that occurs when dry air rises without gain or loss of heat.

dune A mound or ridge of wind-deposited sand.

Earth The planet in the Solar System that is third closest to the Sun and, of the nine major planets, is fifth in order of size.

Earth Summit The June 1992 United Nations Conference on Environment and Development, held in Rio de Janeiro and attended by delegates from 170 nations.

earthflow A flowing mass of fine-grained soil particles mixed with water. Earthflows are less fluid than mudflows.

earthquake A sudden motion or trembling of the Earth caused by the abrupt release of slowly accumulated elastic energy in rocks.

echo sounder An instrument that emits sound waves and then records them after they reflect off the sea floor. The data are then used to record the topography of the sea floor.

eclipse A phenomenon that occurs when a heavenly body is shadowed by another and therefore rendered invisible. When the Moon lies directly between the Earth and the Sun, the Moon blocks our view of the Sun and we observe a *solar eclipse*. When the Earth lies directly between the Sun and the Moon, the Earth's shadow falls on the Moon and we observe a *lunar eclipse*.

ecology The science that studies interactions between living organisms and their environments.

ecosystem An ecologic system composed of organisms and their environment. It is the product of interactions among biological, geological, atmospheric, and hydrospheric systems. An ecosystem is nearly self-contained so that the exchange of nutrients within the system is much greater than exchanges with other systems.

effluent stream A stream that receives water from ground water because its channel lies below the water table. (*syn:* gaining stream)

elastic deformation A type of deformation in which an object returns to its original size and shape when stress is removed.

elastic limit The maximum stress that an object can withstand without permanent deformation.

electromagnetic radiation The transfer of energy by an oscillating electric and magnetic field; it travels as a wave and also behaves as a stream of particles.

electromagnetic spectrum The entire range of electromagnetic radiation, from very long-wavelength (low-frequency) radiation to very short-wavelength (high-frequency) radiation.

electron A fundamental particle which forms a diffuse cloud of negative charge around an atom.

element A substance that cannot be broken down into other substances by ordinary chemical means. An element is made up of a single kind of atom.

emergent coastline A coastline that was recently under water but has been exposed, either because the land has risen or because sea level has fallen.

end moraine A moraine that forms at the end, or terminus, of a glacier.

entisol A poorly developed soil with little vertical profile. Entisols often form in regions with little vegetation, such as deserts, tundra, and areas recently disturbed by glaciers or volcanic eruptions.

environmental geoscience The study of geology as it applies to living organisms and the quality of life on our planet.

eon The longest unit of geologic time. The most recent eon, the Phanerozoic Eon, is further subdivided into eras.

epicenter The point on the Earth's surface directly above the focus of an earthquake.

epidemiology The study of the distribution of sickness in a population.

epoch The smallest unit of geologic time. Periods are divided into epochs.

equinox Either of two times during a year when the Sun shines directly overhead at the equator. The equinoxes are the beginnings of spring and fall, when every portion of the Earth receives 12 hours of daylight and 12 hours of darkness.

era A geologic time unit. Eons are divided into eras, and eras are subdivided into periods.

erosion The removal of weathered rocks by moving water, wind, ice, or gravity.

erratic A boulder that was transported to its present location by a glacier. Usually different from the bedrock in its immediate vicinity.

esker A long, snakelike ridge formed by deposition in a stream that flowed on, within, or beneath a glacier.

estuary A bay that formed when a broad river valley was submerged by rising sea level or a sinking coast.

eustatic Of or pertaining to worldwide changes in sea level.

eutrophic lake A lake characterized by abundant dissolved nitrates, phosphates, and other plant nutrients and by a seasonal deficiency of oxygen in bottom water. Such lakes are commonly shallow.

evaporation The transformation of a liquid into a gas.

evaporite deposit A chemically precipitated sedimentary rock that formed when dissolved ions were concentrated by evaporation of water.

evolution The change in the physical and genetic characteristics of a species over time.

exfoliation Fracturing in which concentric plates or shells split from the main rock mass like the layers of an onion.

exponential growth Growth that occurs at a constant rate of increase over time. Exponential growth plotted versus time produces a characteristic, ever-steepening, J-shaped curve.

extensional stress Tectonic stress in which rocks are pulled apart.

external mold A fossil cavity created in sediment by a shell or other hard body part that bears the impression of the exterior of the original.

externality (environmental) The cost of living in a degraded environment, including direct costs such as medical bills, lost work, and damage to waterways, crops, and livestock. It also includes indirect costs of environmental degradation, such as reduction in tourism and lowered land values.

extinct volcano A volcano that is expected never to erupt again.

extrusive rock An igneous rock formed from material that has erupted onto the surface of the Earth.

fall A type of mass wasting in which unconsolidated material falls freely or bounces down the face of a cliff.

fault A fracture in rock along which displacement has occurred.

fault creep A continuous, slow movement of solid rock along a fault resulting from a constant stress acting over a long time.

fault zone An area of numerous, closely spaced faults.

faunal succession (principle of) See *principle of faunal succession.*

feldspar A common group of aluminum silicate rock-forming minerals that contain potassium, sodium, or calcium.

fetch The distance that the wind has traveled over the ocean without interruption.

fiord A long, deep, narrow arm of the sea bounded by steep walls, generally formed by submergence of a glacially eroded valley.

firn Hard, dense snow that has survived through one summer melt season. Firn is transitional between snow and glacial ice.

firn line The boundary between permanent snow and seasonal snow on a glacier. Above the firn line, winter snow does not melt completely during summer, whereas below the firn line it does.

fissility Fine layering along which a rock splits easily.

fission The spontaneous or induced splitting, by particle collision, of a heavy nucleus into a pair of nearly equal fission fragments plus some neutrons. Large amounts of energy are released during fission (see *fusion*).

flash flood A rapid, intense, local flood of short duration.

flood An overflowing of water onto land that is not usually submerged.

flood basalt Basaltic lava that erupts gently in great volume from cracks on the Earth's surface, to cover large areas of land and form basalt plateaus.

flood plain That portion of a river valley adjacent to the channel; it is built by sediment deposited during floods and is covered by water during a flood.

flow Mass wasting in which individual particles move downslope as a semifluid, not consolidated, mass.

fly ash All solids, including ash, charred paper, cinders, dust, and soot, that are carried in a gas stream, especially in gases escaping from a coal-fired power plant.

focus The initial rupture point of an earthquake.

fog A cloud that forms at or very close to ground level.

fold A bend in rock.

foliation Layering in rock, created by metamorphism.

footwall The rock beneath an inclined fault.

forearc basin A sedimentary basin between the trench and the magmatic arc in either an island arc or an Andean continental margin.

foreshocks Small earthquakes that precede a large quake by an interval ranging from a few seconds to a few weeks.

foreshore The zone that lies between the high and low tides; the intertidal region.

formation A lithologically distinct body of sedimentary, igneous, or metamorphic rock that can be recognized in the field and can be mapped.

fossil The preserved trace, imprint, or remains of a plant or animal.

fossil fuel Fuels formed from the partially decayed remains of plants and animals. The most commonly used fossil fuels are petroleum, coal, and natural gas.

fracture (1) The manner in which minerals break other than along planes of cleavage. (2) A crack, joint, or fault in bedrock.

front In meteorology, a line separating air masses of different temperatures or densities.

frontal weather system A weather system that develops when air masses collide.

frost wedging A process in which water freezes in a crack in rock and the expansion wedges the rock apart.

fuel rod A 2-meter-long column of fuel-grade uranium pellets, used to fuel a nuclear reactor.

fusion The combination of two light nuclei to form a heavier nucleus. Large amounts of energy are released by fusion (see *fission*).

gabbro Igneous rock that is mineralogically identical to basalt but that has a medium- to coarse-grained texture because of its plutonic origin.

gaining stream A stream that receives water from ground water because its channel lies below the water table. (*syn:* effluent stream)

galaxy A large volume of space containing many billions of stars held together by mutual gravitational attraction.

gem A mineral that is prized for its rarity and beauty rather than for industrial use.

geocentric model A model that places the Earth at the center of the celestial bodies.

geologic column A composite columnar diagram that shows the sequence of rocks at a given place or region, arranged to show their positions in the geologic time scale.

geologic resource Any source of raw materials, including soil, surface and ground water, metal ores, coal, petroleum, building stone, sand, gravel, and a host of other rock and mineral products.

geologic time The time extending from the Earth's formation about 4.6 billion years ago to the beginning of human history.

geologic time scale A chronological arrangement of geologic time subdivided into units.

geological structure Any feature formed by deformation or movement of rocks, such as a fold or a fault. Also, the combination of all such features of an area or region.

geology The study of the Earth, the materials of which it is made, the physical and chemical changes that occur on its surface and in its interior, and the history of the planet and its life forms.

geothermal energy Energy derived from the heat of the Earth.

geothermal gradient The rate at which temperature increases with depth in the Earth.

geyser A type of hot spring that intermittently erupts jets of hot water and steam. Geysers occur when ground water comes in contact with hot rock.

glacial polish A smooth polish on bedrock, created when fine particles transported at the base of a glacier abrade the bedrock.

glacial striation Parallel grooves and scratches in bedrock that form as rocks are dragged along at the base of a glacier.

glacier A massive, long-lasting accumulation of compacted snow and ice that forms on land and moves downslope or outward under its own weight.

gneiss A foliated rock with a banded appearance, formed by regional metamorphism.

Gondwanaland The southern part of Wegener's Pangaea, which was the late Paleozoic supercontinent. (syn: Gondwana)

graben A wedge-shaped block of rock that has dropped downward between two normal faults.

graded bedding A type of bedding in which larger particles are at the bottom of each bed, and the particle size decreases toward the top.

graded stream A stream with a smooth, concave profile. A graded stream is in equilibrium with its sediment supply; it transports all the sediment supplied to it without erosion or deposition in the stream bed.

gradient The vertical drop of a stream over a specific distance.

gradualism A theory of evolution that proposes that species change gradually in small increments.

granite A medium- to coarse-grained, sialic, plutonic rock made predominantly of potassium feldspar and quartz.

gravel Unconsolidated sediment consisting mainly of rounded particles with diameters greater than 2 millimeters.

graywacke A poorly sorted sandstone, commonly dark in color and consisting mainly of quartz, feldspar, and rock fragments with considerable quantities of silt and clay in its pores.

Green Revolution The development, beginning in the 1960s, of new crop varieties that produced much higher yields than traditional strains.

greenhouse effect An increase in the temperature of a planet's atmosphere, caused when infrared-absorbing gases are introduced into the atmosphere.

greenhouse gas Any gas, including water vapor, carbon dioxide, methane, chlorofluorocarbons, and nitrogen oxides, that absorbs infrared radiation and may cause rising atmospheric temperature.

groin A narrow wall built perpendicular to the shore to trap sand transported by currents and waves.

ground moraine A moraine formed when a melting glacier deposits till in a relatively thin layer over a broad area.

ground water Water contained in soil and bedrock; all subsurface water.

guyot A flat-topped seamount.

gypsum A mineral with the formula $CaSO_4 \cdot 2H_2O$. It commonly forms in evaporite deposits.

gyre A closed, circular current in either water or air.

Hadean Eon The earliest time in the Earth's history.

half-life The time it takes for half of the nuclei of a radioactive isotope in a sample to decompose.

hanging valley A tributary glacial valley whose mouth lies high above the floor of the main valley.

hanging wall The rock above an inclined fault.

hardness The resistance of the surface of a mineral to scratching.

hazardous waste Defined by the EPA as any waste that is (1) flammable (such as petroleum products or organic solvents); (2) explosive or volatile enough to release toxic fumes; (3) corrosive (such as strong acids and bases); or (4) toxic, carcinogenic, or mutagenic.

headland A coastal point of land that juts out into the sea.

headward erosion The lengthening of a valley in an upstream direction.

heliocentric solar system A model that places the Sun at the center of the Solar System.

Hertzsprung-Russell diagram (H-R diagram) A plot of absolute stellar magnitude (or luminosity) against temperature.

histosol A boggy soil that forms in regions such as the Florida Everglades and the Mississippi River delta.

Homo sapiens Modern humans; the only surviving species of the genus *Homo*.

horn A sharp, pyramid-shaped rock summit where three or more cirques intersect near the summit.

hornblende A rock-forming mineral; the most common member of the amphibole group.

hornfels A fine-grained rock formed by contact metamorphism.

horse latitudes Two regions of the Earth lying at about 30° north and south latitudes, in which air is falling, forming vast areas of high pressure. Generally dry conditions prevail, and steady winds are rare.

horst A block of rock that has moved relatively upward and is bounded by two faults.

hot spot A persistent volcanic center thought to be located directly above a rising plume of hot mantle rock.

hot spring A spring formed where hot ground water flows to the surface.

Hubble's Law A law that states that the velocity of a galaxy is proportional to its distance from Earth. Thus the most distant galaxies are traveling at the highest velocities.

humidity A measure of the amount of water vapor in the air. *Absolute humidity* is the amount of water vapor in a given volume of air. *Relative humidity* is the ratio of the amount of water vapor in a given volume of air divided by the maximum amount of water vapor that can be held by that air at a given temperature.

humus The dark organic component of soil, composed of litter that has decomposed sufficiently so that the origins of the individual pieces cannot be determined.

hurricane See *tropical cyclone*.

hydration The chemical combination of water with another substance.

hydraulic action The mechanical loosening and removal of material by flowing water.

hydroelectric energy Electricity produced by turbines that harness the energy of water dropping downward through a dam.

hydrologic cycle The constant circulation of water among the sea, the atmosphere, and the land.

hydrolysis A decomposition reaction involving water.

hydrosphere All of the Earth's surface and near-surface water, including water in streams, lakes, and oceans; water in the atmosphere; water frozen in glaciers; and ground water, which soaks soil and rock to a depth of 2 or 3 kilometers.

hydrothermal metamorphism Changes in rock that are primarily caused by migrating hot water and by ions dissolved in the hot water.

hydrothermal vein A sheetlike mineral deposit that fills a fault or other fracture, precipitated from hot-water solutions.

ice age A time of extensive glacial activity, when alpine glaciers descended into lowland valleys and continental glaciers spread over the higher latitudes.

ice sheet A glacier that forms a continuous cover of ice over areas of 50,000 square kilometers or more and spreads outward under the influence of its own weight. (*syn:* continental glacier, ice cap)

iceberg A large chunk of ice that breaks from a glacier into a body of water.

igneous rock Rock that solidified from magma.

inceptisol A very young soil with weakly developed layering and little mineral alteration from the parent rock.

incised meander A stream meander that is cut below the level at which it originally formed, usually caused by rejuvenation.

index fossil A fossil that identifies and dates the layers in which it is found. Index fossils are abundantly preserved in rocks, widespread geographically, and existed as a species or genus for only a relatively short time.

Industrial Revolution The social and economic changes that occurred in England from the mid-eighteenth to the mid-nineteenth century, during the rise of modern industrialism.

industrialized country A nation where industry and technology are well developed, characterized by a low fertility rate, a low infant mortality rate, and high per-capita income. Examples are the United States, Japan, Canada, and Germany.

influent stream A stream that lies above the water table. Water percolates from the stream channel downward into the saturated zone. (*syn:* losing stream)

injection well A well for storage or disposal of injected fluid.

intensity (of an earthquake) A measure of the effects of an earthquake at a particular place on buildings and people.

intermediate rocks Igneous rocks with chemical and mineral compositions between those of granite and basalt.

internal mold A type of fossil that forms when the inside of a shell fills with sediment or precipitated minerals.

internal processes Earth processes that are initiated by movements within the Earth and those internal movements themselves; for example, formation of magma, earthquakes, mountain building, and tectonic plate movement.

intertidal zone The part of a beach that lies between the high and low tide lines.

intracratonic basin A sedimentary basin located within a craton.

intrusive rock A rock formed when magma solidifies within bodies of preexisting rock.

inversion (atmospheric) A meteorological condition in which the lower layers of air are cooler than those at higher altitudes. This cool air can remain relatively stagnant and allows air pollutants to concentrate in urban areas.

ion An atom with an electrical charge.

ionic substitution The replacement of one or more kinds of ions in a mineral by other kinds of ions of similar size and charge.

island arc A gently curving chain of volcanic islands in the ocean, formed by convergence of two plates (each bearing ocean crust), and the resulting subduction of one plate beneath the other.

isostasy The condition in which the lithosphere floats on the asthenosphere as an iceberg floats on water.

isostatic adjustment The rising and settling of portions of the lithosphere to maintain equilibrium as they float on the plastic asthenosphere.

isotherm A line on a weather map connecting points of equal temperature.

isotopes Atoms of the same element that have the same number of protons but different numbers of neutrons.

jet stream A relatively narrow, high-altitude, fast-moving air current.

joint A fracture that occurs without movement of rock on either side of the break.

Jovian planets The outer planets—Jupiter, Saturn, Uranus, and Neptune—which are massive, with a high proportion of the lighter elements.

kame A small mound or ridge of layered sediment deposited by a stream that flows on top of, within, or beneath a glacier.

kaolinite A common clay mineral, $Al_2Si_2O_5(OH)_4$.

karst topography A type of topography formed over limestone or other soluble rock and characterized by caverns, sinkholes, and underground drainage.

kerogen The solid bitumen in oil shales that yields oil when the shales are heated and distilled; the precursor of liquid petroleum.

kettle A depression in glacial drift, created by the melting of a large chunk of ice left buried in the drift by a receding glacier. The ice prevents sediment from collecting; when the ice melts, a lake or swamp may fill the depression.

key bed A thin, widespread, easily recognized sedimentary layer that can be used for correlation.

L wave An earthquake wave that travels along the surface of the Earth or along a boundary between layers within the Earth. (syn: surface wave)

lagoon A protected body of water separated from the sea by a reef or barrier island.

lake A large, inland body of standing water that occupies a depression in the land surface.

laminar flow A type of flow in which water moves in straight, even paths without turbulence.

landform Any of the features that make up the Earth's surface, including large-scale features such as mountain ranges, plains, and plateaus, and small features such as hills, stream valleys, and gullies.

landslide A general term for the downslope movement of rock and regolith under the influence of gravity.

latent heat The heat released or absorbed by a substance during a change in state (i.e., melting, freezing, vaporization, condensation, or sublimation).

lateral moraine A ridgelike moraine that forms on or adjacent to the sides of a mountain glacier.

laterite A highly weathered soil, rich in oxides of iron and aluminum, that usually develops in warm, moist tropical or temperate regions.

Laurasia The northern part of Wegener's Pangaea, which was the late Paleozoic supercontinent.

lava Fluid magma that flows onto the Earth's surface from a volcano or fissure. Also, the rock formed by solidification of the same material.

law (scientific) A formal statement of the way in which events always occur under given conditions. It is considered to be factual and always correct. A law is the most certain of scientific statements.

less developed country A nation with little or no industry, characterized by a high fertility rate, a high infant mortality rate, and low per-capita income. Examples include many countries in Africa, Asia, and Latin America.

levee An artificial or natural embankment along a watercourse or an arm of the sea.

light-year The distance traveled by light in one year, approximately 9.5×10^{12} kilometers.

limb The side of a fold in rock.

limestone A sedimentary rock consisting chiefly of calcium carbonate.

lithification The conversion of loose sediment to solid rock.

lithosphere The cool, rigid outer layer of the Earth, about 100 kilometers thick, which includes the crust and part of the upper mantle.

litter Leaves, twigs, and other plant or animal material that have fallen to the surface of the soil but have not decomposed.

loam Soil that contains a mixture of sand, clay, and silt and a generous amount of organic matter.

loess A homogenous, unlayered deposit of windblown silt, usually of glacial origin.

longitudinal dune A long, symmetrical dune oriented parallel with the direction of the prevailing wind.

longshore current A current, flowing parallel and close to the coast, that is generated when waves strike a shore at an angle.

longshore drift Sediment carried by longshore currents.

losing stream A stream that lies above the water table. Water percolates from the stream channel downward into the saturated zone. (syn: influent stream)

low-velocity layer A region of the upper mantle where seismic waves travel relatively slowly, approximately at the same speed as they travel in the asthenosphere.

luster The quality and intensity of light reflected from the surface of a mineral.

mafic rock Dark-colored igneous rock with high magnesium and iron contents and composed chiefly of iron- and magnesium-rich minerals.

magma Molten rock generated within the Earth.

magmatic arc A narrow elongate band of intrusive and volcanic activity associated with subduction.

magnetic reversal A change in the Earth's magnetic field in which the north magnetic pole becomes the south magnetic pole, and vice versa.

magnetometer An instrument that measures the Earth's magnetic field.

magnitude (of an earthquake) A measure of the strength of an earthquake, determined from seismic recordings (see also *Richter scale* and *moment magnitude*).

main sequence A band running across a Hertzsprung-Russell diagram that contains most of the stars. Hydrogen fusion generates the energy in a main-sequence star.

manganese nodule A manganese-rich, potato-shaped rock found on the ocean floor.

mantle A mostly solid layer of the Earth, lying beneath the crust and above the core. The mantle extends from the base of the crust to a depth of about 2900 kilometers.

mantle convection The convective flow of solid rock in the mantle.

mantle plume A rising vertical column of mantle rock.

map A representation on a plane surface of a part or whole of the Earth's surface or of any other surface.

marble A metamorphic rock consisting of fine- to coarse-grained recrystallized calcite and/or dolomite.

marias Dry, barren, flat expanses of volcanic rock on the Moon. They were first thought to be seas.

mass extinction The extinction of numerous species and higher taxa within a relatively short time.

mass wasting The movement of earth material downslope, primarily under the influence of gravity.

meander One of a series of sinuous curves or loops in the course of a stream.

mechanical weathering The disintegration of rock into smaller pieces by physical processes.

medial moraine A moraine formed in or on the middle of a glacier by the merging of lateral moraines as two glaciers flow together.

Mercalli scale A scale of earthquake intensity that measures the strength of an earthquake in a particular place by its effect on buildings and people. It has been replaced by the Richter scale and the moment magnitude scale.

mesa A flat-topped mountain or a tableland that is smaller than a plateau and larger than a butte.

mesosphere The layer of air that lies above the stratopause and extends from about 55 kilometers upward to 80 kilometers above the Earth's surface. Temperature decreases with elevation in the mesosphere.

Mesozoic Era The part of geologic time from roughly 245 million to 65 million years ago. Dinosaurs rose to prominence and became extinct during this era.

metal Any chemical element characterized by a metallic luster, ductility, and the ability to conduct electricity and heat.

metamorphic facies A set of all the metamorphic rock types that formed under similar temperature and pressure conditions.

metamorphic grade The intensity of metamorphism that formed a rock; the maximum temperature and pressure attained during metamorphism.

metamorphic rock A rock formed when igneous, sedimentary, or other metamorphic rocks recrystallize in response to elevated temperature, increased pressure, chemical change, and/or deformation.

metamorphism The process by which rocks and minerals change in response to changes in temperature, pressure, chemical conditions, and/or deformation.

metasomatism Metamorphism accompanied by introduction of ions from an external source.

meteorite A fallen meteoroid.

meteoroid A small interplanetary body in an irregular orbit. Many meteoroids are fragments of asteroids formed during collisions.

methylhydrate An icelike, frozen, methane-rich solid that is common in sea-floor mud. Methylhydrate deposits may contain as much as twice as much carbon as all conventional coal, oil, and gas reserves combined.

mica A layered silicate mineral with a distinctive platy crystal habit and perfect cleavage. Muscovite and biotite are common micas.

mid-channel bar An elongate lobe of sand and gravel formed in a stream channel.

mid-oceanic ridge A continuous submarine mountain chain that forms at the boundary between divergent tectonic plates within oceanic crust.

migmatite A rock composed of both igneous and metamorphic-looking materials. It forms at very high metamorphic grades when rock begins to melt partially to form magma.

mineral A naturally occurring inorganic solid with a definite chemical composition and a crystalline structure.

mineral deposit A local enrichment of one or more minerals.

mineral reserve The known supply of ore in the ground.

mineralization A process of fossilization in which the organic components of an organism are replaced by minerals.

Mohorovičić discontinuity (Moho) The boundary between the crust and the mantle, identified by a change in the velocity of seismic waves.

Mohs hardness scale A standard, numbered from 1 to 10, to measure and express the hardness of minerals based on a series of ten fairly common minerals, each of which is harder than those lower on the scale.

mollisol A soil with a thick layer of humus that develops on prairies. The grain-growing breadbasket of the Midwest lies on mollisols, as do some forests.

moment magnitude A number based on the amount of rock displacement during an earthquake. Moment magnitude is a more accurate indicator of the energy released by a quake than Richter magnitude and has become the standard scale of earthquake strength.

monocline A fold with only one limb.

monsoon A continental wind system caused by uneven heating of land and sea. Monsoons generally blow from the sea to the land in the summer, when the continents are warmer than the ocean, and bring rain. In winter, when the ocean is warmer than the land, the monsoon winds reverse.

moraine A mound or ridge of till deposited directly by glacial ice.

mountain chain A number of mountain ranges grouped together in an elongate zone.

mountain range A series of mountains or mountain ridges that are closely related in position, direction, age, and mode of formation.

mud Wet silt and clay.

mud cracks Irregular, usually polygonal fractures that develop when mud dries. The patterns may be preserved when the mud is lithified.

mudflow Mass wasting of fine-grained soil particles mixed with a large amount of water.

mudstone A nonfissile rock composed of a mixture of clay and silt.

multiple-use Pertaining to a land management method in which public land is managed for many purposes.

mummification A process in which the remains of an animal are preserved by dehydration.

mutagen A substance or process that causes an inheritable birth defect.

mutation A fundamental, inheritable change that produces individuals that are genetically unlike their parents.

natural gas A mixture of naturally occurring light hydrocarbons composed mainly of methane, CH_4.

natural levee A ridge or embankment of flood-deposited sediment along both banks of a stream channel.

nebula An interstellar cloud of gas and dust. A *planetary nebula* is created when a star the size of our Sun explodes.

neutron A subatomic particle with the mass of a proton but no electrical charge.

neutron star A small, extremely dense star composed almost entirely of neutrons (see also *pulsar*).

nimbo A prefix or suffix added to the name of a cloud type to indicate precipitation.

nitrogen cycle The movement of nitrogen through the atmosphere, hydrosphere, biosphere, and solid Earth. Once nitrogen is converted to biologically usable forms, plants and animals conserve and recycle it within the ecosystem.

nonbiodegradable Incapable of being consumed or otherwise decomposed by decay organisms or other natural processes within a reasonable time. A characteristic of many industrial wastes and elemental pollutants.

nonconformity A type of unconformity in which layered sedimentary rocks lie on an erosion surface cut into igneous or metamorphic rocks.

nonfoliated The lack of layering in metamorphic rock.

nonmetallic resource Any useful material that does not have metallic properties; for example, sand and gravel for construction; limestone for cement; granite, limestone, marble, and other rocks for building stone; and phosphorus and potassium for fertilizer.

nonrenewable resource A resource that is not replenished within human lifetimes because the geological processes that created it are much slower than the span of human history; that is, formation of new deposits occurs much more slowly than consumption. Fossil fuels and metal ores are examples.

normal fault A fault in which the hanging wall has moved downward relative to the footwall.

normal lapse rate The decrease in temperature with elevation in air that is neither rising nor falling.

normal polarity A magnetic orientation that matches the Earth's modern magnetic field.

nucleus The small, dense, central portion of an atom, composed of protons and neutrons. Nearly all of the mass of an atom is concentrated in the nucleus.

nuée ardente A swiftly flowing, often red-hot cloud of gas, volcanic ash, and other pyroclastics, formed by an explosive volcanic eruption. (*syn:* ash flow)

obsidian A black or dark-colored glassy volcanic rock usually of rhyolitic composition.

ocean The huge body of salty water that covers nearly three fourths of Earth's surface. Also, any of the geographic segments of this body (usually listed as the Atlantic, Pacific, Indian, Arctic, and Antarctic oceans).

ocean thermal power Electricity produced by using the heat of seawater.

oceanic crust The 7- to 10-kilometer-thick layer of sediment and basalt that underlies the ocean basins.

oceanic island A seamount, usually of volcanic origin, that rises above sea level.

Ogallala aquifer The aquifer that extends for almost 1000 kilometers from the Rocky Mountains eastward, beneath portions of the Great Plains.

oil A naturally occurring liquid or gas composed of a complex mixture of hydrocarbons. (*syn:* petroleum)

oil shale A kerogen-bearing sedimentary rock that yields liquid or gaseous hydrocarbons when heated.

oil trap Any rock barrier that accumulates oil or gas by preventing its upward movement.

oligotrophic lake A lake characterized by nearly pure water with low concentrations of nitrates, phosphates, and other plant nutrients. Oligotrophic lakes have low productivities, meaning that they sustain relatively few living organisms, although lakes of this type typically contain a few huge trout or similar game fish and are commonly deep.

olivine A common rock-forming mineral in mafic and ultramafic rocks, with a composition that varies between Mg_2SiO_4 and Fe_2SiO_4.

ooid A small, rounded accretionary body in sedimentary rock, generally formed of concentric layers of calcium carbonate around a nucleus such as a sand grain.

open universe The hypothesis that if the gravitational force of the Universe is not sufficient to stop its current expansion, the galaxies will continue to fly apart forever. Stars will eventually consume all their nuclear fuel and stop producing energy, and the galaxies will continue speeding outward into the cold void.

ore A natural material that is sufficiently enriched in one or more minerals to be mined profitably.

ore minerals Minerals from which metals or other elements can be recovered profitably.

organic farming Agriculture that uses only organic fertilizer, such as manure and waste plant products, and applies no synthetic pesticides or herbicides.

organism Any living individual, whether plant or animal.

original horizontality (principle of) See *principle of original horizontality*.

orogeny The process of mountain building; all tectonic processes associated with mountain building.

orographic lifting The lifting of air that occurs when air flows over a mountain.

orthoclase A common rock-forming mineral; a variety of potassium feldspar, $KAlSi_3O_8$.

outwash plain A broad, level surface formed where glacial sediment is deposited in front of or beyond a glacier.

oxbow lake A crescent-shaped lake formed where a meander is cut off from a stream and the ends of the cut-off meander become plugged with sediment.

oxidation The loss of electrons from a compound or element during a chemical reaction. In the weathering of common minerals, oxidation usually occurs when a mineral reacts with molecular oxygen.

oxide One of a group of minerals whose major anion is O^{2-}.

oxisol A tropical soil (laterite) that has been intensely leached; found in tropical rainforests.

ozone O_3, a natural component of the stratosphere that absorbs and filters out some of the Sun's ultraviolet light. Several air pollutants rise into the stratosphere and destroy ozone. Ozone is also produced by reactions involving auto exhaust and sunlight and is a component of smog near the Earth's surface.

ozone hole A zone of abnormally low ozone concentration in the stratosphere.

pahoehoe A basaltic lava flow with a smooth, billowy, or "ropy" surface.

paleoclimatology The study of ancient climates.

paleomagnetism The study of natural remnant magnetism in rocks and of the history of the Earth's magnetic field.

paleontology The study of life that existed in the past.

Paleozoic Era The part of geologic time from 570 million to 245 million years ago. During this era invertebrates, fishes, amphibians, reptiles, ferns, and cone-bearing trees were dominant.

Pangaea A supercontinent, identified and named by Alfred Wegener, that existed from about 300 million to 200 million years ago and included most of the continental crust of the Earth. In this book we refer to three supercontinents, identified by Paul Hoffman as Pangaea I (about 2.0 billion to about 1.3 billion years ago), Pangaea II (1 billion to 700 million years ago), and Pangaea III (300 million to 200 million years ago).

parabolic dune A crescent-shaped dune with its tips pointing into the wind.

parallax The apparent displacement of an object caused by the movement of the observer.

parent rock Any original rock before it is changed by weathering, metamorphism, or other geological processes.

parsec The distance to an object that would have a stellar parallax angle of 1 arc/second. One parsec is about 3.2 light-years or 3×10^{13} kilometers.

partial melting The process in which a silicate rock only partly melts as it is heated, to form magma that is more silica-rich than the original rock.

particles In air pollution terminology, any pollutant larger than a molecule.

passive continental margin A margin, characterized by a firm connection between continental and oceanic crust, where little tectonic activity occurs.

passive solar heating Use of the Sun's heat by designing and orienting a building so that the Sun heats it directly.

paternoster lake One of a series of lakes, strung out like beads and connected by short streams and waterfalls, created by glacial erosion.

peat A loose, unconsolidated, brownish mass of partially decayed plant matter; a precursor to coal.

pebble A sedimentary particle between 2 and 64 millimeters in diameter, larger than sand and smaller than a cobble.

pedalfer A common soil type that forms in humid environments, characterized by abundant iron and aluminum oxides and a concentration of clay in the B horizon.

pediment A gently sloping erosional surface that forms along a mountain front uphill from a bajada, usually covered by a patchy veneer of gravel only a few meters thick.

pedocal A soil formed in arid and semiarid climates, characterized by an accumulation of calcium carbonate.

pegmatite An exceptionally coarse-grained igneous rock, usually with the same mineral content as granite.

pelagic sediment Muddy ocean sediment that consists of a mixture of clay and the skeletons of microscopic marine organisms.

peneplain A low, nearly featureless plain formed by lengthy erosion.

perched water table The top of a localized lens of ground water that lies above the main water table, formed by a layer of impermeable rock or clay.

peridotite A coarse-grained plutonic rock composed mainly of olivine; it may also contain pyroxene, amphibole, or mica but little or no feldspar. The mantle is thought to be made of peridotite.

period A geologic time unit longer than an epoch and shorter than an era.

permafrost A layer of permanently frozen soil or subsoil which lies from about a half meter to a few meters beneath the surface in Arctic environments.

permeability A measure of the speed at which fluid can travel through a porous material.

permineralization Fossilization that occurs when mineral matter is deposited in the pore spaces of the hard parts of an animal or plant.

petrified wood Material formed when silica precipitates in the pores of wood, preserving the wood's original form and structure.

petroleum A naturally occurring liquid composed of a complex mixture of hydrocarbons.

pH A measure of acidity. Numbers lower than 7 represent acidic solutions, and numbers higher than 7 represent basic ones.

Phanerozoic Eon The most recent 570 million years of geologic time, represented by rocks that contain evident and abundant fossils.

phenocryst A large, early-formed crystal in a finer matrix in igneous rock.

phosphorus cycle The movement of phosphorus through the atmosphere, hydrosphere, biosphere, and solid Earth. An ecosystem thrives by conserving and recycling its phosphorus.

photon The smallest particle or packet of electromagnetic energy, such as light, infrared radiation, and so forth.

photosphere The surface of the Sun, visible from Earth.

photosynthesis The biological process used by plants, algae, and some bacteria to combine light energy, carbon dioxide, and water to form energy-rich organic molecules such as glucose. The organisms then consume the molecules for their energy.

phyllite A metamorphic rock with a silky appearance and a commonly wrinkled surface, intermediate in grade between slate and schist.

phytoplankton Plants that float at or near the sea surface, forming the base of the marine food chain.

pillow lava Lava that solidified under water, forming spheroidal lumps like a stack of pillows.

placer deposit A surface mineral deposit formed by the mechanical concentration of mineral particles (usually by water) from weathered debris.

planetesimals Mini-planets, ranging in size from a few kilometers to about 100 kilometers in diameter, that coalesced to form the planets of the Solar System about 4.6 billion years ago.

planimetric map A two-dimensional picture of an area that outlines coastlines, rivers, cities, roads, and other geographic, cultural, and political features.

plastic Capable of being deformed permanently without fracture.

plastic deformation A permanent change in the original shape of a solid that occurs without fracture.

plate A relatively rigid, independent segment of the lithosphere that can move independently of other plates.

plate boundary A boundary between two lithospheric plates.

plate tectonics theory A theory of global tectonics in which the lithosphere is segmented into several plates that move about relative to one another by floating on and gliding over the plastic asthenosphere. Seismic and tectonic activity occurs mainly at the plate boundaries.

plateau A large elevated area of comparatively flat land.

platform The part of a continent covered by a thin layer of nearly horizontal sedimentary rocks overlying older igneous and metamorphic rocks of the craton.

playa A dry desert lake bed.

playa lake An intermittent desert lake.

Pleistocene Epoch A span of time from roughly 2 or 3 million to 8000 years ago, characterized by several advances and retreats of glaciers.

plucking A process in which glacial ice erodes rock by loosening particles and then lifting them and carrying them downslope.

plunging fold A fold with a dipping or plunging axis.

pluton An igneous intrusion.

plutonic rock An igneous rock that forms deep (a kilometer or more) beneath the Earth's surface.

pluvial lake A lake formed during a time of abundant precipitation. Many pluvial lakes formed as continental ice sheets melted.

point bar A stream deposit on the inside of a growing meander.

point-source pollution Pollution which arises from a specific site, such as a septic tank or a factory.

pollution An undesirable change in the quality of the environment that is caused by wastes or by excessive heat, noise, or harmful radiation.

polymer A large molecule formed by the linking of many smaller components.

population I star A relatively young star, formed from material ejected by an older, dying star, composed mainly of hydrogen and helium with 1 percent heavier elements. Our Sun is a population I star.

population II star An old star with lower concentrations of heavy elements than a population I star.

pore space The open space between grains in rock, sediment, or soil.

porosity The proportion of the volume of a material that consists of open spaces.

porphyry Any igneous rock containing larger crystals (phenocrysts) in a relatively fine-grained matrix.

porphyry copper A large body of porphyritic igneous rock that contains disseminated copper sulfide minerals and is usually mined by open-pit methods.

pothole A smooth, rounded depression in bedrock in a stream bed, caused by abrasion when currents circulate stones or coarse sediment.

ppm Parts per million. If a dissolved ion such as calcium is present in water at a concentration of 1 ppm, there is 1 kilogram of calcium for every 1 million kilograms of water.

Precambrian All of geologic time before the Paleozoic Era, encompassing approximately the first 4 billion years of Earth's history. Also, all rocks formed during that time.

precipitation (1) A chemical reaction that produces a solid salt, or precipitate, from a solution. (2) Any form in which atmospheric moisture returns to the Earth's surface—rain, snow, hail, and sleet.

preservation A process in which an entire organism or a part of an organism is preserved with very little chemical or physical change.

pressure gradient The change in air pressure over distance.

pressure relief melting The melting of rock and the resulting formation of magma caused by a drop in pressure at constant temperature.

primary (P) wave A seismic wave formed by alternate compression and expansion of rock. P waves travel faster than any other seismic waves.

primary air pollutants Gases, volatiles, and particulates released directly into air from a pollutant source.

principle of cross-cutting relationships The principle that a rock or feature must first exist before anything can happen to it or another rock cuts across it.

principle of faunal succession The principle that fossil organisms succeed one another in a definite and recognizable sequence, so that sedimentary rocks of different ages contain different fossils, and rocks of the same age contain identical fossils. Therefore, the relative ages of rocks can be determined from their fossils.

principle of original horizontality The principle that most sediment is deposited as nearly horizontal beds, and therefore most sedimentary rocks started out with nearly horizontal layering.

principle of superposition The principle that states that in any undisturbed sequence of sediment or sedimentary rocks, the ages decrease from bottom to top.

prominence A flamelike jet of gas rising from the Sun's corona.

Proterozoic Eon The portion of geologic time from 2.5 billion to 570 million years ago.

proton A dense, massive, positively charged particle found in the nucleus of an atom.

protoplanets The planets in their earliest, incipient stage of formation.

protosun The Sun in its earliest, incipient stage of formation. The protosun was a condensing agglomeration of dust and gas.

pulsar A neutron star that emits a pulsating radio signal.

pumice Frothy, usually rhyolitic magma solidified into a rock so full of gas bubbles that it can float on water.

punctuated evolution A theory that evolution occurs in a series of rapid steps broken by long periods of little or no change.

pyroclastic rock Any rock made of material that was ejected explosively from a volcanic vent.

pyroxene A rock-forming silicate mineral group that consists of many similar minerals. Members of the pyroxene group are major constituents of basalt and gabbro.

quartz A rock-forming silicate mineral, SiO_2. Quartz is a widespread and abundant component of continental rocks but is rare in the oceanic crust and mantle.

quartz sandstone Sandstone containing more than 90 percent quartz.

quartzite A metamorphic rock composed mostly of quartz and formed by recrystallization of sandstone.

quasar An object, less than 1 light-year in diameter and very distant from Earth, that emits an extremely large quantity of energy.

radial drainage pattern A drainage pattern formed when a number of streams originate on a mountain and flow outward like the spokes of a wheel.

radioactivity The natural spontaneous decay of unstable nuclei.

radiometric age dating The process of measuring the absolute age of geologic material by measuring the concentrations of radioactive isotopes and their decay products.

radon A radioactive gas, formed by radioactive decay of uranium, that commonly accumulates in some igneous and sedimentary rocks.

rain-shadow desert A desert formed on the lee side of a mountain range.

rainwash The washing away of soil or regolith by rainwater before the water has concentrated into channels or streams. (syn: sheet erosion)

recessional moraine A moraine that forms at the terminus of a glacier as the glacier stabilizes temporarily during retreat.

recharge The replenishment of an aquifer by the addition of water.

recrystallization (in fossils) A process of fossil development in which the fossil retains the shape and features of the original organism or part thereof, but the atoms rearrange to form a new mineral.

red giant A stage in the life of a star when its core is composed of helium that is not undergoing fusion. A hot shell of hydrogen around this core is fusing at a rapid rate, producing enough energy to cause the star to expand.

red shift The frequency shift of light waves that is observed in the spectrum of an object traveling away from an observer. This shift is caused by the Doppler effect.

reef A wave-resistant ridge or mound built by corals or other marine organisms.

reflection The return of a wave that strikes a surface.

refraction The bending of a wave that occurs when the wave changes velocity as it passes from one medium to another.

regional burial metamorphism Metamorphism of a broad area of the Earth's crust, caused by elevated temperatures and pressures resulting from simple burial.

regional dynamothermal metamorphism Metamorphism accompanied by deformation that affects an extensive region of the Earth's crust.

regional metamorphism Metamorphism that is broadly regional in extent, involving very large areas and volumes of rock. Includes both regional dynamothermal and regional burial metamorphism.

regolith The loose, unconsolidated, weathered material that overlies bedrock.

rejuvenated stream A stream that has had its gradient steepened and its erosive ability renewed by tectonic uplift or a drop of sea level.

relative humidity See *humidity*.

relative time Time expressed as the order in which rocks formed and geological events occurred, but not measured in years.

relief The vertical distance between a high point and a low point on the Earth's surface.

remote sensing The collection of information about an object by instruments that are not in direct contact with it.

renewable resource A resource that is replenished within a human lifetime. Stream water and trees are examples.

replacement Fossilization in which the original organic material dissolves and is replaced by new minerals.

reserves Known geological deposits that can be extracted profitably under current conditions.

reservoir Any natural or artificial pond or lake or any ground-water source from which water may be withdrawn for irrigation or water supplies. Also, a subterranean source from which oil can be extracted.

reservoir rock Porous and permeable rock in which liquid petroleum or gas accumulates.

residence time The average time spent by all water molecules in a portion of the hydrologic cycle.

respiration Breathing.

reverse fault A fault in which the hanging wall has moved up relative to the footwall.

reverse flow The upstream migration of seawater in a stream channel; usually caused by excessive diversion of the fresh stream water for irrigation or by drought.

reversed polarity Magnetic orientations in rock which are opposite to the present orientation of the Earth's field. Also, the condition in which the Earth's magnetic field is opposite to its present orientation.

revolve To orbit a central point. A satellite revolves around the Earth (see *rotate*).

rhyolite A fine-grained extrusive igneous rock compositionally equivalent to granite.

Richter scale A numerical scale of earthquake magnitude measured by the amplitude of the largest wave on a standardized seismograph.

rift A zone of separation of tectonic plates at a divergent plate boundary.

rift valley An elongate depression that develops at a divergent plate boundary. Examples include continental rift valleys and the rift valley along the center of the mid-oceanic ridge system.

ring of fire The belt of subduction zones and major tectonic activity, including extensive volcanism, that borders the Pacific Ocean along the continental margins of Asia and the Americas.

rip current A current created when water flows back toward the sea after a wave breaks against the shore. (*syn:* undertow)

ripple marks Small parallel ridges and troughs formed in loose sediment by wind or water currents and waves. They may then be preserved when the sediment is lithified.

risk assessment Evaluation of the health effects or environmental damage that is likely to result from a particular hazard or pollutant.

river Any large stream fed by smaller streams.

roche moutonnée An elongate, streamlined bedrock hill sculpted by a glacier.

rock A naturally formed solid that is an aggregate of one or more different minerals.

rock avalanche A type of mass wasting in which a segment of bedrock slides over a tilted bedding plane or fracture. The moving mass usually breaks into fragments. (*syn:* rockslide)

rock cycle The sequence of events in which rocks are formed, destroyed, altered, and reformed by geological processes.

rock flour Finely ground, silt-size rock fragments formed by glacial abrasion.

rock-forming mineral One of nine minerals (or mineral groups) that make up most of the Earth's crust. They are feldspar, pyroxene, amphibole, micas, clay minerals, olivine, quartz, calcite, and dolomite.

rockslide A type of slide in which a segment of bedrock slides along a tilted bedding plane or fracture. The moving mass usually breaks into fragments. (*syn:* rock avalanche)

rotate To turn or spin on an axis. Tops and planets rotate on their axes (see *revolve*).

rounding The sedimentary process in which sharp, angular edges and corners of grains are smoothed.

rubble Angular particles with diameters greater than 2 millimeters.

runoff Water that flows back to the oceans in surface streams.

S wave A seismic wave consisting of a shearing motion in which the oscillation is perpendicular to the direction of wave travel. S waves travel more slowly than P waves. (*syn:* secondary wave)

salinization A process whereby salts accumulate in soil that is irrigated heavily.

salt cracking A weathering process in which salts that are dissolved in water in the pores of rock crystallize. This process widens cracks and pushes grains apart.

saltation Sediment transport in which particles bounce and hop along the surface.

sand Sedimentary grains that range from 1/16 to 2 millimeters in diameter.

sandstone Clastic sedimentary rock comprising primarily lithified sand.

saturated zone The region below the water table where all the pores in rock or regolith are filled with water.

scale (of a map) The ratio of the distance between any two points on a map and the actual distance on the Earth's surface (e.g., 1 centimeter = 1 kilometer).

scarp A line of cliffs created by faulting or by erosion.

schist A strongly foliated metamorphic rock that has a well-developed parallelism of minerals such as micas.

sea arch An opening that is created when a cave is eroded all the way through a narrow headland.

sea stack A pillar of rock that remains when a sea arch collapses or when the inshore portion of a headland erodes faster than the tip.

sea-floor spreading The hypothesis that segments of oceanic crust are separating at the mid-oceanic ridge.

seamount A submarine mountain, usually of volcanic origin, that rises 1 kilometer or more above the surrounding sea floor.

secondary air pollutant A pollutant generated by chemical reactions within the atmosphere.

secondary recovery Production of oil or gas by artificially augmenting the reservoir energy, as by injection of water, detergent, or other fluid.

secondary wave A seismic wave consisting of a shearing motion in which the oscillation is perpendicular to the direction of wave travel. Secondary waves travel more slowly than primary waves. (*syn:* S wave)

sediment Solid rock or mineral fragments transported and deposited by wind, water, gravity, or ice, precipitated by chemical reactions or secreted by organisms, and that accumulate as layers in loose, unconsolidated form.

sedimentary rock A rock formed when sediment is lithified.

sedimentary structure Any structure formed in sedimentary rock during deposition or by later sedimentary processes; for example, bedding.

seismic gap An immobile region of a fault bounded by moving segments.

seismic profiler A device used to construct a topographic profile of the ocean floor and to reveal layering in sediment and rock beneath the sea floor.

seismic tomography A technique whereby seismic data from many earthquakes and recording stations are analyzed to provide a three-dimensional view of the interior of the Earth.

seismic wave Any elastic wave that travels through rock and is produced by an earthquake or explosion.

seismogram The record made by a seismograph.

seismograph An instrument that records seismic waves.

seismology The study of earthquake waves and the interpretation of the resultant data to elucidate the structure of the interior of the Earth.

selective cutting A logging method in which loggers harvest only the valuable mature trees and leave the rest of the forest untouched.

shale A fine-grained clastic sedimentary rock with finely layered structure, composed predominantly of clay minerals.

shear strength A material's resistance to being pulled apart along its cross section.

sheet flood A broad, thin sheet of flowing water that is not concentrated into channels, typically in arid regions.

shield volcano A large, gently sloping volcanic mountain formed by successive flows of basaltic magma.

sialic rock A rock, such as granite or rhyolite, that contains large proportions of silicon and aluminum.

silica Silicon dioxide, SiO_2. Includes quartz, opal, chert, and many other varieties.

silicate All minerals whose crystal structures contain silicate tetrahedra. All rocks composed principally of silicate minerals.

silicate tetrahedron A pyramid-shaped molecule consisting of a silicon ion bonded to four oxygen ions, SiO_4^{4-}. The silicate tetrahedron is the fundamental building block of all silicate minerals, which are the most abundant minerals in the Earth's crust.

sill A tabular or sheetlike igneous intrusion oriented parallel to the grain or layering of country rock.

silt All sedimentary particles from 1/256 to 1/16 millimeter in size.

siltstone Lithified silt.

sinkhole A circular depression in karst topography, caused by the collapse of a cavern roof or by dissolution of surface rocks.

slate A compact, fine-grained, low-grade metamorphic rock with slaty cleavage that can be split into slabs and thin plates. Slate is intermediate in grade between shale and phyllite.

slaty cleavage A parallel metamorphic foliation in a plane perpendicular to the direction of tectonic compression.

slide Any type of mass wasting in which the rock or regolith initially moves as a consolidated unit over a fracture surface.

slip face The steep lee side of a dune that is at the angle of repose for loose sand, so that the sand slides or slips.

slump A type of mass wasting in which the rock and regolith move as a consolidated unit with a backward rotation along a concave fracture.

smectite A type of clay mineral that contains the abundant elements weathered from feldspar and silicate rocks.

smog Smoky fog. The word is used loosely to define visible air pollution.

snowline The altitude above which there is permanent snow.

soil nutrients The 16 chemical elements needed by plants.

soil Soil scientists define soil as the upper layers of regolith that support plant growth. Engineers define soil as synonymous with regolith.

soil horizon A layer of soil that is distinguished from other horizons by its differences in appearance and in physical and chemical properties.

soil-moisture belt The relatively thin, moist surface layer of soil above the unsaturated zone beneath it.

solar cell A device that produces electricity directly from sunlight.

solar energy Energy derived from the Sun. Current technologies enable us to use solar energy in three ways: passive solar heating, active solar heating, and electricity production by solar cells.

solar wind A stream of atomic particles shot into space by violent storms occurring in the outer regions of the Sun's atmosphere.

solifluction A type of mass wasting that occurs when water-saturated soil moves slowly over permafrost.

solstice Either of two times per year when the Sun shines directly overhead farthest from the equator. The solstices mark the beginnings of summer and winter. One solstice occurs on or about June 21 and marks the longest day of the year in the Northern Hemisphere and the shortest day in the Southern Hemisphere. The other solstice occurs on or about December 22, marking the longest day in the Southern Hemisphere and shortest day in the Northern Hemisphere.

soot Carcinogenic particles, consisting mostly of carbon, that are released whenever fossil fuel is burned. Soot is a common type of air pollutant.

sorting A process in which flowing water or wind separates sediment according to particle size, shape, or density.

source rock The geologic formation in which oil or gas originates.

specific gravity The weight of a substance relative to the weight of an equal volume of water.

specific heat The heat required to raise the temperature of 1 gram of a substance 1° Celsius.

spectrum (electromagnetic) A pattern of wavelengths into which a beam of light or other electromagnetic radiation is separated. The spectrum is seen as colors, or is photographed, or is detected by an electronic device. An *emission spectrum* is obtained from radiation emitted from a source. An *absorption spectrum* is obtained after radiation from a source has passed through a substance that absorbs some of the wavelengths.

speleothem Any mineral deposit formed in a cave by the action of water.

spheroidal weathering Weathering in which the edges and corners of a rock weather more rapidly than the flat faces, giving rise to a rounded shape.

spit A small point of sand or gravel extending from shore into a body of water.

spodosol An acidic soil that forms in cool, moderately moist climates. The A horizon is generally sandy, and the B horizon contains leached organic matter and clay.

spoil Waste rock and soil removed during mining, quarrying, excavating, or dredging.

spreading center The boundary or zone where lithospheric plates rift, or separate from each other. (syn: divergent plate boundary)

spring A place where ground water flows out of the Earth to form a small stream or pool.

stalactite An icicle-like dripstone, deposited from drops of water, that hangs from the ceiling of a cavern.

stalagmite A deposit of mineral matter that is formed on the floor of a cavern by the action of dripping water.

stock An igneous intrusion with an exposed surface area of less than 100 square kilometers.

storm surge A temporary rise in sea level caused by strong onshore winds that push the surface of the ocean against the shore.

strain The deformation (change in size or shape) that results from stress.

strategic mineral A mineral that is deemed necessary for national defense.

stratification The arrangement of sedimentary rocks in strata or beds.

stratified drift Sediment that was transported by a glacier and then sorted, deposited, and layered by glacial meltwater.

stratopause The boundary between the stratosphere and the mesosphere.

stratosphere The layer of air extending from the tropopause to about 55 kilometers. Temperature remains constant and then increases with elevation in the stratosphere.

stratovolcano A steep-sided volcano formed by an alternating series of lava flows and pyroclastic eruptions. (syn: composite volcano)

stratus cloud A horizontally layered, sheetlike cloud.

streak The color of a fine powder of a mineral, usually obtained by rubbing the mineral on an unglazed porcelain streak plate.

stream A moving body of water, confined in a channel and flowing downslope.

stream piracy The natural diversion of the headwaters of one stream into the channel of another.

stream terrace An abandoned flood plain above the level of the present stream.

stress The force per unit area exerted against an object.

striations Parallel scratches in bedrock, caused by rocks embedded in the base of a flowing glacier.

strike The direction of a tilted rock surface (e.g., a bedding plane or fault) as it intersects a horizontal plane. Strike is measured as a compass direction.

strike-slip fault A fault on which the motion is parallel with the fault's strike and is primarily horizontal.

subduction The process in which a lithospheric plate descends beneath another plate and dives into the asthenosphere.

subduction zone (or subduction boundary) The region or boundary where a lithospheric plate descends into the asthenosphere.

sublimation The process by which a solid transforms directly into a vapor or a vapor transforms directly into a solid without passing through the liquid phase.

submarine canyon A deep, V-shaped, steep-walled trough eroded into a continental shelf and slope.

submarine fan A large fan-shaped accumulation of sediment deposited at the bases of many submarine canyons adjacent to the deep-sea floor. (syn: abyssal fan)

submergent coastline A coastline that was recently above sea level but has been drowned, either because the land has sunk or because sea level has risen.

subsidence Settling of the Earth's surface, which can occur as either petroleum or ground water is removed by natural processes.

subsidy (agricultural) A payment to farmers for agricultural practices deemed beneficial to the nation.

sunspot A cool, dark region on the Sun's surface, formed by an intense magnetic disturbance.

supercooling A condition in which water droplets in air do not freeze even when the air cools below the freezing point.

supernova An exploding star that is releasing massive amounts of energy.

superposed stream A stream that has downcut through several rock units and maintained its course as it encountered older geologic structures and rocks.

superposition (principle of) See *principle of superposition*.

supersaturation A condition in which the relative humidity of air rises above 100 percent.

surf The chaotic turbulence created when a wave breaks near the beach.

surface process Any process that acts on the Earth's surface to create or modify landforms, including weathering, erosion, and transport and deposition of sediment.

surface water All water at the Earth's surface, including that in streams, lakes, wetlands, and reservoirs, in contrast with ground water.

surface wave An earthquake wave that travels along the surface of the Earth or along a boundary between layers within the Earth. (syn: L wave)

suspended load That portion of a stream's load that is carried for a considerable time in suspension, free from contact with the stream bed.

suture The junction created when two continents or other masses of crust collide and weld into a single mass of continental crust.

syncline A fold that arches downward and whose center contains the youngest rocks.

synergistic effects Interactions between several independent factors that combine to produce a larger effect.

synfuel Any synthetic fuel derived from coal or other sources.

system A collection of objects bonded together in some way so that the collection is more than an assemblage of independent parts.

tactite A rock formed by contact metamorphism of carbonate rocks. It is typically coarse grained and rich in garnet.

talus slope An accumulation of loose angular rocks at the base of a cliff from which they have been cleared by mass wasting.

tarn A small lake at the base of a cirque.

tectonics A branch of geology dealing with the broad architecture of the outer part of the Earth; specifically the relationships, origins, and histories of major structural and deformational features.

tephra A collective term for all solid material ejected from a volcano and carried through the air. It includes volcanic dust, ash, cinders, bombs, and blocks. (syn: pyroclastic material)

terminal moraine An end moraine that forms when a glacier is at its greatest advance.

terminus The end, or foot, of a glacier.

terrestrial planets The four Earth-like planets closest to the Sun—Mercury, Venus, Earth, and Mars—which are composed primarily of rocky and metallic substances.

terrigenous sediment Sea-floor sediment derived directly from land.

thermal pollution The addition of heat to an aquatic ecosystem.

thermocline A layer of ocean water between 0.5 and 2 kilometers deep where the temperature drops rapidly with depth; the boundary between the warm and cool layers of lake water.

thermoremanent magnetism The permanent magnetism of rocks and minerals that results from cooling through the Curie point.

thermosphere An extremely high and diffuse region of the atmosphere lying above the mesosphere. Temperature remains constant and then rises rapidly with elevation in the thermosphere.

threshold effect A situation in which a sequence of disturbances has little or no perceptible effect, but a small additional disturbance causes a large change.

thrust fault A type of reverse fault with a dip of 45° or less over most of its extent.

tidal current A current caused by the tides.

tidal energy Electricity produced by harnessing tidal currents with a tidal dam and turbine.

tide The cyclic rise and fall of ocean water caused by the gravitational force of the Moon and, to a lesser extent, the Sun.

tidewater glacier A glacier that flows directly into the sea.

till Sediment that was deposited directly by glacial ice and has not been resorted by a stream.

tillage The way in which a farmer prepares a field for planting of crops and control of weeds.

tillite A sedimentary rock formed of lithified till.

tombolo A sand or gravel bar that connects an island to the mainland or to another island.

topographic map A map showing the topography of the land—the shapes of its mountains and valleys.

topsoil The upper, usually organic-rich zone of soil including both O and A horizons.

tornado A small, intense, short-lived, funnel-shaped storm that protrudes from the base of a cumulonimbus cloud.

total fertility rate (TFR) The total number of children a woman in a given population is expected to bear during the course of her life if the birth rate remains constant for at least one generation.

toxic Poisonous or pertaining to poisons.

trace fossil A sedimentary structure consisting of tracks, burrows, or other marks made by an organism.

traction Sediment transport in which particles are dragged or rolled along a stream bed, beach, or desert surface.

trade winds The winds that blow steadily from the northeast in the Northern Hemisphere and from the southeast in the Southern Hemisphere, between 5° and 30° north and south latitudes.

transform fault A strike-slip fault between two offset segments of a mid-oceanic ridge.

transform plate boundary A boundary between two lithospheric plates where the plates are sliding horizontally past each other.

transpiration Direct evaporation from the leaf surfaces of plants.

transport The movement of sediment by flowing water, ice, wind, or gravity.

transverse dune A relatively long, straight dune with a gently sloping windward side and a steep lee face that is oriented perpendicular to the prevailing wind.

trellis drainage pattern A drainage pattern characterized by a series of fairly straight, parallel streams joined at right angles by tributaries.

trench A long, narrow depression of the sea floor, formed where a subducting plate sinks into the mantle.

tributary Any stream that contributes water to another stream.

tropical cyclone A broad, circular storm with intense low pressure that forms over warm oceans (also called a hurricane, typhoon, or cyclone).

tropopause The boundary between the troposphere and the stratosphere.

troposphere The layer of air that lies closest to the Earth's surface and extends upward to about 17 kilometers. Temperature generally decreases with elevation in the troposphere.

trough The lowest part of a wave.

truncated spur A triangular-shaped rock face that forms when a valley glacier cuts off the lower portion of an arête.

tsunami A large sea wave, produced by a submarine earthquake or a volcano, characterized by long wavelength and great speed.

tuff A general term for all rocks formed from consolidated tephra or pyroclastic material.

turbidity current A rapidly flowing submarine current, laden with suspended sediment, that results from mass wasting on the continental shelf or slope.

turbulent flow A pattern in which water flows in an irregular and chaotic manner. It is typical of stream flow.

turnover A process, usually occurring in fall and spring in temperate-climate lakes, in which convection mixes lake water so that it becomes of uniform temperature from the top to the bottom of the lake.

U-shaped valley A glacially eroded valley with a characteristic U-shaped cross section.

ultimate base level The lowest possible level of downcutting of a stream, usually sea level.

ultisol A soil formed under warm, moist conditions but not as extreme as those of oxisols; found in the southeastern United States.

ultramafic rock Rock composed mostly of minerals containing iron and magnesium; for example, peridotite.

unconformity A gap in the geological record, such as an interruption of deposition of sediments or a break between eroded igneous and overlying sedimentary strata; usually of long duration.

undertow A current created by water flowing back toward the sea after a wave breaks. (*syn:* rip current)

uniformitarianism The principle that states that geologic processes and scientific laws operating today also operated in the past and that past geologic events can be explained by forces observable today: "The present is the key to the past." The principle does not imply that geologic change goes on at a constant rate; nor does it exclude catastrophes such as impacts of large meteorites.

unit cell The smallest group of atoms that perfectly describes the arrangement of all atoms in a crystal and repeats itself to form the crystal structure.

unsaturated zone A subsurface zone above the water table that may be moist but is not saturated; it lies above the zone of saturation. (*syn:* zone of aeration)

upper mantle The part of the mantle that extends from the base of the crust downward to about 670 kilometers beneath the surface.

upwelling A rising ocean current that transports water from the depths to the surface.

urban heat island effect A local change in climate caused by a city.

urban sprawl The growth of suburbs and the highway systems linking them to cities.

valley A low-lying area of land, usually traversed by a stream, that receives the drainage of surrounding heights.

valley train A long and relatively narrow strip of outwash deposited in a mountain valley by the streams flowing from an alpine glacier.

varve A pair of light and dark layers that was deposited in a year's time as sediment settled out of a body of still water. Most commonly formed in sediment deposited in a glacial lake.

vent A volcanic opening through which lava and rock fragments erupt.

ventifact A desert cobble or boulder with one or more of its faces flattened and polished by windblown sand.

vertisol A clay-rich soil that expands and contracts with moisture content.

vesicle A bubble formed by expanding gases in volcanic rocks.

volatile A compound that evaporates rapidly and therefore easily escapes into the atmosphere.

volcanic bomb A small blob of molten lava that was hurled out of a volcanic vent and acquired a rounded shape while in flight.

volcanic neck A vertical, pipelike intrusion formed by the solidification of magma in the vent of a volcano.

volcanic rock A rock that formed when magma erupted, cooled, and solidified within a kilometer or less of the Earth's surface.

volcano A hill or mountain that formed from lava and rock fragments ejected through a volcanic vent.

wash An intermittent stream channel found in a desert.

water table The upper surface of a body of ground water at the top of the zone of saturation and below the zone of aeration.

Watt, James Designer of the first practical steam engine (1769), which gave rise to the Industrial Revolution.

wave An oscillatory rising and falling of the surface of a body of water.

wave energy Electricity produced by using the force generated by rising and falling sea waves.

wave height The vertical distance from the crest to the trough of a wave.

wave period The time interval between two crests (or two troughs) as a wave passes a stationary observer.

wave-cut cliff A cliff created when a rocky coast is eroded by waves.

wave-cut platform A flat or gently sloping platform created by erosion of a rocky shoreline.

wavelength The distance between successive wave crests (or troughs).

weather The state of the atmosphere on a particular day, as characterized by temperature, wind, cloudiness, humidity, and precipitation (see *climate*).

weathering The decomposition and disintegration of rocks and minerals at the Earth's surface by mechanical and chemical processes.

welded tuff A hard, tough, glass-rich pyroclastic rock formed by cooling of an ash flow that was hot enough to deform plastically and partly melt after it stopped moving; it often appears layered or streaky.

wet adiabatic lapse rate The rate of cooling when moist air rises and condensation occurs without gain or loss of heat.

wetlands Known variously as swamps, bogs, marshes, sloughs, mud flats, flood plains, and many other names, wetlands are the boundaries between land and water. Some are water-soaked or flooded throughout the entire year; others are dry for much of the year and wet only during times of high water. Some are wet only during exceptionally wet years and may be dry for several years at a time.

white dwarf A stage in the life of a star when fusion has halted and the star glows solely from the residual heat produced during past eras. White dwarfs are very small stars.

withdrawn water Water that is extracted from a surface- or groundwater source and then returned to the Earth near the place from which it was taken. Most of the water used by industry and used domestically is returned locally (see *consumed water*).

xeriscaping A type of landscaping which utilizes indigenous plants that can survive on semiarid and desert rainfall.

zone of aeration A subsurface zone above the water table that may be moist but is not saturated; it lies above the zone of saturation. (*syn:* unsaturated zone)

zone of saturation A subsurface zone below the water table in which the soil and bedrock are completely saturated with water.

zone of wastage The lower portion of a glacier where more snow melts in summer than accumulates in winter, so there is a net loss of glacial ice. (*syn:* ablation area)

Boldface page numbers indicate where key terms are highlighted; italicized page numbers indicate figures; t indicates a table.

A

Ablation area, glacial, **322**
Abrasion, **159**, *159*
Accumulation area, glacial, **322**
Acid mine drainage, **365**
Acid precipitation, **159**
Agriculture
   and flooding on Nile River, 213
   nineteenth-century, 177, *177*
   and soil fertility, 170, 170–172, *171*
   in southern California, water supply and, 253–255
   water consumption by, 255
Air, natural dry, gaseous composition of, 55, 55t
Air pollution, 56, *56*
   due to landfills, 280
   due to trash incinerators, 423
   Teakettle Mountain, 367
Alaska pipeline, 306, *307*
Alligators, salvation of swamp ecosystem by, 231–232, *232*
Alpine glacier(s), *318*, **318**–319, *319*
   erosional landforms created by, 322–325, *324*, *325*, *326*, *327*
Aluminum, 366–369
   and conservation, 368
   and dams, 368
   mining and refining of, environmental impacts of, 366
   production of, and electrical consumption, 366
   recycling of, 368–369
Alvarez, Walter and Luis, 32
ALVIN submarine, 60, 61–62
Amazon River, 208
Ammonoids, **28**, *30*, *33*
Amphibole, **43**, *44*, *45*
Andean margin, **84**
Andes Mountains, development, 83–84, *84*
Andesite, **48**
Angle of repose, **191**, *192*
Animals
   behavior of, and earthquakes, 120
   of desert, 337
   marine, destruction of habitat of, 306, *306*
Aqueduct, Los Angeles, 253, *254*
Aquifer, **218**
   artesian, **220**, *220*
Archean Eon, **26**
Arches National Park, Utah, 338

Arctic National Wildlife Refuge, Alaska, 404–405, *405*
Area, systems of measurement of, A-3
Arête, **324**, *325*
Artesian aquifer, **220**, *220*
Artesian well(s), **220**, *220*
Asbestos, **46**
   and cancer, 45–46
Aseismic ridges, **95**
Ash, volcanic, **133**, 140, *140*
Asthenosphere, **23**, **67**, **70**
Aswan Dam, Egypt, 213, 245–246
Atlantic coast, coastal erosion on, 298, 298t
Atmosphere, **24**, 55–56, *57*
   computer modeling of, 151
   evolution of, 26
   and life, co-evolution of, 24–26
   temperature of, clouds and, 151–152
Atoll, **310**, *311*
Atom, **40**, *40*
Atomic bomb, 392
Atomic energy, uranium and, 391–399
Atomic Weights, International Table of, A-2

B

Bacteria
   anaerobic, **264**, *264*
   denitrifying, **167**
Banded iron formations, **362**, *362*
Barrier island(s), **296**–298, *297*
   Miami, Florida, built on, 304
Basal slip, **320**
Basalt, **46**, **48**
Basalt plateau, **48**, *48*
Base level of stream, **209**, *209*
   temporary, **209**, *209*
   ultimate, **209**
Basement rock, **57**
Bauxite, **163**, **362**, *363*, 366
   mining of, 366, *367*
Beach(es), **295**, *295*
   growth and shrinkage of, *297*
Becquerel, Henri, 391
Bedrock, **57**
Benioff zone, **74**, *74–75*
Big bang, **19**
Big Thompson River, Colorado, 213, *213*
Bioclastic limestone, **52**, *52*

I–1

Biodegradable, **6**
Biodiversity, **8–9**
Biogeochemical cycle, **166**
Biological oxygen demand, **264**
Biomagnification, **267**, *268*
Biomass energy, **422–423**, *423*
Bioremediation, **273**
Biosphere, **24**
Bird, Eocene, *Diatryma*, 32, *32*
Bitumen, **387**
Black lung disease, **383**
Black smokers, 61, **360**, *361*
Bladder cancer, 12
Body waves, **102**
    primary, *103*, **103–104**, 108, *109*
    shear, *103*, **103–104**, 108, *109*
Bonneville Salt Flats, Utah, *52*
Branching chain reaction, 392, *392*
Breakwater(s), **298**, *299*
Breeder reactors, **396**
Bugaboo Mountains, British Columbia, *325*
Bush, George, *13*
Bushveldt intrusion, South Africa, 357

C

Calcite, **43**, *44*, **51**
Caldera(s), *136*, *140*, **141**
    in United States, *142*, 142–143, *143*
Caliche, **163**
Cancer
    asbestos and, 45–46
    bladder, 12
Cap rock, **384**
Capillary action, **162–163**, *163*
Carbonate(s), **42**, **51**
Carbonate rocks, 51–52
Carbonic acid, 159
Carcinogen, **5**
Cars, electric, 405, 415, *416*
Cavern(s), **220**
Cenozoic Era, **32**
Cenozoic life, 32–33
Central Valley, California, cotton field in, 241, *241*
CERCLA, 262, **272**
Chain reaction, branching, 392, *392*
Channel(s), 3, **207**
    banks of, **207**
    bed of, **207**
Channeled scablands, **333**
Channelization, of river, 215
Chemical fertilizers, 171
Chemicals, pollution associated with, 271
    human health and, 270

    toxic, in coastal waters, 305–306, *306*
Chert, 50, *50*
Chesapeake Bay, pollution of, estuaries and, 308–310, *309*
China syndrome, **396**
Chrysotile, **45**
Cinder cone, **133**, *133*
Cinders, volcanic, **133**
Cirque, **323**, *324*, *326*
Clastic sedimentary rocks, 49t, 49–50
Clasts, **49**
Clay, instability of, 180
Clay minerals, **43**, *44*
Clean Water Act, **262**, 267
    surface water in United States after, 268–270
Climate
    changes in, indications of, 345, *345*
    and environment, 150–151
    and soils, 162–164
    volcanoes and, 145–148, *147*
Clouds, atmospheric temperature and, 151–152
Coal, **51**, *51*, 380, 381, *381*
    burning of, 384
    deposits of, in United States, *383*, 383–384
    formation of, 381–382, *382*, 402
    liquefaction of, 389
    reserves of, 401, 402
    types of, 382
Coal mine(s), underground, *382*, 382–383
Coastal erosion
    on Long Island, New York, 298–301, *299*, *300*
    in Louisiana, 302, *302*
    in San Diego County, 303, *303*
Coastal plains, 291
Coastline(s)
    Atlantic, coastal erosion on, 298, 298t, *300*
    developed, protection of, 298, *299*, *300*, 301, 313, *313*
    development along, and public policy, 303–304, *304*
    emergent, **292**, *293*, *294*, 295
    erosion of, 298–303
    and transport along, 296–298
    Gulf, erosion of, 301, 301t
    sediment-poor, **295**, *295*, 296
    sediment-rich, **295**, *295*
    submergent, **292–293**, *294*
    of world, 292–293
Cold fusion, 426
Colorado River, 252, 252–253, 341, *341*

Composite cone volcanoes, **133**, *134*, *135*
Compost, **284**
Composting, as alternative to landfill, 284
Computer modeling, of atmosphere, 151
Conglomerate, **49**, *49*
Conservation
    as alternative energy resource, 411–414
    aluminum and, 368
    of metals, 375
Conservation tillage, 173, *173*
Construction, earthquake-resistant, 110–111, *111*
Consumption, reduction of, as alternative to waste disposal, 285
Continental crust, convergence of oceanic crust with, 68, **74–75**, *74–75*
    divergent plate boundaries in, 73
    two plates carrying, convergence of, 76
Continental drift, theory of, 80
Continental glaciers, **319**, *319*
    and Great Lakes, 332, 332–333
    Pleistocene, effects of, *330*, 330–333, *331*, *332*
Continental interior deserts, 340
Continental margin(s), **89–91**
    active, **91**, *93*
    passive, **89–91**, *92*
Continental rifting, **73**
Continental rise, **91**, *93*
Continental shelf, **91**, *93*
Continental slope, **91**, *93*
Contour plowing, 172–173, *173*
Control rods, **393**
Convection, 77, *77*
Convergent plate boundary(ies), *72*, **73–76**
Copper deposits, porphyry, **359–360**, *360*
Copper mine, open pit, Bingham Canyon, Utah, *364*
Coral, castle, *311*
Core of Earth, 71
Cost–benefit analysis, **7–9**
Country(ies)
    developed, 14
    developing, 14
    industrial, 14
    less developed, 14
Country rock, **51**
Courthouse Wash, Utah, 219, *341*
Cover-cropping, **174**

Crandall, Dwight, 144–145
Crater, volcanic, **130**
Crater Lake, Oregon, 135–136, *136*, 141, 223, *225*
Creep
　fault, **114**
　in mass wasting, **194**, *194–195*, *195*
Crevasses, glacial, 320, *320*
Crop rotation, **174**
Cropping patterns, 173–174
Crude oil, 380, 381, *381*
Crust, **23**, 68, **69**, 69t
　continental. *See* Continental crust.
　depiction of, colors and style for, A-7
　oceanic. *See* Oceanic crust.
Crust–mantle boundary, of Earth, 108
Crystal(s), 41, *41*
Crystal faces, 41, *41*
Crystal settling, 357, *357*
Current, **296**
　longshore, **296**, *296*, **298**, *298*
Cuyahoga River, Cleveland, fire on, 262, *262*
Cyanide
　stability of, 369
　toxicity of, 369–370
Cyanide heap leaching, 369

D

Dam(s), **243**
　aluminum and, 368
　ecological disruptions associated with, 245–246
　recreational and aesthetic losses associated with, 245, *245*
Darwin, Charles, 310
Debris flows
　in flash floods, 342
　in mass wasting, **195–196**, *196*
Decay organisms, aerobic, **264**
Decker strip coal mine, Montana, 383, *383*
Deep injection wells and ground-water pollution, **275**, *276*, *277*
Deflation, **343**, *343*
Delta, **301**
deMello, Collor, 14
Desert(s), **337**, *339*, *339*
　animals of, 337
　continental interior, 340
　dunes on, *344*, 344–345
　effect of latitude on, 338, 339, *339*
　expansion and contraction of, 345, *345*, *347*
　formed by cold ocean currents, 341
　Great American, 250–252, *252*

　land deterioration in, **346**, *346–347*, *347*
　major, of world, *338*, 339–341
　plants of, 337
　rain-shadow, 339, *340*
　water in, 341–342
　wind erosion on, 343, *343*
Desert pavement, **343**, *343*
Desert stream, **219**, *219*
Desertification, **346**, *346*
Detergents, eutrophication caused by, 264, *265*
Deuterium, 426
Developed countries, 14
Developing countries, 14
Development, sustainable, **342**
Dinosaurs, 28, *30*
　in Montana, 31, *31*, *33*
Diorite, **48**
Discharge, of stream, **208**, *208*
Disease organisms, as water pollutant, 263
Divergent plate boundary(ies), **72–73**, 73t, *74*
　in continental crust, 73
　in oceanic crust, **72–73**
Divides, **211**
Dolomite, **43**, **44**, 51
Dolostone, **51**
Downcutting, **209**
　by stream, 209
Drainage basins, **211**, *211*
Drift, **326–327**
　stratified, **326–327**
　　landforms composed of, *328*, 329
Dunes, **344**, *344*
　migration of, *344*, 344–345
Dust bowl, 176, *176*

E

Earth, 3–15
　age of, 3
　core of, 71
　crust–mantle boundary of, 108
　as dynamic planet, 4–6
　early, 21–23
　as layered planet, 68, *69*
　layers of, 23, *25*, 69t, 69–71
　lithosphere–asthenosphere boundary of, 108–109
　modern, 23–26
　orbit of, variations in, 333–334, *334*
　solid, 23, 40–45, 46–54
　as system, 57
　in time, 17–37

Earth flows, in mass wasting, **195–196**
Earth materials, 39–59
Earth science, 153
Earth Summit, **13–14**
　Agenda 21 of, **14**
Earthquake(s), **23**, 99–123
　animal behavior and, 120
　areas of potential damage from, in U.S., 121, *121*
　control of, 121
　damage done by, 106–113
　　construction design and, *110*, 110–111, *111*
　　fire and, 111, *112*
　　rock and soil as influence on, 107–110, *110*
　danger zones, in North America, 113–119
　definition of, 101
　elastic energy of, **101**, *101*
　epicenter of, **103**
　faults and, 113, 113–115, *114*, *115*
　focus of, **102**, *102–103*
　ground motion during, 106–111
　hazards of, and human habitation, 121–122, *122*
　injection wells and, 11, 12, *12*
　landslides associated with, 112
　magnitude of, **105**
　major historical, 100t
　major zones of, *102*
　mass wasting associated with, 192, 197–200, *199*, *200*, *201*
　moment magnitude of, **105**
　in Pacific Northwest, 117–118, *118*
　at plate interiors, 118–119, *119*
　prediction of
　　long-term, 119–120
　　short-term, 120, 120–121
　in San Francisco Bay area, *114*, 114–115, *115*
　in southern California, *115*, 115–116, *116*
　in southern California, *115*, 115–116, *116*
　strength of, measurement of, 105–106, 106t
　subduction and, 74, 74–75
　at subduction zones, *117*, 117
　at transform plate boundaries, *113*, 113–114
　triggered by human activity, 119
　tsunamis created by, *112*, **112–113**
　waves of, *102*, 102–106
　　and Earth's interior structure, 108–109
Eccentricity, **333**
Ecology, **6**

Economics and environment. *See* Environment, economics and.
Economy, growth of, and destruction of environment, 153
Ecosystems, **6–7**, *7*
El Parícutin, 133
Electric cars, 405, 415, 416
Electricity
　generation of, from trash, 423
　use of, to reduce oil consumption, 405
　and work, conversion of heat from fuels to, *412*, 412–413, *413*
Elements, **40**, 40t
　concentrations of, in Earth's crust, 357
　Table of, A-1
　used in industry, future supplies of, 374, 375t
　and uses of, 354, 355t
Energy
　atomic, uranium and, 391–399
　biomass, 422–423, *423*
　commercially produced, use and waste of, *410*, *411*
　conservation of, 405, 409
　　in buildings and homes, 414, *414*, *415*
　　in industry, 415
　　social solutions for, 411, *414*
　　strategies for, 414–415
　　technical solutions for, 414
　　in transportation, 411, *414*, 415
　consumption of
　　in less developed countries, 400, *400*, *401*
　　in United States, 399–400, *400*
　　with water diversion projects, 246
　elastic, of earthquake, **101**, *101*
　geothermal, **420**–422, *421*
　hydroelectric, **419**–420, *420*
　from seas, **423**, 423–424, *424*
　solar. *See* Solar energy.
　sources of, in United States, 380, *380*
　supply of, future of, 399–401, *402*
　tidal, **423**, 423–424
　units of, systems of measurement of, A-4
　use of, in United States, 380, *380*
　wave, **424**, *424*
　from wind, 422, *422*
Energy-consuming devices, efficiencies of, 411, *411*
Energy-consuming systems, end-use efficiencies of, 410, 411, *411*
Energy resource(s), 354

alternative, **409**–428
　conservation as, 411–414
　for United States, 403–405, *404*, *405*
Environment
　climate and, 150–151
　destruction of, and growth of economy, 153
　economics and, 7–13
　　cost–benefit analysis and, 7–9
　　long-term versus short-term considerations in, 9
　　risk assessment and, 9–13
　　tragedy of commons and, 9
　problems of, near nuclear reactors, 393–398
Environmental degradation, **7**, 7–8
Environmental geoscience, **5**–6
Eocene bird *Diatryma*, 32, *32*
Eons, **26**
Epicenter of earthquake, **103**
Epochs, **26**
Eras, **26**
Erosion, **48**
　coastal, 298–303
　　Long Island, New York, 298–301, *299*, *300*
　　in Louisiana, 302, *302*
　　in San Diego County, 303, *303*
　　and transport along coastlines, 298–303
　by glaciers, 322–325
　reservoirs and, 244, *244*
　soil. *See* Soil erosion.
　by stream, 208
　by wind, on desert, 343, *343*
Erosional landforms created by alpine glaciers, 322–325, *324*, *325*, *326*, *327*
Esker(s), **328**, *329*
Estuary(ies), **308**
Ethanol, 423
Eustatic changes, **293**
Eutrophication, cultural, **264**, *265*
Evaporation, **206**
Evaporite deposits, **342**, *342*, *362*
Evaporites, **51**
Excavation for reduction of mass wasting, 202, *202*
Exponential notation, A-6
Externalities, **8**
Exxon Valdez, 306–308, *307*

F

Fall in mass wasting, **192**, *193*, 194t, 197, *198*
Fall turnover, **226**
Farm subsidies, 174

government, and soil in United States, 174
Farming, organic, 172
Farmland and precipitation, in United States, 238–239, *239*
Fault(s), **67**, **82**, *83*
　earthquakes and, *113*, 113–115, *114*, *115*
Fault creep, **114**
Feldspar, **43**, *44*
Fertilizers
　chemical, 171
　crop yields and, *178*
　eutrophication caused by, 264, *265*
　plant/animal matter, 171, *171*
　production of, 178, *179*
Firn, **318**
Firn line, **322**, *322*
Fish, catch of, worldwide, 306, *306*, 308
Fissility, **50**, *50*
Fission products, **275**
Fissures, volcanic, **130**
Fjords, **325**, *327*
Flaming Gorge Dam, Utah, *420*
Flash floods, **342**
Flathead Lake, Montana, *223*
Flood(s), 4, 5, *5*, **207**, 211–213
　costs associated with, 211–212
　factors augmenting, 212
　flash, **342**
　frequency of, 213–214
　and stream size, 213, *213*
Flood basalts, **130**, *131*
Flood control and Mississippi River floods, *214*, 214–215
Flood plain, **207**
　management of, 215
Flooding
　and agriculture, on Nile River, 213
　continental, 293–294
　of Mississippi River, 213
Florida Everglades, *227*, 230–233, *231*, *232*
Flow in mass wasting, **192**, *193*, 194t, 194–197, *195*, *196*
Fluoride pollution, 367
Fold(s), **82**, *82*
Foliation, **54**
Forbes, Stephen, 6
Foreshocks, **120**
Fossil fuels, **354**, 380–381
　advantages and disadvantages of, 379
　consumption of, 5
　depletion of, 399
　formation of, 390
　geopolitics of, 402–403

heat from, conversion of, to work and electricity, *412*, 412–413, *413*
important applications of, 409
as nonrenewable resources, **387**
Fossil water. *See* Ground water.
Fossils, 24, *25*
Franklin, Benjamin, 237
Fresh Kills Landfill, New York City, 282–284, *284*
Fuel(s), fossil. *See* Fossil fuels.
hydrogen gas as, 425
Fuel rod, **393**, *393*
Fusion reactions, 426

G

Gabbro, **48**
Ganges River, India, *269*
Gas, natural, 387, 390
future availability of, 401, *402*
Gas hydrates, 389–391
Gem, **45**
Geologic resource(s), **5**, 354–357
Geologic structure, **82**
Geologic time, 4
Geologic time scale, 26, 27t
Geology, **4**
Geoscience, environmental, 5–6
Geothermal energy, 420–422, *421*
Geothermal energy plant, Fenton Hill, New Mexico, *421*, 421
Glacial epoch. *See* Ice age(s).
Glacial Lake Missoula, 333
Glacial striations, **322**, *323*
Glacier(s), 317, *318*
alpine, *318*, 318–319, *326*
erosional landforms created by, 322–325, *324*, *325*, *326*, *327*
continental, 293, *294*, **319**, *319*
and Great Lakes, *332*, 332–333
Pleistocene, effects of, *330*, 330–333, *331*, *332*
creation of lakes by, 223, *223*
deposits made by, 326–329
erosion by, 322–325
formation of, 318
mass balance of, 322, *322*, *323*
movement of, 319–322, *320*
types of, 318–319
Glen Canyon Dam, Colorado, 243, *243*, *244*, 244, 245, *245*
Global warming and sea-level rise, 312, 312–313, *313*
Gneiss, **53**
Gold, mining of, in South America, 370, *370*
in United States, 369–370

panning for, 360, *361*
prices of, 369
in world circulation, 369
Gold, Thomas, 390
Gondwanaland, **79**, 83, 85, 86–87
Graded stream, **209**, 209–211, *210*
Gradient of stream, **207**, *207*
Grains, consumption of, 179
production of, Green Revolution and, 177–178, *178*, 179
total global, 179, *179*
Granite, **46**, 47, *47*
Gravitational sliding as cause of plate movement, 78, *78–79*
Great Lakes, continental glaciers and, *332*, 332–333
Great Salt Lake, Utah, *225*
Green Revolution, **177**, 178
Greenland ice sheet, ice core from, 13, *13*
Groin(s), **298**, *299*
Gross national product, 375
Ground water, 56, **206**, 216–217, *217*
characteristics of, 216–217, *217*
consumption of, and renewable supply, 251, *251*
desert, 341
diversion of, and environmental impact, 246–249
heated, uses of, 420–421, *421*
heated by processes, 420
industrial wastes in, 272–273, *273*
movement of, 218, 218–219
and nuclear waste disposal, 275–277
pollution of, 270–275, *271*
areas of subsidence and, *247*
deep injection wells and, **275**, *276*, 277
by landfills, 280, 283
San Joaquin Valley, 274–275
in San Joaquin Valley, 274–275
recharge of, **218**
removal of
saltwater intrusion following, 249, *249*
subsidence in, 249, *249*
sewage in, 272, *272*
Gulf coast, erosion of, 301, 301t
Gypsum, **51**

H

Hadean Eon, **26**
Half-life, **275–276**
Halite, 41, *41*, **51**
Hardin, Garrett, "The Tragedy of the Commons" by, 9, *9*

Hawaiian Island–Emperor Seamount chain, 95, *95*
Hazardous waste, **266–267**
Health, human, water pollution and, 269, 269–270
Heat from fuels, conversion of, to work and electricity, *412*, 412–413, *413*
Himalayas, 4, *4*, 83, 84–85, *85*
Hoffman, Paul, 80–82
Horn, **323**–324, *325*
Horses, evolution of, 32–33, *34*
Hot spot(s), **95**, **129**
magma production at, 129, *129*
Hot-spot volcanoes, 137
Hubbard Brook Experimental Forest, 167–168, *168*
Hubbard Glacier, 320–321, *321*
Hubble, Edwin, 19
Human beings
evolution of, 33, *35*
health of, water pollution and, 269, 269–270
Humus, **160**
Hurricanes, 298
damage caused by, 303–304, *304*
Hussain, King, 256
Hydrides, **425**
Hydrocarbons, 390
Hydroelectric energy, 419–420, *420*
Hydrogen gas, as fuel, 425
Hydrologic cycle, **206**, *206*
residence time in, **206**, 207t
Hydrolysis, 159–**160**
Hydrosphere, 24, 56–57
Hydrothermal ore deposits, 359, 359–360, *360*
Hydrothermal processes, 359–360
Hydrothermal solution(s), **359**, 360, *361*
Hydrothermal vein deposit, **359**
Hydrothermal vents, deep-sea, 61
Hypothesis, **10**
forming of, 10
testing of, 10

I

Ice age(s), **329**, 329–330
causes of, 333–334
major, and plate tectonics, 333
Pleistocene, 329, 330, 333
cycles of, and orbital variations, 333–334, *334*
Ice sheet, **319**, *319*
Igneous rocks, 46–48
extrusive, **46**
intrusive, **46**
textures of, 47t

Incineration, **267**, 268
　as alternative to landfill, 284
Incinerators, modern, 284, *284*
India, 85, 86–87
Industrial countries, 14
Industrial wastes, 266–268
　destruction of, 267, *268*
　dispersal of, 267, *267*
　in ground water, 272–273, *273*
Industry, conservation of energy by, 415
Injection wells, deep
　and ground-water pollution, **275**, *276*, *277*
　and earthquakes, 11, 12, *12*
Inorganic compounds, toxic, as water pollutants, 263
Integrated Farm Management Program Option, 174
Intercropping, **174**
Internal process, **4**
International politics, water and, 256–257
Iron, in soil, 5
Irrigation
　for agribusinesses, water consumption by, 255–256
　of croplands, 241, *242*, 242t
　drip, 256, *256*
　flood, 255–256, *256*
　sprinkler, 255–256, *256*
Island arc(s), 91–**94**, *94*
Isostasy, 70, **70**–71
Isostatic adjustment, **71**, *71*

J

Jackson, David, 120
Joint(s), **82**, *83*
Joule, James, 412
Jovian planets, **21**, *21*
Jupiter, *21*

K

Kagan, Yan, 120
Kame(s), *328*, **329**
Karst topography, **221**, *222*
Kerogen, **388**
Kimberlite(s), **358**, *358*
　pipes of, **358**
Kissimmee River, Florida, 230, *231*, 232
Krakatoa, 146

L

L waves, **104**, *104*
Lago Poopó, Bolivia, 338, *343*
Lagoon(s), **296**, 301

Lake(s), 56, 221–**222**, *222*
　Beartooth Mountain, Montana, *57*
　eutrophic, 223–**225**, *224*, *225*, **264**
　formation of, 223, *223*
　life cycle of, 222–225
　oligotrophic, 223–**225**, *224*
　oxbow, 210, *210*
　playa, **342**
　pothole, *223*
　salty, fresh water and, 225, *225*
　temperature layering and turnover in, 226, *226*–227
Lake Baikal, Russia, 223
Lake Nube, Bolivia, *222*
Lake Nyos, Cameroon, 227
Lake Okeechobee, Florida, 232
Land degradation
　in desert, *346*, 346–347, *347*
　human-induced, 169, *170*, 170t
Landfill(s)
　air pollution due to, 280
　composting as alternative to, 284
　incineration as alternative to, 284
　pollution of ground water by, 280, *283*
　wastes buried in, 262, *273*, 273–274
Landslide(s)
　avoidance of, engineering for, *192*, 201
　earthquake-related, 112
　recent, affecting humans, 189, *190*
　Yungay, Peru, 199–200, *200*
Latitude, effect of, on desert, 338, *339*, 339
Laurasia, **79**, 83, 85, 86–87
Lava, 46–47
Lava plateau, **130**, *131*
Law, **10**
Leaching, **162**
Length, systems of measurement of, A-3
Less developed countries, 14
Levee(s), natural, 212, *213*, *214*, **214**–215, *216*
Life
　and atmosphere, co-evolution of, 24–26
　"A Brief History of," 26–33
Lighting, energy consumption and, 414, *414*, *415*
Limestone, **48**, **51**
　bioclastic, **52**, *52*, *54*
　folds in, *82*
Lipari Landfill, Pitman, New Jersey, *273*, 273–274
Liquefaction during earthquake, **109**–110, *110*

Lithification, **48**
Lithium minerals, 358
Lithosphere, 23, **67**, **70**
Lithosphere–asthenosphere boundary of Earth, 108–109
Litter, 160, *161*
　plastic, in coastal waters, 306, *306*
Loam, **160**
Loess, **175**, *175*
Longshore current, **296**, *296*, 298, *298*
Los Angeles Aqueduct, 253, *254*
Los Angeles Water Project, 253–255, *254*
Love Canal, New York, 262, *262*, 272

M

Magma, **23**, **46**
　behavior of, effects of pressure and water on, 130
　　effects of silica on, 130
　depiction of, colors and style for, A-7
　explosive, violent volcanoes and, 137–140
　granitic and basaltic, formation of, 129
　production of
　　at hot spot, 129, *129*
　　pressure-relief melting and, 128, *128*, *129*
　　in spreading center, 128–129, *129*
　　in subduction zone, *127*, 127–128
Magmatic processes, 357
Malthus, Reverend Thomas, 177
Mammals, Miocene, 33
Manganese nodules on sea floor, **363**, *363*
Mantle, **23**, 68, 69t, **69**–71
　depiction of, colors and style for, A-7
Mantle convection, **77**
　as cause of plate movement, 77, *78*
　plate movement as cause of, 77
Mantle plume, **82**
Mantle-plume volcanoes, 137
Marble, **53**, *54*
Marine animals, habitat of, destruction of, 306, *306*
Marine evaporites, **362**
Mass, systems of measurement of, A-3
Mass extinction(s), **28**, 32
Mass wasting, **190**
　earthquakes and volcanoes in, 192
　factors controlling, 190–192

fall in, **192**, *193*, 194t, 197, *198*
flow in, **192**, *193*, 194, 194t, 194–197, *195*, *196*
prediction of, 201
reduction of, excavation for, **202**, *202*
   retaining wall for, **202**
   terraces for, **202**
rock type and orientation of layers and, 191, *191*
slide in, **192**, *193*, 194t, *196*, 196–197, *197*
steepness of slope and, 190
types of, 192–197, *193*, 194t, *195*, *196*, *197*, *198*
unconsolidated materials and, 191, *192*
water and vegetation as influence on, 192, *192*
Matterhorn, 323–324, *325*
McCann, William, 119
Measurement, systems of, A-3–A-6
   conversion factors for, A-5
Melting, pressure-relief, and magma production, 128, *128*, *129*
Mercury, *21*
   contamination from gold mining, 370
   toxicity of, 370
Mesozoic life, 28–32
Metal(s), **354**
   conservation of, 375
   deposits of, 356–357
   mining of, 364–365
   recycling of, 375
   refining of, 365
   reserves of, modern geopolitics of, 372–374, *373*
   substitution of, 375
   supply of, forecasting future of, 374–375
Metamorphic processes, 362–363
Metamorphic rocks, **52**–53
Metamorphism, **52**, *54*
Methane, 387, 390
Methyl hydrate(s), **389**–391
Metric System, basic units of, prefixes for use with, A-4
Mica, **43**, *44*
Mid-oceanic ridge, **72**–73, 88, 88–89, *89*, *90*
Milankovitch, Milutin, 334
Milky Way, 19
Mine
   open pit, **364**, *364*
   surface, **364**, *364*
   tunnel, **364**
Mine spoil(s), 364, 365, 383

Butte–Anaconda area, 365, 365–366, *366*
   Clark Fork River, Montana, 366, *366*
Mineral(s), **40**–45, *41*
   accessory, **45**
   chemical composition of, 40–41
   critical, 373, 374, 375t
   crystalline nature of, **41**
   deposits of, subeconomic, **370**
   future availability of, 370–375
   groups of, 42t
   low-grade resources of, exploitation of, 374–375, *375*
   reserves of, 370–371, *371*
      depletion of, 371–372
   rock-forming, **41**–43, 42t
   strategic, 373–374
Mineral deposit, **356**
Mineral reserves, **357**
Mineral resources, 354
   and mining, 353–377
Mining, in early 1900s, *354*
   gold, in South America, 370, *370*
   in United States, 369–370
   history of, 371–372, *372*
   of metals, 364–365
   mineral resources and, 353–377
Mining Law of 1872, 364
Miocene mammals, *33*
Mississippi River, 208, 301
   delta, 301–302, *302*
   flood control along, *214*, 214–215
   flooding of, 213
Missouri River, Missouri, flooded, *214*
Model, **10**
Moho, **108**
Mohorovičić, Andrija, 108
Mohorovičić discontinuity, **108**
Moisture belt, soil, **217**–218
Moraine(s), **327**, 327t
   end, **327**, *327*
   terminal, **327**, *328*
Mount Adams, 133, *134*
Mount Baker, 200, *201*
Mount Pelée, 125, *126*, 141–144, *145*, *146*
Mount Pinatubo, 136, *136*, 147, *147*
Mount Redoubt, 133
Mount St. Helens, 125, *126*, 133, 137, *138*, 138–139, *139*, 141, 146, 200
Mount Tambora, 144, *145*
Mount Vesuvius, 146, *146*
Mountain chains, two, building of, 83–85
Mountains, building of, plate tectonics and, 82–85
   subduction and, 75

Mudflow(s), in mass wasting, **195**–196
   Nevado del Ruiz, Colombia, 199, *199*
Mudstone, petroleum and, 384
Mulholland, William, 254
Mullineaux, Don, 144–145
Municipal waste, in coastal waters, 305, *305*
Mutagen, **5**
Mutations, **5**

N

National Flood Insurance Program, 304
Neutrons, **40**
Niagara Falls, 209
Niger, North Africa, nomads in, *346*
Nile River, 341
   flooding and agriculture on, 213
Nitrogen cycle, 167, *167*
Nitrogen fixing, **167**
Nonbiodegradable, **6**
Nonmetallic resources, **354**, *355*, 355–356
Nonrenewable resources, **6**
North America, earthquake danger zones in, 113–119
Nuclear fission, **391**
Nuclear fuel cycle, 394–395, *395*
Nuclear fusion, 19, 409, **426**, *426*
Nuclear power industry, present and future status of, 398–399, *399*
Nuclear power plant(s), 391, 393, *393*, *394*, *396*
   accidents at, 396–398, *397*, *398*
   Chernobyl, former Soviet Union, 397, 397–398, *398*
   decommissioning of, **395**
   dismantlement of, **395**
   radioactive emissions from, 393–394
   Three Mile Island, Pennsylvania, 396, *396*
   Washington, 399
Nuclear reactor(s)
   environmental problems near, 393–398
   isolation of, **395**
   operation of, 391–393
Nuclear reactor site, Hanford, Washington, 277, *277*, *278*
Nuclear waste
   disposal of, ground water and, 275–277
   problems associated with, 394–395, *395*
Nutrients in soil, 166, 166t

O

Ocean(s), 56
  central, pollution of, 310
  cold currents of, deserts formed by, 341
  evolution of, 24
  thermal power from, 424
Ocean thermal power plant, 425
Oceanic crust
  convergence of, with continental crust, 68, 74–75, *74–75*
  divergent plate boundaries in, **72–73**
  two plates carrying, convergence of, 76
Oceanic island(s), **94–95**
Oceanic trenches, **74–75**, *75*
Ogallala aquifer, high plains and, *248*, 248–249
Oil, availability of, in future, 401
  consumption of, use of electricity to reduce, 405
  crude, *380*, **381**, *381*
Oil drilling platform, 385–386, *386*
  Mississippi delta, 301, *302*
Oil pipeline, Alaska, 306, *307*
Oil refinery(ies), 386, *386*
Oil shale, 388, *388*, *389*
  Rifle, Colorado, *389*
Oil spill(s), 386
  Prince William Sound, Alaska, 306–308, *307*
Oil trap(s), **384–385**, *385*
Oil well(s), drilling of, 385, 385–386, *386*
  secondary recovery from, 387
Olivine, **43**, *44*
Ollagüe volcano, near Chile, *75*
Open pit mine, 364, *364*
Orbital variations, cycles of Pleistocene ice age and, 333–334, *334*
Ore(s), **356**
  deposits, disseminated, **359**
    precipitation in formation of, 362
    sedimentary sorting in formation of, 360, *361*
  formation of, 357
  hydrothermal deposits of, 359, 359–360, *360*
Ore minerals, **43**
Organic compounds, industrial, as water pollutants, 263
Organic farming, 172
Organic material, development of, 162, *163*
Organic wastes, industrial. *See* Industrial wastes.

Orthoclase, **43**
Owens Valley, California, 253–255, *254*, *255*
Oxbow lake, 210, *210*
Oxidation, **159**
Oxides, **42**
Oxygen demand, biological, **264**

P

P waves, **103**, **103–104**, 108, *109*
Pacific Coast, erosion of, 302, 302t
Paleozoic Era, **28**, *30*
Paleozoic life, 28
Pangaea, **79–80**, *80*, *81*
Pangaea I, 81–82
Pangaea II, **82**
Pangaea III, **82**, *83*
Parent rock, 51, *52*
Peat, **51**, **381**, *382*
Pegmatite(s), **358**, *358*
Periods, **26**
Permafrost, **184**, **196**
Permeability, **216**
Pesticides, biological impact of, 267, *267*, *268*
Petroleum, **381**, *381*
  consumption of, in United States, 399–400, *400*
  extraction, transport, and refining of, 385, 385–386
  formation of, 384, 384–385
  imported, cost of defending, 403
  imports of, by United States, 403
  reserves of, and consumption of, 402–403, *403*
    in United States, 404–405, *405*
  reservoir, **384**, *384*, 387
  source rock, **384**
  used in United States, sources of, 403, *404*
Phanerozoic Eon, 28–33
Phosphorus cycle, *166*, 166–167
Photosynthesis, 24–26, *25*
Phytoplankton, **28**, *30*, *33*, 305
Placer deposit(s), **360**, *361*
Plagioclase, **43**
Plains, high, and Ogallala aquifer, *248*, 248–249
Planetesimals, **19**
Planets
  formation of, 19
  Jovian, **21**, *21*
  terrestrial, 19–**21**
Plant nutrients, as water pollutants, 263
Plants, of desert, 337
  prairie, root systems of, *176*, *176*

Plastic flow, **320**
Plate(s), **67**, *68*
  major, of world, *23*
  movement of, as cause of mantle convection, 77
    gravitational sliding as cause of, 78, 78–79
    mantle convection as cause of, 77, *78*
    possible mechanisms of, 77–79
    push-pull model of, **79**
  and plate tectonics, 71–72
  tectonic, **4**, **76**, 76–77
  two, carrying continental crust, convergence of, 76
    carrying oceanic crust, convergence of, 76
Plate boundary(ies), **67**, 71–72, *72*
  convergent, **72**, *72*, 73t, **73–76**
  divergent, **72–73**, 73t, *74*
  transform, **72**, *72*, 73t, **76**
Plate tectonics, **23**, *23*, 67–97
  major ice ages and, 333
  and mountain building, 82–85
  plates and, 71–72
  and sea floor, 88
  theory, **67**
  volcanoes and, 133–137
Playa(s), **342**, *342*
Playa lake, **342**
Pleistocene glaciers and Great Lakes, 332, 332–333
Pleistocene Ice Age, **329**, *330*, 333
  cycles of, and orbital variations, 333–334, *334*
Plowing, contour, 172–173, *173*
Pluton, **46**, *47*
Plutonic rocks, **46**
Plutonium-239, 396
Pollution, **5**
  air. *See* Air pollution.
  burning coal causing, 384
  control of
    destruction in, 264, 267
    dispersal in, 263, 267
    isolation in, 264, 268
    recycling in, 264
  external and human costs of, 7–8, *8*
  fluoride, 367
  ground-water. *See* Ground water, pollution of.
  ocean, *305*, 305–310. *See also* Water pollution.
  oil refineries causing, 386
  oil spills causing, 386
  thermal, **270**
  water. *See* Water pollution.
Polymers, **130**

Porosity, **216**, *217*
Porphyry, 359
Pothole lakes, *223*
Powell, John Wesley, 250–251
Precautionary principle, **12**
Precession, **333**
Precipitation
   and farmland in United States, 238–239, *239*
   in formation of ore deposits, 362
Pressure and magma production, 130
Proterozoic Eon, **26**
Protium, 426
Protons, **40**
Protosun, **19**
Pyroclastic material, **133**
Pyroxene, **43**, *44*

Q

Quality of life, **8**
Quarry(ies), stone mined from, 356, *356*
Quartz, **43**, *44*, 52–53
Quartzite, **53**

R

Radiation, exposure to, health effects of, 393–394, *394*, 397–398
Radioactive fallout from Chernobyl disaster, 397–398, *398*
Radioactive materials as water pollutants, 263
Radioactive wastes, disposal of, 275–277, *276*
   at Hanford, Washington, 277, *277*, *278*
   problems associated with, 394–395, *395*
Radioactivity, **391**, *391*
Radon and rocks, 53
Rainfall in California, 339, *340*
Rainwash, *169*
Rangeland, deterioration of, 347, *347*
Recharge of ground water, **218**
Recreational resources, rivers as, 8, *8*
Recycling
   as alternative to waste disposal, 285, 285–286, *286*, 287t, *288*
   of aluminum, 368–369
   of metals, 375
   in pollution control, 264
Reef(s), 310–311, *311*
Refining of metals, 365
Regolith, **160**
Reilly, Bernard, 268

Renewable resources, **6**
Reservoir(s), **243–244**
   erosion due to, 244, *244*
   risk of disaster associated with, 244
   silting due to, 243–244
   water loss due to, 243
Residence time in hydrologic cycle, **206**, 207t
Residual deposits, **362**
Resource(s), geologic, **5**
   nonrenewable, **6**
   renewable, **6**
Retaining wall, for reduction of mass wasting, 202
Reverse flow of water, **246**
Rhyolite, **47**, *47*
Richter, Charles, 105
Richter scale, **105**
Rift, **72**, *74*
Rift valleys, **73**
Rift zone volcanoes, 136–137
Rifting, continental, **73**
Rio Declaration, **13**
Risk assessment
   economics and environment and, 9–13
   experiment in, 11, *11*
   first observation in, 11, *12*
   second observation in, 11–12
River(s), **207**
   channelization of, 215
   as recreational resources, 8, *8*
Rock(s), **40**, **57**
   basement, **57**
   carbonate, 51–52
   country, **57**
   earthquake damage related to, 107–110, *110*
   igneous. See Igneous rocks.
   impermeable, **384**
   metamorphic, 52–53
   parent, **51**, *52*, *162*
   permeable, **384**
   plutonic, **46**
   radon and, 53
   sedimentary. See Sedimentary rocks.
   source, for petroleum, **384**
   type(s) of
      and orientation of layers of, mass wasting and, *191*, *191*
      symbols for, A-7
   volcanic, **47–48**, *48*
Rock cycle, **54–55**, *55*
Rockslide
   Kelly, Wyoming, 197
   Madison River, Montana, 199, *199*
   in mass wasting, 197
Runoff, **206**

S

S waves, *103*, **103–104**, 108, *109*
Sadat, Anwar, 256
Safe Drinking Water Act, 269, 270
Sahara Desert, 346–347, *347*
Sahel, 346, *346*
Salinization, **250**, *250*
Salt in Colorado River, 253
Salt dome, **384**
Saltwater, intrusion of, **249**, *249*
   following removal of ground water, 249, *249*
San Andreas fault
   earthquakes and, *113*, 113–114, 115, *115*
   types of motion along, 114
Sand and gravel, mining of, 355, *355*
Sandstone, **48**, **49**, *49*
   joints in, 83
Sanitary landfills, **280–284**, *282*. See also Landfill(s).
Scablands, channeled, **333**
Scarp, **106**, *107*
Schiaparelli, Giovanni, 3
Schist, **54**, *54*
Schneider, Stephen, conservation with, 150–153
Scientific notation, A-6
Sea floor
   manganese nodules on, *363*, *363*
   plate tectonics and, 88
Sea level, global
   changes in, 293–294, *331*
   rise in, global warming and, *312*, 312–313, *313*
Sea wall, 299, *300*
Seamount(s), **94**, *94*, *95*
Seas. See also Ocean(s).
   energy from, *423*, 423–424, *424*
Seashore, 291
Sediment, **48**
   at emergent coastline, 295, *295*
   at submergent coastline, 295, *295*
   transport and deposit of, **48**
   transport of, by stream, 208
   as water pollutant, 263
Sedimentary processes, 360–362, *361*
Sedimentary rocks, **48–52**
   chemical, **51**, *51*
   clastic, 49t, **49–50**
   fault in, 83
   organic, **50**, *50–51*
Sedimentary sorting, in formation of ore deposits, 360, *361*
Seismic gap hypothesis, **119–120**
Seismic waves, **102**
   measurement of, 104–105

Seismogram, **105**
Seismograph, **104**–**105**, *105*
Seismology, **102**
Selway River, Idaho, *207*, *208*, *208*
Semiarid zones, **337**
   land degradation in, *346*, 346–347, *347*
Septic system, **272**
Sewage
   in coastal waters, 305
   in ground water, 272, *272*
   as water pollutant, 263
Sewage treatment
   municipal, 265–266, *266*
   primary, **265**, *266*
   secondary, **265**, *266*
   tertiary, **265**–**266**, *266*
Shale(s), **48**, **50**, **54**
   instability of, 180
   mining of, Colorado River and, 253
   oil, 388, *388*, 389
   petroleum and, 384
Shield volcano, **132**, *132*
SI system of measurement, A-3
Silicate, **42**
Silicate tetrahedron, **43**, *43*
Silting, reservoirs and, 243–244
Sinkhole(s), **220**, *221*, *222*
Slate, **54**
Slide in mass wasting, **192**, *193*, 194t, 196–197, *197*
Slip face, **344**, *344*
Slope, undermining of, 191, *191*
Slump in mass wasting, **196**, 196–197, *197*
Snow, and glacial ice, 318, *322*
Snowline, **322**, *322*
Soil(s), **4**, *4*, 157–186
   arctic, and Alaska pipeline, 184, *184*
   climate and, 162–164
   components of, 160
   degradation of, causes of, 170, *170*
   desert, wind erosion of, 343, *343*
   development of organic material and, 162
   earthquake damage related to, 107–110, *110*
   fertility of, agriculture and, 170, 170–172, *171*
   formation of, rates of, 166
   laterite, **163**, *164*
   nutrients, formation of, rates of, 166, 166t
   parent rock and, 162
   pedalfer, **163**, *164*
   pedocal, **163**, *164*
   slope angle and aspect and, 162
   soil-forming factors in, 162–166

   zone of accumulation of, 162
   zone of leaching of, **162**
Soil engineering, 179
   building foundation beneath river, New York City, *181*, 181–182, *182*, *183*
   case histories of, 180, 180–184, *181*, *182*, *183*, *184*
   construction in Mexico City, *180*, 180–181, *181*
   at housing subdivision, Colorado, 180, *180*
Soil erosion, **159**, *164*, *165*, *169*, 169–170
   case histories of, *175*, 175–176
   in central China, 175
   cropland, reduction of, 172–174
   human activity and, *169*, 169–170, *171*
   off-site impact of, **171**–172
   on-site impact of, **171**
Soil horizons, **160**–162, *161*
Soil moisture belt, **217**–218
Soil profiles, 160–162
Solar cells, **418**–419, *419*
   electricity production by, 419
   on rooftop, 419, *419*
Solar collectors, **418**
Solar energy, *416*, 416–419
Solar heating, active, **418**, *418*
   passive, **416**–418, *417*
Solar house, passive, *417*, 417–418
Solar system
   formation of, 19, *20*
   schematic view of, *22*
Solifluction in mass wasting, **196**, *196*
Spent reactor fuel repository at Yucca Mountain, Nevada, 278–279, *280*
Sphinx, climate changes and, 345, *345*
Spit, **296**, *297*
Spreading center, **72**, *74*
Spring(s), 219, **219**–220
Spring turnover, **226–227**
Spruce Tree House, Mesa Verde National Park, 416, *417*
Stone mined from quarries, 356, *356*
Stratovolcanoes, *133*, *134*, *135*
Stream(s), **207**
   base level of, **209**, *209*
   braided, **332**
   capacity of, **208**
   competence of, **208**
   desert, 219, *219*, 341, *341*
   discharge of, **208**, *208*
   downcutting by, 209
   erosion and transport of sediment by, 208

   flow of, 207–211
   graded, **209**, 209–211, *210*
   gradient of, **207**, *207*
   meandering, 210, *210*
   size of, flooding and, 213, *213*
Strip-cropping, **174**
Subduction, **74**
   and earthquakes, 74, *74–75*
   and mountain building, 75
   and volcanoes, 74, *74–75*
Subduction zone(s), **74**, *74–75*
   earthquakes at, *117*, 117
   volcanoes in, 133–136, *135*
Subsidence, **180**, **249**, *249*
   in removal of ground water, 249, *249*
Subsidies, farm, **174**
Substitution of metals, 375
Sulfur, 358, *359*
Sulfur dioxide, 159
Sulfuric acid, 383, 384
Sun, formation of, 19
Supercontinent(s), **80**–82
Superfund, **262**, 272
Superfund site, Butte, Montana, 365, 365–366
Surface mine, **364**, *364*
Surface processes, **4**
Surface water, 205, **206**
Surface waves, **104**, *104*
Sustainable development, **13**
Synfuels, **387**
Systems, **6**

T

Talus, **191**
Tar sands, 387, *388*
Tectonic plate(s), **4**
   anatomy of, *76*, 76–77
Temperature, atmospheric
   clouds and, 151–152
   changes in, after volcanic eruptions, 147
   systems of measurement of, A-3
Temperature layering and turnover, in lakes, *226*, 226–227
Tephra, **133**
Terminus, glacial, **322**, *323*
Terraces for reduction of mass wasting, **202**
Terrestrial planets, 19–21
Teton Dam, Idaho, disaster at, 244–245, *245*
Theory, **10**
   forming of, 10
Thermal pollution, **270**
Thermal power from oceans, 424

Thermocline, **226**, *226*
Thermodynamics
  First Law of, 412
  Second Law of, 411, 412
Threshold effects, **13**
Thundercloud, 56
Tibet, 85, 86–87
Tidal energy, *423*, 423–424
Tide, migration of, 298t, 301t
Till, **326**
  landforms composed of, 327–329
Tillage, **172**
  conservation, 173, *173*
Tilt, **333**
Time, systems of measurement of, A-3
Tomak fusion test reactor, Princeton University, New Jersey, 426
Topography, effect of, on desert, 339, *340*
Toxic chemicals in coastal waters, 305–306, *306*
Toxic substances, **5**
"Tragedy of the Commons, The," by Garrett Hardin, 9, *9*
Transform faults, **89**, *90*
Transpiration, *167*, **206**
Transportation, conservation of energy in, 411, *414*, *415*
Trash, generation of energy from, 423
Trash incinerators, 423
Trenches, oceanic, *74*–*75*, **75**
Tributaries, **207**
Tritium, 426
Tsunami(s), *112*, **112–113**, 146
Tuff, **140**, *140*
Tunnel mine, **364**
Turbine, **412**, *412*
Turnover, temperature layering and, in lakes, *226*, 226–227

U

Underthrusting, **85**
United States, energy strategies for, 403–405, *404*, *405*
Universe
  closed, **19**
  evolution of, *20*
  open, **18**
  origins of, 19
Uranium, 54
  and atomic energy, 391–399
  energy emitted by, 391
  fuel-grade, 393
  in nuclear fission, 391
  reserves of, 401, *402*
Uranium-235, 396
Uranium-238, 396

Uranium fuels, 379
Uranium ore, natural, 392–393

V

Valley(s)
  hanging, **324–325**, *325*, *326*
  rift, **73**
  U-shaped, **323**, *324*, *325*
  V-shaped, 323
Van Dover, Cindy Lee, conversation with, 60–63
Vent, volcanic, **130**
Volcanic ash, **133**, 140, *140*
Volcanic cinders, **133**
Volcanic eruptions, **23**
Volcanic fissures, **130**
Volcanic rocks, **47**–**48**, *48*
Volcanic vent deposits, *358*, *359*
Volcano(es), 125–149, **130**
  active, **141**
  ash flows from, 140, *140*
  cinder cone, **133**, *133*
  and climate, 145–148, *147*
  composite cone, **133**, *134*, *135*
  danger zones for, in U.S., 144, *144*
  death and damage caused by, 125–126
  disasters associated with, 141t
  disastrous, historical, *146*, *146*
  dormant, **141**
  education and public policy concerning, 145
  El Parícutin, 133
  eruption of, Long Valley Caldera, California, *143*, *143*
  extinct, **141**
  future, assessment of danger of, 141–145
  hot-spot, 137
  Krakatoa, 146
  magma and, 127–130
  mantle-plume, 137
  mass wasting associated with, 192, 197–200, *199*, *200*, *201*
  Mount Adams, *133*, *134*
  Mount Baker, 200, *201*
  Mount Pelée, 125, *126*, 141–144, 145, 146
  Mount Pinatubo, *136*, *136*, 147, *147*
  Mount Redoubt, 133
  Mount St. Helens, 125, *126*, 133, 137, *138*, 138–139, *139*, 141, 146, 200
  Mount Tambora, 144, 145
  Mount Vesuvius, 146, *146*
  and plate tectonics, 133–137

  prediction of, long-term, 141–144, *144*
    short-term, 144–145
  and related landforms, 130–133
  rift-zone, 136–137
  shield, **132**, *132*
  subduction and, 74, *74*–*75*
  subduction zone, 133–136, *135*
  temperature changes after, *147*
  types of, characteristics of, 131t
  violent, explosive magma and, 137–140
Volume, systems of measurement of, A-3

W

Wash, **341**, *341*
Waste(s)
  hazardous, **266–267**
  municipal, disposal of, 280–284
  organic, industrial. *See* Industrial wastes.
  solid, average household in United States
    amount of, 280, *281*
    composition of, *283*
Waste disposal
  alternatives to, 285, 285–286, *286*, *287*
  and fresh-water pollution, 261–289
  of municipal wastes, 280–284
Waste Isolation Pilot Plant at Carlsbad, New Mexico, *278*, *279*
Water
  conservation of, 255–256
  consumed, **242**, *242*
  demands for
    agricultural, **239**, *240*, 240–241, *241*
    domestic, **239**–**240**, *240*
    industrial, **239**, *240*, *240*, *241*
  in desert, 341–342
  deterioration of quality of, with diversion, 246
  diversion, *243*, 243–246, *244*, *245*
  fresh, 205
    and salty lakes, *225*, *225*
  ground. *See* Ground water.
  human use of, 237–259
  international code for, principles of, 257
  and international politics, 256–257
  loss of, reservoirs and, 243
  and magma production, 128, 130
  oil shale development and, *388*, *389*

Water (*continued*)
  problems associated with, in Los Angeles area, 253–254, *254*, *255*
  reverse flow of, **246**
  salinity of, in Colorado River, 253
  surface, 205, **206**
    diversion projects, *243*, 243–246, *244*, *245*
    in United States, 268–270, *269*
    in United States, shortages of, 237, *238*
    supply of and demand for, *238*, 238–242
    use of, in western United States, 250–252, *251*
  withdrawn, **242**, *242*
Water cycle, 206–207
  in forest ecosystem, *168*
Water diversion projects, energy consumption with, 246
Water pollutants
  categories of, 263, *263*
  control of, 263–264
Water pollution
  by biodegradable wastes, 264
  detection of, 269
  and human health, 269, 269–270
  public awareness of, 262
  waste disposal and, 261–289
Water project, Los Angeles, 253–255, *254*
Water table, **217**–220, *218*

Waterway(s), pristine, 268, 269
  synergistic effects in, **269**
Watt, James, 412
Wave(s), **296**
  body, **102**, *103*, **103**–104, 108, *109*
  breaks of, **296**
  of earthquake, *102*, 102–106
  seismic, **102**, 104–105
  surface, **104**, *104*
Wave energy, 424, *424*
Weathering, **48**, 159–160
  chemical, *158*, **159**, 160
  mechanical, *158*, **159**, 160, *160*
Weathering processes, 362
Wegener, Alfred, 79–80, *80*, *81*
Well(s), **218**
  artesian, **220**, *220*
  excessive pumping and, reservoir depletion in, 247, *247*
  formation of, cone of depression and, *246*, 246–247
Wetlands, *227*, **227**–233, *228*
  abundance of, in United States, 228, *230*
  and humans, 228–229
  importance of, 229
  losses of, in United States, 228, *229*
Whale(s), beached, 305–306, *306*
Wildlife, water pollution and, 305–306, *306*, *307*
Wind, development of, 56, *56*
  energy from, 422, *422*
  erosion by, on desert, 343, *343*

Wind farm(s), 422, *422*
Windmill(s), 422, *422*
Wood as biofuel, 422–423, *423*
World Trade Center, New York City, *181*, 181–182, *182*, *183*

X

Xeriscaping, **255**

Y

Yellow River, China, 215
Yellow River Plain, China, 175
Yellowstone National Park, volcanic eruption and, 142–143, *143*
Yellowstone River, valley formed by, 210, *210*
Yosemite National Park, *325*
Young, John, 366
Yucca Mountain Repository for spent reactor fuel, 278–279, *280*

Z

Zion National Park, Utah, lithified dune in, 345, *345*
Zone of aeration, water table and, **217**, *218*
Zone of saturation, water table and, **217**, *218*

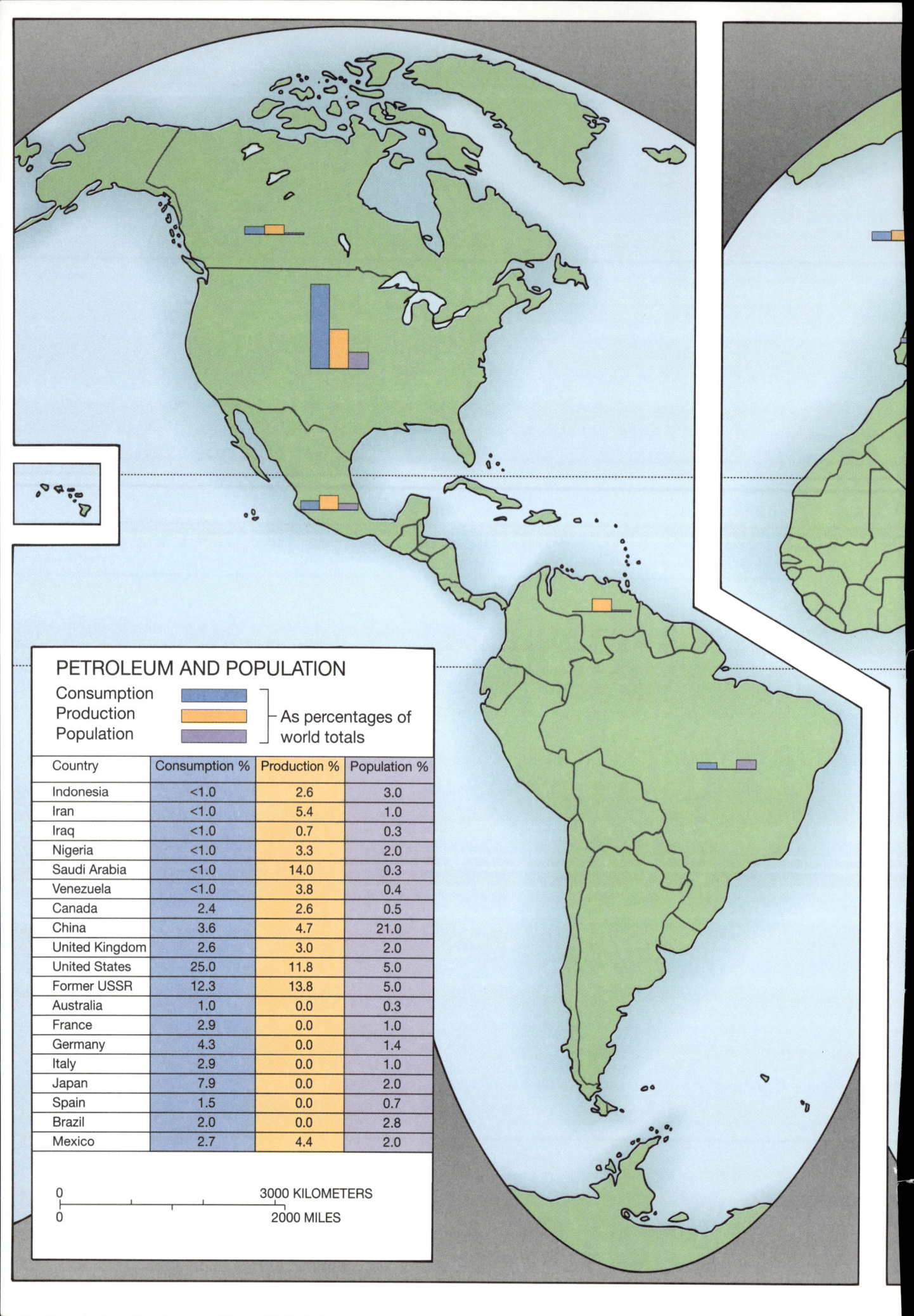

PETROLEUM AND POPULATION

Consumption
Production
Population
— As percentages of world totals

Country	Consumption %	Production %	Population %
Indonesia	<1.0	2.6	3.0
Iran	<1.0	5.4	1.0
Iraq	<1.0	0.7	0.3
Nigeria	<1.0	3.3	2.0
Saudi Arabia	<1.0	14.0	0.3
Venezuela	<1.0	3.8	0.4
Canada	2.4	2.6	0.5
China	3.6	4.7	21.0
United Kingdom	2.6	3.0	2.0
United States	25.0	11.8	5.0
Former USSR	12.3	13.8	5.0
Australia	1.0	0.0	0.3
France	2.9	0.0	1.0
Germany	4.3	0.0	1.4
Italy	2.9	0.0	1.0
Japan	7.9	0.0	2.0
Spain	1.5	0.0	0.7
Brazil	2.0	0.0	2.8
Mexico	2.7	4.4	2.0

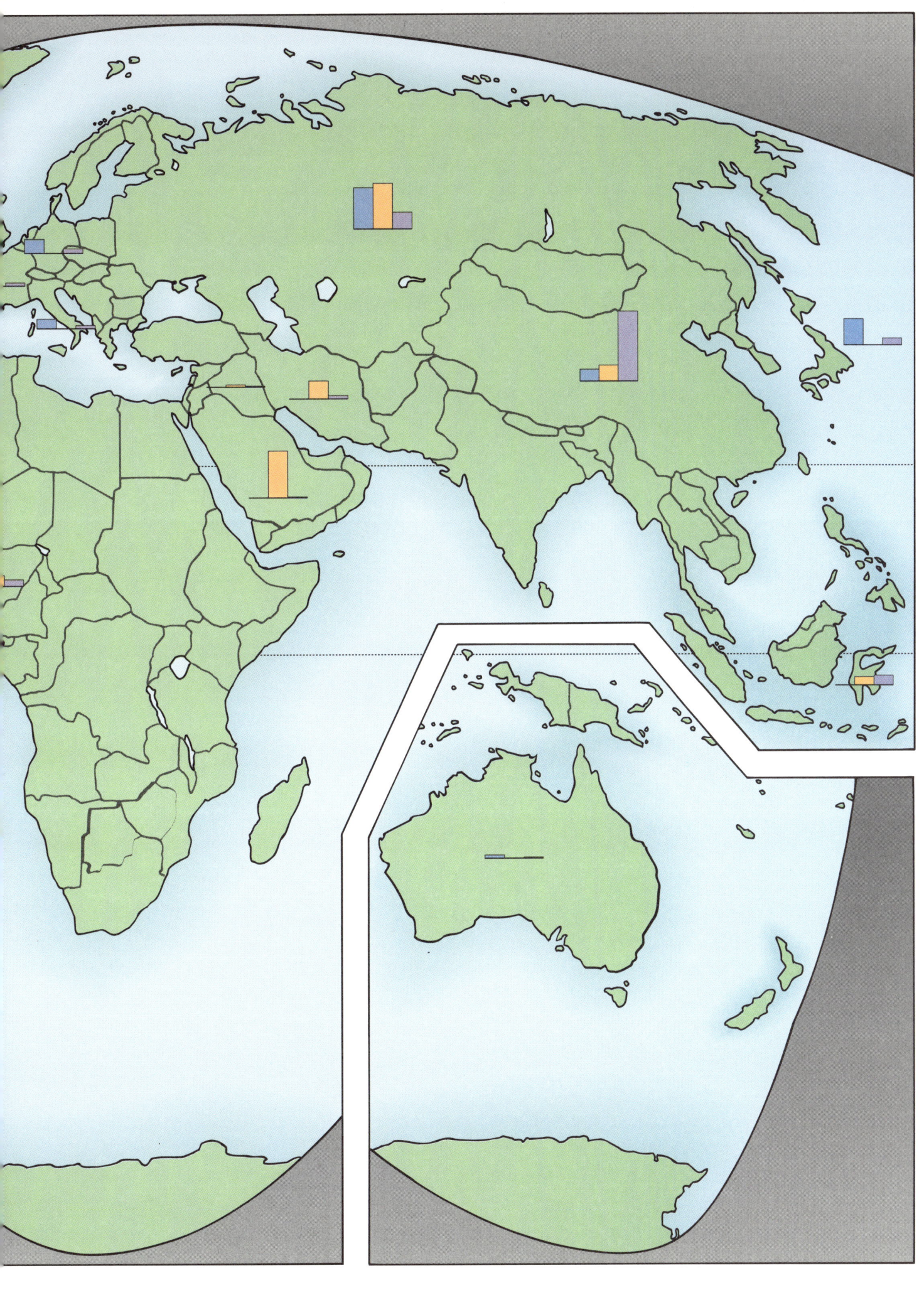